Lecture Notes in Computer Science

Edited by G. Goos, J. Hartmanis and J. van Leeuwen

Mariangiola Dezani-Ciancaglini
Gordon Plotkin (Eds.)

Typed Lambda Calculi and Applications

Second International Conference
on Typed Lambda Calculi and Applications, TLCA '95
Edinburgh, United Kingdom, April 10-12, 1995
Proceedings

Springer

Series Editors

Gerhard Goos
Universität Karlsruhe
Vincenz-Priessnitz-Straße 3, D-76128 Karlsruhe, Germany

Juris Hartmanis
Department of Computer Science, Cornell University
4130 Upson Hall, Ithaca, NY 14853, USA

Jan van Leeuwen
Department of Computer Science, Utrecht University
Padualaan 14, 3584 CH Utrecht, The Netherlands

Volume Editors

Mariangiola Dezani-Ciancaglini
Department of Computer Science, University of Turin
Corso Svizzera, 185, I-10149 Turin, Italy

Gordon Plotkin
Department of Computer Science, University of Edinburgh
The King's Buildings, Mayfield Road, Edinburgh EH9 3JZ, United Kingdom

CR Subject Classification (1991): F.4.1, F.3.0, D.1.1

ISBN 3-540-59048-X Springer-Verlag Berlin Heidelberg New York

CIP data applied for

Typesetting: Camera-ready by author
SPIN: 10485422 45/3140-543210 - Printed on acid-free paper

Preface

This volume is the proceedings of the Second International Conference on Typed Lambda Calculi and Applications, TLCA'95, held in Edinburgh, Scotland, from April 10 to April 12, 1995. There are 29 papers concerning the following topics:

- Proof theory of type systems;
- Logic and type systems;
- Typed λ-calculi as models of (higher–order) computation;
- Semantics of type systems;
- Proof verification via type systems;
- Type systems of programming languages;
- Typed term rewriting systems.

These were selected from a total of 58 submissions, of generally high quality.

We wish to express our gratitude to all the members of the Program Committee, and to the many referees who assisted them. Moreover, we would like to thank the members of the Organizing Committee and all those who submitted papers. Finally, we greatly appreciate the excellent cooperation with Springer-Verlag.

The lambda calculus has served as a source of ideas, problems and applications in computer science for over thirty years. This volume demonstrates its continuing vitality.

Turin, January 1995

Mariangiola Dezani-Ciancaglini Gordon Plotkin

Program Committee

H. Barendregt	(Catholic University of Nijmegen)
M. Dezani	(Chairperson, University of Turin)
J-Y. Girard	(University of Marseilles)
R. Hindley	(University of Swansea)
F. Honsell	(University of Udine)
J. W. Klop	(CWI)
G. Longo	(ENS)
A. Meyer	(MIT)
G. Plotkin	(University of Edinburgh)
P. Scott	(University of Ottawa)
J. Smith	(University of Gothenburg/Chalmers)
J. Tiuryn	(University of Warsaw)

Referees

F. Alessi	T. Altenkirch	G. Amiot
F. Barbanera	E. Barendsen	G. Barthe
L.S. van Benthem Jutting	S. Berardi	C. Berline
M. Bezem	M. Boffa	G. Boudol
R. Blute	F. Cardone	G. Castagna
J.R.B. Cockett	M. Coppo	T. Coquand
A. Compagnoni	R.L. Crole	D. Cubric
P.L. Curien	W. Dekkers	R. Di Cosmo
P. Di Gianantonio	G. Dowek	P. Dybjer
T. Ehrhard	D. Fridlender	H. Geuvers
P. Giannini	G. Ghelli	M. Grabowski
P. de Groote	T. Hurkens	B. Intrigila
B. Jacobs	J.P. Jouannaud	R. Hasegawa
M. Hofmann	D. Kesner	Y. Lafont
M. Lenisa	U. de' Liguoro	P. Lincoln
P. Malacaria	S. Martini	J. Mason
E. Meijer	M. Miculan	E. Moggi
R. Nederpelt	V. van Oostrom	J. Otto
C. Paulin-Mohring	A. Piperno	E. Poll
R. Pollack	F. van Raamsdonk	L. Regnier
J.G. Riecke	S. Ronchi della Rocca	P. Rosolini
L. Roversi	M. Ruys	A. Salibra
A. Schubert	R.A.G. Seeely	R. Statman
P. Urzyczyn	R.C. de Vrijer	B. Werner

Organizing Committee

G. Cleland, P. Gardner, M. Lekuse, G. Plotkin (University of Edinburgh)

Table of Contents

Comparing λ-calculus translations in Sharing Graphs *

Andrea Asperti[1] and Cosimo Laneve[2]

[1] Dip. di Matematica, P.za di Porta S. Donato, 5, 40127 Bologna, Italy.
[2] INRIA Sophia Antipolis, 2004 Route des Lucioles BP 93, 06902 Valbonne, France.

Abstract. Since Lamping's seminal work [Lam90] on optimal graph reduction techniques for the λ-calculus, several different translations based on the same set of control operators (sharing graphs) have been proposed in the literature [GAL92a, GAL92b, AL93a, As94]. In this paper we clarify the correspondence between all these translations, passing through the so called bus-notation [GAL92a]. Indeed all the sharing graph encodings turn out to be equivalent modulo the way of counting bus levels.

1 Introduction

In [Lam90], Lamping proposed a complex graph reduction technique for the λ-calculus that was optimal in the sense of Lévy [Le78]. Lamping's approach was revisited in [GAL92a], where a restricted set of control nodes and reduction rules was proved sufficient for the implementation. In [GAL92b] Gonthier-Abadi-Lévy pointed out a strong analogy between optimal reductions and Linear Logic [Gi86]. In particular, the optimal implementation of λ-calculus is actually a refinement of Girard's proof-net representation. This refinement just provides a *local implementation* of the non-linear operations over shared data (nets into *boxes*), which in proof-nets are performed as a single, global step. By their encoding of Linear Logic in sharing graphs, and some encoding of λ-calculus in Linear Logic, a third (slightly different) translation was implicitly obtained. This encoding has been used (and explicited) in [AL93a], in the much more general case of Interaction Systems.

At the same time of [GAL92b], Asperti proposed yet another and sensibly different translation of λ-calculus into sharing graphs [As94]. This latter translation is strongly connected to the dynamic algebra interpretation of Linear Logic [Re92, DR93] (a relation that has been recently formalized in [ADLR94]).

Up to now, no correspondence has been established in the literature between all these different translations. In this paper we prove that such differences may be explained, up to minor syntactic details, by the way of counting the nesting of boxes into proof-nets. Precisely, the innermost box into a proof-net may be considered as a box at "depth 0" or a box at "depth n", if there are n outer boxes. In order to formalize this intuition, we shall use a further sharing graph implementation of λ-calculus: the so-called *bus-notation* [GAL92a]. The bus notation

* Partially supported by the ESPRIT Basic Research Project 6454 - CONFER.

was introduced with the aim of reducing the number of interaction rules between control nodes. This is achieved by considering edges as set of wires (buses) and an n-indexed control node as affecting the n-th wire only. The above conventions for counting nestings of boxes are reflected into the two ways of counting wires: "from the left" or "from the right". So, if we have $k+1$ wires in the bus, and a node on a wire, this node can be seen as an operator at level n (assuming n wires on the right) or $k-n$ (when counting from the left).

The two different choices underlie respectively the translations in [As94] and [GAL92b], which therefore are mapped onto the same sharing graph in the bus notation. This also explains why exactly the same set of rewriting rules works in both cases.

We remark that this work could have been written in the framework of proof nets [Gi86]. Indeed the results presented here could be transposed to proof nets without any difficulty. We choose to stick to λ-calculus for simplicity sake.

We warn the reader that some general knowledge of the literature (in particular [GAL92a] and [As94]) is a prerequisite for understanding this paper.

2 Sharing Graphs

In [GAL92a] it was proved that Lamping's original set of operators (nodes) needed for the optimal implementation of λ-calculus could be reduced to only five (indexed) *control nodes*, described in the following picture:

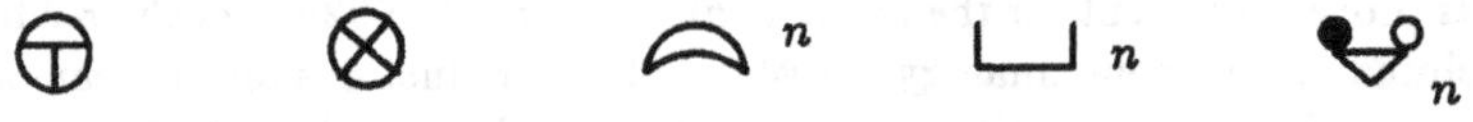

(root) (erasing) (croissant) (bracket) (fan)

Actually, only three of them (fan, croissant and bracket) are really important. The fan node represents sharing (or duplication), while croissant and brackets are needed for the correct interaction of fans during the reduction. In Lamping's original intuition, brackets and croissants must surround fans, implicitly defining their scope. Two different kinds of "brackets" are needed, since they propagate in opposite directions. This is essentially equivalent to consider brackets and croissants as delimiting the extent of the box of Linear Logic (recall that the box is just a sort of global bracket surrounding a datum that can be duplicated or erased) (see [GAL92a, GAL92b]). This relation was furtherly stressed in [As94], where a tight relation between croissant, bracket and the two categorical operations associated with the comonad "!" of Linear Logic was proved.

Erasing nodes are needed in the case of lambda abstractions over variables not appearing in the body of the function; until we do not consider garbage collection rules, these nodes have no operational effect.

In [AL93a] it was shown that all these control nodes could be considered as an abstract set of operators for implementing sharing in virtually every class of higher order rewriting systems. This suggested the name of "sharing graphs", that we shall adopt in the following.

The rules governing the interactions between control operators are drawn in Figure 1. We shall also explicitly consider two "proper nodes", for application and lambda abstraction.

(abstraction) (application)

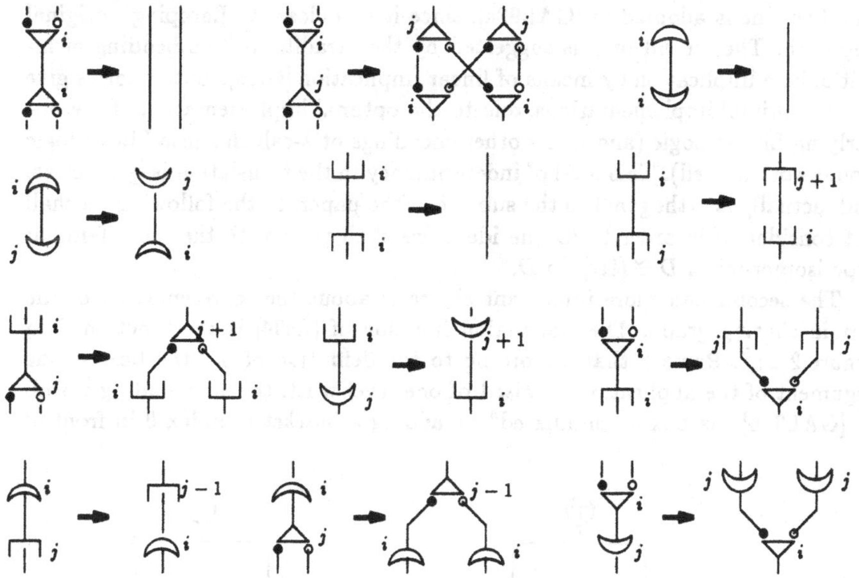

Fig. 1. The control rules ($0 \leq i < j$)

The abstraction and application nodes are *ternary* nodes, i.e. they are connected to three edges. In particular, in the case of abstraction nodes, one edge points to the root, one to the body of the function, and the third one to the bound variable. Thus, bound variables are supposed to be explicitly connected to their binders (an idea that goes back 7to Bourbaki).

As shown in [GAL92a], application and lambda nodes are not actually necessary, since from the operational point of view they can be assimilated to fan. Note in particular that the interaction between an application and an abstraction amounts to connect the root of the application to the root of the body of the function, and the bound variable to the argument of the application.

For the sake of clarity, we prefer to keep the distinction; however we shall not explicitly draw the interaction rules between proper operators and control nodes, that are exactly the same as for fans.

3 The box representation

The somewhat puzzling number of different translations of λ-calculus in sharing graphs is actually due to a double level of indeterminacy. Accepting the fact that the main problem is the optimal encoding of the box of Linear Logic [GAL92b], the first and evident choice is about different encodings of λ-calculus in Linear Logic. Up to now, two possible encodings have been investigated in the literature, respectively based on the type isomorphisms $D \cong !(D \multimap D)$ and $D \cong (!D) \multimap D$. The first one is adopted in [GAL92a], since it was closer to Lamping's original approach. The second one, is suggested by the "traditional" embedding of intuitionistic implication by means of linear implication [Gi86]. Both choices give rise to optimal implementations, due to the optimal implementation of the underlying Linear Logic (and many other encodings of λ-calculus into Linear logic would work as well). This level of indeterminacy in the translation is quite clear, and, actually is orthogonal to the subject of the paper. In the following we shall not consider it: in order to fix the ideas, we shall work with the more familiar type isomorphism $D \cong (!D) \multimap D$.

The second and more important choice is about the representation of the box in sharing graph. The translation function of [As94] is the function $\mathcal{G}$ in Figure 2(a) [3]. Remark that, according to the definition of $\mathcal{G}_n$, the box on the argument of the application is raised of one level w.r.t. the surrounding world. In [GAL92b] the box is "simulated" by adding a bracket of index 0 in front of

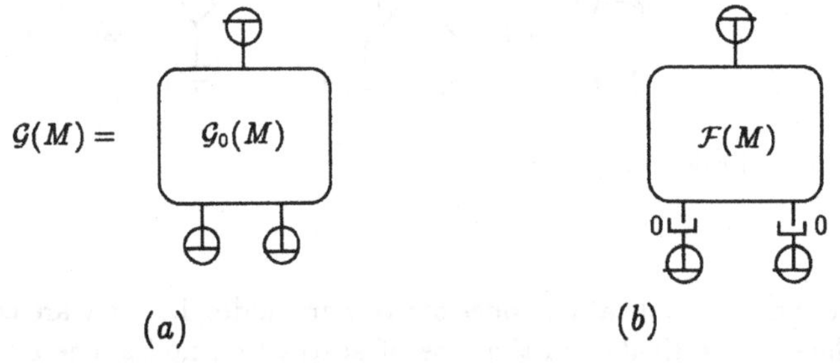

Fig. 2. The translation functions $\mathcal{G}$ and $\mathcal{F}$

its principal port (leaving the box at the same level of the context). This choice is described by the translation function $\mathcal{F}$ of Figure 4.

Note that the operational behaviour is sensibly different in the two cases. In particular, in [As94], the (external) control operators which are propagating inside the box are at a lower level than the (operators inside the) box. The contrary happens in [GAL92b] where the square bracket of index 0 on the principal

[3] Warning: in Figures 3 and 4, when the implementation of $\lambda x\, M$ is defined, we have assumed that $x \in \mathsf{fv}(M)$. The case $x \notin \mathsf{fv}(M)$ follows by connecting an erasing node to the port of the abstraction representing the bound variable.

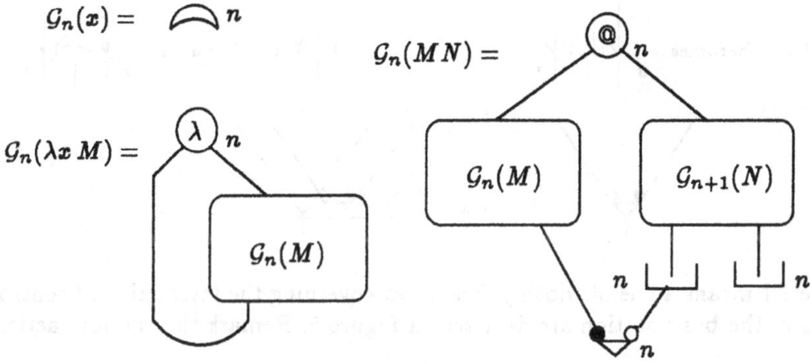

Fig. 3. The implementation of [As94]

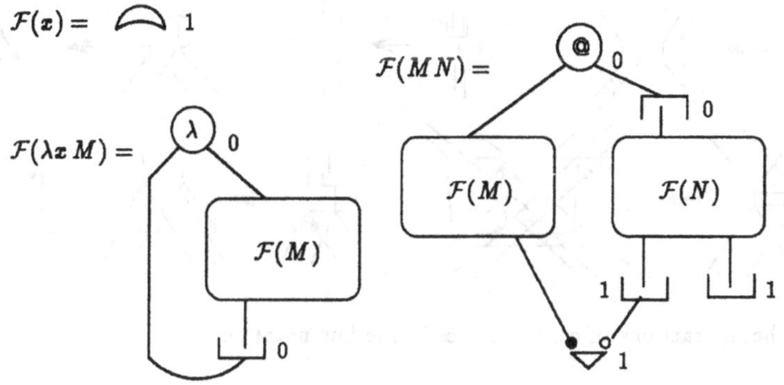

Fig. 4. The implementation of [GAL92b]

port of the box has the effect to raise the level of incoming control operators, avoiding their conflict with inner operators proper to the box. Due to this consideration, it can appear a bit surprising that *exactly the same set of rewriting rules* works in both cases. Moreover, apart the above informal discussion on the idea underlying the two box representations (i.e. their different ways to avoid conflicts), there is a priori no much evidence of a formal correspondence between them.

This is the main problem we shall address in this paper; to this purpose, we need to recall an interesting and odd notation, introduced in [GAL92a]: the *bus notation*.

4 The bus notation

A more explicit graph notation is obtained by interpreting edges in the systems $\mathcal{G}(M)$ and $\mathcal{F}(M)$ as buses, namely set of wires. In this view, an i-indexed node is considered as an operator acting on the wire at depth i. For instance:

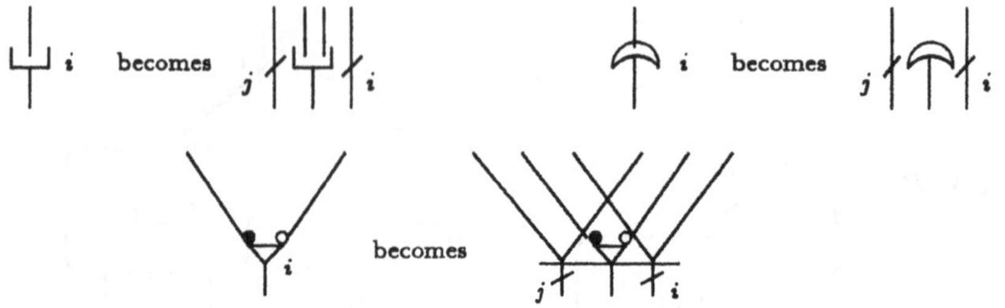

where $_j\dagger$ means a bus of width j. The rules governing the interaction of control nodes in the bus notation are depicted in Figure 5. Remark that no interaction

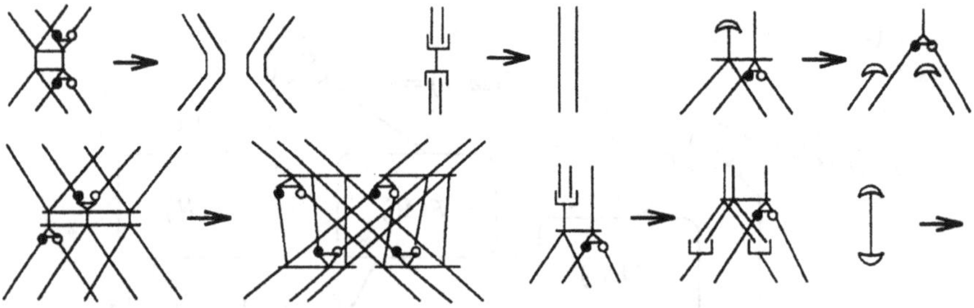

Fig. 5. The interactions of control nodes in the bus notation

between croissants or brackets on different wires is defined. This because the bus notation eliminates such commutations. Nevertheless it is not clear how much it is gained in efficiency with buses, since interactions of fans are no more local but concern every wire of the bus.

5 The bus notation of the translation $\mathcal{F}$

Trying an encoding *ad literam* through buses of the translation $\mathcal{F}$, we immediately realize that we cannot fix a priori the dimension of the bus. This is in contrast with what happens in [GAL92a] where the bus has always dimension 3; we shall come back soon to this point. For the moment, after having fixed some wire "0", we shall suppose to have an infinite numbers of wires at its left. The resulting translation is defined by the function $\mathsf{B}^{\mathcal{F}}$ in Figure 6 (note that $\mathsf{B}^{\mathcal{F}}(M)$ is exactly the bus-counterpart of $\mathcal{F}(M)$). Remark that $\mathsf{B}^{\mathcal{F}}(M)$ uses a *finite number* of wires. This number depends on the syntactical structure of the term; precisely we exactly need an extra-wire at the left every time we have a (nested) application. Alternatively we may add an index to the translation $\mathcal{F}$, mimicking the definition of $\mathcal{G}$. This index expresses now the number of extra-wires required to the left of the term (see Figure 10).

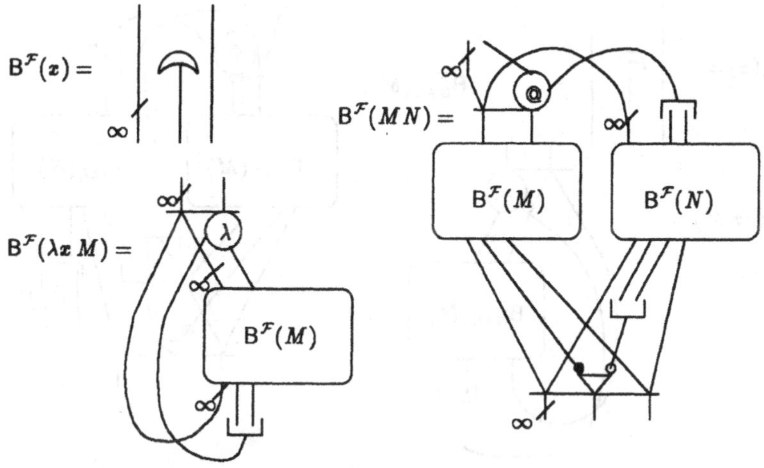

Fig. 6. The bus encoding of the function $\mathcal{F}$

The reader will have probably noted another difference w.r.t. to [GAL92a]: the dimension of the bus at the root of the term is different from its dimension at free variables (in particular, it augments of 1). However, this is merely a consequence of the different encoding of λ-calculus in Linear Logic and we shall not furtherly discuss this point.

It is possible to avoid the infinite (or, in any case, growing) number of wires needed during the translation by introducing suitable operators (square brackets) on the leftmost wire. The resulting translation $\mathsf{B}^{\mathcal{F}}_{\mathrm{GAL}}$ is described in Figure 7. Remark that, in $\mathsf{B}^{\mathcal{F}}_{\mathrm{GAL}}$, the bus has always dimension two at the root of a (sub) term, and 3 at its free variables. This is actually the way followed in [GAL92a]. The main difference with [GAL92a] is in the rule for the application, since $\mathsf{B}^{\mathcal{F}}_{\mathrm{GAL}}$ puts a box in the second argument of the application, whilst in [GAL92b] it was on the rule for abstraction. The purpose of the two brackets added on the leftmost wire is exactly that of "absorbing" and "recreating" the extra-wire required by the box.

The correspondence between $\mathsf{B}^{\mathcal{F}}$ and $\mathsf{B}^{\mathcal{F}}_{\mathrm{GAL}}$ is established by Theorem 10 below. Few preliminary properties about $\mathsf{B}^{\mathcal{F}}_{\mathrm{GAL}}$ are required. We say that a node is *at level n* if, once fixed the direction for counting wires, the node is located on the n-th wire.

Proposition 1. *Let us fix a right-to-left numbering of wires. In the translation* $\mathsf{B}^{\mathcal{F}}_{\mathrm{GAL}}$*, the nodes located on the leftmost wire never change the level of other nodes through interactions.*

This is an immediate consequence of the shape of the rules in Figure 5.

Proposition 2. *Let G be a graph obtained by evaluating λ-terms translated according to* $\mathsf{B}^{\mathcal{F}}_{\mathrm{GAL}}$. *Only brackets may appear on the leftmost wires of G. Moreover such nodes will always remain on the leftmost wire till they disappear.*

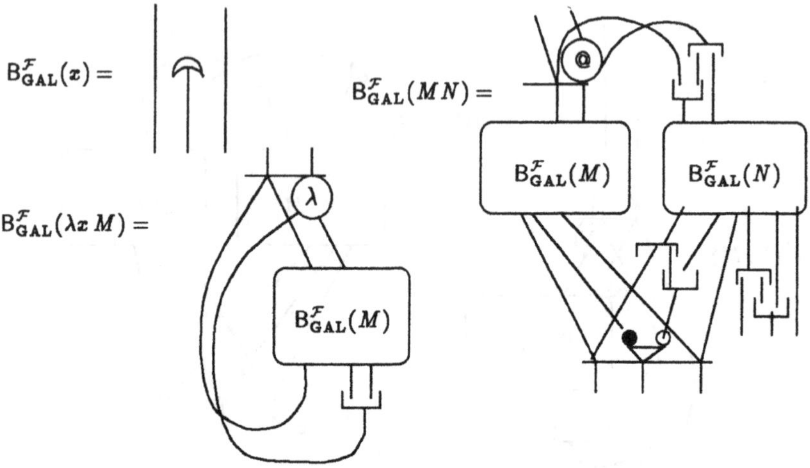

Fig. 7. The translation function $B^{\mathcal{F}}_{GAL}$

Also this property is immediate because it is true initially, and it is obviously preserved by the reduction rules in Figure 5.

Let a *deadlock* be two nodes, one in front of the other, that do not interact, namely no rule of Figure 5 applies. Remark that Proposition 2 guarantees the absence of deadlocks on the leftmost wires of $B^{\mathcal{F}}_{GAL}(M)$ or graphs obtained by its evaluation. Indeed, the nodes on leftmost wires are brackets and two brackets, one in front of the other, can be simplified by the second rule of Figure 5.

Definition 3. The graphs obtained by evaluating $B^{\mathcal{F}}(M)$ and $B^{\mathcal{F}}_{GAL}(M)$ are related by the following function $\imath$:

- $\imath : B^{\mathcal{F}}(M) \rightarrow B^{\mathcal{F}}_{GAL}(M)$ is defined by structural induction. It is immediate when M is a variable or an abstraction. When $M = M'M''$, let $\imath'$ and $\imath''$ the mappings on the graphs for M' and M''. Let us define $\imath$ for the nodes and the wires which aren't in the domains of $\imath'$ and $\imath''$. The application node, the control nodes at level 0 or 1 and the wires at level 0 of $B^{\mathcal{F}}(M)$ are mapped into the corresponding nodes of $B^{\mathcal{F}}_{GAL}(M)$. The wire at level k, $k \geq 1$, of the bus departing from the second argument of the application node of $B^{\mathcal{F}}(M)$ is mapped onto a path of $B^{\mathcal{F}}_{GAL}(M)$ traversing the bracket at level 1. The wires at level 2 representing free variables of $B^{\mathcal{F}}(M'')$ are mapped onto the path traversing the bracket and outgoing from the right-branch. Those at level k, $k \geq 3$ are mapped onto the path traversing the bracket and outgoing the left-branch;
- Suppose to have established a correspondence $\imath$ between a graph G obtained by evaluating $B^{\mathcal{F}}(M)$ and a graph G' resulting from $B^{\mathcal{F}}_{GAL}(M)$ and let $G' \Rightarrow F'$. If this reduction involves a bracket on the leftmost wire, then the same correspondence $\imath$ relates G and F'. Otherwise the redex contracting $G' \Rightarrow F'$ has a counter-image in G (namely it is a redex of the same type). Let $G \Rightarrow F$

be the contraction of such a redex. The mapping $\imath'$ between F and F' is defined by changing $\imath$ according to the redex fired.

Lemma 4. *The correspondence $\imath$ of Definition 3 between the evaluations of $\mathsf{B}^{\mathcal{F}}(M)$ and of $\mathsf{B}^{\mathcal{F}}_{\mathrm{GAL}}(M)$ is well-defined.*

PROOF: Suppose to have a correspondence $\imath$ between G and G' obtained from $\mathsf{B}^{\mathcal{F}}(M)$ and $\mathsf{B}^{\mathcal{F}}_{\mathrm{GAL}}(M)$, respectively, and such that there is a bijection between redexes not involving brackets on the leftmost wires of G'.

Therefore a rewriting not involving the leftmost wire may be performed both in G and G'. A rewriting, let us say r, involving the leftmost wire concerns only G'. But r does not modify $\imath$ for the following reasons:

1. r does not change the level of nodes located on wires different from the leftmost (by Proposition 1);
2. r does not duplicate the nodes on wires different from the leftmost (since, by Proposition 2, there are only brackets on leftmost wires). ∎

Remark. We recall that, in sharing graphs, brackets and croissants are used only with the purpose to guarantee the correct matching of fans and abstractions/applications (namely guaranteeing their correct interaction). Observe that Lemma 4 implies that such matchings are performed in the same way both in $\mathsf{B}^{\mathcal{F}}$ and $\mathsf{B}^{\mathcal{F}}_{\mathrm{GAL}}$.

An immediate consequence of the foregoing lemma is stated by the following corollary.

Corollary 5. *The implementations $\mathsf{B}^{\mathcal{F}}(M)$ and $\mathsf{B}^{\mathcal{F}}_{\mathrm{GAL}}(M)$ always reduce interactions of application-abstraction nodes and between fans which are in corresponding positions of the graphs, where the correspondence is fixed by the function $\imath$ of Definition 3.*

Therefore the interactions of fans and applications/abstractions is completely unsensible to the removal of leftmost brackets in $\mathsf{B}^{\mathcal{F}}_{\mathrm{GAL}}(M)$. Corollary 5 fixes a syntactic correspondence between $\mathsf{B}^{\mathcal{F}}(M)$ and $\mathsf{B}^{\mathcal{F}}_{\mathrm{GAL}}(M)$ which is quite strong, since it ensures that the amount of sharing performed by fans is the same both in $\mathsf{B}^{\mathcal{F}}(M)$ and $\mathsf{B}^{\mathcal{F}}_{\mathrm{GAL}}(M)$. Below we show that this correspondence is also semantics, where with "semantics" we mean the interpretation of a sharing graph G, namely the λ-term associated to G. To this aim we recall from [GAL92a] the notion of *consistent path*, which is the basic brick for the semantics of sharing graphs.

Definition 6. Let a *context* be a term generated by the following grammar:

$$a \ ::= \ \Box \mid \circ \cdot a \mid \bullet \cdot a \mid \langle a, b \rangle$$

The semantics of control nodes is given by relations on sequences of contexts as described by the following picture:
The semantics of abstraction and application nodes may be defined in terms of fans, by using the equations in Figure 9.

Let φ be a path into a sharing graph (with the bus notation). We say that φ is *consistent* if

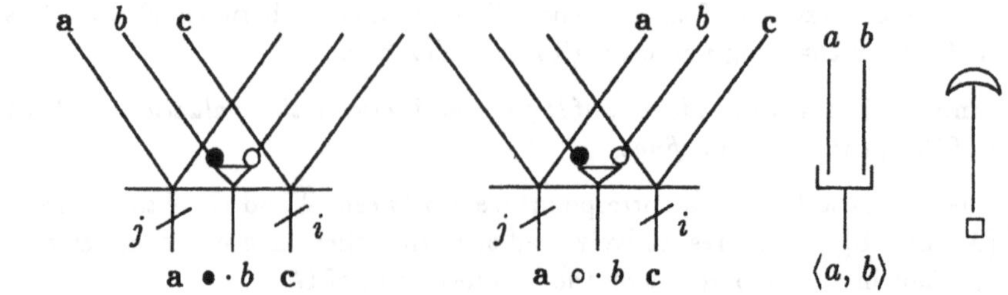

Fig. 8. The semantics of control nodes

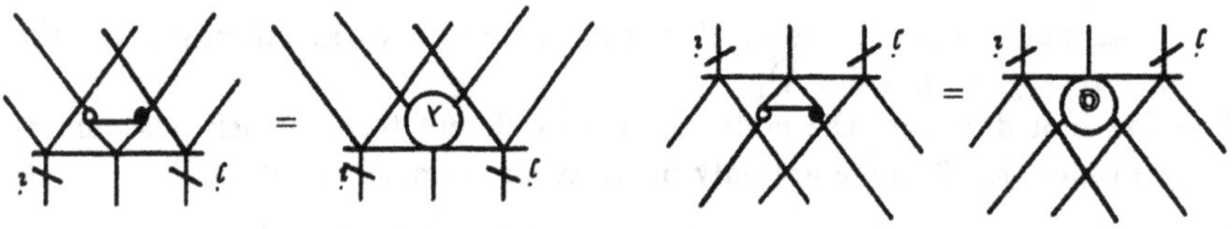

Fig. 9. The interpretation of application and abstraction nodes in terms of fans

1. every (edge of every) wire of φ is labeled by a context;
2. every pair of consecutive wires satisfy the constraints in Figure 8.

Definition 7. Let G be a sharing graph (in the bus notation). Let R_{ij} be the relation on pairs $(\overline{d}, \overline{d'})$, where $\overline{d}$ and $\overline{d'}$ are sequence of contexts, and $(\overline{d}, \overline{d'}) \in R_{ij}$ if and only if there exists a consistent path φ connecting the *root nodes* i and j and such that the context at i is $\overline{d}$ and the context at j is $\overline{d'}$. The *context semantics* of G, denoted $\mathcal{C}(G)$, is the set of relations R_{ij}, where i and j are root nodes.

Theorem 8. *The context semantics of* $\mathsf{B}^{\mathcal{F}}(M)$ *and* $\mathsf{B}^{\mathcal{F}}_{\mathrm{GAL}}(M)$ *is invariant by sharing graph reductions. Namely, let* G *be a sharing graph obtained by evaluating* $\mathsf{B}^{\mathcal{F}}(M)$ *or* $\mathsf{B}^{\mathcal{F}}_{\mathrm{GAL}}(M)$*, if* $G \Rightarrow G'$ *then* $\mathcal{C}(G) = \mathcal{C}(G')$.

PROOF: Because the rewriting rules of Figure 5 do not change the initial and final contexts of consistent paths traversing the redexes. ■

In [GAL92a] the semantics of a sharing graph is defined by reading-back the Böhm-tree of a λ-term. This tree is obtained by taking only the consistent paths root-to-root [4]. The following lemma states that these paths are the same both in $\mathsf{B}^{\mathcal{F}}(M)$ and $\mathsf{B}^{\mathcal{F}}_{\mathrm{GAL}}(M)$, up-to some brackets on leftmost wires of paths in $\mathsf{B}^{\mathcal{F}}_{\mathrm{GAL}}(M)$, which nevertheless do not affect consistency.

---

[4] In the more general setting of proof-nets, we may take the *execution formula* as the semantics of a proof net. This is obtained by the consistent paths root-to-root (see [DR93, ADLR94]).

Lemma 9. *For every root-to-root consistent path in* $\mathsf{B}^{\mathcal{F}}_{\mathrm{GAL}}(M)$, *the corresponding path in* $\mathsf{B}^{\mathcal{F}}(M)$ *– where the "correspondence" is fixed by* $\imath$ *in Definition 3 – is root-to-root and consistent, and* vice versa.

Proof: Let φ be a consistent path in $\mathsf{B}^{\mathcal{F}}_{\mathrm{GAL}}(M)$ and ψ be its counter-image according to the embedding of Lemma 4. Remark that we are considering *every* consistent path, not only those starting and finishing at root-nodes. The case of root-to-root consistent paths follows easily. We assume that, for every section of the bus in $\mathsf{B}^{\mathcal{F}}_{\mathrm{GAL}}(M)$, the corresponding section in $\mathsf{B}^{\mathcal{F}}(M)$ has a strictly larger width. This may be easily obtained by taking enough wires "on the left" in the encoding $\mathsf{B}^{\mathcal{F}}(M)$.

We prove, by induction on the length of ψ, that φ is consistent. In particular, if the width of φ at a given section is k and the width of ψ at the corresponding section is n ($n \geq k$) then

1. the context of the i-th wire, $0 \leq i < k-1$ is the same both in φ and ψ;
2. the context of φ of the $(k-1)$-th wire is $\langle \cdots \langle a_{n-1}, a_{n-2}\rangle \cdots a_{k-1}\rangle$ and the context of the i-th wire of ψ, $k-1 \leq i \leq n-1$, is a_i.

Remark: since we are taking paths φ which are images, according to $\imath$, of paths ψ in $\mathsf{B}^{\mathcal{F}}(M)$, φ never terminates at a bracket on the leftmost wire. The basic case has two subcases: (a) when φ is a bus in between two nodes which aren't brackets on the leftmost wires (and φ does not traverse any other node) and (b) when φ traverses exactly one bracket on the leftmost wire. These two cases are easy and left to the reader.

When $\psi = \psi' \cdot u$, let φ' be the (consistent) path corresponding to ψ' and w be the image of u. If w is a single edge (it does not traverse a bracket on the leftmost wire) u terminates at a node which is not on the leftmost wire, the statement of the induction is easy. Indeed the decision how to lengthen φ' depends on the context at the wire where it is placed the final node of φ' (which cannot be a bracket on the leftmost wire, by definition of $\imath$ – see Definition 3 –).

Otherwise $w = w' \cdot w''$ and let b be the bracket on the leftmost wire traversed by w. There are two subcases:

1. w' enters from the principal port of the bracket. Since the width of the bus u is greater than the corresponding width of w lengthening ψ', the context labeling the leftmost edge of w', let us say the k-th, must have the shape $\langle a, b\rangle$. Hence, by hypothesis, b must mark the k-th wire of u and a stores the contexts of u at wires $k+1$, $k+2$, etc. It is clear that $\varphi' \cdot w'$ may be consistently lengthened with w''.
2. w' enters from an auxiliary port of the bracket. This case is an easy consequence of the inductive hypothesis.

The *vice versa* may be proved in the same way. ■

Theorem 10. *The implementations* $\mathsf{B}^{\mathcal{F}}_{\mathrm{GAL}}(M)$ *and* $\mathsf{B}^{\mathcal{F}}$ *are syntactically and semantically equivalent. This equivalence is up-to brackets on leftmost wires, which are inessential.*

Proof: immediate consequence of Lemma 4 and Lemma 9. ■

6 The correspondence of $\mathcal{F}$ and $\mathcal{G}$

The correspondence between $\mathcal{F}$ and $\mathcal{G}$ will be proved by reasoning again on the bus-notation. The main idea underlying our proof is: ***instead of counting levels from right to left, we shall count them from left to right.*** The application of this criterion to the function $\mathsf{B}^{\mathcal{F}}$ gives $\mathsf{B}^{\mathcal{F},r}$, which is described in Figure 10 (we assume $\mathsf{B}^{\mathcal{F},r}(M) = \mathsf{B}_0^{\mathcal{F},r}(M)$).

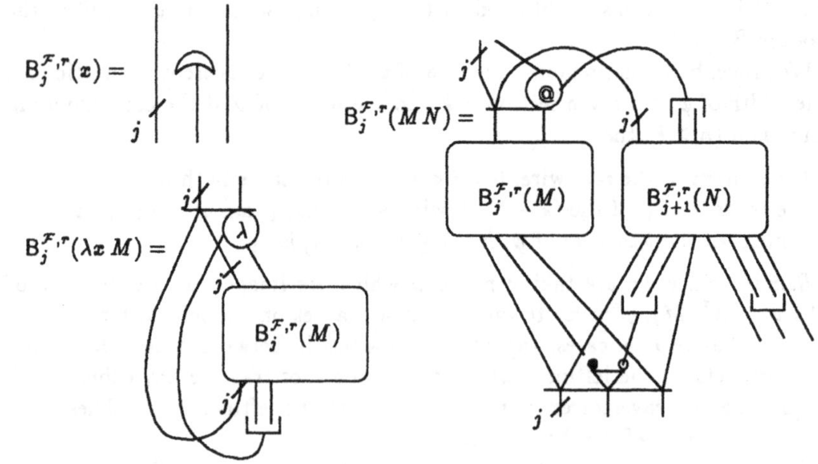

Fig. 10. The translation $\mathsf{B}^{\mathcal{F},r}$

On the other hand, the bus interpretation of $\mathcal{G}$, namely $\mathsf{B}^{\mathcal{G}}$, is illustrated in Figure 11 (as usual $\mathsf{B}^{\mathcal{G}}(M) = \mathsf{B}_0^{\mathcal{G}}(M)$). Observe that, in the definition of $\mathsf{B}^{\mathcal{G}}$, wires are actually counted from the left to the right. We mean that, as far as buses are concerned, the translation $\mathcal{G}$ implicitly uses a dual criterion w.r.t. $\mathcal{F}$ for counting levels.

Remark also that $\mathsf{B}^{\mathcal{F},r}$ differs from $\mathsf{B}^{\mathcal{G}}$ only for the presence of some extra-brackets at the root of the argument of an application and on bound variables. The relation between $\mathsf{B}^{\mathcal{F},r}$ and $\mathsf{B}^{\mathcal{G}}$ may be fixed in a way similar to the correspondence fixed in Section 5. By mimicking Definition 3, it is possible to define a mapping $\jmath$ from the evaluation of $\mathsf{B}^{\mathcal{G}}(M)$ to the evaluation of $\mathsf{B}^{\mathcal{F},r}(M)$ which forgets about the brackets on the rightmost wires of the implementation $\mathsf{B}^{\mathcal{F},r}$ and of their interaction (we leave to the reader the formalization of the details).

Lemma 11. *The above correspondence $\jmath$ between the evaluations of $\mathsf{B}^{\mathcal{G}}(M)$ and of $\mathsf{B}^{\mathcal{F},r}(M)$ is well-defined.*

Establishing the above lemma is less obvious than for Lemma 4, because of the presence of applications and abstractions on rightmost wires (in the situation of Lemma 4, there were only brackets on leftmost wires). In particular, what

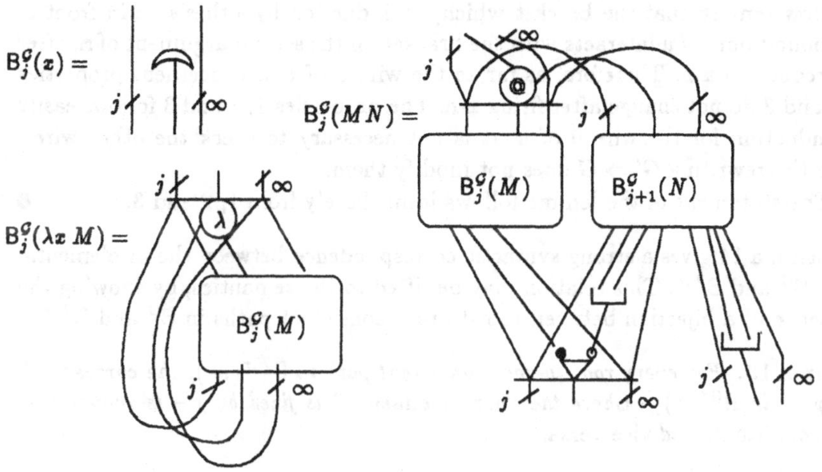

Fig. 11. The bus interpretation of the translation $\mathcal{G}$

could invalidate the lemma is the eventual presence of a deadlock between a bracket and an abstraction (or application) node on the rightmost wire. Such a configuration is forbidden by the following proposition.

Proposition 12. ***Never a bracket is in front of an abstraction or an application on the rightmost wires of graphs obtained evaluating*** $\mathsf{B}^{\mathcal{F},r}(M)$.

Remark. This property has been already proved for the general setting of Interaction Systems (see Lemma 8.13 in [AL94]).

PROOF: We prove by induction on the length of the derivation $\mathsf{B}^{\mathcal{F},r}(M) \Rightarrow^* G$ that

1. on the rightmost wires there are only abstraction, applications or brackets;
2. there is always a bracket in front of a bound port of an abstraction or the second argument of an application;
3. if on a 0-level wire there is a sequence of brackets, such a sequence may be simplified to a single bracket (by firing instances of the second rule of Figure 5) which is in front of a bound port of an abstraction or the second argument of an application.

Items 1, 2 and 3 are clearly true for $\mathsf{B}^{\mathcal{F},r}(M)$. Let us check the inductive step. That is $\mathsf{B}^{\mathcal{F},r}(M) \Rightarrow^* G' \Rightarrow G$. The interesting case is when $G' \Rightarrow G$ is due to the interaction of an application $\mathtt{m}$ and an abstraction $\mathtt{n}$. Then, according to the definition of the rewriting rule, the bus entering on the output port of $\mathtt{m}$ is connected with the body of $\mathtt{n}$ and the bus entering into the bound port of $\mathtt{n}$ is connected with the second argument of $\mathtt{m}$. Let u and w be these new connections, respectively.

Now remark that the bracket which, by induction hypothesis, is in front of the bound port of n interacts with the bracket on the second argument of m, after the reduction m-n. Therefore, as far as the wire 0 of w is concerned, properties 1, 2 and 3 do not change after firing m-n. The properties 1, 2 and 3 follows easily by induction for the wire 0 of u. It is not necessary to check the other wires, since the rewriting $G' \Rightarrow G$ does not modify them.

The statement of the lemma follows immediately from 1, 2 and 3. ∎

Lemma 14 gives a strong syntactic correspondence between the implementations $B^{\mathcal{G}}$ and $B^{\mathcal{F},r}$. This relation may be lifted to the semantics, by showing the existence of a bijection between root-to-root consistent paths in $B^{\mathcal{G}}$ and $B^{\mathcal{F},r}$.

Lemma 13. *For every root-to-root consistent path in* $B^{\mathcal{F},r}(M)$*, the corresponding path in* $B^{\mathcal{G}}(M)$ *– where the "correspondence" is fixed by* $\jmath$ *– is root-to-root and consistent, and* vice versa.

The proof of this lemma is omitted, in order to meet the space requirements. Anyway the proof is similar to the one of Lemma 9, apart some difficulties due to the presence of abstractions and applications on rightmost wires. These problems are accommodated by means of Proposition 12.

Lemma 11 and Lemma 13 imply:

Theorem 14. *The implementations* $B^{\mathcal{G}}$ *and* $B^{\mathcal{F},r}$ *are syntactically and semantically equivalent. This equivalence is up-to brackets on rightmost wires, which are inessential.*

Remark. The above correspondence and the bus notation definitely clarify that the propagation of a control operator inside a box is the same both for $\mathcal{F}$ and for $\mathcal{G}$. The criterion adopted for counting wires is the unique reason for the operators that are propagating inside a box are at a lower level than the (nodes of) the box in $\mathcal{G}$, whilst it is the contrary in $\mathcal{F}$.

7 Conclusions

In this paper we have shown the correspondence between several different translations of λ-calculus into sharing graphs that have been proposed in the literature. We did it by passing through the so called *bus notation* [GAL92a], and showing that, apart from some syntactical details, the main difference is in the different way of reading levels of wires in the bus: from left-to-right, or from right-to-left.

Another source of difference of the translations is due to different encodings of λ-calculus into Linear Logic, based on different type isomorphisms (this topic looks quite clear, and it has not been really addressed in this paper). Then, the several translations can be essentially classified as follows.

By this work, we get some evidence in favour of Asperti's translation, since it uses the minimal number of operators. However, it seems possible that the additional information provided by the extra-operators of other translations could

isomorphism	right-to-left	bus	left-to-right
$D \cong (!D) \multimap D$	[AL93a]	this paper	[As94]
$D \cong !(D \multimap D)$	[GAL92b]	[GAL92a]	

Fig. 12. Classification of translations

be of some theoretical and even practical use (for instance in trying to solve the well known and crucial problem of accumulation of control operators).

References

[As94] A. Asperti. *Linear Logic, Comonads, and Optimal Reductions.* To appear in Fundamenta Informaticae, Special Issue devoted to Categories in Computer Science. 1994.

[AL93a] A. Asperti, C. Laneve. *Optimal Reductions in Interaction Systems.* Proc. of the 4th Joint Conference on the Theory and Practice of Software Development, TAPSOFT'93, Orsay (France). April 1993.

[AL94] A. Asperti, C. Laneve. *Interaction System 2: The practice of optimal reductions.* May 1993. A revised version of this paper may be got by anonymous ftp at cma.cma.fr as file pub/papers/cosimo/newIS2.ps.Z.

[ADLR94] A. Asperti, V. Danos, C. Laneve, L. Regnier . *Paths in the λ-calculus: three years of communications without understandings.* Proc. of the 9th Annual Symposium on Logic in Computer Science (LICS 94), Paris 1994.

[DR93] V. Danos, L. Regnier. *Local and asynchronous beta-reduction.* Proc. of the 8th Annual Symposium on Logic in Computer Science (LICS 93), Montreal. 1993.

[Gi86] J. Y. Girard. *Linear Logic.* Theoretical Computer Science, 50. 1986.

[GAL92a] G. Gonthier, M. Abadi, J.J. Lévy. *The geometry of optimal lambda reduction.* Proc. of the 19th Symposium on Principles of Programming Languages (POPL 92). 1992.

[GAL92b] G. Gonthier, M. Abadi, J.J. Lévy. *Linear Logic without boxes.* Proc. of the 7th Annual Symposium on Logic in Computer Science (LICS'92). 1992.

[Lam90] J. Lamping. *An algorithm for optimal lambda calculus reductions.* Proc. of the 17th Symposium on Principles of Programming Languages (POPL 90). San Francisco. 1990.

[Le78] J.J.Levy. *Réductions correctes et optimales dans le lambda-calcul.* Thèse de doctorat d'état, Université de Paris VII. 1978.

[Re92] L. Regnier. *Lambda Calcul et Réseaux.* Thèse de doctorat, Université Paris VII. 1992.

Extensions of Pure Type Systems

Gilles Barthe

Faculty of Mathematics and Informatics
University of Nijmegen, The Netherlands
email: gillesb@cs.kun.nl

Abstract. *We extend pure type systems with quotient types and subset types and establish an equivalence between four strong normalisation problems: subset types, quotient types, definitions and the so-called K-rules. As a corollary, we get strong normalisation of ECC with definitions, subset and quotient types.*

1 Introduction

The theory of pure type systems provide a general framework for the introduction and study of a wide class of typed λ-calculi ([1, 7]). Pure type systems are minimal calculi, in the sense that they only have one type constructor, the dependent product Π. Hence the use of pure type systems for software or program verification on the one hand and formal mathematics on the other is limited: pure type systems have to be used as frameworks in the spirit of the AUTOMATH project ([16]). For example, one cannot define in a pure type system the type of groups; instead one must work in a context corresponding to the type of groups. If one wants to use type theory in the same way as naive set theory, one must seek for richer, more expressive type systems in which standard mathematical constructions can be performed. These type systems should be expressive enough to allow for

(i) the definition of mathematical structures such as groups or vector spaces;
(ii) manipulations on those mathematical structures; in the case of groups, such manipulations include formation of products, substructures, quotients ...

One can address the problem of extending pure type systems in two different ways:

- *the minimal way.* One can seek for the weakest extension of pure type systems in which the requirements (i)-(ii) are met. This approach attempts to restrict the number of type constructors to be introduced by providing suitable encodings for them. A typical example is *ECC* ([14]): there are only two type constructors, the dependent product Π and the dependent sum Σ. There are no constructor to form subtypes or quotient types; instead, sets are encoded as setoids for which subsetoids and quotient setoids can be constructed.
- *the liberal way.* One tries to get as close as possible to a typed naive set theory by introducing a new type constructor for each standard mathematical construction. The intention is to avoid all unecessary encoding and thus simplify

the task of formalising mathematics or verifying programs. Two particularly important type constructors besides Σ-types are subset and quotients, for which various type-theoretical formulations can be found in the literature (e.g. [5, 9, 10, 12, 15] for quotient types and [5, 11, 15, 17, 20] for subset types).

This paper introduces pure type systems with quotient types and subset types and studies their proof-theoretical properties. We show that for a large class of pure type systems (including the Calculus of Constructions and ECC[1]), strong normalisation is preserved by these extensions. The result is in fact obtained as a corollary of the

Main Theorem. *Let λS be a hol system*[2]*. Then the following are equivalent:*

(i) the extension λS_δ of λS with definitions[3] *is strongly normalising,*
(ii) the extension λS_χ of λS with quotient types is strongly normalising,
(iii) the extension λS_σ of λS with subset types is strongly normalising,
(iv) the extension λS_κ of λS with K-rules[4] *is strongly normalising.*

The proof is purely combinatorial and uses reduction-preserving translations between the different type systems.

The paper is organised as follows: in section 2, we introduce a syntax for subset and quotient types. In section 3, we prove the main theorem. In section 4, we discuss in more detail the syntax and meta-theoretical properties of subset and quotient types. Section 2 has been kept brief quite deliberately: subset types and quotient types are very useful constructions and one cannot hope to cover the range of their applications in a few words. We refer the reader to the above mentioned literature for an introduction to subset and quotient types and some examples of their applications in the formalisation of mathematics. Our syntax for quotient types generalises the one of [12], which provides a particularly relevant introduction to the material on quotient types (including several examples).
Terminology As usual, $\rightarrow$ is used to denote one-step reduction. $\twoheadrightarrow^+$, $\twoheadrightarrow$ and $=$ respectively denote the transitive, reflexive-transitive and reflexive-symmetric-transitive closures of $\rightarrow$.

2 Extensions of pure type systems

The framework of pure type systems is extremely useful because of its generality and conceptual simplicity ([1, 7]). Yet pure type systems do not have enough structure to model standard mathematical constructions and mathematics have to be formalised in suitable extensions of pure type systems. Here we consider

[1] Although ECC is not a pure type system, our proof will apply.
[2] That is a pure type system with a distinguished sort of propositions and enough rules to encode logic, equality and relations.
[3] Pure type systems with definitions have been introduced and studied in [18].
[4] λS_κ is essentially an extension of λS with a K-combinator.

extensions of pure type systems with subset and quotient types (strong sums are treated in the appendix). We restrict ourselves to those pure type systems with a logic and definable relations, for which subset and quotient types are well understood.

Definition 1 *A pure type system* $\lambda S = (S, A, R)$ *is a* higher order logical system *(hol for short) if it has two distinguished sorts* $*$ *and* $\Box$ *such that*

(i) $* : \Box \in A$,
(ii) $\forall s \in S.(s, *, *) \in R$,
(iii) $\forall s \in S.\exists s', s'' \in S.(s, \Box, s'), (s, s', s'') \in R$.

(ii) enables us to define universal quantification and logic whereas (iii) makes it possible to define for every type A Leibniz equality $=_A$ and a type $A \to A \to *$ of binary relations.
Remark. Examples of hol systems include the Calculus of Constructions λC and its extension λC^∞ with an infinite hierarchy of universes. However, Definition 1 is not the most adequate definition of a higher-order logical system: for example, $\lambda PRED\omega$ is not a hol system. We chose this definition for its simplicity; see [4] for a more appropriate definition of a hol system (all the proofs and results can be adapted to this alternative definition).

2.1 Pure type systems with subsets

In a pure type system, it is not possible to form subsets: if A is a type and $P : A \to *$, we are not able to form the type of elements a of A such that Pa. We extend pure type systems with a new construct to build subsets. Several proposals to define subset types can be found in the literature ([5, 11, 15, 17, 20]). They can be classified into two categories: weak subset types and strong subset types. In the latter case, $a : \{x : A|B\}$ is taken as an evidence that $B[a/x]$. This is the case of the rules of ([5, 11]). Weak subset types do not take $a : \{x : A|B\}$ as a proof of $B[a/x]$ (see [15, 20]). Here, we focus on strong subset types, as they seem to be more appropriate for the formalisation of mathematics.

Given a hol system λS, we show how to extend it to a hol system λS_σ with subset types. The pseudo-terms are extended with the constructions $\{V : T|T\}$, $\mathsf{in}_T\ T$, $\mathsf{out}\ T$ and $\mathsf{proof}\ T$. New reduction rules are added:

$$\mathsf{in}_{\{x:A|B\}}(\mathsf{out}\ t) \to_\sigma t$$
$$\mathsf{out}\ (\mathsf{in}_{\{x:A|B\}} t) \to_\sigma t$$

Note that x is bound in $\{x : A|B\}$. The definition of substitution is extended in the obvious way.

σ-reduction is both Church-Rosser and strongly normalising. β-reduction is also Church-Rosser on the extended set of pseudo-terms and commutes with σ-reduction. By the Hindley-Rosen lemma, it follows that $\beta\sigma$-reduction is Church-Rosser.

The notion of derivation is defined by the axioms and rules for PTSs and the following rules:

Subset types

$$\frac{\Gamma \vdash A : s \quad \Gamma, x : A \vdash B : *}{\Gamma \vdash \{x : A|B\} : s} \quad \text{Formation}$$

$$\frac{\Gamma \vdash t : A \quad \Gamma \vdash p : B[t/x] \quad \Gamma \vdash \{x : A|B\} : s}{\Gamma \vdash \mathsf{in}_{\{x:A|B\}}\ t : \{x : A|B\}} \quad \text{Introduction}$$

$$\frac{\Gamma \vdash t : \{x : A|B\}}{\Gamma \vdash \mathsf{out}\ t : A} \quad \text{Elimination}$$

$$\frac{\Gamma \vdash t : \{x : A|B\}}{\Gamma \vdash \mathsf{proof}\ t : B[\mathsf{out}\ t/x]} \quad \text{Proof}$$

$$\frac{\Gamma \vdash t : A \quad \Gamma \vdash A' : s \quad A =_\sigma A'}{\Gamma \vdash t : A'} \quad \sigma\text{-conversion}$$

Remarks.

- In some syntaxes ([5, 17, 20]), one has $a : A \wedge B[a/x] \Rightarrow a : \{x : A|B\}$. Our syntax is loaded to preserve unicity of typing in functional type systems.
- The introduction rule is rather unusual for type theory, because information is lost when applying it, as the proof in the second premise disappears (see subsection 4.1 for some meta-theoretical consequences of this rule). The proof rule reflects our view that $a : \{x : A.B\}$ is a proof of $B[\mathsf{out}\ a/x]$.
- We could have made the rules slightly more general by letting the possibility for $\{x : A|B\}$ to live in a different sort than A.

2.2 Pure type systems with quotients

Another limitation of pure type systems is the absence of construct to form quotients. One can introduce a new construct to form the quotient of a type A by a binary relation on A. It has been done in several ways in the literature ([5, 9, 10, 12, 15]). Our syntax is inspired from [12], where Jacobs gives a syntax for quotient types in a simple type theory. We show how to extend a hol system λS into a hol system with quotients λS_χ. First, extend the pseudo-terms with the constructions T/T, $[T]_T$, $\mathsf{identify}\ T$, $\mathsf{lift}\ T$ and $\mathsf{pick}_T\ V : T\ \mathsf{from}\ T\ \mathsf{in}\ T$.Note that x is bound in $\mathsf{pick}_p\ x : A\ \mathsf{from}\ M\ \mathsf{in}\ N$. The definition of substitution is extended in the obvious way. Second, add the reduction rules

$$\mathsf{pick}_p\ x : A\ \mathsf{from}\ [M]_R\ \mathsf{in}\ N \rightarrow_\chi N[M/x]$$
$$\mathsf{pick}_p\ x : A\ \mathsf{from}\ M\ \mathsf{in}\ N[[x]_R/w] \rightarrow_\chi N[M/w]$$

In the second rule, it is assumed that x is not free in N. Note that the second rule is an η-like rule. There are no reduction rules for lift and identify, whose meaning is purely logical.

The notion of derivation is extended with the following rules:

Quotient types	
$$\frac{\Gamma \vdash A : s \quad \Gamma \vdash R : A \to A \to *}{\Gamma \vdash A/R : s}$$	Formation
$$\frac{\Gamma \vdash a : A \quad \Gamma \vdash A/R : s}{\Gamma \vdash [a]_R : A/R}$$	Introduction
$$\frac{\Gamma \vdash p : R\ a\ a' \quad \Gamma \vdash A/R : s}{\Gamma \vdash \mathsf{identify}\ p : [a]_R =_{A/R} [a']_R}$$	Identify
$$\frac{\Gamma \vdash B : s \quad \Gamma, w : A \vdash t : B \quad \Gamma \vdash a : A/R \quad \Gamma \vdash p : \Pi x : A.\Pi y : A.(R\ x\ y) \to t[x/w] =_B t[y/w]}{\Gamma \vdash \mathsf{pick}_p\ w : A\ \mathsf{from}\ \ a\ \mathsf{in}\ \ t : B}$$	Elimination
$$\frac{\Gamma, w : A/R \vdash B : * \quad \Gamma \vdash p : \Pi x : A.B[[x]_R/w]}{\Gamma \vdash \mathsf{lift}\ p : \Pi w : A/R.B}$$	No junk
$$\frac{\Gamma \vdash t : A \quad \Gamma \vdash A' : s \quad A =_\chi A'}{\Gamma \vdash t : A'}$$	χ-conversion

The introduction rule gives a canonical map from A to A/R which assigns to every element of A its 'equivalence class'. Note that $=_{A/R}$ in the identify rule is not a new connective but Leibniz equality. The no junk rule ensures that the canonical map $[.]$ is surjective. The elimination rule captures the universal property of quotients (as coequalisers): if $f : A \to B$ identifies elements related by R, then f factorises through $[.]_R$. Note that the elimination rule is non-dependent.

Remarks.

- The syntax we give here is not PTS-like but exploits the fact that the original λS is a hol system. The main reason to choose such a syntax is its convenience and conciseness. It is easy to turn these rules into PTS-like ones. Yet the syntax becomes awkward.
- In a hol system, if $A : *$ and $R : A \to A \to *$, we can form the term $\mathsf{lift}\ (\lambda x : A.x) : A/R \to A$ whose computational content is unclear. The status of the 'no junk' rule is clearer for those pure type systems where types and propositions are kept separate: it is a lógical assertion.
- The rules for quotients introduce a dependency of objects on proofs. One might want to overcome this problem by forgetting the proof in the elimination rule of quotient types but it leads to undecidable type-checking. Fortunately, this dependency is inessential:

  Lemma 2 *If* $\mathsf{pick}_\xi\ x : A\ \mathsf{from}\ \ M\ \mathsf{in}\ \ N$ *(for* $\xi = p, p'$*) are terms of type* B*, then* $\mathsf{pick}_p\ x : A\ \mathsf{from}\ \ M\ \mathsf{in}\ \ N =_B \mathsf{pick}_{p'}\ x : A\ \mathsf{from}\ \ M\ \mathsf{in}\ \ N$.

  Proof: assume $\Gamma, M : A/R \vdash \mathsf{pick}_\xi\ x : A\ \mathsf{from}\ \ M\ \mathsf{in}\ \ N : B$ for $\xi = p, p'$. For every $a : A$, we have

  $$(\mathsf{pick}\ \ x : A\ \mathsf{from}\ \ M\ \mathsf{in}\ \ N)\ [[a]/M] \equiv \mathsf{pick}\ \ x : A\ \mathsf{from}\ \ [a]\ \mathsf{in}\ \ N[[a]/M]$$

$$\to_\chi N[a/x,[a]/M]$$

Hence

$$(\mathsf{pick}_p\ x : A \text{ from } M \text{ in } N)\,[[a]/M] =_B (\mathsf{pick}_{p'}\ x : A \text{ from } M \text{ in } N)\,[[a]/M]$$

The lemma follows from the 'no junk' rule.

- As for subset types, we could have given a slightly more general presentation by letting A/R live in a different sort from A.

χ-reduction is Church-Rosser. Indeed, the only problematic critical pair converges, as seen in the diagram below (subscripts are omitted for the sake of readability):

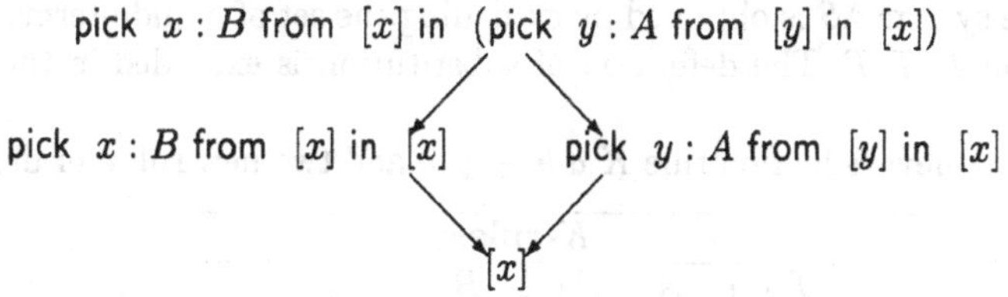

We can apply the Hindley-Rosen lemma to get confluence of $\beta\chi$-reduction.

2.3 Pure type systems with definitions

In [18], Poll and Severi have introduced pure type systems with definitions. A pure type system λS can be extended to a pure type system with definitions λS_δ as follows: first, extend the set of pseudo-terms with the constructions $V = T : T$ (global definition) and $V = T : T \text{ in } T$ (local definition). By convention, x is bound in $x = a : A$ and $x = a : A \text{ in } t$. The definition of substitution is extended in the obvious way.

The notion of derivation is extended with the following rules:

Definitions		
$\dfrac{\Gamma \vdash a : A}{\Gamma, x = a : A \vdash x : A}$	δ-start	where x fresh
$\dfrac{\Gamma \vdash a : A \quad \Gamma \vdash b : B}{\Gamma, x = a : A \vdash b : B}$	δ-weakening	where x fresh
$\dfrac{\Gamma, x = a : A \vdash B : s}{\Gamma \vdash x = a : A \text{ in } B : s}$	δ-formation	
$\dfrac{\Gamma, x = a : A \vdash b : B \quad \Gamma \vdash x = a : A \text{ in } B : s}{\Gamma \vdash (x = a : A \text{ in } b) : (x = a : A \text{ in } B)}$	δ-introduction	
$\dfrac{\Gamma \vdash a : A \quad \Gamma \vdash A' : s \quad \Gamma \vdash A =_\delta A'}{\Gamma \vdash a : A'}$	δ-conversion	

The rules are self-explanatory. In the δ-introduction, we need to require that $x = a : A \text{ in } B$ is legal in order for the system to behave well with respect to reduction. The notion of δ-reduction is used to unfold a definition or remove it when it is vacuous. It is defined relative to a pseudo-context Γ:

$$\Gamma_1, x = a : A, \Gamma_2 \vdash x \rightarrow_\delta a$$
$$\Gamma \vdash x = a : A \text{ in } b \rightarrow_\delta b \quad \text{if } x \notin FV(b)$$

with the obvious compatibility rules. It can be shown that both δ and $\beta\delta$-reduction are confluent.

2.4 Pure type systems with K-rules

Extending a pure type system with the K-rules is hardly interesting per se. However, it is very useful to relate the problem of strong normalisation of pure type systems with definitions, quotients and subset types. The extension λS_κ of a pure type system λS is obtained by extending the set of pseudo-terms with the construction $K\ T\ T$. The definition of substitution is extended in the obvious way.

There is a new reduction rule $K\ a\ b\ \rightarrow_\kappa\ a$ and two new rules for derivations

K-rules	
$\dfrac{\Gamma \vdash a : A \quad \Gamma \vdash b : B}{\Gamma \vdash K\ a\ b : A}$	Introduction
$\dfrac{\Gamma \vdash t : A \quad \Gamma \vdash A' : s \quad A =_\kappa A'}{\Gamma \vdash t : A'}$	κ-conversion

$\beta\kappa$-reduction is Church-Rosser: indeed, κ-reduction is Church-Rosser and commutes with β-reduction. The result follows from the Hindley-Rosen lemma. Although the extension of pure type systems is rather harmless, we have not succeeded in proving that the extension preserves strong normalisation.

3 The main result

Main Theorem. *Let λS be a hol system. Then the following are equivalent*[5]*:*

(i) λS_δ is strongly normalising,
(ii) λS_χ is strongly normalising,
(iii) λS_σ is strongly normalising,
(iv) λS_κ is strongly normalising.

Convention For $\zeta \in \{\delta, \kappa, \sigma, \chi\}$, we let T_ζ be the set of pseudo-terms of λS_ζ and $\vdash_\zeta$ be the entailment relation of λS_ζ. Besides, it is convenient to extend slightly the syntax with terms of the form $\lambda - : A.t$, $- = a : A$ in b and $\text{pick}_Z\ - : A$ from M in N. The idea is that $-$ is a dummy variable which does not occur anywhere; by introducing a new identifier $-$, we avoid clashes of variables.

[5] (iv) implies (i) is already implicit in [18].

3.1 The proof

We give five translations $\lceil . \rceil_\delta, \lceil \cdot \rceil_\chi, \lceil \cdot \rceil_\kappa, \lceil \cdot \rceil_\sigma$ and $\lceil . \rceil_\epsilon$ as in the following diagram

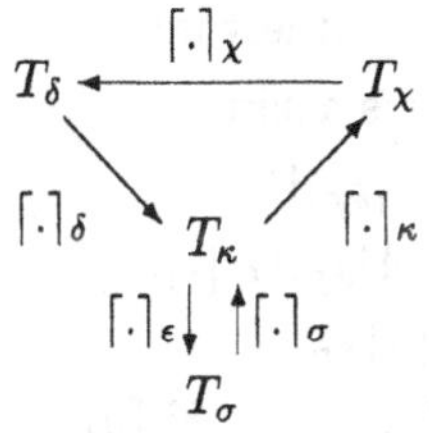

preserving reductions and satisfying

$$\begin{array}{l}
\Gamma \vdash_\delta t : A \Rightarrow \lceil \Gamma \rceil_\delta \vdash_\kappa \lceil t \rceil_\delta : \lceil A \rceil_\delta \\
\Gamma \vdash_\chi t : A \Rightarrow \mathsf{absurd} : \bot, \lceil \Gamma \rceil_\chi \vdash_\delta \lceil t \rceil_\chi : \lceil A \rceil_\chi \\
\Gamma \vdash_\kappa t : A \Rightarrow \lceil \Gamma \rceil_\kappa \vdash_\chi \lceil t \rceil_\kappa : \lceil A \rceil_\kappa \\
\Gamma \vdash_\sigma t : A \Rightarrow \mathsf{absurd} : \bot, \lceil \Gamma \rceil_\sigma \vdash_\kappa \lceil t \rceil_\sigma : \lceil A \rceil_\sigma \\
\Gamma \vdash_\kappa t : A \Rightarrow \lceil \Gamma \rceil_\epsilon \vdash_\sigma \lceil t \rceil_\epsilon : \lceil A \rceil_\epsilon
\end{array}$$

Strictly speaking, the translations $\lceil . \rceil_\chi$, $\lceil \cdot \rceil_\kappa$, $\lceil \cdot \rceil_\sigma$ and $\lceil . \rceil_\epsilon$[6] are not maps but relations. These relations have the property that for every reduction $t \to u$ and t' related to t by $\lceil . \rceil$ there exists u' related to u by $\lceil . \rceil$ such that $t' \twoheadrightarrow^+ u'$, as shown in the following diagram

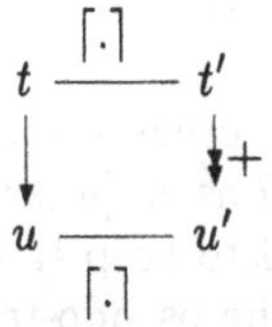

Hence they preserve reductions. Fortunately, the non-functionality of the translations is unproblematic for derivations as well, in the sense that $\Gamma \vdash t : A \Rightarrow \Gamma' \vdash t' : A'$ for some Γ', t', A' related to Γ, t, A by $\lceil . \rceil$. For the sake of simplicity, we shall treat all translations as functions. Note that the non-functionality could be avoided altogether by defining the translation on derivations rather than pseudo-terms. We have chosen not to do so for the clarity of the presentation. It is immediate to infer from the proofs below what the translations on derivations should be. Finally, we do not require $\Gamma \vdash a : A \quad \Rightarrow \quad \lceil \Gamma \rceil \vdash \lceil a \rceil : \lceil A \rceil$ for the translations $\lceil . \rceil_\chi$ and $\lceil . \rceil_\sigma$ as it would imply the conservativity of λS_σ and λS_χ over λS.

The translation $\lceil . \rceil_\delta$ It is essentially the one used by Poll and Severi in [18] to prove that λS_δ is strongly normalising whenever a completion of λS is. $\lceil . \rceil_\delta$ is

[6] We shall omit subscripts in the rest of the section.

defined relative to a pseudo-context Γ and maps pseudo-terms of $\lambda\mathcal{S}_\delta$ to pseudo-terms of $\lambda\mathcal{S}_\kappa$.

$$\lceil x\rceil_{\delta\Gamma} = \begin{cases} \lceil a\rceil_{\delta\Gamma_1} & \text{if } x \text{ is a variable and } \Gamma \equiv \Gamma_1, x = a : A, \Gamma_2 \\ x & \text{otherwise} \end{cases}$$

$$\lceil s\rceil_{\delta\Gamma} = s \quad \text{if } s \text{ is a sort}$$

$$\lceil \lambda x : A.t\rceil_{\delta\Gamma} = \lambda x : \lceil A\rceil_{\delta\Gamma}.\lceil t\rceil_{\delta\Gamma}$$

$$\lceil \Pi x : A.B\rceil_{\delta\Gamma} = \Pi x : \lceil A\rceil_{\delta\Gamma}.\lceil B\rceil_{\delta\Gamma}$$

$$\lceil tu\rceil_{\delta\Gamma} = \lceil t\rceil_{\delta\Gamma}\lceil u\rceil_{\delta\Gamma}$$

$$\lceil x = a : A \text{ in } b\rceil_{\delta\Gamma} = K\ \lceil b\rceil_{\delta\Gamma, x=a:A}\ (K\ \lceil a\rceil_{\delta\Gamma}\ \lceil A\rceil_{\delta\Gamma})$$

It can be proved that

Lemma 3 ([18]) - *δ-reduction is strongly normalising,*
- $a \rightarrow_\beta b \quad \Rightarrow \quad \lceil a\rceil_{\delta\Gamma} \twoheadrightarrow^{+}_{\beta\kappa} \lceil b\rceil_{\delta\Gamma}$,
- $\Gamma \vdash a : A \quad \wedge \quad \Gamma \vdash a \rightarrow_\delta b \quad \Rightarrow \quad \lceil a\rceil_{\delta\Gamma} \twoheadrightarrow_{\beta\kappa} \lceil b\rceil_{\delta\Gamma}$;
- $\Gamma \vdash a : A \quad \Rightarrow \quad \lceil\Gamma\rceil_\delta \vdash \lceil a\rceil_{\delta\Gamma} : \lceil A\rceil_{\delta\Gamma}$.

Strong normalisation of $\lambda\mathcal{S}_\delta$ follows from the above lemma and strong normalisation of $\lambda\mathcal{S}_\kappa$. Indeed, assume there is a derivation $\Gamma \vdash a_0 : A$ with an infinite reduction sequence $\Gamma \vdash a_0 \rightarrow a_1 \rightarrow \ldots$ δ-reduction is strongly normalising, hence the reduction sequence contains infinitely many β-reductions. For each β-reduction $a_i \rightarrow_\beta a_{i+1}$, we have $\lceil a_i\rceil_{\delta\Gamma} \twoheadrightarrow^{+}_{\beta\kappa} \lceil a_{i+1}\rceil_{\delta\Gamma}$. Thus the sequence $\lceil a_0\rceil_{\delta\Gamma} \twoheadrightarrow \lceil a_1\rceil_{\delta\Gamma} \twoheadrightarrow \ldots$ contains infinitely many reduction steps. As $\lceil\Gamma\rceil_\delta \vdash \lceil a_0\rceil_{\delta\Gamma} : \lceil A\rceil_{\delta\Gamma}$, we get a contradiction.

The translation $\lceil . \rceil_\chi$ It maps pseudo-terms of $\lambda\mathcal{S}_\chi$ to pseudo-terms of $\lambda\mathcal{S}_\delta$. The intuition is to translate A/R as A, $[a]_R$ as a and $\mathsf{pick}_p\ w : A$ from M in N as $N[M/w]$. Definitions are used to keep track of all reductions. $\lceil . \rceil_\chi$ is defined inductively on the structure of the pseudo-terms:

$$\lceil x\rceil_\chi = x \quad \text{for variables and constants } x$$

$$\lceil \lambda x : A.t\rceil_\chi = \lambda x : \lceil A\rceil_\chi.\lceil t\rceil_\chi$$

$$\lceil \Pi x : A.B\rceil_\chi = \Pi x : \lceil A\rceil_\chi.\lceil B\rceil_\chi$$

$$\lceil tu\rceil_\chi = \lceil t\rceil_\chi\lceil u\rceil_\chi$$

$$\lceil A/R\rceil_\chi = (- = \lceil R\rceil_\chi) \text{ in } \lceil A\rceil_\chi$$

$$\lceil [a]_R\rceil_\chi = (- = \lceil R\rceil_\chi) \text{ in } \lceil a\rceil_\chi$$

$$\lceil \mathsf{identify}\ p\rceil_\chi = (- = \lceil p\rceil_\chi) \text{ in absurd } (\lceil a\rceil_\chi =_{\lceil A\rceil_\chi} \lceil a'\rceil_\chi)$$

$$\lceil \mathsf{lift}\ p\rceil_\chi = \lceil p\rceil_\chi$$

$$\lceil \mathsf{pick}_p\ w : A \text{ from } M \text{ in } N\rceil_\chi = (w = \lceil M\rceil_\chi : \lceil A\rceil_\chi)\ (- = \lceil p\rceil_\chi) \text{ in } \lceil N\rceil_\chi$$

Note that the non-functionality of $\lceil . \rceil_\chi$ is due to the translation of identify p, where a, a', A are introduced. Of course, the idea is $p : a =_A a'$. The translation satisfies:

Lemma 4 - $t \rightarrow_{\beta\chi} t' \quad \Rightarrow \quad \lceil t \rceil_\chi \twoheadrightarrow^+_{\beta\delta} \lceil t' \rceil_\chi$,
- $\Gamma \vdash t : A \quad \Rightarrow \quad \mathsf{absurd} : \bot, \lceil \Gamma \rceil_\chi \vdash \lceil t \rceil_\chi : \lceil A \rceil_\chi$.

In the first part of the lemma, $\beta\delta$ denotes reduction in the empty context (or any context without global definitions): the translation does not introduce global definitions, hence the context is not important. Besides, it follows from the definition of δ-reduction in a context without global definitions that χ-reductions induce a non-empty sequence of δ-reductions.

The second part of the lemma is proved by induction on the length of the derivation. The key fact is that $\lceil t[u/x] \rceil_\chi = \lceil t \rceil_\chi [\lceil u \rceil_\chi / x]$ for every pseudo-terms t, u. Note that absurd plays a crucial role in the translation because $\lceil . \rceil_\chi$ is not sound for the identify rule. Strong normalisation of $\lambda\mathcal{S}_\chi$ is an immediate consequence of the above lemma and strong normalisation of $\lambda\mathcal{S}_\delta$.

The translation $\lceil . \rceil_\kappa$ It maps pseudo-terms of $\lambda\mathcal{S}_\kappa$ to pseudo-terms of $\lambda\mathcal{S}_\chi$. $\lceil . \rceil_\kappa$ is defined inductively on the structure of the pseudo-terms:

$$\begin{aligned}
\lceil x \rceil_\kappa &= x \quad \text{for variables and constants } x \\
\lceil \lambda x : A.t \rceil_\kappa &= \lambda x : \lceil A \rceil_\kappa . \lceil t \rceil_\kappa \\
\lceil \Pi x : A.B \rceil_\kappa &= \Pi x : \lceil A \rceil_\kappa . \lceil B \rceil_\kappa \\
\lceil tu \rceil_\kappa &= \lceil t \rceil_\kappa \lceil u \rceil_\kappa \\
\lceil K\ t\ u \rceil_\kappa &= \mathsf{pick}_p - : \lceil U \rceil_\kappa \ \mathsf{from}\ [\lceil u \rceil_\kappa]_{(\lambda x : \lceil U \rceil_\kappa . \lambda y : \lceil U \rceil_\kappa . \bot)} \ \mathsf{in}\ \lceil t \rceil_\kappa
\end{aligned}$$

where $p \equiv \lambda x : \lceil U \rceil_\kappa . \lambda y : \lceil U \rceil_\kappa . \lambda z : \bot . z \ (\lceil t \rceil_\kappa =_{\lceil T \rceil_\kappa} \lceil t \rceil_\kappa)$.

The non-functionality of $\lceil . \rceil_\kappa$ is due to the fact that U, T are introduced in the translation (intuitively, $u : U$ and $t : T$); yet $\lceil . \rceil_\kappa$ is functional up to conversion. The translation satisfies:

Lemma 5 - $t \rightarrow_{\beta\kappa} t' \quad \Rightarrow \quad \lceil t \rceil_\kappa \twoheadrightarrow^+_{\beta\chi} \lceil t' \rceil_\kappa$,
- $\Gamma \vdash_\kappa t : A \quad \Rightarrow \quad \lceil \Gamma \rceil_\kappa \vdash_\chi \lceil t \rceil_\kappa : \lceil A \rceil_\kappa$.

By the above lemma, strong normalisation of $\lambda\mathcal{S}_\chi$ implies strong normalisation of $\lambda\mathcal{S}_\kappa$.

The translation $\lceil . \rceil_\sigma$ It maps pseudo-terms of $\lambda\mathcal{S}_\sigma$ to pseudo-terms of $\lambda\mathcal{S}_\kappa$. The intuition is to translate $\{x : A|B\}$ as A and both in and out as identities. The K-rules are used to keep track of reductions. The translation is defined inductively as follows:

$$\begin{aligned}
\lceil x \rceil_\sigma &= x \quad \text{for variables and constants } x \\
\lceil \lambda x : A.t \rceil_\sigma &= \lambda x : \lceil A \rceil_\sigma . \lceil t \rceil_\sigma \\
\lceil \Pi x : A.B \rceil_\sigma &= \Pi x : \lceil A \rceil_\sigma . \lceil B \rceil_\sigma \\
\lceil tu \rceil_\sigma &= \lceil t \rceil_\sigma \lceil u \rceil_\sigma \\
\lceil \{x : A|B\} \rceil_\sigma &= K\ \lceil A \rceil_\sigma\ (\Pi x : \lceil A \rceil_\sigma . \lceil B \rceil_\sigma)
\end{aligned}$$

$$\lceil \mathsf{in}_{\{x:A|B\}}\ a \rceil_\sigma = K\ \lceil a \rceil_\sigma\ \lceil \{x : A|B\} \rceil_\sigma$$
$$\lceil \mathsf{out}\ a \rceil_\sigma = \lceil a \rceil_\sigma$$
$$\lceil \mathsf{proof}\ a \rceil_\sigma = \mathsf{absurd}\ \lceil B \rceil_\sigma [\lceil a \rceil_\sigma / x]$$

Note that the non-functionality of $\lceil . \rceil_\sigma$ is due to the translation of proof a (the intuition is that $a : \{x : A|B\}$). Strong normalisation of λS_σ follows from strong normalisation of λS_κ and the following lemma:

Lemma 6 - $t \rightarrow_{\beta\sigma} t' \quad \Rightarrow \quad \lceil t \rceil_\sigma \twoheadrightarrow^+_{\beta\kappa} \lceil t' \rceil_\sigma$,
- $\Gamma \vdash_\sigma t : A \quad \Rightarrow \quad \mathsf{absurd} : \bot, \lceil \Gamma \rceil_\sigma \vdash_\kappa \lceil t \rceil_\sigma : \lceil A \rceil_\sigma$.

Note that absurd plays a key role as the translation is not sound for the introduction and proof rules.

The translation $\lceil . \rceil_\epsilon$ It maps pseudo-terms of λS_κ to pseudo-terms of λS_σ. $\lceil . \rceil_\epsilon$ is defined inductively on pseudo-terms:

$$\lceil x \rceil_\epsilon = x \quad \text{for variables and constants } x$$
$$\lceil \lambda x : A.t \rceil_\epsilon = \lambda x : \lceil A \rceil_\epsilon . \lceil t \rceil_\epsilon$$
$$\lceil \Pi x : A.B \rceil_\epsilon = \Pi x : \lceil A \rceil_\epsilon . \lceil B \rceil_\epsilon$$
$$\lceil tu \rceil_\epsilon = \lceil t \rceil_\epsilon \lceil u \rceil_\epsilon$$
$$\lceil K\ t\ u \rceil_\epsilon = \mathsf{out}\ (\mathsf{in}_{\{x:\lceil T \rceil_\epsilon | \lceil u \rceil_\epsilon = _{\lceil U \rceil_\epsilon} \lceil u \rceil_\epsilon\}}\ \lceil t \rceil_\epsilon)$$

Note that the non-functionality of the translation comes from the definition of $\lceil K\ t\ u \rceil_\epsilon$ (intuitively $t : T$ and $u : U$). Still $\lceil . \rceil_\epsilon$ is functional up to conversion. Strong normalisation of λS_κ is a consequence of strong normalisation of λS_σ and the following lemma:

Lemma 7 - $t \rightarrow_{\beta\kappa} t' \quad \Rightarrow \quad \lceil t \rceil_\epsilon \twoheadrightarrow^+_{\beta\sigma} \lceil t' \rceil_\epsilon$,
- $\Gamma \vdash_\kappa t : A \quad \Rightarrow \quad \lceil \Gamma \rceil_\epsilon \vdash_\sigma \lceil t \rceil_\epsilon : \lceil A \rceil_\epsilon$.

3.2 Significance of the theorem

The main theorem establishes the equivalence between four strong normalisation problems for hol systems. In fact, it can be strengthened into:

Proposition 8 *Let λS be a hol system. Let Δ, Δ' be non-empty subsets of $\{\delta, \kappa, \sigma, \chi\}$. Then λS_Δ is strongly normalising if and only if $\lambda S_{\Delta'}$ is, where λS_Δ is the obvious extension of λS.*

Besides, the proof can be carried over to pure type systems with $\beta\eta$-reduction ([7]). It follows

Corollary 9 *The Calculus of Constructions λC (with β and $\beta\eta$-reduction) and ECC, extended with definitions, subset types and quotient types are strongly normalising.*

Proof: K-rules are definable in ECC, so ECC_κ and $ECC_{\delta,\sigma,\chi}$ are strongly normalising. λC with K-rules (with β and $\beta\eta$-reduction) is strongly normalising (see for example [7]), hence $\lambda C_{\delta,\sigma,\chi}$ (with β and $\beta\eta$-reduction) is strongly normalising.□
It is immediate to conclude (via a standard normalisation argument) that the extended systems are consistent.
Remarks.

- The equivalence between strong normalisation of λS_δ and strong normalisation of λS_κ holds in an arbitrary pure type system.
- The proof method can be adapted to prove preservation of strong normalisation of quotient and subset types for other type theories than pure type systems, e.g for higher-order logic ([11, 12]).
- The translation $\lceil . \rceil_\chi$ shows how the semantical proof of strong normalisation of λC given in [8] can be extended to a proof of strong normalisation of λC_χ.

4 Discussion

In this section, we focus on alternative syntaxes for quotient types and on the logical and meta-theoretic properties of subset types and quotient types.
A general remark. Basic properties of pure type systems, such as the transitivity, substitution, thinning, generation and subject reduction lemmas can be extended to the new type systems. Yet, decidability of type-checking and the strengthening lemma fail for type systems with subset types.

4.1 Subset types via quotient types

Subset types alter quite dramatically the proof-theoretic properties of the type system:

- type-checking is undecidable in presence of subset types: we cannot decide whether $\Gamma \vdash \mathsf{in}_{\{x:A|\phi\}}\ a : \{x : A|\phi\}$ as proofs are not recorded in the introduction rule,
- the strengthening lemma ([1]) is false. In a hol system with subset types, $\vdash a : A$ implies $\mathsf{absurd} : \bot \vdash \mathsf{in}_{\{x:A|\bot\}}\ a : \{x : A|\bot\}$. Thus $\mathsf{absurd} : \bot \vdash \mathsf{proof}\ a : \bot$. If the strengthening lemma would be true, then $\vdash \mathsf{proof}\ a : \bot$. This would imply that the pure type system with subsets is inconsistent. Yet λC_σ is consistent.

Subset types as defined above are very problematic. Fortunately, they are definable in every hol system λS_χ with quotients and $(s, *, s)$ strong sums[7]. Assume $\Gamma \vdash A : s$ and $\Gamma, x : A \vdash B : *$. Then $\Gamma \vdash \{x : A|B\}$ where

$$\{x : A|B\} \equiv \Sigma x : A.B/(\lambda w : \Sigma x : A.B.\lambda w' : \Sigma x : A.B.w_1 =_A w'_1)$$

[7] Pure type systems with strong sums are introduced in the appendix.

Furthermore we can define for every pseudo-terms a, p, the pseudo-terms

$$\begin{aligned} \mathsf{out}\ a &\equiv \mathsf{pick}_{(\lambda w:\Sigma x:A.B.\lambda w':\Sigma x:A.B.\lambda z:w_1=_A w'_1.z)}\ w : \Sigma x : A.B \text{ from } a \text{ in } w_1 \\ \mathsf{proof}\ a &\equiv (\mathsf{lift}\ (\lambda w : \Sigma x : A.B.w_2))a \\ \mathsf{in}_p\ a &\equiv [\langle a, p\rangle]_{(\lambda w:\Sigma x:A.B.\lambda w':\Sigma x:A.B.w_1=_A w'_1)} \end{aligned}$$

If $\Gamma \vdash a : \{x : A|B\}$, then $\Gamma \vdash \mathsf{out}\ a : A$ and $\Gamma \vdash \mathsf{proof}\ a : B[a/x]$. Moreover, if $\Gamma \vdash a : A$, $\Gamma \vdash p : B[a/x]$ and $\Gamma \vdash p : B[a'/x]$, we have $\Gamma \vdash \mathsf{in}_p\ a : \{x : A|B\}$ and $\mathsf{in}_p\ a =_{\{x:A|B\}} \mathsf{in}_{p'} a$ (is inhabited in context Γ).

Our construction is similar to the one of [15]; yet Mendler can only define a weak notion of subset, because his syntax does not have the 'no junk' rule.

4.2 Variations on quotient types

Quotients with explicit representatives In [2, 4], the author gives a syntax for quotients with a choice operator rep picking a representative for each equivalence class[8]. We briefly describe the new syntax. In the sequel, λS is a hol system. We extend it into a hol system λS_{χ_r} with quotients with explicit representatives. Pseudo-terms are extended with constructions T/T, $[T]_T$, $\mathsf{identify}\ T$, $\mathsf{rep}\ T$ and $\mathsf{ad}(T, T)$. The reduction rules are

$$\begin{aligned} \mathsf{ad}(\mathsf{identify}\ p, q) &\to_{\chi_r} q\ a\ b\ p \\ \mathsf{ad}(p, \lambda x : A.\lambda y : A.\lambda z : R\ x\ y.\mathsf{identify}\ z) &\to_{\chi_r} p \\ [\mathsf{rep}\ a]_R &\to_{\chi_r} a \end{aligned}$$

The notion of derivation is extended with the formation, introduction and identify rules of Section 2.2, a χ_r-conversion rule and the new rules:

Quotient types with explicit representatives	
$\dfrac{\Gamma, w : A \vdash t : B \qquad \Gamma \vdash p : [a]_R =_{A/R} [b]_R \qquad \Gamma \vdash q : \Pi x : A.\Pi y : A.(R\ x\ y) \to t[x/w] =_B t[y/w]}{\Gamma \vdash \mathsf{ad}(p, q) : t[a/w] =_B t[b/w]}$	Adequacy
$\dfrac{\Gamma \vdash a : A/R}{\Gamma \vdash \mathsf{rep}\ a : A}$	Representative

Remarks:

- The syntax does not introduce any dependency of objects on proofs.
- The choice operator rep is rather weak, in the sense that it does not imply the axiom of choice for types. In fact, one can only prove that the axiom of unique choice implies the axiom of choice.

[8] M.Hofmann has also considered the possibility of adding a choice operator for quotients; see [4, 10] for a discussion on the use of explicit representatives to formalise mathematics.

Effectiveness In a hol system, every relation R on A has a reflexive, symmetric, transitive closure R^+. We say the quotient A/R is *effective*[9] if for all $a, a' : A$, $[a]_R =_{A/R} [a']_R$ and $R^+ \; a \; a'$ are logically equivalent. It is possible to force all quotients to be effective by introducing a rule

$$\frac{\Gamma \vdash p : [a]_R =_{A/R} [a']_R}{\Gamma \vdash \mathsf{noconf}\; p : R^+ \; a \; a'} \text{ No confusion}$$

There are new reduction rules

$$\mathsf{noconf}\; p \; R^\star \; r \; s \; t \; i \to_{\chi_e} p$$
$$\mathsf{noconf}\; (\mathsf{identify}\; p) \; R \; r \; s \; t \; i \to_{\chi_e} p$$

where $R^\star \equiv \lambda x : A.\lambda y : A.[x]_R =_{A/R} [y]_R$. The resulting type system is called $\lambda\mathcal{S}_{\chi_e}$. Note that effectiveness has very strong logical consequences:

Lemma 10 *[4] In ECC_{χ_e}, the axiom of choice*[10] *implies excluded middle and proof irrelevance.*

Proof: the first part is a straightforward adaptation of [13]. The second part follows from [19], see also [6].

A strong normalisation result The method of the main theorem can be used to prove a strong normalisation result for these variants of quotient types. However, we have been unable to prove that strong normalisation of $\lambda\mathcal{S}_{\chi_e}$ or $\lambda\mathcal{S}_{\chi_r}$ is a consequence of strong normalisation of $\lambda\mathcal{S}_{\kappa}$. Instead, we have to consider a mild extension of $\lambda\mathcal{S}_{\kappa}$.

Let $\lambda\mathcal{S}_{\mathsf{exfalso}}$ be the extension of $\lambda\mathcal{S}$ with a new propositional constant exfalso of type $\Pi x : *.\bot \to x$ and two new reduction rules:

$$\mathsf{exfalso}\; (\phi \to \phi) \; x \; p \to p$$
$$\mathsf{exfalso}\; (\phi \to \psi) \; x \; (\mathsf{exfalso}\; (\psi \to \phi) \; y \; p) \to p$$

The reduction rules are to be understood as some form of cut-elimination. We can prove

Theorem 11 *If $\lambda\mathcal{S}_{\kappa,\mathsf{exfalso}}$ is strongly normalising, so are $\lambda\mathcal{S}_{\chi_r}$ and $\lambda\mathcal{S}_{\chi_e}$.*

Note that exfalso is used to interpret the adequacy and no confusion rules. See [4] for a full proof.

---

[9] This is a generalisation of the notion of [12].

[10] The axiom of choice is given by two inhabitants *make*, *check* of respective types

$$\Pi A, B : \mathsf{Type}.\Pi R : A \to B \to \mathsf{Prop}.(\forall a : A.\exists b : B.R \; a \; b) \to A \to B$$
$$\forall A, B : \mathsf{Type}.\forall R : A \to B \to \mathsf{Prop}.(\forall a : A.\exists b : B.R \; a \; b) \to \forall x : el \; A.R \; x \; (choice \; R \; x)$$

Congruence types Quotient types do not capture, when it exists, the computational content of the equivalence relation. In [2, 4], the author introduces a variant of quotient types, congruence types, which preserve the computational content of the quotienting relation. If T is an inductive type and R a confluent noetherian rewrite system on T, we can define T/R as a congruence type. Here, the identify rule is replaced by some reductions. Examples of types which can be defined as congruence types include the integers and the free group over an arbitrary set. Congruence types are potentially very useful for equational reasoning in type theory.

5 Conclusion

We have introduced pure type systems with quotient and subset types and linked the problem of strong normalisation of these extensions with the problem of strong normalisation of pure type systems with K-rules and definitions. The technique of the proof is widely applicable, for example to pure type systems with β and $\beta\eta$-reduction and to extensions of pure type systems (with strong sums, inductive types...).

The main theorem is somewhat unsatisfactory, as one would like to prove that $\lambda S_{\delta,\sigma,\chi}$ is strongly normalising for every strongly normalising hol system λS. Note that it follows from

Conjecture 12 *If λS is a strongly normalising pure type system, then λS_κ is strongly normalising.*

Despite its apparent simplicity, Conjecture 12 does not seem to have any obvious combinatorial proof. A possibility to prove the conjecture would be to give a saturated set semantics for an arbitrary strongly normalising pure type system λS and derive strong normalisation of λS_κ from the semantics (see [8, 21] for some related work). Yet our failure to prove the conjecture is in fact part of a more general problem, namely the lack of techniques to prove generic normalisation results for type theory.

Acknowledgements I am indebted to Erik Barendsen and Herman Geuvers for making some useful comments on an earlier version of the paper and to Bart Jacobs for his inspiring work on quotient types ([12]).

References

1. H.Barendregt. *Typed λ-calculi*, Handbook of logic in computer science, Abramsky and *al* eds, OUP 1992.
2. G.Barthe. *Towards a mathematical vernacular*, submitted.
3. G.Barthe. *Formalizing mathematics in type theory: fundamentals and case studies*, submitted.
4. G.Barthe. *An introduction to quotient and congruence types*, manuscript, University of Nijmegen, November 1994.

5. R.Constable and *al. Implementing Mathematics with the NuPrl Proof Development System*, Prenctice Hall, 1986.
6. T.Coquand. *A new paradox in type theory*, in the proceedings of the 9th Congress of Logic, Methodology and Philosophy of Science.
7. H.Geuvers. *Logics and type systems*, Ph.D thesis, University of Nijmegen, 1993.
8. H.Geuvers. *A short and flexible proof of strong normalisation for the Calculus of Constructions*, submitted.
9. M.Hofmann. *Extensional concepts in intensional type theory*, Ph.D thesis, University of Edinburgh, forthcoming.
10. M.Hofmann. *A simple model for quotient types*, in these proceedings.
11. B.Jacobs. *Categorical Logic and Type Theory*, in preparation.
12. B.Jacobs. *Quotients in simple Type Theory*, submitted.
13. J.Lambek and P.J.Scott. *Introduction to higher-order categorical logic*, CUP, 1986.
14. Z.Luo. *Computation and reasoning: a type theory for computer science*, OUP, 1994.
15. N.Mendler. *Quotient types via coequalisers in Martin-Lof's type theory*, in the informal proceedings of the workshop on logical frameworks, Antibes, May 1990.
16. R.Nederpelt and *al* (eds). *Selected Papers on Automath*, North-Holland, 1994.
17. B.Nordstrom, K.Petersson and J.Smith. *Programming in Martin-Lof's type theory*, OUP, 1990.
18. E.Poll and P.Severi. *PTS with definitions*, in proceedings of LFCS'94, LNCS 813.
19. G.Pottinger. *Definite descriptions and excluded middle in the theory of constructions*, TYPES mailing list, November 1989.
20. A.Salvesen and J.Smith. *The strength of the subset type in Martin-Lof's type theory*, proceedings of LICS'88, 1988.
21. J.Terlouw. *Strong normalisation in type systems: a model-theoretical approach*, Dirk van Dalen Festschrift, Utrecht, 1993.

Appendix: pure type systems with strong sums

We sketch how to extend pure type systems with strong sums. Let λS be a pure type system. Let $R_\Sigma \subseteq S^3$. We can extend λS into a pure type system with strong sums λS_Σ as follows. First, extend the pseudo-terms with the constructs $\Sigma V : T.T$, $\langle T, T\rangle$ and T_i (i=1,2). Add the reduction rules $\langle t_1, t_2\rangle_i \rightarrow_\Sigma t_i$ where $i = 1, 2$. The notion of derivation is extended with the following rules:

Sum Types	
$\dfrac{\Gamma \vdash A : s_1 \quad \Gamma, x : A \vdash B : s_2}{\Gamma \vdash \Sigma x : A.B : s_3}$	Formation $(s_1, s_2, s_3) \in R_\Sigma$
$\dfrac{\Gamma \vdash t_1 : A \quad \Gamma \vdash t_2 : B[t_1/x] \quad \Gamma \vdash \Sigma x : A.B : s}{\Gamma \vdash \langle t_1, t_2\rangle : \Sigma x : A.B}$	Pairing
$\dfrac{\Gamma \vdash t : \Sigma x : A.B}{\Gamma \vdash t_1 : A}$	First projection
$\dfrac{\Gamma \vdash t : \Sigma x : A.B}{\Gamma \vdash t_2 : B[t_1/x]}$	Second projection
$\dfrac{\Gamma \vdash t : A \quad \Gamma \vdash A' : s \quad A =_\Sigma A'}{\Gamma \vdash t : A'}$	Σ-conversion

A Model for Formal Parametric Polymorphism: A PER Interpretation for System $\mathcal{R}$

Roberto Bellucci[1,3], Martín Abadi[2], Pierre-Louis Curien[1]

[1] LIENS, CNRS - Département de Mathématiques et Informatique de l'Ecole Normale Supérieure 45, rue d'Ulm, 75005 Paris, France

[2] Digital Equipment Corporation, Systems Research Center, 130 Lytton Avenue, Palo Alto, California 94301, USA

[3] Dipartimento di Matematica, Via del Capitano 15, 53100 Siena, Italy

Abstract. System $\mathcal{R}$ is an extension of system F that formalizes Reynolds' notion of relational parametricity. In system $\mathcal{R}$, considerably more λ-terms can be proved equal than in system F: for example, the encoded weak products of F are strong products in $\mathcal{R}$. Also, many "theorems for free" à la Wadler can be proved formally in $\mathcal{R}$. In this paper we describe a semantics for system $\mathcal{R}$. As a first step, we give a precise and general reconstruction of the per model of system F of Bainbridge et al., presenting it as a categorical model in the sense of Seely. Then we interpret system $\mathcal{R}$ in this model.

1 From Models of F to Models of $\mathcal{R}$

The principle of parametricity has gone through many avatars. First Strachey distinguished parametric polymorphism and ad hoc polymorphism [Str67]. Strachey described parametric polymorphism as the pure polymorphism of functions like *append*, which works on lists of any type uniformly. In contrast, a function like *print* examines and branches on the types of its arguments, and hence Strachey deemed its polymorphism ad hoc.

Reynolds formalized Strachey's notion of parametricity [Rey83], in his attempt to define a set-theoretic model for Girard's system F [Gir72]. According to Reynolds' semantic definition, a polymorphic function is parametric if its instances at related types are related. For example, let us take a polymorphic function f of type $\forall(X).X \rightarrow X$ (the type of the identity function). If X is instantiated to two types A and B with a relation R between them, and a has type A, b has type B, and aRb, then we must obtain $f(a)Rf(b)$. All of the definable functions of system F are parametric, but system F admits models with non-definable, non-parametric elements.

Since Reynolds' work, there have been many studies of parametricity. In particular, Bainbridge et al. introduced a view of parametricity based on dinaturality [BFSS90]. They also developed Reynolds' ideas in a variant of the partial-equivalence-relation (per) model. It seems still unknown whether the standard per model is parametric without modification.

System $\mathcal{R}$ is an extension of system F that formalizes a parametricity requirement. The intent is to capture Reynolds' notion of relational parametricity in a formal system without reference to a particular model. Other formal systems with similar features exist, serving related purposes; see [ACC93] for a comparison.

In a preliminary version of system $\mathcal{R}$, quantification over pers and quantification over relations were equated. This equation was rather attractive but too daring; Hasegawa ingeniously exploited it to derive an inconsistency [ACC93]. The inconsistency stimulated our interest in semantics, and particularly in the problem of harmonizing the use of pers and the use of general relations.

Both pers and relations have a place in the semantics and in the logic of parametricity. Pers are important as the denotations of types. From the point of view of formal reasoning, pers are the basis of equational reasoning. Relations enforce the requirement of parametricity in the construction of types, as in Reynolds' original (non-existent) set-theoretic model and in the model of Bainbridge et al. Logically, relations correspond to predicates, and many useful ones can be defined from the graphs of definable functions. Thus, both pers and relations play a role in system $\mathcal{R}$, and a semantics for system $\mathcal{R}$ should shed some light on their interaction.

In this paper we show how a parametric model of system F can be extended to a parametric model of system $\mathcal{R}$. This extension might be possible for any parametric model of system F, but we carry it out for the modified per model of Bainbridge et al. Along the way, we give a precise and general reconstruction of this per model. We present it as a categorical model in the sense of Seely [See87].

The next two sections introduce the syntax of system $\mathcal{R}$ and the necessary categories of pers and relations. Section 4 defines two parametric semantics of system F; then section 5 extends one of these semantics to system $\mathcal{R}$.

2 System $\mathcal{R}$

This section is an introduction to system $\mathcal{R}$, adapted from [ACC93]. System $\mathcal{R}$ is a formal system with judgements and rules in the style of those of F. In order to deal explicitly with relational parametricity, the judgements of $\mathcal{R}$ generalize those of F; they are:

$$\vdash_{\mathcal{R}} E \qquad\qquad E \vdash_{\mathcal{R}} \begin{array}{c}\sigma\\ \mathcal{S}\\ \tau\end{array} \qquad\qquad E \vdash_{\mathcal{R}} \begin{array}{c}M : \sigma\\ \mathcal{S}\\ N : \tau\end{array}$$

which mean, respectively, E is a legal environment; $\mathcal{S}$ is a relation between types σ and τ in E; and $\mathcal{S}$ relates M of type σ and N of type τ in E.

A built-in equality judgement on values is not necessary. Instead of writing that M and N are equal in σ, we can turn the type σ into a relation σ^* (intuitively, the identity relation on σ) and write that σ^* relates M and N. Similarly, there is no need for a built-in typing judgement. We write:

$$E \vdash_{\mathcal{R}} M : \sigma \text{ for } E \vdash_{\mathcal{R}} \begin{array}{c}M : \sigma\\ \sigma^*\\ M : \sigma\end{array} \qquad \text{and} \qquad E \vdash_{\mathcal{R}} M = N : \sigma \text{ for } E \vdash_{\mathcal{R}} \begin{array}{c}M : \sigma\\ \sigma^*\\ N : \sigma\end{array}$$

The environments of $\mathcal{R}$ are lists of components of two sorts, directly inspired by the corresponding ones for F environments:

$$\begin{array}{c}X\\ \mathcal{W}\\ Y\end{array} \qquad \text{and} \qquad \begin{array}{c}x : \sigma\\ \mathcal{S}\\ y : \tau\end{array}$$

which mean, respectively, $\mathcal{W}$ is a relation variable between type variables X (domain) and Y (codomain); and variables x and y have types σ and τ, respectively, and are related by $\mathcal{S}$

With this notation, we now explain some of the rules for related values judgements of $\mathcal{R}$. We write $x \notin M$ to mean that x is not a free variable of M. We start with rules that imitate those of F for $\rightarrow$ and $\forall$.

The introduction and elimination rules for $\rightarrow$ are, respectively:

$$\frac{E,\ \begin{array}{c}x : \sigma_1\\ \mathcal{R}\\ y : \sigma_2\end{array} \vdash_{\mathcal{R}} \begin{array}{c}M : \tau_1\\ \mathcal{S}\\ N : \tau_2\end{array} \qquad \begin{array}{c}x \notin N, \mathcal{S}\\ y \notin M, \mathcal{S}\end{array} \qquad E \vdash_{\mathcal{R}} \begin{array}{c}\tau_1\\ \mathcal{S}\\ \tau_2\end{array}}{E \vdash_{\mathcal{R}} \begin{array}{c}\lambda(x : \sigma_1).M : \sigma_1 \rightarrow \tau_1\\ \mathcal{R} \rightarrow \mathcal{S}\\ \lambda(x : \sigma_2).N : \sigma_2 \rightarrow \tau_2\end{array}}$$

$$\frac{E \vdash_{\mathcal{R}} \begin{array}{c}M_1 : \sigma_1 \rightarrow \tau_1\\ \mathcal{R} \rightarrow \mathcal{S}\\ M_2 : \sigma_2 \rightarrow \tau_2\end{array} \qquad E \vdash_{\mathcal{R}} \begin{array}{c}N_1 : \sigma_1\\ \mathcal{R}\\ N_2 : \sigma_1\end{array}}{E \vdash_{\mathcal{R}} \begin{array}{c}M_1 N_1 : \tau_1\\ \mathcal{S}\\ M_2 N_2 : \tau_2\end{array}}$$

These rules follow the same pattern as the F rules:

$$\frac{E, x:\sigma \vdash_F M:\tau}{E \vdash_F \lambda(x:\sigma).M : \sigma\to\tau} \qquad \frac{E\vdash_F M:\sigma\to\tau \qquad E\vdash_F N:\sigma}{E\vdash_F MN:\tau}$$

The introduction rule says: Assume that if $\mathcal{R}$ relates x of type σ_1 and y of type σ_2, then $\mathcal{S}$ relates M of type τ_1 and N of type τ_2. For technical reasons, assume also that $\mathcal{S}$ is a relation between τ_1 and τ_2, as one would expect. Then $\mathcal{R}\to\mathcal{S}$, a relation between $\sigma_1\to\tau_1$ and $\sigma_2\to\tau_2$, relates the functions $\lambda(x:\sigma_1).M$ of type $\sigma_1\to\tau_1$ and $\lambda(y:\sigma_2).N$ of type $\sigma_2\to\tau_2$. The elimination rule works in the opposite direction, applying related functions to related inputs and yielding related outputs.

The introduction and elimination rules for $\forall$ are:

$$\frac{E,\begin{array}{c}X\\ \mathcal{W}\\ Y\end{array}\vdash_{\mathcal{R}}\begin{array}{c}M:\sigma_1\\ \mathcal{S}\\ N:\sigma_2\end{array}\quad\begin{array}{c}X\notin\mathcal{S},N,\sigma_2\\ Y\notin\mathcal{S},M,\sigma_1\end{array}}{E\vdash_{\mathcal{R}}\begin{array}{c}\lambda(X).M:\forall(X).\sigma_1\\ \forall(\mathcal{W}).\mathcal{S}\\ \lambda(Y).N:\forall(Y).\sigma_2\end{array}} \qquad \frac{E\vdash_{\mathcal{R}}\begin{array}{c}M:\forall(X).\sigma_1\\ \forall(\mathcal{W}).\mathcal{S}\\ N:\forall(Y).\sigma_2\end{array}\quad E\vdash_{\mathcal{R}}\begin{array}{c}\tau_1\\ \mathcal{R}\\ \tau_2\end{array}}{E\vdash_{\mathcal{R}}\begin{array}{c}M\tau_1:\sigma_1[\tau_1/X]\\ \mathcal{S}[\mathcal{R}/\mathcal{W}]\\ N\tau_2:\sigma_2[\tau_2/Y]\end{array}}$$

These rules follow the same pattern as the F rules:

$$\frac{E, X \vdash_F M:\sigma}{E \vdash_F \lambda(X).M : \forall(X).\sigma} \qquad \frac{E\vdash_F M:\forall(X).\sigma \qquad E\vdash_F \tau}{E\vdash_F M\tau:\sigma[\tau/X]}$$

The introduction rule says: Assume that if $\mathcal{W}$ is a relation between types X and Y, then $\mathcal{S}$ relates M of type σ_1 and N of type σ_2. Then $\forall(\mathcal{W}).\mathcal{S}$, a relation between $\forall(X).\sigma_1$ and $\forall(Y).\sigma_2$, relates the polymorphic terms $\lambda(X).M$ of type $\forall(X).\sigma_1$ and $\lambda(Y).N$ of type $\forall(Y).\sigma_2$. Again, the elimination rule works in the opposite direction.

Since the relation constructions parallel the type constructions, we can easily define a relation σ^* for every type σ: we replace all quantifiers over types with corresponding quantifiers over relations. This $(\ )^*$ operation is used in the rules for variables:

(Rel Val $x\mathcal{R}y$)

$$\frac{\vdash_{\mathcal{R}} E',\begin{array}{c}x:\sigma_1\\ \mathcal{R}\\ y:\sigma_2\end{array},E''}{E',\begin{array}{c}x:\sigma_1\\ \mathcal{R}\\ y:\sigma_2\end{array},E''\vdash_{\mathcal{R}}\begin{array}{c}x:\sigma_1\\ \mathcal{R}\\ y:\sigma_2\end{array}}$$

(Rel Val $\mathcal{R}x$)

$$\frac{\vdash_{\mathcal{R}} E',\begin{array}{c}x:\sigma_1\\ \mathcal{R}\\ y:\sigma_2\end{array},E''}{E',\begin{array}{c}x:\sigma_1\\ \mathcal{R}\\ y:\sigma_2\end{array},E''\vdash_{\mathcal{R}}\begin{array}{c}x:\sigma_1\\ \sigma_1{}^*\\ x:\sigma_1\end{array}}$$

(Rel Val $\mathcal{R}y$)

$$\frac{\vdash_{\mathcal{R}} E',\begin{array}{c}x:\sigma_1\\ \mathcal{R}\\ y:\sigma_2\end{array},E''}{E',\begin{array}{c}x:\sigma_1\\ \mathcal{R}\\ y:\sigma_2\end{array},E''\vdash_{\mathcal{R}}\begin{array}{c}y:\sigma_2\\ \sigma_2{}^*\\ y:\sigma_2\end{array}}$$

The first rule is straightforward. The other two formalize the parametricity condition. Basically, they assert that the relation σ^* relates to itself any element of a type σ.

The preceding rules, together with the rules of β and η conversion, form the core of the part of $\mathcal{R}$ that deals with relations built from variables, $\to$, and $\forall$. This part of $\mathcal{R}$ is not a very powerful proof system on its own, but it suffices to encode F. In particular:

$$\text{if } E\vdash_F M=N:\sigma \text{ then } E\vdash_{\mathcal{R}}\begin{array}{c}M:\sigma\\ \sigma^*\\ N:\sigma\end{array}$$

In the second sequent we use E as an abbreviation of an $\mathcal{R}$ environment; see [ACC93].

In addition, $\mathcal{R}$ has rules for defining relations from functions:

$$\frac{E \vdash_{\mathcal{R}} M : \sigma_1 \to \sigma_2 \qquad E \vdash_{\mathcal{R}} N : \sigma_1}{E \vdash_{\mathcal{R}} \begin{array}{c} N : \sigma_1 \\ < M > \\ MN : \sigma_2 \end{array}} \qquad \frac{E \vdash_{\mathcal{R}} \begin{array}{c} N_1 : \sigma_1 \\ < M > \\ N_2 : \sigma_2 \end{array}}{E \vdash_{\mathcal{R}} \begin{array}{c} MN_1 : \sigma_2 \\ \sigma_2^* \\ N_2 : \sigma_2 \end{array}}$$

According to these rules, a function M from σ_1 to σ_2 can be viewed as a relation $< M >$ between σ_1 and σ_2 (intuitively, the graph of M).

Functional relations are essential to the power of $\mathcal{R}$. They are often useful for obtaining "free theorems" as in Wadler's work [Wad89]. They have no analogue in F.

In $\mathcal{R}$, free theorems can be stated and proved in a logical framework and without any reference to particular classes of models. An easy "theorem for free" asserts that the type $\forall(X).X \to Bool$ contains only constant functions; this is not provable in F. There are many more substantial examples: for example, encoded products, sums, algebras, quantifications are strong. Various metatheorems can also be obtained, such as syntactic versions of Reynolds' abstraction theorem and identity extension lemma.

The full sets of rules for systems F and $\mathcal{R}$ can be found in [ACC93]. Below we list only the rules for the related types judgements.

(Rel $\mathcal{W}$)

$$\frac{\vdash_{\mathcal{R}} E', \begin{array}{c} X \\ \mathcal{W} \\ Y \end{array}, E''}{E', \begin{array}{c} X \\ \mathcal{W} \\ Y \end{array}, E'' \vdash_{\mathcal{R}} \begin{array}{c} X \\ \mathcal{W} \\ Y \end{array}}$$

(Rel $\mathcal{W}X$)

$$\frac{\vdash_{\mathcal{R}} E', \begin{array}{c} X \\ \mathcal{W} \\ Y \end{array}, E''}{E', \begin{array}{c} X \\ \mathcal{W} \\ Y \end{array}, E'' \vdash_{\mathcal{R}} \begin{array}{c} X \\ X \\ X \end{array}}$$

(Rel $\mathcal{W}Y$)

$$\frac{\vdash_{\mathcal{R}} E', \begin{array}{c} X \\ \mathcal{W} \\ Y \end{array}, E''}{E', \begin{array}{c} X \\ \mathcal{W} \\ Y \end{array}, E'' \vdash_{\mathcal{R}} \begin{array}{c} Y \\ Y \\ Y \end{array}}$$

(Rel Arrow)

$$\frac{E \vdash_{\mathcal{R}} \begin{array}{c} \sigma_1 \\ \mathcal{R} \\ \sigma_2 \end{array} \qquad E \vdash_{\mathcal{R}} \begin{array}{c} \tau_1 \\ \mathcal{S} \\ \tau_2 \end{array}}{E \vdash_{\mathcal{R}} \begin{array}{c} \sigma_1 \to \tau_1 \\ \mathcal{R} \to \mathcal{S} \\ \sigma_2 \to \tau_2 \end{array}}$$

(Rel FRel)

$$\frac{E \vdash_{\mathcal{R}} M : \sigma \to \tau}{E \vdash_{\mathcal{R}} \begin{array}{c} \sigma \\ < M > \\ \tau \end{array}}$$

(Rel Forall)

$$\frac{E, \begin{array}{c} X \\ \mathcal{W} \\ Y \end{array} \vdash_{\mathcal{R}} \begin{array}{c} \sigma_1 \\ \mathcal{S} \\ \sigma_2 \end{array} \qquad \begin{array}{c} X \notin \mathcal{S}, \sigma_2 \\ Y \notin \mathcal{S}, \sigma_1 \end{array}}{E \vdash_{\mathcal{R}} \begin{array}{c} \forall(X).\sigma_1 \\ \forall(\mathcal{W}).\mathcal{S} \\ \forall(Y).\sigma_2 \end{array}}$$

3 Categories of Pers and Categories of Relations

In this section we introduce the main notions and tools that we use for the definition of our model of $\mathcal{R}$. For more details, see for example [Bar84].

First we recall the definition of interpretation (over a combinatory algebra) for the untyped λ-calculus. Then we consider the category of partial equivalence relations (pers) and some related categorical constructions. Finally, we introduce the category of saturated relations between pers, extending the categorical constructions previously defined for pers. Our models are based on this last category.

It is a well-known result about combinatory logic that abstraction is internally definable. This means that we can associate a *CL*-term $[x].M$ with every *CL*-term M so that for all *CL*-terms N:

$$([x].M)N \triangleright_{CL} M[N/x]$$

where $\triangleright_{CL}$ is the reduction relation on *CL*-terms obtained by orienting the two equations for k and s from left to right.

This enables us to translate λ-terms to *CL*-terms. Given a λ-term M, we denote its translation by $(M)_{CL}$. Using the translation, we can define the semantics of a λ-term M by $[\![M]\!]_\rho = [\![(M)_{CL}]\!]_\rho$. This interpretation satisfies the rule of β-conversion. The difference between the usual interpretation of the λ-calculus based on λ-models and the combinatory-algebra interpretation regards the ξ rule (if $M = N$ then $\lambda(x).M = \lambda(x).N$), which is not valid in arbitrary combinatory algebras.

In a partial combinatory algebra, the application $\cdot$ is partial and for all a, b, c it satisfies: $(k \cdot a) \cdot b = a$ and $((s \cdot a) \cdot b) \cdot c \simeq (a \cdot c) \cdot (b \cdot c)$; this implies that we cannot define the meaning of general λ-terms. However, we can still define the meaning of some terms, useful in later results:

Lemma 1. *For every λ-term M and variable x, the semantics of $(\lambda x.M)_{CL}$ is defined in every partial combinatory algebra.*

Starting from an arbitrary partial combinatory algebra, we define categories of pers and categories of relations.

Definition 2. The category **PER** of partial equivalence relations (pers) on a partial combinatory algebra D is defined by:

- $A \in Obj(\mathbf{PER})$ iff it is a symmetric and transitive partial relation on D; its domain of definition is denoted $dom(A)$ and the partial quotient of D by A is denoted $Q(A)$;
- a morphism $(f : A \to B) \in Mor(\mathbf{PER})$ is a function $f : Q(A) \to Q(B)$ such that there is an element $n \in D$ that realizes f, that is, for all $a \in dom(A) : f([a]_A) = [n \cdot a]_B$.

Definition 3. The category **SAT** of saturated relations on a partial combinatory algebra D is defined by:

- a saturated relation $(R : A \not\rightarrow B) \in Obj(\mathbf{SAT})$ is given by two pers A and B over D and a relation $R \subseteq dom(A) \times dom(B)$ such that $R = A; R; B$ or, equivalently:
  - R is a relation between $dom(A)$ and $dom(B)$, that is, if aRb then aAa and bBb;
  - R is saturated, that is, if aAb, bRc, and cBd, then aRd.
- a morphism $f : (R : A \not\rightarrow A') \to (S : B \not\rightarrow B') \in Mor(\mathbf{SAT})$ between two saturated relations consists of a couple $(f' : A \to B, f'' : A' \to B')$ of per morphisms such that:

$$\text{for all } a \in dom(A),\ b \in dom(A'),\ \text{if } aRb \text{ then } f'(a)Sf''(b)$$

We often write R for $R : A \not\rightarrow B$, and sometimes refer to a saturated relation simply as a relation. We call A and B the domain and the codomain of $R : A \not\rightarrow B$, respectively, and write $A = dom(R)$ and $B = cod(R)$. More generally, given a tuple $\overline{S}$ of saturated relations, we write $dom(\overline{S})$ and $cod(\overline{S})$ for the tuples of domains and codomains of the relations in $\overline{S}$.

The saturation property allows us to see a saturated relation indistinctly as a relation between equivalence classes or as a relation between elements of such classes.

Moreover, since any per can be seen as the identity saturated relation on itself, and every morphism between pers (seen as saturated relations) in **SAT** must be a couple of equal **PER** morphisms, we find **PER** as a full subcategory of **SAT**.

Next, we generalize the usual categorical constructions of product ($\times$), exponentiation ($\Rightarrow$), and intersection ($\bigcap$) from **PER** to **SAT**. Products do not appear explicitly in any of our formal systems, but they are part of the necessary categorical structure. In order to define products, we encode the projections $(\text{-})_1, (\text{-})_2 : D \to D$ in D as the functions realized by the interpretations of $\lambda z.z(\lambda x.\lambda y.x)$ and $\lambda z.z(\lambda x.\lambda y.y)$, respectively. (By Lemma 1 their semantics is defined even in a partial combinatory algebra.)

We are interested in functions on relations $F : \mathit{Obj}(\mathbf{SAT})^k \to \mathit{Obj}(\mathbf{SAT})$ with the property:

(SAT-FUNC) if $\overline{S} : \overline{A} \nrightarrow \overline{B}$ then $F(\overline{S}) : F(\overline{A}) \nrightarrow F(\overline{B})$

Note that $F(\overline{S}) : F(\overline{A}) \nrightarrow F(\overline{B})$ implies that $F(\overline{A})$ and $F(\overline{B})$ are pers. We stipulate that a function $F : \mathit{Obj}(\mathbf{SAT})^0 \to \mathit{Obj}(\mathbf{SAT})$ satisfies (SAT-FUNC) iff it is constantly equal to a per.

Next we describe how to define functions on relations by product, exponentiation, and intersection with parameters. All of these constructions preserve property (SAT-FUNC).

Recall the definition of product and exponentiation in **PER**:

$$a(A \times B)b \text{ iff } a_1 A b_1 \text{ and } a_2 B b_2$$
$$a(A \Rightarrow B)b \text{ iff for all } c, d, \text{ if } cAd \text{ then } (a \cdot c)B(b \cdot d)$$

Product and exponentiation in **SAT** are similar:

Definition 4 (Product). Given $R : A \nrightarrow B$ and $S : C \nrightarrow D$, their product $R \times S : (A \times C) \nrightarrow (B \times D)$ is defined by:

$$a(R \times S)b \quad \text{iff} \quad \begin{cases} a(A \times C)a \\ a_1 R b_1 \text{ and } a_2 S b_2 \\ b(B \times D)b \end{cases}$$

Definition 5 (Exponentiation). Given $R : A \nrightarrow B$ and $S : C \nrightarrow D$ their exponentiation $R \Rightarrow S : (A \Rightarrow C) \nrightarrow (B \Rightarrow D)$ is defined by:

$$a(R \Rightarrow S)b \quad \text{iff} \quad \begin{cases} a(A \Rightarrow C)a \\ \text{for all } c, d, \text{if } cRd \text{ then } (a \cdot c)S(b \cdot d) \\ b(B \Rightarrow D)b \end{cases}$$

It is easy to prove that these constructions are the categorical product and exponentiation in **SAT**, and extend those defined in **PER**.

Like pers, saturated relations also support infinite intersections:

Definition 6 (Intersection with Parameters). Given a function

$$F : \mathit{Obj}(\mathbf{SAT})^{k+1} \to \mathit{Obj}(\mathbf{SAT})$$

with property (SAT-FUNC), we define the intersection of F on its last argument

$$\bigcap_{R:A \nrightarrow B}^{SAT} F[-, R] : \mathit{Obj}(\mathbf{SAT})^k \to \mathit{Obj}(\mathbf{SAT})$$

with

$$\left(\bigcap_{R:A \nrightarrow B}^{SAT} F[-, R]\right)(\overline{S}) : \left(\bigcap_{R:A \nrightarrow B} F(dom(\overline{S}), R)\right) \nrightarrow \left(\bigcap_{R:A \nrightarrow B} F(cod(\overline{S}), R)\right)$$

by:

$$a\left(\left(\bigcap_{R:A\not\rightarrow B}^{SAT} F[-,R]\right)(\overline{S})\right)b \text{ iff } \begin{cases} a\left(\bigcap_{R:A\not\rightarrow B} F(dom(\overline{S}),R)\right)a \\ a\left(\bigcap_{R:A\not\rightarrow B} F(\overline{S},R)\right)b \\ b\left(\bigcap_{R:A\not\rightarrow B} F(cod(\overline{S}),R)\right)b \end{cases}$$

for all $\overline{S}:\overline{A}\not\rightarrow\overline{B}\in Obj(\mathbf{SAT})^k$.

Note that in the right-hand side of the previous definition we used set-theoretical intersection. In section 4.2 we show that infinite intersections are categorical products in **SAT**.

Proposition 7. *Intersection with parameters is well defined, that is, it maps tuples of saturated relations to saturated relations. Moreover it satisfies (SAT-FUNC), that is, if* $\overline{S}:\overline{A}\not\rightarrow\overline{B}$ *then*

$$\left(\bigcap_{R:A\not\rightarrow B}^{SAT} F[-,R](\overline{S})\right):\left(\bigcap_{R:A\not\rightarrow B}^{SAT} F[-,R](\overline{A})\right)\not\rightarrow\left(\bigcap_{R:A\not\rightarrow B}^{SAT} F[-,R](\overline{B})\right)$$

In particular, this proposition says that even if $F[\overline{A},R]$ is in general a saturated relation, its intersection over all saturated relations R is a per. This is the crucial property that allows us, in the next section, to interpret quantification over type variables as ranging over all saturated relations and not only over pers.

We now have the ingredients for our parametric interpretation of systems F and $\mathcal{R}$, described in sections 4 and 5, respectively.

4 Parametric Semantics of System *F*

We define an interpretation of system F, which we call the parametric per model of F, or **SAT** model of F. In the first part of this section, we define the interpretation concretely, by directly interpreting the judgements of the syntax of F in **SAT**. This is a reconstruction of the model sketched in [BFSS90]. The second part of the section gives a more general categorical description of the **SAT** model. The main difference in outcome is that the second construction requires less of the underlying structure—namely, only a partial combinatory algebra is needed.

4.1 Concrete Models

Throughout this subsection, we assume that D is at least a total combinatory algebra.

Types. In the parametric per model, quantification over types is understood as quantification over all relations. In order to interpret the quantification on a type variable as a quantification over all relations, we must be able to instantiate the variable with arbitrary relations. Therefore, we must be able to provide the semantics of types in a general context where type variables are interpreted as relations.

The type expressions of F are obtained from type variables by the $\rightarrow$ and $\forall$ constructors. We interpret a type expression σ by a function

$$[\sigma]: RelAssign \rightarrow Obj(\mathbf{SAT})$$

where *RelAssign* is the domain of (total) *relation assignments* which map type variables to relations. If a relation assignment maps every type variable to a per, then we call it a *per assignment*. The subset of per assignments is *PerAssign*. Typically, ε denotes a relation assignment and η a per assignment. The operators $\uparrow$ and $\downarrow$ transform an arbitrary relation assignment ε into a per assignment:

$$\varepsilon\uparrow(X) = dom(\varepsilon(X)) \qquad \varepsilon\downarrow(X) = cod(\varepsilon(X))$$

Definition 8 (Semantics of Type Expressions). The semantics of type expressions is defined by:

$$[\![X]\!]_\varepsilon = \varepsilon(X) \qquad [\![\sigma_1 \to \sigma_2]\!]_\varepsilon = [\![\sigma_1]\!]_\varepsilon \Rightarrow [\![\sigma_2]\!]_\varepsilon \qquad [\![\forall(X)\sigma]\!]_\varepsilon = \bigcap^{SAT}_{R:A \nrightarrow B} [\![\sigma]\!]_{\varepsilon[R/X]}$$

Proposition 9. *The previous definition is proper, that is:*

1. *for all type expressions σ and all per environments η:* $\quad [\![\sigma]\!]_\eta \in Obj(\mathbf{PER})$
2. *for all type expressions σ and all relation environments ε:* $\quad [\![\sigma]\!]_\varepsilon : [\![\sigma]\!]_{\varepsilon\uparrow} \nrightarrow [\![\sigma]\!]_{\varepsilon\downarrow}$

This proposition easily follows from Proposition 7 together with the remark that product and exponentiation in **SAT** extend those defined in **PER**. Part (1) amounts to what is known in the literature as the Identity Extension Lemma. It is stated in [Rey83] for an hypothetical set-theoretic semantics, and is proved in [BFSS90] for the **SAT** model.

We close this section with a standard substitution lemma:

Lemma 10 (Substitution Lemma for Types). *For all type expressions σ and τ and all relation assignments ε:*

$$[\![\sigma[\tau/X]]\!]_\varepsilon = [\![\sigma]\!]_{\varepsilon[[\![\tau]\!]_\varepsilon/X]}$$

Typings. In what follows, we consider mostly derivable judgements, and often write judgement instead of derivable judgement. We interpret the typing judgements $E \vdash_F M : \sigma$ by functions

$$[\![E \vdash_F M : \sigma]\!] : PerAssign \to ValAssign \to \bigcup_{A \in Obj(PER)} Q(A)$$

where *ValAssign* is the domain of *value assignments*, which map each term variable to an equivalence class of some per. Typically ρ denotes a value assignment.

Definition 11.

- A value assignment ρ satisfies an environment E with respect to a per assignment η iff, for all $(x : \sigma) \in E$, $\rho(x) \in Q([\![\sigma]\!]_\eta)$. This is denoted $\rho \models_\eta E$.
- Two value assignments ρ_1 and ρ_2 satisfy an environment E with respect to a relation assignment ε iff, for all $(x : \sigma) \in E$, $\rho_1(x)\ [\![\sigma]\!]_\varepsilon\ \rho_2(x)$. This is denoted $\rho_1, \rho_2 \models_\varepsilon E$.

Definition 12. Given a value assignment ρ, we can obtain from it a new (untyped) value assignment $\rho^\star$ such that $\rho^\star(x) \in \rho(x)$, that is, $\rho^\star$ maps every term variable x to an (arbitrary) element of the equivalence class $\rho(x)$.

The *erase* function associates to every typed term M the untyped term *erase*(M) obtained by erasing all type decorations present in M. What follows is a version in our relational setting of a well-known semantical property of erasures of typed terms in **PER** models (see [Mit90]). It is usually stated for **PER** models over a λ-model D, but the proof actually uses only the total combinatory algebra structure: the totality is used to make sure that the interpretation of untyped terms in D is always defined.

Theorem 13. *For all typing judgements $E \vdash_F M : \sigma$ and all assignments η and ρ such that $\rho \models_\eta E$:*

$$\left(\llbracket erase(M) \rrbracket_{\rho^*}\right) \llbracket \sigma \rrbracket_\eta \left(\llbracket erase(M) \rrbracket^D_{\rho^*}\right)$$

Moreover, this is independent from the specific choice of ρ^*.

To prove this result we need a more general statement.

Theorem 14 (Abstraction Theorem for Erasures). *For all typing judgements $E \vdash_F M : \sigma$ and all assignments ε, ρ_1, and ρ_2 such that $\rho_1, \rho_2 \models_\varepsilon E$:*

$$\left(\llbracket erase(M) \rrbracket_{\rho_1^*}\right) \llbracket \sigma \rrbracket_\varepsilon \left(\llbracket erase(M) \rrbracket_{\rho_2^*}\right)$$

Moreover, this is independent from the specific choice of ρ_1^* and ρ_2^*.

The previous theorem allows us to define the semantics of typed terms from the semantics of their erasures, provided that the **SAT** model is based on a total combinatory algebra:

Definition 15 (Semantics of Typing Judgements). Given a per assignment η and a value assignment ρ such that $\rho \models_\eta E$, we can define the semantics of the judgement $E \vdash_F M : \sigma$ by:

$$\llbracket E \vdash_F M : \sigma \rrbracket_{\eta\rho} = \left[\llbracket erase(M) \rrbracket_{\rho^*}\right]_{\llbracket \sigma \rrbracket_\eta}$$

Using the Abstraction Theorem for Erasures it is easy to verify that the previous definition is proper. We obtain:

Corollary 16 (Abstraction Theorem). *For all typing judgements $E \vdash_F M : \sigma$ and all assignments ε, ρ_1, and ρ_2 such that $\rho_1, \rho_2 \models_\varepsilon E$:*

$$\left(\llbracket E \vdash_F M : \sigma \rrbracket_{\varepsilon\uparrow\rho_1}\right) \llbracket \sigma \rrbracket_\varepsilon \left(\llbracket E \vdash_F M : \sigma \rrbracket_{\varepsilon\downarrow\rho_2}\right)$$

To close this section, we state a standard substitution lemma.

Lemma 17 (Substitution Lemma for Terms). *For all typing judgements $E, \overline{X} \vdash_F M : \sigma$ all type expressions $\overline{\tau}$ whose free variables are in the domain of E, and all assignments η and ρ such that $\rho \models_\eta E$, we have:*

$$\llbracket E \vdash_F M[\overline{\tau}/\overline{X}] : \sigma[\overline{\tau}/\overline{X}] \rrbracket_{\eta\rho} = \llbracket E, \overline{X} \vdash_F M : \sigma \rrbracket_{\eta[\overline{\llbracket \tau \rrbracket_\eta}/\overline{X}]\rho}$$

Equalities. The **SAT** model is sound with respect to the equality rules of F.

Theorem 18 (Soundness for F Equalities). *Given an equality judgement $E \vdash_F M = N : \sigma$, a per assignment η and a value assignment ρ such that $\rho \models_\eta E$, we have:*

$$\llbracket E \vdash_F M : \sigma \rrbracket_{\eta\rho} \llbracket \rho \rrbracket_\eta \llbracket E \vdash_F N : \sigma \rrbracket_{\eta\rho}$$

4.2 Categorical Models

In the previous subsection we have assumed a total combinatory algebra in order to define the semantics of typed terms from the semantics of their erasures. In this section we overcome this limitation. We build a **SAT** model for system F starting from an arbitrary partial combinatory algebra. This will be possible by moving from an untyped semantics for typed terms (that is, a semantics based on erasures) to a typed semantics. The typed semantics is presented as a categorical model

Categorical models of system F are based on the quantifiers-as-adjoints paradigm, which goes back to Lawvere [Law69]; Seely has defined them under the name of PL-categories [See87].

PL-categories. PL-categories are an algebraic generalization of the models of simply typed λ-calculus in a bi-dimensional universe of cartesian closed categories indexed over a global category. PL-categories are sometimes referred to as *external* models of F in contrast with the *internal* ones which use the internal category theory.

We assume some acquaintance with the notion of indexed category and of categorical models of system F, but provide the main definitions. The definition of external model is based on that of indexed category. A model is given via a contravariant functor $\mathbf{G}$ from a category $\mathbf{E}$ to $\mathbf{CCCat}$, the category of all (small) cartesian closed categories and cartesian closed functors between them. The category $\mathbf{E}$ is cartesian and has a distinguished object Ω, interpreting the collection of types. Products Ω^p (denoted by p from now) are used to give meanings to environments E declaring p type variables. Types legal in E are interpreted by arrows in $\mathbf{E}[\Omega^p, \Omega]$. The functor $\mathbf{G} : \mathbf{E}^{op} \to \mathbf{CCCat}$ takes Ω^p in $\mathbf{E}$ to a (local) category $\mathbf{G}(\Omega^p)$ whose objects are the types legal in E. Thus, types appear both as arrows in $\mathbf{E}$ and as objects in the local category $\mathbf{G}(\Omega^p)$ and we require:

$$Obj((\mathbf{G}(\Omega^p)) = \mathbf{E}[\Omega^p, \Omega]$$

The arrows of the local category $\mathbf{G}(\Omega^p)$ interpret the terms of system F whose free type variables are in E. Every local category is required to be a model of the simply typed λ-calculus, that is, a cartesian closed category. The abstraction on type variables is described as the right adjoint to the diagonal functor.

SAT as a PL-category. We now recast the **SAT** model as a PL-category.

Definition 19 (Global Category). The objects of the global category $\mathbf{E}$ are the set $Obj(\mathbf{SAT})$ and its powers. The set of morphisms is defined in two steps: first we define

$$\mathbf{E}[p, 1] = \{F : Obj(\mathbf{SAT})^p \to Obj(\mathbf{SAT}) \mid F \text{ satisfies } (SAT\text{-}FUNC)\}$$

and then

$$\mathbf{E}[p, q] = \mathbf{E}[p, 1]^q$$

Definition 20 (Indexed Category). The collection of objects of the indexed category $\mathbf{G}(p)$ is $\mathbf{E}[p, 1]$, by the definition of PL-categories. The morphisms in $\mathbf{G}(p)$ are the uniformly realized arrows between objects, that is, the arrows $\Phi : F \to H$ such that:

- $F, H \in Obj(\mathbf{G}(p))$;
- $\Phi : p \to Mor(\mathbf{SAT})$ and for all $\overline{R} \in p$, $\Phi(\overline{R}) : F(\overline{R}) \to H(\overline{R})$ in $Mor(\mathbf{SAT})$;
- there exists $n \in D$ that realizes uniformly Φ in the sense that for all $\overline{R} \in p$, (n, n) realizes $\Phi(\overline{R})$ in $Mor(\mathbf{SAT})$.

Given $L \in \mathbf{E}[p, q]$, we define $\mathbf{G}(L)$ as the functor from $\mathbf{G}(q)$ to $\mathbf{G}(p)$ that acts on both objects and morphisms as the pre-composition with L.

Product and exponentiation in the fibers $\mathbf{G}(p)$ are defined componentwise using those of **SAT** as follows. For all $F, G \in \mathbf{G}(p)$ and $\overline{R} \in p$:

- $(F \times G)(\overline{R}) = F(\overline{R}) \times G(\overline{R})$
- $F^G(\overline{R}) = F(\overline{R})^{G(\overline{R})}$

Note that $\mathbf{G}(0)$ is isomorphic to **PER**. This corresponds to the fact that closed type expressions are interpreted as pers.

We interpret quantification by the intersection operator introduced in Definition 6.

Definition 21 (Indexed Adjunction). Given an object L of $\mathbf{G}(p+1)$, we define

$$\forall(p)(L) = \bigcap_{R:A \not\leftrightarrow B}^{SAT} L[-,R]$$

The behavior of $\forall(p)$ on morphisms is that if n realizes the morphism $\Phi : F \to H$ in $\mathbf{G}(p+1)$ then $\forall(p)(\Phi : F \to H)$ is the arrow from $\forall(p)(F)$ to $\forall(p)(H)$ in $\mathbf{G}(p)$ realized by n. For every p, $F \in Obj(\mathbf{G}(p))$ and $H \in Obj(\mathbf{G}(p+1))$, we define the isomorphism

$$\Delta(p) : \mathbf{G}(p+1)[\mathbf{Fst}(p)(F), H] \xrightarrow{\cong} \mathbf{G}(p)[F, \forall(p)(H)]$$

so that it sends a morphism $\Phi : \mathbf{Fst}(p)(F) \to H$ realized by n to the unique morphism from F to $\forall(p)(H)$ in $\mathbf{G}(p)$ realized by n.

For any $F \in \mathbf{G}(p+1)$ we denote by $\mathbf{Proj}^F(p)$ the counit of the indexed adjunction, that is, the following arrow in $\mathbf{G}(p+1)$:

$$\mathbf{Proj}^F(p) = \Delta^{-1}(p)(id_{\forall(F)}) : \mathbf{Fst}(p) \circ \forall(p)(F) \to F$$

Categorical Interpretation of Types and Terms. Using the standard machinery of categorical semantics over an indexed category it is straightforward to define the interpretation of the judgements of system F. For more details see [AM92].

Given a type judgement $E \vdash_F \sigma$ with p type variables declared in E, the semantics of σ in E, denoted by $[\sigma]_p$, is an object of the fiber $\mathbf{G}(p)$, that is, an arrow $Obj(\mathbf{SAT})^p \to Obj(\mathbf{SAT})$. We do the same for environments, so if $\vdash_F E$ is an environment judgment with p type variables declared in E, then its semantics, denoted by $[\vdash_F E]_p$, is an element of the fiber $\mathbf{G}(p)$, namely the product of the semantics of the types of the term variables declared in it. Finally, the semantics of a typing judgement $E \vdash_F M : \sigma$ with p type variables declared in E, denoted by $[E \vdash_F M : \sigma]_p$, is a morphism in the fiber $\mathbf{G}(p)$ from $[\vdash_F E]_p$ to $[\sigma]_p$.

With this kind of semantics the Abstraction Theorem comes "for free" since we have that

$$[E \vdash_F M : \sigma]_p : [\vdash_F E]_p \to [\sigma]_p$$

This means, in particular, that the semantics of M maps related values for its free term variables to related values.

Proposition 22. *In the case that the combinatory algebra D is total the categorical model coincides with the concrete one.*

5 Semantics of System $\mathcal{R}$

In this section we extend the first **SAT** model of F to a model of $\mathcal{R}$. We believe that an analogous extension is possible for the categorical **SAT** model of F. We prefer treating the first model because the structure of $\mathcal{R}$ complicates the notations for categorical models even further. The complications arise from the dependence of relation expressions upon term variables. Because of this dependence, the semantics of relation expressions is defined only under correct assignments to term variables; this is hard to express in a categorical style.

5.1 Related Types

Since type and relation variables are both present explicitly in system $\mathcal{R}$, we define relation assignments on both type and relation variables in such a way that type variables are mapped to pers and relation variables are mapped to relations. The domain *RelAssign* of relation assignments becomes:

$$RelVar \cup TypeVar \to Obj(\mathbf{SAT})$$

We denote $\eta \updownarrow$ the restriction of a relation assignment η to the set of type variables. Value assignments remain the same as in system F.

We interpret the related types judgements by functions of type:

$$\left[\!\!\left[E \vdash_{\mathcal{R}} \begin{matrix} \sigma \\ \mathcal{S} \\ \tau \end{matrix} \right]\!\!\right] : RelAssign \to ValAssign \to Obj(\mathbf{SAT})$$

Defining the semantics of relation expressions is a little more difficult than defining the semantics of type expressions. One technical reason for this is that relation expressions may contain term variables, so we are forced to make their meaning depend upon term variable assignments. Moreover, the presence of term variables inside relation expressions implies that not all relation expressions are meaningful, since we must add some hypothesis on the interpretation of term variables which, in turn, depend on the particular (syntactic) environment considered. In order to define the semantics of relation expressions, then, we would like to know in which environment we are working and that the relation expression is well-formed (that is, derivable) in this environment. For these reasons, it seems best to define the semantics of entire related types judgements, as we do next.

To interpret functional-relation expressions, we use an auxiliary function:

Definition 23. Let $FRel : Mor(\mathbf{PER}) \to Obj(\mathbf{SAT})$ be the function that maps an $f : A \to B$ to the relation $FRel(f) : A \not\leftrightarrow B$ such that:

$$a(FRel(f))b \text{ iff } b \in f([a]_A)$$

Note that, by construction, $FRel(f)$ is always saturated.

Definition 24. Given a derivable environment judgement $\vdash_{\mathcal{R}} E$ and a relation assignment η, we say that η is an assignment for E iff:

$$\text{for all} \begin{pmatrix} X \\ \mathcal{W} \\ Y \end{pmatrix} \in E, \quad \eta(\mathcal{W}) : \eta(X) \not\leftrightarrow \eta(Y).$$

Definition 25. We define the satisfaction of an environment judgement by a relation assignment and a value assignment, and the meaning of a related types judgement, with a joint inductive definition:

- Given an environment judgement $\vdash_{\mathcal{R}} E$, a relation assignment η for it, and a value assignment ρ, we say that ρ satisfies E with respect to η iff:

$$\textit{for all} \begin{pmatrix} x : \sigma_1 \\ \mathcal{R} \\ y : \sigma_2 \end{pmatrix} \in E, \quad \rho(x) \left[\!\!\left[E' \vdash_{\mathcal{R}} \begin{matrix} \sigma_1 \\ \mathcal{R} \\ \sigma_2 \end{matrix} \right]\!\!\right]_{\eta\rho} \rho(y)$$

where E' is such that $E = E', \begin{matrix} x : \sigma_1 \\ \mathcal{R} \\ y : \sigma_2 \end{matrix}, E''$. We write this $\rho \models_\eta E$.

– Given assignments η and ρ such that $\rho \models_\eta E$, we define the semantics of the related types judgement $E \vdash_{\mathcal{R}} \begin{array}{c}\sigma_1\\ \mathcal{S}\\ \sigma_2\end{array}$ by induction on its derivation:

$$\left[\!\!\left[E', \mathcal{W}, E'' \vdash_{\mathcal{R}} \begin{array}{c}X\\ \mathcal{W}\\ Y\end{array} \right]\!\!\right]_{\eta\rho} = \eta(\mathcal{W})$$

$$\left[\!\!\left[E', \mathcal{W}, E'' \vdash_{\mathcal{R}} \begin{array}{c}X\\ X\\ Y\end{array} \right]\!\!\right]_{\eta\rho} = \eta(X) \qquad \left[\!\!\left[E', \mathcal{W}, E'' \vdash_{\mathcal{R}} \begin{array}{c}X\\ Y\\ Y\end{array} \right]\!\!\right]_{\eta\rho} = \eta(Y)$$

$$\left[\!\!\left[E \vdash_{\mathcal{R}} \begin{array}{c}\sigma_1 \to \tau_1\\ \mathcal{S}_1 \to \mathcal{S}_2\\ \sigma_2 \to \tau_2\end{array} \right]\!\!\right]_{\eta\rho} = \left[\!\!\left[E \vdash_{\mathcal{R}} \begin{array}{c}\sigma_1\\ \mathcal{S}_1\\ \sigma_2\end{array} \right]\!\!\right]_{\eta\rho} \Rightarrow \left[\!\!\left[E \vdash_{\mathcal{R}} \begin{array}{c}\tau_1\\ \mathcal{S}_2\\ \tau_2\end{array} \right]\!\!\right]_{\eta\rho}$$

$$a \left[\!\!\left[E \vdash_{\mathcal{R}} \begin{array}{c}\forall(X).\sigma_1\\ \forall(\mathcal{W}).\mathcal{S}\\ \forall(Y).\sigma_2\end{array} \right]\!\!\right]_{\eta\rho} b \text{ iff } \begin{cases} a\, [\![\forall(X).\sigma_1]\!]^F_{\eta\natural}\, a \\ \text{for all } R : A \nrightarrow B \qquad a \left[\!\!\left[E, \mathcal{W} \vdash_{\mathcal{R}} \begin{array}{c}X \quad \sigma_1\\ \mathcal{S}\\ Y \quad \sigma_2\end{array} \right]\!\!\right]_{\tilde{\eta}\rho} b \\ \quad (\text{where } \tilde{\eta} = \eta[R, A, B/\mathcal{W}, X, Y]) \\ b\, [\![\forall(Y).\sigma_2]\!]^F_{\eta\natural}\, b \end{cases}$$

$$\left[\!\!\left[E \vdash_{\mathcal{R}} \begin{array}{c}\sigma_1\\ < M >\\ \sigma_2\end{array} \right]\!\!\right]_{\eta\rho} = \mathit{FRel}([\![E_F \vdash_F M : \sigma_1 \to \sigma_2]\!]^F_{\eta\natural,\rho})$$

Note that for the semantics of quantification and functional relations we have used the parametric semantics of system F (denoted $[\![-]\!]^F$). Thus, the semantics of the corresponding related types judgements do not depend upon their derivations. Moreover, the derivation of a related types judgement always follows exactly the structure of the relation except in the case of functional relations. Therefore, the semantics of a related types judgement is always independent of its derivation.

In the judgement $E_F \vdash_F M : \sigma_1 \to \sigma_2$, E_F stands for the F flattening of E, obtained by retaining the type variables and the term variables of E, and removing the relations. In [ACC93] appears a lemma, called (Flattened F derivations from $\mathcal{R}$ derivations), that asserts that if $E \vdash_{\mathcal{R}} M : \sigma_1 \to \sigma_2$ is provable in $\mathcal{R}$, then $E_F \vdash_F M : \sigma_1 \to \sigma_2$ is provable in F.

The presence of functional relations prevents us from using the intersection operator of Definition 6 since it is no longer true that $F(\overline{S}) : F(\overline{A}) \nrightarrow F(\overline{B})$. Indeed, consider the case where the function F is constantly equal to a functional relation: it does not map pers to pers. Instead, we have given a direct, pointwise definition of the meaning of intersection.

Theorem 26 (Soundness for Related Types Judgements).
For every related types judgement $E \vdash_{\mathcal{R}} \begin{array}{c}\sigma_1\\ \mathcal{S}\\ \sigma_2\end{array}$ *and assignments* η *and* ρ *such that* $\rho \models_\eta E$:

$$\left[\!\!\left[E \vdash_{\mathcal{R}} \begin{array}{c}\sigma_1\\ \mathcal{S}\\ \sigma_2\end{array} \right]\!\!\right]_{\eta\rho} : [\![\sigma_1]\!]^F_{\eta\natural} \nrightarrow [\![\sigma_2]\!]^F_{\eta\natural}$$

Given a type expression σ, we expect its F semantics to be equal to the $\mathcal{R}$ semantics of σ^*. This turns out to be an essential property.

Proposition 27. *If $E \vdash_{\mathcal{R}} \sigma$ then for all assignments η and ρ such that $\rho \models_\eta E$ we have*

$$\left[\!\!\left[E \vdash_{\mathcal{R}} \begin{matrix} \sigma \\ \sigma^* \\ \sigma \end{matrix} \right]\!\!\right]_{\eta\rho} = [\sigma]^F_{\eta\ddagger}$$

This proposition is a corollary of a more general statement:

Lemma 28. *If $E, \begin{matrix} \overline{X'} \\ \overline{W} \\ \overline{X''} \end{matrix} \vdash_{\mathcal{R}} \begin{matrix} \sigma[\overline{X'}/\overline{X}] \\ \sigma^*[\overline{W}/\overline{X}] \\ \sigma[\overline{X''}/\overline{X}] \end{matrix}$ then for all assignments η and ρ such that $\rho \models_\eta E$, and for all tuples $\overline{S} : \overline{A} \not\leftrightarrow \overline{B}$ of relations, we have:*

$$\left[\!\!\left[E, \begin{matrix} \overline{X'} \\ \overline{W} \\ \overline{X''} \end{matrix} \vdash_{\mathcal{R}} \begin{matrix} \sigma[\overline{X'}/\overline{X}] \\ \sigma^*[\overline{W}/\overline{X}] \\ \sigma[\overline{X''}/\overline{X}] \end{matrix} \right]\!\!\right]_{\eta[\overline{S},\overline{A},\overline{B}/\overline{W},\overline{X'},\overline{X''}],\rho} = [\sigma]^F_{\eta\ddagger[\overline{S}/\overline{X}]}$$

As usual, we have also a substitution lemma:

Lemma 29 (Substitution Lemma for Relations).

For all related types judgements $E, \begin{matrix} \overline{X} \\ \overline{W} \\ \overline{Y} \end{matrix} \vdash_{\mathcal{R}} \begin{matrix} \sigma_1 \\ \mathcal{R} \\ \sigma_2 \end{matrix}$ and $E \vdash_{\mathcal{R}} \begin{matrix} \overline{\tau_1} \\ \overline{S} \\ \overline{\tau_2} \end{matrix}$ and all assignments η and ρ such that $\rho \models_\eta E$, we have:

$$\left[\!\!\left[E \vdash_{\mathcal{R}} \begin{matrix} \sigma_1[\overline{\tau_1}, \overline{\tau_2}/\overline{X}, \overline{Y}] \\ \mathcal{R}[\overline{S}, \overline{\tau_1}, \overline{\tau_2}/\overline{W}, \overline{X}, \overline{Y}] \\ \sigma_2[\overline{\tau_1}, \overline{\tau_2}/\overline{X}, \overline{Y}] \end{matrix} \right]\!\!\right]_{\eta,\rho} = \left[\!\!\left[E, \begin{matrix} \overline{X} \\ \overline{W} \\ \overline{Y} \end{matrix} \vdash_{\mathcal{R}} \begin{matrix} \sigma_1 \\ \mathcal{R} \\ \sigma_2 \end{matrix} \right]\!\!\right]_{\eta[\overline{S},\overline{A},\overline{B}/\overline{W},\overline{X},\overline{Y}],\rho}$$

where

$$\left[\!\!\left[E \vdash_{\mathcal{R}} \begin{matrix} \overline{\tau_1} \\ \overline{S} \\ \overline{\tau_2} \end{matrix} \right]\!\!\right]_{\eta,\rho} = \overline{S} : \overline{A} \not\leftrightarrow \overline{B}$$

5.2 Related Values

In order to check the validity of a related values judgement, we interpret the two terms of the judgement in the parametric per model of F, and we prove that these interpretations are related by the semantics of the relation expression of the judgement.

Thus we first interpret a related values judgement by a function:

$$\left[\!\!\left[E \vdash_{\mathcal{R}} \begin{matrix} M : \sigma_1 \\ S \\ N : \sigma_2 \end{matrix} \right]\!\!\right] : RelAssign \to ValAssign \to \bigcup_{A,B \in Obj(PER)} Q(A) \times Q(B)$$

$$\left[\!\!\left[E \vdash_{\mathcal{R}} \begin{matrix} M : \sigma_1 \\ S \\ N : \sigma_2 \end{matrix} \right]\!\!\right]_{\eta\rho} = ([E_F \vdash_F M : \sigma_1]^F_{\eta\ddagger\rho}, [E_F \vdash_F N : \sigma_1]^F_{\eta\ddagger\rho})$$

Finally, we obtain that the two components of this interpretation are related:

Theorem 30 (Soundness for Related Values Judgements).

Given a related values judgement $E \vdash_{\mathcal{R}} \begin{matrix} M : \sigma_1 \\ S \\ N : \sigma_2 \end{matrix}$ and assignments η and ρ such that $\rho \models_\eta E$, we have:

$$[E_F \vdash_F M : \sigma_1]^F_{\eta\ddagger\rho} \left[\!\!\left[E \vdash_{\mathcal{R}} \begin{matrix} \sigma_1 \\ S \\ \sigma_2 \end{matrix} \right]\!\!\right]_{\eta\rho} [E_F \vdash_F N : \sigma_2]^F_{\eta\ddagger\rho}$$

6 Conclusion

We have defined two parametric models of system F and used one of them as a basis for an interpretation of system $\mathcal{R}$. In hindsight, our results may not seem surprising. However, the definitions include a number of tricky, "obvious" details. Details of this sort were left implicit in the work of Bainbridge et al. [BFSS90], and misunderstood in the first, inconsistent version of $\mathcal{R}$. The concrete interpretation of $\mathcal{R}$ carried out here establishes the soundness of $\mathcal{R}$. It would be interesting to have an abstract characterization of the notion of model for $\mathcal{R}$, and then to recast our proof in more abstract terms.

The interpretation has been helpful both in understanding $\mathcal{R}$ and in thinking about other formal systems for reasoning about polymorphic programs. Several other formal systems come to mind. Following a suggestion of [PA93], we have started to consider a formal system with relations of arities other than 2. Reynolds discussed relations of all arities in his original work, but binary relations have been preferred more recently (e.g., in [BFSS90]), in part arbitrarily. It seems interesting to extend the model presented here to support relations of all arities.

Acknowledgements

Ryu Hasegawa and Eugenio Moggi both made useful suggestions.

References

[ACC93] M. Abadi, L. Cardelli, and P.-L. Curien. Formal parametric polymorphism. In *A Collection of Contributions in Honour of Corrado Böhm on the Occasion of his 70th Birthday*, volume 121 of *Theoretical Computer Science*, pages 9–58, 1993. An early version appeared in the Proceedings of the *20th Ann. ACM Symp. on Principles of Programming Languages.*

[AM92] A. Asperti and S. Martini. Categorical models of polymorphism. *Information and Computation*, 99:1–79, 1992.

[Bar84] H.P. Barendregt. *The Lambda Calculus. Its Syntax and Semantics.* Number 103 in Studies in Logic and the Foundations of Mathematics. North-Holland, 1984. Revised edition.

[BFSS90] E.S. Bainbridge, P. Freyd, A. Scedrov, and P.J. Scott. Functorial polymorphism. *Theoretical Computer Science*, 70:35–64, 1990.

[Gir72] J.-Y. Girard. *Interprétation fonctionnelle et élimination des coupures dans l'arithmétique d'ordre supérieur.* Thèse de Doctorat d'Etat, Université Paris VII, 1972.

[Law69] F.W. Lawvere. Adjointness in foundations. *Dialectica*, 23(3-4):281–296, 1969.

[Mit90] J.C. Mitchell. A type-inference approach to reduction properties and semantics of polymorphic expressions. In G. Huet, editor, *Logical Foundations of Functional Programming*, Reading, MA, USA, pages 195–212. Addison-Wesley, 1990.

[PA93] G. Plotkin and M. Abadi. A logic for parametric polymorphism. In *Proceedings of the International Conference on Typed Lambda Calculi and Applications*, March 1993, Utrecht, NL, number 664 in Lecture Notes in Computer Science, pages 361–375. Springer-Verlag, 1993.

[Rey83] J.C. Reynolds. Types, abstraction and parametric polymorphism. In R.E.A. Mason, editor, *INFORMATION PROCESSING '83*, pages 513–523. Elsevier Science Publishers B.V.North-Holland, 1983.

[See87] R.A.G. Seely. Categorical semantics for higher order polymorphic lambda calculus. *The Journal of Symbolic Logic*, 52(4):969–989, December 1987.

[Str67] C. Strachey. Fundamental concepts in programming languages. Lecture Notes, International Summer School in Programming Languages, Copenhagen, Denmark, Unpublished, August 1967.

[Wad89] P. Wadler. Theorems for free! In *Proceedings of the Fourth International Conference on Functional Programming Languages and Computer Architecture*, pages 347–359. ACM press, 1989.

A realization of the negative interpretation of the Axiom of Choice

Stefano Berardi
Chalmers
Torino University *

Marc Bezem
Utrecht University †

Thierry Coquand
Chalmers
University of Gothenburg ‡

Abstract

We present a possible computational content of the negative translation of classical analysis with the Axiom of Choice. Interestingly, this interpretation uses a refinement of the realizibility semantics of the absurdity proposition, which is *not* interpreted as the empty type here. We also show how to compute witnesses from proofs in classical analysis, and how to extract algorithms from proofs of ∀∃ statements.

1 Introduction

It is well-known that the Axiom of Choice [13] is validated by the Brouwer–Heyting–Kolmogoroff explanation of the logical constants [3]. In view of the negative interpretation of classical arithmetic into intuitionistic arithmetic [6], one would expect that it is possible to make constructive sense of the Axiom of Choice as used in informal mathematics, for instance in the form of Zorn's Lemma.

This, however, appears to be non-trivial. The combination of the Axiom of Choice *and* the Excluded Middle turns out to be extremely problematic from a constructive point of view. To make constructive sense of such a combination can actually be seen as one the main aims of Hilbert's programme [8, 9, 1]. We address here the more modest question of the analysis of the computational content of the Axiom of Choice, by giving a novel realizability interpretation of the negative translation of the Axiom of Choice. This interpretation is due to the third author, motivated by [5].

*Dip. Informatica, C.so Svizzera 185, 10149 Torino, Italy, e-mail `stefano@di.unito.it`.
†Department of Philosophy, P.O. Box 80126, 3508 TC Utrecht, The Netherlands, e-mail `bezem@phil.ruu.nl`.
‡Department of Computer Sciences, S-41296, Gothenburg, Sweden, e-mail `coquand@cs.chalmers.se`.

Most of the work cited above has been inspired by metamathematical questions (consistency proofs, proof theoretic strength). Quite a different motivation arises from the computer science point of view, namely the extraction of algorithms from proofs. Here one encounters the same problem of the combination of the Axiom of Choice with classical logic. Upto now there were only two possibilities: (i) to use the bar recursive Dialectica interpretation of the Axiom of Choice due to Spector [10, 15]; (ii) to avoid the Axiom of Choice whenever possible, for example by encoding functions as relations such as done by Murthy in [14].

We improve on (i) since our interpretation is computationally much more direct than Gödel's Dialectica interpretation, and the resulting algorithm is more intuitive than bar recursion.

We improve on [14, 4], where the Axiom of Choice is avoided at the cost of encoding functions as relations. This encoding is often unnatural (see the discussion of Higman's lemma in [14]). Moreover, the encoding relies on the definability of a function value $f(x)$ as the smallest y satisfying $\phi(x,y)$, and this only works well if y ranges over an effective well-ordering such as the natural numbers. The approach here allows to interpret directly the first informal proof of Higman's lemma presented in [14], and applies also to the case where y ranges over an arbitrary simple type over N.

Our paper is organized as follows. In the first section we present the formal system under consideration and we state the essential difficulty of making constructive sense of the combination of the Axiom of Choice and the Excluded Middle. In the next section we present the programming language in which the realizing objects live. The central and longer part, Section 4, is then the description of a realizability interpretation, with a precise and detailed proof of correctness. This proof of correctness is non-constructive. (We use an intuitionistic meta-theory throughout this paper, unless explicitely stated otherwise.) We end with a conclusion and a discussion of further research.

2 Presentation of HA^ω

2.1 Types

The types of HA^ω are N and with τ, τ' also $\tau \rightarrow \tau'$. Here and below types will be denoted by lower case greek letters $\tau, \tau', \ldots$.

2.2 Terms

The terms of HA^ω are built from (typed) variables and constants using lambda abstraction and (well typed) application. There are countably many variables $x, y, z, \ldots$ for each type τ. The constants are: $0 : \mathsf{N}$, $\mathsf{s} : \mathsf{N} \rightarrow \mathsf{N}$ and $\mathsf{R}_\tau : \tau \rightarrow (\mathsf{N} \rightarrow \tau \rightarrow \tau) \rightarrow \mathsf{N} \rightarrow \tau$ for every type τ. Terms are denoted by $M, M', N, \ldots$, and $M : \tau$ expresses that the term M has type τ.

2.3 Formulae

Prime formulae are equations of the form $M = M'$, with $M, M' : \mathsf{N}$. Higher type equations, say $M = M'$ with $M, M' : \mathsf{N} \rightarrow \mathsf{N}$, are abbreviations of equations of lowest type, such as $Mx = M'x$ with x fresh. The set of formulae of HA^ω is generated in the usual way from the prime formulae by the boolean connectives $\wedge, \Rightarrow, \bot$ and the quantifiers $\forall, \exists$. We use $\phi, \phi', \ldots$ to denote formulae. We abbreviate $\phi \Rightarrow \bot$ by $\neg\phi$.

2.4 Theory

The theory HA^ω, intuitionistic higher-order arithmetic, is built up from three parts: (i) axioms and rules for first order many-sorted intuitionistic predicate logic; (ii) equality axioms and the axiom schema of induction; (iii) lambda calculus axioms and rules and the defining equations of the constants R_τ, $\mathsf{R}_\tau xy0 = x$, $\mathsf{R}_\tau xy(\mathsf{s}z) = yz(\mathsf{R}_\tau xyz)$. Thus our theory HA^ω essentially coincides with HA^ω from [18], the only difference being that we consider $\vee$ as defined and use the lambda version instead of the combinator version. The theory HA^ω_c is HA^ω with classical logic; HA^ω_-, *minimal higher-order arithmetic*, is HA^ω without the axiom schema $\bot \Rightarrow \phi$.

2.5 The Axiom of Choice

Theories of classical (intuitionistic) analysis can be obtained from HA^ω_c (HA^ω) by adding the Axiom of Choice. The Axiom of Choice of types τ, τ', denoted by $\mathsf{AC}(\tau, \tau')$, is the axiom schema

$$[\forall x : \tau\ \exists y : \tau'\ \phi(x, y)] \Rightarrow \exists f : \tau \rightarrow \tau'\ \forall x : \tau\ \phi(x, fx)$$

(schematic in formula ϕ). Here we will mainly consider $\mathsf{AC}(\mathsf{N}, \tau)$ (schematic in τ), that we may abbreviate to AC. The axiom AC is sufficiently strong to formalize a large part of classical analysis. Intuitionistically, AC is not a strong axiom, as may be expected from the Brouwer–Heyting–Kolmogorov interpretation of a $\forall\exists$ prefix. More formally, it follows from results of Goodman [7] that adding $\mathsf{AC}(\mathsf{N}, \mathsf{N})$ and $\mathsf{AC}(\mathsf{N}, \mathsf{N} \rightarrow \mathsf{N})$ to HA^ω is conservative over Heyting Arithmetic.

2.6 A negative interpretation

We will use the notation $\nabla\ \phi$ for $\neg\neg\phi$. This notation is justified by the fact that $\neg\neg$ can be thought of as a modal operator on formulae; we can prove indeed that $\nabla\ \phi$ follows from ϕ, and $\nabla\ \psi$ from $\nabla\ \phi$ and $\phi \Rightarrow \nabla\ \psi$.

Since absurdity is not interpreted by the empty type, we cannot realize $\bot \Rightarrow \phi$ for all ϕ. We overcome this problem by exercising some care in the negative interpretation. The idea is essentially due to Kolmogorov [11]. We employ the fact that $\bot \Rightarrow\ \nabla\ \phi$ can be proved for all ϕ without using the axiom schema

$\bot \Rightarrow \phi$. Although our prime formulae are decidable, the negative interpretation of a prime formula ϕ will be $\nabla\ \phi$.

As negative interpretation we use a standard version of the double negation translation, i.e. prefixing prime formulae and $\exists$ by ∇. The negative interpretation of a formula ϕ, denoted by G ϕ, is inductively defined by (here and below, $\equiv$ denotes syntactical identity):

- $\mathsf{G}\ \bot \equiv \bot$
- $\mathsf{G}\ \phi \equiv \nabla\ \phi$ if ϕ is a prime formula
- $\mathsf{G}\ [\phi \Rightarrow \psi] \equiv [\mathsf{G}\ \phi] \Rightarrow \mathsf{G}\ \psi$
- $\mathsf{G}\ [\phi \wedge \psi] \equiv \mathsf{G}\ \phi \wedge \mathsf{G}\ \psi$
- $\mathsf{G}\ \forall x : \tau\ \phi \equiv \forall x : \tau\ \mathsf{G}\ \phi$
- $\mathsf{G}\ \exists x : \tau\ \phi \equiv \nabla\ \exists x : \tau\ \mathsf{G}\ \phi$

The negative interpretation satisfies the following preservation property.

2.1. Fact. If ϕ is provable in HA^{ω}_{c}, then G ϕ is provable in HA^{ω}_{-}.

In the presence of AC, one cannot expect such a simple preservation result since, as we shall see below, AC is classically much stronger than intuitionistically.

The negative interpretation of AC, G AC, reads:

$$[\forall x : \mathsf{N}\ \nabla\ \exists y : \tau\ \mathsf{G}\ \phi(x,y)] \Rightarrow \nabla\ \exists f : \mathsf{N} \to \tau\ \forall x : \mathsf{N}\ \mathsf{G}\ \phi(x, fx).$$

By the stability[1] of formulae after the negative interpretation, G AC is subsumed by the following axiom schema:

$$[\forall x : \mathsf{N}\ \nabla\ \exists y : \tau\ \neg\phi(x,y)] \Rightarrow \nabla\ \exists f : \mathsf{N} \to \tau\ \forall x : \mathsf{N}\ \neg\phi(x, fx)$$

Par abus de langage we will from now on denote this schema by G AC. The theory HA^{ω}_{-} + G AC will be called *negative analysis*.

We can now extend the preservation property above.

2.2. Fact. If ϕ is provable in classical analysis, then G ϕ is provable in negative analysis.

The proof goes by inspection of the standard preservation proof, taking care that $\bot \Rightarrow \phi$ is avoided.

The following fact makes clear that the (straightforward) realizability interpretation of intuitionistic analysis does not suffice.

2.3. Fact. G AC fails to be an intuitionistic consequence of the Axiom of Choice[2]

[1] A formula is stable iff it is equivalent to the negation of another formula. Equivalently: ϕ is stable iff $\nabla\ \phi$ is equivalent to ϕ.

[2] This is to be contrasted with the induction schema over integers, whose negative interpretation is an instance of the induction schema itself.

As a consequence we cannot escape from realizing G AC. The chances for recursive realizability of G AC seem particularly bad in view of the following fact.

2.4. FACT. $\mathsf{HA}^{\omega}_{=}$ + G AC refutes Church's Thesis, stating that every function is recursive.

For exactly this reason, and because the semantics of the system NuPrl is based on recursive realizability, the work [14, 4] restricts itself to a fragment of classical logic that does not include the Axiom of Choice.

3 Presentation of the programming language $\mathcal{P}$

The programming language $\mathcal{P}$ extends the types, terms and equations of HA^{ω} with type constants for interpreting $\perp$ and equations, type constructors for lists and pairs, term constants associated to the new types, general recursion and infinite terms to form choice sequences.

3.1 Types of $\mathcal{P}$

The types of $\mathcal{P}$ are N, Unit, Abs and with τ, τ' also $\tau \rightarrow \tau'$, $\tau \times \tau'$ (cartesian product) and $[\tau]$ (lists over type τ). The type Unit serves to interpret prime formulae, and the type Abs to interpret $\perp$. The type Abs will not be empty. Like in the case of HA^{ω}, types will be denoted by lower case greek letters $\tau, \tau', \ldots$. The types of HA^{ω} will be called N-types.

3.2 Terms of $\mathcal{P}$

The terms of $\mathcal{P}$ are built from (typed) variables and constants using lambda abstraction, (well typed) application, and the formation of infinite terms: if $M_0, M_1, \ldots$ is an infinite sequence of terms of type τ, then $(\lambda\!\!\lambda x.M_x)$ is a term of type $\mathsf{N} \rightarrow \tau$. (The infinite terms are *not* for computational purposes, they only play a role in the termination proof.) There are countably many variables $x, y, z, \ldots$ for each type τ. The set of constants of $\mathcal{P}$ extends that of HA^{ω} with constants R_{τ} for types τ that are not N-types, () : Unit, Dummy : Abs, $\mathsf{Axiom}_1, \mathsf{Axiom}_2 : \mathsf{N} \rightarrow \mathsf{Abs}$, constants for general recursion (fixpoint combinators of all appropriate types) and constants for pairing and projection and list construction and destruction. Terms are again denoted by $M, M', N, \ldots$, and $M : \tau$ expresses that the term M has type τ.

3.3 Equations of $\mathcal{P}$

The only formulae of $\mathcal{P}$ are equations of the form $M = M'$, with M, M' terms of $\mathcal{P}$ of the same type, not necessarily N.

3.4 Theory of $\mathcal{P}$

The theory of $\mathcal{P}$ is equational, built up from the usual lambda calculus axioms and rules, defining equations for the constants R_τ as in HA^ω, but now for all types τ of $\mathcal{P}$, pairing axioms and list axioms and axioms for general recursion as usual, and the following axiom schema for infinite terms:

$$(\beta\beta) \qquad (\lambda\lambda x.M_x)\underline{k} = M_k$$

(schematic in τ, $M_0, M_1, \ldots$ of type τ and natural number k).

3.5 Pragmatics of $\mathcal{P}$

We allow ourselves a liberal use of $\mathcal{P}$. We will assume that all terms are well typed and we will reduce type information to a minimum that is required to reconstruct the type of a well typed term from the context. For every natural number k, we abbreviate $\mathsf{s}(\ldots(\mathsf{s}\ 0)\ldots)$ (with k times s) by $\underline{k}$, called the numeral $\underline{k}$. The term $\underline{k}_\tau$ of N-type τ is defined inductively by: $\underline{k}_{\mathsf{N}} \equiv \underline{k}$; $\underline{k}_{\tau\to\tau'} \equiv \lambda x : \tau.\underline{k}_{\tau'}$. We write pairs as (M, N) and triples as (M, N, P) instead of $(M, (N, P))$. We even write $\lambda(x, y).M$ instead of $\lambda z : \tau{\times}\tau'.M'$, where M' is obtained from M by replacing x by the first projection of z and y by the second. Lists are denoted by $[M_1, \ldots, M_n]$ ($M_i : \tau$ for $1 \leq i \leq n$). Adding a term H at the beginning of a list L will be denoted by $H : L$.

Instead of the explicit use of fixpoint combinators we define terms by giving the recursion equations. As an example we define a term which will play a role in the sequel, $\mathsf{get} : \mathsf{N}\to[\mathsf{N}{\times}\tau{\times}\tau']\to(\tau\to\tau'\to\tau'')\to\tau''\to\tau''$. The term $(\mathsf{get}\ x\ l\ f\ a)$ searches the list l for the first triple whose first component matches x; if such a triple is found, then the output is f applied to the second and third component of the triple, otherwise the output is a. Formally,

$$\begin{array}{lcl} \mathsf{get}\ x\ []\ f\ a & = & a \\ \mathsf{get}\ x\ ((x', y, y') : l)\ f\ a & = & \mathsf{if}\ (x = x')\ \mathsf{then}\ (f\ y\ y')\ \mathsf{else}\ (\mathsf{get}\ x\ l\ f\ a) \end{array}$$

Here and below $\mathsf{if}\ (M = M')\ \mathsf{then} \ldots \mathsf{else} \ldots$ (with M, M' of type N) is a sugared version of a well-known primitive recursive term.

3.6 Known facts about $\mathcal{P}$

There exists a reduction relation $\longrightarrow$ on the terms of $\mathcal{P}$ such that the reflexive, symmetric and transitive closure of $\longrightarrow$ coincides with $=$ (convertibility) on the terms of $\mathcal{P}$. Moreover, $\longrightarrow$ satisfies:

(i) the Church Rosser Theorem;

(ii) every closed normal form of type N is a numeral $\underline{k}$;

(iii) every closed normal form of type Unit is ();

(iv) every closed normal form of type Abs is either Dummy or of the form $\mathsf{Axiom}_1\underline{k}$ or $\mathsf{Axiom}_2\underline{k}$;

(v) the Continuity Lemma: let $M : (\mathsf{N}{\rightarrow}\tau){\rightarrow}\tau'$ and $N : \mathsf{N}{\rightarrow}\tau$ be such that MN has a closed normal form. Then there exists a natural number m such that for all $N' : \mathsf{N}{\rightarrow}\tau$ with $N\underline{i} = N'\underline{i}$ for all $i < m$ we have $MN = MN'$. In particular we have extensionality: if $N\underline{i} = N'\underline{i}$ for all natural numbers i, then $MN = MN'$.

4 Realizability

Realizability, due to Kleene, aims at formalizing the notion of constructive truth, see [19] for an overview. A realizability interpretation interprets a logic, usually an extension of Heyting Arithmetic, in a programming language. More specifically, to each formula ϕ of the logic is associated a type $|\phi|$ of the programming language. One then defines by formula induction when a program of type $|\phi|$ realizes the formula ϕ. Intuitively, it means that this program is a constructive justification of the formula ϕ. Finally, to establish soundness, to each proof of a closed formula ϕ is associated a program of type $|\phi|$ which realizes the formula ϕ.

In this section a realizability interpretation of HA^ω + G AC in $\mathcal{P}$ will be given. It consists of a mapping of formulas of HA^ω to types of $\mathcal{P}$ together with a realizability relation between programs in $\mathcal{P}$ and formulas of HA^ω, where the program has the type to which the formula is mapped. The main result will be that every theorem of HA^ω + G AC can be realized in $\mathcal{P}$. The difficult step in proving this result is the realization of G AC.

4.1 Mapping formulas of HA^ω to types of $\mathcal{P}$

The idea behind this mapping is usually referred to as "forgetting dependencies", due to Martin-Löf. By formula induction we define a type $|\phi|$ of $\mathcal{P}$ for every formula ϕ of HA^ω:

- $|M = M'| \equiv \mathsf{Unit}$
- $|\bot| \equiv \mathsf{Abs}$
- $|\phi \Rightarrow \psi| \equiv |\phi|{\rightarrow}|\psi|$
- $|\phi \wedge \psi| \equiv |\phi| \times |\psi|$
- $|\forall x : \tau\ \phi| \equiv \tau{\rightarrow}|\phi|$
- $|\exists x : \tau\ \phi| \equiv \tau \times |\phi|$

Note that the domains of quantification in HA^ω are types of HA^ω, and hence of $\mathcal{P}$, so that the mapping $|\ |$ is well defined.

4.2 Reducibility in $\mathcal{P}$

In order to define the realizability relation we need a notion of reducibility for closed terms of $\mathcal{P}$ of N-type. By induction on the N-type we define:

- $M : \mathsf{N}$ is reducible iff M reduces to a numeral
- $M : \tau \to \tau'$ is reducible iff MN is reducible for every reducible $N : \tau$

In the sequel, we shall need the following technicalities.

4.1. DEFINITION. Two expressions E and E' (terms or formulae) are called *related* if they are syntactically identical up to the indices of the constants Axiom_i. Note that related terms are of the same type.

4.2. LEMMA. *If M and M' are two related terms of type* N, *then $M = \underline{n}$ iff $M' = \underline{n}$. If M and M' are two related terms of type* Unit, *then $M = ()$ iff $M' = ()$.*

PROOF. By induction on the length of reduction sequences. □

4.3. LEMMA. *If M and M' are two related terms of* N*-type, then M is reducible iff M' is reducible.*

PROOF. By an easy induction on the common N-type of M and M', using Lemma 4.2 for the base case N. □

4.3 Realizability relation

What follows is essentially the notion of modified realizability, due to Kreisel, with realizing objects from the programming language $\mathcal{P}$. We give an inductive definition of "M realizes ϕ", where ϕ is a closed formula of HA^ω with possibly closed reducible terms of $\mathcal{P}$ occurring in the prime constituents of ϕ, and $M : |\phi|$ a closed term of $\mathcal{P}$. Let $\underline{k}$ denote any numeral. Then

$M : \mathsf{Abs}$ realizes $\bot$	iff	$M = \mathsf{Axiom}_i\underline{k}$ for some $i = 1, 2$ and $\underline{k}$
$M : \mathsf{Unit}$ realizes $M_1 = M_2$	iff	$M = ()$ and M_1, M_2 reduce to the same $\underline{k}$
$M : \|\phi\| \to \|\psi\|$ realizes $\phi \Rightarrow \psi$	iff	MN realizes ψ whenever N realizes ϕ
$M : \|\phi \wedge \psi\|$ realizes $\phi \wedge \psi$	iff	$M = (N, P)$ with N realizes ϕ and P realizes ψ
$M : \tau \to \|\phi\|$ realizes $\forall x : \tau\ \phi$	iff	MN realizes $\phi[x := N]$ for all *reducible* $N : \tau$
$M : \tau \times \|\phi\|$ realizes $\exists x : \tau\ \phi$	iff	$M = (N, P)$ with $N : \tau$ *reducible* and P realizes $\phi[x := N]$

Note that the above definition uses reducibility for N-types only. In the case $M_1 = M_2$ above, the equation is of type N. The terms M_1 and M_2 come from (possibly) open terms of HA^ω in which closed reducible terms of $\mathcal{P}$ are substituted for the variables. Thus M_1 and M_2 are closed and reducible, since all constants of HA^ω are reducible constants of $\mathcal{P}$. Hence we can verify $M_1 = M_2$ in $\mathcal{P}$, relying on Fact (i) and (ii) from 3.6.

In the sequel, we shall need the following technical lemma.

4.4. LEMMA. *If ϕ and ϕ' are two related formulae, and M, M' two related terms, then M realizes ϕ iff M' realizes ϕ'.*

PROOF. By an easy induction on the realization relation, using Lemma 4.2 and Lemma 4.3. □

4.4 Main Result and applications

In this subsection we formulate the main result, sketch a proof and give two applications of the main result. The essential and difficult step in the proof of the main result, the realization of G AC is given in the next subsection.

4.5. THEOREM. *Every theorem of* $\mathsf{HA}^\omega_- + \mathsf{G\ AC}$ *can be realized by a term in $\mathcal{P}$ in which no constants* Axiom_i *occur ($i = 1, 2$).*

PROOF. Apart from the realization of G AC, the proof is more or less standard. For example, the axiom $\forall x : \mathsf{N}\ \neg(\mathsf{s}x = 0)$ is realized by $M \equiv \lambda x : \mathsf{N}\ \lambda h : \mathsf{Unit}.\mathsf{Dummy}$. Indeed, $M\underline{n}$ realizes $\neg(\mathsf{s}\underline{n} = 0)$ for every natural number n, since nothing realizes $\mathsf{s}\underline{n} = 0$ (here we use Fact (i) and (ii) from 3.6). □

4.4.1 Application 1: the consistency of analysis

The main result immediately implies the consistency of analysis. Assume $\bot$ is provable in $\mathsf{HA}^\omega_c + \mathsf{AC}$. Then G $\bot$, i.e. $\bot$, is provable in $\mathsf{HA}^\omega_- + \mathsf{G\ AC}$. Hence $\bot$ is realizable by a term of $\mathcal{P}$ in which no constants Axiom_i occur ($i = 1, 2$). This is impossible by the definition of realization.

4.4.2 Application 2: how to compute witnesses with AC and classical logic

Assume a formula $\phi(x)$ of HA^ω, with $x : \mathsf{N}$, is decidable, i.e. of the form $M_\phi x = 0$ for suitable closed term M_ϕ of HA^ω. We will freely write $\phi(x)$ instead of $M_\phi x = 0$. Assume $\exists x : \mathsf{N}\ \phi(x)$ is a theorem of $\mathsf{HA}^\omega_c + \mathsf{AC}$. Then $\nabla\ \exists x : \mathsf{N}\ \mathsf{G}\ \phi(x)$ is a theorem of $\mathsf{HA}^\omega_- + \mathsf{G\ AC}$. Using $\neg\exists x\ \phi \iff \forall x\ \neg\phi$ and the stability of $\neg\phi$, we have that $\nabla\ \exists x : \mathsf{N}\ \phi(x)$ is a theorem of $\mathsf{HA}^\omega_- + \mathsf{G\ AC}$, and hence realizable by a term not containing constants Axiom_i ($i = 1, 2$), say by M. We have that

$$N \equiv \lambda(x, h) : \mathsf{N}\times\mathsf{Unit}.\mathsf{if}\ \phi(x)\ \mathsf{then}\ (\mathsf{Axiom}_1 x)\ \mathsf{else}\ \mathsf{Dummy}$$

realizes $\neg\exists x : \mathsf{N}\ \phi(x)$. So MN realizes $\bot$ and must hence be convertible to $\mathsf{Axiom}_1\underline{n}$ for some numeral $\underline{n}$. We claim $\phi(\underline{n})$, i.e. $\underline{n}$ is a witness. Consider the following extensionally equal terms:

$$F \equiv \lambda x : \mathsf{N}\lambda h : \mathsf{Unit}.\mathsf{if}\ \phi(x)\ \mathsf{then}\ (\mathsf{Axiom}_1 x)\ \mathsf{else}\ \mathsf{Dummy}$$

$$F' \equiv \lambda x : \mathsf{N}\lambda h : \mathsf{Unit}.\mathsf{if}\ \phi(x)\ \mathsf{then}\ (\mathsf{if}\ \phi(x)\ \mathsf{then}\ (\mathsf{Axiom}_1 x)\ \mathsf{else}\ \mathsf{Dummy})\ \mathsf{else}\ \mathsf{Dummy}$$

Since F and F' are extensionally equal we have by extensionality:

$$\mathsf{Axiom}_1\underline{n} = MN = M(\lambda(x,h).Fxh) = M(\lambda(x,h).F'xh)$$

Note that F' can be obtained from F by replacing Axiom_1 by $\lambda x.\mathsf{if}\ \phi(x)\ \mathsf{then}\ (\mathsf{Axiom}_1 x)\ \mathsf{else\ Dummy}$.
Since M does not contain the constant Axiom_1, it follows that

$$\mathsf{Axiom}_1\underline{n} = \mathsf{if}\ \phi(\underline{n})\ \mathsf{then}\ (\mathsf{Axiom}_1\underline{n})\ \mathsf{else\ Dummy}.$$

Using Fact (iv) from 3.6 we get $\phi(\underline{n})$.

4.4.3 Application 3: extraction of algorithms

Let $\phi(y,x)$ be $M_\phi yx = 0$ with $x : \mathsf{N}$ and $y : \tau$. Assume $\forall y : \tau\ \exists x : \mathsf{N}\ \phi(y,x)$ is a theorem of HA_c^ω+AC. Like in Application 2 above, there exists an Axiom_i-free realizer M of $\forall y : \tau\ \nabla\ \exists x : \mathsf{N}\ \phi(y,x)$ in $\mathcal{P}$. Then we have for every reducible, Axiom_i-free $Y : \tau$, that MY is an Axiom_i-free realizer of $\nabla\ \exists x : \mathsf{N}\ \phi(Y,x)$. Define

$$N \equiv \lambda y \lambda(x,h) : \mathsf{N}{\times}\mathsf{Unit}.\mathsf{if}\ \phi(y,x)\ \mathsf{then}\ (\mathsf{Axiom}_1 x)\ \mathsf{else\ Dummy},$$

then NY realizes $\neg\exists x : \mathsf{N}\ \phi(Y,x)$ as above. Hence $MY(NY)$ realizes $\bot$ and hence reduces to $\mathsf{Axiom}_1\underline{n}$ for some numeral $\underline{n}$. Like above, we have $\phi(Y,\underline{n})$. At this point, observe that in the reduction of $MY(NY)$ to $\mathsf{Axiom}_1\underline{n}$ no special features of the constants Abs, Axiom_i, Dummy are used. As a consequence, they may be considered as variables as well, and hence

$$MY(NY)[\mathsf{Abs} := \mathsf{N}, \mathsf{Axiom}_1 := \lambda x : \mathsf{N}.x, \mathsf{Dummy} := 0]$$

is a well-typed term which reduces to $\underline{n}$. Thus the term

$$F \equiv \lambda y : \tau.My(Ny)[\mathsf{Abs} := \mathsf{N}, \mathsf{Axiom}_1 := \lambda x : \mathsf{N}.x, \mathsf{Dummy} := 0]$$

is a well-typed term of $\mathcal{P}$ with the property that $\phi(Y, FY)$ for every reducible, Axiom_i-free $Y : \tau$. In particular, for $\tau \equiv \mathsf{N}$ we have $\phi(\underline{n}, F\underline{n})$ for every numeral $\underline{n}$.

4.5 Realization of G AC

Recall that G AC is the following schema:

$$[\forall x : \mathsf{N}\ \nabla\ \exists y : \tau\ \neg\phi(x,y)] \Rightarrow \nabla\ \exists f : \mathsf{N}{\to}\tau\ \forall x : \mathsf{N}\ \neg\phi(x, fx)$$

We start with some preliminary calculations:

$$|\forall x : \mathsf{N}\ \nabla\ \exists y : \tau\ \neg\phi(x,y)| \equiv \mathsf{N}{\to}((\tau{\times}(|\phi|{\to}\mathsf{Abs})){\to}\mathsf{Abs}){\to}\mathsf{Abs}$$

$$|\neg\exists f : \mathsf{N}{\to}\tau\ \forall x : \mathsf{N}\ \neg\phi(x,fx)| \equiv ((\mathsf{N}{\to}\tau){\times}(\mathsf{N}{\to}|\phi|{\to}\mathsf{Abs})){\to}\mathsf{Abs}$$

A realizer of G AC should be a term M such that MHP realizes $\bot$ whenever H realizes $\forall x : \mathsf{N}\ \nabla\ \exists y : \tau\ \neg\phi(x,y)$ and P realizes $\neg\exists f : \mathsf{N}{\to}\tau\ \forall x : \mathsf{N}\ \neg\phi(x, fx)$.

Moreover, M should not contain constants Axiom_i $(i = 1,2)$. The general idea is to approximate a function witnessing $\exists f : \mathsf{N}\to\tau\ \forall x : \mathsf{N}\ \neg\phi(x, fx)$ by means of a list L of triples (X,Y,Z), where $X : \mathsf{N}$ and $Y : \tau$ are reducible, and Z realizes $\neg\phi(X,Y)$. Given such a list $L = [(X_1,Y_1,Z_1),\ldots,(X_n,Y_n,Z_n)]$ with all the X_i's distinct, we consider a function $\mathsf{fun}\ L : \mathsf{N}\to\tau$ which maps X_i to Y_i $(1 \le i \le n)$ and takes function values $\underline{0}_\tau$ in arguments different from all X_i's. Formally:

$$\mathsf{fun}\ l\ x = \mathsf{get}\ x\ l\ (\lambda y : \tau\lambda z : |\phi|\to\mathsf{Abs}.y)\ \underline{0}_\tau$$

Furthermore, we consider a function $\lambda x : \mathsf{N}.\mathsf{rea}\ L\ x\ A : \mathsf{N}\to|\phi|\to\mathsf{Abs}$ which maps X_i to the realizer Z_i $(1 \le i \le n)$ and takes values A (to be specified below) in arguments different from all X_i's. Formally:

$$\mathsf{rea}\ l\ x\ a = \mathsf{get}\ x\ l\ (\lambda y : \tau\lambda z : |\phi|\to\mathsf{Abs}.z)\ a$$

Consider

$$P(\mathsf{fun}\ L, \lambda x : \mathsf{N}.\mathsf{rea}\ L\ x\ A).$$

If $\lambda x : \mathsf{N}.\mathsf{rea}\ L\ x\ A$ realizes $\forall x\ \neg\phi(x, \mathsf{fun}\ L\ x)$, then we would have that $P(\mathsf{fun}\ L, \lambda x : \mathsf{N}.\mathsf{rea}\ L\ x\ A)$ realizes $\bot$ and we would be done. However, this is in general not the case since we cannot choose A such that A realizes $\neg\phi(x,\underline{0}_\tau)$ for all x different from all X_i's. We claim that, as A may depend on x, there is a possibility to construct A in such a way that it allows us to compute a better approximation of the function witnessing $\exists f : \mathsf{N}\to\tau\ \forall x : \mathsf{N}\ \neg\phi(x, fx)$. The type of A must be $|\phi|\to\mathsf{Abs}$, so we must have

$$A \equiv \lambda x' : |\phi|.\cdots,$$

where $\cdots$ is of type Abs. It is tempting to fill in $\cdots$ with $(\mathsf{Axiom}_i x)$. The resulting term

$$P(\mathsf{fun}\ L, \lambda x.\mathsf{rea}\ L\ x\ (\lambda x' : |\phi|.\mathsf{Axiom}_i x))$$

of type Abs is not a solution, since it contains Axiom_i, but it will play an important role in the discussion below. Note that $\lambda x : \mathsf{N}.\mathsf{rea}\ L\ x\ A$ only accesses A in case x does not occur as first component of a triple in L. The basic intuition is that, if the above term reduces to $\mathsf{Axiom}_i\underline{k}$, then this tells us that P needs more information about its arguments, in particular it needs a function value and a realizer for the argument k.

Observe that, so far, H realizing $\forall x : \mathsf{N}\ \nabla\ \exists y : \tau\ \neg\phi(x,y)$ has not been used. For filling in $\cdots$ we use H. Recall that the type of H is $\mathsf{N}\to((\tau\times(|\phi|\to\mathsf{Abs}))\to\mathsf{Abs})\to\mathsf{Abs}$. The obvious way to continue is putting

$$A \equiv \lambda x' : |\phi|.H\ x\ \cdots.$$

Now $\cdots$ is of type $(\tau\times(|\phi|\to\mathsf{Abs}))\to\mathsf{Abs}$ and is hence of the form $\lambda(y,z).\cdots$, so that we have

$$A \equiv \lambda x' : |\phi|.H\ x\ (\lambda(y,z).\cdots),$$

where $\cdots$ is again of type Abs. The crucial idea is now to put

$$A \equiv \lambda x' : |\phi|.H\ x\ (\lambda(y,z).\cdots((x,y,z):L)),$$

where $\cdots$ stands for a recursive call of the whole procedure described above.

This informal discussion motivates the following recursive definition:

$$\Phi\ p\ h\ l\ =\ p\,(\mathsf{fun}\ l, \lambda x.\mathsf{rea}\ l\ x\ (\lambda x'.h\ x\ (\lambda(y,z).\Phi\ p\ h\ ((x,y,z){:}l))))$$

We shall prove that, given H and P as above, $\Phi\ P\ H\ []$ realizes $\bot$. Thus $\lambda h \lambda p.\Phi\ p\ h\ []$ realizes G AC.

The first step will be, in the next lemma, to check that each recursive call to Φ on a "good argument" indeed corresponds to an extension of the list L approximating a witness f such that $\forall x : \mathsf{N}.\neg\phi(x, fx)$.

4.6. Lemma. *Let H, P, Φ be as above. Furthermore, let $L = [(X_1, Y_1, Z_1), \ldots, (X_n, Y_n, Z_n)]$ be such that L does not contain* Axiom_1 *and, for all $1 \leq i \leq n$, $X_i : \mathsf{N}, Y_i : \tau$ are reducible and $Z_i : |\phi| \rightarrow \mathsf{Abs}$ realizes $\neg\phi(X_i, Y_i)$. If $\Phi\ P\ H\ L$ does not realize $\bot$, then there exist $X : \mathsf{N}, Y : \tau$ and $Z : |\phi| \rightarrow \mathsf{Abs}$ not containing* Axiom_1 *such that:*

(i) X and Y are reducible;

(ii) X is different from all X_i's;

(iii) Z realizes $\neg\phi(X, Y)$;

(iv) $\Phi\ P\ H\ ((X,Y,Z){:}L)$ does not realize $\bot$.

Proof. Let conditions be as above. By Lemma 4.4 we may assume that Axiom_1 does not occur in P, H. Recall

$$\Phi\ P\ H\ L\ =\ P\,(\mathsf{fun}\ L, \lambda x.\mathsf{rea}\ L\ x\ (\lambda x'.H\ x\ (\lambda(y,z).\Phi\ P\ H\ ((x,y,z){:}L)))).$$

Let $\Phi'\ P\ H\ L$ be the term discussed above (not depending on H):

$$\Phi'\ P\ H\ L\ =\ P(\mathsf{fun}\ L, \lambda x.\mathsf{rea}\ L\ x\ (\lambda x' : |\phi|.\mathsf{Axiom}_1 x)).$$

By our assumption on L we have that

$$\lambda x : \mathsf{N}.\mathsf{rea}\ L\ x\ (\lambda x' : |\phi|.\mathsf{Axiom}_1 x)$$

realizes $\forall x : \mathsf{N}.\neg\phi(x, \mathsf{fun}\ L\ x)$. Consequently, $\Phi'\ P\ H\ L$ realizes $\bot$, and hence reduces to $\mathsf{Axiom}_i \underline{k}$ for some numeral $\underline{k}$ and $i = 1, 2$. Since the only occurrence of Axiom_1 in $(\Phi'\ P\ H\ L)$ is the one which is explicitly shown, we have that

$$\Phi\ P\ H\ L\ =\ (\Phi'\ P\ H\ L)[\mathsf{Axiom}_1 := \lambda x.H\ x\ (\lambda(y,z).\Phi\ P\ H\ ((x,y,z){:}L))]$$

By our assumption that $\Phi\ P\ H\ L$ does not realize $\bot$, it follows that $i = 1$, and hence

$$\begin{aligned}\Phi\ P\ H\ L\ &=\ (\mathsf{Axiom}_1 \underline{k})[\mathsf{Axiom}_1 := \lambda x.H\ x\ (\lambda(y,z).\Phi\ P\ H\ ((x,y,z){:}L))]\\ &=\ H\ \underline{k}\ (\lambda(y,z).\Phi\ P\ H\ ((\underline{k},y,z){:}L)).\end{aligned}$$

Since H realizes $\forall x : \mathsf{N}\ \nabla\ \exists y : \tau\ \neg\phi(x,y)$ and $\Phi\ P\ H\ L$ does not realize $\bot$, it follows that $\lambda(y,z).\Phi\ P\ H\ ((\underline{k},y,z){:}L)$ does not realize $\neg\exists y : \tau\ \neg\phi(\underline{k}, y)$. By the definition of realizability and using classical logic it follows that there exists a reducible $Y : \tau$ and a $Z : |\phi| \rightarrow \mathsf{Abs}$ realizing $\neg\phi(\underline{k}, Y)$ such that $\Phi\ P\ H\ ((\underline{k}, Y, Z) : L)$ does not realize $\bot$. By Lemma 4.4 we may assume that Axiom_1 does not occur in Y, Z. Now we are done with (i), (iii) and (iv); it remains to prove (ii). To this end, we reason similarly to the argument used in Application 2 above. Consider the following extensionally equal terms:

$$F \equiv \lambda x.\mathsf{rea}\ L\ x\ (\lambda x' : |\phi|.\mathsf{Axiom}_1 x)$$
$$F' \equiv \lambda x.\mathsf{rea}\ L\ x\ (\lambda x' : |\phi|.\mathsf{if}\ (\mathsf{member}\ x\ L)\ \mathsf{then}\ \mathsf{Dummy}\ \mathsf{else}\ (\mathsf{Axiom}_1 x))$$

Here $(\mathsf{member}\ x\ L)$ tests if x occurs as first component of a triple in L. Since the only occurrence of Axiom_1 in F is the one which is explicitly shown, we have that

$$F' = F[\mathsf{Axiom}_1 := \lambda x.\mathsf{if}\ (\mathsf{member}\ x\ L)\ \mathsf{then}\ \mathsf{Dummy}\ \mathsf{else}\ (\mathsf{Axiom}_1 x)]$$

Recall that $P(\mathsf{fun}\ L, F) = \Phi'\ P\ H\ L = \mathsf{Axiom}_1\underline{k}$. By extensionality we have $P(\mathsf{fun}\ L, F') = P(\mathsf{fun}\ L, F) = \mathsf{Axiom}_1\underline{k}$. By substitution we get

$$\begin{aligned} P(\mathsf{fun}\ L, F') &= (\mathsf{Axiom}_1\underline{k})[\mathsf{Axiom}_1 := \lambda x.\mathsf{if}\ (\mathsf{member}\ x\ L)\ \mathsf{then}\ \mathsf{Dummy}\ \mathsf{else}\ (\mathsf{Axiom}_1 x)] \\ &= \mathsf{if}\ (\mathsf{member}\ \underline{k}\ L)\ \mathsf{then}\ \mathsf{Dummy}\ \mathsf{else}\ (\mathsf{Axiom}_1\underline{k}) \end{aligned}$$

By Fact (i) (Church Rosser Theorem) from 3.6 we get $\neg(\mathsf{member}\ \underline{k}\ L)$. This is (ii) and completes the proof of the lemma. □

Using the lemma above we can prove by contradiction that $\Phi\ P\ H\ []$ realizes $\bot$. We give an informal argument, which can easily be formalized using the axiom of Dependent Choice and classical logic. The argument is similar to the argument used by Tait in [17]. Suppose $\Phi\ P\ H\ []$ does not realize $\bot$. Then, by the lemma above, there exist X_1, Y_1, Z_1 such that $[(X_1, Y_1, Z_1)]$ satisfies the conditions of the lemma, in particular $\Phi\ P\ H\ [(X_1, Y_1, Z_1)]$ does not realize $\bot$. Applying the lemma again yields X_2, Y_2, Z_2 such that ... Iterating this argument infinitely many times yields an infinite sequence of triples (X_i, Y_i, Z_i) $(i = 1, 2, \ldots)$ such that each finite initial subsequence satisfies the conditions of the lemma. Note that the X_i's all convert to different numerals. Define (classically!) $(\underline{n}, Y'_n, Z'_n)$ to be (X_i, Y_i, Z_i) if $X_i = \underline{n}$ for some i, and $(\underline{n}, \underline{0}_\tau, \mathsf{Axiom}_1\underline{n})$ otherwise. This results in an infinite sequence $\lambda\!\!\lambda n.(\underline{n}, Y'_n, Z'_n)$ such that $\lambda\!\!\lambda n.Y'_n$ is reducible and $\lambda\!\!\lambda n.Z'_n$ realizes $\forall x : \mathsf{N}\ \neg\phi(x, (\lambda\!\!\lambda n.Y'_n)x)$. It follows that $P(\lambda\!\!\lambda n.Y'_n, \lambda\!\!\lambda n.Z'_n)$ realizes $\bot$ and hence reduces to a term of the form $\mathsf{Axiom}_i\underline{k}$. By Fact (v) from 3.6, the Continuity Lemma, $P(\lambda\!\!\lambda n.Y'_n, \lambda\!\!\lambda n.Z'_n)$ only depends on a finite initial subsequence of $\lambda\!\!\lambda n.(\underline{n}, Y'_n, Z'_n)$. This conflicts with the above construction of $[(X_1, Y_1, Z_1), \ldots, (X_n, Y_n, Z_n)]$ for n large enough.

5 Conclusion and further research

We gave a quite direct computational interpretation of the Axiom of Choice. For instance, it allows to interpret directly the first informal proof of Higman's lemma presented in [14], and avoids in that case the encoding of a function as a relation. Observe furthermore that the computation in the previous subsection is demand-driven in the sense that the list of triples only contains function values and realizers that are really needed for the computation of a realizer of $\perp$.

It should be noted that our interpretation works for $\mathsf{AC}(\tau, \tau')$ where $\tau = \mathsf{N}$ but τ' is arbitrary. As it is given, it uses in an essential way the restriction $\tau = \mathsf{N}$, but we hope to be able to extend it to arbitrary types τ. By analogous techniques we can realize the Axiom of Dependent Choice and Double Negation Shift (see [15]).

There are many directions in which this work can be improved, and some potential connections seem worthwhile to analyse.

One is the formulation of this work in the framework of sequential algorithms, which is indeed the natural framework in which one can observe "intensional behaviour" of functionals, which has been our main tool in the realization of the negative interpretation of the Axiom of Choice. We expect from such a formulation a much more transparent proof of termination (and it will be interesting to see what becomes in this formalism the difficulty of finding "fresh" constants). We also think that in such a framework, the generalization to higher types, that is, a computational interpretation of $\mathsf{AC}(\tau, \tau')$ for arbitrary τ and τ', should be straightforward. An elegant form would be an interpretation of the notion of dependent types in the framework of sequential algorithms, together with a direct interpretation of $\nabla\ \phi$, and the inference of $\nabla\ \forall x : \tau.\phi(x)$ from $\forall x : \tau.\nabla\ \phi(x)$.

Closely related should be the question of a constructive formulation of our proof of realizability. Can we adapt the method of [2] and avoid the use of infinite terms? Our hope is that our interpretation, computationally more direct than the one of Spector [15, 10], may provide help for a constructive understanding of classical analysis with the Axiom of Choice.

Experiments with some examples, in particular Higman's Lemma, have revealed a computational inefficiency of our interpretation. Intuitively, our algorithms proceed by trial and errors and may "forget too many things" of previous trials. It seems possible to design improved algorithms, that "remember" and can use all previous trials. This problem is closely connected to the problem of the computational interpretation of the implication $\nabla\ \phi \Rightarrow \nabla\ \psi \Rightarrow \nabla\ (\phi \wedge \psi)$ so that its solution should have consequences even for the problem of the computational content of propositional classical logic.

The previous remark, and metaphors such as "trials and errors", "conjecturing laws that are refined by experimentation", are immediately suggested by trying to understand the computational behaviour of our interpretation. They hint at possible connections with learning theories, such as the one described in [16], where the learning agent may benefit from negative information.

Yet another natural connection is with the work of Hilbert [8, 9] and Ackermann [1]. Our interpretation can be seen as a variation of Hilbert's epsilon method [8, 9], with a (classical) proof of termination. It will be interesting to compare our algorithm with the one described in Ackermann's paper [1] [3].

[3] According to [12], if it is well-known that the proof of convergence of Ackermann's algorithm was defective as presented in [1], it is not known yet, not even non-constructively, if the method converges at all.

Acknowledgement

We thank Anne Troelstra for helpful comments and answers to various questions that arose during the preparation of this paper, Loïc Colson for interesting discussions on the axiom of choice and impredicativity at the beginning of this work, and Marco Hollenberg and Jan Springintveld for carefully reading successive draft versions.

References

[1] W. Ackermann. Begründung des Tertium non datur mittels der Hilbertschen Theorie der Widerspruchsfreiheit. Mathematische Annalen, 93, 1924, p. 1-36

[2] M. Bezem. Strong Normalization of Barrecursive Terms Without Using Infinite Terms. Archiv für mathematische Logik und Grundlagenforschung, 25, 1985, p. 175-182

[3] E. Bishop. *Foundations of Constructive Analysis.* New York, McGraw-Hill, 1967.

[4] R. Constable and C. Murthy. Finding Computational Content in Classical Proofs. In G.Huet and G. Plotkin, editors, *Logical Frameworks*, 341 - 362, (1991), Cambridge University Press.

[5] Th. Coquand. A semantics of evidence for classical arithmetic. Journal of Symbolic Logic, to appear, 1994.

[6] K. Gödel. *Collected Work.* Volumes I and II, S. Feferman, J. W. Dawson, S.C. Kleene, G.H. Moore, R. M. Solovay, J. van Heijenoort, editors, Oxford, 1986.

[7] N. Goodman. *Intuitionistic arithmetic as a theory of constructions.* PhD thesis, Stanford University, 1968.

[8] D. Hilbert. Die logischen Grundlagen der Mathematik. Mathematische Annalen, 88, 1923, p. 151-165

[9] D. Hilbert. The foundations of mathematics. In van Heijenoort ed., *From Frege to Gödel*, p. 465-479.

[10] W.A. Howard. Functional interpretation of bar induction by bar recursion. Compos. Mathematica 20 (1968), 107-124.

[11] A.N. Kolmogorov. On the principle of the excluded middle. In *From Frege to Gödel*, J. van Heijenoort, editor, Harvard University Press, Cambridge MA, 1971, pp. 414-437.

[12] G. Kreisel. Mathematical Logic. In *Lectures on Modern Mathematics*, vol. III, ed. Saaty, Wiley, N.Y., 1965, p. 95 - 195.

[13] G.E. Moore. *Zermelo's Axiom of Choice: Its Origins, Development and Influence.* Springer-Verlag, 1982

[14] C. Murthy. *Extracting Constructive Content from Classical Proofs.* Ph. D. Thesis, 1990.

[15] C. Spector. Provably recursive functionals of analysis: a consistency proof of analysis by an extension of principles formulated in current intuitionistic mathematics. *Recursive Function Theory*, J.C.E. Dekker ed., Proceedings of Symposia in Pure Mathematics V, AMS, p. 1 - 27, 1961.

[16] D.N. Osherbon, M. Stob and S. Weinstein. *Systems That Learn: An Introduction to Learning Theory for Cognitive and Computer Scientists.* MIT Press, 1986.

[17] W.W. Tait. Normal form theorem for bar recursive functions of finite type. In *Proceedings of the second Scandinavian Logic Symposium*, J.E. Fenstad ed., North Holland, Amsterdam, 1971, pp. 353-367.

[18] A.S. Troelstra. *Metamathematical Investigation of Intuitionistic Arithmetic and Analysis.* Lecture Notes in Mathematics 344, Springer-Verlag, Berlin, 1973.

[19] A.S. Troelstra. Realizability. ILLC Prepublication Series for Mathematical Logic and Foundations ML-92-09.

Using Subtyping in Program Optimization

S. Berardi[1] and L. Boerio[2]

[1] Dipartimento di Informatica, Universita' di Torino
Corso Svizzera 185, 10149 Torino, Italy
stefano@di.unito.it

[2] Dipartimento di Informatica, Universita' di Torino
Corso Svizzera 185, 10149 Torino, Italy
lucab@di.unito.it

Abstract. Constructive logics can be used to write the specifications of programs as logic formulas to be proved. By using the Curry-Howard isomorphism, we can automatically extract executable code from constructive proofs. Normally, programs automatically generated are ineffcient. They contains parts useless to compute the final result. In this paper, we show a technique to erase such useless parts from programs. Our technique is essentially an extension of another method, developed by S. Berardi, optimizing simply typed λ-terms. By using the notion of subtyping, we can overcome some intrinsecal limitations of the original method. We prove that optimized terms are equivalent to the original ones and we give an algorithm to find such optimizations

1 Introduction

In our work, we have studied an algorithm to simplify programs represented by λ-terms of first order typed λ-calculus. In this way, we can obtain simpler programs, operationally equivalent to the original ones, but requiring less time and space to run. A main motivation of our work, is to produce better code for program extracted from formal proofs. From a formal proof in a Logical Framework, by using the Curry-Howard isomorphism or the notion of realizability ([1] and [7]), we can automatically extract executable code in form of typed λ-terms. Generally, such λ-terms extracted from proofs are ineffcent. They contain parts that are useless for the computation of the final result. Several approaches have been followed to erase useless code from programs (see e.g. [1], [5], [6]). One of these is due to S. Berardi ([3]). By building on the top of the algorithm of C. Mohring and Y. Takayama, Stefano Berardi developed a term simplification technique that he called "pruning". In Berardi's technique, useless subterms are found by analyzing the types of the terms. Berardi's original technique, however, had some weakness. Some simplifications were not possible because they would lead to ill typed applications of functions. To overcome these limitations, in this paper we introduce the notion of subtypes. As a consequence, an application is now well typed if the argument has a type not only equal, but more in general included in the input type of the corresponding function. In this way more simplifications are now possible, and we have developed a method that encompasses original pruning overcoming some of its limits.

This is the plan of the paper. In section 2, we introduce the typed system used to write our programs. In section 3, we introduce the notion of subtype and we describe how to use it in order to discover useless subterms in expressions. In section 4, we define the simplifications relations used to compare the original programs to their simplified version. In section 5, we prove that the optimized programs are observationally equivalent to the original ones, if we preserve type and context. In section 6, we will prove that there exist minimum simplifications and, in section 7, we will introduce an algorithm to find such minimum simplifications and give some examples. In section 8, the conclusions.

2 The System

In this section, we define a simply typed lambda calculus, $\mathcal{T}_\Omega$, used to write the terms representing the programs to be simplified. $\mathcal{T}_\Omega$ is an extension of Gödel's system $\mathcal{T}$, obtained by adding subtyping and a special atomic type Ω to $\mathcal{T}$. The notion of subtyping we consider involves set theoretical inclusion and object identification. We consider types paired with an equality notion on them, hence, with "A is a subtype of B" we mean, at the same time, that each inhabitant of A is an inhabitant of B and two inhabitants equated in A are equated in B. Hence B may contains more objects than A, and it may equate objects kept distinct in A.

For sake of simplicity, we introduce only a minimum of atomic types and inclusions between them. The atomic types we consider are N, the data type of natural number equipped with standard equality, and Ω, to be thought as the type of "useless" (to be removed) integers, and consisting of the natural numbers all identified together. With the definition we gave, the type N is included in Ω. The system $\mathcal{T}_\Omega$ without Ω will correspond exactly to Gödel system $\mathcal{T}$. The intuition behind the extension of $\mathcal{T}$ with Ω, and the use of Ω in removing useless parts of programs in $\mathcal{T}_\Omega$ will be discussed in the next section.
In the remaining of this section, the reader may simply consider Ω as it were any atomic type including N (like the type of relative integers).
We now complete the description of $\mathcal{T}_\Omega$.

Definition 1.

i) The types of $\mathcal{T}_\Omega$ are inductively defined from the atomic types N and Ω by repeatedly using the constructors $\rightarrow$ and $\times$.
ii) Let A, A_1, A_2, B, B_1, B_2 and C be types. We define the inclusion relation,

$\subseteq_\Omega$[3], over the set of types of $\mathcal{T}_\Omega$ with the following deduction system:

$(\subseteq_\Omega - \Omega)\ N \subseteq_\Omega \Omega$

$(\subseteq_\Omega - \text{Ref}_N)\ \ N \subseteq_\Omega N \qquad (\subseteq_\Omega - \text{Ref}_\Omega)\ \ \Omega \subseteq_\Omega \Omega$

$$(\subseteq_\Omega - \rightarrow)\ \frac{B_1 \subseteq_\Omega A_1 \quad A_2 \subseteq_\Omega B_2}{A_1 \rightarrow A_2 \subseteq_\Omega B_1 \rightarrow B_2} \quad (\subseteq_\Omega - \times)\ \frac{A_1 \subseteq_\Omega B_1 \quad A_2 \subseteq_\Omega B_2}{A_1 \times A_2 \subseteq_\Omega B_1 \times B_2}$$

Remark the controvariance of the arrow in the first argument. We write $A \supseteq_\Omega B$ when $B \subseteq_\Omega A$.

iii) Now we define the notion of Ω-type, whose relevance will be pointed out in the next section. A type T is said an *Ω-type* if it is defined by the following syntax:

$$O ::= \Omega \mid A \rightarrow O \mid O \times O$$

where A can be any type. In other words, a term has a Ω-type iff we may obtain out of it, by repeated applications and projections, some terms of type Ω but no terms of type N. Consequently, when in the next section Ω will be interpreted as the type of "useless" integers, any term t having an Ω-type will be interpreted as a "useless" program, since out of it we may only get "useless" integers as output.

It's immediate to see that the relation of inclusion among types is a partial order relation.

Example 1. It's easy to verify that:

- $\Omega \rightarrow N \subseteq_\Omega \Omega \rightarrow \Omega$, $N \rightarrow N \subseteq_\Omega N \rightarrow \Omega$. We note that $\Omega \rightarrow \Omega$ and $N \rightarrow N$ are not comparable.
- $N \times N \subseteq_\Omega \Omega \times N$, $N \times \Omega \subseteq_\Omega \Omega \times \Omega$. Again, we note that $\Omega \times N$ and $N \times \Omega$ are not comparable.
- Each type is included in some Ω-type. Ω-types are the "largest" type.

Now, we examine the language of terms of $\mathcal{T}_\Omega$. The term formation operators are: abstraction, application, pair formation and pair selection.

Definition 2. The set of pre-terms of $\mathcal{T}_\Omega$ is defined by the following syntax:

$$M ::= x^T \mid c \mid \lambda x^T.M \mid (MM) \mid \langle M, M\rangle \mid \pi_1(M) \mid \pi_2(M)$$

where x^T is a typed variable, T a type, c a constant and $\langle .,. \rangle$ and $\pi_i(.)$ the constructors and selectors of the ordered pairs.

[3] To formalize the notion of inclusion, we use the symbol of $\subseteq_\Omega$ instead of the more traditional $\leq$ to avoid confusion with another relation, defined in a section 4

The basic constants of the language are: an infinite set of dummy constants, $\mathbf{d}^O$ (for each Ω-type O), denoting one "canonical" inhabitant for each Ω-type, whose interpretation will be given in the next section, $\mathbf{0}^N$, $\mathbf{succ}^{N\to N}$ and a family of recursion operators over the integers, $\mathbf{rec}^{N\to A\to(N\to A\to A)\to A}$ (or just $\mathbf{rec}_A$ for short) for each type A. We call Σ_Ω such a set of constants. We call *atom* any atomic type and *symbol* any term constant and variable. We use α, β, γ, ... to denote atoms and s, s', s", ... to denote symbols.

Definition 3.

i) Let T be any type. With T_ω we denote the type obtained from T by replacing each atom N by an Ω.
ii) A context is any finite set of typed variables, having pairwise distinct names. If Γ is a context then with dom(Γ) we indicate the set of "names" of the typed variables of Γ and with $\Gamma_{/x}$ we denote the context obtained from Γ by erasing x. Contexts are ranged over by Γ, Δ, ...
iii) If Γ is a context, t a term and T a type, we write $t : T$ if t is a well formed term of type T and $\Gamma \vdash_{\text{pr}} t : T$, if t is a well formed term of principal type T under (a part of) the assumptions contained in Γ. If Γ is the empty set we just write $\vdash_{\text{pr}} t : T$

Definition 4. Let Γ be a context, t and u be terms, x a typed variable, c^A a constant of Σ_Ω, A, A_1, A_2 and B be types. The type inference system for terms of $\mathcal{T}_\Omega$ and principal types is:

$$\Gamma \vdash_{\text{pr}} c^A : A \qquad\qquad \Gamma.x^A \vdash_{\text{pr}} x^A : A$$

$$\frac{\Gamma.x^A \vdash_{\text{pr}} t : B}{\Gamma \vdash_{\text{pr}} \lambda x^A.t : A \to B} \qquad \frac{\Gamma \vdash_{\text{pr}} t : A \to B \quad \Gamma \vdash_{\text{pr}} u : A' \quad A' \subseteq_\Omega A}{\Gamma \vdash_{\text{pr}} (t\,u) : B}$$

$$\frac{\Gamma \vdash_{\text{pr}} t : A_1 \quad \Gamma \vdash_{\text{pr}} u : A_2}{\Gamma \vdash_{\text{pr}} \langle t, u\rangle : A_1 \times A_2} \quad \frac{\Gamma \vdash_{\text{pr}} t : A_1 \times A_2}{\Gamma \vdash_{\text{pr}} \pi_j(t) : A_j} \quad (\text{for } j = 1, 2)$$

We call the set of well formed terms of this system Λ_Ω.

Note that, if t is a well formed term then its principal type is unique. We call (general) type of a term t, each type including the principal type of t. Formally we write $\Gamma \vdash t : B$ if $\Gamma \vdash_{\text{pr}} t : A$ and $A \subseteq_\Omega B$.

Equivalently we could define the type judgement by a deduction system using the same system used to define the principal type judgement, but replacing the rule for term application with the rules

$$\frac{\Gamma \vdash t : A \to B \quad \Gamma \vdash u : A}{\Gamma \vdash (t\,u) : B} \qquad (\text{sub}) \quad \frac{\Gamma \vdash t : A \quad A \subseteq_\Omega B}{\Gamma \vdash t : B}$$

Then we could define the principal type of a term t as the minimum (w.r.t. $\subseteq_\Omega$) type that we can assign to t.

Types (and principal types) are ranged over by A, B, C, etc. Terms and pre-terms are ranged over by t, u, v etc. Generally we use upper cases for types and lower cases for terms. We assume the standard associativity and precedence rules of operators. For simplicity, in the rest of the paper we omit type decorations over symbols if the type is clear from the context. We assume the standard definition of *free* and *bound* variable, *closed term*, *substitution* and *closed substitution*. We denote the set of the free variables of a term t with FV(t) and substitutions with σ, τ, ...In the rest of the paper we suppose that, for each substitution σ, the names of the variables in the domain of the substitution, dom(σ), are pairwise disjoint. As usual, a typed context, $\mathcal{C}[.]$ is a w.f. term of the language with a hole. Since we are in a typed setting, also holes in contexts have types.

We remark that $\mathcal{T}_\Omega$ extends the traditional system $\mathcal{T}$ of Gödel. We can say that $\mathcal{T}$ is the fragment Ω-free, namely without the atomic type Ω and constants $\mathbf{d}^O$, of $\mathcal{T}_\Omega$.

The system has the conventional rules of β, η and π reduction and conversion. We also have reduction rules for the constant **rec**. We suppose α rule implicit.

Definition 5.

i) Let t, t_1, t_2, u and f terms, A a type, x a variable, n a term of type N. The elementary contraction rules of the system are:

$$
\begin{array}{lll}
(\beta) & (\lambda x^A.t)\,u & \rightarrow_\beta \; t[x := u] \\
(\pi) & \pi_i(\langle t_1, t_2\rangle) & \rightarrow_\pi \; t_i \;\; (\text{for } i = 1, 2) \\
(\eta) & \lambda x^A.(f\,x) & \rightarrow_\eta \; f \;\; (\text{with } x \text{ not free in } f) \\
(\text{rec}0) & \mathbf{rec}\;\mathbf{0}\;a\;f & \rightarrow_{rec0} \; a \\
(\text{recS}) & \mathbf{rec}\;(\mathbf{succ}\;n)\;a\;f & \rightarrow_{recS} \; f\;n\;(\mathbf{rec}\;n\;a\;f)
\end{array}
$$

We write $\rightarrow_{rec}$ for $\rightarrow_{rec0} \cup \rightarrow_{recS}$

ii) Let $r \in \{\beta, \pi, \eta, \text{rec}\}$. With $\rightarrow_r$ we denote also the contextual closure of the elementary contraction rule $\rightarrow_r$. We write $\rightarrow_r^*$ to indicate any finite number of steps (included 0) of the reduction r, while $\rightarrow^*$ denotes any finite number of steps of any reduction. We use $=_r$ for conversion w.r.t. the reduction r, while we use $=_\mathcal{T}$ to denote conversion under β, π and **rec** conversion rules together. We use '$\equiv$' to denote sintactic equality.

We assume the usual definitions of redex and normal form.

To formalize the notion of program equivalence, we use an equivalence relation over terms of our system known as "observational equivalence".

Definition 6. Let A be any type. Let t and u be terms of type A. We say that t and u are observationally equivalent in A, writing $t =_{obs} u : A$ iff for each closed typed context of type N that closes t and u, with a hole of type A, we have $\mathcal{C}[t] =_\mathcal{T} \mathcal{C}[u]$

Observational equivalence is a kind of operational equivalence, since two observationally equivalent programs behave in the same way. We note that observational equivalence implies extensional equivalence for programs over natural

numbers. In fact, we have that if t and u are closed terms of type $N \to N$:
$t =_{obs} u : N \to N \quad \Rightarrow \quad \forall closed\, n : N.(t\, n) \; =_{\mathcal{T}} \; (u\, n)$.

We just recall that, an *equational theory* or just a *theory* for a typed system, is an equivalence relation over terms of the system, that is compatible with term formation, types of terms and reductions.

Definition 7. Let T be a theory over the system $\mathcal{T}$. If t and u are terms:

i) we write $t =_T u$ if $(t, u) \in T$;
ii) with $t \neq_T u$ we mean that $t =_T u$ doesn't hold;
iii) a theory T is consistent if there exist two terms t and u such that $t \neq_T u$;
iv) a theory T is a maximum theory if it is consistent and for each consistent theory T' and for each pair of term t and u, we have $t =_{T'} u \Rightarrow t =_T u$.

Lemma 8. *(Statman 86) Observational equivalence is the maximum equational theory for the system $\mathcal{T}$.*

We note that the principal types weakly decrease with reduction. For instance $((\lambda x^A.x)\, y^B)$ has principal type A, if $B \subseteq_\Omega A$. If we reduce the redex, we obtain just y^B that has principal type B.
So we have the following properties:

Proposition 9.

i) Let t be a term of principal type A and σ a substitution. Then $\sigma(t)$ is a w.f. term of principal type A' for some type $A' \subseteq_\Omega A$.
ii) Let $(\lambda x^A.t)\, u$ be a w.f. term of principal type T. Then $t[x := u]$ is a w.f. term of principal type $T' \subseteq_\Omega T$.
iii) In $\mathcal{T}_\Omega$ holds a "weak" subject reduction for principal types. If t and u are terms of principal type A and B respectively, s.t. $t \to^* u$ then $B \subseteq_\Omega A$. By our definition of general type, we see that the subject reduction still holds for types.
iv) If t and u are terms s.t. $t =_{obs} u : A$ then, $t =_{obs} u : A'$ for each type A' s.t. $A \subseteq_\Omega A'$.
v) The system $\mathcal{T}_\Omega$ is Church-Rosser and strongly normalizing[4].

3 The Semantics

In this section we will discuss the intuition behind the extension of system $\mathcal{T}$ with Ω and the use of Ω in removing useless parts of programs of $\mathcal{T}_\Omega$.

We first explain the meaning of Ω and of the inclusion $N \subseteq_\Omega \Omega$. We said that N is the type of integers with standard equality, while Ω is the type of

[4] In order to see it, just replace in any term each Ω by N and each $\mathbf{d}^O$, with O a Ω-type, by a fresh variable of type O', where O' is the type obtained from O by replacing each Ω with a N. Then use the strong normalization for system $\mathcal{T}$

integers all equated together, and represents the "useless" numerals. Indeed, if a term has the type Ω, then we know that it is a numeral, but nothing more, because in Ω 0, 1, 2, ..., are all identified together, hence we cannot distinguish which is which. If a term has the type N, then it still brings the information "I am a numerical value" but now, since we have in N standard equality, we may distinguish it from all other (different) numerals. The condition to be an element of the Ω-set of terms is weaker than the corresponding one to be an element of the N-set of terms and this fact is formalized be the $N \subseteq_\Omega \Omega$.

What we said should justify our claim of Ω being the type of useless numeral. We now may explain through examples the use of Ω to express the fact that the input or the ouput of a program, or the program itself, or a part of it are useless.

If we consider the identity function over numerals, $\lambda x^N.x$, the types we can infer are $N \to N$ and, by using the inclusion $\subseteq_\Omega$, $N \to \Omega$. The first type expresses the fact that we may input the identity with a numeral that we use, receiving back a numeral we may still use. The second one, instead, expresses the fact that we may choose not to use the output of the identity (by changing the output type from N to Ω). In this case the entire program for the identity becomes useless: there is no use of a function if we do not use its ouput. The type $N \to \Omega$ may be consider as a type of useless functions. In particular, a function of principal type $N \to \Omega$ like $\lambda x^N.\mathbf{d}^\Omega$ is useless, because it may only returns useless integers as outputs. More in general, any function of principal type $A_1 \to \ldots \to A_n \to \Omega$ is useless.

In the same way, we may argue that a pair, having principal type $\Omega \times \Omega$, instead of $N \times N$ is of no use. Indeed, the type $\Omega \times \Omega$ expresses the fact that both components of the pair are useless integers, and therefore there is no use for the pair itself, either. By combining the last two remarks, we see that a function having principal type $N \to (\Omega \times \Omega)$ is useless. The same reasoning apply to all programs having an Ω-type as principal type. In particular $\mathbf{d}^O$ always denotes a useless program.

By using the interpretation we gave to Ω and $\mathbf{d}^O$, we may now explain what we mean by simplification of a program in our setting. A set of subtypes and subterms in a given term is redundant and may be simplified if we may replace the subtypes with Ω (a useless type) and the subterms with some $\mathbf{d}^O$ (a useless term), *while preserving the well-formation of the term.* In the next section we will formalize this notion of simplification. In the remaining of this section we will explain how a formal semantics of our types can be given using the notion of partial equivalence relation (see for more details [2]).

Definition 10.

i) We call Terms the set of type free terms of pure λ-calculus built by using the constants **d**, **0**, **succ** and **rec** and the term constructors for abstaction, application, pairing and projection.
ii) We call Terms^0 the set of closed terms of Terms
iii) We call Pure the type free λ-calculus, having as set of terms Terms and as reduction rules the type free versions of the rules seen in section 2.

iv) We indicate $=_{\mathcal{P}}$ the convertibility in Pure under β, π and rec rules together.
v) We call *erasure*, the function from Λ_Ω to Terms that remove from a typed term each type annotation. We denote $|.|$ such a function. If t is a term of Λ_Ω, we call erasure of t, the term in Terms obtained by applying $|.|$ to t

Definition 11.

i) A partial equivalence relation (PER) over a set A is a simmetric and transitive binary relation over A.
ii) If R is a PER over a set A, we call domain of R the set defined as: $\mathrm{dom}(R) = \{t \mid t \in A \text{ and } \langle t, t\rangle \in R\}$.
iii) A PER over a type T is a PER over the set of Terms^0.
iv) To each type A in T_Ω, we can associate a PER over Terms^0. Let A_1, A_2, A and B be types, and t, u, v_1 and v_2 be terms. We inductively define the semantics of the types of T_Ω:

$$
\begin{aligned}
[\![N]\!] &= \{\langle t, u\rangle \mid t, u \in \mathrm{Terms}^0 \text{ and } t =_{\mathcal{P}} u =_{\mathcal{P}} \mathbf{succ}^k(\mathbf{0}) \text{ (for some } k)\} \\
[\![\Omega]\!] &= \{\langle t, u\rangle \mid t, u \in \mathrm{Terms}^0\} \\
[\![A_1 \times A_2]\!] &= \{\langle t, u\rangle \mid t, u \in \mathrm{Terms}^0 \text{ and } \langle \pi_i(t), \pi_i(u)\rangle \in [\![A_i]\!] \ (i = 1,\ 2)\} \\
[\![A \to B]\!] &= \{\langle t, u\rangle \mid t, u \in \mathrm{Terms}^0 \text{ and} \\
&\qquad \langle (t\, v_1), (u\, v_2)\rangle \in [\![B]\!] \text{ for all } \langle v_1, v_2\rangle \in [\![A]\!]\}
\end{aligned}
$$

The informal discussion about the types N, Ω and Ω-types may now be translated by formal properties in the semantics:

Proposition 12. *Let O be any Ω-type.*

i) $[\![N]\!] \subseteq [\![\Omega]\!]$
ii) $[\![N \to \Omega]\!] = [\![\Omega \to \Omega]\!] = [\![O]\!] =$ *the trivial PER, identifying all terms* $t, u \in$ *Terms*^{0}
iii) $[\![\Omega \to N]\!] = \{\langle f, g\rangle \mid f, g$ *constant and returning the same output*$\}$
iv) $[\![N \to N]\!] = \{\langle f, g\rangle \mid f(n) =_{\mathcal{P}} g(n)$, *for each* $n \in dom([\![N]\!])\}$

The last property formally expresses the fact that Ω-types are types of "useless" programs.

Our syntactic definition reflects the semantic properties above remarked. In particular about the relation of $\subseteq_\Omega$, we have a double informal interpretation. If A and a are types then $A \subseteq_\Omega B$ has the meaning of "A carries more information then B". The other meaning is "the set of values denoted by A is a subset (from an insiemistic point of view) of the set denoted by B".

The syntactic $\subseteq_\Omega$ is reflected in the semantics by the inclusion relation.

Proposition 13. *Let A and B be types, then $A \subseteq_\Omega B \Rightarrow [\![A]\!] \subseteq [\![B]\!]$*

4 Simplification Relations

In order to formalize the notion of "simplifying a term", introduced in the previous section, in this section we define three relations of $\leq$, one on types, one over (well formed) terms of the system and one over contexts.

Definition 14. Let t, t_1, t_2, u, u_1, u_2 be any terms, A, A_1, A_2, B, B_1, B_2, T any types, c the costant **0** or **succ**, x, y, any term variables, Γ and Γ' any contexts.

i) The simplification relation for types, $\leq_T$, is the smallest relation including $\Omega \leq_T A$ and compatible by type formation, namely $N \leq_T N$ and if $A_1 \leq_T A_2$ and $B_1 \leq_T B_2$ then $A_1 \to B_1 \leq_T A_2 \to B_2$ and $A_1 \times B_1 \leq_T A_2 \times B_2$.
ii) The simplification relation for contexts, $\Gamma \leq_{ctx} \Gamma'$, is defined by: $\Gamma \leq_{ctx} \Gamma'$ iff $(\forall x^A)\ ((x^A \in \Gamma) \Rightarrow (\exists B)\ (x^B \in \Gamma'$ and $A \leq_T B))$.
iii) The simplification relation for terms, $\leq_t$, is defined by the following deduction system.

$$\frac{O \leq_T U}{\mathbf{d}^O \leq_t u}\ (\text{with u:U}) \qquad c^A \leq_t c^A \qquad \frac{A_1 \leq_T A_2}{\mathbf{rec}_{A_1} \leq_t \mathbf{rec}_{A_2}}$$

$$\frac{A \leq_T B}{x^A \leq_t x^B} \qquad \frac{A \leq_T B \quad t_1[x^A := z^A] \leq_t t_2[y^B := z^B]}{\lambda x^A.t_1 \leq_t \lambda y^B.t_2}\ (*)$$

$$\frac{t_1 \leq_t t_2 \quad u_1 \leq_t u_2}{t_1 u_1 \leq_t t_2 u_2} \qquad \frac{t_1 \leq_t u_1 \quad t_2 \leq_t u_2}{\langle t_1, t_2\rangle \leq_t \langle u_1, u_2\rangle} \qquad \frac{t \leq_t u}{\pi_i(t) \leq_t \pi_i(u)}\ (i = 1, 2)$$

(*) In expressions involving binders we have to introduce a fresh variable z in order to have the relation invariant up to α conversion. Without this caution we would have for example, $\lambda x^A.x \leq_t \lambda x^A.x$, but not $\lambda x^A.x \leq_t \lambda y^A.y$.

To simplify the notation, in the rest of the paper we often write only $\leq$, both for $\leq_T$, for $\leq_{ctx}$ and for $\leq_t$.

Example 2.

- We can compare the $\leq_T$ with $\subseteq_\Omega$, by remembering Example 1. We have $\Omega \leq_T \Omega \to \Omega \leq_T N \to \Omega$, $\Omega \to N \leq_T N \to N$. We remark here the covariance of the arrow.
- $\mathbf{d}^\Omega \leq_t \lambda x^N.\mathbf{d}^\Omega \leq_t \lambda x^N.x$

Remember always that the relation $\leq$ is defined on well formed expressions.

The intended meaning of these two relations is that if a term t is less than or equal another term u than t is "simpler" than "u" in the sense that it requires less time and space to evaluate it. On these relations hold some properties.

Proposition 15.

i) The three simplification relations are order relations.
ii) Let t and u be terms having principal type A and B, respectively. If $t \leq_t u$ then $A \leq_T B$ and $FV(t) \leq_{ctx} FV(u)$
iii) Let t and u be terms of type A s.t. $t \leq u$. Let $C[.]$ a typed context with a hole of type A. Then $C[t] \leq C[u]$

5 Simplifications and Observational Equivalence

In this section, we want to prove that the simplification relation '$\leq_t$' and preserve the observational equivalence if it preserves type and context; in this case our simplification is in fact an optimization. To this aim we define a family of binary relations over well formed closed terms of Λ_Ω and another family over substitutions. By using these relations, we can prove that, if we simplify a term, preserving its type and its context, we obtain a term operationally equivalent to the original one.

Definition 16.

i) For each pair of types A and B s.t. $A \leq_T B$ we inductively define the relation $\mathrm{R}_{A,B}(t,u)$, where t and u are closed terms of type A and B respectively, with the following deduction system:

$$(\mathrm{AX}_\Omega)\ \ \mathrm{R}_{\Omega,A}(t,u) \qquad (\mathrm{AX}_N)\ \ \frac{t =_T u}{\mathrm{R}_{N,N}(t,u)}$$

$$(\rightarrow)\ \ \frac{\forall v_1, v_2 . \mathrm{R}_{A_1,B_1}(v_1,v_2) \Rightarrow \mathrm{R}_{A_2,B_2}(tv_1, uv_2)}{\mathrm{R}_{A_1\rightarrow A_2, B_1\rightarrow B_2}(t,u)}$$

$$(\times)\ \ \frac{\mathrm{R}_{A_1,B_1}(\pi_1(t),\pi_1(u)) \quad \mathrm{R}_{A_2,B_2}(\pi_2(t),\pi_2(u))}{\mathrm{R}_{A_1\times A_2, B_1\times B_2}(t,u)}$$

ii) If Γ and Δ are two contexts s.t. $\Gamma \leq \Delta$ and σ and τ are two closed substitutions s.t. $\mathrm{dom}(\sigma) = \Gamma$ and $\mathrm{dom}(\tau) = \Delta$ we define:

$$\mathrm{R}_{\Gamma,\Delta}(\sigma,\tau) \text{ iff for all x s.t. } x^A \in \Gamma \text{ and } x^B \in \Delta : \mathrm{R}_{A,B}(\sigma(x^A), \tau(x^B))$$

By induction over t we may prove:

Theorem 17. *Let t and u be terms, of types A and B, s.t. $t \leq u$ and $A \leq_T B$. Let $\Gamma = FV(t)$ and $\Delta = FV(u)$. If σ and τ are closed substitutions that close t and u s.t. $dom(\sigma) = \Gamma$, $dom(\tau) = \Delta$ and $R_{\Gamma,\Delta}(\sigma,\tau)$, then $R_{A,B}(\sigma(t), \tau(u))$.*

At this point we can bind the simplification relations to the observational equivalence. By remarking that $\mathrm{R}_{A,A}(t,u)$ implies $t =_{obs} u$, from the previous theorem in the case $A = B$ and $\sigma = \tau =$ identity, we may conclude:

Theorem 18. *Let t and u be terms s.t. $\Gamma \vdash t : A$, $\Gamma \vdash u : A$. If $t \leq_t u$ then $t =_{obs} u : A$*

Example 3. Let us consider the term $(\lambda x^N.\lambda y^N.x)\,u\,v$ of type N, where u and v are some terms of type N. It is an obvious consideration that the bound variable y is useless to compute the final result of the expression, and the same holds for v. In fact we have $(\lambda x^N.\lambda y^\Omega.x)\,u\,\mathbf{d}^\Omega \leq_t (\lambda x^N.\lambda y^N.x)\,u\,v$ with same type and context. By the previous theorem these two terms are observationally equivalent.

In this section we have proved that given a term t, if we simplify it by replacing some parts with the dummy constants, without altering its type, we obtain a new equivalent term. In the next section we'll see that there exists a "simplest" term equivalent to the one given in input.

6 Minimum Simplifications

In this section we show that there exist minimum or best simplifications of terms among those preserving type and context.
We have seen that $\leq$ is an order relation. We can prove that:

Proposition 19.

i) There exists the gratest lower bound w.r.t. $\leq$, of each pair of types and of terms. We denote the g.l.b. of e_1, e_2 by $\inf(e_1, e_2)$.
ii) For each pair of contexts there exists their g.l.b., denoted by infctx and, if Γ and Δ are contexts, we have:
$infctx(\Gamma, \Delta) = \{x^C \mid x^A \in \Gamma,\ x^B \in \Delta$ and $C = \inf(A, B)\}$
iii) If t and u are terms s.t. $\Gamma \vdash t : A$ and $\Delta \vdash u : B$ and there exists a term z s.t. $t \leq z$ and $u \leq z$, then $infctx(\Gamma, \Delta) \vdash \inf(t, u) : \inf(A, B)$

At this point, we introduce two structures useful for the optimization algorithm that we will define in the next section.

Definition 20. For each term t, s. t. $\Gamma \vdash t : T$ we define:

i) $\mathrm{LE}(t) = \{t' \mid t' \leq t\}$
ii) $\mathrm{CLE}(t) = \{t' \mid t' \leq t$ and $\Gamma' \vdash t' : T$ and $\Gamma' \subseteq \Gamma\}$

If t is a term, the set LE(t) (*less equal* t) is the set of all the terms that we can obtain from t by replacing some parts with the dummy constants, while the set CLE(t) (*contestual less equal*) is the set of simplified version of t with same type and context included in that of t. By using the proposition 19 we can prove:

Proposition 21. *Given any term t we have:*

i) LE(t) is a finite complete lower semilattice w.r.t. $\leq$
ii) CLE(t) is a sub semilattice of LE(t)

Definition 22. Let t any term. We denote the minimum element of CLE(t) as FL(t)[5].

By the results of section 5, FL(t) $=_{obs}$ t. Given a term t, FL(t) is the best simplification of t that we can obtain without altering its operational meaning. At this point, our aim is to define an algorithm to find such a minimum term.

[5] FL shortens "flow algorithm". The choice of the name refers to the fact that the bottom of CLE(t) may be computed by a kind of "data flow algorithm"

7 The Optimization Algorithm

In this section we show an algorithm to compute the minimum element of the set CLE(t), namely FL(t), defined in the previous section. This algorithm takes in input a term t and gives in output the "best optimization" of t, namely the simplest well formed term (w.r.t. $\leq_t$) observationally equivalent to t. In this setting, more simple means that it takes less space and time to return a result. We note that, because of type constraints, the problem of finding a minimum simplification is not closed by subproblems. If we have a function applied to an argument, we cannot simply optimize the function and the argument separately, since the application of the two simplified terms may not be consistent.

In order to overcome this problem, we have defined a more general algorithm. It takes in input two type judgements, $\Gamma \vdash_{\text{pr}} t : A$ and $\Gamma_0 \vdash_{\text{pr}} t_0 : A_0$, such that $\Gamma_0 \leq_{ctx} \Gamma$, $t_0 \leq t$ and $A_0 \leq A$ and it takes also a context Γ' and a type A' s.t. $\Gamma_0 \leq_{ctx} \Gamma' \leq_{ctx} \Gamma$ and $A_0 \leq A' \leq A$.

In output the function gives the minimum term u (w.r.t. $\leq_t$), the minimum context Δ and the minimum type B, s.t. $\Delta \vdash_{\text{pr}} u : B$ and $t_0 \leq u \leq t$, $\Gamma' \leq_{ctx} \Delta \leq_{ctx} \Gamma$, $A' \leq B \leq A$. This kind of problem is closed by subproblems and it is more general of the original one. So, if we can solve the former we can solve the latter. Now we are ready to show the algorithm. It is inductively defined.

Definition 23.
The optimization algorithm
Input: $\Gamma \vdash_{\text{pr}} t : A$, $\Gamma_0 \vdash_{\text{pr}} t_0 : A_0$, Γ', A' s.t. $t_0 \leq t$, $\Gamma_0 \leq_{ctx} \Gamma' \leq_{ctx} \Gamma$ and $A_0 \leq A' \leq A$
Output: The minimum u, Δ and B s.t. $\Delta \vdash_{\text{pr}} u : B$, $t_0 \leq u \leq t$, $\Gamma' \leq_{ctx} \Delta \leq_{ctx} \Gamma$ and $A' \leq B \leq A$

If $\Gamma' = \Gamma_0$ and $A' = A_0$ then set $u = t_0$, $\Delta = \Gamma_0$ and $B = A_0$
else
if A_0 and A' are Ω-types and $t_0 = \mathbf{d}^{A_0}$ then set $u = \mathbf{d}^{A'}$, $\Delta = \Gamma'$ and $B = A'$
else

- if $t = c^A$, with $c = \mathbf{0}$ or $c = \mathbf{succ}$, then set $u = c^A$, $\Delta = \Gamma'$ and $B = A'$
- if $t = \mathbf{rec}_T$, with $A = N \to T \to (N \to T \to T) \to T$ then set $u = \mathbf{rec}_{T_1}$, $\Delta = \Gamma'$ and $B = N \to T_1 \to (N \to T_1 \to T_1) \to T_1$, where T_1 is the minimum element of the set $\{C \mid C' \equiv N \to C \to (N \to C \to C) \to C \text{ and } A' \leq C' \leq A\}$
- if $t = x^A$, with $\Gamma = \Gamma_1.\{x^A\}$, then set $u = x^{A'''}$, $\Delta = \Gamma'_{/x}.\{x^{A'''}\}$ and $B = A'''$, where $A''' = sup(A', A'')$ and if $x \in dom(\Gamma')$ then $\Gamma' \equiv \Gamma'_{/x}.\{x^{A''}\}$ else $A'' = \Omega$
- if $t = \lambda x^{T'}.t'$, with $A \equiv T' \to T''$ and $A' \equiv C' \to C''$ then
  - if t_0 is an abstraction, we suppose $t_0 \equiv \lambda x^{T'_0}.t'_0$, $A_0 \equiv T'_0 \to T''_0$ and $t'_0 : T''_0$
  - if t_0 is not an abstraction but A_0 is a functional type, we suppose $t_0 \equiv \mathbf{d}^{A_0}$ and $A_0 \equiv T'_0 \to T''_0$

- if t_0 is not an abstraction and A_0 is not a functional type, we have $t_0 \equiv \mathbf{d}^{\Omega}$ and we set $T'_0 = T''_0 = \Omega$
- apply recursively the algorithm on $\Gamma.\{x^{T'}\} \vdash_{\text{pr}} t' : T''$, $\Gamma_0.\{x^{T'_0}\} \vdash_{\text{pr}} t'_0 : T''_0$, $\Gamma'.\{x^{C'}\}$ and C'', obtaining $\Delta'.\{x^{B'}\} \vdash_{\text{pr}} u' : B''$, with $t'_0 \leq u' \leq t'$, $\Gamma'.\{x^{C'}\} \leq_{ctx} \Delta'.\{x^{B'}\} \leq_{ctx} \Gamma.\{x^{T'}\}$ and $C'' \leq B'' \leq T''$
- set $u = \lambda x^{B'}.u'$, $\Delta = \Delta'$ and $B = B' \to B''$

if $t = (f\,a)$, with $f : T \to A$, $a : S$ and $S \subseteq_{\Omega} T$, then

- if t_0 is an application then, we suppose $t_0 \equiv (f_0\,a_0)$, $f_0 : T_0 \to A_0$, $a_0 : S_0$ and $S_0 \subseteq_{\Omega} T_0$
- if t_0 is not an application then, we suppose $t_0 \equiv \mathbf{d}^{A_0}$, we set $T_0 = S_0 = \Omega$ and $f_0 = \mathbf{d}^{\Omega \to A_0}$, $a_0 = \mathbf{d}^{\Omega}$
- apply recursively the algorithm on $\Gamma \vdash_{\text{pr}} f : T \to A$, $\Gamma_0 \vdash_{\text{pr}} f_0 : T_0 \to A_0$, Γ' and $T_0 \to A'$, obtaining $\Delta_1 \vdash_{\text{pr}} f_1 : T_1 \to A_1$, with $f_0 \leq f_1 \leq f$, $\Gamma' \leq_{ctx} \Delta_1 \leq_{ctx} \Gamma$ and $T_0 \to A' \leq T_1 \to A_1 \leq T \to A$
- let S_1 the minimum element of the set $\{V \mid V \subseteq_{\Omega} T_1 \text{ and } S_0 \leq V \leq S\}$
- apply recursively the algorithm on $\Gamma \vdash_{\text{pr}} a : S$, $\Gamma_0 \vdash_{\text{pr}} a_0 : S_0$, Δ_1 and S_1, obtaining $\Delta_2 \vdash_{\text{pr}} a_2 : S_2$, with $\alpha_0 \leq a_2 \leq a$, $\Delta_1 \leq_{ctx} \Delta_2 \leq_{ctx} \Gamma$ and $S_0 \leq S_2 \leq S$
- let $i = 2$
  while $\Delta_i \neq \Delta_{i-1}$ and ((i is even $\Rightarrow S_i \not\subseteq_{\Omega} T_{i-1}$) or ($i$ is odd $\Rightarrow S_{i-1} \not\subseteq_{\Omega} T_i$)) **do**
  - increment i
  - apply recursively the algorithm alternatively on the function and on the argument using the results of the previous step, obtaining new $\Delta_i \vdash_{\text{pr}} f_i : T_i \to A_i$ or $\Delta_i \vdash_{\text{pr}} a_i : S_i$
- if, at the end of the iteration, i is even then set $u = f_{i-1}a_i$, $\Delta = \Delta_i$ and $B = A_{i-1}$ else set $u = f_i a_{i-1}$, $\Delta = \Delta_i$ and $B = A_i$

if $t = \langle t', t'' \rangle$, with $A = T' \times T''$ then

- if t_0 is a pair then we suppose $t_0 \equiv \langle t'_0, t''_0 \rangle$ and $A_0 \equiv A'_0 \times A''_0$
- if t_0 is not a pair but A_0 is a cartesian product then we suppose $A_0 \equiv A'_0 \times A''_0$ and we set $t'_0 = \mathbf{d}^{A'_0}$, $t''_0 = \mathbf{d}^{A''_0}$
- if t_0 is not a pair and A_0 is not a cartesian product then $t_0 \equiv \mathbf{d}^{\Omega}$, so we set $A'_0 = A''_0 = \Omega$ and $t'_0 = t''_0 = \mathbf{d}^{\Omega}$
- we suppose $A' \equiv C' \times C''$
- apply recursively the algorithm on $\Gamma \vdash_{\text{pr}} t' : T'$, $\Gamma_0 \vdash_{\text{pr}} t'_0 : A'_0$, Γ' and C', obtaining $\Delta_1 \vdash_{\text{pr}} u_1 : B_1$, with $\Gamma' \leq \Delta_1 \leq \Gamma$ and $C' \leq B_1 \leq T'$
- apply recursively the algorithm on $\Gamma \vdash_{\text{pr}} t'' : T''$, $\Gamma_0 \vdash_{\text{pr}} t''_0 : A''_0$, Δ_1 and C'', obtaining $\Delta_2 \vdash_{\text{pr}} u_2 : B_2$, with $\Delta_1 \leq \Delta_2 \leq \Gamma$ and $C'' \leq B_2 \leq T''$
- let $i = 2$
  while $\Delta_i \neq \Delta_{i-1}$ **do**
  - increment i
  - apply recursively the algorithm alternatively on the first and the second components of the terms using the results of the previous step, obtaining new $\Delta_i \vdash_{\text{pr}} u_i : B_i$

  - if, at the end of the iteration, i is even then set $u = \langle u_{i-1}, u_i \rangle$, $\Delta = \Delta_i$, $B = B_{i-1} \times B_i$ else set $u = \langle u_i, u_{i-1} \rangle$, $\Delta = \Delta_i$, $B = B_i \times B_{i-1}$
- if $t = \pi_i(t')$, with $t' : A^1 \times A^2$ and $A \equiv A^i$ $(i = 1, 2)$, then
  - if t_0 is a projection, we suppose $t_0 \equiv \pi_i(t'_0)$ and $t'_0 : C_0 \equiv A^1_0 \times A^2_0$, with $A_0 \equiv A^i_0$, and if $i = 1$ then we set $C = A' \times A^2_0$ else we set $C = A^1_0 \times A'$
  - if t_0 is not a projection, we suppose $t_0 \equiv \mathbf{d}^{A_0}$ and if $i = 1$ then we set $C = A' \times \Omega$ and $t'_0 = \mathbf{d}^{A_0 \times \Omega}$ of type $C_0 \equiv A_0 \times \Omega$ else we set $C = \Omega \times A'$ and $t'_0 = \mathbf{d}^{\Omega \times A_0}$ of type $C_0 \equiv \Omega \times A_0$
  - apply recursively the algorithm on $\Gamma \vdash_{\mathrm{pr}} t' : A^1 \times A^2$, $\Gamma_0 \vdash_{\mathrm{pr}} t'_0 : C_0$, Γ' and C, obtaining $\Delta' \vdash_{\mathrm{pr}} u' : B^1 \times B^2$, with $t'_0 \leq u' \leq t'$, $\Gamma' \leq_{ctx} \Delta \leq_{ctx} \Gamma$ and $C \leq B^1 \times B^2 \leq A^1 \times A^2$
  - set $u = \pi_i(u')$, $\Delta = \Delta'$ and $B = B^i$

Now we can come back to the original problem, how to optimize a term.

Definition 24. We call Simp the function from Λ_Ω to Λ_Ω that, given a term t, returns the minimum element of CLE(t). It is formally defined by the following algorithm:

The optimization function

Input: A term t, s.t. $\Gamma \vdash_{\mathrm{pr}} t : T$

Output: The minimum element of CLE(t)

- Run the optimization algorithm on $\Gamma \vdash_{\mathrm{pr}} t : T$, $\vdash_{\mathrm{pr}} \mathbf{d}^\Omega : \Omega$, Γ and T', where T' is the minimum type w.r.t. $\leq_T$ included in T, obtaining $\Gamma \vdash_{\mathrm{pr}} u : T''$, for some T''.
- Return u

We have the following soundness result:

Theorem 25. *Given a term t the function Simp computes the simplest term t' equivalent to t, namely Simp(t) = FL(t).*

We conclude the paper with two interesting examples. In the first one, both the old and the new simplification algorithm obtain the same result. Let $t \equiv \lambda x^N.\pi_1(\mathbf{rec}_{N \times N}\ x\ a\ F)$ of type $N \to N$ where we have $a \equiv \langle n_1, n_2 \rangle$ and $F \equiv \lambda n^N.\lambda w^{N \times N}.\langle (f\ \pi_1(w)), (g\ \pi_1(w)\ \pi_2(w)) \rangle$, with f and g free variables of types $N \to N$ and $N \to N \to N$, respectively. By applying any of the two algorithms we have the simplification $t \equiv \lambda x^N.\pi_1(\mathbf{rec}_{N \times \Omega}\ x\ a'\ F')$ of type $N \to N$ where $a \equiv \langle n_1, \mathbf{d}^\Omega \rangle$ and $F \equiv \lambda n^N.\lambda w^{N \times \Omega}.\langle (f\ \pi_1(w)), \mathbf{d}^\Omega \rangle$.

In this second example, we can see how the new technique can allow us to reduce computational space and time in a considerable way.

Let's suppose to extend the set of constants Σ_Ω, by adding the set of constants: $\mathbf{it}_A : N \to A \to (A \to A) \to A$ (for each type A). These new constants implement iteration over natural numbers. The corresponding contraction rules are: $\mathbf{it}_A\ 0\ a\ f \to_{\mathrm{it}} a$ and $\mathbf{it}_A\ (\mathbf{succ}\ n)\ a\ f \to_{\mathrm{it}} f\ (\mathbf{it}_A\ n\ a\ f)$, where $a : A$ and $f : A \to A$. Now consider the following term:

$$t \equiv \lambda n^N.\lambda v^{N \times N}.\pi_1(\mathbf{it}_{N \times N}\ n\ v\ F) \quad \text{of type } N \to N \times N \to N \text{ where}$$

$$F \equiv \lambda w^{N\times N}.\langle (f\,\pi_1(w)), (g\,\pi_2(w))\rangle$$

with f, g free variables of type $N \to N$. The original pruning is unable to simplify t. Instead, by applying the new algorithm we obtain:

$$t' \equiv \lambda n^N.\lambda v^{N\times\Omega}.\pi_1(\mathbf{it}_{N\times\Omega}\ n\ v\ F') \quad \text{of type } N \to N \times \Omega \to N \text{ where}$$
$$F' \equiv \lambda w^{N\times\Omega}.\langle (f\,\pi_1(w)), \mathbf{d}^{\Omega}\rangle$$

8 Conclusions

In this paper we have described a conservative extension of the Berardi's term simplification technique known as "pruning". In our method we look for parts of expressions that are never evaluated to compute the final result. We replace such useless parts by dummy constants, maintaining the type consistency of the expression. We have weakened the definition of type consistency so to introduce a notion of inclusion polymorphism our system. We have proved that, by replacing redundant parts with the dummy constant without altering consistency and Input-Output relation, we can obtain simpler term that are operationally equivalent to the original ones. We have also proved that there exist simplest terms, i.e. best simplifications and we have presented an algorithm to compute best simplifications.

Acknowledgments

The authors want to thank M. Coppo for his valuable remarks and suggestions about this paper, and all his help during their Ph. D. thesis.

References

1. M. Beeson, *Foundations of Constructive Mathematics*, Berlin, Springer-Verlag, 1985
2. S. Berardi, *An Application of PER Models to Program Extraction*, Technical Report, Turin University, 1992.
3. S. Berardi, *Pruning Simply Typed λ-terms*, Technical Report, Turin University, 1993.
4. L. Boerio, *Extending Pruning Techniques to Polymorphic Second Order λ-Calculus*, Proceedings of ESOP '94, Edinburgh, April 1994, LNCS 788, D. Sannella (ed.), Springer-Verlag, pp. 120-134.
5. C. Paulin-Mohring, *Extracting F_ω's Programs from Proofs in the Calculus of Constructions*, In: Association for Computing Machinery, editor, Sixteenth Annual ACM Symposium on Priciples of Programming Languages, 1989.
6. Y. Takayama, *Extraction of Redundancy-free Programs from Constructive Natural Deduction Proofs*, Journal of Symbolic Computation, 1991, 12, 29-69
7. A. S. Troelstra, *Mathematical Investigation of Intuitionistic Arithmetic and Analysis*, Lecture Notes in Mathematics, 344, Springer-Verlag, 1973

What is a Categorical Model of Intuitionistic Linear Logic?

G.M. Bierman
University of Cambridge Computer Laboratory

Abstract. This paper re-addresses the old problem of providing a categorical model for Intuitionistic Linear Logic (**ILL**). In particular we compare the now standard model proposed by Seely to the lesser known one proposed by Benton, Bierman, Hyland and de Paiva. Surprisingly we find that Seely's model is *unsound* in that it does not preserve equality of proofs. We shall propose how to adapt Seely's definition so as to correct this problem and consider how this compares with the model due to Benton *et al.*

1 Intuitionistic Linear Logic

For the first part we shall consider only the multiplicative, exponential fragment of Intuitionistic Linear Logic (**MELL**). Rather than give a detailed description of the logic and associated term calculus we assume that the reader is familiar with other work [15, 5]. The sequent calculus formulation is originally due to Girard [9] and is given below.

$$\frac{}{A \vdash A}\ \textit{Identity}$$

$$\frac{\Gamma \vdash B \qquad B, \Delta \vdash C}{\Gamma, \Delta \vdash C}\ \textit{Cut}$$

$$\frac{\Gamma \vdash A}{\Gamma, I \vdash A}\ (I_{\mathcal{L}}) \qquad \frac{}{\vdash I}\ (I_{\mathcal{R}})$$

$$\frac{\Gamma, A, B \vdash C}{\Gamma, A \otimes B \vdash C}\ (\otimes_{\mathcal{L}}) \qquad \frac{\Gamma \vdash A \qquad \Delta \vdash B}{\Gamma, \Delta \vdash A \otimes B}\ (\otimes_{\mathcal{R}})$$

$$\frac{\Gamma \vdash A \qquad \Delta, B \vdash C}{\Gamma, \Delta, A \multimap B \vdash C}\ (\multimap_{\mathcal{L}}) \qquad \frac{\Gamma, A \vdash B}{\Gamma \vdash A \multimap B}\ (\multimap_{\mathcal{R}})$$

$$\frac{\Gamma \vdash B}{\Gamma, !A \vdash B}\ \textit{Weakening} \qquad \frac{\Gamma, !A, !A \vdash B}{\Gamma, !A \vdash B}\ \textit{Contraction}$$

$$\frac{\Gamma, A \vdash B}{\Gamma, !A \vdash B}\ \textit{Dereliction} \qquad \frac{!\Gamma \vdash A}{!\Gamma \vdash !A}\ \textit{Promotion}$$

Sequents are written as $\Gamma \vdash A$, where A, B represent formulae and Γ, Δ represent multisets of formulae. Where Γ represents the multiset $A_1, \ldots, A_n$, then $!\Gamma$ is taken to represent the multiset $!A_1, \ldots, !A_n$.

The natural deduction presentation proved harder to formalize and early proposals [1, 15] failed to have the vital property of *closure under substitution.* A natural deduction system which has this property was given by Benton *et al.* [4] and is given below.

$$\dfrac{\begin{matrix}[A^x]\\ \vdots \\ B\end{matrix}}{A \multimap B}\,(\multimap_{\mathcal{I}})_x \qquad \dfrac{\begin{matrix}\vdots \\ A \multimap B\end{matrix} \quad \begin{matrix}\vdots \\ A\end{matrix}}{B}\,(\multimap_{\mathcal{E}})$$

$$\dfrac{}{I}\,(I_{\mathcal{I}}) \qquad \dfrac{\begin{matrix}\vdots \\ A\end{matrix} \quad \begin{matrix}\vdots \\ I\end{matrix}}{A}\,(I_{\mathcal{E}})$$

$$\dfrac{\begin{matrix}\vdots \\ A\end{matrix} \quad \begin{matrix}\vdots \\ B\end{matrix}}{A \otimes B}\,(\otimes_{\mathcal{I}}) \qquad \dfrac{\begin{matrix}\vdots \\ A \otimes B\end{matrix} \quad \begin{matrix}[A^x]\ [B^y]\\ \vdots \\ C\end{matrix}}{C}\,(\otimes_{\mathcal{E}})_{x,y}$$

$$\dfrac{\begin{matrix}\vdots \\ !B\end{matrix} \quad \begin{matrix}\vdots \\ C\end{matrix}}{C}\,\textit{Weakening} \qquad \dfrac{\begin{matrix}\vdots \\ !B\end{matrix} \quad \begin{matrix}[!B^x]\ [!B^y]\\ \vdots \\ C\end{matrix}}{C}\,\textit{Contraction}_{x,y}$$

$$\dfrac{\begin{matrix}\vdots \\ !B\end{matrix}}{B}\,\textit{Dereliction} \qquad \dfrac{\begin{matrix}\vdots \\ !A_1\end{matrix} \ \ldots \ \begin{matrix}\vdots \\ !A_n\end{matrix} \quad \begin{matrix}[\![!A_1^{x_1} \ \cdots \ !A_n^{x_n}]\!]\\ \vdots \\ B\end{matrix}}{!B}\,\textit{Promotion}_{x1,\ldots,xn}$$

The main difference between this and earlier presentations is in the *Promotion* rule where here substitutions are 'built-in'.

The Curry-Howard correspondence [10] provides a systematic process for attaching names, or *terms*, to proof trees from the natural deduction formulation of a given constructive logic (a clear description is given by Gallier [7]). We can apply it to get the following term assignment system for **MELL**, which rather than presenting in a tree-like fashion, we choose to present in a sequent style.

$$\frac{}{x\colon A \triangleright x\colon A}\ \textit{Identity}$$

$$\frac{\Gamma, x\colon A \triangleright M\colon B}{\Gamma \triangleright \lambda x\colon A.M\colon A \multimap B}\ (\multimap_{\mathcal{I}}) \qquad \frac{\Gamma \triangleright M\colon A \multimap B \qquad \Delta \triangleright N\colon A}{\Gamma, \Delta \triangleright MN\colon B}\ (\multimap_{\mathcal{E}})$$

$$\frac{}{\triangleright *\colon I}\ (I_{\mathcal{I}}) \qquad \frac{\Gamma \triangleright M\colon A \qquad \Delta \triangleright N\colon I}{\Gamma, \Delta \triangleright \mathsf{let}\ N\ \mathsf{be}\ *\ \mathsf{in}\ N\colon A}\ (I_{\mathcal{E}})$$

$$\frac{\Gamma \triangleright M\colon A \qquad \Delta \triangleright N\colon B}{\Gamma, \Delta \triangleright M \otimes N\colon A \otimes B}\ (\otimes_{\mathcal{I}}) \quad \frac{\Delta \triangleright M\colon A \otimes B \qquad \Gamma, x\colon A, y\colon B \triangleright N\colon C}{\Gamma, \Delta \triangleright \mathsf{let}\ M\ \mathsf{be}\ x \otimes y\ \mathsf{in}\ N\colon C}\ (\otimes_{\mathcal{E}})$$

$$\frac{\Gamma_1 \triangleright M_1\colon !A_1 \quad \cdots \quad \Gamma_n \triangleright M_n\colon !A_n \qquad x_1\colon !A_1, \ldots, x_n\colon !A_n \triangleright N\colon B}{\Gamma_1, \ldots, \Gamma_n \triangleright \mathsf{promote}\ M_1, \ldots, M_n\ \mathsf{for}\ x_1, \ldots, x_n\ \mathsf{in}\ N\colon !B}\ \textit{Promotion}$$

$$\frac{\Gamma \triangleright M\colon !A \qquad \Delta \triangleright N\colon B}{\Gamma, \Delta \triangleright \mathsf{discard}\ M\ \mathsf{in}\ N\colon B}\ \textit{Weakening}$$

$$\frac{\Delta \triangleright M\colon !A \qquad \Gamma, x\colon !A, y\colon !A \triangleright N\colon B}{\Gamma, \Delta \triangleright \mathsf{copy}\ M\ \mathsf{as}\ x, y\ \mathsf{in}\ N\colon B}\ \textit{Contraction}$$

$$\frac{\Gamma \triangleright M\colon !A}{\Gamma \triangleright \mathsf{derelict}(M)\colon A}\ \textit{Dereliction}$$

Normalization is the process of removing 'detours' from a proof in natural deduction. At the level of terms it can be seen as providing a set of reduction rules, which are known as β-rules. For **MELL** there are six β-rules which are given below.

1. $(\lambda x\colon A.M)\ N \leadsto_\beta M[x := N]$
2. $\mathsf{let}\ *\ \mathsf{be}\ *\ \mathsf{in}\ M \leadsto_\beta M$
3. $\mathsf{let}\ M \otimes N\ \mathsf{be}\ x \otimes y\ \mathsf{in}\ P \leadsto_\beta P[x := M, y := N]$
4. $\mathsf{derelict}(\mathsf{promote}\ \vec{M}\ \mathsf{for}\ \vec{x}\ \mathsf{in}\ N) \leadsto_\beta N[\vec{x} := \vec{M}]$
5. $\mathsf{discard}\ (\mathsf{promote}\ \vec{M}\ \mathsf{for}\ \vec{x}\ \mathsf{in}\ N)\ \mathsf{in}\ P \leadsto_\beta \mathsf{discard}\ \vec{M}\ \mathsf{in}\ P$
6. $\mathsf{copy}\ (\mathsf{promote}\ \vec{M}\ \mathsf{for}\ \vec{x}\ \mathsf{in}\ N)\ \mathsf{as}\ y, z\ \mathsf{in}\ P \leadsto_\beta \mathsf{copy}\ \vec{M}\ \mathsf{as}\ \vec{u}, \vec{v}\ \mathsf{in}$ $P\,[y := \mathsf{promote}\ \vec{u}\ \mathsf{for}\ \vec{x}\ \mathsf{in}\ N,$ $z := \mathsf{promote}\ \vec{v}\ \mathsf{for}\ \vec{x}\ \mathsf{in}\ N]$

In addition there are other term equalities: *commuting conversions*, which arise from consideration of the subformula property, as well as those suggested by the process of cut elimination for the sequent calculus formulation.[1] For the purposes of this paper these need not be considered here. The interested reader is again referred to other work [6, 4].

[1] In fact there are other term equalities due to the interaction between our formulation of the *Promotion* rule and the fact that we are suppressing the *Exchange* rule.

2 Two Categorical Models

The fundamental idea of a categorical treatment of proof theory is that propositions should be interpreted as the objects of the category and proofs should be interpreted as morphisms. The proof rules correspond to natural transformations between appropriate hom-functors. As mentioned above, the proof theoretic setting will reveal a number of reduction rules, which can be viewed as equalities between proofs. In particular, these equalities should hold in the categorical model.

Let us fix some notation. The interpretation of a proof is represented using semantic braces, $[\![-]\!]$, making the usual simplification of using the same letter to represent a proposition as its interpretation. Given a term $\Gamma \triangleright M\colon A$ where $M \rightsquigarrow_\beta N$, we shall write $\Gamma \triangleright M = N\colon A$.

Definition 1. A category, $\mathbb{C}$, is said to be a *categorical model* of a given logic, $\mathcal{L}$, iff

1. For all proofs $\Gamma \triangleright_{\mathcal{L}} M\colon A$, there is a morphism $[\![M]\!]\colon \Gamma \to A$ in $\mathbb{C}$,
2. For all equalities $\Gamma \triangleright_{\mathcal{L}} M = N\colon A$ it is the case that $[\![M]\!] =_{\mathbb{C}} [\![N]\!]$ (where $=_{\mathbb{C}}$ represents equality of morphisms in the category $\mathbb{C}$).

The second condition is often referred to as 'soundness'. Given this definition we shall now consider in detail two proposals for a categorical model of Linear Logic. Firstly that proposed by Seely [14] and secondly that of Benton *et al.* [4]. First we recall Seely's definition (where for clarity we have named the natural isomorphisms relating the tensor and categorical products).

Definition 2 (Seely). A *Seely category*, $\mathbb{C}$, consists of

1. A symmetric monoidal closed category (SMCC) with finite products, together with a comonad $(!, \varepsilon, \delta)$, such that
2. For each object A of $\mathbb{C}$, $(!A, d_A, e_A)$ is a comonoid with respect to the tensor product,
3. There exists natural isomorphisms $\mathsf{n}\colon !A \otimes !B \xrightarrow{\sim} !(A \times B)$ and $\mathsf{p}\colon I \xrightarrow{\sim} !1$,
4. The functor ! takes the comonoid structure of the cartesian product to the comonoid structure of the tensor product.

It is instructive to consider this definition in more detail. The naturality of n amounts to the following diagram commuting for morphisms $f\colon A \to C$ and $g\colon B \to D$.

$$\begin{array}{ccc}
!A \otimes !B & \xrightarrow{\ \mathsf{n}\ } & !(A \times B) \\
{\scriptstyle !f \otimes !g}\downarrow & & \downarrow{\scriptstyle !(f \times g)} \\
!C \otimes !D & \xrightarrow[\ \mathsf{n}\ ]{} & !(C \times D)
\end{array}$$

Condition 4 (which seems to have been overlooked by Barr [2] and Troelstra [15]) amounts to requiring that the following two diagrams commute.

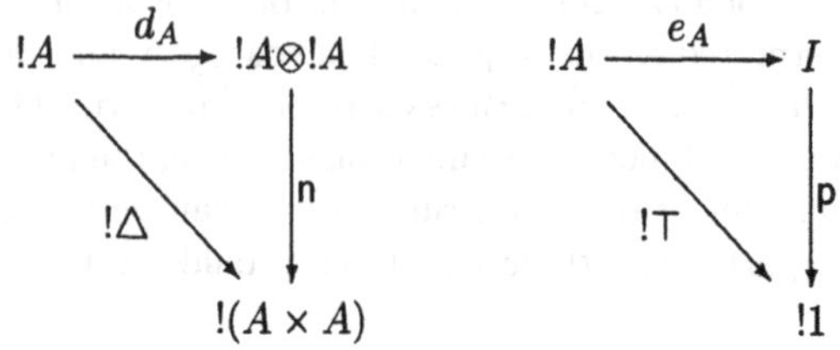

Now let us consider the model proposed by Benton *et al.* (the version given here is taken from my thesis [6] and is a slight adaptation from the original definition [4]).

Definition 3. A *Linear category*, $\mathbb{C}$, consists of

1. A SMCC, $\mathbb{C}$, together with
2. A symmetric monoidal comonad $(!, \varepsilon, \delta, \mathsf{m}_{A,B}, \mathsf{m}_I)$ such that
   (a) For every free !-coalgebra $(!A, \delta_A)$ there are two distinguished monoidal natural transformations with components $e_A\colon !A \to I$ and $d_A\colon !A \to !A \otimes !A$[2] which form a commutative comonoid and are coalgebra morphisms,
   (b) Whenever $f\colon (!A, \delta_A) \to (!B, \delta_B)$ is a coalgebra morphism between free coalgebras, then it is also a comonoid morphism.

Let us consider in detail the conditions in this definition. Firstly requiring that $(!, \mathsf{m}_{A,B}, \mathsf{m}_I)$ is a symmetric monoidal functor amounts to the following diagrams commuting.

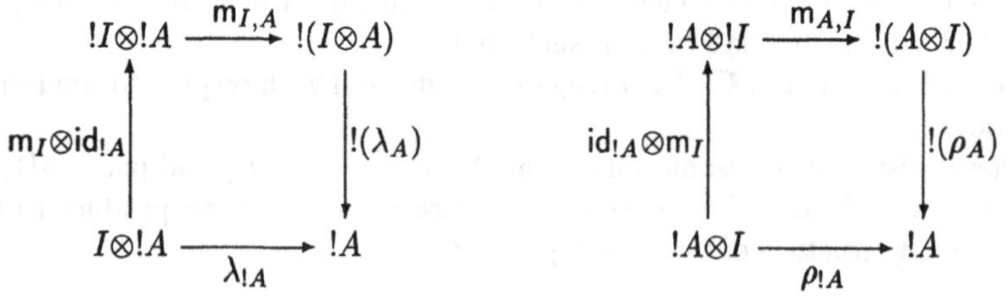

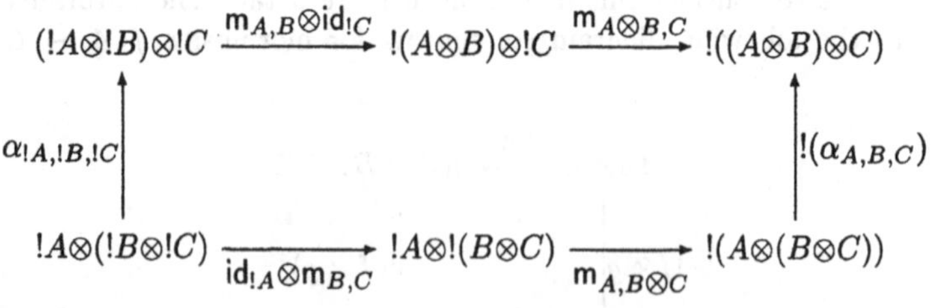

[2] This necessitates showing that $!\otimes!$ and I are monoidal functors, but this is trivial and omitted.

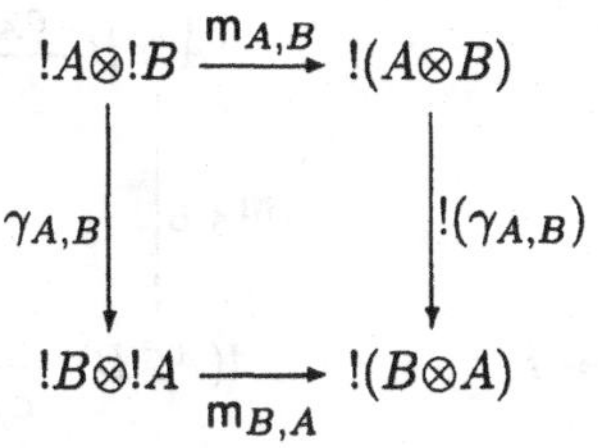

Requiring that ε is a monoidal natural transformation amounts to the following two commuting diagrams.

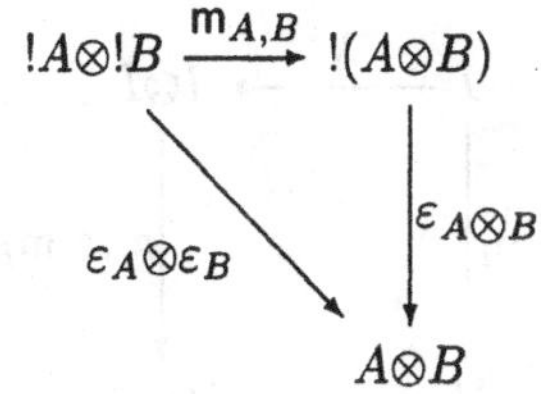

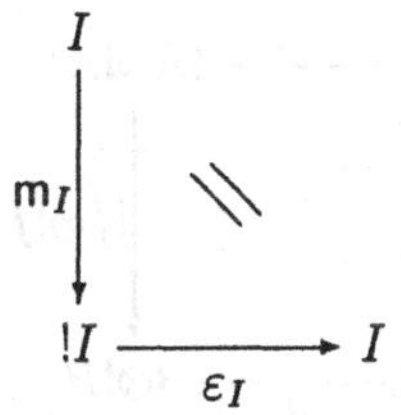

Requiring that δ is a monoidal natural transformation amounts to the following two commuting diagrams.

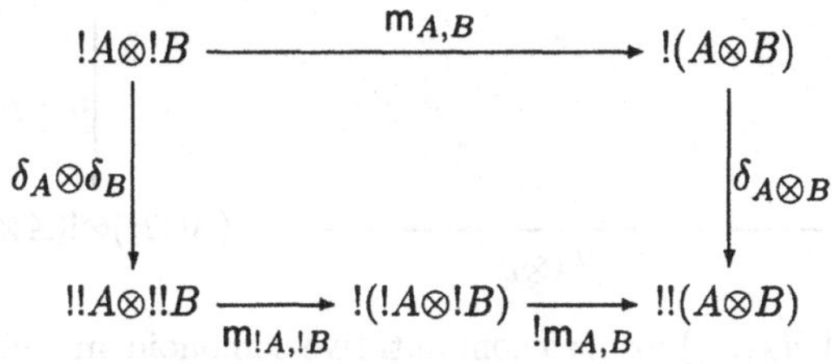

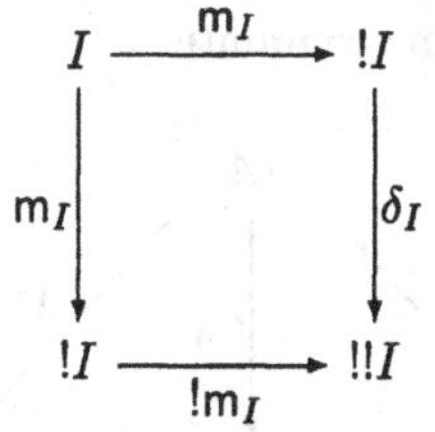

Requiring that $e_A\colon !A \to I$ is a monoidal natural transformation amounts to requiring that the following three diagrams commute, for any morphism $f\colon A \to B$.

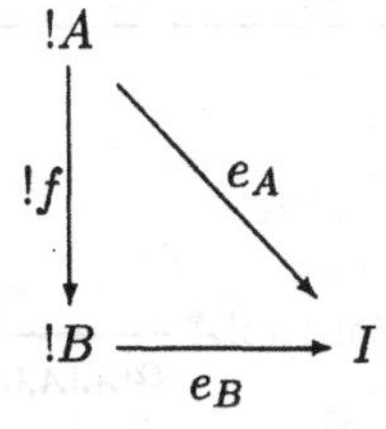

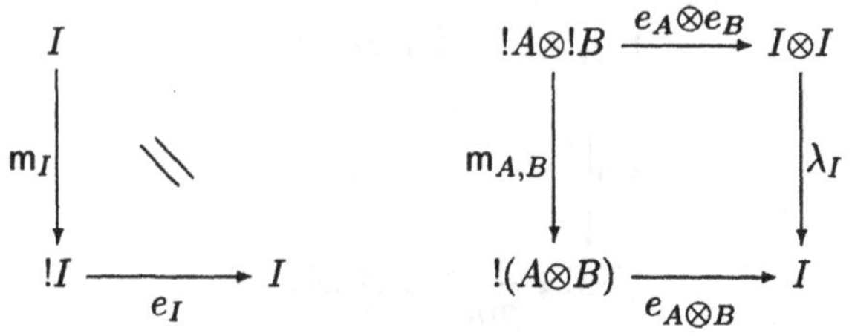

Requiring that $d_A\colon !A \to !A\otimes !A$ is a monoidal natural transformation amounts to requiring that the following three diagrams commute, for all $f\colon A \to B$.

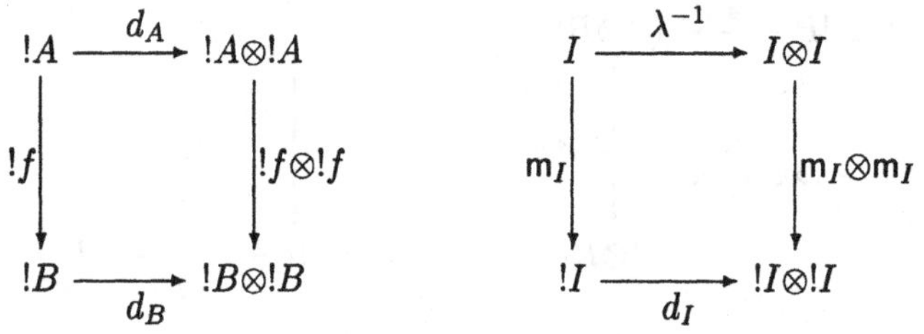

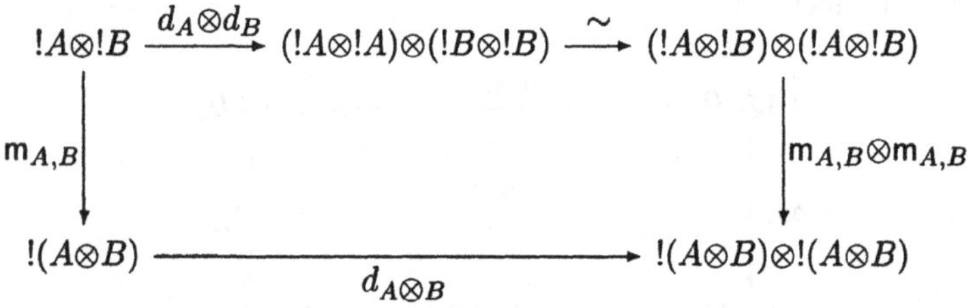

Requiring that $(!A, d_A, e_A)$ forms a commutative comonoid amounts to requiring that the following three diagrams commute.

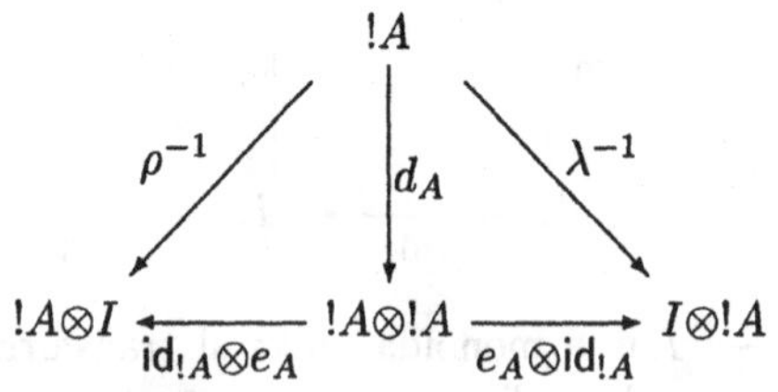

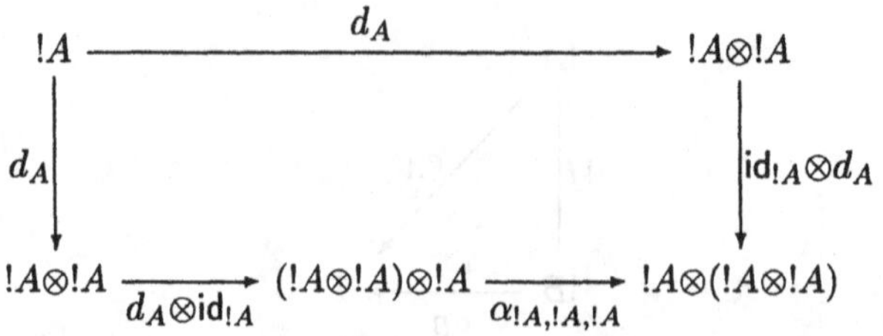

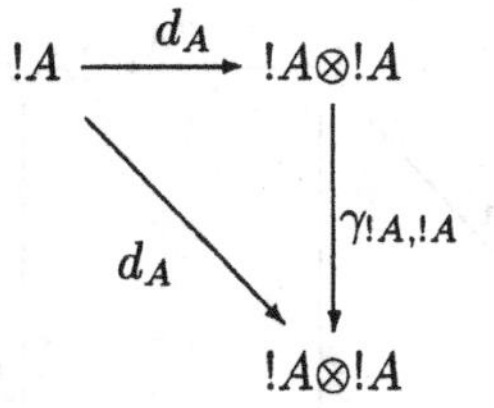

Requiring that e_A is a coalgebra morphism amounts to requiring that the following diagram commutes.

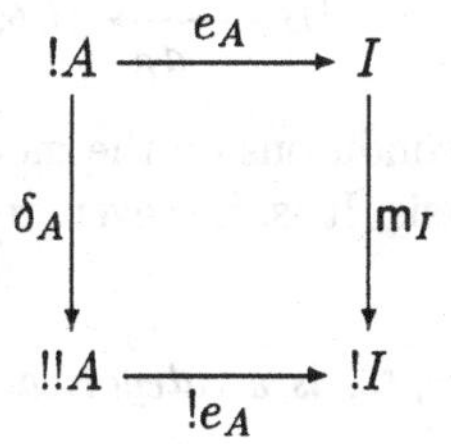

Requiring that d_A is a coalgebra morphism amounts to requiring that the following diagram commutes.

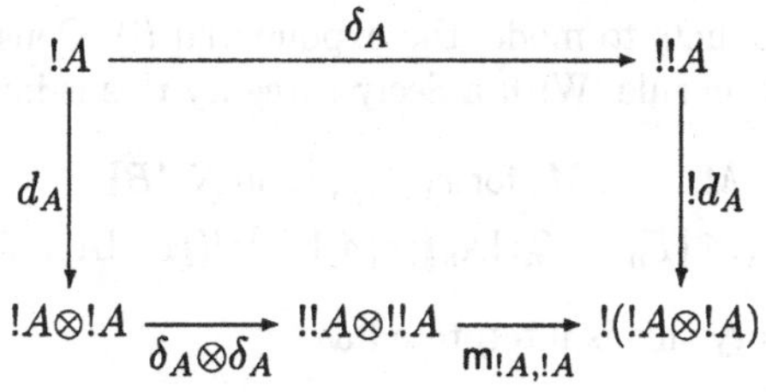

Finally all coalgebra morphisms between *free* coalgebras are also comonoid morphisms. Thus given a coalgebra morphism f, between the free coalgebras $(!A, \delta_A)$ and $(!B, \delta_B)$, i.e. which makes the following diagram commute.

$$
\begin{array}{ccc}
!A & \xrightarrow{\ f\ } & !B \\
{\scriptstyle \delta_A}\downarrow & & \downarrow{\scriptstyle \delta_B} \\
!!A & \xrightarrow[\ !f\ ]{} & !!B
\end{array}
$$

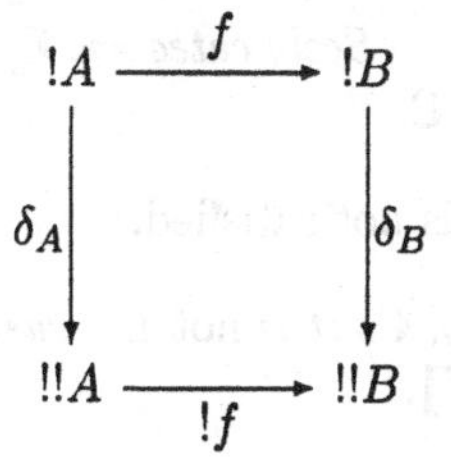

Then it is also a comonoid morphism between the comonoids $(!A, e_A, d_A)$ and $(!B, e_B, d_B)$, i.e. it makes the following diagram commute.

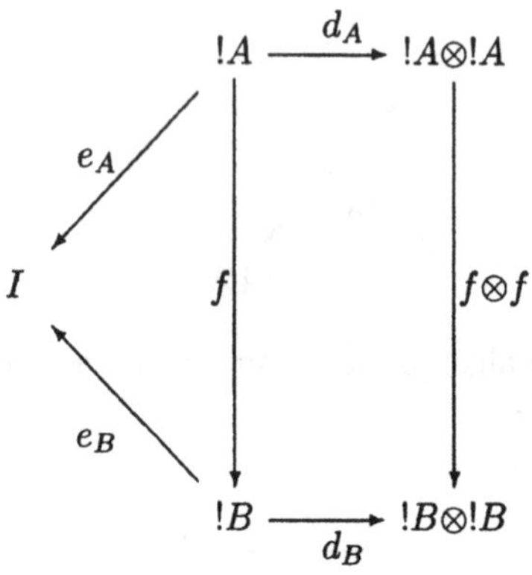

These amount to some strong conditions on the model and some of their consequences are explored in my thesis. It is, however, reasonably straightforward to show the following.

Theorem 1. *A Linear category,* $\mathbb{C}$, *is a categorical model for* **MELL**.

Proof. The first condition is proved by a trivial induction on the structure of the proof $\Gamma \triangleright M: A$. The second condition is proved by checking the six β-rules from earlier.

The main difference between these two models is that a Seely category critically needs categorical products to model the exponential (!). Consider the interpretation of the *Promotion* rule. With a Seely category this is interpreted as

$$[\![\Gamma_1, \ldots, \Gamma_n \triangleright \mathsf{promote}\ M_1, \ldots, M_n\ \mathsf{for}\ x_1, \ldots, x_n\ \mathsf{in}\ N: !B]\!]$$
$$\stackrel{\text{def}}{=} [\![\Gamma_1 \triangleright M_1: !A_1]\!] \otimes \ldots \otimes [\![\Gamma_n \triangleright M_n: !A_n]\!]; \mathsf{n}; \delta; !\mathsf{n}^{-1}; !([\![x_1: !A_1, \ldots, x_n: !A_n \triangleright N: B]\!]).$$

With a Linear category this is interpreted as

$$[\![\Gamma_1, \ldots, \Gamma_n \triangleright \mathsf{promote}\ M_1, \ldots, M_n\ \mathsf{for}\ x_1, \ldots, x_n\ \mathsf{in}\ N: !B]\!]$$
$$\stackrel{\text{def}}{=} [\![\Gamma_1 \triangleright M_1: !A_1]\!] \otimes \ldots \otimes [\![\Gamma_n \triangleright M_n: !A_n]\!]; \delta \otimes \ldots \otimes \delta; \mathsf{m}; !([\![x_1: !A_1, \ldots, x_n: !A_n \triangleright N: B]\!]).$$

Let us consider whether a Seely category is a categorical model for **MELL**. Seely showed that the first requirement is satisfied.

Proposition 1 (Seely). *Given a Seely category,* $\mathbb{C}$, *for all proofs* $\Gamma \triangleright M: A$ *there is a morphism* $[\![M]\!]: \Gamma \to A$ *in* $\mathbb{C}$.

However the second condition is not satisfied.

Fact 1. *Given a Seely category,* $\mathbb{C}$, *it is* not *the case that for all term equalities* $\Gamma \triangleright M = N: A$ *that* $[\![M]\!] =_{\mathbb{C}} [\![N]\!]$.[3]

[3] It should be noted that the term equalities were not generally known when Seely proposed his model.

One counter-example is the sixth β-rule from earlier. In fact we only need use a simplified version where the promoted term, N, has only one free variable, i.e.

$$\begin{array}{l}\Gamma \triangleright \mathsf{copy}\ (\mathsf{promote}\ M\ \mathsf{for}\ x\ \mathsf{in}\ N)\ \mathsf{as}\ y,z\ \mathsf{in}\ P \\ \quad = \mathsf{copy}\ M\ \mathsf{as}\ x',x''\ \mathsf{in}\ P[y := \mathsf{promote}\ x'\ \mathsf{for}\ x\ \mathsf{in}\ N, z := \mathsf{promote}\ x''\ \mathsf{for}\ x\ \mathsf{in}\ N]\colon C.\end{array}$$

This term equality implies the *same commuting diagram* for a Linear category as for a Seely category,

$$\begin{array}{ccccccccc} \Gamma & \xrightarrow{m} & !A & \xrightarrow{\delta} & !!A & \xrightarrow{!n} & !B & & \\ & & \downarrow{\scriptstyle d} & & & & \downarrow{\scriptstyle d} & & \\ & & !A\otimes !A & \xrightarrow[\delta\otimes\delta]{} & !!A\otimes !!A & \xrightarrow[!n\otimes !n]{} & !B\otimes !B & \xrightarrow[p]{} & C. \end{array} \qquad (1)$$

For a Linear category we can complete the diagram as

$$\begin{array}{ccccccccc} \Gamma & \xrightarrow{m} & !A & \xrightarrow{\delta} & !!A & \xrightarrow{!n} & !B & & \\ & & \downarrow{\scriptstyle d} & & \downarrow{\scriptstyle d} & & \downarrow{\scriptstyle d} & & \\ & & !A\otimes !A & \xrightarrow[\delta\otimes\delta]{} & !!A\otimes !!A & \xrightarrow[!n\otimes !n]{} & !B\otimes !B & \xrightarrow[p]{} & C. \end{array}$$

The left hand square commutes by the condition that all free coalgebra morphisms are comonoid morphisms. The right hand square commutes by naturality of d. Unfortunately it is not clear how to make diagram 1 commute for a Seely category. Indeed it is straightforward to see how a term model can be constructed from Seely's original definition such that this diagram does *not* commute.

At this stage we might try adding the condition that all (free) coalgebra morphisms are comonoid morphisms to Seely's definition (and hence add extra equations to the term model). This proves still to be incomplete as we find that neither the sixth nor the fifth term equalities are modelled correctly in the cases when the promoted term, N, has zero or more than one free variable. One might be further tempted to add additional *ad-hoc* conditions to make a Seely category a model for **MELL**. However, as shown in my thesis, this is by no means simple and rather it would seem more prudent to consider a more abstract view. Rather we consider some of the motivation behind the Seely construction.

First we shall recall a construction, the dual of which (i.e. that generated by a monad) is known as the "Kleisli category" [13, Page 143].

Definition 4. Given a comonad $(!,\varepsilon,\delta)$ on a category $\mathbb{C}$, we take all the objects A in $\mathbb{C}$ and for each morphism $f\colon !A \to B$ in $\mathbb{C}$ we take a new morphism $\hat{f}\colon A \to B$. The objects and morphisms form the *co-Kleisli category* $\mathbb{C}_!$, where the composition of the morphisms $\hat{f}\colon A \to B$ and $\hat{g}\colon B \to C$ is defined as $\hat{f};\hat{g} \stackrel{\text{def}}{=} \widehat{(\delta_A; !f; g)}$.

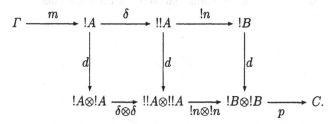

The interest in this construction is that it has strong similarities with the Girard translation [8] of Intuitionistic Logic (**IL**) into **ILL** where the intuitionistic implication is decomposed as $(A \supset B)^{\circ} \stackrel{\text{def}}{=} !(A^{\circ}) \multimap B^{\circ}$. In fact, as first shown by Seely [14], the co-Kleisli construction can be thought of as a categorical equivalent of the Girard translation in the following sense.

Proposition 2 (Seely). *Given a Seely category, $\mathbb{C}$, the co-Kleisli category, $\mathbb{C}_!$, is cartesian closed.*

Proof. (Sketch) Given two objects A and B their exponent is defined to be $!A \multimap B$. Then we have the following sequence of isomorphisms.

$$\begin{array}{rll} \mathbb{C}_!(A \times B, C) & \cong \mathbb{C}(!(A \times B), C) & \text{By definition,} \\ & \cong \mathbb{C}(!A \otimes !B, C) & \text{By use of the } \mathsf{n} \text{ isomorphism,} \\ & \cong \mathbb{C}(!A, !B \multimap C) & \text{By } \mathbb{C} \text{ having a closed structure,} \\ & \cong \mathbb{C}_!(A, !B \multimap C) & \text{By definition.} \end{array}$$

We know from Kleisli's construction that we have the adjunction

$$\begin{array}{c} \mathbb{C}_! \\ G \uparrow \dashv \downarrow F \\ \mathbb{C} \end{array}$$

where G is the functor defined by $g\colon A \to B \mapsto \widehat{(\varepsilon; g)}$ and F is the functor defined by $\hat{f}\colon A \to B \mapsto \delta_A; !f$.

Seely's model arises from at least the desire to make the co-Kleisli category a cartesian closed category (CCC), which is achieved by including the n and p natural isomorphisms. This means that there is an adjunction between a SMCC ($\mathbb{C}$) and a CCC ($\mathbb{C}_!$). As a CCC is trivially a SMCC, there is then an adjunction between two SMCCs. We might expect that this is a *monoidal adjunction*.

Definition 5. An adjunction $\langle F, G, \eta, \epsilon \rangle\colon \mathbb{C} \rightharpoonup \mathbb{D}$ is said to be a *monoidal* adjunction when F and G are monoidal functors and η and ϵ are monoidal natural transformations.

We now state a new definition for a Seely-style category and then investigate some of its properties.

Definition 6. A *new-Seely category*, $\mathbb{C}$, consists of

1. a SMCC, $\mathbb{C}$, with finite products, together with
2. a comonad, $(!, \varepsilon, \delta)$, and
3. two natural isomorphisms, $\mathsf{n}\colon !A \otimes !B \xrightarrow{\sim} !(A \times B)$ and $\mathsf{p}\colon I \xrightarrow{\sim} !1$

such that the adjunction, $\langle F, G, \eta, \varepsilon \rangle$, between $\mathbb{C}$ and $\mathbb{C}_!$ is a *monoidal* adjunction.

Assuming that F is monoidal gives us the morphism and natural transformation

$$\mathsf{m}_I : I \to F1,$$
$$\mathsf{m}_{A,B} : FA \otimes FB \to F(A \times B).$$

Assuming that G is monoidal gives us the morphism and natural transformation

$$\mathsf{m}'_1 : 1 \to GI,$$
$$\mathsf{m}'_{A,B} : GA \times GB \to G(A \otimes B).$$

By assumption ε and η are monoidal natural transformations.

It is easy to see that m_I is Seely's morphism p and $\mathsf{m}_{A,B}$ is Seely's natural transformation n. In fact, we can define their inverses

$$\mathsf{m}_I^{-1} \stackrel{\text{def}}{=} F\mathsf{m}'_1 ; \varepsilon_I : F1 \to I,$$
$$\mathsf{m}_{A,B}^{-1} \stackrel{\text{def}}{=} F(\eta_A \times \eta_B) ; F\mathsf{m}'_{FA,FB} ; \varepsilon_{FA \otimes FB} : F(A \times B) \to FA \otimes FB.$$

Hence the monoidal adjunction itself provides the isomorphisms $!A \otimes !B \cong !(A \times B)$ and $I \cong !1$. As the co-Kleisli category is a CCC it has a trivial commutative comonoid structure, $(A, \Delta, \top)$, on all objects A. We can use this and the natural transformations arising from the monoidal adjunction to define a comonoid structure, $(F(A), d, e)$, on the objects of $\mathbb{C}$, with the structure maps defined as

$$d \stackrel{\text{def}}{=} F(\Delta) ; \mathsf{m}_{A,A}^{-1} : F(A) \to F(A) \otimes F(A),$$
$$e \stackrel{\text{def}}{=} F(\top) ; \mathsf{m}_I^{-1} : F(A) \to I.$$

It is easy to see that these definitions amount to condition 4 of Seely's original definition. Thus there is at least as much structure as in Seely's original definition but with the extra structure of the monoidal adjunction. Some consequences of this adjunction are given in the following lemma.

Lemma 1. *Given a new-Seely category, $\mathbb{C}$, the following facts hold:*

1. *The induced comonad $(FG, F\eta_G, \varepsilon)$ on $\mathbb{C}$ is a* monoidal comonad $(FG, F\eta_G, \varepsilon, m_{A,B}, m_I)$.
2. *The comonoid morphisms $e : FG(A) \to I$ and $d : FG(A) \to FG(A) \otimes FG(A)$ are monoidal natural transformations.*
3. *The comonoid morphisms $e : FG(A) \to I$ and $d : FG(A) \to FG(A) \otimes FG(A)$ are coalgebra morphisms.*
4. *If $f : (FG(A), F\eta_{GA}) \to (FG(B), F\eta_{GB})$ is a coalgebra morphism then it is also a comonoid morphism.*

Proof. For part 1 we take the definitions

$$m_I \stackrel{\text{def}}{=} \mathsf{m}_I ; F\mathsf{m}'_1 : I \to FG(I),$$
$$m_{A,B} \stackrel{\text{def}}{=} \mathsf{m}_{GA,GB} ; F\mathsf{m}'_{A,B} : FG(A) \otimes FG(A) \to FG(A \otimes B).$$

The rest of the lemma holds by construction.

Corollary 1. *Every new-Seely category is a Linear category.*

(It is clear that the converse is not true, as the Linear category need not have finite products.) We can hence show that a new-Seely category is a sound model for the **MELL**.

Theorem 2. *A new-Seely category,* $\mathbb{C}$, *is a categorical model for* **MELL**

3 Including the additives

Now we shall consider the whole of **ILL** by adding the additive connectives to **MELL**. Logically these are given by the following sequent calculus rules (we shall ignore the additive units).

$$\frac{\Gamma, A \vdash C}{\Gamma, A \times B \vdash C}\,(\times_{\mathcal{L}-1}) \qquad \frac{\Gamma, B \vdash C}{\Gamma, A \times B \vdash C}\,(\times_{\mathcal{L}-2})$$

$$\frac{\Gamma \vdash A \qquad \Gamma \vdash B}{\Gamma \vdash A \times B}\,(\times_{\mathcal{R}})$$

$$\frac{\Gamma, A \vdash C \qquad \Gamma, B \vdash C}{\Gamma, A \oplus B \vdash C}\,(\oplus_{\mathcal{L}})$$

$$\frac{\Gamma \vdash A}{\Gamma \vdash A \oplus B}\,(\oplus_{\mathcal{R}-1}) \qquad \frac{\Gamma \vdash B}{\Gamma \vdash A \oplus B}\,(\oplus_{\mathcal{R}-2})$$

There are a number of ways of formulating the additives in a natural deduction system which are discussed in my thesis. However, for now we shall simply take the term assignment system which is familiar from that of the λ-calculus. The term assignment rules as well as the β-rules for the additives are given below.

$$\frac{\Gamma \triangleright M : A \qquad \Gamma \triangleright N : B}{\Gamma \triangleright \langle M, N \rangle : A \times B}\,(\times_{\mathcal{I}})$$

$$\frac{\Gamma \triangleright M : A \times B}{\Gamma \triangleright \mathsf{fst}(M) : A}\,(\times_{\mathcal{E}-1}) \qquad \frac{\Gamma \triangleright M : A \times B}{\Gamma \triangleright \mathsf{snd}(M) : B}\,(\times_{\mathcal{E}-2})$$

$$\frac{\Gamma \triangleright M : A}{\Gamma \triangleright \mathsf{inl}(M) : A \oplus B}\,(\oplus_{\mathcal{I}-1}) \qquad \frac{\Gamma \triangleright M : B}{\Gamma \triangleright \mathsf{inr}(M) : A \oplus B}\,(\oplus_{\mathcal{I}-2})$$

$$\frac{\Delta \triangleright M : A \oplus B \qquad \Gamma, x : A \triangleright N : C \qquad \Gamma, y : B \triangleright P : C}{\Gamma, \Delta \triangleright \mathsf{case}\ M\ \mathsf{of}\ \mathsf{inl}(x) \to N \parallel \mathsf{inr}(y) \to P : C}\,(\oplus_{\mathcal{E}})$$

$$\begin{array}{rl}
\mathsf{fst}(\langle M,N\rangle) & \leadsto_\beta M\\
\mathsf{snd}(\langle M,N\rangle) & \leadsto_\beta N\\
\mathsf{case}\,(\mathsf{inl}(M))\ \mathsf{of}\ \mathsf{inl}(x)\ \to\ N \,\|\, \mathsf{inr}(y)\ \to\ P & \leadsto_\beta N[x:=M]\\
\mathsf{case}\,(\mathsf{inr}(M))\ \mathsf{of}\ \mathsf{inl}(x)\ \to\ N \,\|\, \mathsf{inr}(y)\ \to\ P & \leadsto_\beta P[y:=M]
\end{array}$$

To model these additive connectives we shall add finite products and coproducts to a Linear category and finite coproducts to a new-Seely category. As might be expected both models are sound.

Theorem 3. *Both a new-Seely category with finite coproducts and a Linear category with finite products and coproducts, are models for* **ILL**.

Somewhat surprisingly, we find that the so-called Seely isomorphisms (n and p) exist in a Linear category with products.

Lemma 2. *Given a Linear category with finite products we can define the natural isomorphisms*

$$\begin{aligned}
n &\stackrel{\text{def}}{=} \delta\otimes\delta; \mathsf{m}_{!A,!B}; !(\Delta); !((\mathsf{id}\otimes e_B)\times(e_A\otimes\mathsf{id})); !(\rho\times\lambda); !(\varepsilon\times\varepsilon)\colon !A\otimes!B \to !(A\times B)\\
n^{-1} &\stackrel{\text{def}}{=} d_{A\times B}; !\mathsf{fst}\otimes!\mathsf{snd}\colon !(A\times B)\to !A\otimes!B,\\
p &\stackrel{\text{def}}{=} \mathsf{m}_I; !\top\colon I \to !1,\\
p^{-1} &\stackrel{\text{def}}{=} e_1\colon !1 \to I.
\end{aligned}$$

Thus the co-Kleisli category associated with a Linear category is also a CCC. Given our earlier calculations we might consider the adjunction between a Linear category and its co-Kleisli category, where we find the following holds.

Lemma 3. *The adjunction between a Linear category, $\mathbb{C}$, with finite products and its co-Kleisli category, $\mathbb{C}_!$, is a* monoidal *adjunction.*

Thus when considering the complete intuitionistic fragment, the new-Seely and Linear categories are equivalent. It is easy to check that common models such as coherent spaces, dI-domains and pointed cpos and strict maps are all examples of new-Seely/Linear categories.

An interesting question is whether the co-Kleisli category $\mathbb{C}_!$ has an induced coproduct structure given a coproduct structure on $\mathbb{C}$. Seely [14] showed that $\mathbb{C}_!$ does *not* have a coproduct structure, but in fact it is possible to identify a *weak* coproduct structure. We use the following well-known fact about the co-Kleisli category [12, Corollary 6.9].

Fact 2. *The co-Kleisli category of a comonad is equivalent to the full subcategory of the category of coalgebras consisting of the free coalgebras.*

Lemma 4. *Given two free coalgebras $(!A,\delta_A)$ and $(!B,\delta_B)$, we define their coproduct to be $(!(!A\oplus!B),\delta_{!A\oplus!B})$. We define the injection morphisms to be* inl $\stackrel{\text{def}}{=}$

δ_A; !inl: $!A \to !(!A \oplus !B)$ and inr $\stackrel{\text{def}}{=} \delta_B$; !inr: $!B \to !(!A \oplus !B)$, which are (free) coalgebra morphisms. Given two (free) coalgebra morphisms $f\colon !A \to !C$ and $g\colon !B \to !C$, then the morphism $(![f,g]; !\varepsilon_C)\colon !(!A \oplus !B) \to !C$ is a (free) coalgebra morphism and makes a coproduct diagram commute.

So far we have followed others [14, 11] and only considered whether the co-Kleisli category $\mathbb{C}_!$ generated by the comonad is cartesian closed. It is should be noted that alternatively one can consider the full Eilenberg-Moore category of coalgebras ($\mathbb{C}^!$) instead. In other work [4, 6], various subcategories of $\mathbb{C}^!$ are shown to be cartesian closed. An important feature of these (sub)categories is that the underlying category $\mathbb{C}$ need *not* necessarily have products, in contrast to the situation for $\mathbb{C}_!$. The interested reader is referred to these other works.

4 Conclusions

In this paper we have considered the definition of a categorical model for **ILL**. Surprisingly, Seely's now standard definition [14] was shown to be unsound, in that it does not model all equal proofs with equal morphisms. A model given in our earlier work [4] was shown to be sound. We have also considered a method for improving Seely's original definition so as to be sound. In fact both (sound) models turn out to be equivalent. It is worth pointing out that these models are sound with respect to the equalities arising from the commuting conversions.

Lafont [11] also proposed a categorical model for **ILL**, which amounts to requiring an adjunction between a SMCC and a category of commutative comonoids. In my thesis [6] it is shown that this model is a categorical model of **ILL** by demonstrating that every Lafont category is a Linear category.

In Lemma 1 it was proved that a monoidal adjunction between a particular SMCC (a new-Seely category) and CCC (its co-Kleisli category) yielded the structure of a Linear category. Lemma 3 shows that a Linear category also has the structure of a monoidal adjunction between it (a SMCC) and its associated co-Kleisli category (a CCC). Thus the notion of a Linear category is in some senses equivalent to the existence of a monoidal adjunction between a SMCC and a CCC. This observation has been used by Benton [3] to derive the syntax of a mixed linear and non-linear term calculus.

Categorically, most models proposed for Classical Linear Logic (**CLL**) are extensions of Seely's model for **ILL** to $\star$-autonomous categories [14, 2]. Thus the problems identified with Seely's model in this paper apply to these models. Extending a Linear category with a dualizing object gives a (sound) model of **CLL**, although the categorical import of this construction is work in progress.

Acknowledgements

This paper is taken in part from my PhD thesis. I should like to thank Valeria de Paiva, Martin Hyland and Nick Benton for their help and useful comments on my work. I have used Paul Taylor's LaTeX macros in this paper. I am currently funded with research fellowships from the EPSRC and Wolfson College, Cambridge.

References

1. S. Abramsky. Computational interpretations of linear logic. *Theoretical Computer Science*, 111(1–2):3–57, 1993. Previously Available as Department of Computing, Imperial College Technical Report 90/20, 1990.
2. M. Barr. $\star$-autonomous categories and linear logic. *Mathematical Structures in Computer Science*, 1:159–178, 1991.
3. P.N. Benton. A mixed linear and non-linear logic: Proofs, terms and models. Technical Report 352, Computer Laboratory, University of Cambridge, 1994.
4. P.N. Benton, G.M. Bierman, V.C.V. de Paiva, and J.M.E. Hyland. Term assignment for intuitionistic linear logic. Technical Report 262, Computer Laboratory, University of Cambridge, August 1992.
5. P.N. Benton, G.M. Bierman, V.C.V. de Paiva, and J.M.E. Hyland. A term calculus for intuitionistic linear logic. In M. Bezem and J.F. Groote, editors, *Proceedings of Conference on Typed Lambda Calculi and Applications*, volume 664 of *Lecture Notes in Computer Science*, pages 75–90, 1993.
6. G.M. Bierman. *On Intuitionistic Linear Logic*. PhD thesis, Computer Laboratory, University of Cambridge, December 1993. Available as Computer Laboratory Technical Report 346. August 1994.
7. J. Gallier. Constructive logics part I: A tutorial on proof systems and typed λ-calculi. *Theoretical Computer Science*, 110(2):249–339, March 1993.
8. J.-Y. Girard. Linear logic. *Theoretical Computer Science*, 50:1–101, 1987.
9. J.-Y. Girard and Y. Lafont. Linear logic and lazy computation. In *Proceedings of TAPSOFT 87*, volume 250 of *Lecture Notes in Computer Science*, pages 52–66, 1987. Previously Available as INRIA Report 588, 1986.
10. W.A. Howard. The formulae-as-types notion of construction. In J.R. Hindley and J.P. Seldin, editors, *To H.B. Curry: Essays on combinatory logic, lambda calculus and formalism*. Academic Press, 1980.
11. Y. Lafont. The linear abstract machine. *Theoretical Computer Science*, 59:157–180, 1988. Corrections *ibid.* 62:327–328, 1988.
12. J. Lambek and P.J. Scott. *Introduction to higher order categorical logic*, volume 7 of *Cambridge studies in advanced mathematics*. Cambridge University Press, 1987.
13. S. Mac Lane. *Categories for the Working Mathematican*, volume 5 of *Graduate Texts in Mathematics*. Springer Verlag, 1971.
14. R.A.G. Seely. Linear logic, *-autonomous categories and cofree algebras. In *Conference on Categories in Computer Science and Logic*, volume 92 of *AMS Contemporary Mathematics*, pages 371–382, June 1989.
15. A.S. Troelstra. *Lectures on Linear Logic*, volume 29 of *Lecture Notes*. CSLI, 1992.

An explicit *Eta* rewrite rule

Daniel BRIAUD

Centre de Recherche en Informatique de Nancy (CNRS)
and INRIA-Lorraine
Campus Scientifique, BP 239,
F54506 Vandœuvre-lès-Nancy, France

email: Daniel.Briaud@loria.fr

Abstract. In this paper, we extend λ-calculi of explicit substitutions by an *Eta* rule. The previous definition of *Eta* is due to Hardin (1992) and Ríos (1993). Their definition is a conditional rewrite rule and does not stick fully to the philosophy of explicit substitutions. Our main result is making the η-contraction explicit by means of an unconditional, *generic* *Eta* rewrite rule and of an extension of the substitution calculus. We do this in the framework of λv, a calculus introduced by Lescanne (1994). We prove the correction, confluence and strong normalization (on typed terms) of $\lambda v\eta$. We also show how this explicit *Eta* leads to η', a very general alternative to the classical η, that allows confluent contractions not captured by η.

Introduction

The main feature of a λ-calculus of explicit substitutions is that the classical β-contraction is expressed by a first-order term rewriting system. In this way, such λ-calculi are a step towards achieving Curry's program [CF58]. Indeed, one aim was to find a fully formalized prelogic, consisting of a theory of substitution (not a metatheory) and a theory of types. According to Curry, there are two forms of such a prelogic: Combinatory Logic which is the ultimate and λ-conversion which is intermediate. We think λ-calculus of explicit substitutions is as good as Combinatory Logic, but more intuitive. Historically, explicit substitutions were first designed by De Bruijn [dB78] but with a crude terminology. They have been made popular more recently by Abadi, Cardelli, Curien and Lévy [ACCL91].

These authors present $\lambda\sigma$, a λ-calculus of explicit substitutions which is a result of almost a decade of research work. The starting point is CCL, categorical combinatory logic [Cur83, Cur86b, Cur86a], a combinatory logic more intuitive than the classical one. In particular, it is based on λ-calculus with cartesian products and keeps its structure. Hardin [Har87, Har89] studied confluence on open terms of this calculus. An important contribution toward explicit substitutions is the $\lambda\rho$-calculus [Cur86b], suited for weak reduction. It has been extended to $\lambda\sigma$ [ACCL91]. Hardin and Levy designed $\lambda\sigma_{\Uparrow}$[HL89], thus achieving an important goal: confluence on open terms. Guided by implementation grounds, this family of calculi rests fundamentally on the concept of composition of substitutions.

In this paper, we will work in $\lambda\upsilon$, a recent λ-calculus of explicit substitutions introduced in [Les94] and thoroughly studied in [LRD94a]. What characterizes this calculus is that the substitution calculus υ is an orthogonal rewriting system which does not manage composition of substitutions. Like most λ-calculi of explicit substitutions, $\lambda\upsilon$ uses De Bruijn indices for terms [dB72], beginning at $\underline{1}$. We recall the syntax of λ-terms with De Bruijn indices:

Definition 1 Λ**.** The set Λ is the set generated by:

$$\begin{array}{ll} \textbf{Terms} & a ::= \underline{n} \mid aa \mid \lambda a \\ \textbf{Naturals} & n ::= n+1 \mid 1. \end{array}$$

Terms of Λ are called pure terms.

How does $\lambda\upsilon$ break the β-contraction? First, the rule *Beta* creates a substitution stored in a closure denoted by [ ]:

$$Beta \quad (\lambda a)b \rightarrow a[b/]$$

Then, rules of υ (Cf figure 1) distribute this substitution and apply it to indices. The rules $\{Beta\} \cup \upsilon$ are defined on the following set of terms $\Lambda\upsilon$:

$$\begin{array}{ll} \textbf{Terms} & a ::= \underline{n} \mid aa \mid \lambda a \mid a[s] \\ \textbf{Substitutions} & s ::= a/ \mid \Uparrow(s) \mid \uparrow \\ \textbf{Naturals} & n ::= n+1 \mid 1. \end{array}$$

For example, $(\lambda x.\lambda y.xy)z$, denoted $(\lambda(\lambda(\underline{2}\ \underline{1})))\underline{1}$ in De Bruijn notation, is con-

(Beta)	$(\lambda a)b \rightarrow a[b/]$
(App)	$(ab)[s] \rightarrow a[s]b[s]$
(Lambda)	$(\lambda a)[s] \rightarrow \lambda(a[\Uparrow(s)])$
(FVar)	$\underline{1}\,[a/] \rightarrow a$
(RVar)	$\underline{n+1}[a/] \rightarrow \underline{n}$
(FVarLift)	$\underline{1}[\Uparrow(s)] \rightarrow \underline{1}$
(RVarLift)	$\underline{n+1}\,[\Uparrow(s)] \rightarrow \underline{n}\,[s][\uparrow]$
(VarShift)	$\underline{n}\,[\uparrow] \rightarrow \underline{n+1}$

Fig. 1. The rewrite system $\lambda\upsilon$

tracted by υ as follows:

$$
\begin{aligned}
(\lambda(\lambda(\underline{2}\ \underline{1})))\underline{1} &\underset{Beta}{\longrightarrow} (\lambda(\underline{2}\ \underline{1}))[\underline{1}\ /] \\
&\underset{v}{\longrightarrow} \lambda((\underline{2}\ \underline{1})[\Uparrow(\underline{1}\ /)]) \qquad \text{rule } \mathit{Lambda} \\
&\underset{v}{\longrightarrow} \lambda(\underline{2}\ [\Uparrow(\underline{1}\ /)]\ \underline{1}\ [\Uparrow(\underline{1}\ /)]) \qquad \text{rule } \mathit{App} \\
&\overset{+}{\underset{v}{\longrightarrow}} \lambda(\underline{1}\ [\underline{1}\ /][\uparrow])\ \underline{1}) \qquad \text{rules } \mathit{RVarLift},\ \mathit{FVarLift} \\
&\underset{v}{\longrightarrow} \lambda(\underline{1}\ [\uparrow])\ \underline{1}) \qquad \text{rule } \mathit{FVar} \\
&\underset{v}{\longrightarrow} \lambda(\underline{2}\ \underline{1}) \qquad \text{rule } \mathit{VarShift}
\end{aligned}
$$

Its v-normal form, namely $\lambda(\underline{2}\ \underline{1})$, is equivalent to $\lambda y.z\ y$. We may η-reduce it to z. Indeed, the practical reason for defining η-reduction comes from the natural equality $(\lambda x.fx)a =_\beta fa$ if x has no free occurences in f. It comes from the wish to make equal two functions which behave the same, that is which return the same result when applied to the same parameter (extensional equality). This is the role of η-contraction :

$$(\lambda x.fx) \underset{\eta}{\longrightarrow} f \text{ if } x \text{ has no free occurences in } f.$$

Accordingly, we would like to η-reduce $\lambda(\underline{2}\ \underline{1}\ )$ to $\underline{1}$. We observe that in De Bruijn notation, η-contraction is not as trivial as in the classical formalism: there is some work to do to compute the η-reduct. Our aim is now to find a first order rewrite rule which makes explicit the substitution process involved in the η-contraction and hidden at the meta-level in all the other approaches [Har92, Río93]. In the following, we make the η-reduction explicit by means of an unconditional *Eta* rewrite rule and of an extension of the substitution calculus, v. We show that the rewriting system obtained, denoted by $\lambda v\eta$, provides a correct implementation of the η-contraction. Moreover, *Eta* leads to a new definition of the classical η-contraction. We especially study consequences of this new definition w.r.t. ground confluence. Next, we prove some properties of the $\lambda v\eta$-calculus, namely confluence on ground terms[1] and strong normalization on typed terms. Finally, we compare our *Eta* rule to previous approaches.

1 Definition of η and *Eta*

We now work in De Bruijn notation. Ríos [Río93] gives an operational definition of η on the set Λ of pure terms. Specifically, he proves :

$$\lambda(a\ \underline{1}) \underset{\eta}{\longrightarrow} a_1 \textit{ if } a_1 \textit{ is defined,}$$

with the partial function a_n, defined at the meta-level:

[1] In the following, we write confluent instead of ground confluent

$$(ab)_n = a_n b_n \qquad \underline{m}_n = \begin{cases} \underline{m-1} \text{ if } m > n \\ undefined \text{ if } m = n \\ \underline{m} \text{ if } m < n \end{cases}$$
$$(\lambda a)_n = \lambda a_{n+1}$$

Lemma 2 Ríos. *η is the rewrite extension on Λ of $\lambda(a\ \underline{1}) \underset{\eta}{\longrightarrow} a_1$ if a_1 is defined.*

Starting from this and unlike Ríos, we try to define a primitive *Eta* rule. Our reasoning is based on two facts. First, a_n is an effective partial function and should be computed by a rewrite system. Second, the function a_n should look like a / substitution according to the following similarities.

a_{n+1} leaves unchanged the indices from $\underline{1}$ to $\underline{n}$, decreases the indices after $\underline{n+2}$ and rules out $\underline{n+1}$.

$a[\Uparrow^n(b/)]$ leaves unchanged the indices from $\underline{1}$ to $\underline{n}$, decreases the indices after $\underline{n+2}$ and replaces $\underline{n+1}$ by $b[\uparrow^n]$.

We see that a_{n+1} and $a[\Uparrow^n(b/)]$ have the same effect if the term a does not contain the index $\underline{n+1}$. If a contains $\underline{n+1}$, its presence is remembered by a special term, namely $\perp$.

Let us define the set of terms $\Lambda\upsilon_\perp$.

Definition 3. The set $\Lambda\upsilon_\perp$ is the set generated by:

$$\begin{array}{ll} \textbf{Terms} & a ::= \underline{n} \mid aa \mid \lambda a \mid a[s] \mid \perp \\ \textbf{Subst} & s ::= a/ \mid \Uparrow(s) \mid \uparrow \\ \textbf{Naturals} & n ::= n+1 \mid 1. \end{array}$$

Now, we are able to define a rule on $\Lambda\upsilon_\perp$ which we call *Eta*:

Definition 4. Let $a, b \in \Lambda\upsilon_\perp$. *Eta* is the rewrite extension on $\Lambda\upsilon_\perp$ of :

$$\lambda(a\ \underline{1}) \underset{Eta}{\longrightarrow} a[\perp/]$$

As $\perp$ is a constant, we add the rule:

Definition 5. $\upsilon_\perp$ is: $\upsilon \cup \{\perp[s] \to \perp\}$.

In [Bri94], we extend the properties of $\lambda\upsilon$ proved in [LRD94a] to a $\lambda\upsilon$-calculus with a finite set of constants. A consequence is that $a[\Uparrow^n(\perp/)]$ has the expected behaviour:

Lemma 6. *Let $m \geq 1$ and $n \geq 0$.*

1. $\underline{m}[\Uparrow^n(\perp/)] \overset{\bullet}{\underset{\upsilon}{\longrightarrow}} \underline{m}$ *if* $1 \leq m \leq n$
2. $\underline{n+1}[\Uparrow^n(\perp/)] \overset{\bullet}{\underset{\upsilon}{\longrightarrow}} \perp[\uparrow^n] \overset{\bullet}{\underset{\upsilon_\perp}{\longrightarrow}} \perp$

3. $\underline{m}\,[\Uparrow^n(\bot/)] \xrightarrow[\upsilon]{\bullet} \underline{m-1}$ *if* $m \geq n+2$.

We show that *Eta* is correct on pure terms w.r.t. η, that is to say, an *Eta*-rewrite followed by υ-normalization is equivalent to an η-contraction. We introduce another relation η' defined on Λ and we prove this relation is actually η:

Definition 7. Let $a \in \Lambda$. $\lambda(a\ \underline{1}) \xrightarrow[\eta']{} b$ iff $\lambda(a\ \underline{1}) \xrightarrow[Eta]{} a[\bot/]$ and $\upsilon(a[\bot/]) = b \in \Lambda$. We extend η' by rewrite extension on Λ.

Lemma 8. *Let* $a \in \Lambda$, a_{n+1} *is defined if and only if* $\upsilon(a[\Uparrow^n(\bot/)]) \in \Lambda$ *and in that case* $a_{n+1} = \upsilon(a[\Uparrow^n(\bot/)])$.

Proof. 1. Let $a \in \Lambda$ and suppose that a_{n+1} is defined. We proceed by structural induction on a:
   (a) $a = \underline{m}$
   By hypothesis, a_{n+1} exists, so $a = \underline{m} \neq \underline{n+1}$.
      i. $m > n+1$, $\underline{m}_{n+1} = \underline{m-1}$
      By lemma 6, $\upsilon(\underline{m}\,[\Uparrow^n(\bot/)]) = \underline{m-1} \in \Lambda, = \underline{m}_{n+1}$
      ii. $m < n+1, \underline{m}_{n+1} = \underline{m}$
      By lemma 6, $\upsilon(\underline{m}\,[\Uparrow^n(\bot/)]) = \underline{m} \in \Lambda, = \underline{m}_{n+1}$
   (b) $a = \lambda b$
   As a_{n+1} is defined and $a_{n+1} = (\lambda b)_{n+1} = \lambda b_{n+2}$, b_{n+2} is defined. Then, by the induction hypothesis, $\upsilon(b[\Uparrow^{n+1}(\bot/)]) \in \Lambda$ and equals b_{n+2}.

$$
\begin{array}{rcl}
a[\Uparrow^n(\bot/)] & = & (\lambda b)[\Uparrow^n(\bot/)] \\
& \xrightarrow[\upsilon]{} & \lambda(b[\Uparrow^{n+1}(\bot/)]) \\
& \xrightarrow[\upsilon]{\bullet} & \lambda(b_{n+2}) \\
& \in & \Lambda \\
& = & (\lambda b)_{n+1} \\
& = & a_{n+1}
\end{array}
$$

   (c) $a = bc$. Immediate by application of the induction hypothesis to b and c.
2. Let $a \in \Lambda$ and suppose $\upsilon(a[\Uparrow^n(\bot/)]) \in \Lambda$. a_{n+1} is defined. Specifically, by structural induction on a:
   (a) $a = \underline{m}$
   Suppose a_{n+1} is not defined. Therefore $a = \underline{n+1}$ and by lemma 6, $\upsilon(\underline{n+1}\,[\Uparrow^n(\bot/)]) = \bot[\uparrow^n] \notin \Lambda$. This contradicts the hypothesis, so a_{n+1} is defined.
   (b) $a = \lambda b$
   $\upsilon((\lambda b)[\Uparrow^n(\bot/)]) = \lambda\upsilon(b[\Uparrow^{n+1}(\bot/)]) \in \Lambda$. So $\upsilon(b[\Uparrow^{n+1}(\bot/)]) \in \Lambda$, and by the induction hypothesis, b_{n+2} is defined. As $a_{n+1} = (\lambda b)_{n+1} = \lambda b_{n+2}$, a_{n+1} is defined.
   (c) $a = bc$. Immediate by application of the induction hypothesis to b and c.

We now prove equivalence of η and η' contractions for redexes located at the head of terms:

Lemma 9. *Let* $a \in \Lambda$, $\lambda(a\ \underline{1}) \underset{\eta}{\longrightarrow} a_1$ *if and only if* $\lambda(a\ \underline{1}) \underset{\eta'}{\longrightarrow} \upsilon(a[\bot/]) = a_1$.

Proof. 1. Suppose $\lambda(a\ \underline{1}) \underset{\eta}{\longrightarrow} a_1$. By the previous lemma, since a_1 is defined, we know: $\upsilon(a[\bot/]) = a_1 \in \Lambda$. So $\lambda(a\ \underline{1}) \underset{Eta}{\longrightarrow} a[\bot/] \overset{\bullet}{\underset{\upsilon}{\longrightarrow}} a_1$. That is, $\lambda(a\ \underline{1}) \underset{\eta'}{\longrightarrow} \upsilon(a[\bot/]) = a_1 \in \Lambda$.

2. Suppose $\lambda(a\underline{1}) \underset{\eta'}{\longrightarrow} \upsilon(a[\bot/]) \in \Lambda$. By the previous lemma, a_1 is defined and equals $\upsilon(a[\bot/])$. So, $\lambda(a\ \underline{1}) \underset{\eta}{\longrightarrow} \upsilon(a[\bot/])$.

More generally, by rewrite extension on Λ, we get the following proposition: *Eta* followed by the υ-normalization[2] correctly implements the classical η-reduction.

Proposition 10 Correction. *Let* $a, b \in \Lambda$, $a \underset{\eta}{\longrightarrow} b$ *if and only if* $a \underset{\eta'}{\longrightarrow} b$.

We achieve our aim: we compute the η-reduct by a first-order term rewriting system, as this computation is expressed through explicit substitutions.

2 Confluence of $\beta\eta'$

The *Eta* rule leads us to a new view of the classical η-contraction, i.e. η expressed in the classical formalism (without De Bruijn notation). Indeed, we define a new η-contraction in the classical λ-calculus, denoted by η' as:

$$\lambda x.(a\ x) \underset{\eta'}{\longrightarrow} a\{\bot/x\}$$

Notice that η' is unconditional. As shown by correction, this rule coincides with the classical η in the case x is not a free variable of a. In the other case, the classical η is not allowed, but η' is. In some cases, it makes sense to forget the precondition on the application of classical η. The following examples show that some η'-contractions, classically forbidden, can in fact be allowed and keep the entire calculus confluent.

1. $(\lambda xy.x)(\lambda x.x)(\lambda x.x\ x) \underset{\eta'}{\longrightarrow} (\lambda xy.x)(\lambda x.x)\bot \overset{\bullet}{\underset{\beta}{\longrightarrow}} (\lambda x.x)$
   $(\lambda xy.x)(\lambda x.x)(\lambda x.x\ x) \overset{\bullet}{\underset{\beta}{\longrightarrow}} (\lambda x.x)$
2. $\lambda x.((\lambda y.z)x\ x) \underset{\eta'}{\longrightarrow} (\lambda y.z)\bot \underset{\beta}{\longrightarrow} z$
   $\lambda x.((\lambda y.z)x\ x) \underset{\beta}{\longrightarrow} \lambda x.(z\ x) \underset{\eta'}{\longrightarrow} z$
3. $(\lambda u.(\lambda x.uxx))(\lambda y.z) \underset{\eta'}{\longrightarrow} (\lambda u.u\bot)(\lambda y.z) \underset{\beta}{\longrightarrow} (\lambda y.z)\bot \underset{\beta}{\longrightarrow} z$
   $(\lambda u.(\lambda x.uxx))(\lambda y.z) \underset{\beta}{\longrightarrow} (\lambda x.(\lambda y.z)xx) \underset{\beta}{\longrightarrow} (\lambda x.zx) \underset{\eta'}{\longrightarrow} z$

[2] We do not need the constant rule $\bot[s] \rightarrow \bot$

The rest of this section is devoted to the study of this unconditional η'. To be consistent with our older definitions, we go back to De Bruijn notation, but the following statements hold in the classical λ-calculus. First, we extend Λ to $\Lambda_\perp$, the set of pure terms that may contain $\perp$ constant occurences. We define unconditional η' on this set. Next, we look for a set Λm on which the relation $\beta\eta'$ is confluent. To prove this confluence, we first show the postponement of η'-steps w.r.t. β-steps.

We define β' and η' on the set $\Lambda_\perp$:

Definition 11 $\Lambda_\perp$. The set $\Lambda_\perp$ is the set generated by :

$$\begin{aligned}
\textbf{Terms} \quad & a ::= \underline{n} \mid aa \mid \lambda a \mid \perp \\
\textbf{Naturals} \quad & n ::= n+1 \mid 1.
\end{aligned}$$

Definition 12. Let $a, b \in \Lambda_\perp$.

1. β' is the rewrite extension on $\Lambda_\perp$ of : $(\lambda a)b \xrightarrow[\beta']{} \upsilon_\perp(a[b/])$

2. η' is the rewrite extension on $\Lambda_\perp$ of : $\lambda(a\ \underline{1}) \xrightarrow[\eta']{} \upsilon_\perp(a[\perp/])$

In [Bri94], we have shown that β' and classical β coincide on the set $\Lambda_\perp$, as $\perp$ is a constant. Hence, we will write β for both.

$\beta\eta'$ is clearly not confluent on $\Lambda_\perp$, i.e. on the set of terms containing $\perp$ occurences, as shown by the critical pair between β and η' :

$$ab \xleftarrow[\beta']{} (\lambda(a\ \underline{1}))b \xrightarrow[\eta']{} a[\perp/]b$$

In $\Lambda_\perp$, we have the counter-example :

$$\underline{1}\ \underline{1} \xleftarrow[\beta']{} (\lambda(\underline{1}\ \underline{1}))\ \underline{1} \xrightarrow[\eta']{} \perp\ \underline{1}$$

But by restricting η' to a reasonably large subset of $\Lambda_\perp$, we can make $\beta\eta'$ confluent. This set is Λm :

Definition 13.

$$\Lambda m = \{a \in \Lambda_\perp \mid \exists b \in \Lambda : a \xrightarrow[\beta\eta']{\bullet} b\}$$

This set is larger than both Λ and the set used by Hardin and Ríos, in the sense that η' contains more reductions and Λm contains terms with occurences of $\perp$. For example, the terms $(\lambda(\underline{1}\ \underline{1}))\ \underline{1}$ and $\underline{1}\ \underline{1}$ belong to Λm, but $\perp\ \underline{1}$ does not.

As already noticed, an η'-contraction does not necessarily coincide with a classical η-contraction. We call such an η'-contraction biased. In the following, we show that on the set Λm, we can either postpone a biased η'-contraction and thereby transform it in a classical η one, or eliminate it. From this, confluence of $\beta\eta'$ follows immediately. Consider the reduction :

$$a \xrightarrow[\eta']{} b \xrightarrow[\beta]{} c$$

We successively prove :

1. the "below case" : when the η'-redex is strictly contained in the β-redex, it can be postponed and may be duplicated (or eliminated).
2. the "upon" case : when the η'-redex strictly contains the β-redex, it can be postponed.
3. the "critical" case : when there is a critical pair, the η'-contraction can be eliminated.

The first two cases hold in $\Lambda_\perp$, the last one holds only in Λm. Due to potential duplication, there is added complexity for the below case.

First, we show the below case. To treat it formally and according to the philosophy of explicit substitution, we need what are called projection lemmas. In the case of *Eta*, these lemmas say that an *Eta*-rewrite between terms of $\Lambda\upsilon_\perp$ maps to a η'-reduction between their $\upsilon_\perp$-normal forms. If the *Eta*-rewrite takes place inside a closure [ ] we call it internal, if it takes place outside each closure, we call it external. External and internal are abbreviated by the superscripts *ext* and *int*.

Lemma 14 Projection lemmas on $\Lambda_\perp$.

1. If $a, b \in \Lambda\upsilon_\perp$ such that $a \xrightarrow[Eta]{}^{ext} b$ then $\upsilon_\perp(a) \xrightarrow[\eta']{} \upsilon_\perp(b)$.
2. If $a, b \in \Lambda\upsilon_\perp$ such that $a \xrightarrow[Eta]{}^{int} b$ then $\upsilon_\perp(a) \overset{\bullet}{\underset{\eta'}{\longrightarrow}} \upsilon_\perp(b)$

The proof of this lemma is in [Bri94]. The *Beta* projection lemma is similar, its proof extends the one given in [LRD94a].

Lemma 15 The case η' below β. *Let $a, b, c \in \Lambda$.*

1. If $(\lambda a)b \xrightarrow[\eta']{} (\lambda c)b \xrightarrow[\beta]{} \upsilon_\perp(c[b/])$ then $(\lambda a)b \xrightarrow[\beta]{} \upsilon_\perp(a[b/]) \xrightarrow[\eta']{} \upsilon_\perp(c[b/])$.
2. If $(\lambda b)a \xrightarrow[\eta']{} (\lambda b)c \xrightarrow[\beta]{} \upsilon_\perp(b[c/])$ then $(\lambda b)a \xrightarrow[\beta]{} \upsilon_\perp(b[a/]) \overset{\bullet}{\underset{\eta'}{\longrightarrow}} \upsilon_\perp(b[c/])$.

The proof comes directly from the projection lemma 14.

Proof. 1. We assume:

$$(\lambda\lambda(a\ \underline{1}))b \xrightarrow[\eta']{} (\lambda\upsilon_\perp(a[\perp/]))b \xrightarrow[\beta]{} \upsilon_\perp(\upsilon_\perp(a[\perp/][b/])) = \upsilon_\perp(a[\perp/][b/])$$

We observe :

$$(\lambda\lambda(a\ \underline{1}))b \xrightarrow[Beta]{}^{ext} (\lambda(a\ \underline{1}))[b/] \xrightarrow[Eta]{}^{ext} a[\perp/][b/]$$

By external projection lemmas, we get :

$$(\lambda\lambda(a\ \underline{1}))b \xrightarrow[\beta]{} \upsilon_\perp((\lambda(a\ 1))[b/]) \xrightarrow[\eta']{} \upsilon_\perp(a[\perp/][b/])$$

The key part, i.e. Barendregt style substitution lemma [Bar84], is inside the projection lemma:

$$a[\Uparrow(b/)][\perp/] \overset{\bullet}{\underset{\upsilon_\perp}{\longleftrightarrow}} a[\Uparrow(b/)][\perp[b/]/] \overset{\bullet}{\underset{\upsilon_\perp}{\longleftrightarrow}} a[\perp/][b/]$$

2. We assume:

$$(\lambda b)(\lambda(a\ \underline{1})) \underset{\eta'}{\longrightarrow} (\lambda b)(v_{\perp}(a[\perp/])) \underset{\beta}{\longrightarrow} v_{\perp}(b[v_{\perp}(a[\perp/])/]) = v_{\perp}(b[a[\perp/]/])$$

We observe:

$$(\lambda b)(\lambda(a\ \underline{1})) \underset{Beta}{\longrightarrow}^{ext} b[(\lambda(a\ \underline{1}))/] \underset{Eta}{\longrightarrow}^{int} b[a[\perp/]/]$$

By external and internal projection lemmas, we get:

$$(\lambda b)(\lambda(a\ \underline{1})) \underset{\beta}{\longrightarrow} v_{\perp}(b[(\lambda(a\ \underline{1}))/]) \underset{\eta'}{\overset{\bullet}{\longrightarrow}} v_{\perp}(b[a[\perp/]/])$$

Lemma 16 The case η' above β. *Let $a, c \in \Lambda_{\perp}$. If $\lambda(a\ \underline{1}) \underset{\eta'}{\longrightarrow} v_{\perp}(a[\perp/]) \underset{\beta}{\longrightarrow} c$ then there exists $d \in \Lambda_{\perp}$ such that $\lambda(a\ \underline{1}) \underset{\beta}{\longrightarrow} d \underset{\eta'}{\longrightarrow} c$.*

Proof. We translate the hypothesis

$$\lambda(a\ \underline{1}) \underset{\eta'}{\longrightarrow} v_{\perp}(a[\perp/]) \underset{\beta}{\longrightarrow} c$$

into a β-reduction:

$$(\lambda\ a)\perp \underset{\beta}{\longrightarrow} v_{\perp}(a[\perp/]) \underset{\beta}{\longrightarrow} c$$

Since $\perp$ is not an abstraction, the first β-contraction does not create a β-redex in $v_{\perp}(a[\perp/])$. So, the structure of the β-redex in $v_{\perp}(a[\perp/])$ is already in a. Hence, there exists a $d \in \Lambda_{\perp}$ such that:

$$a \underset{\beta}{\longrightarrow} d$$

Since we have:

$$(\lambda\ a)\perp \underset{\beta}{\longrightarrow} v_{\perp}(a[\perp/]) \underset{\beta}{\longrightarrow} c$$

$$(\lambda\ a)\perp \underset{\beta}{\longrightarrow} (\lambda\ d)\perp \underset{\beta}{\longrightarrow} v_{\perp}(d[\perp/])$$

By the substitution lemma for β [HS86], we get:

$$c = v_{\perp}(d[\perp/])$$

Returning to η'-contraction, we get the postponement:

$$\lambda(a\ \underline{1}) \underset{\beta}{\longrightarrow} \lambda(d\ \underline{1}) \underset{\eta'}{\longrightarrow} v_{\perp}(d[\perp/]) = c$$

The previous lemmas imply:

Lemma 17. *Let $a, b, c \in \Lambda_{\perp}$. If $a \underset{\eta'}{\overset{\bullet}{\longrightarrow}} b \underset{\beta}{\overset{\bullet}{\longrightarrow}} c$ when η' is below or above β then there exists $d \in \Lambda_{\perp}$ such that: $a \underset{\beta}{\overset{\bullet}{\longrightarrow}} d \underset{\eta'}{\overset{\bullet}{\longrightarrow}} b$.*

The proof is by a double induction on first, the length of the η'-reduction, and second, the length of the β-reduction.

The following lemma shows what happens with the classical commutation of classical β and η.

Lemma 18 The critical case. *Let* $M, N \in \Lambda_\perp$.
If $(\lambda(M\ \underline{1}))N \underset{\eta'}{\longrightarrow} (\upsilon_\perp(M[\perp/]]))N \overset{\bullet}{\underset{\beta}{\longrightarrow}} P \in \Lambda$ *then* $(\lambda(M \cdot \underline{1}))N \overset{\bullet}{\underset{\beta}{\longrightarrow}} P$.

Proof. We prove it with the classical formalism. Here, we treat $\perp$ as a free variable and we rename $\perp$ by $\top$, a fresh variable, in M and N:
$N' \equiv N[\top/\perp]$ and $N' \equiv N[\top/\perp]$. We have:

$$(\lambda x.M'\ x)N' \underset{\eta'}{\longrightarrow} (M'[\perp/x])N' \overset{\bullet}{\underset{\beta}{\longrightarrow}} P \in \Lambda$$

By stability of β (or substitution lemma [HS86]):

$$((M'[\perp/x])N')[N'/\perp] \overset{\bullet}{\underset{\beta}{\longrightarrow}} P[N'/\perp]$$

As $\perp \notin FV(N'P)$, $P[N'/\perp] = P$ and $N'[N'/\perp] = N'$.

$$(M'[\perp/x][N'/\perp])N' \overset{\bullet}{\underset{\beta}{\longrightarrow}} P$$

And $M'[\perp/x][N'/\perp] = M'[N'/x]$,

$$(M'[N'/x])N' \overset{\bullet}{\underset{\beta}{\longrightarrow}} P$$

By β-expansion:

$$(\lambda x.M'\ x)N' \overset{\bullet}{\underset{\beta}{\longrightarrow}} P$$

By renaming, as $\top \notin FV(MN)$,

$$(\lambda x.M\ x)N \overset{\bullet}{\underset{\beta}{\longrightarrow}} P$$

Lemma 19 Postponement of η'-contractions. *Let* $a \in \Lambda m$ *and* $c \in \Lambda$ *such that:* $a \overset{\bullet}{\underset{\beta\eta'}{\longrightarrow}} c$. *Then there exists* $d \in \Lambda : a \overset{\bullet}{\underset{\beta}{\longrightarrow}} d \overset{\bullet}{\underset{\eta}{\longrightarrow}} c$.

Proof. We proceed by induction on the number of η'-steps. We consider the last η' step. If there is no η'-step, we are done. If this step is a critical case, by lemma 18, we eliminate it. If this step is an upon or below case, by lemma 17, it is postponed and may be duplicated (or eliminated). As only β-steps eliminate $\perp$ occurences, final η'-steps are in fact classical η-steps : $a \overset{\bullet}{\underset{\beta\eta'}{\longrightarrow}} b \overset{\bullet}{\underset{\eta}{\longrightarrow}} c$ and $b \in \Lambda$, so that we apply the induction hypothesis on $a \overset{\bullet}{\underset{\beta\eta'}{\longrightarrow}} b$.

Theorem 20. $\beta\eta'$ *is confluent on* Λm.

Proof. Let $a, b, c \in \Lambda m$ such that $a \xrightarrow[\beta\eta']{*} b$ and $a \xrightarrow[\beta\eta']{*} c$. As $b, c \in \Lambda m$, there exists $b', c' \in \Lambda$ such that $b \xrightarrow[\beta\eta']{*} b'$ and $c \xrightarrow[\beta\eta']{*} c'$. So:

$$a \xrightarrow[\beta\eta']{*} b' \quad a \xrightarrow[\beta\eta']{*} c'$$

By the previous lemma, we associate with these two reductions two classical ones:

$$a \xrightarrow[\beta\eta]{*} b' \quad a \xrightarrow[\beta\eta]{*} c'$$

Since the classical $\beta\eta$ is confluent on Λ plus a constant, there exists $d \in \Lambda$ ($d \notin \Lambda_\perp$ because η-steps are correct and $b', c' \in \Lambda$):

$$b' \xrightarrow[\beta\eta]{*} d \quad c' \xrightarrow[\beta\eta]{*} d$$

By correction, η' can simulate η:

$$b' \xrightarrow[\beta\eta']{*} d \quad c' \xrightarrow[\beta\eta']{*} d$$

The definition of Λm is very general and we do not know a syntactic characterization for it (we conjecture there is none). If one wants to implement an η' reduction strategy, one may wish to know if one stays in Λm. The absence of a syntactic characterization seems to prevent providing such a criterion. Even a smaller set, like,

$$\{a \in \Lambda_\perp | \exists b \in \Lambda : a \xrightarrow[\beta]{*} b\}$$

is of no more help. Nevertheless, η' sheds more light on the relation between η and β. Concerning $\lambda\upsilon\eta$, that is to say the rewrite system $\{Beta, Eta\} \cup \upsilon_\perp$, one may find in [Bri94] a proof of its ground confluence on the set $\Lambda\upsilon\eta$:

$$\Lambda\upsilon\eta = \{a \in \Lambda\upsilon_\perp | \exists b \in \Lambda\upsilon : a \xrightarrow[\upsilon_\perp]{*} b\}$$

3 $\lambda\upsilon\eta$ and Strong Normalization

In this section, we study the preservation of strong normalization of $\lambda\upsilon\eta$ on $\Lambda\upsilon_\perp$ terms. Then, we deduce strong normalization of $\lambda\upsilon\eta$ on simply typed terms. The proofs of these properties are straightforward extensions of the $\lambda\upsilon$ ones, thus we just sketch them.

We adapt the $\lambda\upsilon$ strong normalization preservation proof [BBLRD94] and discuss it. The main ideas are: use the strong normalization of $\beta\eta'$ and the fact that $b/$ substitutions, with $b \neq \perp$, come from *Beta* rewrites. We emphasize the fact that this last property of υ is essential to this proof of strong normalization of $\lambda\upsilon$ and $\lambda\upsilon\eta$. The following lemma formalizes this property.

Lemma 21. *Let $a_0 \in \Lambda_\perp$ such that $a_0 \xrightarrow[\lambda\upsilon\eta]{*} a_n \equiv C\{d[\Uparrow^i(c/)]\}$ with $c \neq \perp$. Then there exists a_i, $0 \leq i \leq n$ such that: $a_i \equiv D\{(\lambda e)b\}$ and $b \xrightarrow[\lambda\upsilon\eta]{*} c$.*

The proof can be found in [BBLRD94]. This lemma does not hold in the σ calculus because of the rule $(a \cdot s) \circ t \rightarrow a[t] \cdot (s \circ t)$. In particular, one observes that it creates a closure $[\,]$ and so it may create $[b \cdot id]$, the equivalent in $\lambda\sigma$ of $[b/]$ [Mel95].

We now state the second key point of this normalization proof: this lemma isolates the sources of all potentially infinite derivations in closures $[\,]$.

Lemma 22. *Let $a \in \Lambda\upsilon_\perp$ such that $\upsilon_\perp(a)$ is strongly normalizing. In a $\lambda\upsilon\eta$ derivation starting from a, there exists a rank N such that each $\lambda\upsilon\eta$-step following N is internal.*

That property, proved in [BBLRD94], depends only on υ, not on *Beta*, $\perp$ or *Eta*. It is not shared by the σ substitution calculus [ACCL91] because of the same rule. Indeed, in the σ-derivation:

$$\underline{1}[(a \cdot s) \circ t] \rightarrow^{int} \underline{1}[a[t] \cdot (s \circ t)] \rightarrow^{ext} a[t]$$

the external redex $\underline{1}[\cdot]$ produced by the internal σ-rewrite $(a \cdot s) \circ t \rightarrow a[t] \cdot (s \circ t)$ can not be moved earlier in the derivation.

By the two previous lemmas, we get:

Theorem 23. *Let $a \in \Lambda\upsilon_\perp$ such that $\upsilon_\perp(a)$ is $\beta\eta'$-strongly normalizing. Then, a is $\lambda\upsilon_\perp$-strongly normalizing.*

The proof follows closely the $\lambda\upsilon$ case, described in [BBLRD94]. It is based on a minimal counter-example, more precisely a minimal $\lambda\upsilon\eta$-derivation.

Thus $\lambda\upsilon\eta$ preserves strong normalization. As a consequence, we derive strong normalization of a simply typed version of $\lambda\upsilon\eta$ on $\Lambda\upsilon_\perp$. For this, first, we need a typing system, second, we have to show $\beta\eta'$ is strongly normalizing on simply typed pure terms.

Therefore, we enlarge $\lambda\upsilon$ simply typed terms [LRD94a] to $\lambda\upsilon\eta$ ones. This part relies heavily on the simply typed version of $\lambda\upsilon$-calculus described in [LRD94a]. To introduce a typed *Eta* rule, we have to type terms of $\Lambda\upsilon_\perp$. For this, instead of a single constant $\perp$, we need, for each simple type A, a typed constant $\perp_A$ and a rule $\perp_A[s] \rightarrow \perp_A$. We just add an axiom scheme to the typing system of $\lambda\upsilon$ in order to type $\perp_A$ and terms containing occurences of $\perp_A$. The grammar of the pseudo-terms is:

Type	$A ::= \mathsf{A}_1 \mid \ldots \mid \mathsf{A}_n \mid A \Rightarrow B$
Naturals	$n ::= 1 \mid n+1$
Terms	$a ::= \underline{n} \mid ab \mid \lambda A.a \mid a[s] \mid \perp_A$
Substitutions	$s ::= a/ \mid \uparrow \mid \Uparrow(s)$
Context	$\Gamma ::= [\,] \mid A \cdot \Gamma$

where $\mathsf{A}_1 \ldots \mathsf{A}_n$ is a family of atomic types. The set of $\lambda\upsilon\eta$-simply typed terms is noted $\Lambda\upsilon_\perp^\rightarrow$ and is produced by the typing system:

Terms

$$\frac{}{\Gamma \vdash \bot_A : A}$$

$$\frac{\Gamma \vdash a : A \Rightarrow B \quad \Gamma \vdash b : A}{\Gamma \vdash ab : B} \quad \frac{A \cdot \Gamma \vdash a : B}{\Gamma \vdash \lambda A.a : A \Rightarrow B}$$

$$\frac{\Gamma \vdash a : A \quad \Delta \vdash s : \Gamma}{\Delta \vdash a[s] : A} \quad \frac{}{A \cdot \Gamma \vdash \underline{1} : A} \quad \frac{\Gamma \vdash \underline{n} : A}{B \cdot \Gamma \vdash \underline{n+1} : A}$$

Substitutions

$$\frac{\Gamma \vdash a : A}{\Gamma \vdash a/ : A \cdot \Gamma} \quad \frac{}{A \cdot \Gamma \vdash \uparrow : \Gamma} \quad \frac{\Gamma \vdash s : \Delta}{A \cdot \Gamma \vdash \Uparrow(s) : A \cdot \Delta}$$

We define a typed *Eta* rule accordingly:

Lemma 24 Subject reduction theorem. *Let $a \in \Lambda\upsilon_{\bot}^{\rightarrow}$. The rewrite extension on $\Lambda\upsilon_{\bot}^{\rightarrow}$ of $\lambda A.(a\ \underline{1}) \underset{Eta}{\longrightarrow} a[\bot_A/]$, i.e. typed Eta, preserves types.*

Proof.

$$\frac{\dfrac{A \cdot \Gamma \vdash a : A \Rightarrow B \quad A \cdot \Gamma \vdash \underline{1} : A}{A \cdot \Gamma \vdash a\,\underline{1} : B}}{\Gamma \vdash \lambda A.(a\ \underline{1}) : A \Rightarrow B} \quad \frac{A \cdot \Gamma \vdash a : A \Rightarrow B \quad \dfrac{\Gamma \vdash \bot_A : A}{\Gamma \vdash \bot_A/ : A \cdot \Gamma}}{\Gamma \vdash a[\bot_A/] : A \Rightarrow B}$$

Having defined simply typed terms, it remains to show $\beta\eta'$ strong normalization on simply typed pure terms. We denote the set of simply typed pure terms $\Lambda_{\bot}^{\rightarrow}$; it is a subset of $\Lambda\upsilon_{\bot}^{\rightarrow}$.

Lemma 25. *Let $a \in \Lambda_{\bot}^{\rightarrow}$, a is $\beta\eta'$-strongly normalizing.*

The adaptation of the $\beta\eta$ case [HS86] is quite straightforward. It relies on an easy "reverse" substitution lemma: let M, L be pure terms and x, y be variables. If $M\{y/x\} \underset{\beta\eta'}{\longrightarrow} L$, then there is a pure term N such that: $M \underset{\beta\eta'}{\longrightarrow} N$ and $L \equiv N\{y/x\}$.

The conditions of the preservation theorem 23 are fulfilled, thus:

Corollary 26. *Let $a \in \Lambda\upsilon_{\bot}^{\rightarrow}$, a is $\lambda\upsilon\eta$-strongly normalizing.*

Conclusion

Hardin's and Ríos' definition[3] of *Eta* [Har92, Río93] in the framework of $\lambda\upsilon$, is:

$$\lambda(a\ \underline{1}) \underset{Eta}{\longrightarrow} b \text{ if } \upsilon(a) = \upsilon(b[\uparrow]) \qquad (HR)$$

[3] To be consistent with our notations, we take Ríos' syntax.

That rule does not make the η-reduct computation explicit since it uses an υ-matching [JK91] instead of our $\upsilon_\perp$-normalization. More precisely, imagine *Eta* applied to a term $\lambda(a\ \underline{1})$. To apply rule (HR), one must solve, modulo the theory υ, the equation $a =_\upsilon b[\uparrow]$ where b, the *Eta*-reduct, is the unknown (the variable to instantiate). This computation is what we call υ-matching. Clearly, υ-matching is more complex than $\upsilon_\perp$-normalization: we do not even know whether υ-matching is decidable or not and since υ-matching may produce several solutions, rule (HR) may produce several reducts for the same *Eta*-redex, among them the classical η-reduct. Consequently, rule (HR) is less operational than our *Eta* rule. Moreover, our *Eta* rule is generic. Indeed, in this paper we apply our definition to $\lambda\upsilon$. But all we need in order to define *Eta* is a term rewriting system that computes β-contraction; for instance we do not use renaming operators like $\uparrow$. Hence, *Eta* can be adapted to every λ-calculus of explicit substitution, with explicit names or not. For instance, in $\lambda\sigma_{\Uparrow}$ [HL89], we would write:

$$\lambda(a\ \underline{1}) \underset{Eta}{\longrightarrow} a[\perp \cdot id]$$

and in $\lambda\chi$ [LRD94b]:

$$\lambda x_i.(a\ x_i) \underset{Eta}{\longrightarrow} a[\perp / x_i]_0$$

Lastly, our rule is unconditional. We have seen that this led to a very general alternative of the classical η, namely η' which does not require De Bruijn notation. The condition of application of the classical η rule is too strong, and as we have shown, there are other confluent reductions which are not captured by this rule. This confirms our conviction that explicit substitutions help in a deeper understanding of λ-calculus, not only of its β-reduction aspect but also of other aspects like η-reduction. In that manner, according to Curry [CF58], explicit substitutions solidify the fundamentals of logic.

Acknowledgments. We would like to thank Pierre Lescanne and Jocelyne Rouyer-Degli for their constant support, and Philippe de Groote and Roberto Amadio for their remarks.

References

[ACCL91] M. Abadi, L. Cardelli, P.-L. Curien, and J.-J. Lévy. Explicit substitutions. *Journal of Functional Programming*, 1(4):375–416, 1991.

[Bar84] H. P. Barendregt. *The Lambda-Calculus, its syntax and semantics*. Studies in Logic and the Foundation of Mathematics. Elsevier Science Publishers B. V. (North-Holland), Amsterdam, 1984. Second edition.

[BBLRD94] Z. Benaissa, D. Briaud, P. Lescanne, and J. Rouyer-Degli. $\lambda\upsilon$, a calculus of explicit substitutions which preserves strong normalisation. Submitted, December 1994.

[Bri94] D. Briaud. An explicit *Eta* rewrite rule. Rapport de Recherche 2417, INRIA, 1994.

[CF58] H. B. Curry and Feys. *Combinatory Logic*, volume 1. Elsevier Science Publishers B. V. (North-Holland), Amsterdam, 1958.

[Cur83] P.-L. Curien. *Combinateurs catégoriques, algorithmes séquentiels et programmation applicative.* Thèse de Doctorat d'Etat, Univ. Paris 7, 1983.

[Cur86a] P.-L. Curien. Categorical combinators. *Information and Control*, 69:188–254, 1986.

[Cur86b] P.-L. Curien. *Categorical Combinators, Sequential Algorithms and Functional Programming.* Pitman, 1986.

[dB72] N. G. de Bruijn. Lambda calculus with nameless dummies, a tool for automatic formula manipulation. *Indag. Mat.*, 34:381-392, 1972.

[dB78] N. G. de Bruijn. A namefree lambda calculus with facilities for internal definition of expressions and segments. TH-Report 78-WSK-03, Technological University Eindhoven, Netherlands, Department of Math., 1978.

[Har87] Th. Hardin. *Résultats de confluence pour les règles fortes de la logique combinatoire catégorique et liens avec les lambda-calculs.* Thèse de Doctorat d'Etat, Univ. Paris 7, 1987.

[Har89] Th. Hardin. Confluence results for the pure strong categorical combinatory logic CCL: λ-calculi as subsystems of CCL. In *TCS*, 65:291–342, 1989.

[Har92] Th. Hardin. Eta-conversion for the languages of explicit substitutions. In *3rd ALP, LNCS 632*, Volterra, Italy, 1992.

[HL89] Th. Hardin and J.-J. Lévy. A confluent calculus of substitutions. In *France-Japan Artificial Intelligence and Computer Science Symposium*, Izu, 1989.

[HS86] J. Roger Hindley and Johnathan P. Seldin. *Introduction to Combinators and Lambda-calculus.* Cambridge University, 1986.

[JK91] J.-P. Jouannaud and Claude Kirchner. Solving equations in abstract algebras: a rule-based survey of unification. In J.-L. Lassez and G. Plotkin, editors, *Computational Logic. Essays in honor of Alan Robinson*, chapter 8, pages 257–321. The MIT press, Cambridge (MA, USA), 1991.

[Les94] P. Lescanne. From $\lambda\sigma$ to $\lambda\upsilon$, a journey through calculi of explicit substitutions. In *21st POPL, Portland (Or., USA)*, pages 60–69. ACM, 1994.

[LRD94a] P. Lescanne and J. Rouyer-Degli. The calculus of explicit substitutions $\lambda\upsilon$. Technical Report RR-2222, INRIA-Lorraine, January 1994.

[LRD94b] P. Lescanne and J. Rouyer-Degli. Explicit substitutions with de Bruijn's levels, August 1994.

[Mel95] P.-A. Melliès. Typed λ-calculi with explicit substitutions may not terminate. In M. Dezani, editor, *TLCA*, 1995.

[Río93] A. Ríos. *Contributions à l'étude des λ-calculs avec des substitutions explicites.* Thèse de Doctorat d'Université, U. Paris VII, 1993.

Extracting Text from Proofs

Yann Coscoy & Gilles Kahn & Laurent Théry *

INRIA Sophia Antipolis, France.
{coscoy,kahn,thery}@inria.fr

Abstract. In this paper, we propose a method for presenting formal proofs in an intelligible form. We describe a transducer from proof objects (λ-terms in the Calculus of Constructions) to pseudo natural language that has been implemented for the Coq system.

1 Introduction

Almost all computer proof assistants today are used in the following manner. The user states a theorem to prove. Then using a variety of commands that are recorded in a *proof script*, the user brings the interactive system to a state indicating that the theorem has been proved. The theorem is then archived for later reuse, and the corresponding proof script is kept *preciously*.

There are two basic reasons for safekeeping the proof script. First, a proof must often be verified again in a slightly different context, or a different version of the theorem may be needed. Thus, the script is kept as a model for constructing later variants of a proof. We remark in passing that the command language of some systems does not facilitate this task.

Second, the proof script is used as tangible evidence that the proof was actually carried out, and as a means of communicating its intellectual content. In the context of program verification, such evidence must be presented to industrial auditors. As a vehicle for the communication of proofs, we feel that proof scripts *alone* are largely *inadequate*. Decoding what a proof script actually does is the province of expert users of a given proof assistant. A proof script invokes sophisticated tactics that are system specific, that attempt to do things that fail and are irrelevant for the final proof. Furthermore, these tactics may be refined as the proof assistant is being improved. On the other hand, a proof script contains invaluable information on the level of abstraction at which the proof is carried out. In her work, A. Cohn ([Cohn88]) tries to produce intelligible text from the proof script.

By contrast, in this paper, we will be concerned with proof assistants that construct a *proof object*, i.e. a data structure that explicitly represents the proof of facts established with the system. Proof objects are built by a number of modern proof assistants ([Coq91, Hol92, Lego92]), but they are rarely used as they are considered to be exceedingly large and difficult to understand. On the basis of experiments carried out in the last three years with several computer

* Part of this research was done when the third author was at ATT Bell Laboratories.

proof assistants, we take exception with this commonly held view and find proof objects useful and important in many respects.

First it is possible to make good sense out of proof objects, and this is what we show in this paper. Second, proof objects are far more independent of the proof assistant than proof scripts and they form a better basis for understanding and displaying the intellectual content of a proof. As a result, they are useful in debugging automatic proof tactics. Last, if they can be built incrementally as in ALF [Alf93], proof objects provide a useful interactive feedback on what is going on in the proof.

2 Extracting Text from λ-terms

A number of authors ([Chester76, Ep93, Felty88, Huang94]) have investigated the possibility of producing text out of formal proofs. We approach the problem in a somewhat different fashion and use techniques that are familiar in code generation. We do not produce very fluid natural language and we handle only the logical structure of the proofs, in the spirit of [Gentzen69].

Our work is applicable to any representation of natural deduction proofs, but type theoretic proofs give the most concise and elegant definition. For the rest of this paper, we use a presentation based on the Calculus of Constructions as in [Coq91]. We deal with a λ-calculus with three predefined types: ***Type*** for types, ***Prop*** for propositions and ***Set*** for sets, plus constants representing axioms and theorems. At this stage, we restrict our extraction to λ-terms that inhabit ***Prop***, i.e. proofs of propositions. In what follows, we first define a basic transducer by giving simple translation rules and then present some further optimizations.

2.1 Basic Transducer

To construct a transducer from λ-terms to text, we must address some problems.

Type information Trying to obtain a textual presentation from the raw λ-term is ***not*** possible: we need additional type information. To see this, consider the tautology $A \supset A$. A proof of this is the identity function $\lambda h{:}\, A.\ h$. If we decorate this term with type information, we obtain:

$$(\lambda h{:}\, A.\ h_A)_{A \supset A}$$

and it becomes clear that the text of the proof needs to make reference at least in two places to type information that is not structurally part of the λ-term:

$$(\lambda h{:}\, A.\ h_{\underline{A}})_{\underline{A \supset A}} \quad \triangleright \quad \begin{array}{l} \text{Assume } A\ (h) \\ \quad\quad \text{By } (h) \text{ we have } \underline{A} \\ \text{We have proved } \underline{A \supset A} \end{array}$$

Textual variants In fact, the proof above presumes that A is a proposition. The complete proof, including this assumption, is $\lambda A{:}\,Prop.\ \lambda h{:}\,A.\ h$, where *Prop* is a distinguished type that represents propositions. So we use the word "Assume" because the type of A is *Prop*:

$$(\lambda A{:}\,Prop_{\underline{Type}}.(\lambda h{:}\,A_{\underline{Prop}}.\ h_A)_{A\supset A})_{\forall A:Prop.A\supset A} \ \triangleright$$

<u>Let</u> A : *Prop*
<u>Assume</u> A (h)
  By (h) we have A
We have proved $A \supset A$
We have proved $\forall A{:}\,Prop.\,A \supset A$

while we use "Let" for the outer λ-abstraction because the type of *Prop* is *Type*. In other circumstances, we will want to leave the λ-notation as it is, when the λ-abstraction denotes an authentic function. For example the successor function:

$$(\lambda h{:}\,nat.\ h+1)_{\mathrm{N}\to\mathrm{N}}$$

This discussion shows that we need to use fairly different textual variants for a given construct (such as the λ-abstraction) depending on the type of its arguments. Type expressions are another example. A most general (dependent) product will be written $\Pi x{:}\,P.\ Q$. If Q is of type *Prop*, we write more conventionally $\forall x{:}\,P.\ Q$. If the product is non-dependent, i.e. x does not occur free in Q, one notes $P \to Q$. When additionally P and Q are of type *Prop*, we prefer $P \supset Q$.

Consider now an application MN and examine the type of M. We understand $M_{\forall x:P.Q}\ N$ as a specialization, while $M_{P\supset Q}\ N$ is an instance of *modus ponens* and $M_{P\to Q}\ N$ is a function application. The typed λ-term is a very compact notation, but examining the types brings out the logical structure of the proof.

To sum up, we propose a first attempt at transduction rules. In these rules, τ is assumed of type *Prop* and metavariable l may denote a list of bound variables. Expressions that don't match any rule are left unchanged.

Rules for abstraction

$(\lambda l{:}\,A_{Type}.\ M)_\tau$	$\triangleright$	Let l: A M We have proved τ
$(\lambda h{:}\,A_{Prop}.\ M)_\tau$	$\triangleright$	Assume A (h) M We have proved τ
$(\lambda x{:}\,A_{Set}.\ M)_\tau$	$\triangleright$	Consider an arbitrary x in A M We have τ, since x is arbitrary

Rules for application

$(M_{\forall x:P.\,Q}\, N)_\tau$ ▷	M In particular τ
$(M_{P\supset Q}\, N)_\tau$ ▷	- N - M We deduce τ

For identifiers, we make a slight distinction between assumptions appearing in the proof term (metavariable h) and theorems (metavariable T) found in the context.

Rules for identifiers

h_τ ▷	By h we have τ
T_τ ▷	Using T we get τ

Example:
With the rules above, the proof (S) $\lambda A, B, C{:}\,Prop.\ \lambda h{:}\,A \supset B \supset C.\ \lambda h_0{:}\,A \supset B.\ \lambda h_1{:}\,A.\ (h\,h_1(h_0\,h_1))$ reads

Let $A, B, C{:}\,Prop$
Assume $A \supset B \supset C$ (h)
    Assume $A \supset B$ (h_0)
        Assume A (h_1)
                -By h_1 we have A
                -By h_0 we have $A \supset B$
              -We deduce B
                -By h_1 we have A
                -By h we have $A \supset B \supset C$
              -We deduce $B \supset C$
            We deduce C
        We have proved $A \supset C$
    We have proved $(A \supset B) \supset A \supset C$
We have proved $(A \supset B \supset C) \supset (A \supset B) \supset A \supset C$
We have proved $\forall A, B, C{:}\,Prop.\ (A \supset B \supset C) \supset (A \supset B) \supset A \supset C$

While the text above is very clear, it is also painfully lengthy. One reason is immediately apparent. Assumptions, corresponding to abstractions, are introduced (and discharged) one by one. Indeed, if the outer λ-abstraction had been decomposed as a succession of elementary bindings, the result might be even more verbose.

Repeated constructs To improve the density of the proof text, we should

- ignore inessential intermediate results
- reduce the drift toward the right margin of the page caused by repeated indentations.

An effective technique of achieving this is to rewrite the rules for abstraction and application in the case where a given rule is being applied repeatedly.

Rules for iterated abstraction

$(\lambda l^1 : A^1_{Type} \cdots \lambda l^k : A^k_{Type} . M)_\tau$	$\triangleright$	Let $l^1 : A^1$ $\vdots$ Let $l^k : A^k$ M We have proved τ
$(\lambda h^1 : A^1_{Prop} \cdots \lambda h^k : A^k_{Prop}\ M)_\tau$	$\triangleright$	Assume A^1 (h^1) and ... and A^k (h^k) M We have proved τ
$(\lambda x^1 : A^1_{Set} \cdots \lambda x^k : A^k_{Set}\ M)_\tau$	$\triangleright$	Choose arbitrarily x^1 in A^1, ..., x^k in A^k M Thus we have τ

Rules for iterated application

$(M_{\forall x^1 : P^1 . \cdots \forall x^k : P^k . Q}\ N^1 \cdots N^k)_\tau$	$\triangleright$	M In particular τ
$(M_{P^1 \supset \cdots \supset P^k \supset Q}\ N^1 \cdots N^k)_\tau$	$\triangleright$	-N^1 $\vdots$ -N^k -M We deduce τ

Remarks:

1. The second rule of abstraction uses the connective "and" in a non-commutative manner: due to the dependence between types, a later assumption might refer to an earlier one [Ranta94]. The third rule uses the comma in the same way.
2. The rules for iterated application are a refinement of the familiar rule for representing curried applications with less parentheses:

   $$((\cdots(M_{P^1 \to \cdots P^k \to Q}\ N^1)\cdots)N^k) \text{ is rather written } (M_{P^1 \to \cdots P^k \to Q}\ N^1 \cdots N^k)$$

3. Due to the rule of repeated implications, subproofs will occur in a more natural order.
4. The format of the rules involving implications uses a dash "-", to emphasize the argument structure in the deduction. Because our style is generally postfix, if we write this symbol at the beginning of a subproof, we might have inaesthetic sequences of dashes. To avoid this, we require the dash to be printed in front of the *conclusion* of a subproof. The corresponding subproof itself, appearing above, is indented slightly to the right.

References to assumptions To alleviate the presentation further, we choose a shorter rule to refer to assumptions introduced in the proof:

$$h_{\tau} \triangleright \text{ We have } h$$

Additionally, we make a special case for (iterated) applications where the operator is a variable f (be it a local hypothesis or a theorem) and for applications where all operands are variables.

To progress further, we consider theorems related with the method used for defining new symbols, including the usual logical connectives.

Introduction theorems New symbols are defined, possibly inductively, with the help of introduction theorems. Our concern being the proof structure, the only introduction that interests us are those which create propositions. Such introduction theorems are of the form:

$$\mathtt{C}intro : \forall x^1 \colon A^1 . \ldots \forall x^n \colon A^n . \Phi^1 \supset \cdots \supset \Phi^i \supset (\mathtt{C}\, u^1 \ldots u^k)$$

where the subterms u^j are bound variables. For example, the disjunction is defined by two introduction theorems:

$$\vee intro_l \colon \forall A, B \colon Prop.\, A \supset A \vee B$$
$$\vee intro_r \colon \forall A, B \colon Prop.\, B \supset A \vee B$$

and the transitive closure R^* by the theorems:

$$R^*_0 \colon \forall U \colon Type. \forall R \colon (Relation\ U). \forall x \colon U. (R^*\ U\ R\ x\ x)$$
$$R^*_n \colon \forall U \colon Type. \forall R \colon (Relation\ U). \forall x,y,z \colon U. (R\ x\ y) \supset (R^*\ U\ R\ y\ z) \supset (R^*\ U\ R\ x\ z)$$

Consider now a proof involving such theorems:

$$\lambda A, B, C \colon Prop.\, \lambda h \colon A.\, (\vee intro_r\ B\ (A \vee C)\ (\vee intro_l\ A\ C\ h))$$

Let A, B, C: *Prop*
Assume A (h)
- Applying $\vee intro_l$ with h we get $A \vee C$
Applying $\vee intro_r$ we get $B \vee (A \vee C)$
We have proved $A \supset B \vee (A \vee C)$
We have proved $\forall A, B, C \colon Prop.\, A \supset B \vee (A \vee C)$

Phrasing the proof as using two theorems is quite pedantic, since these facts constitute the definition of $\vee$. To make this distinction more apparent, we propose the following presentation rules:

Rules for introduction theorems

	$(\mathsf{C}intro\, M^1 \cdots M^n\, N^1 \cdots N^i)_\tau$	▷	$-N^1$ $\vdots$ $-N^i$ So by definition of C we have τ
$i = 0$	$(\mathsf{C}intro\, M^1 \cdots M^n)_\tau$	▷	By definition of C we have τ
$i = 1$	$(\mathsf{C}intro\, M^1 \cdots M^n\, N)_\tau$	▷	N By definition of C we have τ

Remarks:

1. Just as we did in the case of applications, we can use textual variants when applying an introduction theorem to variables. Such rules are easy to define, and we leave them to the imagination of the reader.
2. It is an advantage not to see the names of the introduction theorems in the proof text, because we know of no good principle of naming for them.
3. We could simply say "By definition", rather than "By definition of C" since the symbol C is the leading operator of τ.

Elimination theorems Elimination theorems express how one uses new symbols. Elimination theorems have the following forms, depending on whether the symbol C is used to create a proposition or not:

$$\mathsf{C}elim : \forall x^1{:}\, A^1. \ldots \forall x^n{:}\, A^n. \Phi^1 \supset \cdots \supset \Phi^i \supset (\mathsf{C}\, u^1 \ldots u^k) \supset B$$
$$\mathsf{C}elim : \forall x^1{:}\, A^1. \ldots \forall x^n{:}\, A^n. \Phi^1 \supset \cdots \supset \Phi^i \supset \forall x{:}\, (\mathsf{C}\, u^1 \ldots u^k). B$$

The elimination of disjunction is an example of elimination of the first kind:

$$\vee elim{:}\ \forall A, B, P{:}\, Prop. (A \supset P) \supset (B \supset P) \supset (A \vee B) \supset P$$

and the principle of induction over the naturals is an example for the second kind:

$$\mathrm{N}elim{:} \forall P{:}\, \mathrm{N} \rightarrow Prop. (P\, 0) \supset (\forall n{:}\, \mathrm{N}. (P\, n) \supset (P\, (\mathrm{Suc}\, n))) \supset \forall n{:}\, \mathrm{N}. (P\, n)$$

Note that both i and k may be 0, as in the following:

$$\bot elim{:}\ \forall P{:}\, Prop. \bot \supset P$$

Consider now the following proof of the commutativity of disjunction:

$\lambda A, B{:}\, Prop. \lambda h{:}\, A \vee B.$
$\quad (\vee elim\ A\ B\ (B \vee A)\ (\lambda i : A.\ \vee intro_r\ B\ A\ i)\ (\lambda j : B.\ \vee intro_l\ B\ A\ j)\ h)$

Let A, B : *Prop*
Assume $A \vee B$ (h)
    Assume A (i)
        From i and the definition of $\vee$, we have $B \vee A$
    -We have proved $A \supset B \vee A$
    Assume B (j)
        From j and the definition of $\vee$, we have $B \vee A$
    -We have proved $B \supset B \vee A$
    -We have h
    Applying $\vee elim$ we get $B \vee A$
We have proved $A \vee B \supset B \vee A$
We have proved $\forall A, B{:}\, Prop.\, A \vee B \supset B \vee A$

The layout of the arguments in repeated applications has the unhappy consequence that the argument to eliminate appears *last*. But this argument is the one that drives the reasoning. It is much more appropriate to give it first, and then to announce the possible cases. So we propose the following rules for elimination theorems of the first kind:

$(\mathtt{C}elim\, M^1 \cdots M^n N^1 \cdots N^i P)_\tau$	$\triangleright$	P
		Therefore by definition of $\mathtt{C}$, to prove τ we have i cases:
		Case$_1$:
		    N^1
		⋮
		Case$_i$:
		    N^i
		So we have τ

$i = 0$ $(\mathtt{C}elim\, M^1 \cdots M^n P)_\tau$	$\triangleright$	P, by definition of $\mathtt{C}$ there is a contradiction So we can assert τ
$i = 1$ $(\mathtt{C}elim\, M^1 \cdots M^n\, N\, P)_\tau$	$\triangleright$	P Therefore by definition of $\mathtt{C}$ to prove τ N So we have τ

With these rules, we obtain results that are longer, but more perspicuous:

Let A, B : *Prop*
Assume $A \vee B$ (h)
    We have h
    Therefore by definition of $\vee$ to prove $B \vee A$, we have two cases:
    Case$_1$:
        Assume A (i)
            From i and the definition of $\vee$, we have $B \vee A$
        We have proved $A \supset B \vee A$
    Case$_2$:
        Assume B (j)
            From j and the definition of $\vee$, we have $B \vee A$
        We have proved $B \supset B \vee A$
    So we have $B \vee A$
We have proved $A \vee B \supset B \vee A$
We have proved $\forall A, B{:}\, Prop.\, A \vee B \supset B \vee A$

Slightly different rules are necessary for elimination theorems of the second kind.

$(\texttt{C}elim\ M^1 \cdots M^n\ N^1 \cdots N^i)_\tau$	▷	By definition of C, to prove τ we have i cases: Case$_1$: $\quad N^1$ $\vdots$ Case$_i$: $\quad N^i$ So we have τ
$i = 0$ $(\texttt{C}elim\ M^1 \cdots M^n)_\tau$	▷	C is empty, so τ
$i = 1$ $(\texttt{C}elim\ M^1 \cdots M^n\ N)_\tau$	▷	By definition of C to prove τ, $\quad N$ So we have τ

To illustrate these rules, we look at a little inductive proof in Peano arithmetic. Assume $\leq$ is defined inductively by the following introduction theorems:

$$\leq\!intro_b\colon \forall n\colon \mathbb{N}.\, n \leq n \quad \text{and} \quad \leq\!intro_r\colon \forall n, m\colon \mathbb{N}.\, n \leq m \supset n \leq (\mathrm{Suc}\, m)$$

Here is a proof of the fact $\forall m\colon \mathbb{N}.\, (0 \leq m)$:

$$(\mathbb{N}elim\ (\lambda x.\, 0 \leq x)\ (\leq\!intro_b\, 0)\ \ \lambda m\colon \mathbb{N}.\, \lambda h\colon (0 \leq m).\, (\leq intro_r\, 0\, m\, h))$$

and the way it is layed out now, without any rule that is specific of $\mathbb{N}elim$:

By definition of $\mathbb{N}$ to prove $\forall n\colon \mathbb{N}.\, 0 \leq n$, we have two cases:
Case$_1$:
    By definition of $\leq$ we have $0 \leq 0$
Case$_2$:
    Let $m\colon \mathbb{N}$
    Assume $0 \leq m$ (h)
        From h and the definition of $\leq$, we have $0 \leq (\mathrm{Suc}\, m)$
    We have proved $0 \leq m \supset 0 \leq (\mathrm{Suc}\, m)$
    We have proved $\forall m\colon \mathbb{N}.\, 0 \leq m \supset 0 \leq (\mathrm{Suc}\, m)$
So we have $\forall n\colon \mathbb{N}.\, 0 \leq n$

Omitting the conclusion of a proof The two proofs above show that the application of elimination theorems accounts elegantly for two forms of reasoning:

- the commutativity of $\vee$ is proved *by cases*,
- the theorem on $\leq$ uses a proof *by induction.*

If the proof is by cases, the constructor that is eliminated doesn't occur in the contexts of the subordinate cases; otherwise the proof is by induction. But in both situations, the subordinate proofs introduce a context that they will finally discharge. Being told explicitly about the discharge of this local context seems inessential. In both examples above, it seems we could omit statements such as

"We have proved $A \supset B \vee A$" or "We have proved $\forall m{:}\, \mathrm{N}.\, 0 \leq m \supset 0 \leq (\mathrm{Suc}\, m)$" without jeopardizing the clarity of the text. We have to be careful, however, not to lose the line of reasoning. When arguing by induction, we feel it useful to recall the goal of each subordinate case.

To specify this, we define a special (transducer) context called $*$ where the transduction *forgets* the conclusion of a series of λ-abstractions. As a bonus, we can use the $*$ context at the top level of the proof object and announce the statement of the theorem to prove. Revisiting one of our examples, we get improved explanations:

Theorem: $\forall A, B{:}\, \mathit{Prop}.\, A \vee B \supset B \vee A$
Let $A, B : \mathit{Prop}$
Assume $A \vee B$ (h)
We have h
Therefore by definition of $\vee$ to prove $B \vee A$, we have two cases:
Case_1:
    Assume A (i)
    From i and the definition of $\vee$, we have $B \vee A$
Case_2:
    Assume B (j)
    From j and the definition of $\vee$, we have $B \vee A$
So we have $B \vee A$

Remarks:

1. Cases may be imbedded within cases. It is preferable to number cases absolutely, within a given proof, using a Dewey notation like "Case 1.2.1:".
2. Some axioms or theorems are of such frequent and quasi-implicit use that they do deserve specific linguistic wording. This is the case for theorems concerning usual logical constants, or the theorems used when reasoning classically, such as the axiom of the Excluded Middle. In fact, it is useful, for any given theory, to be able to associate specific wording and layout to any axiom or theorem. So, as an example we show in Fig. 1 and Fig. 2 rules we propose for the logical connectives. Notice that some rules demand their arguments to be λ-abstractions. However, the applicability of these rules is not reduced if we accept the η rule: $M \equiv \lambda x.\, M x$ (x does not appear free in M).

2.2 Further optimizations

Looking at the text obtained for the commutativity of $\vee$, we see that one aspect is less than idiomatic: we often give a name to an assumption, like h, i or j, although this assumption is used only once, and immediately. It is easy to check, on the body of a λ-term, that an assumption occurs only once. In that case, we can omit the name of the assumption and replace it, at its unique occurrence, by "the assumption". It is bit more tricky to check that its use occurs *immediately after* the assumption, because we have to track that fact through all of our rules.

Assume we have been able to establish that an occurrence of h is unique and appears immediately after the introduction of h. Then we may drop the reference to h entirely. Also the text for $(\lambda h\colon A.\ h)_\tau$ becomes now too terse, so we write "Trivially τ". And for $(\lambda h\colon A.\ M)_\tau$ where h doesn't occur in M we prefer:

M
A fortiori τ

A λ-term where x occurs only once and has the form;

$$\lambda x\colon (\mathtt{C}\, u^1 \ldots u^k).(\mathit{Celim}\ M^1 \cdots M^n\ N^1 \cdots N^i\ x)$$

also deserves special treatment. Other possibilities, based on an analysis of dependence are approached in [Coscoy94]. Figure 3 shows the kind of results one obtains.

3 Implementation

The rules discussed in this paper have been implemented ([Massol93, Coscoy94]) as a back end for the Coq system [Coq91]. The implementation is faithful to the rules in this paper and works in two parts.

1. First, a program implemented within the Coq proof engine traverses the proof, recognizes patterns and equality of some subterms, checks occurrences of bound variables, decides what type annotations will appear in the proof text and computes Dewey numbers for all cases.
2. Then, an annotated proof term is produced as input for a rule-driven pretty-printer. This tool applies the rules given in this paper, taking into account font selection and the limited width of the page.

4 Conclusion

The transducer is part of the freely available user-interface package [CtCoq94]. User reaction confirms the usefulness of proof texts and shows that our approach remains practical for moderate size proofs. Clearly though, much work remains to obtain proof texts that are both compact and perspicuous. Most attractive is the fact that the system takes automatically into account the user's own inductive definitions. The use of normal definitions is not even recorded in the Coq (V5.8) proof object, so some proofs look mysteriously quick. As a consequence, one may develop a style of axiomatisation where inductive definitions are prevalent.

The exact wording proposed by our transduction rules may not be to everyone's taste. For example, instead of "Assume A", one may prefer "Assume A holds". This is very easy to change, and doesn't alter the basic argument of this paper: with a careful analysis of types, and an understanding of the role of specific theorems, it is possible to extract a good text from a typed λ-term. We have presented the basic elements of this analysis, but it is likely that it can be refined much further. We have left object-language formulae alone, but the work of [Ranta94] shows that much can be done in this area. Indeed a first area where this would be pleasant is for expressions of type *Type*, that may be rendered as substantives ([deBruijn87]). For other expressions, in the context of proof assistants, it seems more important to be able to introduce specialized mathematical notation for a given theory. We plan to implement a mechanism that loads pretty-printing rules for a given theory when the proof engine adds this theory to the current context. Beautifully, type theory allows one to load specific rules for the basic theorems of that theory with exactly the same mechanism.

Should the proof still be to verbose, it is possible to ask interactively for eliding subterms. We plan to keep a connection with the proof script, which might help in automating elision. But we will not try to discover ways to make the proof more abstract. In our view, it is the responsability of the user to introduce the right mathematical abstractions in the proof development, and if the user is unhappy with the proof text in that respect, then the proof itself should be improved.

Finally, users often ask whether such texts could also be used as input. We have not considered the issue, although we try to use distinctive phrases. The idea seems natural but non trivial. However, we have positive experience ([Théry94, Alf93]) with using partial proof text rather than sequents to represent the current state of the proof assistant. Such an approach introduces an entirely different concern in the production of text: from one step to the next, unless backtracking is specifically required, the text should grow monotonically for the user to feel comfortable. This precludes certain global optimizations, which may then be applied only when the proof (or a subordinate proof) is complete.

References

[Alf93] L. Magnusson, B. Nordström, *The ALF proof editor and its proof engine*, in Proceedings of the 1993 Types Worshop, Nijmegen, LNCS 806, 1994.

[Chester76] D. Chester, *The translation of Formal Proofs into English*, Artificial Intelligence 7 (1976), 261-278, 1976.

[Cohn88] A. Cohn, *Proof Accounts in HOL*, unpublished draft.

[Coscoy94] Y. Coscoy, *Traductions de preuves en langage naturel pour le systeme Coq*, Rapport de Stage, Ecole Polytechnique, 1994.

[Coq91] G. Dowek, A. Felty, H. Herbelin, G. Huet, C. Paulin-Mohring, B. Werner, *The Coq Proof Assistant User's Guide*, INRIA Technical Report no. 134, 1991.

[CtCoq94] Y. Bertot, *the CtCoq Interface*, available by ftp at `babar.inria.fr:/pub/centaur/ctcoq`.

[deBruijn87] N.G. de Bruijn, *The Mathematical Vernacular, a language for Mathematics with typed sets* in Proceedings from the Workshop on Programming Logic, P. Dybjer *et al.*, Programming Methodology Group, Volume 37, University of Göteborg, 1987.

[Ep93] A. Edgar, F.J. Pelletier, *Natural language explanation of Natural Deduction proofs*, in Proceedings of the First Conference of the Pacific Association for Computational Linguistics, Simon Fraser University, 1993.

[Felty88] A. Felty, *Proof explanation and revision*, Technical Report, University of Pennsylvania MS-CIS-88-17, 1988.

[Gentzen69] G. Gentzen, *Investigations into logical deduction*, in "The collected papers of Gerhard Gentzen", M.E. Szabo (Ed.), pp. 69-131, North Holland, 1969.

[Hol92] M.J.C. Gordon, T.F. Melham, *HOL: a proof generating system for higher-order logic*, Cambridge University Press, 1992.

[Huang94] X. Huang, *Reconstructing Proofs at the Assertion Level*, in Proceedings of CADE-12, Nancy, LNAI 814, Springer Verlag, 1994.

[Lego92] Z. Luo, R. Pollack, *LEGO proof development system : user's manual*, Technical Report, University of Edinburgh, 1992.

[Massol93] A. Massol, *Présentation de Preuves en Langue Naturelle pour le Système Coq*, Rapport de DEA, Université de Nice-Sophia-Antipolis, 1993.

[Ranta94] A. Ranta, *Type Theory and the Informal Language of Mathematics*, in Proceedings of the 1993 Types Worshop, Nijmegen, LNCS 806, 1994.

[Théry94] L. Théry, *Une méthode distribuée de création d'interfaces et ses applications aux démonstrateurs de théorèmes*, PhD, Université Denis Diderot, Paris, 1994.

$\underline{\wedge intro}: \forall A, B: Prop.\, A \supset B \supset A \wedge B$

$(\wedge intro\; A\, B\, P\, Q)_\tau$ ▷
- P
- Q

Altogether we have τ

$\underline{\vee intro_l}: \forall A, B: Prop.\, A \supset A \vee B$

$(\vee intro_l\; A\, B\, P)_\tau$ ▷ P

Obviously τ

$\underline{\vee intro_r}: \forall A, B: Prop.\, B \supset A \vee B$

$(\vee intro_r\; A\, B\, P)_\tau$ ▷ P

Obviously τ

$\underline{\exists intro}: \forall A: Set. \forall P: A \rightarrow Prop. \forall y: A.\, (P y) \supset \exists x: A.\, (P\, x)$

$(\exists intro\; A\, P\, c\, Q)_\tau$ ▷ Q

The element c proves τ

$\underline{\neg intro}: \forall A: Prop.\, (A \supset \bot) \supset \neg A$

$(\neg intro\; A\, (\lambda i: A.P))_\tau$ ▷ Suppose $A\,(i)$

P

So we have $\neg A$ (negating i)

$\underline{\wedge elim}: \forall A, B, C: Prop.\, (A \supset B \supset C) \supset (A \wedge B) \supset C$

$(\wedge elim\; A\, B\; C\, (\lambda i: A.\, \lambda j: B.\, P)\; Q)_\tau$ ▷ Q

We know A (i) and B (j)

P

Fig. 1. Text for familiar logical connectives

$\underline{\vee elim}: \forall A, B, C: Prop. (A \supset C) \supset (B \supset C) \supset (A \vee B) \supset C$

		R
		So we have two cases
		Case$_1$:
		Suppose A (i)
$(\vee elim\, A\, B\, C\, (\lambda i: A. P)\, (\lambda j: B. Q)\, R)_\tau$	$\triangleright$	P
		Case$_2$:
		Suppose B (j)
		Q
		We have τ in both cases

$\underline{\exists elim}: \forall A: Set. \forall P: A \rightarrow Prop. \forall C: Prop. (\forall y: A. (P\, y) \supset C) \supset (\exists x: (P\, x)) \supset C$

		R
		So there exists a y in A such that $T(i)$
$(\exists elim\, A\, P\, C\, (\lambda y: A. \lambda i: T. Q)\, R)_\tau$	$\triangleright$	Q
		We have τ (independently of y)

$\underline{\neg elim}: \forall A: Prop. A \supset \neg A \supset \bot$

		- P
$(\neg elim\, A\, P\, Q)_\tau$	$\triangleright$	- Q
		We have a contradiction

$\underline{\bot elim}: \forall P: Prop. \bot \supset P$

		Q
$(\bot elim\, P\, Q)_\tau$	$\triangleright$	So we can assert τ

Classical Reasoning:

$\underline{ExMid}: \forall A: Prop. A \vee \neg A$

		If we have A (i)
		P
		Otherwise $\neg A$ (j)
$(\vee elim\, A\, \neg A\, C\, (\lambda i{:}A.P)(\lambda j{:}\neg A.Q)(ExMid\, A))_\tau$	$\triangleright$	Q
		So (classically) τ

$\underline{Contrad}: \forall A: Prop. (\neg A \supset \bot) \supset A$

		Suppose $\neg A$ (i)
$(Contrad\, A\, (\lambda i: A.P))_\tau$	$\triangleright$	P
		So (classically) we have A (negating i)

$\underline{ClassicQuantif}: \forall A: Set. \forall P: A \rightarrow Prop. \neg(\forall x: A. (P\, x)) \supset (\exists x: A. \neg(P\, x))$

$(ClassicQuantif\, M)_\tau$	$\triangleright$	M
		Therefore (classically) we have τ

Fig. 2. Text for familiar logical connectives and classical reasoning

Theorem: ***præclarum***
Statement
$\forall x, y, z, t\colon Prop.\,(x \supset z) \wedge (y \supset t) \supset x \wedge y \supset z \wedge t$
Proof
Let x, y, z, t be propositions
Assume we know $x \supset z$ (i) and $y \supset t$ (j)
Assume we know x (k) and y (l)
- Using i with k we deduce z
- Using j with l we deduce t
Altogether we have $z \wedge t$

Theorem: ***Symmetric and transitive relation is nearly reflexive.***
Statement
$\forall A\colon Set.\,\forall R\colon A \to A \to Prop.\,(\forall x, y\colon A.\,(R\,x\,y) \supset (R\,y\,x)) \supset$
$(\forall x, y, z\colon A,\,(R\,x\,y) \supset (R\,y\,z) \supset (R\,x\,z)) \supset \forall x\colon A.\,(\exists y_0\colon A.\,(R\,x\,y_0)) \supset (R\,x\,x)$
Proof
Let A be a set
Let $R\colon A \to A \to Prop$
Assume $\forall x, y\colon A.\,(R\,x\,y) \supset (R\,y\,x)$ (*symmetry*)
and $\forall x, y, z\colon A.\,(R\,x\,y) \supset (R\,y\,z) \supset (R\,x\,z)$ (*transitivity*)
Consider an arbitrary x in A
Assume there exists an element y_0 in A such that $R\,x\,y_0$
Using *symmetry* we deduce $R\,y_0\,x$
Using *transitivity* we deduce $R\,x\,x$

Theorem: ***Drinker's Principle***
Statement
$\forall A\colon Set.\,\forall P\colon A \to Prop.\,\forall y\colon A.\exists x\colon A.\,(P\,x) \supset \forall z\colon A.\,(P\,z).$
Proof
Let A be a set
Let $P\colon A \to Prop$
Consider an arbitrary y in A
If we have $\forall z\colon A.\,(P\,z)$
A fortiori $(P\,y) \supset \forall z\colon A.\,(P\,z)$
The element y proves $\exists x\colon A.\,(P\,x) \supset \forall z\colon A.\,(Pz)$
Otherwise $\neg(\forall z\colon A.\,(P\,z))$
Therefore (classically) we have $\exists z\colon A.\,\neg(P\,z)$
So there exists a t in A such that $\neg(P\,t)$ (i)
Assume $(P\,t)$
From i, we have a contradiction
So we can assert $\forall z\colon A.\,(P\,z)$
We have proved $(P\,t) \supset \forall z\colon A.\,(P\,z)$
The element t proves $\exists x\colon A.\,(P\,x) \supset \forall z\colon A.\,(P\,z)$
We have $\exists x\colon A.\,(P\,x) \supset \forall z\colon A.\,(P\,z)$ (independently of t)
So (classically) $\exists x\colon A.\,(P\,x) \supset \forall z\colon A.\,(P\,z)$

Fig. 3. Three examples

Higher-Order Abstract Syntax in Coq

Joëlle Despeyroux[1], Amy Felty[2], André Hirschowitz[3]

[1] INRIA, Sophia-Antipolis, 2004 Route des Lucioles, F-06565 Valbonne Cedex, France
joelle.despeyroux@sophia.inria.fr
[2] AT&T Bell Laboratories, 600 Mountain Ave., Murray Hill, NJ 07974, USA
felty@research.att.com
[3] CNRS URA 168, University of Nice, 06108 Nice Cedex 2, France
andre.hirschowitz@sophia.inria.fr

Abstract. The terms of the simply-typed λ-calculus can be used to express the *higher-order abstract syntax* of objects such as logical formulas, proofs, and programs. Support for the manipulation of such objects is provided in several programming languages (e.g. λProlog, Elf). Such languages also provide embedded implication, a tool which is widely used for expressing *hypothetical judgments* in natural deduction. In this paper, we show how a restricted form of second-order syntax and embedded implication can be used together with induction in the Coq Proof Development system. We specify typing rules and evaluation for a simple functional language containing only function abstraction and application, and we fully formalize a proof of type soundness in the system. One difficulty we encountered is that expressing the higher-order syntax of an object-language as an inductive type in Coq generates a class of terms that contains more than just those that directly represent objects in the language. We overcome this difficulty by defining a predicate in Coq that holds only for those terms that correspond to programs. We use this predicate to express and prove the adequacy for our syntax.

1 Introduction

Abstraction in the λ-calculus can be used to represent various binding operators such as quantification in formulas or abstraction in functional programs. By making use of the implementation of the λ-calculus in programming languages that support it, the programmer is freed from such concerns as implementing substitution algorithms and correctly handling the scope and names of bound variables. Many examples exist and illustrate the usefulness of higher-order syntax in programming and theorem proving. For example, the Logical Framework (LF) [10] provides a uniform framework for specifying a large class of languages and inference systems. A variety of logics and typed λ-calculi have been specified using it [2]. Theorem provers for several of these logics have been specified and implemented in the logic programming language λProlog, which provides support for the manipulation of objects expressed in higher-order syntax [5, 6]. The λProlog language has also been used to specify program evaluators and transformers [7, 8]. Elf, a logic programming implementation of LF, has been used to specify and verify properties of inference systems [11, 14] and compilers [9]. In many of these examples, embedded implication (*i.e.*, an implication on the left of an implication) is used, providing an elegant mechanism for handling scoping of sets of assumptions during proof construction, or of contexts during program evaluation.

Higher-order syntax and hypothetical judgments can be expressed in many theorem provers. However, there is little experience using them in proofs. In this paper, we illustrate the use of a restricted form of second-order syntax and embedded implication in the Coq Proof Development system [4] by defining typing rules and evaluation for a simple functional language containing only function abstraction and application. We prove type soundness for this language, *i.e.*, that evaluating a term preserves its type. By using this syntax much of the details of proofs, in particular those concerning substitution and names and scopes of variables, are greatly simplified. In addition, this work represents a step towards the goal of providing support for higher-order abstract syntax and allowing both programming and program verification in a unified setting.

We have chosen the Coq Proof Development System because it implements the Calculus of Inductive Constructions (CIC) [13], a type theory which provides a notion of inductive definitions. Defining a type inductively provides a principle of structural induction and an operator for defining functions recursively over the type. These operators can be used directly and there are no requirements placed on the user to prove their correctness. However, in order to use the built-in support for induction, we had to overcome two obstacles.

The first obstacle is that negative occurrences of the type being defined are not allowed in inductive definitions. If L is the type of terms of the language being defined, the usual way to express the higher-order syntax of an abstraction operator such as function abstraction in our example is to introduce a constant such as *Lam* and assign it the type $(L \rightarrow L) \rightarrow L$. That is, *Lam* takes one argument of functional type. Thus function abstraction in the object-language is expressed using λ-abstraction in the meta-language. As a result, bound variables in the object-language are identified with bound variables in the meta-language. In inductive types in Coq, negative occurrences such as the first occurrence of L in the above type are disallowed. As in [3], we get around this problem by introducing a separate type *var* for variables and giving *Lam* the type $(var \rightarrow L) \rightarrow L$. We must then add a constructor for injecting variables into terms of L. Thus, in our restricted form of higher-order syntax, we still define function abstraction using λ-abstraction in Coq and it is still the case that α-convertible terms in our object-language map to α-convertible terms in Coq, but we cannot directly define object-level substitution using Coq's β-reduction. Instead we define substitution as an inductive predicate. Its definition is simple and similar to the one found in [12].

The second obstacle is that defining the type L as an inductive type with the usual constructors for application and abstraction plus the special constructor for variables gives a set of terms in Coq that is "too large". That is, there are more terms in L than those that correspond to objects in the object-language. To solve this problem for our functional language, we succeeded in the task of defining an object-level predicate, which we call *valid*, that is true only for those terms that correspond to programs. This predicate, however, does allow some terms that do not directly represent programs, namely, those that are extensionally equal to terms that do. We define extensional equality for the type L in Coq, and consider that each term of the object-language is in fact represented by an equivalence class of terms determined by this equality relation.

This work extends two related projects where higher-order syntax is used in formal proof. In [9], Elf is used in compiler verification. In Elf, there is no quantification over predicates, and thus induction principles cannot be expressed inside the language. As a result, much of the detail of proofs must be done outside the system. Tools such as schema-checking [14] have been developed to help with this task. In [3], a different approach to higher-order syntax in Coq is adopted. There, like here, a separate type *var* for variables is introduced and *Lam* is defined as above. However, instead of directly representing (closed) terms of the object-language by terms of type L, closed and open terms of the object-language are implemented together as functions from lists of arguments (of type L) to terms of type L. Semantics are given on these functional terms. A predicate on these terms is introduced which defines valid terms to be the expected ones. Induction over terms is carried out by using the induction principle for this predicate. Here, we instead define typing and evaluation directly on (closed) terms of type L. We succeed in reasoning about them by directly using the induction principles generated by their definitions.

The rest of the paper is organized as follows. In Section 2, we define the higher-order syntax of our functional language in Coq. In Sections 3, 4, 5, and 6, we show that our syntax adequately represents our object-language. In Section 3, we give a definition of our object-language in LF for which adequacy has already been proved, and in Section 4, we express a translation between the LF and Coq syntaxes. In Section 5, we explain which kind of terms should be ruled out and the need for extensionality. In Section 6, we implement the predicate *valid* which selects terms that represent terms of the object-language and prove the correctness of this implementation. In Section 7, we define and implement substitution in Coq and prove its correctness. Although the definitions of *valid* and substitution are simple, finding them was one of the main challenges of this work. In defining them, and in the proofs in this paper, we succeeded in avoiding any need to refer to variable names or occurrences of variables in terms, or a notion of *fresh* variables not occurring in terms. Section 8 presents the Coq definitions for typing and evaluation in the object-language, which specify the usual natural semantics style presentation of these judgments, and discusses the Coq proof of type soundness. In Section 9, we conclude and discuss future work.

Notation. In proving the adequacy of our representation and correctness of substitution, we will often use notation close to the syntax of Coq. To make the distinction, for those definitions or statements not intended to be Coq or LF definitions, we will use (*) as a superscript on Coq keywords.

2 Specifying Provisional Syntax

We assume the reader is familiar with the Calculus of Inductive Constructions. We simply note the notation used in this paper, much of which is taken from the Coq system. Let M and N represent terms of CIC. The syntax of terms is as follows.

$$\begin{gathered} Prop \mid Set \mid Type \mid MN \mid \lambda x : M.N \mid \forall x : M.N \mid M \rightarrow N \mid \\ M \wedge N \mid M \vee N \mid \exists x : M.N \mid \neg M \mid M = N \mid Rec\, M\, N \mid \\ Case\ x : M\ of\ M_1 \Rightarrow N_1, \ldots, M_n \Rightarrow N_n \end{gathered}$$

Here $\forall$ is the dependent type constructor and the arrow ($\rightarrow$) is the usual abbreviation when the bound variable does not occur in the body. Of the remaining constants, *Prop*, *Set*, *Type*, λ, *Rec*, and *Case* are primitive, while the others are defined. *Rec* and *Case* are the operators for defining inductive (*Case*) and recursive (*Rec*) functions over inductive types. Equality on *Set* ($=$) is Leibnitz equality.

A parameter is introduced using the Parameter keyword and inductive types are introduced with an Inductive Set or Inductive Definition declaration where each constructor is given with its type, separated by vertical bars.

We specify a provisional syntax for our object-language, the λ-calculus, by introducing a type for variables and defining terms and types inductively.

Parameter $var : Set.$
Inductive Set L $=$
  $Var : var \rightarrow L \mid App : L \rightarrow L \rightarrow L \mid Lam : (var \rightarrow L) \rightarrow L.$
Inductive Set tL $= TVar : var \rightarrow tL \mid Arrow : tL \rightarrow tL \rightarrow tL.$

For instance, $(Lam\ (\lambda x : var.\ (App\ (Var\ x)\ (Var\ x))))$ encodes the function $\lambda x.(x\ x)$. This syntax is provisional since, although it is clear how to encode each term of the object-language as a term of type L, for most instantiations of the type var, the type L contains *exotic* terms, that is, terms that do not encode any λ-term. Describing these terms and finding a way to rule them out is the subject of the next few sections.

The following induction principle for L is generated by the system and proven automatically. It illustrates a general form of induction over higher-order syntax.

$$\begin{aligned} &\forall P : L \rightarrow Prop. \\ &\quad (\forall v : var.(P\ (Var\ v))) \rightarrow \\ &\quad (\forall m, n : L.(P\ m) \rightarrow (P\ n) \rightarrow (P\ (App\ m\ n))) \rightarrow \\ &\quad (\forall E : var \rightarrow L.(\forall v : var.(P\ (E\ v))) \rightarrow (P\ (Lam\ E))) \rightarrow \\ &\forall e : L.(P\ e). \end{aligned}$$

By asserting var as a parameter, the theorems we prove will hold for any instantiation of this type. The important ones to consider will be those that satisfy any axioms we assert which make assumptions about this type. All those that we will need here should follow from the var_nat assumption below which asserts a surjective mapping from var to the natural numbers.

Inductive Set $nat = 0 : nat \mid S : nat \rightarrow nat$
Axiom $var\_nat : \exists s : var \rightarrow nat.\ \forall n : nat.\ \exists v : var.\ (s\ v) = n.$

3 Specifying Syntax in LF

To prove that our syntax adequately represents our object-language, we begin with an LF signature for the λ-calculus, for which adequacy has already been proven [2]. In LF, the syntax is introduced simply by declaring the type l_0 for terms and two constructors for application and abstraction.

$l_0 : Type$
$app_0 : l_0 \rightarrow l_0 \rightarrow l_0$
$lam_0 : (l_0 \rightarrow l_0) \rightarrow l_0.$

The set of ($\alpha\beta\eta$-equivalence classes of) LF terms generated by this signature has three important properties, which we state informally as follows:

1. All terms of the object-language can be represented (as trees) using only the two constructors (induction principle).
2. Each term has a unique representation (injection principle).
3. Any two terms that are extensionally equal are also equal (extensionality principle).

The formulation of these principles, which we will not give here, involves LF terms of type l_0, $l_0 \rightarrow l_0$, $l_0 \rightarrow l_0 \rightarrow l_0$, etc. We will use this sequence of types in the translation of our object-language from the LF representation to the Coq representation in the next section. Thus we give a formal definition:

Definition* $l_n := if\ n = 0\ then\ l_0\ else\ l_0 \rightarrow l_{n-1}.$

In the context of this sequence of types, instead of the original two constructors, the induction and injection principles involve what we call the higher-order constructors, which are defined as follows:

Definition* $ref = \lambda n : nat.\lambda i \in [0..n-1].\lambda x_{n-1}, \ldots, x_0 : l_0.x_i$
Definition* $app = \lambda n : nat.$
$\lambda e, e' : l_n.\lambda x_{n-1}, \ldots, x_0 : l_0.(app_0\ (e\ x_{n-1}\ \ldots\ x_0)\ (e'\ x_{n-1}\ \ldots\ x_0))$
Definition* $lam = \lambda n : nat.Case\ n\ of$
$0 \Rightarrow \lambda e : l_1.\ (lam_0\ e)$
$(S\ m) \Rightarrow \lambda e : l_{m+2}.\ \lambda x_m, \ldots, x_0 : l_0.(lam_0\ (e\ x_m\ \ldots\ x_0)).$

These higher-order constructors will allow us to give a very simple translation from the above LF syntax into Coq (see below). Note that they are not LF terms. However, for each $n \geq 0$ and $i < n$, the terms $(ref\ n\ i)$, $(app\ n)$, and $(lam\ n)$, which we abbreviate as $ref_{n,i}$, app_n, and lam_n, are LF terms. These three families of higher-order λ-terms have natural interpretations in any Cartesian closed category with reflexive objects (cf [1] definition 9.3.1 page 219). The interpretation there is highly semantic in nature. Our purpose here is syntax, and our motivation for the above definitions is not the use of object-level β-reduction.

In [3], our provisional syntax was used to yield and manipulate an implementation of l_0, and in fact of the l_n's. There, adequacy caused no problem, but the final syntax and semantics were invaded by (object-level) lists. Here we will succeed in implementing syntax and semantics without object-level lists.

4 Translation Between LF and Coq Syntaxes

In order to express our translation we define the Coq counterpart of the types l_n and the corresponding higher-order constructors.

Definition* $L_n := if\ n = 0\ then\ L\ else\ var \rightarrow L_{n-1}$.
Definition* $\mathcal{R}ef = \lambda n : nat.\lambda i \in [0..n-1].\lambda x_{n-1}, \ldots, x_0 : var.(Var\ x_i)$
Definition* $\mathcal{A}pp = \lambda n : nat.\lambda e, e' : L_n.\lambda x_{n-1}, \ldots, x_0 : var.$
$(App\ (e\ x_{n-1}\ \ldots\ x_0)\ (e'\ x_{n-1}\ \ldots\ x_0))$
Definition* $\mathcal{L}am = \lambda n : nat.Case\ n\ of$
$0 \Rightarrow \lambda e : L_1.\ (Lam\ e)$
$(S\ m) \Rightarrow \lambda e : L_{m+2}.\lambda x_m, \ldots, x_0 : var.(Lam\ (e\ x_m\ \ldots\ x_0))$.

As before, we use abbreviations $\mathcal{R}ef_{n,i}$, $\mathcal{A}pp_n$, and $\mathcal{L}am_n$.

To show the correspondence between Coq terms of type L and LF terms of type l_0, we begin by defining the following natural translation $Trans$ from the l_n's into the L_n's.

Inductive Definition* $Trans : \forall n : nat.\ l_n \rightarrow L_n \rightarrow Prop$
$= Trans\_ref : \forall n : nat.\forall i \in [0..n-1].(Trans\ n\ ref_{n,i}\ \mathcal{R}ef_{n,i})$
$\mid Trans\_app : \forall n : nat.\forall a, b : l_n.\forall a', b' : L_n.\ (Trans\ n\ a\ a') \rightarrow$
$(Trans\ n\ b\ b') \rightarrow (Trans\ n\ (app_n\ a\ b)\ (\mathcal{A}pp_n\ a'\ b'))$
$\mid Trans\_lam : \forall n : nat.\forall e :\ l_{n+1}.\forall f :\ L_{n+1}.\ (Trans\ (n+1)\ e\ f) \rightarrow$
$(Trans\ n\ (lam_n\ e)\ (\mathcal{L}am_n\ f))$.

Proposition*: The above definition $Trans$ defines, for each n, an injective map from l_n to L_n.

The proof of this proposition relies heavily on the induction and injection principles mentioned earlier. For a similar statement and a fully mechanized proof of it, see [3]. Our next task is to characterize the image of this translation, which in fact is the subset of terms in L_n that specify λ-terms, and thus are the terms we are interested in. The natural definition that selects this subset is the following one (see [3]):

Inductive Definition* $Valid : \forall n : nat.\ L_n \rightarrow Prop$
$= Valid\_ref : \forall n, i : nat.(i < n) \rightarrow (Valid\ n\ \mathcal{R}ef_{n,i})$
$\mid Valid\_app : \forall n : nat.\forall e, e' : L_n.$
$(Valid\ n\ e) \rightarrow (Valid\ n\ e') \rightarrow (Valid\ n\ (\mathcal{A}pp_n\ e\ e'))$
$\mid Valid\_lam : \forall n : nat.\forall e :\ L_{n+1}.\ (Valid\ (n+1)\ e) \rightarrow (Valid\ n\ (\mathcal{L}am_n\ e))$.

Indeed, we have the following result, whose proof is straightforward.

Theorem* : For each integer n, *Trans* yields a bijection between terms of type l_n and terms of type L_n satisfying $(Valid\ n)$.

Note that *Valid* is not a Coq predicate. Furthermore, it is not clear how to define a Coq predicate for each instance of n without using the definition for $n+1$ (see the *Valid_lam* case). This is precisely the task we will turn to now, at least for $n = 0$. We will succeed only modulo extensionality (see Section 6). Note that the *Valid_ref* case becomes irrelevant when $n = 0$ and thus only closed terms satisfy $(Valid\ 0)$. Our solution will replace the proposition $(Valid\ (n+1)\ e)$ with an equivalent one that depends on n instead of $n+1$, thus allowing us to drop n altogether. One obvious candidate is $\forall v : var.(Valid\ n\ (e\ v))$. However, as we will see, this is not sufficient and does not rule out all the necessary terms.

5 Exotic Terms, Extensionality and Extended Validity

There are three kinds of exotic terms that the type L may contain. We illustrate by instantiating *var* to *nat*. In this case, we have irreducible functional terms of type $nat \to L$ that use a *Case* operator. Exotic terms of type L are generated through the *Lam* constructor.

The first kind of exotic term is illustrated by the following term:

$$exot_1 = (Lam\ \lambda x : nat.Case\ x : nat\ of\ 0 \Rightarrow (Var\ 0)\ \ (S\ n) \Rightarrow (Var\ (S\ n))).$$

The above term is extensionally equal to the term $(Lam\ \lambda x : nat.(Var\ x))$, and thus we could accept it as a well-formed term. Extensional equality is defined as the following Coq definition.

$$
\begin{array}{l}
\textsf{Inductive Definition}\ eq_L : L \to L \to Prop \\
\quad = eq_L\_var : \forall x : var(eq_L\ (Var\ x)\ (Var\ x)) \\
\quad \mid eq_L\_app : \forall m_1, m_2, n_1, n_2 : L. \\
\qquad (eq_L\ m_1\ n_1) \to (eq_L\ n_2\ m_2) \to (eq_L\ (App\ m_1\ m_2)\ (App\ n_1\ n_2)) \\
\quad \mid eq_L\_lam : \forall M, N : var \to L. \\
\qquad (\forall x : var.(eq_L\ (M\ x)\ (N\ x))) \to (eq_L\ (Lam\ M)\ (Lam\ N)).
\end{array}
$$

It will be difficult to define predicates which are able to distinguish *Valid*-terms from terms extensionally equal to them. For instance our predicate *subst* (see below) does not. We circumvent this problem by considering that a λ-term is in fact represented by an equivalence class of terms for this eq_L relation.

The second kind of exotic terms are the *open* ones, namely those with "free variables" such as $exot_2 = (Var\ (S\ 0))$. These open terms will play a role in our approach; we will first introduce a meta-level predicate *Valid_v*, a slight modification of *Valid* allowing open terms; we will then succeed in defining a Coq predicate *valid* which implements $(Valid\_v\ 0)$ up to extensionality.

The third kind of exotic term is more problematic and we definitely want our *valid* predicate to discard them. These are terms that contain functions that are not "uniform" in their argument. For example, let $exot_3 :=$

$$\lambda f, x : nat.Case\ x : nat\ of\ 0 \Rightarrow (Var\ x)\ \ (S\ n) \Rightarrow (App\ (Var\ f)\ (Var\ n)).$$

The term $(Lam\ \lambda f : nat.(Lam\ \lambda x : nat.(exot_3\ f\ x)))$ does not represent a λ-term.

In order to integrate the first kind of exotic term, we define the meta-level predicate $Valid\_ext$ which selects terms extensionally equal to $Valid$ ones. For this, we have to extend eq_L to the sequence of types L_n. Note that for each integer n, eq_{L_n} is a Coq term.

Definition* $eq_{L_0} = eq_L$.
Definition* $eq_{L_{n+1}} = \lambda e, f : L_{n+1}.\forall v : var.(eq_{L_n}\ (e\ v)\ (f\ v))$.
Definition* $Valid\_ext = \lambda n : nat.\lambda e : L_n.\exists e' : L_n.((eq_{L_n}\ e\ e') \wedge (Valid\ n\ e'))$.

In order to integrate exotic terms of the second kind, we start by introducing a fourth higher-order constructor $\mathcal{V}$.

Definition* $\mathcal{V} = \lambda n : nat.\lambda v : var.\lambda x_1 : var. \ldots \lambda x_n : var.(Var\ v)$.

Note that although $\mathcal{V}$ is not a Coq term, for each n, $(\mathcal{V}\ n)$ is a Coq term in L_{n+1}, which we denote by $\mathcal{V}_n$. We are now in a position to mimic our characterization of well-formed closed terms through $Valid$ and $Valid\_ext$ to obtain the following predicates, $Valid\_v$ and $Valid\_v\_ext$, which characterize our open terms.

Inductive Definition* $Valid\_v : \forall n : nat.\ L_n \rightarrow Prop$
$= Valid\_v\_var : \forall n : nat.\forall v : var.(Valid\_v\ n\ (\mathcal{V}_n\ v))$
$\mid Valid\_v\_ref : \forall n, i : nat.(i < n) \rightarrow (Valid\_v\ n\ \mathcal{R}ef_{n,i})$
$\mid Valid\_v\_app : \forall n : nat.\forall e, e' : L_n.$
$(Valid\_v\ n\ e) \rightarrow (Valid\_v\ n\ e') \rightarrow (Valid\_v\ n\ (\mathcal{A}pp_n\ e\ e'))$
$\mid Valid\_v\_lam : \forall n : nat.\forall e :\ L_{n+1}.$
$(Valid\_v\ (n+1)\ e) \rightarrow (Valid\_v\ n\ (\mathcal{L}am_n\ e))$.

Definition* $Valid\_v\_ext =$
$\lambda n : nat.\lambda e : L_n.\exists e' : L_n.((eq_{L_n}\ e\ e') \wedge (Valid\_v\ n\ e'))$.

In the next section, we will implement $(Valid\_v\_ext\ 0)$ and $(Valid\_ext\ 0)$.

6 Implementing Validity

As stated earlier, for any n, the challenge of defining a Coq predicate implementing $(Valid\ n)$ is to remove the dependence of the Lam case on $(n+1)$. We show here how we successfully overcome this difficulty for the case when $n = 1$ and define the Coq predicate $valid_1$ implementing $(Valid\ 1)$ modulo extensionality. Since we only need $(Valid\ 0)$, we can then implement it directly (modulo extensionality) using $valid_1$.

We denote by VL_n the subset of $Valid\_v\_ext$-terms in L_n and by VL the union of the VL_n's.

The basis of our implementation of $(Valid\ 1)$ is the following fact which shows that we are able to express quite simply $(Valid\ 2)$ in terms of $(Valid\ 1)$ (modulo extensionality).

Proposition* $Separate\_val : \forall e : var \rightarrow var \rightarrow L.$
$(\forall v : var.(Valid\_v\_ext\ 1\ (e\ v))) \rightarrow$
$(\forall v : var.(Valid\_v\_ext\ 1\ \lambda u : var.(e\ u\ v))) \rightarrow (Valid\_v\_ext\ 2\ e).$

Proof: It follows from the induction and injection principles that equivalence classes (modulo eq_{L_m}) of terms of VL_m are in one-one correspondence with (second-order abstract syntax) trees built from the higher-order constructors $\mathcal{R}ef_{m,i}$, $\mathcal{A}pp_m$, $\mathcal{L}am_m$, and $\mathcal{V}_{m,v}$. Here a tree is a set of (abstract) paths together with a map from this set to our set of constructors.

Let e be a term satisfying the assumptions. We pick three values u,v and w in var. (Note that by the var_nat axiom, we know there are infinitely many terms of type var. Here, we require at least three). By the first assumption, both $(e\ u\ v)$ and $(e\ u\ w)$ are values of the $(Valid\_v\_ext\ 1)$ function $(e\ u)$, thus their associated trees differ at most in some leaves, where they both have a $\mathcal{V}_m$, with different arguments. By the second assumption, a similar statement holds for the trees associated with $(e\ w\ v)$ and $(e\ u\ v)$. By transitivity, we infer that for any four-tuple (u, v, w, x) in var, the trees associated with $(e\ u\ v)$ and $(e\ w\ x)$ differ at most in some leaves, where they both have a $\mathcal{V}_m$ with different arguments. We denote by P the set of paths p where the constructor associated with $(e\ u\ v)$ is $(\mathcal{V}_{m_p}\ (\phi_p\ u\ v))$. We have to prove that for each p in P, ϕ_p is either one of the two projections or a constant function. We know that for any u, $\lambda v.(\phi_p\ u\ v)$ is either constant or the identity. Similarly, for any u, $\lambda v.(\phi_p\ v\ u)$ is either constant or the identity. The following lemma will complete our proof and illustrate why at least three distinct variables are needed.

Lemma*: Let var be a set with at least three elements. Let ϕ be a function from $var \times var$ into var satisfying the property that for any u, $\lambda v.(\phi\ u\ v)$ and $\lambda v.(\phi\ v\ u)$ are either constant functions or the identity. Then ϕ is either a constant function or one of the two projections.

Proof: First suppose that for some u, $\lambda x.(\phi\ u\ x)$ is a constant w different from u. Then for any v different from w, $\lambda x.(\phi\ x\ v)$ has to be constant and equal to w. Now choose u'. For x different from w, $(\phi\ u'\ x)$ is equal to w. Since there are at least two such x's, $\lambda x.(\phi\ u'\ x)$ has to be constant and equal to w.

The same argument applies when for some u, $\lambda x.(\phi\ x\ u)$ is a constant w different from u. In the remaining cases, $(\phi\ u\ v)$ can only be u or v.

Now suppose that for some u, $\lambda x.(\phi\ u\ x)$ is the constant function $\lambda x.u$. Using the previous assumption, we deduce that for any v different from u, $\lambda x.(\phi\ x\ v)$ is the identity. Now for any u', $\lambda x.(\phi\ u'\ x)$ takes the value u' at least twice, and hence is the constant function $\lambda x.u'$, and we are done.

In the remaining case, $\lambda x.(\phi\ u\ x)$ is the identity for any u, thus ϕ is the second projection.

The above lemma is crucial since it has the corollary below, which concerns the set of terms WL_1 also defined below, and allows a direct implementation of $(Valid\_v\_ext\ 1)$. We denote by VVL_1 the set of $(Valid\_v\ 1)$-terms. Note that VL_1 is the set of terms extensionally equal to terms in VVL_1.

Definition*: We define WL_1 as the smallest among the subsets W of L_1 satisfying the following conditions:

1. W contains $\lambda x.(Var\ x)$.
2. W contains $\lambda x.(Var\ u)$ for any u in *var*.
3. W contains $\lambda x.(App\ (a\ x)\ (b\ x))$ for any pair (a, b) of terms in W.
4. W contains $\lambda x.(Lam\ (e\ x))$ for any e of type $var \to var \to L$ satisfying the two conditions:
   (a) for any u in *var*, $(e\ u)$ is in W;
   (b) for any u in *var*, $\lambda x.(e\ x\ u)$ is in W.

Corollary*: If type *var* has at least three terms, then WL_1 contains VVL_1 and is contained in VL_1.

Proof: We first check that VL_1 is a subset of L_1 satisfying the above four conditions. It is clear for the first three. For the fourth one, if e is such that for any u in *var*, $(e\ u)$ and $\lambda x.(e\ x\ u)$ are in VL_1, then by the *Separate_val* proposition, e satisfies $(Valid\_v\_ext\ 2)$ and thus is eq_{L_2} equivalent to some *Valid_v*-term e'. Thus $\lambda x.(Lam\ (e\ x))$ is eq_{L_1} equivalent to $\lambda x.(Lam\ (e'\ x))$ which satisfies $(Valid\_v\ 1)$ by *Valid_v_lam*. It follows that $\lambda x.(Lam\ (e\ x))$ satisfies $(Valid\_v\_ext\ 1)$. Since WL_1 is the smallest set satisfying the above conditions, WL_1 is contained in VL_1.

We now check that VVL_1 is contained in any such W. We choose such a W and we prove by induction that any term t satisfying $(Valid\_v\ 1)$ is in W. Induction is on the length of the proof of $(Valid\_v\ 1\ t)$. If t has a height of 1, then t is of the form $\lambda x.(Var\ x)$ or $\lambda x.(Var\ u)$, and hence it is in W. If t has bigger height, then the head constructor is either *App* or *Lam*. If $t = \lambda x.(App\ (a\ x)\ (b\ x))$, then by inversion of the definition of *Valid_v*, a and b both satisfy $(Valid\_v\ 1)$. By the induction hypothesis, they are in W, and thus so is t. If $t = \lambda x.(Lam\ (e\ x))$, then by inversion of the definition of *Valid_v*, e satisfies $(Valid\_v\ 2)$. This implies that for any u in *var*, $(e\ u)$ and $\lambda x.(e\ x\ u)$ are in VVL_1, and thus in W since the height of these terms is smaller than the height of t.

We conjecture that WL_1 is in fact equal to VVL_1 but have not yet proved it.

Now we state our Coq definitions. The definition of $valid_1$ is derived directly from the definition of WL_1 above and selects exactly the terms of type $var \to L$ that we want.

Inductive Definition $valid_1 : (var \to L) \to \text{Prop}$
$= valid_1\_var : \forall v : var.(valid_1\ \lambda x : var.(Var\ v))$
$\mid valid_1\_ref : (valid_1\ \lambda x : var.(Var\ x))$
$\mid valid_1\_app : \forall e, e' : var \to L.$
$(valid_1\ e) \to (valid_1\ e') \to (valid_1\ \lambda x : var.(App\ (e\ x)\ (e'\ x)))$
$\mid valid_1\_lam : \forall e : var \to var \to L.$
$(\forall u : var.(valid_1\ \lambda v : var.(e\ u\ v)) \wedge (valid_1\ \lambda v : var.(e\ v\ u))) \to$
$(valid_1\ \lambda x : var.(Lam\ (e\ x)))$.

Inductive Definition $valid_0 : L \rightarrow$ Prop
$= valid_0\_var : \forall v : var.(valid_0\ (Var\ v))$
$\mid valid_0\_app : \forall a, b : L.(valid_0\ a) \rightarrow\ (valid_0\ b) \rightarrow (valid_0\ (App\ a\ b))$
$\mid valid_0\_lam : \forall e : var \rightarrow L.(valid_1\ e) \rightarrow (valid_0\ (Lam\ e)).$

Definition $valid = \lambda e : L.\exists e' : L.((eq_L\ e\ e') \wedge (valid_0\ e')).$

Definition $closed = \lambda t : L.((valid_0\ t) \wedge$
$\forall e : var \rightarrow L.(valid_1\ e) \rightarrow \forall x : var.(t = (e\ x)) \rightarrow \forall y : var.(eq_L\ (e\ y)\ t)).$

It follows easily from the above statements that the *valid* predicate implements $(Valid\_v\_ext\ 0)$ and that *closed* implements $(Valid\_ext\ 0)$.

Note that $(valid_0\ exot_1)$ does not hold, but $(valid\ exot_1)$ does if we take e' in the definition of *valid* to be $(Lam\ \lambda x : nat.(Var\ x))$. Note also that in order for $(Lam\ \lambda f : nat.(Lam\ \lambda x : nat.(exot_3\ f\ x)))$ to satisfy $valid_0$, $\lambda f : nat.(Lam\ \lambda x : nat.(exot_3\ f\ x))$ must satisfy $valid_1$. Although it is the case that $\forall u : var.(valid_1\ \lambda v : var.(exot_3\ v\ u))$ holds, $\forall u : var.(valid_1\ \lambda v : var.(exot_3\ u\ v))$ does not. In fact, no term with a *Case* operator will satisfy $valid_0$. However, we must include those that are extensionally equal to terms with no *Case* operator in order to correctly implement substitution, which we define in the next section.

7 Substitution

In order to specify evaluation for our language, we must specify β-reduction which for our representation is the operation that, given a redex of the form $(App\ (Lam\ \lambda x : var.M)\ N)$, replaces all occurrences of $(Var\ x)$ in M by N. To do so, we define a Coq predicate *subst*. We proceed as we did to define *valid*, here starting with a definition of *Subst* of type $\forall n : nat.L_{n+1} \rightarrow L_0 \rightarrow L_n \rightarrow Prop$. The proposition $(Subst\ n\ E\ p\ r)$ holds if E has the form $\lambda x : var.F$ and r is the term obtained by replacing all occurrences of $(Var\ x)$ in F by p. Although we define it relationally, *Subst* is functional on the first three arguments.

Inductive Definition* $Subst : \forall n : nat.L_{n+1} \rightarrow L_0 \rightarrow L_n \rightarrow Prop$
$= Subst\_ref\_rename : \forall n : nat.\forall p : L_0(Subst\ n\ \mathcal{R}ef_{n+1,n}\ p\ \lambda x_1.\ldots.\lambda x_n.p)$
$\mid Subst\_ref\_keep : \forall n : nat.\forall p : L_0.\forall i \in [0..n-1].$
$(Subst\ n\ \mathcal{R}ef_{n+1,i}\ p\ \mathcal{R}ef_{n,i})$
$\mid Subst\_var : \forall n : nat.\forall v : var.\forall p : L_0.(Subst\ n\ (\mathcal{V}_{n+1}\ v)\ p\ (\mathcal{V}_n\ v))$
$\mid Subst\_app : \forall n : nat.\forall p : L_0.\forall A, A' : L_{n+1}\forall B, B' : L_n.$
$(Subst\ n\ A\ p\ B) \rightarrow (Subst\ n\ A'\ p\ B') \rightarrow$
$(Subst\ n\ (\mathcal{A}pp_{n+1}\ A\ A')\ p\ (\mathcal{A}pp_n\ B\ B'))$
$\mid Subst\_lam : \forall n : nat.\forall p : L_0.\forall A : L_{n+2}.\forall B : L_{n+1}.$
$(Subst\ (n+1)\ A\ p\ B) \rightarrow (Subst\ n\ (\mathcal{L}am_{n+1}\ A)\ p\ (\mathcal{L}am_n\ B)).$

As before this definition cannot be directly transformed into a Coq definition because it requires an infinite series of definitions where for each n $(Subst\ n)$ requires that of $(Subst\ (n+1))$. As before, we must implement $(Subst\ 0)$. As for $(Valid\ 0)$, we cannot do it directly, but instead must work modulo extensionality. In particular, we implement:

Definition* $Subst\_ext = \lambda n : nat.\lambda e : L_{n+1}.\lambda p : L_0.\lambda r : L_n.\exists e' : L_{n+1}.\exists p' : L_0.$
$\exists r' : L_n.(eq_{L_{n+1}}\ e\ e') \to (eq_{L_0}\ p\ p') \to (eq_{L_n}\ r\ r') \to (Subst\ n\ e'\ p'\ r')$.

Note that if $(Subst\ n\ E\ p\ r)$ holds, then E, p and r are $Valid\_v$-terms, and if $(Subst\_ext\ n\ E\ p\ r)$ holds, then E, p and r are $Valid\_v\_ext$-terms. The property below is crucial and reduces $(Subst\_ext\ (n+1))$ to $(Subst\_ext\ n)$. Here $t@x$ denotes the term $\lambda x_1, \ldots, x_n : var.(t\ x_1 \ldots x_n\ x)$, where the value of n can be determined from context.

Lemma* : $\forall n : nat.\forall E : L_{n+2}.\forall p : L_0.\forall r : L_{n+1}.$
$(Valid\_v\_ext\ (n+2)\ E) \to (Valid\_v\_ext\ 0\ p) \to (Valid\_v\_ext\ (n+1)\ r) \to$
$(\forall v : var(Subst\_ext\ n\ E@v\ p\ r@v)) \to (Subst\_ext\ (n+1)\ E\ p\ r)$

Proof: Let r' be a term such that $(Subst\_ext\ (n+1)\ E\ p\ r')$. Using the fact that replacing all the arguments (except the variable being substituted) by a value before or after the substitution leads to the same result, we have that $\forall v : var.(Subst\_ext\ n\ E@v\ p\ r'@v)$ holds. From the fact that $Subst$ is functional, we know that for all v of type var, $r@v$ and $r'@v$ are eq_{L_n} equivalent. This directly implies that r and r' are $eq_{L_{n+1}}$ equivalent.

Note that this lemma would not hold without extensionality.
This property leads to the following implementation of $(Subst\_ext\ 0)$.

Inductive Definition $subst : (var \to L) \to L \to L \to Prop$
$= subst\_ref : \forall p : L.(subst\ \lambda x : var.(Var\ x)\ p\ p)$
$\mid subst\_var : \forall v : var.\forall p : L.(subst\ \lambda x : var.(Var\ v)\ p\ (Var\ v))$
$\mid subst\_app : \forall p : L.\forall A, A' : (var \to L).\forall B, B' : L.$
$(subst\ A\ p\ B) \to (subst\ A'\ p\ B') \to$
$(subst\ \lambda v : var.(App\ (A\ v)\ (A'\ v))\ p\ (App\ B\ B'))$
$\mid subst\_lam : \forall p : L.\forall A : var \to var \to L.\forall B : var \to L.$
$(\forall v : var.(subst\ (\lambda x : var.(A\ x\ v))\ p\ (B\ v))) \to$
$(subst\ (\lambda x : var.(Lam\ (A\ x)))\ p\ (Lam\ B))$.

Definition $subst\_ext = \lambda e : var \to L.\lambda p : L.\lambda r : L.\exists e' : var \to L.\exists p' : L.\exists r' : L.$
$(\forall v : var.(eq_L\ (e\ v)\ (e'\ v))) \to (eq_L\ p\ p') \to (eq_L\ r\ r') \to (subst\ e'\ p'\ r')$.

The correctness of these definitions is expressed by the following statement whose proof is straightforward. (The second part follows directly from the lemma above).

Proposition* : $\forall E : var \to L.\forall p, r : L.(Subst\_ext\ 0\ E\ p\ r) \to (subst\_ext\ E\ p\ r)$.
Conversely, $\forall E : var \to L.\forall p, r : L.(valid_1\ E) \to (valid\ p) \to (valid\ r) \to (subst\_ext\ E\ p\ r) \to (Subst\_ext\ 0\ E\ p\ r)$.

In proving properties of our object-language, we may choose to use either $subst$ or $subst\_ext$. In the next section, we choose the former.

8 An Example Proof: the Subject Reduction Theorem

In this section, we specify type assignment and evaluation for our object-language by introducing inductive types for each. We then outline the proof of type soundness (also called subject reduction) which we have fully formalized in Coq.

For typing, we first introduce a predicate for assigning variables to types along with two assumptions about it stating that each variable has a unique type and that there is a variable at every type

Parameter $typvar : var \rightarrow tL \rightarrow Prop.$
Axiom $uniq\_var\_type : \forall x : var.\forall t, s : tL.(typvar\ x\ t) \rightarrow (typvar\ x\ s) \rightarrow (s = t).$
Axiom $exists\_new\_var : \forall t : tL.\exists x : var.(typvar\ x\ t).$

Like *var*, *typvar* is introduced as a parameter, and thus the theorems we prove will hold for any instantiation. Here, the important ones to consider will be those for which we can prove the above axioms. Note, for example, that these two axioms hold trivially if we take *var* to be *tL* and *typvar* to be equality on *tL*.

The usual natural deduction style inference rules for assigning simple types to untyped terms is specified by the following inductive type.

Inductive Definition $type : L \rightarrow tL \rightarrow Prop$
$= type\_Var : \forall x : var.\forall s : tL.(typvar\ x\ s) \rightarrow (type\ (Var\ x)\ s)$
$\mid type\_App : \forall e, e' : L.\forall t', t : tL.$
$(type\ e\ (Arrow\ t'\ t)) \rightarrow (type\ e'\ t') \rightarrow (type\ (App\ e\ e')\ t)$
$\mid type\_Lam : \forall E : var \rightarrow L.\forall t, t' : tL.$
$(\forall x : var.(typvar\ x\ t) \rightarrow (type\ (E\ x)\ t')) \rightarrow (type\ (Lam\ E)\ (Arrow\ t\ t')).$

The third clause in this definition uses a hypothetical judgment with an embedded arrow for typing λ-abstractions. It asserts the fact that $(Lam\ E)$ has functional type $(Arrow\ t\ t')$ if under the assumption that for arbitrary variable x of type t, it can be shown that the expression $(E\ x)$ (the expression obtained by replacing all occurrences of the variable bound by the λ-abstraction at the head of E with x) has type t'.

Similar definitions of *type* have been given in LF and λProlog where the predicate defining typing appears on the left of the embedded arrow. Here this would mean replacing $(typvar\ x\ t)$ by $(type\ (Var\ x)\ t)$. Note that this change results in a negative occurrence of *type* which is disallowed in Coq. For this reason, we need a separate *typvar* predicate, just as we needed a separate type *var* in the definition of L.

The following induction principle for this definition illustrates the general form of induction over inductively defined predicates, in particular when they involve embedded universal quantification and implication.

$\forall P.L \rightarrow tL \rightarrow Prop.$
$(\forall x : var.\forall s : tL.(typvar\ x\ s) \rightarrow (P\ (Var\ x)\ s)) \rightarrow$
$(\forall E : var \rightarrow L.\forall t, t' : tL.$
$(\forall x : var.(typvar\ x\ t) \rightarrow (type\ (E\ x)\ t')) \rightarrow$
$(\forall x : var.(typvar\ x\ t) \rightarrow (P\ (E\ x)\ t')) \rightarrow (P\ (Lam\ E)\ (Arrow\ t\ t'))) \rightarrow$
$(\forall e, e' : L.\forall t', t : tL.$
$(type\ e\ (Arrow\ t'\ t)) \rightarrow (P\ e\ (Arrow\ t'\ t)) \rightarrow$
$(type\ e'\ t') \rightarrow (P\ e'\ t') \rightarrow (P\ (App\ e\ e')\ t)) \rightarrow$
$\forall e : L.\forall t : tL.(type\ e\ t) \rightarrow (P\ e\ t)$

Call by value semantics for our simple functional language is defined by the following inductive definition. Note the use of substitution in the β-redex case.

Inductive Definition $eval : L \rightarrow L \rightarrow Prop$
$= eval\_Lam : \forall E : var \rightarrow L.(eval\ (Lam\ E)\ (Lam\ E))$
$\mid eval\_App : \forall E : var \rightarrow L.\forall e_1, e_2, e_3, v_1, v_2 : L.(eval\ e_1\ (Lam\ E)) \rightarrow$
$(eval\ e_2\ v_2) \rightarrow (subst\ E\ v_2\ e_3) \rightarrow (eval\ e_3\ v_1) \rightarrow (eval\ (App\ e_1\ e_2)\ v_1).$

The proof of type soundness is quite simple and follows naturally from the definitions and axioms presented in this section, the definitions in Section 2, and the definition of *subst*. The main lemma needed for this theorem is that the predicate *subst* preserves typing. For this lemma, we need to define the notion of two terms having the same type. This definition, lemma, and the main theorem are stated as follows.

Definition $same\_type = \lambda m, n : L.\exists t : tL.(type\ m\ t) \wedge (type\ n\ t).$
Lemma $subst\_sr : \forall E : var \rightarrow L.\forall p, n : L.(subst\ E\ p\ n) \rightarrow$
$\forall x : var.(same\_type\ (Var\ x)\ p) \rightarrow \forall t : tL.(type\ (E\ x)\ t) \rightarrow (type\ n\ t).$
Theorem $subj\_reduction : \forall e, v : L.(eval\ e\ v) \rightarrow \forall t : tL.(type\ e\ t) \rightarrow (type\ v\ t).$

The proof of the lemma proceeds by induction on $(subst\ E\ p\ n)$, while the proof of subject reduction proceeds by induction on $(eval\ e\ v)$.

9 Conclusion and Future Work

We have shown by example how higher-order syntax can be used in formal proof. Our method of specification of syntax is in fact quite general. Although we have not yet done it, we plan to generalize it formally as is done in [3]. For any object-language that can be expressed in second-order syntax, it is easy to see how to define the corresponding *valid* and *subst* predicates. Proofs of adequacy follow similarly. In doing proofs, the user is then freed from concerns of α-conversion, and substitution is greatly simplified. In fact, it is possible to automate generation of these definitions and to automate certain aspects of proof search that occur repeatedly in such proofs. Although our proof is already simple (500 lines of Coq script), it would be further simplified by such automated tools.

In addition to type soundness as presented here, several other examples are in progress including a proof of correctness of a λProlog program that computes the negation normal form of formulas in first-order logic and a proof of the Church-Rosser property for the λ-calculus.

Finally, we have not considered here the question of adequacy for our semantic definitions. This, together with the correctness of the Coq theorems with respect to the corresponding 'meta' theorems will be the subject of future work. The latter will follow naturally (see. Theorem 1. in Section 3.5 in [3]).

References

1. A. Asperti and G. Longo. *Categories, Types, and Structures.* MIT Press, Foundations of Computing Series, London, England, 1991.
2. A. Avron, F. Honsell, I. A. Mason, and R. Pollack. Using typed lambda calculus to implement formal systems on a machine. *Journal of Automated Reasoning*, 9(3):309–354, Dec. 1992.
3. J. Despeyroux and A. Hirschowitz. Higher-order syntax and induction in coq. In *Proceedings of the fifth Int. Conf. on Logic Programming and Automated Reasoning (LPAR 94), Kiev, Ukraine, July 16–21, 1994*, 1994. Also available as an INRIA Research Report RR-2292, Inria-Sophia-Antipolis, France, June 1994.
4. G. Dowek, A. Felty, H. Herbelin, G. Huet, C. Murthy, C. Parent, C. Paulin-Mohring, and B. Werner. The coq proof assistant user's guide. Technical Report 154, INRIA, 1993.
5. A. Felty. A logic programming approach to implementing higher-order term rewriting. In L.-H. Eriksson, L. Hallnäs, and P. Schroeder-Heister, editors, *Proceedings of the January 1991 Workshop on Extensions to Logic Programming*, pages 135–161. Springer-Verlag LNCS, 1992.
6. A. Felty. Implementing tactics and tacticals in a higher-order logic programming language. *Journal of Automated Reasoning*, 11(1):43–81, Aug. 1993.
7. J. Hannan. *Investigating a Proof-Theoretic Meta-Language for Functional Programs.* PhD thesis, University of Pennsylvania, Technical Report MS-CIS-91-09, Jan. 1991.
8. J. Hannan and D. Miller. From operational semantics to abstract machines. *Mathematical Structures in Computer Science*, 2:415–459, 1992.
9. J. Hannan and F. Pfenning. Compiler verification in LF. In *Seventh Annual Symposium on Logic in Computer Science*, pages 407–418, 1992.
10. R. Harper, F. Honsell, and G. Plotkin. A framework for defining logics. *Journal of the ACM*, 40(1):143–184, Jan. 1993.
11. S. Michaylov and F. Pfenning. Natural semantics and some of its meta-theory in elf. In L.-H. Eriksson, L. Hallnäs, and P. Schroeder-Heister, editors, *Proceedings of the January 1991 Workshop on Extensions to Logic Programming*, pages 299–344. Springer-Verlag LNCS, 1992.
12. D. Miller. Unification of simply typed lambda-terms as logic programming. In *Eighth International Logic Programming Conference.* MIT Press, 1991.
13. C. Paulin-Mohring. Inductive definitions in the system Coq; rules and properties. In M. Bezem and J. F. Groote, editors, *Proceedings of the International Conference on Typed Lambda Calculi and Applications*, volume 664, pages 328–345. Springer Verlag Lecture Notes in Computer Science, 1993.
14. F. Pfenning and E. Rohwedder. Implementing the meta-theory of deductive systems. In *Eleventh International Conference on Automated Deduction*, pages 537–551. Springer-Verlag LNCS, 1992.

Expanding Extensional Polymorphism*

Roberto Di Cosmo[1] *Adolfo Piperno*[2]

[1] DMI-LIENS - Ecole Normale Supérieure
45, Rue d'Ulm - 75230 Paris (France)
dicosmo@dmi.ens.fr

[2] Dipartimento di Scienze dell'Informazione - Università di Roma "La Sapienza"
Via Salaria 113 - 00198 Roma (Italy)
piperno@dsi.uniroma1.it

Abstract. We prove the confluence and strong normalization properties for second order lambda calculus equipped with an expansive version of η-reduction. Our proof technique, based on a simple abstract lemma and a labelled λ-calculus, can also be successfully used to simplify the proofs of confluence and normalization for first order calculi, and can be applied to various extensions of the calculus presented here.

1 Introduction

The typed lambda calculus provides a convenient framework for studying functional programming and offers a natural formalism to deal with proofs in intuitionistic logic. It comes traditionally equipped with the β equality $(\lambda x.M)N = M[N/x]$ as fundamental computational mechanism, and with the η *(extensional)* equality $\lambda x.Mx = M$ as a tool for reasoning about programs. This basic calculus can then be extended by adding further types, like products, unit and second order types, each coming with its own computational mechanism and/or its extensional equalities.

To reason about programs and the proofs that they represent, one has to be able to orient each equality into a rewriting rule, and to prove that the resulting rewriting system is indeed confluent and strongly normalizing: these properties guarantee that to each program (or proof) P we can associate an equivalent canonical representative which is unique and can be found in finite time by applying the reduction rules to P in whatever order we choose. The β equality, for example, is always turned into the reduction rule $(\lambda x.M)N \longrightarrow M[N/x]$.

Traditionally, the extensional equalities are turned into *contraction* reduction rules, the most known example being the η rule $\lambda x.Mx \longrightarrow M$, but this approach raises a number of difficult problems when trying to add other rules to the system. For example the extensional first order lambda calculus associated to Cartesian Closed Categories, where one needs a special *unit* type T with an axiom $M{:}T = *{:}T$ (see [CDC91] and especially [DCK94b] for a longer discussion and references) is no longer confluent. Another example is the extensional first order lambda calculus enriched with a confluent algebraic rewriting system, where confluence is also broken [DCK94a].

This inconvenient can be fortunately overcome, as proposed in several recent works[Aka93, Dou93, DCK94b, Cub92, JG92], by turning the extensional equalities into *expansion* rules: η becomes then

$$M{:}A \to B \longrightarrow \lambda x.Mx.$$

These expansions are suitably restricted to ensure termination [3], and several *first*

* This work has been partially supported by grants from HCM "Typed Lambda Calculus" and CNR-CNRS projects

[3] We refer the interested reader to[DCK93, DCK94b] for a more detailed discussion of these restrictions.

order systems incorporating both the expansive η rule and an expansive version of the Surjective Pairing extensional rule for products can be proven confluent and strongly normalizing. In[DCK94b] Delia Kesner and the first author even proved that a system with expansions for Surjective Pairing is confluent in the presence of a fixpoint combinator, while it is known that confluence does not hold with the contractive version of Surjective Pairing[Nes89].

These recent works raise a natural question: is it possible to carry on this approach to extensional equalities via expansion rules to the *second order* typed lambda calculus? The answer is not obvious: for an expansion rule to be applicable on a given subterm, we need to look at the type of that subterm, and when we add second order quantification a *subterm* can *change* its type during evaluation. As we will see, this fact rules out a whole class of modular proof techniques that would easily establish the result, and makes the study of expansion rules more problematic.

In this paper we focus on the *second order* typed lambda calculus and extensionality axioms for the arrow type: this system corresponds to the Intuitionistic Positive Calculus with implication, and quantification over propositions.

For this calculus we provide a reduction system based on expansion rules that is confluent and strongly normalizing, by means of an interpretation into a normalizing fragment of the untyped lambda calculus.

This result gives a natural justification of the notion of η-long normal forms used in higher order unification and resolution: they can be now defined simply as the normal forms w.r.t. our extensional rewriting system.

1.1 Survey

The restrictions imposed on the expansion rules in order to insure termination make several usual properties of the λ-calculus fail, most notably η-postponement, that would allow a very simple proof of normalization for the calculus[4], but several proof techniques have been developed over the past years to show that the expansionary interpretation of the extensional equalities yields a confluent and normalizing system in the first order case. One idea is to try to separate the expansion rules from the rest of the reduction, and then try to show some kind of modularity of the reduction systems. One traditional technique for confluence that comes to mind is the well known

Lemma 1.1 (Hindley-Rosen ([Bar84], §3)) *If R and S are confluent, and commute with each other, then $R \cup S$ is confluent.*

Unfortunately, this technique does not work in the presence of restricted expansion rules, because β can destroy expansion redexes, but in[Aka93] Akama gives a modular proof using the following property, requiring some additional conditions on R and S:

Lemma 1.2 *Let S and R be confluent and strongly normalizing reductions, s.t.*

$$\forall M, N \quad (M \xrightarrow{S} N) \quad \textit{implies} \quad (M^R \overset{S}{\underset{+}{\twoheadrightarrow}} N^R),$$

where M^R and N^R are the R-normal forms of M and N, respectively; then $S \cup R$ is also confluent and strongly normalizing.

In[Aka93] R is taken to be the expansionary system alone and S is the usual non extensional reduction relation.

In[DCK94b], confluence and strong normalization of the full expansionary system is reduced to that of the traditional one without expansions using the following:

[4] For a very broad presentation of the properties that fail in presence of restricted expansions, see[DCK94b].

Proposition 1.3 *Let $\mathcal{R}_1$ and $\mathcal{R}_2$ be two reduction systems and $\mathcal{T}$ a translation from $\mathcal{R}_1$-terms to $\mathcal{R}_2$-terms.*

(i) If for every reduction $M_1 \xrightarrow{\mathcal{R}_1} M_2$ there is a non empty sequence $P_1 \xrightarrow[+]{\mathcal{R}_2} P_2$ such that $\mathcal{T}(M_i) = P_i$, for $i = 1, 2$ (simulation property), then the strong normalization of $\mathcal{R}_2$ implies that of $\mathcal{R}_1$.

(ii) If in addition the translation is the identity on $\mathcal{R}_1$ normal forms, and these normal forms are included in the $\mathcal{R}_2$ normal forms, then the confluence of $\mathcal{R}_2$ also implies the confluence of the full system.

The translation used in[DCK94b] inserts in all positions where an expansion could take place a special term Δ_A (called an *expansor*) depending on the type A of the expansion redex, and then all that one is left to prove is the simulation property.

A different non-modular approach is taken in [JG92] and [Dou93], where the proofs of strong normalization are based on an extension of the traditional techniques of reducibility and allow to handle also the peculiarity of the expansion rules. But that is not all, since one is left to prove weak confluence separately, which is not an easy task in the presence of expansion rules (see[DCK94b] for details).

Finally, an even different technique is used in [Cub92], where confluence is shown by a careful study of the residuals in the reduction.

As suggested in the introduction, in the presence of second order quantification, the type of a subterm can evolve during evaluation, and this fact allows us to build very simple examples suggesting that the modular approaches[Aka93, DCK94b] cannot be extended to the second order case.

Expansions and polymorphism are not modular

The following simple example shows that we cannot use the modular techniques developed up to now to separate the complexities introduced by expansion rules and polymorphic typing by singling out the expansions in a separate rewriting system.

Example 1.4 Let $M = (\Lambda\sigma.\lambda x{:}\,(\forall\mu.\mu \to A).(x[\sigma \to \sigma])(\lambda y{:}\,\sigma.y))[A \to B]$. Then, the term M is a normal form w.r.t expansion rules, but its immediate β^2 reduct is not:

$$M' = \lambda x{:}\,(\forall\mu.\mu \to A).(x[(A \to B) \to (A \to B)])(\lambda y{:}\,A \to B.y)$$

In fact, M' reduces to the term

$$M'' = \lambda x{:}\,(\forall\mu.\mu \to A).(x[(A \to B) \to (A \to B)])(\lambda y{:}\,A \to B.\lambda z{:}\,A.yz)$$

Now, there is no way to reduce M to M'' without expansions, so the hypothesis of lemma 1.2 are not satisfied.

This very same example can be used to show how the use of expansor terms is neither viable[5].

Notice that Akama's lemma fails also if we put β^2 together with η in the reduction relation R, *because β does not preserve β^2 normal forms.*

As for the reducibility technique, it can be adapted as in[JG92] for the first order calculi with expansion rules, but there is a fundamental difference between the proof

[5] For readers acquainted with the techniques used in[DCK94b], it is easy to see that the term

$$M = (\Lambda\sigma.\lambda x{:}\,(\forall\mu.\mu \to A).(x[\sigma \to \sigma])(\lambda y{:}\,\sigma.y))[A \to B]$$

is the translation of itself, as there is no expansion redex, but its β^2 reduct

$$\lambda x{:}\,(\forall\mu.\mu \to A).(x[(A \to B) \to (A \to B)])(\lambda y{:}\,A \to B.y)$$

gets translated to $\lambda x{:}\,(\forall\mu.\mu \to A).(x[(A \to B) \to (A \to B)])(\lambda y{:}\,A \to B.\Delta_{A\to B}y)$ and there is no way for M to reduce to it, so the hypothesis of proposition 1.3 are not satisfied.

for the simply typed and the proof for polymorphic lambda calculus: one does not work with just one reducibility candidate, but with all reducibility candidates at once. This requires to deal with many subtle points in the second order case that do not appear at all for the simply typed calculi: for example, in the second order setting one has to show that the set of all strongly normalizing terms is indeed a reducibility candidate: this is straightforward using Girard's original definition of reducibility, but the modifications of Girard's **(CR3)** property imposed by the extensional rules make even this task extremely difficult. Up to now no such proof is known.

1.2 Our approach

Since the previous modular techniques are not viable as used traditionally, and the reducibility properties modified as in[JG92] do not extend nicely to second order, we had to look for something new, and since all the problems come up immediately as soon as we add β^2 to the first order expansionary system, we focused on a simple system with β, expansive η and β^2 first (weak confluence for this calculus is quite straightforward). Here, we first observed that:

- an infinite reduction path in the typed calculus implies the existence of an infinite reduction path containing infinite β steps in the untyped calculus, each untyped term being the erasure of the corresponding typed one (this is the case because β^2 alone, that leaves the erasures unchanged, always terminates, and because β^2 and η together are easily seen to be strongly normalizing);
- an untyped term that is typable in the second order lambda calculus cannot have an infinite β reduction sequence;
- η-postponement holds if we lift the restrictions on expansions, and we can use it on the erasure of the reduction, obtaining a reduction sequence that contains all the β steps, consecutive and at the beginning.

Now, this would clearly suffice to prove strong normalization, if η-postponement could be done without deleting some β or η steps from the original reduction, as then from an infinite typed reduction we could get an infinite β reduction starting from a typable term, that is impossible.

Unfortunately, η-postponement with unrestricted expansions *does* delete some η and β steps, as in the following loops that are the very motivation for imposing restrictions on the η expansions:

$$MN \xrightarrow{\eta} (\lambda x.Mx)N \xrightarrow{\beta} MN \qquad \text{becomes} \qquad MN = MN$$
$$\lambda x.M \xrightarrow{\eta} \lambda z.(\lambda x.M)z \xrightarrow{\beta} \lambda x.M \qquad \text{becomes} \qquad \lambda x.M = \lambda x.M$$

We carefully analysed these deletions during the η postponement, showing that the only β steps that get erased are the ones that are created by expansions which violate the restrictions. To study such reductions, we work in a labelled calculus where abstractions introduced by η expansions are marked, with unrestricted η turned into $M{:}A \to B \xrightarrow{\eta} \lambda^* x.Mx$.

In this labelled calculus, it is easy to identify the β steps that get erased during postponement, and we singled them out as the following β_* rule:

$$(\beta_*) \quad \begin{cases} (\lambda^* x.M)N \xrightarrow{\beta_*} M[N/x] \\ \lambda^* y.D[(\lambda x.M)Y] \xrightarrow{\beta_*} \lambda y.D[M[Y/x]], \ \text{if } y \xrightarrow{\eta}\!\!\!\!\rightarrow Y \wedge [\,] \xrightarrow{\eta}\!\!\!\!\rightarrow D[\,]. \end{cases}$$

$$(\beta) \quad (\lambda x.M)N \xrightarrow{\beta} M[N/x], \quad \text{if } (\lambda x.M)N \text{ is not in a context where } (\beta_*) \text{ applies.}$$

Thus, when a β_*-redex is also a β-redex, we assume that its contraction constitutes a β_*-step.

Now, this is the key step: β_*, unlike the full β, is well behaved wrt β^2, as it preserves β^2 normal forms, and one can apply Akama's lemma to show that β^2, η and β_* together are confluent and normalizing.

This means that an infinite typed reduction must contain infinite β steps that are not β_* steps, and we are done because these β steps are not deleted by the η postponement that we perform on the erasure of the typed reduction: we can finally build an infinite β reduction leaving from a typable term, a contradiction, as we wanted.

What is particularly satisfactory in this proof technique is the fact that we really show that the only source of danger are the real β reductions, and not the β_* redexes produced by η expansion, which are really harmless. This proof technique can be applied successfully also to various extensions of the simple calculus presented here, as we will detail in the Conclusions.

1.3 Structure of the paper

The two main technical points in the paper will then be to present the η postponement in the unrestricted case and to prove the hypothesis of Akama's Lemma for $\beta_* \cup \eta \cup \beta^2$, but there is something else.

Indeed, we found that applying Akama's Lemma is hard: one has to show commutation of one reduction relation wrt the reduction to normal form for the other reduction relation, and this is done in Akama's paper in an ad hoc fashion for a specific calculus by a difficult technical analysis of η normal forms. We wanted a more easily applicable technique, that we found, and that we decided, for its generality, to present in a section by its own.

So, we will first expose formally the results on the untyped calculus, in the next section, then we will show how to drastically simplify the proofs involved in applying Akama's Lemma, in the following section, and finally we will apply this simpler technique to the typed reduction $\beta^2 \cup \eta \cup \beta_*$ in order to obtain the proof of strong normalization for the full reduction. Confluence will follow by Newman's Lemma.

We will then conclude with an overview of the applicability of our technique, and with some ideas for further work.

2 The Untyped Case

In this Section we introduce a λ-calculus with markers, which enable us to keep under control variables introduced by applications of the expansive version of the η rule. We characterize a relevant class of terminating reductions in such calculus.

Terms of the *untyped marked λ-calculus* are defined by the following syntax

$$M ::= x \mid MM \mid \lambda x.M \mid \lambda^* x.M, \tag{1}$$

where x ranges over a denumerable set Var of term variables; $FV(M)$ denotes the set of variables occurring free in M. We call Λ^* the set of terms resulting from (1), while Λ is the set of unmarked terms.

As usual, terms are considered modulo α-conversion, i.e. modulo names of bound variables.

One step η-reduction is the least binary relation on Λ^* which passes contexts and satisfies

$$(\eta) \quad M \xrightarrow{\eta} \lambda^* x.Mx, \qquad \text{if } x \notin FV(M).$$

$\xrightarrow{\eta}\!\!\!\!\rightarrow$ denotes the reflexive and transitive closure of the one step η-reduction relation, while $\xrightarrow[n]{\eta}\!\!\!\!\rightarrow$ denotes the n times composition of $\xrightarrow{\eta}$ with itself.

Remark 2.1 For simplicity of notation, we will always use η to denote *expansion* rules. In the untyped system, such reduction is applicable in any context; in the typed system, some restrictions will be introduced on the applicability of η-reduction. Nevertheless, the same notation will be adopted for both (either untyped and unrestricted or typed and restricted) η-reductions. Indeed, a special notation will be used in the typed case for unrestricted η-reduction.

One step β_* and β-reductions (notation: $\xrightarrow{\beta_*}$ and $\xrightarrow{\beta}$, respectively) are defined as the least binary relations on Λ^* which pass contexts and respectively satisfy:

$$(\beta_*) \quad \begin{cases} (\lambda^* x.M)N \xrightarrow{\beta_*} M[N/x] \\ \lambda^* y.D[(\lambda x.M)Y] \xrightarrow{\beta_*} \lambda y.D[M[Y/x]], \text{ if } y \xrightarrow{\eta}\!\!\!\!\rightarrow Y \wedge [\,] \xrightarrow{\eta}\!\!\!\!\rightarrow D[\,]. \end{cases}$$

$$(\beta) \quad (\lambda x.M)N \xrightarrow{\beta} M[N/x], \quad \text{if } (\lambda x.M)N \text{ is not in a context where } (\beta_*) \text{ applies}$$

Let $\rho = \beta \cup \beta_* \cup \eta$. Finally, let $\xrightarrow{\beta}\!\!\!\!\rightarrow$ ($\xrightarrow{\beta_*}\!\!\!\!\rightarrow$, $\xrightarrow{\rho}\!\!\!\!\rightarrow$) denote the reflexive and transitive closure of one step the β- (β_*, ρ) reduction relation.

Definition 2.2 We define Λ^*_η to be the subset of Λ^* whose elements are obtained from unmarked terms via η-reduction.

$$\Lambda^*_\eta = \{M \in \Lambda^* \mid \exists M' \in \Lambda.\, M' \xrightarrow{\eta}\!\!\!\!\rightarrow M\}.$$

Recall that a context $C[\,](\in \Lambda[\,], \Lambda^*[\,], \Lambda^*_\eta[\,])$ is a term (belonging to $\Lambda, \Lambda^*, \Lambda^*_\eta$) with one hole in it.

Property 2.3 *(i)* $M \in \Lambda^*_\eta \Leftrightarrow (\forall C[\,] \in \Lambda^*_\eta[\,].\, M \equiv C[\lambda^* x.N] \Rightarrow N \equiv D[N'X])$, *where* $N' \in \Lambda^*_\eta \wedge FV(N') \not\ni x \xrightarrow{\eta}\!\!\!\!\rightarrow X \wedge [\,] \xrightarrow{\eta}\!\!\!\!\rightarrow D[\,]$.

(ii) $(M \in \Lambda^*_\eta \wedge M \xrightarrow{\rho} N) \Rightarrow N \in \Lambda^*_\eta$;

Proof. *(i)* ($\Leftarrow$) is trivial. To prove ($\Rightarrow$), observe that if $M \in \Lambda^*_\eta$, then $\exists M' \in \Lambda$ and $n \in \mathbb{N}$, s.t. $M' \xrightarrow[n]{\eta}\!\!\!\!\rightarrow M$. We will reason by induction on n.

$n = 0$. Vacuously true.

$n = m + 1$. Then there is an $M' \in \Lambda$ s.t. $M' \xrightarrow[m]{\eta}\!\!\!\!\rightarrow M'' \xrightarrow{\eta} M$. We know by induction that the property holds for M''. If $M \equiv C[\lambda^* x.N]$, then, two cases:

(a) $\lambda^* x.N$ was already in M'', that is, for some $C'[\,] \in \Lambda^*_\eta[\,], M'' \equiv C'[\lambda^* x.N]$, and we are done by induction hypothesis.

(b) $\lambda^* x.N$ is the result of the last step $M'' \xrightarrow{\eta} M$, and we have two cases:

(b.1) $M'' \equiv C[N']$ and $M \equiv C[\lambda^* x.N'x]$, so $N = N'x$ with $x \notin FV(N')$, and we are done.

(b.2) $M'' \equiv C[\lambda^* x.N']$ and $M \equiv C[\lambda^* x.N]$, with $N' \xrightarrow{\eta} N$. By induction hyp. we know that $N' \equiv D[N''X]$, where $FV(N'') \not\ni x \xrightarrow{\eta}\!\!\!\!\rightarrow X \wedge [\,] \xrightarrow{\eta}\!\!\!\!\rightarrow D[\,]$. Now, the last expansion can be either $N'' \xrightarrow{\eta} N'''$, and then $M \equiv C[\lambda^* x.D[N'''X]]$,
or $X \xrightarrow{\eta} X'$, and then $M \equiv C[\lambda^* x.D[N''X']]$,
or $D[N''X] \xrightarrow{\eta} D'[N''X]$, and then $M \equiv C[\lambda^* x.D'[N''X]]$,
with $D[\,] \xrightarrow{\eta} D'[\,]$. In all cases, the conditions are satisfied and we are done.

(ii) If the ρ reduction is an η, then the property holds by the very definition of Λ^*_η. In the other cases (β and β_*), the proof is deferred after Lemma 2.7 from which it follows immediately. □

The following facts can be shown by simple calculations

Fact 2.4 *If* $H \in \Lambda^*_\eta$ *and* $x \in \mathsf{Var}$ *then* $x \stackrel{\eta}{\twoheadrightarrow} X \Rightarrow H \stackrel{\eta}{\twoheadrightarrow} X[H/x]$.

Fact 2.5 *If* $H, J \in \Lambda^*_\eta$ *and* $x \in \mathsf{Var}$ *then*

$$(H \stackrel{\eta}{\twoheadrightarrow} H', J \stackrel{\eta}{\twoheadrightarrow} J') \Rightarrow H[J/x] \stackrel{\eta}{\twoheadrightarrow} H'[J'/x].$$

Fact 2.6 *If* $x \in \mathsf{Var}$ *and* $x \stackrel{\eta}{\twoheadrightarrow} X \stackrel{\eta}{\twoheadrightarrow} X'$ *then* $X'[X/x] \stackrel{\beta_*}{\twoheadrightarrow} X'$.

Lemma 2.7 *Let* $P \in \Lambda$ *and* $M, N \in \Lambda^*_\eta$.

(i) If $P \stackrel{\eta}{\twoheadrightarrow} M \xrightarrow{\beta_*} N$, *then* $P \stackrel{\eta}{\twoheadrightarrow} N$.

(ii) If $P \stackrel{\eta}{\twoheadrightarrow} M \xrightarrow{\beta} N$, *then there exists* $Q \in \Lambda$ *such that* $P \xrightarrow{\beta} Q \stackrel{\eta}{\twoheadrightarrow} N$.

(iii) Let $M, N \in \Lambda^*_\eta$ *and* $\tau = \beta_* \cup \eta$. *If* $M \stackrel{\tau}{\twoheadrightarrow} N$, *then there exists* $Q \in \Lambda^*_\eta$ *such that* $M \stackrel{\eta}{\twoheadrightarrow} Q \stackrel{\beta_*}{\twoheadrightarrow} N$.

Proof. *(i)* We distinguish the two cases for β_*:

(a) If $M \equiv C[(\lambda^* x.S)R]$, then by Property 2.3 we have $S \equiv D[QX]$ where $Q \in \Lambda^*_\eta$ and $x \stackrel{\eta}{\twoheadrightarrow} X$, and then

$$M \equiv C[(\lambda^* x.D[QX])R] \xrightarrow{\beta_*} C[D[QX[R/x]]] \equiv N.$$

Hence $P \equiv C'[Q'R']$, where $Q' \stackrel{\eta}{\twoheadrightarrow} Q$, $R' \stackrel{\eta}{\twoheadrightarrow} R$ and $C'[] \stackrel{\eta}{\twoheadrightarrow} C[]$. This case is settled using Fact 2.4.

(b) $M \equiv C[\lambda^* y.D[(\lambda x.Q)Y]] \xrightarrow{\beta_*} C[\lambda y.D[Q[Y/x]]] \equiv N$, with $y \stackrel{\eta}{\twoheadrightarrow} Y$. Hence $P \equiv C'[\lambda x.Q']$, where $Q' \stackrel{\eta}{\twoheadrightarrow} Q$ and $C'[] \stackrel{\eta}{\twoheadrightarrow} C[]$. This case is settled using Fact 2.4.

(ii) $M \equiv C[(\lambda x.Q)R] \xrightarrow{\beta} C[Q[R/x]] \equiv N$. Hence

$$P \equiv C'[(\lambda x.Q')R'],$$

where $Q' \stackrel{\eta}{\twoheadrightarrow} Q$, $R' \stackrel{\eta}{\twoheadrightarrow} R$ and $C'[] \stackrel{\eta}{\twoheadrightarrow} C[]$. This case is settled using Fact 2.5.

(iii) We distinguish the two cases for β_* and we observe that:

(a)

$$\begin{array}{ccc} C[(\lambda^* x.D[QX])R] & \xrightarrow{\beta_*} & C[D[QX[R/x]]] \\ \downarrow \eta & & \downarrow \eta \\ C'[(\lambda^* x.D'[Q'X'])R'] & \xrightarrow{\beta_*} & C'[D'[Q'X'[R'/x]]] \end{array}$$

where $C[](D[]Q, X, R$, resp.$) \stackrel{\eta}{\twoheadrightarrow} C'[](D'[]Q', X', R'$, resp.$)$;

(b)

$$\begin{array}{ccc} C[\lambda^* y.D[(\lambda x.Q)Y]] & \xrightarrow{\beta_*} & C[\lambda y.D[Q[Y/x]]] \\ \downarrow \eta & & \downarrow \eta \\ C'[\lambda^* y.D'[(\lambda x.Q'[Y'_i/x^{(i)}]_{i=1\ldots n})Y]] & \underset{(2.6)}{\stackrel{\beta_*}{\twoheadrightarrow}} & C'[\lambda y.D'[Q'[Y'_i/x^{(i)}]_{i=1,\ldots,n}]] \end{array}$$

where $x^{(1)}, \ldots, x^{(n)}$ denote the occurrences of the free variable x in Q and

$$C[](D[], Q, \text{resp.}) \stackrel{\eta}{\twoheadrightarrow} C'[](D'[], Q, \text{resp.}), y \stackrel{\eta}{\twoheadrightarrow} Y \stackrel{\eta}{\twoheadrightarrow} Y'_i,$$

for $i = 1, \ldots, n$.

Let now $M \stackrel{\tau}{\twoheadrightarrow} N$. The lemma follows by an easy induction on the number of β_* steps which are followed by an η step in the reduction from M to N. □

Definition 2.8 Let $M_0 \in \Lambda^*_\eta$. A ρ-reduction path

$$\Pi\colon M_0 \xrightarrow{\rho} M_1 \xrightarrow{\rho} M_2 \xrightarrow{\rho} \cdots$$

starting from M_0 is called *fair* iff either it is finite or, for any $i \in \mathbb{N}$, it satisfies the following conditions

(i) $M_i \xrightarrow{\beta} M_{i+1} \Rightarrow \exists k > 0.\neg(M_{i+k} \xrightarrow{\beta} M_{i+k+1})$;
(ii) $M_i \xrightarrow{\beta_*} M_{i+1} \Rightarrow [\exists k > 0.\neg(M_{i+k} \xrightarrow{\beta_*} M_{i+k+1})] \wedge \neg(M_{i+1} \xrightarrow{\eta} M_{i+2})$;
(iii) $M_i \xrightarrow{\eta} M_{i+1} \Rightarrow \exists k > 0.\neg(M_{i+k} \xrightarrow{\eta} M_{i+k+1})$.

Lemma 2.9 (Main Lemma) *Let $M \in \Lambda$ be a β-strongly normalizing term and let Π be a ρ-reduction path starting from M. Π is finite iff it is fair.*

Proof. Assume the existence of an infinite fair ρ-reduction path starting from M. By definition, an infinite fair ρ-reduction path contains an infinite amount of β steps. Indeed, it does not contain infinite subpaths constituted by all β (β_*, η, repectively) steps, and also it does not contain any infinite subpath in which β steps do not appear, since by Definition 2.8.*(ii)* a β_* step is never followed by an η step.

Using Lemma 2.7, we can build an infinite β-reduction starting from M, which is absurd. Indeed, take a fair reduction sequence starting from a term $M \in \Lambda$ and containing an infinite number of β steps. Consider now the first β step in the sequence.

By Lemma 2.7, we can assume that all reduction steps from M to this first β are η steps: if not, these steps must be a sequence of η followed by a sequence of β_*, by definition of fair reduction sequence, and then we can apply Lemma 2.7.*(i)* and get rid of the β_* sequence, obtaining a reduction sequence that is still fair. Then, from $\Lambda \ni M \xrightarrow{\eta}\!\!\!\!\rightarrow M' \xrightarrow{\beta} M'' \longrightarrow\!\!\!\!\rightarrow \cdots$ we get, using Lemma 2.7.*(iii)* a new fair sequence $\Lambda \ni M \xrightarrow{\beta} M''' \xrightarrow{\eta}\!\!\!\!\rightarrow M'' \longrightarrow\!\!\!\!\rightarrow \cdots$. Now, it suffices to notice that M''' is still in Λ, and that the sequence starting from M''' is again fair and contains an infinite number of β steps, so we can iterate this pumping process and build an infinite β reduction sequence starting from M. □

3 Simplifying Akama's Lemma

It is now time to turn to Akama's Lemma: applying it directly is hard just like the usual Hindley-Rosen's Lemma 1.1, as one has to handle a multi-step reduction.

But for the Hindley-Rosen's Lemma to be applicable, there is a well known sufficient condition; this just asks us to verify that any divergent diagram $M' \xleftarrow{S} M \xrightarrow{R} M''$ can be closed using as many R steps as we want, but no more than one S step. This sufficient condition is what is always used, for its simplicity (see for example [Bar84]).

Along our investigation, we had to devise a similar sufficient condition for Akama's Lemma, to simplify the otherwise extremely difficult proof of the Lemma's hypothesis. This sufficient condition is so general and nice to prove, that even the results in Akama's original paper can be obtained in a few lines, without the complex syntactic analysis used there.

Notation 3.1 Let $\langle \mathcal{A}, \longrightarrow \rangle$ be an Abstract Reduction System. We denote by

$\xrightarrow[=]{}$ the reflexive closure of $\longrightarrow$;
$\xrightarrow[+]{}\!\!\!\!\rightarrow$ the transitive closure of $\longrightarrow$;
$\longrightarrow\!\!\!\!\rightarrow$ the reflexive and transitive closure of $\longrightarrow$.

Lemma 3.2 *Let $\langle \mathcal{A}, \xrightarrow{R}, \xrightarrow{S} \rangle$ be an Abstract Reduction System, where R-reduction is strongly normalizing. Let the following commutation hold*

$$\forall a,b,c,d \in \mathcal{A} \quad \begin{array}{ccc} a & \xrightarrow{R} & c \\ {\scriptstyle S}\downarrow & & \downarrow{\scriptstyle S} \\ b & \overset{R}{\underset{+}{\dashrightarrow}}\!\!\!\!\rightarrow & d \end{array}$$

Then we have

(i) $\xrightarrow{\;R\;}$ and $\xrightarrow{\;S\;}$ commute.

(ii) if R preserves S normal forms (let $S\downarrow$ denote reduction to S normal form), then

$$\forall a,b,c,d \in \mathcal{A} \quad \begin{array}{ccc} a & \xrightarrow{\;R\;} & c \\ {\scriptstyle S\downarrow}\big\downarrow & & \big\downarrow{\scriptstyle S\downarrow} \\ b & \xdashrightarrow{\;R\;} & d \end{array}$$

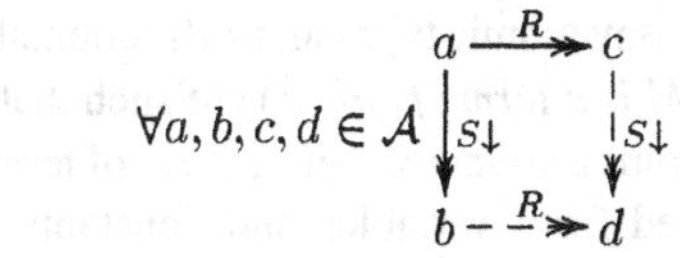

(iii) if S is also confluent and R preserves S normal forms, then

$$\forall a,b,c,d \in \mathcal{A} \quad \begin{array}{ccc} a & \xrightarrow{\;R\;} & c \\ {\scriptstyle S\downarrow}\big\downarrow & & \big\downarrow{\scriptstyle S\downarrow} \\ b & \xdashrightarrow[+]{\;R\;} & d \end{array}$$

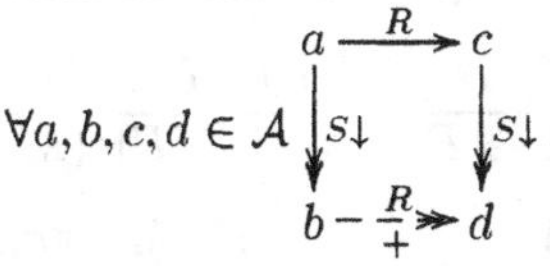

Proof. We just prove the first result, as the others are very simple consequences of it. Such result has been independently obtained by Alfons Geser in his PhD Thesis [Ges]. If $a_1, a_2 \in \mathcal{A}$, then denote $deg(a_1)$ the length of the longest R-reduction path out of a_1 and $dist(a_1, a_2)$ the length of a S-reduction sequence from a_1 to a_2. The proof is by induction on pairs $(deg(b), dist(a,b))$, ordered lexicographically. Indeed, if $deg(b) = 0$ or $dist(a,b) = 0$, then the lemma trivially holds. Otherwise, by hypothesis, there exist a', a'', a''' as in the following diagram.

$$\begin{array}{ccccc} a & \xrightarrow{\;R\;} & a' & \xrightarrow{\;R\;} & c \\ {\scriptstyle S}\big\downarrow & & {\scriptstyle S}\big\downarrow & & \big| \\ a'' & \xrightarrow{R} \to & a''' & & \big| \\ & & & & {\scriptstyle S}\big| \\ {\scriptstyle S}\big| & D_1 & {\scriptstyle S}\big| & D_2 & \big| \\ b & \dashrightarrow{R}\!\!\!\twoheadrightarrow & b' & \dashrightarrow{R}\!\!\!\twoheadrightarrow & d \end{array}$$

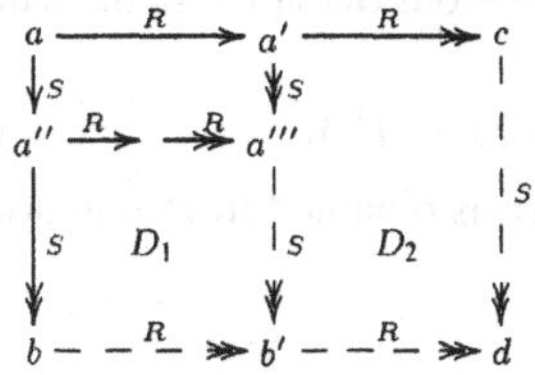

We can now apply the inductive hypothesis to the diagram D_1, since

$$(deg(b), dist(a'', b)) <_{lex} (deg(b), dist(a,b)).$$

Finally, we observe that $b \xrightarrow[+]{\;R\;} b'$, just composing the diagram in the hypothesis down from a.

Hence we can apply the inductive hypothesis to the diagram D_2, since

$$(deg(b'), dist(a', b')) <_{lex} (deg(b), dist(a,b)),$$

and we are done. $\square$

Lemma 3.2.*(iii)* tells us that in using Akama's Lemma, before trying to prove directly the commutation between R and $S\downarrow$ we should better check the one step commutation between R and S, and verify if R preserves S normal forms, which can be boring, but usually simple tasks.

A simple proof of confluence and normalization for $\lambda^1 \beta\eta\pi*$

As a simple application, consider the typed lambda calculus $\lambda^1 \beta\eta\pi*$ for Cartesian Closed Categories: this consists of β, η, π, SP and a rule Top that collapses all terms of a special type T into a single constant $*$ (with both η and SP taken as expansions). If we take $R = \beta \cup \pi \cup Top$ and $S = \eta \cup SP$, it is extremely simple to verify our sufficient condition, and then confluence and normalization for the full system are a consequence of the same properties for the two separate subsystems, that can be shown fairly easily with simple traditional tehniques.

4 The calculus $\lambda^2 \beta\eta$

We briefly recall that in the second order λ-calculus $\lambda^2 \beta\eta$

Types are defined by the following grammar:

$$Type ::= At \mid TVar \mid Type \to Type \mid \forall\sigma.Type,$$

where At are countably many atomic types and $TVar$ countably many type variables.

Terms (M:A will stand for *M is a term of type A*) are such that

- the set of terms contains a countable set $x, y, \ldots$ of term variables for each type
- terms are constructed from variables and constants via the following term formation rules (notice the perfect analogy with the introduction and elimination rules for second order logic in natural deduction style)

$$\frac{\Gamma, x : A \vdash M : B}{\Gamma \vdash \lambda x.M : A \to B} \qquad \frac{\Gamma \vdash M : A \to B \quad \Gamma \vdash N : A}{\Gamma \vdash (MN) : B}$$

$$\frac{\Gamma \vdash M : A}{\Gamma \vdash \Lambda\sigma.M : \forall\sigma.A}\,^{6} \qquad \frac{\Gamma \vdash M : \forall\sigma.A}{\Gamma \vdash M[B] : A[B/\sigma]} \text{ for any type B.}$$

Equality is generated by

$$(\beta) \quad (\lambda x.M)N = M[N/x] \qquad (\eta) \quad \lambda x.Mx = M \text{ if } x \notin FV(M)$$

$$(\beta^2) \quad (\Lambda\sigma.M)[A] = M[A/\sigma]$$

Now we can introduce marked abstractions as in the previous Section. We have then again a set of pre-terms Λ^2_* generated by the grammar

$$M ::= x \mid MM \mid \lambda x.M \mid \lambda^* x.M \mid \Lambda\sigma.M \mid M[A] \qquad (2)$$

and from these we define a set of marked second order terms obtained from unmarked terms by means of *unrestricted* expansions.

Definition 4.1 (Marked terms)

$\Lambda^2_{*\eta} = \{M \in \Lambda^2_* \mid \exists M' \in \lambda^2\beta\eta.\ M'\ \eta\text{-expands in an unrestricted way to} M\}$.

These are the terms of the marked typed calculus $\lambda^2\beta\eta_*$, that has the following associated rewriting system:

$$(\beta_*) \quad \begin{cases} (\lambda^* x.M)N \xrightarrow{\beta_*} M[N/x] \\ \lambda^* y.D[(\lambda x.M)Y] \xrightarrow{\beta_*} \lambda y.D[M[Y/x]], \text{ if } y \xrightarrow{\eta}\!\!\!\!\rightarrow Y \wedge [\,] \xrightarrow{\eta}\!\!\!\!\rightarrow D[\,] \end{cases}$$

$$(\beta) \quad (\lambda x{:}\, A.M)N \xrightarrow{\beta} M[N/x]$$

if $(\lambda x{:}\, A.M)N$ is not in a context where (β_*) applies

$$(\eta) \quad M \xrightarrow{\eta} \lambda^* x{:}\, A.Mx \text{ if } \begin{cases} x \notin FV(M) \\ M{:}\, A \to B \\ M \text{ is not applied} \end{cases}$$

$$(\beta^2) \quad (\Lambda\sigma.M)[A] \xrightarrow{\beta^2} M[A/X]$$

Again, we have split (β) into (β) and (β_*).

The one-step reduction relation between terms is defined as the least relation which includes $\beta, \beta_*, \beta^2, \eta$ and is closed for all the contexts *except* in the application case:

$$\text{if } M \Longrightarrow M', \text{ then } MN \Longrightarrow M'N \text{ except in the case } M \xrightarrow{\eta} M';$$

but, for the sake of simplicity, we will avoid using an additional symbol $\Longrightarrow$ to denote it.

[6] With the proviso that the type variable σ is not free in the type of any free variable of the term M.

Notation 4.2 The transitive and the reflexive transitive closure of $\longrightarrow$ are noted $\overset{}{\underset{+}{\longrightarrow\!\!\!\!\rightarrow}}$ and $\longrightarrow\!\!\!\!\rightarrow$ respectively. Furthermore, we denote $\xrightarrow{\eta_{unr}}\!\!\!\!\rightarrow$ the one-step unrestricted η-reduction.

The so obtained *typed* calculus still has the following property:

Property 4.3 (See 2.3)

(i) $M \in \Lambda^2_{*\eta} \Leftrightarrow (\forall C[\,] \in \Lambda^2_{*\eta}[\,].\, M \equiv C[\lambda^* x.N] \Rightarrow N \equiv D[N'X])$,
where $N' \in \Lambda^2_{*\eta} \wedge FV(N') \not\ni x \xrightarrow{\eta}\!\!\!\!\rightarrow X \wedge [\,] \xrightarrow{\eta}\!\!\!\!\rightarrow D[\,]$.

(ii) $(M \in \Lambda^2_{*\eta} \wedge M \longrightarrow N) \Rightarrow N \in \Lambda^2_{*\eta}$;

Proof. Property *(i)* can be shown by induction exactly as in the untyped case. As for property *(ii)*, we just need to focus on and β^2 reduction, as for the other ones one can proceed exactly as in 2.3. For this, it suffices to show that if $M \xrightarrow{\eta_{unr}}\!\!\!\!\rightarrow M' \xrightarrow{\beta^2} N$, then there exists an M'' such that $M \xrightarrow{\beta^2} M'' \xrightarrow{\eta_{unr}}\!\!\!\!\rightarrow N$. This is easy, because we are using the unrestricted η expansion. Then, given any term M in $\Lambda^2_{*\eta}$, we have $M' \xrightarrow{\eta_{unr}}\!\!\!\!\rightarrow M \xrightarrow{\beta^2} M''$ for some $M \in \Lambda^2_{*\eta}$, that can be turned into $M' \xrightarrow{\beta^2} M'''$ $\xrightarrow{\eta_{unr}}\!\!\!\!\rightarrow M''$ for some $M''' \in \Lambda^2_{*\eta}$, so $M'' \in \Lambda^2_{*\eta}$ too. □

4.1 Properties of Reduction

Let γ be a notion of reduction; we denote by $\gamma\!\downarrow$ an exhaustive γ-reduction path.

Remark 4.4 If $Q \xrightarrow{\eta} Q'$, then $Q[A] \xrightarrow{\eta} Q'[A]$.

Proof. It is an easy induction on the structure of Q. □

Remark 4.5 The reductions β^2 and η alone are confluent and strongly normalizing.

Proof. Folklore for β^2, see [Kes93, Cub92, DCK94a, Min79] for η. □

Lemma 4.6 (Commutation of β^2 wrt η) *β^2 commutes (in one step) with η .*

Proof. We consider all possible critical pairs between η and β^2 :

$$\begin{array}{ccc} (\Lambda\sigma.M)[A] & \xrightarrow{\eta} & \lambda y : B.((\Lambda\sigma.M)[A])y \\ \beta^2\!\downarrow & & \beta^2 \\ M[A/\sigma] & \overset{\eta}{\dashrightarrow} & \lambda y : B.(M[A/\sigma])y \end{array} \qquad \begin{array}{ccc} (\Lambda\sigma.M)[A] & \xrightarrow{\eta} & (\Lambda\sigma.\lambda y : B.My)[A] \\ \beta^2\!\downarrow & & \beta^2 \\ M[A/\sigma] & \overset{\eta}{\dashrightarrow} & \lambda y : B[A/\sigma].(M[A/\sigma])y \end{array}$$

In these diagrams, the erasure of $(\Lambda\sigma.M)[A]$ is not an abstraction, because we can apply η ; but the erasure of $M[A/\sigma]$ is the same, so we can still apply an η , and close the diagram in one step. Using these diagrams, the one step commutation property for the general case is shown by induction on the structure of contextual reductions. □

Lemma 4.7 (Commutation of η with reduction to β^2 n.f.)

$$\begin{array}{ccc} M & \xrightarrow{\eta} & N \\ \beta^2\!\downarrow & & \beta^2\!\downarrow \\ M' & \overset{\eta}{\dashrightarrow} & N' \end{array}$$

Proof. Consider the reduction sequence from M to the β^2 normal form M' of M, and the reduction $M \xrightarrow{\eta} N$. We can apply repeatedly Lemma 4.6 to close the diagram, obtaining

$$\begin{array}{ccc} M & \xrightarrow{\eta} & N \\ {\scriptstyle \beta^2}\downarrow & & \downarrow{\scriptstyle \beta^2} \\ M' & \dashrightarrow_{\eta} & N'' \end{array} \qquad \text{hence} \qquad \begin{array}{ccc} M & \xrightarrow{\eta} & N \\ {\scriptstyle \beta^2\downarrow}\downarrow & & \downarrow{\scriptstyle \beta^2\downarrow} \\ M' & \dashrightarrow_{\eta} & N'' \end{array}$$

since η preserves β^2 normal forms. Finally, being β^2 normal forms unique, $N' = N''$ so $M' \xrightarrow{\eta} N'$ as needed. □

Corollary 4.8 *$\beta^2 \cup \eta$ is confluent and strongly normalizing.*

Proof. Using the previous lemma, and knowing that β^2 and η separatley are CR and SN, we obtain the result by Akama's Lemma. □

Property 4.9 *Relationship between:*

(i) β_* and β^2 :
$$\begin{array}{ccc} M & \xrightarrow{\beta_*} & N \\ {\scriptstyle \beta^2}\downarrow & & \downarrow{\scriptstyle \beta^2} \\ Q & \dashrightarrow_{\beta_*} & R \end{array}$$

(ii) β_* and η:
$$\begin{array}{ccc} M & \xrightarrow{\beta_*} & N \\ {\scriptstyle \eta}\downarrow & & \downarrow{\scriptstyle \eta} \\ Q & \dashrightarrow^{+}_{\beta_*} & R \end{array}$$

Proof. *(i)* There are no non-trivial critical pairs between β_* and β^2 and since β^2 is a rewriting rule without restrictions, it is a matter of a simple induction on the derivation of the reductions to prove the commutation. (The fact that we need only one β_* step to close the diagram comes from the fact that β^2 cannot duplicate subterms.)

(ii) We use our knowledge of the structure of a marked abstraction to distinguish two cases:

(a) $M \equiv C[(\lambda^* x.D[PX])T]$, where $x \notin FV(P), x \xrightarrow{\eta} X, [\,] \xrightarrow{\eta} D[\,]$.
We have $M \xrightarrow{\beta_*} C[D[PX[T/x]]] \equiv N$, i.e.

$$Q \xleftarrow{\eta} C[(\lambda^* x.D[PX])T] \equiv M \xrightarrow{\beta_*} N \equiv C[D[PX[T/x]]].$$

Now, four cases are possible:

1. $Q \equiv C'[(\lambda^* x.D[PX])T]$, with $C[\,] \xrightarrow{\eta} C'[\,]$. Then we have two cases:
   - $C'[\,] \equiv C[\lambda^* y.[\,]y]$ and $D[PX[T/x]]$ is an abstraction. This can happen only if $D[\,] \equiv \lambda^* z.D'[\,]z$, but then
   $$\begin{array}{rl} Q & \xrightarrow{\beta_*} C[\lambda^* y.(\lambda^* z.(D'[PX[T/x]])z)y] \\ & \xrightarrow{\beta_*} C[\lambda^* y.D'[PX[T/x]]y] \\ & \equiv \;\; C[D[PX[T/x]]]; \end{array}$$
   - $Q \xrightarrow{\beta_*} C'[D[PX[T/x]]] \equiv R \xleftarrow{\eta} C[D[PX[T/x]]]$, otherwise.
2. $Q \equiv C[(\lambda^* x.D[P'X])T]$, with $P \xrightarrow{\eta} P'$.
   The expansion in P' cannot be at the root (P' is applied) and it can be performed after the β_*, closing the diagram with $R \equiv C[D[P'X[T/x]]]$.
3. $Q \equiv C[(\lambda^* x.D[PX'])T]$, with $X \xrightarrow{\eta} X'$.
   Two cases are possible here: if $N \xrightarrow{\eta} C[D[PX'[T/x]]]$, then we are done. Otherwise, $x \equiv X \xrightarrow{\eta} \lambda^* t.xt \equiv X'$ and T has an initial abstraction. Hence we have the thesis observing that
   $$\begin{array}{l} Q \equiv C[(\lambda^* x.D[P(\lambda^* t.xt)])(\lambda w.T')] \xrightarrow{\beta_*} \\ C[D[P(\lambda^* t.(\lambda w.T')t)]] \xrightarrow{\beta_*} \\ C[D[P(\lambda t.T'[t/w])]] \equiv N. \end{array}$$
   The case where the λ binding the variable w is a marked one is similar.

4. $Q \equiv C[(\lambda^* x.D[PX])T']$, with $T \xrightarrow{\eta} T'$.
Here again, if η is not allowed after the β_* reduction, it is the case that we can perform another β_* step to close the diagram.

(b) $M \equiv C[\lambda^* y.D[(\lambda x.P)Y]]$, where $y \notin FV(P), y \xrightarrow{\eta}\!\!\!\!\rightarrow Y, [\,] \xrightarrow{\eta}\!\!\!\!\rightarrow D[\,]$.
We have :

$$Q \xleftarrow{\eta} C[\lambda^* y.D[(\lambda x.P)Y]] \equiv M \xrightarrow{\beta_*} N \equiv C[\lambda y.D[P[Y/x]]].$$

Now, four cases are possible:

1. $Q \equiv C'[\lambda^* y.D[(\lambda x.P)Y]]$, with $C[\,] \xrightarrow{\eta} C'[\,]$.
Here $Q \xrightarrow{\beta_*} R$ and $N \xrightarrow{\eta} R$, where $R \equiv C'[\lambda y.D[P[Y/x]]]$, and this case is settled.
2. $Q \equiv C[\lambda^* y.D'[(\lambda x.P)Y]]$, with $D[\,] \xrightarrow{\eta} D'[\,]$.
Two cases are possible here: if $N \xrightarrow{\eta} C[\lambda y.D'[P[Y/x]]]$, the thesis follows exactly as in case 1. Otherwise, we are in the case that
$$\neg(N \xrightarrow{\eta} C[\lambda y.D'[P[Y/x]]]).$$
This may only happen when $D \equiv [\,], D' \equiv \lambda^* t.[\,]t$ and P has an external abstraction. Hence we have the thesis observing that
$$\begin{array}{rcl} Q & \equiv & C[\lambda^* y.\lambda^* t.(\lambda x.(\lambda w.P'))Yt] \\ & \xrightarrow{\beta_*} & C[\lambda y.\lambda^* t.(\lambda w.P')[Y/x]t] \\ & \xrightarrow{\beta_*} & C[\lambda y.\lambda t.P'[Y/x, t/w]] \equiv N. \end{array}$$
The case where the λ binding the variable w is a marked one is similar.
3. $Q \equiv C[\lambda^* y.D[(\lambda x.P')Y]]$, with $P \xrightarrow{\eta} P'$.
Similar to case 2, with some small adjustments.
4. $Q \equiv C[\lambda^* y.D[(\lambda x.P)Y']]$, with $Y \xrightarrow{\eta} Y'$.
Two cases are possible here: if $N \xrightarrow{\eta}\!\!\!\!\rightarrow C[\lambda y.D[P[Y'/x]]]$, the thesis follows exactly as in case 1. Otherwise, we are in the case that
$$\neg(N \xrightarrow{\eta}\!\!\!\!\rightarrow C[\lambda y.D[P[Y'/x]]]).$$
This may happen when $y \equiv Y$ and some occurrences of x are P is in functional position in applications. Let us then distinguish such occurrences, denoting them by $\bar{x}$; moreover, let us assume that $P[Y'/x, Y/\bar{x}]$ denotes the term obtained from P substituting Y' for occurrences of x which are not in functional position in P, and Y for those in functional position. Hence we have the thesis observing that
$$Q \xrightarrow{\beta_*} C[\lambda y.D[P[Y'/x]]] \xrightarrow{\beta_*}\!\!\!\!\rightarrow C[\lambda y.D[P[Y'/x, Y/\bar{x}]]] \xleftarrow{\eta}\!\!\!\!\leftarrow N.$$

□

Property 4.10 *β_* preserves β^2 and η-normal forms.*

Proof. We show that if a reduct is not in β^2 (η)-normal form, then the redex is not in β^2 (η)-normal form either.

It is not possible to create η expansion redexes by β-reduction in general, since this reduction preserves the type of all subterms: imagine indeed we have a reduction $C[(\lambda x{:}\, A.M)N] \xrightarrow{\beta} C[M[N/x]]$, where the second term has an η-redex. If the redex is inside N or M or $C[\,]$, then it already exists in the first term. If it is M or $C[\,]$, then again it is already in the first term. If it is one of the new occurrences of N, then notice that these occurrences have the same type as N in the first term, so N in the first term is a redex too.

For β^2 , we use the fact that the substitutions done by β_* always involve terms that are not of quantifed type, and hence cannot create β^2 redexes.

Lemma 4.11 (Commutation of β^2 and η n.f. wrt β_*) *If $M \xrightarrow{\beta_*} N$, then at least one step of β_* can be performed on the $\beta^2 \cup \eta$-n.f. of M to reach the $\beta^2 \cup \eta$-n.f. of N.*

Proof. Just notice that Properties 4.9and 4.10 fulfill the hypothesis of Lemma 3.2.*(iii)*. □

Corollary 4.12 *The reduction $\beta^2 \cup \eta \cup \beta_*$ is strongly normalizing.*

Proof. By the previous lemma, and the separate strong normalization of $\beta^2 \cup \eta$ reduction and β_* reduction. Notice that, since β_* is *not* confluent, one cannot apply here directly Akama's Lemma. Indeed, one can prove that $\beta^2 \cup \eta \cup \beta_*$ is confluent also, but it is not necessary for the general result. □

Theorem 4.13 *The reduction $\beta^2 \cup \eta \cup \beta \cup \beta_*$ is confluent and strongly normalizing.*

Proof. Assume the existence of an infinite reduction in the typed λ -calculus:

$$\Pi: M_0 \longrightarrow M_1 \longrightarrow \cdots$$

We associate to Π a sequence

$$\Pi': M'_0 \xrightarrow{=} M'_1 \xrightarrow{=} \cdots$$

in the untyped λ-calculus, such that, for all i, $M'_i = erasure(M_i)$. We observe that Π' is still infinite, since, by Corollary 4.12, Π must contain an infinite amount of β steps, and

$$\forall M, N \in \Lambda^2_{*\eta}. (M \xrightarrow{\beta} N) \Rightarrow (erasure(M) \xrightarrow{\beta} erasure(N)).$$

By Lemma 2.7.*(iii)*, Π' can be transformed into a fair sequence Π''. Now, we know that M'_0 is strongly β-normalizing, since it is the erasure of a typed term. Hence Π'' contradicts Lemma 2.9, and this proves the strong normalization property.
Finally, the system is weakly confluent (for independent reasons, the diagrams in the previous Lemmas show almost all relevant cases), so confluence follows by Newman's Lemma. □

Corollary 4.14 (Strong normalization and confluence for $\lambda^2\beta\eta$)
The reduction $\beta^2 \cup \eta \cup \beta$ is confluent and strongly normalizing.

Proof. A simple consequence of the previous result, because of the direct correspondence between reduction sequences in the marked and in the unmarked calculi. □

5 Conclusion

In this paper, not only we presented the very first proof that the expansive approach to extensional equalities, most notably η, can be succesfully carried on to the second order typed λ -calculus, but we did it by means of extremely elementary methods, that do not involve reducibility candidates, complex translations or difficult synactic analysis of expansionary normal forms.

This elementarity can be clearly seen by considering the first order case: in the absence of β^2, there is no need to single out a β_* reduction as in the second order case, and using the Lemma in Section 3 one can get a proof much simpler that all the known proofs mentioned in the Introduction.

The key of the success is twofold: on one side, the marking that tracks the β-redexes created because of expansions, and on the other side, the simple Lemma 3.2, whose hypothesis are easy to verify (this last can have, in these authors' opinion, wide applicability in the theory of abstract term rewriting systems).

It is now important to turn towards several extensions of this result: is it possible to handle in the same way extensionality for quantified types (η^2)? What about combinations with algebraic rewriting systems? What about the Top type? All these questions are currently under active investigation.

Acknowledgements

We would like to thank Delia Kesner, for many discussions and her fundamental help with all matters concerning expansion rules.

References

[Aka93] Yohji Akama. On Mints' reductions for ccc-calculus. In *Typed Lambda Calculus and Applications*, number 664 in LNCS, pages 1–12. Springer Verlag, 1993.

[Bar84] Henk Barendregt. *The Lambda Calculus; Its syntax and Semantics (revised edition)*. North Holland, 1984.

[CDC91] Pierre-Louis Curien and Roberto Di Cosmo. A confluent reduction system for the λ-calculus with surjective pairing and terminal object. In Leach, Monien, and Artalejo, editors, *Intern. Conf. on Automata, Languages and Programming (ICALP)*, volume 510 of *Lecture Notes in Computer Science*, pages 291–302. Springer-Verlag, 1991.

[Cub92] Djordje Cubric. On free ccc. Distributed on the `types` mailing list, 1992.

[DCK93] Roberto Di Cosmo and Delia Kesner. Simulating expansions without expansions. Technical Report LIENS-93-11/INRIA 1911, LIENS-DMI and INRIA, 1993.

[DCK94a] Roberto Di Cosmo and Delia Kesner. Modular properties of first order algebraic rewriting systems, recursion and extensional lambda calculi. In *Intern. Conf. on Automata, Languages and Programming (ICALP)*, Lecture Notes in Computer Science. Springer-Verlag, 1994.

[DCK94b] Roberto Di Cosmo and Delia Kesner. Simulating expansions without expansions. *Mathematical Structures in Computer Science*, 1994. A preliminary version is available as Technical Report LIENS-93-11/INRIA 1911.

[Dou93] Daniel J. Dougherty. Some lambda calculi with categorical sums and products. In *Proc. of the Fifth International Conference on Rewriting Techniques and Applications (RTA)*, 1993.

[Ges] Alfons Geser. *Relative termination*. PhD thesis, Dissertation, Fakultät für Mathematik und Informatik, Universität Passau, Germany, 1990. Also available as: Report 91-03, Ulmer Informatik-Berichte, Universität Ulm, 1991.

[JG92] Colin Barry Jay and Neil Ghani. The virtues of eta-expansion. Technical Report ECS-LFCS-92-243, LFCS, 1992. University of Edimburgh.

[Kes93] Delia Kesner. *La définition de fonctions par cas à l'aide de motifs dans des langages applicatifs*. Thèse de doctorat, Université de Paris XI, Orsay, december 1993. To appear.

[Min79] Gregory Mints. Teorija categorii i teoria dokazatelstv.I. *Aktualnye problemy logiki i metodologii nauky*, pages 252–278, 1979.

[Nes89] Dan Nesmith. An application of Klop's counterexample to a higher-order rewrite system. Draft Paper, 1989.

Lambda-calculus, combinators and the comprehension scheme

Gilles Dowek

INRIA-Rocquencourt, B.P. 105, 78153 Le Chesnay Cedex, France
Gilles.Dowek@inria.fr

Abstract. The presentations of type theory based on a comprehension scheme, a skolemized comprehension scheme and λ-calculus are equivalent, both in the sense that each one is a conservative extension of the previous and that each one can be coded in the previous preserving provability. A similar result holds for set theory.

In the presentation of a theory we can either choose to give notations for objects and axioms expressing the properties of these objects, or to give axioms expressing the existence of objects verifying the desired properties. For instance, relations with a maximal element can either be defined by the language $\leq$, M and the axiom $\forall x\ (x \leq M)$ or by the language $\leq$ and the axiom $\exists y\ \forall x\ (x \leq y)$. From an existential formulation we can produce an explicit one by skolemizing the axioms [12, 2]. The skolemized theory is a conservative extension of the non skolemized one. When we have also unicity axioms, we can translate the skolemized language into the non skolemized one preserving provability and thus the two theories are equivalent in a stronger meaning.

Church's type theory (also called higher order logic) [3, 2] can be presented in several ways. A first presentation uses an explicit notation for functions, i.e. a notation $\lambda x\ a$ (λ-calculus) and an axiom scheme (conversion)

$$\forall x\ ((\lambda x\ a)\ x) = a$$

Another merely states an axiom scheme expressing the existence of functions (comprehension)

$$\exists f\ \forall x_1\ ...\ \forall x_n\ ((f\ x_1\ ...\ x_n) = a)$$

For instance, in the first formulation, we have an explicit notation $(\lambda x\ x)$ for the identity function and an axiom $\forall x\ ((\lambda x\ x)\ x) = x$. In the second formulation, we only have an axiom $\exists f\ \forall x\ (f\ x) = x$. The equivalence of the two formulations is informally stated in [2]. In this paper we prove this statement.

When we skolemize the comprehension scheme, we get an explicit presentation of type theory. The language obtained this way is not λ-calculus, but rather a language based on combinators (in the sense of Curry [4]). Indeed, when we skolemize this scheme we introduce for each term a and for each sequence of variables $x_1, ..., x_n$ a primitive symbol $c_{x_1,...,x_n,a}$, also written $x_1, ..., x_n \mapsto a$ and the comprehension scheme becomes

$$\forall x_1\ ...\ \forall x_n\ (((x_1, ..., x_n \mapsto a)\ x_1\ ...\ x_n) = a)$$

which is roughly the conversion scheme [2].

There are however two essential differences between this language and λ-calculus. First, when we skolemize the comprehension scheme, we get symbols $x_1, ..., x_n \mapsto a$ only for terms a that do not contain further abstractions. Then, as abstractions are primitive symbols in the language of the skolemized comprehension scheme, there is nothing like the rule of λ-calculus that permits to substitute under abstractions with renaming.

Thus, we have in fact three presentations of type theory: the presentation with comprehension scheme, the one with the skolemized comprehension scheme and the one with λ-calculus. We show that these three presentations are equivalent. A weak equivalence result is that each presentation is a conservative extension of the previous. In this case, we have also a stronger result, each presentation can be translated into the previous one preserving provability.

A similar equivalence result for second order logic is proved in [8]. There is however a difference between the second order case and the higher order one, as in the latter extensionality seems to play a central role.

There are a few choices in the formulation of the comprehension scheme. First, we may take an n-ary comprehension scheme

$$\exists f \ \forall x_1 \ ... \ \forall x_n \ ((f \ x_1 \ ... \ x_n) = a)$$

or only the unary one

$$\exists f \ \forall x \ ((f \ x) = a)$$

In λ-calculus functions are unary and n-ary functions are coded as unary ones by currification. But, we show that this cannot be done with the comprehension scheme: some instances of the n-ary comprehension scheme cannot be derived from the unary one.

Another choice concerns the free variables of a. In the proposition

$$\exists f \ \forall x_1 \ ... \ \forall x_n \ ((f \ x_1 \ ... \ x_n) = a)$$

we can decide that all the free variables of a have to be among $x_1, ..., x_n$. We get this way the *closed comprehension scheme.* We can also decide that some free variables of a may not be arguments of f, in this case, we quantify universally these variables at the head of the axiom to get a closed axiom. We get this way the *open comprehension scheme*

$$\forall y_1 \ ... \ \forall y_p \ \exists f \ \forall x_1 \ ... \ \forall x_n \ ((f \ x_1 \ ... \ x_n) = a)$$

These two formulations of type theory are equivalent, but they lead to different languages when we skolemize them.

The last point concerns set theory. Like type theory, set theory can be presented either with existence axioms (Zermelo's axioms, or refinements) or with a language for objects including symbols $\mathcal{P}$, $\bigcup$ and $\{,\}$ where $\mathcal{P}(A)$ is the power set of A, $\bigcup(A)$ is the union of the elements of A (also written $\bigcup_{x \in A} x$) and $\{,\}(A, B)$ is the pair containing A and B (also written $\{A, B\}$) and a notation $\{x \in A \mid P\}$ quite similar to λ-calculus, binding the variable x in the proposition

P. When we skolemize the existence axioms, we do not get the language with binders, but again these languages are equivalent.

Presentations of type theory and set theory with existence axioms, skolemized existence axioms or with binders are useful in different situations. First, the presentation with existence axioms or skolemized existence axioms are better suited to express these theories in a first order setting (set theory is a first order theory and type theory can be coded as a first order theory [6]). But theories with an explicit notation (i.e. with skolemized existential axioms or binders) are better when we want to use these theories in practice, as a language for mathematics. It seems that the presentations with binders are easier to use than the ones with skolemized axioms, but this statement still needs to be justified.

Most proof are omited in this extended abstract. They are detailed in [5].

1 Type theory

1.1 Type theory based on λ-calculus

Definition 1. (Types)
Types are inductively defined by:

- ι and o are types,
- if T and U are types then $T \rightarrow U$ is a type.

There is an infinite number of variables of each type. There is an infinite number of primitive symbols $=_T$ of type $T \rightarrow T \rightarrow o$.

Definition 2. (Terms and propositions)
Terms of type T are inductively defined by:

- variables of type T are terms of type T,
- primitive symbols of type T are terms of type T,
- if a is a term of type $T \rightarrow U$ and b is a term of type T then $(a\ b)$ is a term of type U,
- *if a is a term of type U and x is a variable of type T then $\lambda x\ a$ is a term of type $T \rightarrow U$,*
- $\top$ and $\bot$ are terms of type o (resp. *truth* and *falsehood*),
- if A is a term of type o then $\neg A$ is a term of type o,
- if A and B are terms of type o then $A \wedge B$, $A \vee B$, $A \Rightarrow B$, $A \Leftrightarrow B$ are terms of type o,
- if A is a term of type o and x a variable of type T then $\forall x\ A$ and $\exists x\ A$ are terms of type o.

Propositions are terms of type o.

Definition 3. (Substitution)

- $x[x \leftarrow b] = b$,
- $y[x \leftarrow b] = y$,

- $c[x \leftarrow b] = c$, if c is a primitive symbol,
- $(c\ d)[x \leftarrow b] = (c[x \leftarrow b]\ d[x \leftarrow b])$,
- $(\lambda y\ c)[x \leftarrow b] = \lambda z\ (c[y \leftarrow z][x \leftarrow b])$ *where z is a fresh variable, i.e. a variable not occurring in $\lambda y\ c$ or b,*
- $\top[x \leftarrow b] = \top$, $\bot[x \leftarrow b] = \bot$,
- $(\neg A)[x \leftarrow b] = \neg(A[x \leftarrow b])$,
- $(A * B)[x \leftarrow b] = (A[x \leftarrow b]) * (B[x \leftarrow b])$ for $* \in \{\wedge, \vee, \Rightarrow, \Leftrightarrow\}$,
- $(Qy\ A)[x \leftarrow b] = Qz\ (A[y \leftarrow z][x \leftarrow b])$ where z is a fresh variable, i.e. a variable not occurring in $Qx\ A$ or b, for $Q \in \{\forall, \exists\}$.

Because several choices are possible for the variable z, substitution is not defined on terms, but rather on classes of terms equivalent modulo bound variables renaming.

Definition 4. (Axioms)
Conversion:

$$\forall y_1 \ldots \forall y_p\ \forall x\ (((\lambda x\ a)\ x) = a)\ (\beta)$$

Extensionality:

$$\forall f\ \forall g\ ((\forall x\ ((f\ x) = (g\ x))) \Rightarrow (f = g))$$

$$\forall P\ \forall Q\ ((P \Leftrightarrow Q) \Rightarrow (P = Q))$$

Equality:

$$\forall x\ (x = x)$$

$$\forall w_1 \ldots \forall w_p\ \forall x\ \forall y\ ((x = y) \Rightarrow (P[z \leftarrow x] \Rightarrow P[z \leftarrow y]))$$

Remark. In the axiom schemes, the free variables of a (resp. P) different from x (resp. z) are quantified universally at the head of the proposition. This way all axioms are closed.

Remark. In the conversion scheme

$$\forall y_1 \ldots \forall y_p\ \forall x\ ((\lambda x\ a)\ x) = a$$

the variable x is bound twice (once by the quantifier and once by the λ). If we want to avoid this, we can reformulate the axiom, for instance, as

$$\forall y_1 \ldots \forall y_p\ \forall z\ ((\lambda x\ a)\ z) = a[x \leftarrow z]$$

Deduction rules are the usual ones. We take a natural deduction presentation (see, for instance, [7]).

1.2 Hyper-combinators

Definition 5. (Hyper-combinators)
The set of hyper-combinators is the smallest set of λ-terms such that:

- variables are hyper-combinators,
- primitive symbols are hyper-combinators,
- if a and b are terms then $(a\ b)$ is an hyper-combinator,
- *if a is an hyper-combinator, $FV(a) \subset \{x_1, ..., x_n\}$ and no subterm of a is a λ-abstraction, then $\lambda x_1 \ ... \ \lambda x_n \ a$ is an hyper-combinator,*
- $\top$ and $\bot$ are hyper-combinators,
- if A is an hyper-combinator then $\neg A$ is an hyper-combinator,
- if A and B hyper-combinators then $A \wedge B$, $A \vee B$, $A \Rightarrow B$, $A \Leftrightarrow B$ are hyper-combinators,
- if A is an hyper-combinator and x a variable then $\forall x\ A$ and $\exists x\ A$ are hyper-combinators.

Remark. When we relax the rules above to allow abstractions to occur in the body of abstractions, we get super-combinators [9]. Thus, hyper-combinators are super-combinators, but $\lambda f\ \lambda x\ (f\ ((\lambda y\ y)\ x))$ is a super-combinator which is not an hyper-combinator.

Remark. An hyper-combinator which is an abstraction is a closed term thus, if a is an hyper-combinator and an abstraction then $a[x \leftarrow b] = a$.

Proposition 6. *If a and b are hyper-combinators, then $a[x \leftarrow b]$ is an hyper-combinator.*

Definition 7. (λ-lifting)
Let a be a λ-term, we define an hyper-combinator a' as follows:

- $x' = x$, if x is a variable or a primitive symbol,
- $(b\ c)' = (b'\ c')$,
- if $a = (\lambda x_1 \ ... \ \lambda x_n\ b)$ (b not an abstraction) then let $y_1, ..., y_p$ be the free variables of a, let b' be the translation of b, let $c_1, ..., c_q$ be the maximal subterms of b' which are abstractions (these terms are closed hyper-combinators), let b'' be the term obtained by replacing every c_i by a fresh variable z_i, we let $a' = ((\lambda y_1 \ ... \ \lambda y_p\ \lambda z_1 \ ... \ \lambda z_q\ \lambda x_1 \ ... \ \lambda x_n\ b'')\ y_1 \ ... \ y_p\ c_1 \ ... \ c_q)$,
- $\top' = \top$, $\bot' = \bot$,
- $(\neg A)' = (\neg A')$,
- $(A \wedge B)' = (A' \wedge B')$, $(A \vee B)' = (A' \vee B')$, $(A \Rightarrow B)' = (A' \Rightarrow B')$, $(A \Leftrightarrow B)' = (A' \Leftrightarrow B')$,
- $(\forall x\ A)' = \forall x\ A'$, $(\exists x\ A)' = \exists x\ A'$.

Proposition 8. *For every λ-term a, $a = a'$ is a theorem of type theory.*

Corollary 9. *A proposition P is provable in type theory if and only if P' is provable in type theory.*

Remark. Now, we want to restrict type theory in such a way that all the propositions in the proofs are hyper-combinators. It is not obvious that if P is provable in the λ-calculus based type theory then P' is provable in this restriction. For instance, the proposition

$$((\lambda f\ \lambda x\ (f\ x))\ f\ x) = (f\ x)$$

is provable in the λ-calculus based theory: using twice the scheme β we prove that $((\lambda f\ \lambda x\ (f\ x))\ f\ x)$ is equal to $((\lambda x\ (f\ x))\ x)$ and then to $(f\ x)$. As this proposition is an hyper-combinator, it is its own translation, but the proof above does not go through in the restriction, as the intermediate term $((\lambda x\ (f\ x))\ x)$ is not an hyper-combinator. Thus, we replace the scheme β by the scheme β'

$$\forall y_1\ ...\ \forall y_p\ \forall x_1\ ...\ \forall x_n\ ((\lambda x_1\ ...\ \lambda x_n\ a)\ x_1\ ...\ x_n) = a\ (a \text{ not an abstraction})\ (\beta')$$

In type theory based on λ-calculus, the theory obtained by replacing β by β' is obviously equivalent to the initial one, but it seems to be more powerful when the theory is restricted to hyper-combinators, as using this scheme we can prove the proposition

$$((\lambda f\ \lambda x\ (f\ x))\ f\ x) = (f\ x)$$

Definition 10. The *hyper-combinators based type theory* is the restriction of the λ-calculus based type theory such that all the propositions in the proofs are hyper-combinators and the scheme β is replaced by the scheme β'.

Remark. Even with the scheme β', it is still not obvious that every proof in λ-calculus based type theory can be translated as a proof of P' in the hyper-combinators based type theory. Consider for instance the terms

$$a = ((\lambda x\ \lambda y\ \lambda z\ x)\ w\ w) \qquad b = ((\lambda x\ \lambda y\ \lambda z\ y)\ w\ w)$$

The proposition $a = b$ is a theorem in the λ-calculus based type theory, as both a and b are provably equal to $\lambda z\ w$. But this proof does not go through in the hyper-combinators based type theory. In this theory, we need to use the extensionality axiom, then we are reduced to prove that for every u, $(a\ u) = (b\ u)$, i.e.

$$((\lambda x\ \lambda y\ \lambda z\ x)\ w\ w\ u) = ((\lambda x\ \lambda y\ \lambda z\ y)\ w\ w\ u)$$

and these terms are both provably equal to the term w.

Proposition 11. *If A is an axiom of the λ-calculus based type theory then A' is provable in the hyper-combinators based type theory.*

Proposition 12. *Let a and b be terms, the proposition $(a[x \leftarrow b])' = a'[x \leftarrow b']$ is a theorem of the hyper-combinators based type theory.*

Proof. By induction over the structure of a.

- If a is a variable or a primitive symbol then the result is obvious.
- If $a = (c\ d)$ then we apply the induction hypothesis.

- If $a = \lambda y_1 \ldots \lambda y_n\ c$ (c not an abstraction) then if x is not free in a then the result is trivial, otherwise by extensionality we are reduced to prove

$$((a[x \leftarrow b])'\ y_1 \ldots y_n) = (a'[x \leftarrow b']\ y_1 \ldots y_n)$$

We first show that the terms $(a'[x \leftarrow b']\ y_1 \ldots y_n)$ and $c'[x \leftarrow b']$ are provably equal. Let $x, u_1, \ldots, u_p$ be the free variables of a. Let $e_1, \ldots, e_q$ the maximal abstractions in the term c', $v_1, \ldots, v_q$ be fresh variables and c'' the term c' where the terms $e_1, \ldots, e_q$ are replaced by the variables $v_1, \ldots, v_q$. We have

$$a' = ((\lambda x\ \lambda u_1 \ldots \lambda u_p\ \lambda v_1 \ldots \lambda v_q\ \lambda y_1 \ldots \lambda y_n\ c'')\ x\ u_1 \ldots u_p\ e_1 \ldots e_q)$$

$$(a'[x \leftarrow b']\ y_1 \ldots y_n) =$$
$$((\lambda x\ \lambda u_1 \ldots \lambda u_p\ \lambda v_1 \ldots \lambda v_q\ \lambda y_1 \ldots \lambda y_n\ c'')\ b'\ u_1 \ldots u_p\ e_1 \ldots e_q\ y_1 \ldots y_n)$$

And this term is provably equal to

$$c''[x \leftarrow b', v_1 \leftarrow e_1, \ldots, v_q \leftarrow e_q] = c''[v_1 \leftarrow e_1, \ldots, v_q \leftarrow e_q][x \leftarrow b'] = c'[x \leftarrow b']$$

Then, we show that the terms $((a[x \leftarrow b])'\ y_1 \ldots y_n)$ and $(c[x \leftarrow b])'$ are provably equal. Let $\lambda z_1 \ldots \lambda z_m\ d$ (d not an abstraction) be the term $c[x \leftarrow b]$. We need to prove

$$((\lambda y_1 \ldots \lambda y_n\ \lambda z_1 \ldots \lambda z_m\ d)'\ y_1 \ldots y_n) = (\lambda z_1 \ldots \lambda z_m\ d)'$$

By extensionality, we are reduced to prove

$$((\lambda y_1 \ldots \lambda y_n\ \lambda z_1 \ldots \lambda z_m\ d)'\ y_1 \ldots y_n\ z_1 \ldots z_m) = ((\lambda z_1 \ldots \lambda z_m\ d)'\ z_1 \ldots z_m)$$

and these terms are both provably equal to d'.
At last, by induction hypothesis, the terms $c'[x \leftarrow b']$ and $(c[x \leftarrow b])'$ are provably equal. Thus, $(a'[x \leftarrow b']\ y_1 \ldots y_n)$ and $((a[x \leftarrow b])'\ y_1 \ldots y_n)$ are provably equal.
- If $a = \neg A$, $a = A \wedge B$, $a = A \vee B$, $a = A \Rightarrow B$, $a = A \Leftrightarrow B$, we apply the induction hypothesis.
- If $a = Qy\ A$ ($Q \in \{\forall, \exists\}$), by induction hypothesis, we have

$$A'[x \leftarrow b'] = (A[x \leftarrow b])'$$

Thus

$$A'[x \leftarrow b'] \Leftrightarrow (A[x \leftarrow b])'$$
$$\forall y\ (A'[x \leftarrow b'] \Leftrightarrow (A[x \leftarrow b])')$$
$$(Qy\ A'[x \leftarrow b']) \Leftrightarrow (Qy\ (A[x \leftarrow b])')$$
$$((Qy\ A)'[x \leftarrow b']) \Leftrightarrow ((Qy\ A)[x \leftarrow b])'$$

By the second extensionality axiom, we get

$$((Qy\ A)'[x \leftarrow b']) = ((Qy\ A)[x \leftarrow b])'$$

Proposition 13. *A proposition P is provable in the λ-calculus based type theory if and only if P' is provable in the hyper-combinators based type theory.*

1.3 The closed comprehension scheme

In type theory with the comprehension scheme, we do not have the notation $\lambda x\ a$ any more, thus we have the following definitions.

Definition 14. (Type theory with the open comprehension scheme)
Types are defined as in Definition 1, *terms* and *propositions* are defined as in Definition 2 without the clause 4 and *substitution* is defined as in Definition 3 without the clause 5. Equality and extensionality axioms are the same, but we drop the conversion scheme and add the comprehension scheme:

$$\forall y_1 \ldots \forall y_p\ \exists f\ \forall x_1 \ldots \forall x_n\ ((f\ x_1 \ldots x_n) = a)$$

where f is not free in a and $y_1, \ldots, y_p$ are the free variables of a that are not among $x_1, \ldots, x_n$. Deduction rules are the same as in the λ-calculus based presentation.

Definition 15. (Type theory with the closed comprehension scheme)
Type theory with the closed comprehension scheme is the restriction of type theory with the open comprehension scheme, where we take only the instances of the scheme such that all the free variables of a are among $x_1, \ldots, x_n$. In such an instance, $p = 0$ and the comprehension scheme is rephrased

$$\exists f\ \forall x_1 \ldots \forall x_n\ ((f\ x_1 \ldots x_n) = a)$$

Proposition 16. *Type theory with the open or the closed comprehension scheme are equivalent.*

1.4 The skolemized closed comprehension scheme

When we skolemize the closed comprehension scheme, we introduce an infinite number of primitive symbols $c_{x_1,\ldots,x_n,a}$ (also written $x_1, \ldots, x_n \mapsto a$) where all the free variables of a are among $x_1, \ldots, x_n$ and a does not contain symbols of the form $y_1, \ldots, y_p \mapsto b$. We get the axiom

$$\forall x_1 \ldots \forall x_n\ (((x_1, \ldots, x_n \mapsto a)\ x_1 \ldots x_n) = a)$$

Remark. When we define the language of this theory, we first define the set of terms without Skolem's symbols, then for each such term a and sequence of variables $x_1, \ldots, x_n$ where all the free variables of a are among $x_1, \ldots, x_n$ we introduce a Skolem's symbol $x_1, \ldots, x_n \mapsto a$, at last we define the full set of terms.

Proposition 17. *If a proposition P contains no symbol $x_1, \ldots, x_n \mapsto a$, then P is provable in the theory of the comprehension scheme if and only if it is provable in the theory of the skolemized comprehension scheme. Thus, the skolemized theory is a conservative extension of the non skolemized one.*

Definition 18. Let P be a proposition in the language of the skolemized comprehension scheme, we define the proposition P° in the language of the (non skolemized) comprehension scheme as follows. We replace in P each symbol $x_{i,1}, ..., x_{i,n_i} \mapsto a_i$ by a variable f_i, we get a proposition P_1 and we take

$$P^\circ = \exists f_1 \, ... \, \exists f_p \, ((\forall x_{1,1} \, ... \, \forall x_{1,n_1} \, (f_1 \; x_{1,1} \, ... \, x_{1,n_1}) = a_1) \\ \wedge ... \wedge (\forall x_{p,1} \, ... \, \forall x_{p,n_p} \, (f_p \; x_{p,1} \, ... \, x_{p,n_p}) = a_p) \wedge P_1)$$

Proposition 19. *The proposition P° is provable in the skolemized theory if and only if P is provable in the same theory.*

Corollary 20. *The proposition P is provable in the skolemized theory if and only if P° is provable in the non skolemized theory.*

1.5 The skolemized closed comprehension scheme and hyper-combinators

Now, we can show that there is a simple one to one translation from type theory based on hyper-combinators and on the skolemized closed comprehension scheme. In both cases $\lambda x_1 \, ... \, \lambda x_n \; a$ and $x_1, ..., x_n \mapsto a$ are closed terms and a does not contain further abstractions. Thus, the only difference between the hyper-combinators and the language of the skolemized comprehension scheme is that the former provides a syntactical rule to construct abstractions, while the latter provides a new primitive symbol for each abstraction.

Definition 21. Let a be a term in the hyper-combinators based type theory, the translation a^+ of a is a term in type theory with the skolemized closed comprehension scheme defined as follows

- $x^+ = x$,
- $c^+ = c$,
- $(a \; b)^+ = (a^+ \; b^+)$,
- $(\lambda x_1 \, ... \, \lambda x_n \; a)^+ = x_1, ..., x_n \mapsto a^+$ (a not an abstraction),
- $\top^+ = \top$, $\bot^+ = \bot$,
- $(\neg A)^+ = \neg A^+$,
- $(A \wedge B)^+ = A^+ \wedge B^+$, $(A \vee B)^+ = A^+ \vee B^+$, $(A \Rightarrow B)^+ = A^+ \Rightarrow B^+$, $(A \Leftrightarrow B)^+ = A^+ \Leftrightarrow B^+$,
- $(\forall x \; A)^+ = \forall x \; A^+$, $(\exists x \; A)^+ = \exists x \; A^+$.

Proposition 22. *A proposition P is provable in type theory with hyper-combinators if and only if P^+ is provable in type theory with the skolemized comprehension scheme.*

2 Alternatives in type theory with the comprehension scheme

In this section, we discuss alternatives of the type theory with the comprehension scheme. First, we show that the n-ary comprehension scheme cannot be replaced by an unary one. Then, we show that if we drop the extensionality axiom, the equivalence result does not hold any more. At last, we characterize the language obtained when we skolemize the open comprehension scheme. This language lies somewhere between hyper-combinators and λ-calculus.

2.1 Independence of the binary comprehension scheme

In λ-calculus repeated application of the λ-rule permits to form n-ary functions. For instance from the term x we can form the term $\lambda y\ x$ and then $\lambda x\ \lambda y\ x$. In contrast, it is not possible to iterate the use of the unary comprehension scheme to build n-ary functions. Indeed, using the hypotheses $\forall y\ (f\ y) = x$ and $\forall x\ (g\ x) = f$ we can derive the proposition

$$\forall y\ (g\ x\ y) = x$$

But we cannot quantify over x in this proposition as the introduction rule for the universal quantifier requires the variable x to have no free occurrence in the hypotheses. We cannot either eliminate the first hypothesis using the axiom $\exists f\ \forall y\ (f\ y) = x$ as the elimination rule for the existential quantifier requires the variable f to have no free occurrence in the side hypotheses, and we cannot eliminate the second hypothesis using the axiom $\exists g\ \forall x\ (g\ x) = f$ as the elimination rule for the existential quantifier requires the variable g to have no free occurrence in the conclusion. Thus, it seems that the proposition

$$\exists g\ \forall x\ \forall y\ (g\ x\ y) = x$$

cannot be derived from the unary comprehension scheme. We show that this is indeed the case.

Proposition 23. *The binary comprehension scheme is independent.*

Proof. (sketch)
Let T be a type, we define by induction over the structure of T the set of *tails* of T.

- if $T = \iota$ or $T = o$ then $Tails(T) = \{T\}$,
- if $T = U \to V$ then $Tails(T) = \{T\} \cup Tails(V)$.

We define, by induction over the structure of T, a family $\mathcal{M}_T$:

- $\mathcal{M}_\iota$ is any set containing at least two elements,
- $\mathcal{M}_o = \{0, 1\}$,
- if $\iota \in Tails(U)$ and $U \notin Tails(T)$ then $\mathcal{M}_{T \to U}$ is the set of constant functions from $\mathcal{M}_T$ to $\mathcal{M}_U$,
- otherwise, $\mathcal{M}_{T \to U}$ is the set of all functions from $\mathcal{M}_T$ to $\mathcal{M}_U$.

We then show that the unary comprehension scheme, the extensionality axioms and the equality axioms are valid in this model, but not the proposition

$$\exists g\ \forall x\ \forall y\ (g\ x\ y) = x$$

Remark. The axiom of descriptions [3, 2] is also valid in the model above. The axiom of infinity [3, 2] is valid if $\mathcal{M}_\iota$ is infinite. Thus, the binary comprehension scheme is still independent if we add these axioms to type theory.

2.2 Extensionality

Extensionality is used several times in the proof of the equivalence between the presentations of type theory based on λ-calculus and based on the skolemized comprehension scheme. Thus, we may wonder if these theories are still equivalent if we drop the extensionality axioms. When we drop these axioms, the proposition

$$((\lambda x\ \lambda y\ \lambda z\ x)\ w\ w) = ((\lambda x\ \lambda y\ \lambda z\ y)\ w\ w)$$

is still a theorem of the λ-calculus based type theory, but we show that it is not a theorem of the hyper-combinators based type theory, similarly the proposition

$$((x, y, z \mapsto x)\ w\ w) = ((x, y, z \mapsto y)\ w\ w)$$

is not a theorem of type theory with the skolemized comprehension scheme and no extensionality axiom.

Proposition 24. *The proposition*

$$((x, y, z \mapsto x)\ w\ w) = ((x, y, z \mapsto y)\ w\ w)$$

is not a theorem of type theory with the skolemized comprehension scheme and no extensionality axiom.

Proof. (sketch) We define, by induction over the structure of T, a family $\mathcal{M}_T$

- $\mathcal{M}_\iota$ is a non empty set, we consider an element α of $\mathcal{M}_\iota$,
- $\mathcal{M}_o = \{0, 1\}$,
- $\mathcal{M}_{\iota \to \iota} = \mathcal{M}_\iota^{\mathcal{M}_\iota} - \{k_\alpha\} \cup \{K, K'\}$ where k_α is the constant function equal to α and K and K' are two objects not in $\mathcal{M}_\iota^{\mathcal{M}_\iota}$,
- $\mathcal{M}_{T \to U} = \mathcal{M}_U^{\mathcal{M}_T}$ otherwise.

Then we define the denotation of a term in such a way that the skolemized comprehension scheme and the equality axioms are valid in $\mathcal{M}$, but the term $((x, y, z \mapsto x)\ w\ w)$ denotes K and $((x, y, z \mapsto y)\ w\ w)$ denotes K'.

Remark. The axiom of descriptions [3, 2] is also valid in the model above. The axiom of infinity [3, 2] is valid if $\mathcal{M}_\iota$ is infinite. Thus, the proposition above is still independent if we add these axioms to type theory.

Remark. The proposition above is formulated in type theory with the skolemized comprehension scheme. The question of the existence of a proposition in the language of the (non skolemized) comprehension scheme, i.e. without abstractions that would be provable in the presentation with λ-calculus and not

provable in the presentation with the comprehension scheme is left open. The natural candidate given by the translation of definition 18
$\exists f\ \exists g\ (\forall x\ \forall y\ \forall z\ ((f\ x\ y\ z) = x)\ \wedge \forall x\ \forall y\ \forall z\ ((g\ x\ y\ z) = y)$
$\wedge \forall w\ ((f\ w\ w) = (g\ w\ w)))$
is unfortunately provable in type theory with the comprehension scheme. (Notice that the proposition 19 fails when extensionality is dropped.) Indeed, consider f such that $\forall x\ \forall y\ \forall z\ (f\ x\ y\ z) = x$, C such that $\forall u\ \forall x\ \forall y\ (C\ u\ x\ y) = (u\ y\ x)$ and $g = (C\ f)$. We have $(f\ x\ y\ z) = x$, $(g\ x\ y\ z) = (C\ f\ x\ y\ z) = (f\ y\ x\ z) = y$ and $(g\ w\ w) = (C\ f\ w\ w) = (f\ w\ w)$.

Remark. In [8] Henkin presents a proof of the equivalence of a presentation of second order logic based on a rule of substitution of functional variables and a presentation based on a comprehension scheme. As, in the second order logic presented in this paper, there seems to be no equality between functions or predicates, the proof goes through without extensionality axiom. It seems that the extensionality axiom is required to translate the full type theory.

2.3 Skolemizing the open comprehension scheme

Type theory can be presented either with the closed comprehension scheme, or the open one. When we skolemize the closed comprehension scheme we get hyper-combinators. In this section we characterize the language obtained by skolemizing the open comprehension scheme.

When we skolemize an instance of the open comprehension scheme

$$\forall x_1\ ...\ \forall x_n\ \exists f\ \forall y_1\ ...\ \forall y_p\ (f\ y_1\ ...\ y_p) = a$$

then we introduce a function symbol $f_{(x_1,...,x_n),(y_1,...,y_p),a}$ and the axiom

$$\forall x_1\ ...\ \forall x_n\ \forall y_1\ ...\ \forall y_p\ ((f_{(x_1,...,x_n),(y_1,...,y_p),a}\ x_1\ ...\ x_n)\ y_1\ ...\ y_p) = a$$

As remarked in [10, 11], sound skolemization in type theory requires that the symbol $f_{(x_1,...,x_n),(y_1,...,y_p),a}$ alone is not a term but a function symbol, i.e. a symbol such that if $a_1, ..., a_n$ are terms then $(f\ a_1\ ...\ a_n)$ is a term.

Proposition 25. *Let a be a term containing no Skolem's symbols. Let $x_1, ..., x_n$, $y_1, ..., y_p$ be variables such that all the free variables of a are among $x_1, ..., x_n$, $y_1, ..., y_p$. Let $b_1, ..., b_n$ be terms containing no Skolem's symbols. Let $y'_1, ..., y'_p$ be variables not occurring free in these terms and let $x'_1, ..., x'_{n'}$ be variables such that the free variables of $a[y_1 \leftarrow y'_1]...[y_p \leftarrow y'_p][x_1 \leftarrow b_1]...[x_n \leftarrow b_n]$ are among $x'_1, ..., x'_{n'}, y'_1, ..., y'_p$. We have*
$(f_{(x_1,...,x_n),(y_1,...,y_p),a}\ b_1\ ...\ b_n) =$
$(f_{(x'_1,...,x'_{n'}),(y'_1,...,y'_p),a[y_1\leftarrow y'_1]...[y_p\leftarrow y'_p][x_1\leftarrow b_1]...[x_n\leftarrow b_n]}\ x'_1\ ...\ x'_{n'})$

Remark. We consider a restriction of the language above, in such a way that a symbol $f_{(x_1,...,x_n),(y_1,...,y_p),a}$ can only be applied to the variables $x_1, ..., x_n$. We write $y_1, ..., y_p \mapsto a$ for the term $(f_{(x_1,...,x_n),(y_1,...,y_p),a}\ x_1\ ...\ x_n)$ where a is a term containing no Skolem's symbols.

Notice that the free variables of $y_1, ..., y_p \mapsto a$ are $x_1, ..., x_n$, i.e. the free variables of a but $y_1, ..., y_p$. As a corollary of the previous proposition we have

$$(y_1, ..., y_p \mapsto a)[x \leftarrow b] = (y'_1, ..., y'_p \mapsto a[y_1 \leftarrow y'_1]...[y_p \leftarrow y'_p][x \leftarrow b])$$

if a and b are terms containing no Skolem's symbols, and $y'_1, ..., y'_n$ variable not occurring in these terms.

This language is larger than the language of hyper-combinators as it allows free variables in the body of abstractions and substitution with renaming, but it is not λ-calculus, as it does not allow nested abstractions.

Remark. If $a, b_1, ..., b_n$ are terms without Skolem's symbols we have $(f_{(x_1,...,x_n),(y_1,...,y_p),a}\ b_1\ ...\ b_n) =$
$$y'_1, ..., y'_p \mapsto a[y_1 \leftarrow y'_1]...[y_p \leftarrow y'_p][x_1 \leftarrow b_1]...[x_n \leftarrow b_n]$$
We can extend the notation above and write

$$y'_1, ..., y'_p \mapsto a[y_1 \leftarrow y'_1]...[y_p \leftarrow y'_p][x_1 \leftarrow b_1]...[x_n \leftarrow b_n]$$

for the term $(f_{(x_1,...,x_n),(y_1,...,y_p),a}\ b_1\ ...\ b_n)$ even if the terms $b_1, ..., b_n$ contain Skolem's symbols.

This language allows free variables in the body of abstractions, substitution under abstractions with renaming and nested abstractions, but it is not λ-calculus because a variable bound in an abstraction cannot be bound upper in the term in another abstraction. For instance the term $x \mapsto (y \mapsto x)$ cannot be constructed in this language.

Remark. If we skolemize this way the unary comprehension scheme, we get a language with unary functions and no variable binding through abstractions. In this language there is no term for the first projection since we cannot prove the proposition $\exists f\ \forall x\ \forall y\ (f\ x\ y) = x$. Thus to define this function, we need either variable binding through abstractions $x \mapsto (y \mapsto x)$ or n-ary functions $x, y \mapsto x$.

3 Set theory

Like type theory, set theory can be presented either with existence axioms (Zermelo's axioms or refinements), with an explicit language for objects obtained by skolemizing these axioms or with a language with binders, including symbols $\mathcal{P}$, $\bigcup$ and $\{,\}$ where $\mathcal{P}(A)$ is the power set of A, $\bigcup(A)$ is the union of the elements of A and $\{,\}(A, B)$ is the pair containing A and B and a notation $\{x \in A \mid P\}$. In this section show the equivalence of these presentations.

3.1 Set theory with binders

Definition 26. *Terms and Propositions* are inductively defined by:

- variables are terms,
- if a is a term then $\mathcal{P}(a)$ is a term,

- if a is a term then $\bigcup(a)$ is a term,
- if a and b are terms then $\{,\}(a,b)$ (i.e. $\{a,b\}$) are terms,
- if A is a term and P a proposition then $\{z \in A \mid P\}$ is a term,
- if a and b are terms then $a \in b$ is a proposition,
- if a and b are terms then $a = b$ is a proposition,
- $\top$ and $\bot$ are propositions,
- if A is proposition then $\neg A$ is a proposition,
- if A and B are propositions, then $A \wedge B$, $A \vee B$, $A \Rightarrow B$, $A \Leftrightarrow B$ are propositions,
- if A is proposition and x a variable then $\forall x\ A$ and $\exists x\ A$ are propositions.

Substitution is defined as in λ-calculus with renaming to avoid captures in the terms $\{z \in A \mid P\}$.

Definition 27. (Axioms)
Conversion (power set, union, pair and subset scheme):

$$\forall x\ \forall y\ ((y \in \mathcal{P}(x)) \Leftrightarrow \forall z\ ((z \in y) \Rightarrow (z \in x)))$$
$$\forall x\ \forall y\ ((y \in \bigcup(x)) \Leftrightarrow \exists z\ ((y \in z) \wedge (z \in x)))$$
$$\forall x\ \forall y\ \forall z\ ((z \in \{,\}(x,y)) \Leftrightarrow ((z = x) \vee (z = y)))$$
$$\forall x_1\ ...\ \forall x_n\ \forall y\ \forall z\ ((z \in \{z \in y \mid P\}) \Leftrightarrow ((z \in y) \wedge P))$$

Extensionality:

$$\forall x\ \forall y\ ((\forall z(z \in x) \Leftrightarrow (z \in y)) \Rightarrow (x = y))$$

Equality:

$$\forall x\ (x = x)$$
$$\forall w_1\ ...\ \forall w_p\ \forall x\ \forall y\ ((x = y) \Rightarrow (P[z \leftarrow x] \Rightarrow P[z \leftarrow y]))$$

Deduction rules are the usual ones.

3.2 Set theory with existential axioms

Definition 28. (Zermelo's set theory)
Comprehension (power set, union, pair and subset scheme):

$$\forall x\ \exists A\ \forall y\ ((y \in A) \Leftrightarrow \forall z\ ((z \in y) \Rightarrow (z \in x)))$$
$$\forall x\ \exists A\ \forall y\ ((y \in A) \Leftrightarrow \exists z\ ((y \in z) \wedge (z \in x)))$$
$$\forall x\ \forall y\ \exists A\ \forall z\ ((z \in A) \Leftrightarrow ((z = x) \vee (x = y)))$$
$$\forall x_1\ ...\ \forall x_n\ \forall y\ \exists A\ \forall z\ ((z \in A) \Leftrightarrow ((z \in y) \wedge P))$$

where A is not free in P and $x_1, ..., x_n$ are the free variables of P different from z.
Extensionality:

$$\forall x\ \forall y\ (\forall z\ (z \in x) \Leftrightarrow (z \in y)) \Rightarrow (x = y)$$

Equality:

$$\forall x\ (x = x)$$
$$\forall w_1\ ...\ \forall w_p\ \forall x\ \forall y\ ((x = y) \Rightarrow (P[z \leftarrow x] \Rightarrow P[z \leftarrow y]))$$

3.3 Set theory with skolemized axioms

The skolemization of the power set axiom, the union axiom and the pair axiom introduces function symbols $\mathcal{P}$, $\bigcup$ and $\{,\}$. When we skolemize the subset axiom, we introduce, for each proposition P that does not contain Skolem's symbols and sequence of symbols $x_1, ..., x_n, z$ such that all the free variables of P are among $x_1, ..., x_n, z$ a function symbol $f_{x_1,...,x_n,z,P}$ and an axiom

$$\forall x_1 \ ... \ \forall x_n \ \forall y \ \forall z \ (z \in (f_{x_1,...,x_n,z,P} \ x_1 \ ... \ x_n \ y) \Leftrightarrow ((z \in y) \wedge P))$$

By Skolem's theorem, the theory obtained this way is a conservative extension of set theory.

Now, we want to prove a result similar to proposition 19 and corollary 20, i.e. for every proposition P, we want to build a Skolem's symbols free proposition P° that is provable, if and only if P is provable in the skolemized theory. In fact we shall prove such a result for an arbitrary first order theory and then apply this result to set theory.

We consider a theory with axioms of the form $\forall x_1 \ ... \ \forall x_n \ \exists y \ P$ and we skolemize them as $\forall x_1 \ ... \ \forall x_n \ P[y \leftarrow (f \ x_1 \ ... \ x_n)]$.

Definition 29. Let A be a proposition, we define the proposition A° as follows:

- If A is atomic, we replace every subterm of the form $(f \ t_{i,1} \ ... \ t_{i,n})$ by a variable y_i we get a proposition A_1, let
  $A^\circ = \exists y_1 \ ... \ \exists y_n \ ((P[x_1 \leftarrow t_{1,1}, ..., x_n \leftarrow t_{1,n}, y \leftarrow y_1])^\circ$
  $\wedge ... \wedge (P[x_1 \leftarrow t_{p,1}, ..., x_n \leftarrow t_{p,n}, y \leftarrow y_p])^\circ \wedge A_1)$
- $(\neg A)^\circ = \neg A^\circ$, $(A \wedge B)^\circ = A^\circ \wedge B^\circ$, $(A \vee B)^\circ = A^\circ \vee B^\circ$,
  $(A \Rightarrow B)^\circ = A^\circ \Rightarrow B^\circ$, $(A \Leftrightarrow B)^\circ = A^\circ \Leftrightarrow B^\circ$,
- $(\forall x \ A)^\circ = \forall x \ A^\circ$, $(\exists x \ A)^\circ = \exists x \ A^\circ$.

Consider a proposition A, call n the multiset of the sizes of the subterms of A that whose head is a Skolem's symbol, call p the number of occurrences of connectors and quantifiers in A. This definition is by induction over $< n, p >$.

Proposition 30. *If for every axiom of the form*

$$\forall x_1 \ ... \ \forall x_n \ \exists y \ P$$

we can prove the following unicity property

$$\forall x_1 \ ... \ \forall x_n \ \forall y_1 \ \forall y_2 \ ((P[y \leftarrow y_1] \wedge P[y \leftarrow y_2]) \Rightarrow (y_1 = y_2))$$

Then for every proposition A, the proposition $A \Leftrightarrow A^\circ$ is provable in the skolemized theory.

Corollary 31. *The proposition A is provable in the skolemized theory if and only if A° is provable in the non skolemized one.*

Corollary 32. *For every proposition A in the language of the skolemized set theory, we can build a proposition A° in the language of the (non skolemized) set theory such that A is provable in the skolemized set theory if and only if A° is provable in the (non skolemized) set theory.*

3.4 Set theory with binders and the skolemized axioms

We define a translation of set theory with binders to the language of the skolemized theory. This transformation is analogous to λ-lifting.

Definition 33. – $x' = x$,
- $\{z \in A \mid P\}' = (f_{x_1,...,x_n,z,P'^{\circ}}\ x_1\ ...\ x_n\ A')$, where $x_1, ..., x_n$ are the variables free in P'° and different from z,
- $(\bigcup(a))' = \bigcup(a')$, $(\mathcal{P}(a))' = \mathcal{P}(a')$, $\{a, b\}' = \{a', b'\}$,
- $(a = b)' = (a' = b')$, $(a \in b)' = (a' \in b')$,
- $(\neg A)' = \neg A'$, $(A \wedge B)' = (A' \wedge B')$, $(A \vee B)' = (A' \vee B')$, $(A \Rightarrow B)' = (A' \Rightarrow B')$, $(A \Leftrightarrow B)' = (A' \Leftrightarrow B')$,
- $(\forall x\ A)' = \forall x\ A'$, $(\exists x\ A)' = \exists x\ A'$.

Proposition 34. *When we interpret the term $(f_{x_1,...,x_n,z,P}\ a_1\ ...\ a_n\ A)$ as the term $\{z \in A \mid P[x_1 \leftarrow a_1, ..., x_n \leftarrow a_n]\}$, the language of the skolemized set theory can be seen as a sub-language of the one of set theory with binders. If a is a term, the proposition $a = a'$ is provable set theory. If A is a proposition, the proposition $A \Leftrightarrow A'$ is provable set theory with binders.*

Proposition 35. *If P is an axiom of set theory with binders then P' is provable in the skolemized set theory.*

Proposition 36. *If a and b are terms of set theory with binders then the proposition*

$$(a[x \leftarrow b])' = a'[x \leftarrow b']$$

is provable in the skolemized set theory.

If P is a proposition of the set theory with binders and b is a term of the set theory with binders then the proposition $(P[x \leftarrow a])' \Leftrightarrow P'[x \leftarrow a']$ is provable in the skolemized set theory.

Proposition 37. *The proposition P is provable in set theory with binders if and only if P' is provable if the skolemized set theory.*

Remark. The translation from the theory with binders to the skolemized theory is simpler in set theory than in type theory (where it requires the n-ary conversion scheme). This is because P is a proposition in $\{x \in A \mid P\}$ while a is a term in $\lambda x\ a$. Thus in set theory, we can use the fact that every proposition of the skolemized language is equivalent to one without Skolem's symbols, while no such results holds for terms.

Acknowledgements

The author thanks Thérèse Hardin and Gérard Huet for their help in the preparation of this paper and the anonymous referees for many helpful remarks.

References

1. P.B. Andrews, General models, descriptions and choice in type theory, *The Journal of Symbolic Logic*, 37, 2 (1972) pp. 385-394.
2. P.B. Andrews, An introduction to mathematical logic and type theory: to truth through proof, *Academic Press*, Orlando (1986).
3. A. Church, A formulation of the simple theory of types, *The Journal of Symbolic Logic*, 5 (1940), pp. 56-68.
4. H. Curry, An analysis of logical substitution, *American Journal of Mathematics*, 51 (1929), pp. 363-384.
5. G. Dowek, Lambda-calculus, combinators and the comprehension scheme, *manuscript* (1994).
6. G. Dowek, Collections, types and sets, *manuscript* (1994).
7. J.Y. Girard, Y. Lafont, P. Taylor, Types and proofs, *Cambridge University Press* (1989).
8. L. Henkin, Banishing the rule of substitution for functional variables, *The Journal of Symbolic Logic*, 18, 3 (1953), pp. 201-208.
9. R.J.M. Hughes, Super-combinators, a new implementation method for applicative languages, *Proceedings of Lisp and Functional Programming* (1982).
10. D.A. Miller, Proofs in higher order logic, *PhD Thesis*, Carnegie Mellon University (1983).
11. D.A. Miller, A compact representation of proofs, *Studia Logica*, 46, 4 (1987).
12. T. Skolem, Über die mathematische logik, *Norsk Matematisk Tidsskrift*, 10 (1928), pp. 125-142.

$\beta\eta$-Equality for Coproducts

Neil Ghani
LFCS, Department of Computer Science
University of Edinburgh
The King's Buildings, Mayfield Road
Edinburgh, UK, EH9 3JZ
e-mail: ng@dcs.ed.ac.uk

Abstract. Recently several researchers have investigated $\beta\eta$-equality for the simply typed λ-calculus with exponentials, products and unit types. In these works, η-conversion was interpreted as an expansion with syntactic restrictions imposed to prevent the expansion of introduction terms or terms which form the major premise of elimination rules. The resulting rewrite relation was shown confluent and strongly normalising to the long $\beta\eta$-normal forms. Thus reduction to normal form provides a decision procedure for $\beta\eta$-equality.

This paper extends these methods to sum types. Although this extension was originally thought to be straight forward, the proposed η-rule for the sum is substantially more complex than that for the exponent or product and contains features not present in the previous systems. Not only is there a facility for expanding terms of sum type analogous to that for product and exponential, but also the ability to permute the order in which different subterms of sum type are eliminated.

These different aspects of η-conversion for the sum type is reflected in our analysis. The rewrite relation is decomposed into two parts, a strongly normalising and confluent fragment resembling that found in the calculus without coproducts and a relation which generalises the 'commuting conversions' appearing in the literature. This second fragment is proved decidable by constructing for each term its (finite) set of *quasi-normal forms* and, by embedding the whole relation into this conversion relation, decidability, confluence and quasi-normal forms for the full relation are derived.

1 Introduction

Extensional equality for terms of the simply typed λ-calculus requires η-conversion, whose interpretation as a rewrite rule is traditionally as a contraction $\lambda x.fx \Rightarrow f$ where $x \notin \mathrm{FV}(f)$. When combined with the usual β-reduction, the resulting rewrite relation is strongly normalising and confluent, and thus reduction to normal form provides a decision procedure for the associated equality on terms. Unfortunately these properties typically fail if further datatypes are introduced. Even the presence of the unit type (necessary for the definition of types with given constants such as integers and booleans) with η-rewrite rule $t \Rightarrow *$ leads to a loss of confluence [14]. Specifically if f is a variable of type $1 \rightarrow 1$ then the divergence $\lambda x.* \Leftarrow \lambda x.fx \Rightarrow f$ cannot be completed.

Recently several authors [1, 5, 6, 13, 3] have accepted the old proposal [10, 15, 16] that η-conversion be interpreted as an expansion $f \Rightarrow \lambda x.fx$ and the resulting rewrite relation has been shown confluent. In these works infinite reduction sequences such as

$$f \Rightarrow \lambda x.fx \Rightarrow \lambda x.(\lambda y.fy)x \Rightarrow \ldots$$

are prohibited by imposing syntactic restrictions to limit the possibilities for expansion; namely λ-abstractions cannot be expanded and nor can terms which are applied. This restricted expansion relation was shown to be strongly normalising, confluent and to generate the same equational theory as the unrestricted expansionary rewrite relation. Thus $\beta\eta$-equality could be decided by reduction to normal form in the restricted fragment. These normal forms were also shown to be exactly Huet's *long $\beta\eta$-normal forms* [9, 16].

This interpretation of η-conversion as an expansion, and the restrictions required to recover strong normalisation, have a mathematical explanation within categorical models of reduction [12, 17, 18] where types are represented as objects, terms as morphisms and rewrites as 2-cells. In such models introduction and elimination form an adjoint pair of functors whose local unit and counit [11, 8] correspond to η-expansion and β-contraction respectively. These are linked by local triangle laws which when cut give rise to the restrictions mentioned above.

This paper uses the same categorical methods to derive rewrite rules for coproducts and analyses the resulting rewrite relation. Although originally thought to be a straightforward extension of previous results, the η-rule for coproducts turns out to be substantially more complex than that for products and exponentials. Not only is there a facility for expanding terms of sum type analogous to that for expanding terms of product and exponential type, but also the ability to permute the order in which different subterms of sum type may be eliminated.

These different aspects of η-conversion for the sum type are reflected in our analysis. After defining the terms of the calculus, the categorical methods outlined above are used to define an extensional rewrite relation over these terms, and this relation is then decomposed into two fragments. The first part of this decomposition contains β-redexes, commuting conversions and limited possibilities for η-expansion and is strongly normalising and confluent. All subterms of a normal form are either of base type, the major premise of an elimination rule or a *quasi-introduction term* and so normal forms of this relation may be seen as generalising the long $\beta\eta$-normal forms to this calculus.

The second part of the decomposition allows permutation of the order in which subterms of sum type may be eliminated, an example of which are the 'commuting conversions' appearing in the literature [16, 7]. For each term there are a finite set of such permutations, and so in general unique normal forms do not exist for this relation. Instead to each term is associated a (finite) set of *quasi-normal forms* and terms equivalent in the associated equational theory are shown to have the same set of quasi-normal forms. Confluence and decidability of this conversion relation are corollaries to these results. Finally by appropriately embedding the whole relation in the conversion relation, full confluence and decidability may be proved.

Historically the use of expansions for products and exponentials can be traced back to [15], although the proof that they form a strongly normalising relation had to wait a decade for the papers mentioned above. A partial solution to the problem of $\beta\eta$-equality for coproducts was provided by [6] but in this approach confluence can only be proved for terms of ground type. As of writing the author is aware of several other on-going attempts to tackle this problem, but is unaware of any actual solutions.

The rest of this paper is organised as follows. Section 2 contains notation required later, section 3 a definition of the term calculus and section 4 uses categorical methods to derive an expansionary η- and contractive β-rewrite rule for the coproduct. Section 5 defines the conversion relation and proves decidability and confluence for this fragment, while section 6 deals with the strongly normalising fragment and its embedding into the conversion relation. The paper concludes with some remarks on the possible direction of future research.

2 Notation

While basic knowledge of term rewriting is assumed [4, 10], an introduction to *occurrences* is given (a full development of which may be found in [10]). Occurrences are sequences of natural numbers used to index the subterms of a term and whose analysis forms the technical core of this paper. Let $\mathcal{N}^*$ be the set of sequences of natural numbers with the empty sequence denoted ϵ, while $u.v$ denotes the concatenation of u with v. If $u \neq \epsilon$, then u^+ is the sequence obtained by omitting the last element of u, while u^- is the sequence obtained by omitting the first element. The prefix partial ordering is defined $u \leq v$ iff $\exists w.v = u.w$ and then $v/u = w$. These operations on sequences are extended pointwise to sets of sequences e.g. $X/u = \{w | u.w \in X\}$.

Now let $\mathcal{T}$ be the terms of some calculus. Given any $t \in \mathcal{T}$, its set of *occurrences* $\mathsf{O}(t) \subseteq \mathcal{N}^*$, and the subterm indexed at occurrence $\sigma \in \mathsf{O}(t)$, denoted t/σ, is defined

- If $t = x$, where x is a variable, then $\mathsf{O}(t) = \{\epsilon\}$ and $t/\epsilon = t$
- If $t = Ft_0 \ldots t_n$, then $\mathsf{O}(t) = \{\epsilon\} \cup \{i.\sigma | i \leq n, \sigma \in \mathsf{O}(t_i)\}$ and

$$t/\sigma = \begin{cases} t & \text{if } \sigma = \epsilon \\ t_i/\sigma^- & \text{if } \sigma \neq \epsilon \text{ and } \sigma_0 = i \end{cases}$$

When no danger of confusion exists, the distinction between an occurrence and the subterm so indexed may be blurred. The conversion relation is not left linear, i.e. different occurrences in a redex, indexing syntactically equal subterms, may be mapped to the same occurrence in the reduct. Thus define a set X of occurrences to be *consistent* iff $\forall \sigma, \sigma' \in X. t/\sigma = t/\sigma'$ and if X is non-empty, the subterm(s) so indexed are denoted t/X. Finally $t[\sigma_i \leftarrow u_i]_{i \in \mathcal{I}}$ denoted the textual replacement of terms u_i at occurrences σ_i.

The conversion relation, although not strongly normalising, does posses slightly weaker properties which are nevertheless of both theoretical and practical importance. A *quasi-normal* form for a term t is a term t' such that all reducts of t reduce to t' and the set of R-quasi-normal forms of t is denoted $R(t)$. The one-step reducts of a relation R are denoted $x/R = \{x' | (x, x') \in R\}$, the reflexive closure R^+ and the reflexive transitive closure R^*. If R is an equivalence relation, the equivalence class of an element is denoted $x^=$, while if a term t is R-strongly normalising, its R-rank is denoted $|t|_R$. Finally term constructors are often considered operators on sets of terms, i.e. $F(S_1, \ldots, S_n)$ denotes the set of terms $\{F(t_1, \ldots, t_n) | t_i \in S_i\}$.

3 Almost Bicartesian Closed Logic

Although this paper is primarily concerned with the definition and decidability of $\beta\eta$-equality for coproducts, in order to maintain continuity with previous work and to avoid certain trivial simplifications, a calculus which includes products, terminal object and exponentials is studied. This calculus is called "Almost Bicartesian Closed" as it corresponds to the internal language of bicartesian closed categories, without an initial object. Although the details have not been fully verified, it is hoped that the techniques developed here are sufficient to cope with the addition of an initial object.

The *types* of "Almost Bicartesian Logic", denoted BCC^-, are freely generated by the syntax

$$T := B \mid 1 \mid T + T \mid T \rightarrow T \mid T \times T$$

where B is any base type. For each type T, there are constants $Con(T)$ including the special constant $* \in Con(1)$. Now let Var be a set of variables disjoint from the constants. A context is a list of pairs of variables and types, written $x_1 : \sigma_1, \ldots, x_n : \sigma_n$ such that the variables are pairwise distinct. The concatenation of contexts Γ and Δ is written Γ, Δ. The *term judgements* of BCC^- are of the form $\Gamma \vdash t : T$ and are generated by the traditional structural rules of *Weakening*, *Contraction* and *Exchange* and by the following logical rules

$$\frac{x \in Var}{x : A \vdash x : A}\ Ax \qquad \frac{c \in Con(A)}{\vdash c : A}\ Cons$$

$$\frac{\Gamma \vdash e : A \quad \Gamma \vdash e' : B}{\Gamma \vdash \langle e, e' \rangle : A \times B}\ \times R \qquad \frac{\Gamma \vdash t : A_1 \times A_2}{\Gamma \vdash \pi_i t : A_i}\ \times L$$

$$\frac{\Gamma \vdash t : A_i}{\Gamma \vdash \mathbf{in}_i(t) : A_1 + A_2}\ +R \qquad \frac{\Delta \vdash t : A + B \quad \begin{array}{c}\Gamma, x : A \vdash u : C \\ \Gamma, y : B \vdash v : C\end{array}}{\Gamma, \Delta \vdash \mathbf{case}(t, x.u, y.v) : C}\ +L$$

$$\frac{\Gamma, x : A \vdash e : B}{\Gamma \vdash \lambda x.e : A \rightarrow B}\ \rightarrow R \qquad \frac{\Gamma \vdash e : A \rightarrow B \quad \Delta \vdash e' : A}{\Gamma, \Delta \vdash ee' : B}\ \rightarrow L$$

where the variables occuring in Γ and Δ must be disjoint in those rules involving multiple premises. Given any term judgement $\Gamma \vdash t : T$, we say t is a term of type T and this is written $t : T$. The free variables of a term t are denoted $\mathtt{FV}(t)$ and substitution of terms for free variables of the same type is defined as expected. The suggested η-rewrite rule for coproducts considers terms expressed as substitutions, i.e. subterms whose free variables are not bound in their context. Thus we define the variables bound at an occurrence $\sigma \in \mathtt{O}\ (t)$ as

$$\mathtt{BV}(\sigma, t) = \begin{cases} \emptyset & \text{if } \sigma = \epsilon \\ \{x\} \cup \mathtt{BV}(\sigma^-, t') & \text{if } t = \lambda x.t' \text{ and } \sigma \neq \epsilon \\ \{x_1\} \cup \mathtt{BV}(\sigma^-, t') & \text{if } t = \mathbf{case}(u, x_1.v_1, x_2.v_2) \text{ and } \sigma \geq 1 \\ \{x_2\} \cup \mathtt{BV}(\sigma^-, t') & \text{if } t = \mathbf{case}(u, x_1.v_1, x_2.v_2) \text{ and } \sigma \geq 2 \\ \mathtt{BV}(\sigma^-, t/\sigma_0) & \text{otherwise} \end{cases}$$

and then the free occurrences of a term by $\mathtt{FO}(t) = \{\sigma \in \mathtt{O}\ (t) | \mathtt{FV}(t/\sigma) \cap \mathtt{BV}(\sigma, t) = \emptyset\}$

Familiarity with calculi such as that above is assumed [7, 2]. A term e is called an *introduction* term if it is a λ-abstraction, pair, injection or the constant $*$. If a term is not an introduction term, then it is called an *elimination* term. An occurrence $\sigma \in \mathtt{O}\ (t)$ is the *major premise* of an elimination rule if the subterm so indexed is either applied, projected or the first argument of a *case*-expression.

4 A Rewrite Relation for BCC^-

In [13] extensional rewrite relations for the product, unit and exponential were derived by constructing categorical models of reduction and taking introduction and elimination to be (locally) adjoint functors. When applied to coproducts this approach again generates a contractive β-rewrite rule and an expansionary η-rewrite rule. If $\mathcal{C}(\Gamma, X)$ is the category whose objects are judgements of the form $\Gamma \vdash e : X$ and whose morphisms are rewrites between appropriate judgements, then coproduct introduction and elimination are both functors between the categories shown. These functors constitute an adjoint pair

$$\mathcal{C}(\Gamma.x : A, C) \times \mathcal{C}(\Gamma.y : B, C) \underset{(\_[\mathbf{in}_1(x)/z], \_[\mathbf{in}_2(y)/z])}{\overset{\mathbf{case}(z,\, x.\_,\, y.\_)}{\rightleftarrows}} \mathcal{C}(\Gamma.z : A + B, C) \quad (\top)$$

and the associated unit and co-unit are the rewrite rules.

$$\begin{array}{lll} (\beta_{+,1}) & \mathbf{case}(\mathbf{in}_1(x), x.u, y.v) \Rightarrow u & \\ (\beta_{+,2}) & \mathbf{case}(\mathbf{in}_2(x), x.u, y.v) \Rightarrow v & \\ (\eta_+) & e & \Rightarrow \mathbf{case}(z, x.e[\mathbf{in}_1(x)/z], y.e[\mathbf{in}_2(y)/z]) \end{array}$$

These reduction rules, when closed under substitution and taken together with the reduction rules for the exponential, product and unit connectives, generate *expansionary rewrite relation*, denoted $\Rightarrow$, whose analysis is the subject of the rest of the paper.

$$\begin{array}{lrl} (\beta_{\times,1}) & \pi_0\langle a, b\rangle & \Rightarrow a \\ (\beta_{\times,2}) & \pi_1\langle a, b\rangle & \Rightarrow b \\ (\eta_\times) & c & \Rightarrow \langle \pi_0 c, \pi_1 c\rangle \\ (\beta_\rightarrow) & (\lambda x.t)u & \Rightarrow t[u/x] \\ (\eta_\rightarrow) & t & \Rightarrow \lambda x.tx \\ (\eta_1) & a & \Rightarrow * \\ (\beta_{+,1}) & \mathbf{case}(\mathbf{in}_1(t), x.u, y.v) & \Rightarrow u[t/x] \\ (\beta_{+,2}) & \mathbf{case}(\mathbf{in}_2(t), x.u, y.v) & \Rightarrow v[t/y] \\ (\eta_+) & e[e'/z] & \Rightarrow \mathbf{case}(e', x.e[\mathbf{in}_1(x)/z], y.e[\mathbf{in}_2(y)/z]) \end{array}$$

where various restrictions are implicitly imposed to ensure well-typedness. In addition, to avoid the capture of free variables, $x, y \notin \mathtt{FV}(e)$ in the rewrite rule η_+ and in $\eta_\rightarrow$ we assume $x \notin \mathtt{FV}(t)$.

The rewrite rule η_+ is highly non-local in that (free) subterms of sum type may be expanded to the head of the term and as a result the rewrite rule is significantly more complex than $\eta_\times$ and $\eta_\rightarrow$. As terms typically contain many such subterms a unique normal form cannot be given; rather we may associate to each term a set of normal forms, one for each of the different permutations in which subterms may be expanded. For example the term $\langle \mathbf{case}(t, x.x, y.y), \mathbf{case}(t', x'.x', y'.y') \rangle$ has two normal forms

$$\mathbf{case}(t, x.\mathbf{case}(t', x'.\langle x, x' \rangle, y'.\langle x, y' \rangle), y.\mathbf{case}(t', x'.\langle y, x' \rangle, y'.\langle y, y' \rangle))$$

and

$$\mathbf{case}(t', x'.\mathbf{case}(t, x.\langle x', x \rangle, y.\langle x', y \rangle), y'.\mathbf{case}(t, x.\langle y', x \rangle, y.\langle y', y \rangle))$$

depending on the order in which subterms are expanded. To accommodate this feature, the rewrite rule η_+ is decomposed into two parts, the first of which converts subterms of sum type into the major premises of *case*-expressions and the second which permutes the order in which such subterms are eliminated.

A special case of η_+ occasionally mentioned in the literature [7, 16] is the rewrite rule

$$t \Rightarrow \mathbf{case}(t, x.\mathbf{in}_1(x), y.\mathbf{in}_2(y))$$

This specialisation of η_+ is more akin to $\eta_\times$ and $\eta_\rightarrow$ and indeed, once suitable restrictions have been imposed upon the applicability of these expansions, and taken together with the β-redexes and commuting conversions, forms a strongly normalising and confluent relation. The proof of normalisation is essentially an adaptation of that in [13], although a couple of new innovations are required to deal with the non-congruent nature of this relation. The normal forms of this relation generalise the notion of a long $\beta\eta$-normal form to this calculus as all subterms of a normal form are either of base type, the major premise of an elimination rule or a *quasi-introduction term*. As with the calculus without coproducts, these $\beta\eta$-normal forms can be calculated by first contracting all β-redexes and then performing all remaining expansion possibilities.

The second part of the decomposition is a generalisation of the 'commuting conversions' appearing in [16, 7]. A *conversion* is a subterm occuring as the major premise of a sum elimination and the *conversion relation* develops an algebra of these conversions, allowing them to be identified, discarded or expanded to the head of a term, e.g. the two normal forms above are interconvertable in the conversion relation. Although not strongly normalising, each term has a (finite, enumerable) set of quasi-normal forms and terms equivalent in the equational theory generated by the conversion relation have the same set of quasi-normal forms. Confluence and decidability of the conversion relation are corollaries to these results. The conversion relation is fairly complex and the mathematics required for its complete analysis is too detailed to include in full here. However we indicate the intuitions and motivations behind our definitions and theorems.

Finally the whole expansionary rewrite relation is shown confluent and decidable by embedding it into the conversion relation. Formally terms equivalent in the full theory have normal forms under the first part of the decomposition which are equivalent in the conversion relation.

5 The Conversion Relation

The η_+-rewrite rule extracts free subterms of sum type and inserts injections at their occurrences in the redex; if the original subterm was the major premise of a *case*-expression new β_+-redexes are created by this process. The conversion relation restricts application of the η_+-rewrite rule to extract only those (free) subterms which occur as the major premise of a *case*-expression, and then contracts the resulting β-redexes.

Given a term t, its set of *conversions* is defined by

$$\mathsf{C}\,(t) = \{\sigma \in \mathsf{O}\,(t) | \sigma \text{ occurs as the first argument of a } \textit{case}\text{-expression } \}$$

The free conversions of t are simply those occurrences which are both free and conversions $\mathtt{FC}(t) = \mathtt{C}\,(t) \cap \mathtt{FO}(t)$. Every conversion has a *binding* which consists of the variables bound at the *case*-expression associated to the conversion. These variable bindings play an important role in avoiding variable capture but due to lack of space we often gloss over the details of their treatment. We also assume that whenever sets of conversions are considered, the subterms so indexed are all of the same type and have the same binding.

If $X \subseteq \mathtt{C}\,(t)$, then the result of contracting the β-redexes formed upon insertion of injections at these occurrences are called the *first and second residues* and may be calculated

$$t \setminus_i X = \begin{cases} t & \text{if } X = \emptyset \\ v_i \setminus_i X_i & \text{if } 0 \in X \text{ and } t = \mathbf{case}(u, x.v_1, y.v_2) \\ F(t_1 \setminus_i X_1, \cdots, t_n \setminus_i X_n) & \text{if } X \neq \emptyset,\ 0 \notin X \text{ and } t = F(t_1, \cdots, t_n) \end{cases}$$

where $i = 1, 2$ and $X_n = X/n$.

Lemma 5.1 *Given a set of conversions $X \subseteq \mathtt{C}\,(t)$ binding the variables x_1 and x_2, then for $i = 1$ or 2 there is a reduction sequence $t[\sigma \leftarrow \mathbf{in}_i(x_i)]_{\sigma \in X} \Rightarrow^* t \setminus_i X$.*
Proof A straight forward induction over t □

Residues form part of the definition of the conversion relation which is formally given in terms of a calculus for deriving triples of the form $\langle \sigma, X \rangle : t \Rightarrow_c t'$ where σ is the depth in the term where the rewrite occurs and X are conversions to be expanded, i.e. $X \subseteq \mathtt{FC}(t/\sigma)$. These triples are generated by three inference rules.

$$\text{Expansion}\quad \frac{X \subseteq \mathtt{FC}(t) \quad X \text{ consistent} \quad X \neq \emptyset}{\langle \epsilon, X \rangle : t \Rightarrow_c \mathbf{case}(t/X, x.t \setminus_1 X, y.t \setminus_2 X)}$$

$$\text{Weakening}\quad \frac{x \notin FV(t) \quad y \notin FV(t)}{\langle \epsilon, \emptyset \rangle : \mathbf{case}(u, x.t, y.t) \Rightarrow_c t}$$

$$\text{Congruence}\quad \frac{\langle \sigma, X \rangle : t_j \Rightarrow_c t'_j}{\langle j.\sigma, X \rangle : \mathcal{T}(t_0, \ldots, t_n) \Rightarrow_c \mathcal{T}(t_0, \ldots, t_n)[j \leftarrow t'_j]}$$

where in the first clause x, y are the variables bound by each $\sigma \in X$ and to avoid variable capture $x, y \notin \mathtt{FV}(t) \cup \mathtt{BV}(\sigma, t)$. These conditions can always be met, if necessary by a change of bound variables. The label part of a rewrite is sometimes omitted.

The *Expansion* clause requires the set X of conversions to be free and consistent so that the redex may be expressed in a form compatible with the η_+-rewrite rule. In addition this set is required to be non-empty to prevent expansions of the form $u \Rightarrow \mathbf{case}(t, x.u, y.u)$ which would allow terms to grow arbitrary large, new free variables to be introduced and other undesirable features. However to ensure these terms are identified in the equational theory generated by the conversion relation, rewrites of this form have been inverted and included under the *Weakening* clause.

Lemma 5.2 *Given a triple $\langle \sigma, X \rangle : t \Rightarrow_c t'$, then $t = t'$ in the expansionary rewrite relation.*
Proof Assume first that $\sigma = \epsilon$. If the rewrite is of the form $\mathbf{case}(u, x.t, y.t) \Rightarrow_c t$ then given a variable z not free in t

$$t = t[u/z] \Rightarrow_{\eta_+} \mathbf{case}(u, x.t[\mathbf{in}_1(x)/z], y.t[\mathbf{in}_2(y)/z]) = \mathbf{case}(u, x.t, y.t)$$

If however X in non-empty then, since X is non-empty and consists of free consistent conversions,

$$\begin{aligned} t &= t[\sigma \leftarrow z]_{\sigma \in X}[z := t/X] \\ &\Rightarrow_{\eta_+} \mathbf{case}(t/X, x.t[\sigma \leftarrow \mathbf{in}_1(x)]_{\sigma \in X}, y.t[\sigma \leftarrow \mathbf{in}_2(y)]_{\sigma \in X}) \\ &\Rightarrow^* \mathbf{case}(t/X, x.t \setminus_1 X, y.t \setminus_2 X) \end{aligned}$$

Finally if $\sigma \neq \epsilon$ then, as both relations are congruences, the lemma follows by induction. □

The conversion relation is so named because they generalise the *commuting conversions* [7, 16], i.e. redexes formed when *case*-expressions form the major premise of an elimination rule. The reader is invited to check the following commuting conversion (which illustrates many interesting features and is used later) may be derived as a conversion rewrite with label $\langle \epsilon, \{00\}\rangle$.

$$\begin{array}{ll} \beta_{+,+}: & \mathbf{case}(\mathbf{case}(t, x.u, y.v), x'.u', y'.v') \;\Rightarrow \\ & \mathbf{case}(t, x.\mathbf{case}(u, x'.u', y'.v'), y.\mathbf{case}(v, x'.u', y'.v')) \end{array} \quad (1)$$

The core of our analysis of the conversion relation is to use the structure of a rewrite, represented in the associated label, to describe its action on arbitrary conversions. More formally to each rewrite $r : t \Rightarrow t'$ is associated a function $\overline{r} : \mathsf{C}\,(t) \rightarrow \mathcal{P}\mathsf{C}\,(t')$ which relates conversions in the redex, called *ancestors*, to conversions in the reduct, called *descendants*. A single conversion may have more than one descendant and the ordering on conversions is not necessarily preserved, e.g. in reduction 1 the conversion $00 \mapsto \{0\}$ while $0 \mapsto \{10, 20\}$. Another interesting reduction is

$$\langle \epsilon, \{0, 10\}\rangle : \mathbf{case}(t, x.\mathbf{case}(t, x.u, y.v), y.s) \Rightarrow \mathbf{case}(t, x.u, y.s) \quad (2)$$

which shows how a conversion, e.g. any inside v, may have no descendants, and how a conversion in the reduct may have more than one ancestor. The construction of $\overline{r}$ is fairly lengthy and involved and so only a sketch is given. Firstly given (consistent) conversions X the set $\mathsf{C}\,(t)$ may be partitioned into those conversions which are sub-conversions of (unique) members of X, those conversions which have descendants in one or both of the residues and those conversions which fit into neither of these categories and hence have no descendants. If the calculation of a residue is viewed as indicating a kind of path through the term, then conversions of this last type are those which are unreachable, e.g. conversions like those inside v in reduction 2. The unique descendant of a conversion σ in a residue $t \setminus_i X$ is given by the following function, which is undefined if no such descendant exists.

$$\Omega_i(X, \sigma) = \begin{cases} \Omega_i(X/i, \sigma^-) & \text{if } 0 \in X \text{ and } \sigma \geq i \\ \text{undefined} & \text{if } 0 \in X \text{ and } \sigma \not\geq i \\ 0 & \text{if } 0 \notin X \text{ and } \sigma = 0 \\ \sigma_0.\Omega_i(X/\sigma_0, \sigma^-) & \text{otherwise} \end{cases}$$

The subterm indexed by descendant conversions is given by $(t \setminus_i X)/\Omega_i(X, \tau) = (t/\tau) \setminus_i (X/\tau)$. Now if τ a conversion and r is a *Weakening*, its descendents are

$$\overline{r}(\tau) = \begin{cases} \emptyset & \text{if } \tau \geq 0 \\ \{\tau^-\} & \text{otherwise} \end{cases}$$

while if r is an expansion of the non-empty set of conversions X

$$\overline{r}(\tau) = \begin{cases} \{0.\tau/\sigma\} & \text{if there is a } \sigma \in X \text{ with } \tau \geq \sigma \\ \{1.\Omega_1(X, \tau), 2.\Omega_2(X, \tau)\} & \text{otherwise} \end{cases}$$

where σ in the first clause is necessarily unique and those functions undefined in the second clause are deleted. Finally if r is induced by a congruence then $\overline{r}$ is calculated by induction. The functions $\overline{r}$ are extended pointwise to sets of conversions. Rather lengthy, but tedious, inductive arguments show that $\overline{r}(\tau)$ actually consists of conversions, and that $\overline{r}$ is surjective, i.e. the the function

$$r^{-1}(\tau) = \{\sigma \in \mathsf{C}\,(t) | \tau \in \overline{r}(\sigma)\}$$

does not return the empty set, i.e. the conversion relation does not "create" new conversions. This surjectivity result means that the possibilities for further reduction in a reduct may be traced back to the associated redex, i.e. a static analysis of all the possibilities for reduction may be given. In the next two subsections these comments are validated by the construction of the normal forms of a given term.

5.1 A Decidability Result

Given a rewrite $r : t \Rightarrow_c t'$ and a set of conversions $X \subseteq C(t)$, we shall try to find conditions under which r may be decomposed into its action on individual conversions and its action on the residues, i.e. find conditions under which rewrites of the following form exist

$$r/\sigma' : t/\sigma \Rightarrow_c t'/\sigma' \quad \text{and} \quad r \setminus_i X : t \setminus_i X \Rightarrow_c t' \setminus_i \overline{r}(X)$$

where $\sigma \in X$ and $\sigma' \in \overline{r}(\sigma)$. Such decompositions of a rewrite into rewrites on smaller terms will lead to a recursive characterisation of the possibilities of rewriting for a given term and thus eventually to confluence and decidability. The reduction 1 shows that in general a rewrite may not be localised to a conversion as the conversion 0 is mapped to the conversions 10 and 20 but no reduction exists between the corresponding subterms. This problem occurs as the conversion 0 is mapped into the residues while one of its sub-conversions is expanded to the head of the term and is thus 'removed' from the original conversion. The prefix-ordering endows conversions with a layer structure and the key to localising a rewrite to a conversion is to ensure that this layer structure is preserved. Thus a rewrite $\langle \tau, X \rangle : t \Rightarrow t'$ is said to *preserve* a conversion $\sigma \in C(t)$ iff $\forall \omega \in X. \neg(\tau < \sigma < \omega)$, i.e. no subconversions of σ are expanded outside of σ. This generates a subrelation where all conversions are preserved.

$$\Rightarrow_p = \{r : t \Rightarrow t' | r \text{ preserves } C(t)\}$$

Note the commuting conversion 1 represents the breakdown of the conversion layer structure, i.e. any $\Rightarrow_c$-reduction may be expressed as a sequence of $\Rightarrow_p$-reductions and $\Rightarrow_{\beta_+, +}$-reductions.

Given a rewrite r, a set of conversions X is r-closed iff $r^{-1}\overline{r}(X) = X$. Such sets of conversions induce rewrite in the associated residues, the closure requirement is linked to the non-left linearity of the conversion relation. There is also a slight complication regarding the direction of the residual rewrite. For example if r represents a basic expansion of conversions $Y \subseteq C(t)$, then the obvious candidate for $r \setminus_i X$ is the basic expansion of the descendants of Y in the residue $t \setminus_i X$. However this set may be empty, in which case the direction of the rewrite must be reversed.

Lemma 5.3 *Let* $r : t \Rightarrow_c t'$ *preserve* $\sigma \in C(t)$ *and* $\sigma' \in \overline{r}(\sigma)$. *Then there is a rewrite* $r/\sigma' : t/\sigma \Rightarrow t'/\sigma'$ *and if* X *is an* r-*closed set of conversions then there are either rewrites* $r \setminus_i X : t \setminus_i X \Rightarrow t' \setminus_i \overline{r}(X)$ *or rewrites in the reverse direction. In addition if* r *preserves all conversions then so do the localised and residual rewrites.*

Proof The proof is by induction on r. □

We can now prove the conversion relation decidable but, because of the conditions required by the above lemma, this is done in two stages, firstly for the subrelation $\Rightarrow_p$ and then for the whole conversion relation. Decidability of $\Rightarrow_p$ is proved by constructing for each term its finite and enumerable set of $\Rightarrow_p$-quasi normal forms and using Lemma 5.3 to show that $\Rightarrow_p$-equivalent terms have the same set of quasi-normal forms.

If a term is a quasi-normal form containing free conversions, then the term must be a *case*-expression, as otherwise a rewrite to such a term would exist, but not one in the other direction. Thus the construction of normal forms is essentially a process of expanding as many conversions as possible. However as a non-free conversion may have a free descendant, we must consider not just free conversions but also *potentially free conversions*. Secondly as these normal forms are to be $\Rightarrow_p$-reducts, only *maximal* conversions are to be expanded, i.e. those conversions which are not subconversions of other conversions. The construction also performs two other tasks, checking for possible applications of *Weakening* and also ensuring as much identification of conversions as possible.

The set of *maximal conversions* is defined $MC(t) = \{\sigma \in C(t) | \nexists \sigma' \in C(t). \sigma' < \sigma\}$. The construction of quasi-normal forms is presented in terms of an inductively defined relation NF_p

- If t is a variable then $NF_p(t) = \{t\}$.

- If t is not a variable then define the set of ***potentially free conversions***, denoted $\mathtt{PFC}(t)$, as

$$\{\sigma \in \mathtt{C}(t) \mid \forall u \in \mathtt{NF}_p(t/\sigma).\mathtt{BV}(\sigma) \cap \mathtt{FV}(u) = \emptyset\}$$

The set of ***maximal potentially free conversions*** of t, denoted $\mathtt{MPFC}(t)$, consists of those potentially free conversions which are also maximal and this set is equipped with an equivalence relation $\sigma_1 \sim \sigma_2$ iff $\mathtt{NF}_p(t/\sigma_1) = \mathtt{NF}_p(t/\sigma_2)$. There are three inference rules for constructing normal forms

$$\frac{\mathtt{MPFC}(t) = \emptyset \quad t_i \Rightarrow_{\mathtt{NF}_p} \alpha_i}{T(t_1, \ldots, t_n) \Rightarrow_{\mathtt{NF}_p} T(\alpha_1, \ldots, \alpha_n)}$$

$$\frac{\sigma \in \mathtt{MPFC}(t) \quad \mathtt{NF}_p(t \setminus_1 \sigma^=) = \mathtt{NF}_p(t \setminus_2 \sigma^=) \quad t \setminus_1 \sigma^= \Rightarrow_{\mathtt{NF}_p} u}{t \Rightarrow_{\mathtt{NF}_p} u}$$

$$\frac{\sigma \in \mathtt{MPFC}(t) \quad \mathtt{NF}_p(t \setminus_1 \sigma^=) \neq \mathtt{NF}_p(t \setminus_2 \sigma^=) \quad t \setminus_i \sigma^= \Rightarrow_{\mathtt{NF}_p} \beta_i \quad t/\sigma \Rightarrow_{\mathtt{NF}_p} \alpha}{t \Rightarrow_{\mathtt{NF}_p} \mathbf{case}(\alpha, x.\beta_1, y.\beta_2)}$$

where in the last clause x, y are the variables bound by the set of conversions $\sigma^=$. The set of terms $\mathtt{NF}_p(t)$ is clearly non-empty, finite, enumerable and the maximality condition ensures that if $t' \in \mathtt{NF}_p(t)$ then t' is a $\Rightarrow_p$-reduct of t.

Proposition 5.4 *Let* $r : t \Rightarrow_p t'$.

- *If* $\sigma \in \mathtt{C}(t)$ *and* $\sigma' \in \overline{r}(\sigma)$, *then* $\sigma \in \mathtt{MPFC}(t)$ *iff* $\sigma' \in \mathtt{MPFC}(t')$ *and for such a* σ, $\overline{r}(\sigma^=) = \sigma'^=$.
- *The sets* $\mathtt{NF}_p(t) = \mathtt{NF}_p(t')$ *are equal*

Proof The lemma is proved simultaneously by induction on the sum of the sizes of the terms in question.

- That σ is maximal iff σ' is follows by induction on the definition of the function $\overline{r}$. Secondly, by induction $\mathtt{NF}_p(t/\sigma) = \mathtt{NF}_p(t'/\sigma')$ and so for any $u \in \mathtt{NF}_p(t/\sigma)$, there is a reduction $t \Rightarrow t' \Rightarrow t'[\sigma' \leftarrow u]$. We may further show that $\mathtt{BV}(\sigma) \cap \mathtt{FV}(u) = \mathtt{BV}(\sigma') \cap \mathtt{FV}(u)$ and so σ is potentially free iff σ' is. Finally given $\tau' \in (\sigma')^=$, there is a $\tau \in \mathtt{MPFC}(t)$ such that $\tau' \in \overline{r}(\tau)$ and by induction
$$\mathtt{NF}_p(t/\tau) = \mathtt{NF}_p(t'/\tau') = \mathtt{NF}_p(t'/\sigma') = \mathtt{NF}_p(t/\sigma)$$
Thus $\tau \in \sigma^=$ and $\tau' \in \overline{r}(\sigma^=)$. Similar arguments prove the reverse containment.
- There are two possibilities. Firstly $\mathtt{MPFC}(t) = \emptyset$ then by the first part of this lemma $\mathtt{MPFC}(t') = \emptyset$ and so the result follows by induction. If however $\exists\sigma \in \mathtt{MPFC}(t)$ then $\overline{r}(\sigma) \subseteq \mathtt{MPFC}(t')$. As $\sigma^=$ is r-closed, there are rewrites
$$t/\sigma \Rightarrow t'/\sigma' \text{ and } t \setminus_1 \sigma^= \Rightarrow t' \setminus_1 \overline{r}(\sigma^=) \text{ and } t \setminus_2 \sigma^= \Rightarrow t' \setminus_2 \overline{r}(\sigma^=)$$
where $\sigma' \in \overline{r}(\sigma)$ and the second and third reductions may be reversed. For each of these equalities the set of normal forms of the left hand side is the same as those of the right hand side. It's now routine to check that any normal form of t is a normal form of t' and vice versa.

□

Theorem 5.5 *The relation* $\Rightarrow_p$ *is confluent and given a term* t, *the set* $\mathtt{NF}_p(t)$ *consists of* $\Rightarrow_p$-*quasi normal forms*

Proof Given two reducts of t, a completion may be constructed to any element of the non-empty set $\mathtt{NF}_p(t)$. Now consider any reduct t' of t. By Proposition 5.4, $\mathtt{NF}_p(t') = \mathtt{NF}_p(t)$ so t' reduces to every element of $\mathtt{NF}_p(t)$. □

5.2 Decidability of the Conversion relation

Any $\Rightarrow_c$-reduction can be expressed as a sequence of $\Rightarrow_p$-reductions and $\beta_{+,+}$-commuting conversions. Thus the construction of $\Rightarrow_c$-quasi normal forms can be viewed as synthesising the $\Rightarrow_p$-normal forms just defined with the normal forms of the strongly normalising and confluent relation generated by $\beta_{+,+}$. This synthesis is a three stage process. Firstly all maximal conversions are recursively normalised, and then all $\beta_{+,+}$-commuting conversions performed. This produces terms all of whose conversions contain no potentially free conversions, and so all reducts of such terms must be $\Rightarrow_p$-reducts. Thus the construction of normal forms may be finished off by using the normalisation procedure for $\Rightarrow_p$. The relations NF and NF° are defined to be the identity on terms containing no conversions and

$$\mathtt{NF}^\circ(t) = \{\beta_{+,+}(t[\sigma \leftarrow \alpha_\sigma])|\sigma \in \mathtt{MC}(t) \text{ and } \alpha_\sigma \in \mathtt{NF}(t/\sigma)\} \quad \text{and} \quad \mathtt{NF}(t) = \{\mathtt{NF}_p(t')|t' \in \mathtt{NF}^\circ(t)\}$$

The following conditions, although not sufficient, indicate the structure of $\Rightarrow_c$-quasi normal forms and are important in reasoning about them. A term is *stable* iff $\forall\sigma \in \mathtt{C}\,(t).\mathtt{PFC}\,(t/\sigma) = \emptyset$ and *distributed* iff either $0 \in \mathtt{C}\,(t)$ or $\mathtt{MPFC}(t) = \emptyset$ and all subterms are also distributed.

Lemma 5.6 *All $\Rightarrow_p$-reducts of stable terms are stable. Thus NF° terms are stable and NF terms are both stable and distributed.*

Proof If $t \Rightarrow_p t'$ and there is a conversion in t' containing a potentially free sub-conversion, then there is also a conversion in t' containing a maximal potentially free sub-conversion. Thus by Proposition 5.4 there is a conversion in t also containing a maximal potentially free conversion and so the redex can't be stable.

For the second part, NF° terms are shown stable by induction, while stability of NF terms now follows from the first part of the lemma. Finally NF terms are $\Rightarrow_p$-normal forms and so automatically distributed. □

Lemma 5.7 *If $t \Rightarrow_c t'$, then $\mathtt{NF}^\circ(t)$ and $\mathtt{NF}^\circ(t')$ are $\Rightarrow_p$-equivalent while the sets $\mathtt{NF}(t)$ and $\mathtt{NF}(t')$ are equal.*

Proof The second half of the lemma follows from Prop 5.4. For terms $\mathbf{case}(t, x.u, y.u)$ and u

$$\mathtt{NF}^\circ\mathbf{case}(t, x.u, y.u) = \beta_{+,+}\mathbf{case}(\mathtt{NF}t, x.\mathtt{NF}^\circ u, y.\mathtt{NF}^\circ u) =_p \mathtt{NF}^\circ u$$

If the reduction in question is a $\beta_{+,+}$-commuting conversion, the lemma is proved by direct calculation, while a basic expansion of maximal conversion $X \subseteq \mathtt{FC}(t)$ requires induction if $\mathtt{NF}(t/X)$ contains free conversions. Finally if the reduction is a congruence then the lemma follows by induction. □

Theorem 5.8 *The conversion relation is confluent, decidable and the set $\mathtt{NF}(t)$ consists of $\Rightarrow_c$-quasi normal forms of t.*

Proof The proof is as for the relation $\Rightarrow_p$. □

6 A Normalising Fragment

The second half of the decomposition of the expansionary rewrite relation is generated by β-redexes, commuting conversions and η-expansions. Since the normalisation proof shares many similarities with that in [13], we concentrate on the technical innovations required by the presence of coproducts. Because of the presence of commuting conversions, the proof is based on Prawitz notion of validity [16] as opposed to reducibility [7] as used before. The second innovation in the proof is required to cope with the failure of substitutivity. An operator $\mathcal{I}$, short for the identity, is introduced $\mathcal{I}t = (\lambda x.x)t$ and used to alter the traditional β-redexes as follows

Definition 6.1 *The reduction relation $\Rightarrow_{\beta'}$ is generated by the following basic redexes*

$$
\begin{array}{ll}
(\beta_{\times,i}) & \pi_0\langle u_0, u_1\rangle \Rightarrow u_1 \\
(\beta_{+,\times}) & \pi_i\mathbf{case}(t, x.u, y.v) \Rightarrow \mathbf{case}(t, x.\pi_i u, y.\pi_i v) \\
(\beta_{\rightarrow}) & (\lambda x.t)t' \Rightarrow t[t'/x] \\
(\beta_{+,\rightarrow}) & \mathbf{case}(t, x.u, y.v)t' \Rightarrow \mathbf{case}(t, x.ut', y.vt') \\
(\beta'_{+,1}) & \mathbf{case}(\mathbf{in}_1(t), x.u, y.v) \Rightarrow u[\mathcal{I}t/x] \\
(\beta'_{+,2}) & \mathbf{case}(\mathbf{in}_2(t), x.u, y.v) \Rightarrow v[\mathcal{I}t/y]
\end{array}
$$

and

$$
\begin{array}{ll}
(\beta_{+,+}) & \mathbf{case}(\mathbf{case}(t, x.u, y.v), x'.u', y'.v') \Rightarrow \\
& \mathbf{case}(t, x.\mathbf{case}(u, x'.u', y'.v'), y.\mathbf{case}(v, x'.u', y'.v'))
\end{array}
$$

The rewrite relation obtained by replacing the altered β'-redexes with the more traditional β-redexes of the introduction is denoted $\Rightarrow_\beta$

Strong normalisation of $\Rightarrow_{\beta'}$ implies strong normalisation of $\Rightarrow_\beta$ and both relations share the same (unique) normal forms. To this relation is added limited possibilities for η-expansion. A term may be expanded providing it is neither an introduction term, nor a *case*-expressions of sum type and does not occur as the major premise of an elimination rule. The reduct of such an expansion is given by

$$
\eta(t) = \begin{cases}
t & \text{if } t \text{ is of base type} \\
* & \text{if } t \text{ is of unit type} \\
\lambda x.tx & \text{if } t \text{ is of function type} \\
\langle \pi_0 t, \pi_1 t\rangle & \text{if } t \text{ is of product type} \\
\mathbf{case}(t, x.\mathbf{in}_1(x), y.\mathbf{in}_2(y)) & \text{if } t \text{ is of sum type}
\end{cases}
$$

The context-sensitive nature of the restrictions on expansion manifest themselves in the definition of the relation simultaneously with a subrelation

Definition 6.2 *The relations $\Rightarrow_{\mathcal{F}}$ and $\Rightarrow_{\mathcal{I}}$ are defined simultaneously by the following inference rules*

$$\frac{t \text{ expandable, not atomic type}}{t \Rightarrow_{\mathcal{F}} \eta(t)} \qquad \frac{t \Rightarrow_{\beta'} t'}{t \Rightarrow_{\mathcal{I}\cap\mathcal{F}} t'}$$

$$\frac{u \Rightarrow_{\mathcal{I}\cup\mathcal{F}} u'}{\langle u, v\rangle \Rightarrow_{\mathcal{I}\cap\mathcal{F}} \langle u', v\rangle} \qquad \frac{v \Rightarrow_{\mathcal{I}\cup\mathcal{F}} v'}{\langle u, v\rangle \Rightarrow_{\mathcal{I}\cap\mathcal{F}} \langle u, v'\rangle}$$

$$\frac{t \Rightarrow_{\mathcal{I}} t'}{\pi_0 t \Rightarrow_{\mathcal{I}\cap\mathcal{F}} \pi_0 t'} \qquad \frac{t \Rightarrow_{\mathcal{I}} t'}{\pi_1 t \Rightarrow_{\mathcal{I}\cap\mathcal{F}} \pi_1 t'}$$

$$\frac{t \Rightarrow_{\mathcal{I}} t'}{tu \Rightarrow_{\mathcal{I}\cap\mathcal{F}} t'u} \qquad \frac{u \Rightarrow_{\mathcal{I}\cup\mathcal{F}} u'}{tu \Rightarrow_{\mathcal{I}\cap\mathcal{F}} tu'}$$

$$\frac{t \Rightarrow_{\mathcal{I}\cup\mathcal{F}} t'}{\lambda x.t \Rightarrow_{\mathcal{I}\cap\mathcal{F}} \lambda x.t'} \qquad \frac{t \Rightarrow_{\mathcal{I}} t'}{\mathbf{case}(t, x.u, y.v) \Rightarrow_{\mathcal{I}\cap\mathcal{F}} \mathbf{case}(t', x.u, y.v)}$$

$$\frac{u \Rightarrow_{\mathcal{I}\cup\mathcal{F}} u'}{\mathbf{case}(t, x.u, y.v) \Rightarrow_{\mathcal{I}\cap\mathcal{F}} \mathbf{case}(t, x.u', y.v)} \qquad \frac{v \Rightarrow_{\mathcal{I}\cup\mathcal{F}} v'}{\mathbf{case}(t, x.u, y.v) \Rightarrow_{\mathcal{I}\cap\mathcal{F}} \mathbf{case}(t, x.u, y.v')}$$

$$\frac{t \Rightarrow_{\mathcal{I}\cup\mathcal{F}} t'}{\mathbf{in}_1(t) \Rightarrow_{\mathcal{I}\cap\mathcal{F}} \mathbf{in}_1(t')} \qquad \frac{t \Rightarrow_{\mathcal{I}\cup\mathcal{F}} t'}{\mathbf{in}_2(t) \Rightarrow_{\mathcal{I}\cap\mathcal{F}} \mathbf{in}_2(t')}$$

where $\mathcal{R} \in \{\mathcal{I}, \mathcal{F}\}$ and $\Rightarrow_{\mathcal{I}\cup\mathcal{F}}$, $\Rightarrow_{\mathcal{I}\cap\mathcal{F}}$ are the union and intersection of $\Rightarrow_{\mathcal{I}}$ and $\Rightarrow_{\mathcal{F}}$ respectively.

The relation $\Rightarrow_{\mathcal{I}}$ is the subrelation of $\Rightarrow_{\mathcal{F}}$ obtained by deleting top level expansions and so may be safely applied in situations where $\Rightarrow_{\mathcal{F}}$ may threaten to violate the context sensitive restrictions. The relations $\Rightarrow_{\mathcal{I}}$ and $\Rightarrow_{\mathcal{F}}$ are not congruences and this complicates traditional normalisation and confluence proofs. The major difficulty is the failure of substitutivity in that if $t \Rightarrow_{\mathcal{F}} t'$, there is in general no rewrite $t[u/x] \Rightarrow_{\mathcal{F}} t'[u/x]$, e.g. if u is an introduction term and the rewrite is induced by an expansion of a free occurrence of x in t. However the altered $\Rightarrow_{\beta'}$-redex substitutes an application term and so is immune from substitutivity problems. Of course this cannot be extended to function types as infinite reduction sequences would be formed $(\lambda x.x)(t) \Rightarrow x[\mathcal{I}t/x] = (\lambda x.x)(t)$, but fortunately the induction hypothesis in the strong normalisation proof is sufficient to cope with this latter case and so there is no need to alter $\beta_{\rightarrow}$. The relation $\Rightarrow_{\mathcal{F}}$ (although not $\Rightarrow_{\mathcal{I}}$) is locally confluent. This is a straight-forward generalisation of the equivalent proof in the calculus without coproducts and so is omitted

6.1 Strong Normalisation of $\Rightarrow_{\mathcal{F}}$

The relation $\Rightarrow_{\mathcal{F}}$ is proved strongly normalising by adapting the notion of validity [16] to cope with the presence of expansionary η-rewrites. The set of *valid* terms of type X, denoted $V(X)$ is defined to be the least non-empty type-indexed set of terms such that $V(X)/\Rightarrow_{\mathcal{I}} \subseteq V(X)$ and

- $\rightarrow R$: $\lambda x.t \in V(X \rightarrow Y)$ if $u \in V(X)$ then $t[u/x] \in V(Y)$
- $\times R$: $\langle u, v\rangle \in V(X \times Y)$ if $u \in V(X)$ and $v \in V(Y)$
- $+R$: $\mathbf{in}_1(t) \in V(A+B)$ if $t \in V(A)$
- $+R$: $\mathbf{in}_2(t) \in V(A+B)$ if $t \in V(B)$
- $+L$: $\mathbf{case}(t, x.u, y.v) \in V(X)$ if $u, v \in V(X)$
- $+L$: $\mathbf{case}(t, x.u, y.v) \in V(X_0 \times X_1)$ if $\mathbf{case}(t, x.\pi_i u, y.\pi_i v) \in V(X_i)$
- $+L$: $\mathbf{case}(t, x.u, y.v) \in V(X \rightarrow Y)$ if for any $w \in V(X)$ not containing x, y as free variables then $\mathbf{case}(t, x.uw, y.vw) \in V(Y)$

Notice that as variables have no $\Rightarrow_{\mathcal{I}}$-reducts, they are automatically valid. The last two clauses are required to deal with commuting conversions, but play a minor role in the proof. Before showing that valid terms satisfy strong closure properties, the reducts of a variable must be characterised.

Definition 6.3 *Define a function, Δ, which maps variables to sets of terms by induction over type structure*

$\Delta(z) = \{z\}$	*z is of base type*
$\Delta(z) = \{z, *\}$	*z is of unit type*
$\Delta(z) = \{z\} \cup \{\lambda x.t_1[zt_2/y] \mid t_1 \in \Delta(y), t_2 \in \Delta(x)\}$	*z is of exponent type*
$\Delta(z) = \{z\} \cup \{\langle t_0[\pi_0 z/x], t_1[\pi_1 z/y]\rangle \mid t_0 \in \Delta(x), t_1 \in \Delta(y)\}$	*z is of product type*
$\Delta(z) = \{z\} \cup \{\mathbf{case}(z, x.\mathbf{in}_1(t_1), y.\mathbf{in}_2(t_2)) \mid t_1 \in \Delta(x), t_2 \in \Delta(y)\}$	*z is of sum type*

where the variables x and y are of appropriate type. The function Δ is extended to terms by

$$\Delta(t) = \{t_0[t/z] \mid t_0 \in \Delta(z)\}$$

Lemma 6.4 *The function Δ gives the rewrites of a variable, i.e. $\Delta(z) = z/\Rightarrow^*_{\mathcal{F}}$*

Proof The proof is by induction on the type of the variable. That $\Delta(z) \subseteq z/\Rightarrow^*_{\mathcal{F}}$ follows by induction while, as $z \in \Delta(z)$, the reverse containment follows from showing $\Delta(z)/\Rightarrow_{\mathcal{F}} \subseteq \Delta(z)$. □

As usual in such normalisation proofs, valid terms are shown to satisfy certain closure properties

Definition 6.5 *Let P be a set of terms. Define the four predicates*

V1: If $t \in P$ then t is $\Rightarrow_{\mathcal{F}}$-strongly normalising
V2: If $t \in P$ then $t/\Rightarrow_{\mathcal{F}} \subseteq P$
V3: If $t \in P$ then $\Delta(t) \subseteq P$
V4: If $t \in P$ then $\mathcal{I}t \in P$

Although V4 has been included as a separate predicate, if P is the set of valid terms of some type, it is actually a consequence of the first three and so valid terms are shown, by induction over types, to satisfy only the first three predicates. Terms of base type cannot be expanded and so for such terms the relations $\Rightarrow_{\mathcal{I}}$ and $\Rightarrow_{\mathcal{F}}$ coincide. Thus valid terms of base type trivially satisfy the validity predicates. The only expansion of terms of unit type has as reduct the valid constant $*$ and so the validity predicates hold. The mathematics required for product and function types is essentially the same as in [13] and so omitted.

Quasi-introduction terms are generated by the syntax

$$q := \mathbf{in}_1(t) \mid \mathbf{in}_2(t) \mid \mathbf{case}(t, x.q, y.q)$$

and are closed under $\Rightarrow_{\mathcal{F}}$, are non-expandable and also the reduct of an η-expansion is a quasi-introduction term. They thus form a stepping stone in proving the validity predicates for terms of sum type. Let *Arm* be the function which maps a quasi-introduction term to the set of injections at its leafs.

Lemma 6.6 *If $V(X)$ and $V(Y)$ satisfy V1, V2 and V3, then the set of valid quasi-introduction terms of type $X+Y$ satisfies V1 and V2. Also if $u : X$ and $v : Y$ are valid, then so are $\mathbf{in}_1(u)$ and $\mathbf{in}_2(v)$.*
Proof Since quasi-introduction terms are non-expandable and closed under reduction, the first part of the lemma follows by induction on validity. The second half follows by induction on $|u|_{\mathcal{F}}$ and $|v|_{\mathcal{F}}$ □

Lemma 6.7 *Let $V(X)$ and $V(Y)$ satisfy V1, V2 and V3 and t be a valid term of type $X+Y$.*

- *If $t = \mathbf{case}(t_0, x.u, y.v)$, then $\mathbf{case}(t_0, x.\alpha, y.\beta)$ is valid, where $\alpha \in \eta(u)/\Rightarrow^*_{\mathcal{F}}$ and $\beta \in \eta(v)/\Rightarrow^*_{\mathcal{F}}$*
- *All terms $t' \in \Delta(t)$ are valid.*

Proof The proof is by simultaneous induction on the validity of t.

(i) By induction $\eta(u)$ and $\eta(v)$ are valid, quasi-introduction terms and hence so are α and β. Thus $|\alpha|_{\mathcal{F}} + |\beta|_{\mathcal{F}}$ is used as an inner induction rank. We are left to prove that all $\Rightarrow_{\mathcal{I}}$-reducts of $\mathbf{case}(t_0, x.\alpha, y.\beta)$ are valid. Those induced by reductions of proper subterms are valid by induction. This leaves two cases. If t_0 is an introduction term, and say $\mathbf{case}(\mathbf{in}_1(s), x.\alpha, y.\beta) \Rightarrow_{\mathcal{I}} \alpha[\mathcal{I}s/x]$ then by induction $\eta(u[\mathcal{I}s/x]) = \eta(u)[\mathcal{I}s/x]$ is valid quasi-introduction term. As $\alpha[\mathcal{I}s/x]$ is a reduct of this term it is also valid. Similarly if t_0 is a *case*-expression, the result of a commuting conversion is shown valid by applying induction to the reduct of the original term obtained by a commuting conversion.

(ii) We must prove that $\mathbf{case}(t, x.\mathbf{in}_1(u), y.\mathbf{in}_2(v))$ is valid where $u \in \Delta(x)$ and $v \in \Delta(y)$. By induction u and v are valid and strongly normalising and hence so are $\mathbf{in}_1(u)$ and $\mathbf{in}_2(v)$. These normalisation ranks then form an inner induction. Those 1-step $\Rightarrow_{\mathcal{I}}$ reducts induced by reductions of proper subterms are valid by induction, while the result of a top-level commuting conversion is valid by the first half of this lemma. Finally a reduction

$$\mathbf{case}(\mathbf{in}_1(t_0), x.\mathbf{in}_1(u), y.\mathbf{in}_2(v)) \Rightarrow_{\mathcal{F}} \mathbf{in}_1(u)[\mathcal{I}t_0/x]$$

is valid as by assumption t_0 is valid and of smaller type, so by induction so is $\mathcal{I}t_0$. Finally, by V3, so is $u[\mathcal{I}t_0/x]$ and hence $\mathbf{in}_1(u)[\mathcal{I}t_0/x]$

□

Corollary 6.8 *Let $V(X)$ and $V(Y)$ satisfy V1, V2 and V3. Then they hold for the set of terms $V(X+Y)$.*
Proof Let t be a term. The lemma is established by induction on the validity of t. All $\Rightarrow_{\mathcal{I}}$-reducts are valid and, by induction, strongly normalising. The only other reduct is a valid quasi-introduction term which is also strongly normalising. Thus all reducts of t are strongly

normalising and hence so is t. The $\Rightarrow_{\mathcal{I}}$-reducts of a term are valid by definition, while the result of a basic expansion has already been shown valid. Finally V3 has just been established above. □

Having shown valid terms satisfy the validity predicates and are thus strongly normalising the last stage is to prove that all terms are valid. However before this can be done the criteria under which a **case**-expression may be shown to be valid must be simplified.

Lemma 6.9 *The term* $\mathbf{case}(t, x.u, y.v)$ *is valid iff* t *is strongly normalising,* u, v *are valid, and in addition if* $t' \in t/\Rightarrow_{\mathcal{I}}^{*}$ *is a quasi-introduction term and* $\mathbf{in}_1(\alpha) \in \mathrm{Arm}(t')$ *then* $u[\mathcal{I}\alpha/x]$ *is valid, and similarly for right injections.*

Proof The proof is by induction on firstly the type of the *case*-expression and secondly on $|t|_{\mathcal{F}} + |u|_{\mathcal{F}} + |v|_{\mathcal{F}}$. There are two proof obligations. Firstly if the *case*-expression is of sum or function type, the clauses pertaining to commuting conversions are easily established by induction. Secondly, those $\Rightarrow_{\mathcal{I}}$-reducts induced by reductions of subterms are valid by induction with the second part of the induction hypothesis following from substitutivity considerations, while a basic β-reduction has a valid reduct by assumption. Finally if t is a *case* expression then the result of a basic commuting conversion is shown valid by first using induction to prove the arms valid and then once more for the whole term. □

Finally all terms are shown valid in the traditional manner

Lemma 6.10 *Let* t *be a term and* θ *be a valid substitution. Then* $t\theta$ *is valid*

Proof The proof is by induction on t and follows the standard pattern. The only interesting part is for the term $\mathbf{case}(t, x.u, y.v)$. The terms $u\theta$, $v\theta$ and $t\theta$ are valid and thus strongly normalising by induction. Thus so is any $t' \in (t\theta)/\Rightarrow_{\mathcal{I}}^{*}$ and hence any $\mathbf{in}_1(\alpha) \in Arm(t')$. From this we may deduce α is also valid and so $\theta; x \mapsto \mathcal{I}\alpha$ is a valid substitution. Thus

$$(u\theta)[\mathcal{I}\alpha/x] = u(\theta; x \mapsto \mathcal{I}\alpha)$$

is a valid term □

Theorem 6.11 *The relations* $\Rightarrow_{\mathcal{F}}$ *and* $\Rightarrow_{\mathcal{I}}$ *are strongly normalising.*

Proof All terms are valid since the identity substitution is a valid substitution, and all valid terms are $\Rightarrow_{\mathcal{F}}$-strongly normalising. □

Lemma 6.12 *In a* $\Rightarrow_{\mathcal{F}}$*-normal form all subterms are either of base type, (quasi-)introduction terms or occur as the major premise of an elimination rule.* λ*-abstractions or are applied.*

Proof Simple induction over term structure □

Finally the whole of the expansionary rewrite relation embeds into the conversion relation of the previous section.

Theorem 6.13 *If* t *and* t' *are* $\Rightarrow_c$*-equivalent, then so are their* $\Rightarrow_{\mathcal{F}}$*-normal forms. Hence the expansionary rewrite relation is both confluent and decidable.*

Proof The lemma is proved by calculation the effect on conversions of $\Rightarrow_{\mathcal{F}}$-rewrites on t and t'.

Decidability and confluence of the expansionary rewrite relation may now be lifted from the same results for the conversion relation. □

7 Conclusions and Further Work

In this paper an extensional equality for terms of BCC^{-} was given. To each term we can associate a finite set of normal forms, calculable in two stages; firstly by $\beta\eta$-normalisation and secondly by expanding as many conversions as possible. As terms equivalent in the equational theory have the same set of normal forms, comparison of these normal forms provides a decision procedure for equality of terms.

There are two principle directions in which this research may be extended. Firstly the inability to define a unique normal form is closely linked to form of the *case*-expression which permits the elimination of one term at a time. An alternate, parallel elimination, allowing the concurrent elimination of several terms should permit the definition of unique normal forms and this is the subject of current work.

In a different direction, extensionality principles can be applied to other theories, e.g. those with recursive and dependent types. However the nature of the resulting proof theory is still vague and requires considerable research.

8 Acknowledgements

I would like to thank my first supervisor Barry Jay who first introduced me to the applications of category theory in term rewriting and provided many insights and suggestions. I would also like to thank Stefan Kahrs, Don Sanella, Alex Simpson and Christoph Luth for their considerable help and support on many occasions.

References

1. Y.Akama, *On Mints' Reduction for ccc-Calculus*, in *Typed Lambda Calculi and Applications*, 1993.
2. H.P. Barendregt, *The Lambda Calculus Its Syntax and Semantics (Revised Edition)* Studies in Logic and the Foundations of Mathematics 103 (North Holland, 1984).
3. D. Čubrić, *On Free CCC*, manuscript.
4. N.Dershowitz, J.P. Jouannaud, *Rewrite Systems*, in *The Handbook of Theoretical Computer Science*, Elsevier, 1990.
5. R.Di Cosmo and D.Kesner, *A confluent reduction for the extensional typed λ-calculus*, in *Proceedings ICALP '93*
6. D.Dougherty, *Some λ-calculi with categorical sums and products*, in *Rewriting Techniques and Applications* LNCS 690, 137-151.
7. J-Y. Girard, P. Taylor and Y. Lafont, *Proofs and Types*, Cambridge Tracts in Theoretical Computer Science (Cambridge University Press, 1989).
8. J.W. Gray, *Formal category theory: adjointness for 2-categories*, Lecture Notes in Mathematics 391 (Springer-Verlag, 1974).
9. G. Huet, *Résolution d'équations dans des langages d'ordre* 1, 2, ..., ω. Thèse d'Etat, Université de Paris VII, 1976.
10. G. Huet, *Abstract Properties and Applications of Term Rewriting Systems*, in JACM, Vol. 27, No 4, pp. 797-821, 1980.
11. C.B. Jay, *Local adjunctions*, J. Pure and Appl. Alg. 53 (1988) 227–238.
12. C.B. Jay, Modelling reduction in confluent categories, in: *Proceedings of the Durham Symposium on Applications of Categories in Computer Science*, Lecture Note Series 177 (London Mathematics Society, 1992) 143–162.
13. C.B. Jay and N.Ghani, *The Virtues of Eta-expansion*, to appear in *Journal of Functional Programming*.
14. J. Lambek and P. Scott, *Introduction to higher order categorical logic*, Cambridge Studies in Advanced Mathematics 7 (Cambridge Univ. Press, 1986).
15. G.E. Mints, Teorija categorii i teoria dokazatelstv.I., in: *Aktualnye problemy logiki i metodologii nauky , Kiev, 1979* 252-278.
16. D. Prawitz, Ideas and results in proof theory, in: J.E. Fenstad (ed) *Proc. 2nd Scandinavian Logic Symp.* (North-Holland, 1971) 235–307.
17. D.E. Rydeheard & J.G. Stell, *Foundations of equational deduction: A categorical treatment of equational proofs and unification algorithms*, in: Pitt et al, (eds), *Category Theory and Computer Science*, Lecture Notes in Computer Science 283 (Springer, 1987) 114 – 139.
18. R.A.G. Seely, *Modelling computations: a 2-categorical framework*, in: *Proceedings of the Second Annual Symposium on Logic in Computer Science* (1987).

Typed Operational Semantics

Healfdene Goguen

[1] INRIA Sophia-Antipolis, 2004, route des Lucioles, B. P. 93
06902 Sophia-Antipolis Cedex, France
[2] Department of Computer Science, University of Edinburgh
The King's Buildings, Edinburgh, EH9 3JZ, United Kingdom

Abstract. This paper introduces typed operational semantics, a class of formal systems which define a reduction to normal form for the well-typed terms of a particular type theory. These systems lead to a new approach to the metatheory for type theories, which we develop here for the simply typed lambda calculus. A similar approach can be used to study systems with dependent types.

1 Introduction

Untyped reduction is an intuitive operational semantics for type theory and a useful tool for implementations of type theories. However, it fails to give an adequate explanation of computation, because it lacks type information, an explicit reduction strategy and explicit mention of normal forms. We introduce a new style of operational semantics for type theory, which we call typed operational semantics, that define a reduction to normal form for well-typed terms. These systems give both an appealing account of reduction in type theory and an elegant development of the metatheory.

The central result about a typed operational semantics is the soundness of the semantics for the original type theory, for which proofs resemble traditional proofs of normalization. Completeness of the semantics is straightforward. We show how to exploit this equivalence between the declarative and operational presentations of type theory to obtain straightforward proofs of important metatheoretic properties, such as strengthening, subject reduction and strong normalization.

This paper introduces typed operational semantics in the context of the simply typed lambda calculus. This allows us to give a concise presentation of the important characteristics of the systems. However, the most important benefits of typed operational semantics are demonstrated in systems with dependent types, where our development of the metatheoretic properties is considerably simpler than the existing techniques. We discuss this briefly in Sect. 7, but a full treatment can be found elsewhere [7, 8].

This paper is organized as follows. In Sect. 2 we give a brief introduction to the simply typed lambda calculus and present the syntax of the semantically motivated presentation. We introduce the general idea of typed operational semantics by presenting a simple formal system in Sect. 3. In Sect. 4 we introduce the typed operational semantics studied in this paper and briefly discuss the rules of inference of this system. Section 5 contains the basic development of the metatheory. In Sect. 6 we give a proof of soundness of the typed operational semantics for the semantic presentation, closely related to the proof of strong normalization for the simply typed lambda calculus. Section 7 outlines the difficulties in the metatheory of systems with dependent types and gives an intuition for why typed operational semantics yield a considerably simpler treatment of these systems. We discuss related work in Sect. 8 and make concluding remarks in Sect. 9.

2 $\lambda^{\rightarrow}$: The Semantic Presentation

In this section we introduce the syntax of types and terms for the simply typed lambda calculus and present the judgements and rules of inference for the system $\lambda^{\rightarrow}$. Mitchell [14] gives a thorough introduction to this system.

2.1 Types and Terms

We first introduce the types of the simply typed lambda calculus. We assume that there is only one base type o. For any two types A and B, we have the type $A \rightarrow B$ of functions from A to B.

We assume the existence of an infinite set V of variables. We use the usual language of terms for the simply typed lambda calculus: *variables* x if $x \in V$, *abstraction* $\lambda x{:}A.M$, and *application* $M(N)$. We identify terms which are equivalent up to the renaming of bound variables and write $M \equiv N$ if M and N are equal in this way. We write $\mathrm{FV}(M)$ for the free variables in a term M, those variables not bound by abstractions. We denote the substitution of N for the free variable x in M by $[N/x]M$, where this substitution is defined to avoid the capture of free variables.

A *pre-context* is a sequence $\Gamma = x_1{:}A_1, \ldots, x_n{:}A_n$ of pairs of variables $x_i \in V$ and types A_i. We denote the empty pre-context by $()$. We write $dom(\Gamma)$ for the set $\{x_1, \ldots, x_n\}$. A *context* is a pre-context where the x_i are distinct.

2.2 Judgements and Derivations

We have two judgement forms in our presentation of the simply typed lambda calculus: $\Gamma \vdash M : A$, meaning that the term M has type A in context Γ; and $\Gamma \vdash M = N : A$, meaning that the terms M and N are equal and of type A in context Γ. We call the second judgement form *judgemental equality.* The rules of inference for the system are given in Figs. 1 and 2. We write $\Gamma \vdash J$ for an arbitrary judgement in $\lambda^{\rightarrow}$.

We have added rules of inference for thinning and substitution, following Martin-Löf's semantic explanation of type theory [12]. Although the usual presentations of the simply typed lambda calculus do not have these rules, the rules are always admissible: that is, if there are derivations of the premisses then

$$(Var)\ \frac{\Gamma \text{ context} \quad x{:}A \in \Gamma}{\Gamma \vdash x : A} \qquad (Thin)\ \frac{\Gamma_0, \Gamma_1 \vdash M : A \quad z \notin dom(\Gamma_0, \Gamma_1)}{\Gamma_0, z{:}C, \Gamma_1 \vdash M : A}$$

$$(\lambda)\ \frac{\Gamma, x{:}A \vdash M : B}{\Gamma \vdash \lambda x{:}A.M : A \to B} \qquad (App)\ \frac{\Gamma \vdash M : A \to B \quad \Gamma \vdash N : A}{\Gamma \vdash M(N) : B}$$

$$(Subst)\ \frac{\Gamma_0, z{:}C, \Gamma_1 \vdash M : A \quad \Gamma_0 \vdash N : C}{\Gamma_0, \Gamma_1 \vdash [N/z]M : A}$$

Fig. 1. Typing Rules

$$(Refl)\ \frac{\Gamma \vdash M : A}{\Gamma \vdash M = M : A} \qquad (Sym)\ \frac{\Gamma \vdash M = N : A}{\Gamma \vdash N = M : A}$$

$$(Trans)\ \frac{\Gamma \vdash M = N : A \quad \Gamma \vdash N = P : A}{\Gamma \vdash M = P : A}$$

$$(\beta)\ \frac{\Gamma, x{:}A \vdash M : B \quad \Gamma \vdash N : A}{\Gamma \vdash (\lambda x{:}A.M)(N) = [N/x]M : B}$$

$$(\eta)\ \frac{\Gamma \vdash M : A \to B}{\Gamma \vdash \lambda x{:}A.M(x) = M : A \to B}$$

$$(\lambda\text{-}Eq)\ \frac{\Gamma, x{:}A \vdash M = N : B}{\Gamma \vdash \lambda x{:}A.M = \lambda x{:}A.N : A \to B}$$

$$(App\text{-}Eq)\ \frac{\Gamma \vdash M = P : A \to B \quad \Gamma \vdash N = Q : A}{\Gamma \vdash M(N) = P(Q) : B}$$

$$(Subst\ Eq)\ \frac{\Gamma_0, z{:}C, \Gamma_1 \vdash M : A \quad \Gamma_0 \vdash P = Q : C}{\Gamma_0, \Gamma_1 \vdash [P/z]M = [Q/z]M : A}$$

$$(Eq\ Subst)\ \frac{\Gamma_0, z{:}C, \Gamma_1 \vdash M = N : A \quad \Gamma_0 \vdash P : C}{\Gamma_0, \Gamma_1 \vdash [P/z]M = [P/z]N : A}$$

Fig. 2. Judgemental Equality

there is a derivation of the conclusion. Therefore, adding them is not an essential change to the system.

3 Typed Operational Semantics

A typed operational semantics presents type theory from the perspective of computation instead of that of logical inference: we still need the full type information to derive the well-typedness of any term, but we replace the logical rules for application and abstraction by rules which instead express the reduction behavior of these terms in the calculus. As an example of a typed operational semantics, we present the system λ^V in Fig. 3. This system has one judgement form, $\Gamma \vdash^V M \to^{nf} P : A$, by which we mean informally that both M and P

have type A in context Γ, and that M reduces under the call-by-value strategy to normal form P.

$$(V\text{-}Base)\ \frac{\Gamma \vdash^V M_i \to^{\mathrm{nf}} P_i : A_i \quad (1 \leq i \leq n)}{\Gamma \vdash^V x(M_1,\ldots,M_n) \to^{\mathrm{nf}} x(P_1,\ldots,P_n) : B}\ \textit{if } x{:}A_1 \to \ldots \to A_n \to B \in \Gamma$$

$$(V\text{-}\lambda)\ \frac{\Gamma, x{:}A \vdash^V M \to^{\mathrm{nf}} P : B}{\Gamma \vdash^V \lambda x{:}A.M \to^{\mathrm{nf}} \lambda x{:}A.P : A \to B}\ \textit{if } P \equiv Q(x) \textit{ implies } x \in \mathrm{FV}(Q)$$

$$(V\text{-}\eta)\ \frac{\Gamma, x{:}A \vdash^V M \to^{\mathrm{nf}} P(x) : B}{\Gamma \vdash^V \lambda x{:}A.M \to^{\mathrm{nf}} P : A \to B}\ \textit{if } x \notin \mathrm{FV}(P)$$

$$(V\text{-}\beta)\ \frac{\Gamma \vdash^V M \to^{\mathrm{nf}} \lambda x{:}A.P : B \to C \quad \Gamma \vdash^V N \to^{\mathrm{nf}} Q : B \quad \Gamma \vdash^V [Q/x]P \to^{\mathrm{nf}} R : C}{\Gamma \vdash^V M(N) \to^{\mathrm{nf}} R : C}$$

Fig. 3. Call-By-Value Reduction

λ^V is typical of typed operational semantics in being a deterministic reduction to normal form with type information. We can prove strengthening, normalization and the decidability of type-checking for $\lambda^{\to}$ by proving these results in λ^V and transferring them by soundness and completeness. However, λ^V encodes a particular reduction strategy to normal form, so we are unable to use this system to establish properties about arbitrary reductions, such as strong normalization or subject reduction. In the next section, we consider a typed operational semantics that is universal with respect to reductions and therefore allows us to show these results as well.

4 λ^S: A System for Standard Reduction

4.1 Standard Reduction

Standard reduction [2] is an important notion in the untyped lambda calculus: any reduction sequence can be expressed as a standard, or left to right, reduction sequence. In this section we introduce the system λ^S, a synthesis of this reduction and the semantic presentation $\lambda^{\to}$. We shall use this system to study the metatheory of the simply typed lambda calculus. Because the reduction used in λ^S is universal with respect to reductions, the important results about the relationship between typed terms and reduction follow easily by reasoning in this system.

4.2 Judgements and Derivations

We now define the calculus λ^S. The judgement forms for λ^S and their intuitive meanings are:

- $\Gamma \vdash^S M \rightarrow^{\mathrm{nf}} P : A$, meaning that M has normal form P of type A in context Γ.
- $\Gamma \vdash^S M \rightarrow^{\mathrm{wh}} N : A$, meaning that M weak head reduces to N of type A in context Γ.

We write $\Gamma \vdash^S J$ for an arbitrary judgement in λ^S. We also use the abbreviation $\Gamma \vdash^S M : A$ for $\Gamma \vdash^S M \rightarrow^{\mathrm{nf}} P : A$ when the normal form P is not important.

The typed operational semantics λ^S is defined by the rules of inference in Figs. 4 and 5.

$$(S\text{-}Base)\ \frac{\Gamma \vdash^S M_i \rightarrow^{\mathrm{nf}} P_i : A_i \quad (1 \leq i \leq n)}{\Gamma \vdash^S x(M_1, \ldots, M_n) \rightarrow^{\mathrm{nf}} x(P_1, \ldots, P_n) : B}\ \textit{if } x{:}A_1 \rightarrow \ldots \rightarrow A_n \rightarrow B \in \Gamma$$

$$(S\text{-}\lambda)\ \frac{\Gamma, x{:}A \vdash^S M \rightarrow^{\mathrm{nf}} P : B}{\Gamma \vdash^S \lambda x{:}A.M \rightarrow^{\mathrm{nf}} \lambda x{:}A.P : A \rightarrow B}\ \textit{if } P \equiv Q(x) \textit{ implies } x \in \mathrm{FV}(Q)$$

$$(S\text{-}\eta)\ \frac{\Gamma, x{:}A \vdash^S M \rightarrow^{\mathrm{nf}} P(x) : B \quad \Gamma \vdash^S P \rightarrow^{\mathrm{nf}} Q : A \rightarrow B}{\Gamma \vdash^S \lambda x{:}A.M \rightarrow^{\mathrm{nf}} Q : A \rightarrow B}$$

$$(S\text{-}WH)\ \frac{\Gamma \vdash^S M \rightarrow^{\mathrm{wh}} N : A \quad \Gamma \vdash^S N \rightarrow^{\mathrm{nf}} P : A}{\Gamma \vdash^S M \rightarrow^{\mathrm{nf}} P : A}$$

Fig. 4. Canonical Forms

$$(W\text{-}\beta)\ \frac{\Gamma \vdash^S \lambda x{:}A.M : A \rightarrow B \quad \Gamma \vdash^S N : A}{\Gamma \vdash^S (\lambda x{:}A.M)(N) \rightarrow^{\mathrm{wh}} [N/x]M : B}$$

$$(W\text{-}App)\ \frac{\Gamma \vdash^S M \rightarrow^{\mathrm{wh}} P : A \rightarrow B \quad \Gamma \vdash^S N : A}{\Gamma \vdash^S M(N) \rightarrow^{\mathrm{wh}} P(N) : B}$$

Fig. 5. Weak Head Reduction

4.3 Discussion

We discuss the rules of inference of λ^S in order to illustrate the basic idea of the system and to explain some choices in the presentation.

First, a variable applied to a sequence of terms, which we call a *base term*, evaluates by the rule (*S-Base*) to the variable followed by the normal forms of the terms. We need to know that x occurs in the context to ensure well-typedness.

The rules (S-λ) and (S-η) for abstraction implement η-postponement and preserve normality of the right-hand side of the reduction. The rules first fully reduce the body M of $\lambda x{:}A.M$, and by analysis of the resulting normal form either the abstraction is already in normal form or a final η-reduction takes place. This yields a simple factorization of the reduction sequence into β-reductions followed by η-reductions.

The second premiss of the rule (S-η), requiring the normal form of the η-redex to be well-typed the context Γ, is unnecessary from the point of view of our reduction, although it serves as a side condition requiring x to be distinct from the free variables of P. By strengthening the premisses, we have weakened the rule and simplified proofs by induction on derivations of λ^S. Furthermore, using our metatheoretic development, we can prove the admissibility of the more natural η-reduction rule:

$$(S\text{-}\eta')\ \frac{\Gamma, x{:}A \vdash^S M \rightarrow^{\mathrm{nf}} P(x) : B}{\Gamma \vdash^S \lambda x{:}A.M \rightarrow^{\mathrm{nf}} P : A \rightarrow B}\ \textit{if } x \notin \mathrm{FV}(P)$$

Weak head reduction is defined by the rules of inference for the judgement form $\Gamma \vdash^S M \rightarrow^{\mathrm{wh}} N : A$. Intuitively weak head redices are outermost redices: either a β-redex or a weak head redex on the left of an application. By using this judgement form, we also require the reduct to be well-typed, because the only rule incorporating derivations of this judgement into the main judgement $\Gamma \vdash^S M \rightarrow^{\mathrm{nf}} P : A$ is the rule (S-WH).

We do not use the normal forms of either $\lambda x{:}A.M$ or N in the rule (S-β). However, it is important from the point of view of reduction that the two terms are normalizing, and from the point of view of typing that the two terms are well-typed.

5 Metatheory

In this section we prove the essential metatheoretic properties for the simply typed lambda calculus. Our presentation is influenced by Luo's thesis [9] and developments of Pure Type Systems [3, 17], although the formal system is considerably different. Since we shall eventually show the equivalence of λ^S and $\lambda^{\rightarrow}$, all of the results for λ^S will transfer to $\lambda^{\rightarrow}$ as well.

5.1 Basic Definitions

We first introduce the basic notions of reduction and parallel substitution.

The untyped reduction rules β and η are the usual analogues of the corresponding rules for judgemental equality. If R is a relation on terms then we denote its compatible closure by $M \triangleright_R N$, its transitive closure by R^+ and its reflexive and transitive closure by R^*. Untyped reduction, $M \triangleright N$, is the compatible closure of $\beta\eta$.

A term M is *normal* if it has no reductions, that is if there is no N such that $M \rhd N$. A term is *strongly normalizing* if all reduction sequences starting from that term terminate. A term M satisfies the *diamond property* if $M \rhd^* N$ and $M \rhd^* P$ imply that there is a Q such that $N \rhd^* Q$ and $P \rhd^* Q$.

A *pre-substitution* for a finite set of variables S is a function from S to terms. If δ is a pre-substitution for $dom(\Gamma)$ then we write $\widehat{\delta}(M)$ for the result of simultaneously substituting the values for the variables in the domain of δ:

$$\widehat{\delta}(M) =_{\text{df}} [\delta(x_1), \ldots, \delta(x_n)/x_1, \ldots, x_n]M$$

If ϕ is a pre-substitution for Φ then the *composition* of ϕ and δ, $\phi \circ \delta$, is

$$\phi \circ \delta(x) =_{\text{df}} \widehat{\phi}(\delta(x))$$

We write $\delta[x{:=}M]$ for the extended pre-substitution for $dom(\Gamma, x{:}A)$. If Δ has all components of Γ then we define the substitutions $\text{weak}_\Gamma^\Delta(x) =_{\text{df}} x$ for $x \in dom(\Gamma)$, and $\text{id}_\Gamma =_{\text{df}} \text{weak}_\Gamma^\Gamma$.

A *substitution* δ from Δ to Γ is a pre-substitution for $dom(\Gamma)$ such that $\Delta \vdash^S \delta(x) : A$ for each $x{:}A \in \Gamma$. A *renaming* δ from Δ to Γ is a substitution from Δ to Γ such that $\delta(x) = y$, where $y \in V$, for each $x{:}A \in \Gamma$.

An important property of renamings is that if δ is a renaming from Δ to Γ and ϕ is a renaming from Φ to Δ then $\phi \circ \delta$ is a renaming from Φ to Γ. This property is not straightforward for arbitrary substitutions, and we shall only know it as a consequence of our final soundness theorem.

5.2 Basic Metatheory

In this section we establish the basic metatheoretic properties we need for the two calculi presenting the simply typed lambda calculus. Most of the results will be for the system λ^S, because we shall transfer the results from this system to $\lambda^{\rightarrow}$ by soundness and completeness results.

The results in this section follow by straightforward induction on derivations unless otherwise mentioned.

Lemma 1. *If* $\Gamma \vdash M = N : A$ *then* $\Gamma \vdash M : A$ *and* $\Gamma \vdash N : A$.

Lemma 2 Free Variables. *If* $\Gamma \vdash^S M \to^{\text{nf}} P : A$ *then* $\text{FV}(M) \subseteq dom(\Gamma)$ *and* $\text{FV}(P) \subseteq dom(\Gamma)$.

In contrast to the semantic presentation, we need to show Free Variables before Generation to treat the rule of inference for η-redices.

Lemma 3 Generation. *Any judgement which is derivable in* λ^S *has a uniquely determined last rule of inference.*

Lemma 4 Uniqueness of Normal Forms. *If* $\Gamma \vdash^S M \to^{\text{nf}} P : A$ *and* $\Gamma \vdash^S M \to^{\text{nf}} Q : B$ *then* $P \equiv Q$ *and* $A = B$.

This lemma states that the operational semantics is deterministic. The system also has the stronger property of being *syntax directed*, meaning that for each context Γ and term M there are at most one term M and type A such that $\Gamma \vdash^S M \rightarrow^{\mathrm{nf}} P : A$, but we do not need this fact explicitly.

Lemma 5 Context. *If $\Gamma \vdash^S J$ then Γ is a context.*

Lemma 6 Completeness. *If $\Gamma \vdash^S M \rightarrow^{\mathrm{nf}} P : A$ then $\Gamma \vdash M = P : A$.*

We now show that any judgement in a context Γ also holds in any context which has all components of Γ. We first prove a general lemma inspired by McKinna and Pollack's treatment [13], where we allow names of variables in the context to change.

Lemma 7 Renaming. *If δ is a renaming from Δ to Γ and $\Gamma \vdash^S M \rightarrow^{\mathrm{nf}} P : A$ then $\Delta \vdash^S \widehat{\delta}(M) \rightarrow^{\mathrm{nf}} \widehat{\delta}(P) : A$.*

The reasoning for the rules $(S\text{-}\lambda)$ and $(S\text{-}\eta)$ is complicated. We have the usual Thinning lemma as a corollary.

Corollary 8 Thinning. *If $\Gamma \vdash^S J$ and Δ is a context with all components of Γ then $\Delta \vdash^S J$.*

Strengthening for the simply typed lambda calculus is a straightforward result. However, the property can be very difficult to prove for more complicated calculi with dependent types, and the fact that the typed operational semantics is deterministic leads to a simple and elegant proof of this property for such systems.

Lemma 9 Strengthening. *If $\Gamma_0, z{:}C, \Gamma_1 \vdash^S M \rightarrow^{\mathrm{nf}} P : A$ and $z \notin \mathrm{FV}(M)$ then $\Gamma_0, \Gamma_1 \vdash^S M \rightarrow^{\mathrm{nf}} P : A$.*

Proving a substitution or cut lemma for λ^S directly seems to be difficult. Indeed, substitution on the judgements of λ^S is difficult to define, because we need to know the normal form of the substituted term. Substitution is therefore closely related to soundness of λ^S for $\lambda^{\rightarrow}$, which we have emphasized by including the rule (*Subst*) in the rules of inference for $\lambda^{\rightarrow}$.

5.3 λ^S and Reduction

The calculus λ^S is a synthesis of standard reduction and the typing rules from the simply typed lambda calculus, which makes it a powerful tool for showing results about the relationship between typing in λ^S and untyped reduction.

Lemma 10 Adequacy for Reduction.

- *If $\Gamma \vdash^S M \rightarrow^{\mathrm{nf}} P : A$ then $M \rhd^* P$ and P is normal, and furthermore there is an N such that $M \rhd^*_\beta N \rhd^*_\eta P$.*

– *If* $\Gamma \vdash^S M \to^{\rm wh} N : A$ *then* $M \rhd_\beta N$.

Corollary 11. *If* $\Gamma \vdash^S M \to^{\rm nf} P : A$ *and* M *is normal then* $M \equiv P$.

We now prove a general lemma about the relationship between untyped reduction and the system λ^S. This lemma will imply the important properties of strong normalization, subject reduction and the diamond property.

Definition 12. The predicate $\mathcal{S}_\Gamma^{P,A}(M)$ is defined as the least relation such that $\mathcal{S}_\Gamma^{P,A}(M)$ holds if:

- $\Gamma \vdash^S M \to^{\rm nf} P : A$,
- for all N which are subterms of M there are B and Q such that $\Gamma, \Delta \vdash^S N \to^{\rm nf} Q : B$, where Δ is the sequence of binders which N occurs under in M, and
- $M \rhd N$ implies $\mathcal{S}_\Gamma^{P,A}(N)$ for all N.

$\mathcal{C}_\Gamma^{P,A}(M)$ will be the same as $\mathcal{S}_\Gamma^{P,A}(M)$ except that in the last clause instead of $M \rhd N$ we use $M \rhd^+ N$.

This definition is closely related to the definition of strong normalization in Altenkirch's thesis [1], although we have added type information. We have used this definition because of the similarity of the proofs of subject reduction, strong normalization and the diamond property in this setting. The predicate is well-defined because it is strictly positive. We have the obvious generation lemma for $\mathcal{S}_\Gamma^{P,A}(M)$, and we can also show that $\mathcal{S}_\Gamma^{P,A}(M)$ iff $\mathcal{C}_\Gamma^{P,A}(M)$, and that Strengthening holds for the predicates.

We need an auxiliary lemma about η-reduction, which follows using Generation, Adequacy for Reduction and Strengthening.

Lemma 13 Subject Reduction for (η). *If* $\Gamma \vdash^S \lambda x{:}A.M(x) \to^{\rm nf} P : A \to B$ *and* $x \notin {\rm FV}(M)$ *then* $\Gamma \vdash^S M \to^{\rm nf} P : A \to B$.

Lemma 14 Reduction.

- *If* $\Gamma \vdash^S M \to^{\rm nf} P : A$ *then* $\mathcal{S}_\Gamma^{P,A}(M)$.
- *If* $\Gamma \vdash^S M \to^{\rm wh} N : A$ *and* $\mathcal{S}_\Gamma^{P,A}(N)$ *then* $\mathcal{S}_\Gamma^{P,A}(M)$.

Proof. By induction on derivations. We outline two cases:

- (S-λ). We need to show that:
  - $\Gamma \vdash^S \lambda x{:}A.M \to^{\rm nf} \lambda x{:}A.P : A \to B$,
  - there are C and Q such that $\mathcal{S}_{\Gamma,\Delta}^{Q,C}(N)$ for all subterms N of M, and
  - $\lambda x{:}A.M \rhd N$ implies $\mathcal{S}_\Gamma^{\lambda x:A.P, A\to B}(N)$.

  The first follows directly and the second follows by inductive hypothesis. For the last, we know that $M \rhd N$ implies that $\mathcal{S}_{\Gamma,x:A}^{P,B}(N)$ for all N. The result follows by induction on $\lambda x{:}A.M \rhd N$, where we use Strengthening, Lemma 13 and Uniqueness of Normal Forms for (η).

– (W-β). We know that $\mathcal{S}_\Gamma^{P,A\to B}(\lambda x{:}A.M)$ and $\mathcal{S}_\Gamma^{Q,A}(N)$ by inductive hypothesis, and $\mathcal{S}_\Gamma^{R,B}([N/x]M)$ by assumption, so $\mathcal{C}_\Gamma^{R,B}([N/x]M)$. The result follows by induction on the normalization premisses.

□

All of the important results about reduction are corollaries to Adequacy for Reduction and Reduction.

Corollary 15 Strong Normalization for λ^S. *If* $\Gamma \vdash^S M \to^{\mathrm{nf}} P : A$ *then* M *is strongly normalizing.*

Corollary 16 Subject Reduction for λ^S. *If* $\Gamma \vdash^S M \to^{\mathrm{nf}} P : A$ *and* $M \rhd^* N$ *then* $\Gamma \vdash^S N \to^{\mathrm{nf}} P : A$.

Corollary 17 Diamond Property for λ^S. *If* $\Gamma \vdash^S M \to^{\mathrm{nf}} Q : A$, $M \rhd^* N$ *and* $M \rhd^* P$ *then* $N \rhd^* Q$ *and* $P \rhd^* Q$.

Finally, we show the admissibility of the more natural η rule using Adequacy for Reduction, Subject Reduction, Generation, Corollary 11 and Strengthening.

Proposition 18 Admissibility of (S-η'). *If* $\Gamma, x{:}A \vdash^S M \to^{\mathrm{nf}} P(x) : B$ *and* $x \notin \mathrm{FV}(P)$ *then* $\Gamma \vdash^S \lambda x{:}A.M \to^{\mathrm{nf}} P : A \to B$.

We now have a proof of the diamond property for $\beta\eta$-equality. It is well-known that this property fails on non-well-typed terms with type labels: the term $\lambda x{:}A.(\lambda y{:}B.M)(x)$ reduces to either $\lambda x{:}A.[x/y]M$ by (β) or $\lambda y{:}B.M$ by (η), and if A and B are not the same then the two terms have no common reduct. On the other hand, this counterexample does not hold for well-typed terms, because by the rules of inference the type labels A and B must be identical for any well-typed term. The relative simplicity of our approach serves to emphasize the usefulness of having type information in the computational presentation of type theory.

6 Soundness

In this section we prove the fundamental result of soundness, which relates the semantic presentation $\lambda^{\to}$ to the typed operational semantics λ^S. Specifically, we show that if $\Gamma \vdash M : A$ then there is a P such that $\Gamma \vdash^S M \to^{\mathrm{nf}} P : A$. The soundness result is crucial to our treatment of the metatheory of the simply typed lambda calculus, because we use it to transfer the metatheoretic development of λ^S to $\lambda^{\to}$.

An important first step in a proof of normalization is to elicit the reduction behavior of the system being studied. Our typed operational semantics serves this purpose: the rules of inference show that the well-typed terms are exactly those constructed with variables and abstraction and a series of weak head expansions. The technique we use to show soundness is closely related to existing proofs of normalization because these proofs do construct such reduction sequences. A proof of soundness of λ^V for $\lambda^{\rightarrow}$ would be easier than the proof of soundness for λ^S, in the same way that proofs of normalization are easier than proofs of strong normalization.

Our proof follows Coquand and Gallier's technique [5] of a Kripke-style interpretation and uses finite contexts and the general notion of substitutions. We show that if a term is well-typed in $\lambda^{\rightarrow}$ then that term is well-typed in λ^S, relative to any substitution.

6.1 The Interpretation

We use a slight variation of the usual model construction to show normalization.

Definition 19 Semantic Object. A *semantic object* for Δ and A is a term M such that $\Delta \vdash^S M : A$.

We emphasize that semantic objects are always normalizing. Therefore, the fact that an element of the interpretation is normalizing will require no proof, unlike many proofs of normalization [14, 16].

Definition 20 Interpretation. Let Δ be a context. The interpretation of a type A in Δ, $[\![A]\!]\Delta$, is given by induction on the structure of types:

- $[\![o]\!]\Delta =_{\mathrm{df}} \{M \mid M$ is a semantic object for Δ and $o\}$.
- $[\![A \rightarrow B]\!]\Delta$ is the set of semantic objects M for Δ and $A \rightarrow B$ such that for any renaming δ from Δ' to Δ and any $N \in [\![A]\!]\Delta'$ we know that $\widehat{\delta}(M)(N) \in [\![B]\!]\Delta'$.

Definition 21 Interpretation of Contexts. The interpretation of a context Γ in a context Δ, $[\![\Gamma]\!]\Delta$, is a set of substitutions from Δ to Γ defined by induction on the structure of Γ:

- $[\![()]\!]\Delta =_{\mathrm{df}} \{\epsilon\}$.
- $[\![\Gamma, x{:}A]\!]\Delta =_{\mathrm{df}} \{\rho[x{:=}M] \mid \rho \in [\![\Gamma]\!]\Delta$ and $M \in [\![A]\!]\Delta\}$.

6.2 Properties of the Interpretation

In this section we list the basic properties of the interpretation.

Definition 22 Saturated Set. A set S of semantic objects for Δ and A is a *saturated set* for Δ and A if:

(S1) If M is a base term then $M \in S$.
(S2) If $N \in S$ and $\Delta \vdash^S M \rightarrow^{\mathrm{wh}} N : A$ then $M \in S$.

The usual third condition on saturated sets, that if $M \in S$ then M is strongly normalizing, is enforced by the definition of semantic object.

The following properties are standard.

Lemma 23 Saturated Set. *$[\![A]\!]\Delta$ is a saturated set for any Δ and A.*

Lemma 24 Monotonicity. *If $M \in [\![A]\!]\Delta$ and δ is a renaming from Δ' to Δ then $\widehat{\delta}(M) \in [\![A]\!]\Delta'$.*

Lemma 25 Monotonicity for Contexts. *If $\rho \in [\![\Gamma]\!]\Delta$ and δ is a renaming from Δ' to Δ then $\delta \circ \rho \in [\![\Gamma]\!]\Delta'$.*

6.3 Soundness

Finally, we can show the soundness result.

Lemma 26 Soundness. *If $\Gamma \vdash M : A$ and $\rho \in [\![\Gamma]\!]\Delta$ then $\widehat{\rho}(M) \in [\![A]\!]\Delta$.*

Proof. By induction on derivations of $\Gamma \vdash M : A$.

We outline (λ), the one difficult case. We first show that $\lambda x{:}A.M$ is a semantic object for Δ and $A \to B$ by considering the substitution

$$(\mathrm{weak}_{\Delta}^{\Delta,y:A} \circ \rho)[x{:=}y] \in [\![\Gamma, x{:}A]\!]\Delta, y{:}A$$

for y fresh in Δ, using $(S\text{-}\lambda)$ or $(S\text{-}\eta)$ as appropriate. We then need to show that $\widehat{\delta}(\lambda x{:}A.M)(N) \in [\![B]\!]\Delta'$ for any renaming δ from Δ' to Δ and any $N \in [\![A]\!]\Delta'$, which we show using Renaming, $(W\text{-}\beta)$, *(S2)*, and the definition of $[\![A \to B]\!]\Delta$. □

It is straightforward to show that $\mathrm{id}_\Gamma \in [\![\Gamma]\!]\Gamma$, and the main result follows.

Corollary 27. *If $\Gamma \vdash M : A$ then there is a P such that $\Gamma \vdash^S M \to^{\mathrm{nf}} P : A$.*

Corollary 28 Soundness for $\Gamma \vdash M = N : A$. *If $\Gamma \vdash M = N : A$ then there is a P such that $\Gamma \vdash^S M \to^{\mathrm{nf}} P : A$ and $\Gamma \vdash^S N \to^{\mathrm{nf}} P : A$.*

Proof. By induction on derivations that $\Gamma \vdash M = N : A$, using for example Uniqueness of Normal Forms for (*Trans*), and Soundness and Lemma 13 for (η).

The rule (*App-Eq*) follows by Soundness, Adequacy for Reduction, Subject Reduction and Uniqueness of Normal Forms. □

6.4 Consequences of Soundness

We now transfer the metatheoretic study of λ^S in Sect. 5 to the simply typed lambda calculus by using Corollary 27 and Completeness.

Corollary 29 Strong Normalization. *If $\Gamma \vdash M : A$ then M is strongly normalizing.*

Corollary 30 Subject Reduction. *If $\Gamma \vdash M : A$ and $M \rhd N$ then $\Gamma \vdash N : A$.*

Corollary 31 Church–Rosser. *If $\Gamma \vdash M = N : A$ then there is a P such that $M \rhd^* P$ and $N \rhd^* P$.*

We also know the properties of Free Variables, Context, Renaming and Strengthening for $\lambda^{\to}$.

7 Dependent Types

We have demonstrated how typed operational semantics give a new approach to the metatheory for the simply typed lambda calculus, but there is one important aspect of typed operational semantics which is only brought out in the study of systems with dependent types and $\beta\eta$-equality. In these systems, with type equalities that allow a term to have syntactically different but judgementally equal types, there is a complex interaction between the properties of strengthening, subject reduction, Church–Rosser and strong normalization. The counterexample to the diamond property for non-well-typed terms mentioned in Sect. 5.3 now presents further difficulties: the labels on the two abstractions can be judgementally equal but require an arbitrary number of reductions to reach a common reduct.

Salvesen [15] gave the first proof of subject reduction and Church–Rosser for Pure Type Systems, and Geuvers [6] gives another proof. These proofs show strong normalization first, and use this to show strengthening, subject reduction and Church–Rosser, using complicated syntactic arguments on a different term structure. Coquand [4] shows normalization and the other results for Martin-Löf's Logical Framework with $\beta\eta$-equality by proving the completeness of an untyped algorithm for testing conversion. The algorithm comprises a reduction relation to weak head normal form and a binary relation which compares these weak head normal forms for $\beta\eta$-convertibility.

The development using typed operational semantics is very similar to the above treatment of the simply typed lambda calculus: the essential difference is in the definition of the system, which makes use of the normal forms of types in the rules for application, where the application is well-typed only if the normal forms of the types of the subterms agree. Normalization replaces the type equality rule in the semantic presentation, which coerces a term from one type to another, and leads to a system where there is at most one derivation for each judgement. This, together with the separate rule of inference for η, allows us to follow the outline of the development for the simply typed lambda calculus, proving strengthening directly and following by showing the other important metatheoretic properties.

Luo's type theory *UTT* [10] includes an impredicative universe of propositions, predicative universes of types and a general schema for inductive types, formulated in Martin-Löf's Logical Framework with η-equality on the framework function spaces and abstractions with labels. These features lead to the problems in the metatheory which we have mentioned above, and which are the same as those in the Calculus of Constructions with η-conversion. We have used typed operational semantics to develop the metatheory of this system [7, 8].

8 Related Work

Several ideas have been influential in the development of typed operational semantics. Definitions of typed reduction are rare in the literature, but Martin-Löf [11] proves normalization and Church–Rosser for a reduction defined by the rules of inference for judgemental equality excluding symmetry.

Mitchell [14] defines a type closed predicate to be a predicate $\mathcal{P}$ on terms which satisfies the following conditions:

- If $\mathcal{P}(M_1), \ldots, \mathcal{P}(M_n)$ then $\mathcal{P}(x(M_1, \ldots, M_n))$, for x of appropriate type.
- If for all N, if $\mathcal{P}(N)$ then $\mathcal{P}(M(N))$, then $\mathcal{P}(M)$.
- If $\mathcal{P}(N)$ and $\mathcal{P}(([N/x]M)(N_1, \ldots, N_n))$ then $\mathcal{P}((\lambda x{:}A.M)(N, N_1, \ldots, N_n))$.

Typed operational semantics have several advantages over these predicates. Most importantly, our technique is appealing and familiar because of its similarity to the dynamic semantics of programming languages. More technically, typed operational semantics include information about reduction, normal forms and contexts, and furthermore the rules of inference have a finite number of premisses, which is not the case for the second rule for type closed predicates. Each of these differences either makes typed operational semantics simpler to use than type closed predicates or leads to a larger class of results which can be proved using our technique.

However, the similarity of typed operational semantics and type closed predicates makes clear the view of λ^S as an alternative induction principle for well-typed terms in the simply typed lambda calculus: the rules of inference can be seen as conditions on predicates, and reasoning by induction on derivations corresponds to verifying these conditions. One way of thinking of λ^S is as a formal system expressing general properties of predicates which need to be established in order to know that the predicate holds for judgements in the simply typed lambda calculus. The soundness result says that such predicates are a model for the simply typed lambda calculus.

Coquand's algorithm for testing conversion [4], mentioned in Sect. 7, also provided part of the inspiration for typed operational semantics. His algorithm implements a particular strategy of reduction to normal form and also does not include type information. This means that there is no direct induction principle in this setting, which leads to proofs of strengthening and Church–Rosser that are indirect and more complicated than ours. Furthermore, it seems that strong normalization cannot be proved using this technique.

9 Conclusions

We have introduced a new class of formal systems, typed operational semantics, which present type theory from a computational perspective. In doing so, we have arrived at a strategy for studying reduction in type theory opposite to the frequently adopted approach of removing type information from the proof of normalization. The specific typed operational semantics that we study, based on standard reduction, itself gives a precise, coherent description of the relationship between typing and reduction. We have demonstrated that this system gives a new treatment of fundamental results, including strong normalization, Church–Rosser and subject reduction, and we have outlined how this approach simplifies the study of systems with dependent types and η-equality considerably.

Acknowledgments

Zhaohui Luo taught me much of what I know about the metatheory for type theories. The work in this paper was influenced by discussions with him and James McKinna. Rod Burstall suggested presenting typed operational semantics in the context of the simply typed lambda calculus. Adriana Compagnoni and Jan Smith read drafts of this paper and suggested improvements. Finally, the anonymous referees made useful suggestions, including adding the section which discusses the technique for dependent types.

References

1. T. Altenkirch. *Constructions, Inductive Types and Strong Normalization.* PhD thesis, University of Edinburgh, 1993.
2. H. Barendregt. *The Lambda Calculus: its Syntax and Semantics.* North Holland, revised edition, 1984.
3. H. Barendregt. Lambda calculi with types. In S. Abramsky, D. M. Gabbai, and T. S. E. Maibaum, editors, *Handbook of Logic in Computer Science*, volume 2. Oxford University Press, 1991.
4. T. Coquand. An algorithm for testing conversion in type theory. In G. Huet and G. Plotkin, editors, *Logical Frameworks.* Cambridge University Press, 1991.
5. T. Coquand and J. Gallier. A proof of strong normalization for the theory of constructions using a Kripke-like interpretation. In *Workshop on Logical Frameworks–Preliminary Proceedings*, 1990.
6. H. Geuvers. *Logics and Type Systems.* PhD thesis, Katholieke Universiteit Nijmegen, Sept. 1993.
7. H. Goguen. The metatheory of *UTT*. Submitted to Proceedings of BRA Workshop on Types for Proofs and Programs, 1994.
8. H. Goguen. *A Typed Operational Semantics for Type Theory.* PhD thesis, University of Edinburgh, Aug. 1994.
9. Z. Luo. *An Extended Calculus of Constructions.* PhD thesis, University of Edinburgh, Nov. 1990.
10. Z. Luo. *Computation and Reasoning.* Oxford University Press, 1994.
11. P. Martin-Löf. An intuitionistic theory of types: Predicative part. In H. Rose and J. C. Shepherdson, editors, *Logic Colloqium 1973*, pages 73–118, 1975.
12. P. Martin-Löf. *Intuitionistic Type Theory.* Bibliopolis, 1984.
13. J. McKinna and R. Pollack. Pure type systems formalized. In M. Bezem and J. F. Groote, editors, *Proceedings of the International Conference on Typed Lambda Calculi and Applications*, pages 289–305. Springer–Verlag, LNCS 664, Mar. 1993.
14. J. Mitchell. Type systems for programming languages. In J. van Leeuwen, editor, *Handbook of Theoretical Computer Science.* North Holland, 1990.
15. A. Salvesen. The Church-Rosser property for pure type systems with $\beta\eta$-reduction, Nov. 1991. Unpublished manuscript.
16. W. W. Tait. Intensional interpretation of functionals of finite type I. *Journal of Symbolic Logic*, 32, 1967.
17. L. van Benthem Jutting, J. McKinna, and R. Pollack. Typechecking in pure type systems. In H. Barendregt and T. Nipkow, editors, *Types for Proofs and Programming.* Springer–Verlag, 1993.

A Simple Calculus of Exception Handling

Philippe de Groote

INRIA-Lorraine – CRIN – CNRS
615, rue du Jardin Botanique - B.P. 101
54602 Villers lès Nancy Cedex – FRANCE
e-mail: degroote@loria.fr

Abstract. We introduce a simply-typed λ-calculus ($\lambda_{exn}^{\rightarrow}$) featuring an ML-like exception handling mechanism. This calculus, whose type system corresponds to classical logic through the Curry-Howard isomorphism, satifies several interesting properties: among other, Church-Rosser, subject reduction, and strong-normalisation. Moreover, its typing system ensures that the reduction of well-typed expressions cannot give rise to uncaught exceptions.

1 Introduction

During the last four years, several authors have introduced various calculi that extend the Curry-Howard isomorphism to classical logic and therefore provide a computational interpretation of classical proofs [3, 4, 10, 12, 14, 16, 21]. In this paper we propose yet another such calculus that we call $\boldsymbol{\lambda}_{exn}^{\rightarrow}$.

The originality of $\boldsymbol{\lambda}_{exn}^{\rightarrow}$ derives from its basis on an exception handling mechanism à la ML. Moreover, from a programming point of view, $\boldsymbol{\lambda}_{exn}^{\rightarrow}$ satifies the interesting property that well-typed programs cannot give rise to uncaught exceptions. This result is achieved by introducing a conservative extention of ML-like operational semantics.

The paper is organised as follows.

Section 2 performs an informal type-theoretic analysis of ML-like exception handling. This analysis results in an interpretation of the type of exceptions as the absurdity type, and yields a typing system that corresponds to classical logic.

In Section 3, we define formally the syntax of $\boldsymbol{\lambda}_{exn}^{\rightarrow}$ and its typing relation.

The ML-like operational semantics of exceptions does not completely match the typing system of $\boldsymbol{\lambda}_{exn}^{\rightarrow}$. We discuss this issue, which is related to the subject reduction property, in Section 4.

In Section 5, we provide $\boldsymbol{\lambda}_{exn}^{\rightarrow}$ with a modified operational semantics. This modified semantics is such that $\boldsymbol{\lambda}_{exn}^{\rightarrow}$ satisfies the Church-Rosser and subject reduction property. Consequently, when evaluating a well-typed program, any raised exception is eventually handled.

In Section 6, we investigate the relation between the ML-like and the modified semantics. We prove that, for non-exceptional values, the modified semantics is equivalent to the ML-like semantics.

Section 7 introduces a CPS-interpretation of $\boldsymbol{\lambda}_{exn}^{\rightarrow}$ and shows that there exists a logical embedding of $\boldsymbol{\lambda}_{exn}^{\rightarrow}$ into the simply-typed λ-calculus.

In Section 8, we establish the strong normalisation of $\boldsymbol{\lambda}_{exn}^{\rightarrow}$.

We discuss related work and conclusions in Section 9.

2 Informal Analysis of ML-like Exception handling

The notion of exception and the one of data type constructor in Standard ML are unified. This unification, which follows a proposal by MacQueen [1], is based on the special datatype *exn* whose values are exceptions. Indeed the following standard ML declaration:

$$\texttt{exception}\ \mathit{foo}\ \texttt{of}\ \mathit{int};$$

amounts to the declaration of the constructor *foo* whose type is $int \to exn$.

Values of type *exn* are first-class citizens, they may be stored, be passed as parameters, returned as results, etc. In addition, and in contrast to other values, they may also be turned into *packets* by being *raised* (see [13] for details).

The typing and reduction rules of the operator **raise** are the following:

$$\frac{\Gamma \vdash M : exn}{\Gamma \vdash (\texttt{raise}\, M) : \alpha}$$

$$\begin{array}{rcll} V\,(\texttt{raise}\, M) & \to & (\texttt{raise}\, M) & \text{(for } V \text{ a value)} \\ (\texttt{raise}\, M)\, N & \to & (\texttt{raise}\, M) & \text{(for } N \text{ any expression)} \end{array}$$

These rules correspond respectively to the deduction rule and the proof reduction rules that are used in natural deduction for *falsity* [9].

$$\frac{\bot}{\alpha} \qquad\qquad \frac{\begin{array}{c}\vdots\\ \alpha \to \beta\end{array} \quad \begin{array}{c}\vdots\\ \dfrac{\bot}{\alpha}\end{array}}{\beta} \;\to\; \frac{\begin{array}{c}\vdots\\ \bot\end{array}}{\beta} \qquad\qquad \frac{\begin{array}{c}\vdots\\ \dfrac{\bot}{\alpha \to \beta}\end{array} \quad \begin{array}{c}\vdots\\ \alpha\end{array}}{\beta} \;\to\; \frac{\begin{array}{c}\vdots\\ \bot\end{array}}{\beta}$$

Therefore it makes sense to identify, through the Curry-Howard isomorphism, the type of exceptions (exn) with the logical notion of falsity ($\bot$). Let us accept this identification and proceed further with the type-theoretic analysis of exception handling.

Packets, i.e. raised exceptions, are propagated and then possibly handled. The typing rule for exception handlers is akin to the following:

$$\frac{\Gamma \vdash M : \alpha \qquad \Gamma \vdash N : exn \to \alpha}{\Gamma \vdash M\ \texttt{handle}\ N : \alpha}$$

This rule is certainly sound, but not satisfactory. On the one hand, we would like to have a rule that allows exception declarations to be discarded. This is mandatory if we want to preserve the logical consistency of the type system because of the identifification of the type of exception with false. On the other hand, as it is stated, this rule does not properly reflect the standard ML exception handling mechanism. Indeed in standard ML the right hand side of the operator **handle** is not an expression but a *match*.

A solution to these two issues is to use the `let` construct to declare exception constructors locally and to consider the following typing rule:

$$\frac{\Gamma, y : \alpha \to exn \vdash M : \beta \qquad \Gamma, x : \alpha \vdash N : \beta}{\Gamma \vdash \texttt{let exception}\, y\ \texttt{of}\ \alpha\ \texttt{in}\ M\ \texttt{handle}\ (y\, x) \Rightarrow N\ \texttt{end} : \beta}$$

This rule, which is consistent with the definition of standard ML, corresponds to the elimination of the disjunction for the particular case of the excluded middle. Therefore it is sound with respect to classical logic. This is not too surprising because it is known, since Griffin's work [12], that there is a strong connection between classical logic and sequential control.

The fact that the above rule is classical allows us to write *classical programs* in the sense of [12]. This is illustrated by the following example.

Example 2.1 In classical logic, conjunction can be defined in terms of implication and negation as follows: $\alpha \wedge \beta = \neg(\alpha \to \neg\beta)$. This definition allows a *classical* pairing operator, together with the associated projections, to be defined:

```
fun pair (x:int) (y:int) = fn (f:(int->int->exn)) => (f x y);

fun proj1 (p:(int -> int -> exn) -> exn) =
              let exception y of int
              in
              (raise (p (fn x => (raise y x))))
              handle
              (y x) => x
              end;

fun proj2 (p:(int -> int -> exn) -> exn) =
              let exception y of int
              in
              (raise (p (fn x => y)))
              handle
              (y x) => x
              end;
```

The types that a compiler infers for these three pieces of code are the following:

```
val pair = fn : int -> int -> (int -> int -> exn) -> exn
val proj1 = fn : ((int -> int -> exn) -> exn) -> int
val proj2 = fn : ((int -> int -> exn) -> exn) -> int
```

Finally, the execution of the programs yields the expected results:

```
- proj1 (pair 1 2);
val it = 1 : int
- proj2  (pair 1 2);
val it = 2 : int
```

3 Definition of a Simple Calculus of Exception Handling

We define a simple calculus ($\lambda^{\rightarrow}_{exn}$) of exception handling based on the ideas discussed in the previous section.

Definition 3.1 *The* types *of* $\lambda^{\rightarrow}_{exn}$ *are given by the following grammar:*

$$\tau \ ::= \ a \mid \bot \mid (\tau \rightarrow \tau)$$

where "a" range over a finite set of ground types, and $\bot$ *is a distinguished ground type.*

We let the lowercase Greek letters $(\alpha, \beta, \gamma, \ldots)$ range over types, and we write $\neg\alpha$ as an abbreviation for $(\alpha \rightarrow \bot)$.

Definition 3.2 *The* expressions *of* $\lambda^{\rightarrow}_{exn}$ *are built upon two distinct alphabets of variables: the* λ*-variables and the exception variables. The raw syntax of the expressions is the following.*

$$\begin{aligned} T \ ::= \ & c \mid x \mid y \mid \lambda x.T \mid (T\,T) \mid \\ & (\mathtt{raise}\,T) \mid \mathtt{let}\; y : \neg\alpha \;\mathtt{in}\; T \;\mathtt{handle}\; (y\,x) \Rightarrow T \;\mathtt{end} \end{aligned}$$

where c ranges over possible constants, x ranges over λ*-variables, and y over exception variables.*

We use uppercase Roman letters $(A, B, C, \ldots)$ to denote expressions, and we adopt the usual notational conventions [5, pp. 22–23]. The notions of free and bound (occurrences of) variables are defined as usual. In particular, the scoping rule for the **handle** construct is the following: in an expression of the form

$$\texttt{let}\ y : \neg\alpha\ \texttt{in}\ A\ \texttt{handle}\ (y\,x) \Rightarrow B\ \texttt{end},$$

the exception variable y is bound in the subexpression A, and the λ-variable x is bound in the subexpression B. We also assume that some implicit convention (e.g. [5, p. 26]) prevents clashes between free and bound variables, and we let $A[x := B]$ denote the usual capture-avoiding substitution.

When writing proofs, we use a more compact notation for the **raise** and the **handle** constructs. We write

$$(\mathcal{R}\,A) \quad \text{for} \quad (\texttt{raise}\ A).$$

Similarly, we write

$$\langle\, x.A \mid y.B \,\rangle \quad \text{for} \quad \texttt{let}\ x : \neg\alpha\ \texttt{in}\ A\ \texttt{handle}\ (x\,y) \Rightarrow B\ \texttt{end},$$

α being left implicit. We also write

$$\langle\, (x_i).A \mid (y_i.B_i) \,\rangle \quad \text{for} \quad \langle\, x_1.\langle\, x_2.\cdots\langle\, x_n.A \mid y_n.B_n \,\rangle \cdots \mid y_2.B_2 \,\rangle \mid y_1.B_1 \,\rangle,$$

the number n of nested constructs being left implicit.

$\lambda^{\rightarrow}_{exn}$ is a call-by-value language. We thus have to define the notion of value.

Definition 3.3 *The notion of* value *is defined as follows:*

$$V \;::=\; c \mid x \mid y \mid \lambda x.T \mid (y\,V)$$

Note that there is a significant difference between λ-variables and exception variables. The latter act as datatype constructors. Therefore, the application of an exception variable to a value is a value. From now on, the uppercase Roman letter V (with possible subscripts) will range over values.

Finally the typing rules of our calculus are the ones of the simply typed λ-calculus together with the rules discussed in the previous section for the **raise** and the **handle** constructs.

Definition 3.4 *Define a* typing environment *to be a function, undefined almost everywhere, that assigns types to λ-variables and that assigns types of the form $\neg\alpha$ to exception variables. Let $\Gamma, \Delta, \ldots$ range over typing environments, and let σ be a given function that assigns a type to each constant.*

The typing rules *of the calculus are the following:*

$$\Gamma \vdash c : \sigma(c)$$

$$\Gamma \vdash x : \Gamma(x) \qquad\qquad \Gamma \vdash y : \Gamma(y)$$

$$\frac{\Gamma, x : \alpha \vdash B : \beta}{\Gamma \vdash \lambda x.B : \alpha \rightarrow \beta} \qquad\qquad \frac{\Gamma \vdash A : \alpha \rightarrow \beta \qquad \Gamma \vdash B : \alpha}{\Gamma \vdash A B : \beta}$$

$$\frac{\Gamma \vdash A : \bot}{\Gamma \vdash (\texttt{raise}\ A) : \alpha} \qquad\qquad \frac{\Gamma, y : \neg\alpha \vdash A : \beta \qquad \Gamma, x : \alpha \vdash B : \beta}{\Gamma \vdash \texttt{let}\ y : \neg\alpha\ \texttt{in}\ A\ \texttt{handle}\ (y\,x) \Rightarrow B\ \texttt{end} : \beta}$$

As we noted, the typing rules of $\lambda^{\rightarrow}_{exn}$, as a logical system, are consistent with respect to classical logic. It is quite easy to show that the system is also complete. We thus have the following proposition.

Proposition 3.5 *Consider the calculus $\lambda^{\rightarrow}_{exn}$ without constants. Given a type α, there exists an expression A such that $\vdash A : \alpha$ if and only if α, seen as a proposition, is a classical tautology.* □

4 Consistency Problems with ML-like Operational Semantics

Let the function σ that assigns types to constants be consistent, in the sense that its range is a consistent set of propositions. Then Proposition 3.5 ensures that closed expressions of type $\bot$ cannot exist. This property, in turn, ensures that the execution of a program cannot give rise to an uncaught exception, provided that the calculus satisfies the subject reduction property.

Unfortunately this is not the case with an ML-like semantics. Indeed, consider the following example:

Example 4.1 The following piece of code defines another *classical* pairing operator:

```
fun var_pair x y = let exception P of (int -> int -> exn)
                   in P handle (P g) => raise (g x y)
                   end;
```

It is easy to check that the expression

```
proj_1 (var_pair 1 2)
```

is a well-typed expression of type *int*. The execution of this expression, however, gives rise to an uncaught exception.

The dynamic semantics of standard ML is given in a natural semantics style [13]. Table 1 adapts this semantics to $\lambda^{\rightarrow}_{exn}$, in terms of reduction rules. The problem with subject

$(\lambda x.\, M)\, V \;\rightarrow\; M[x{:=}V]$	(β_V)
$V_1\,(\mathbf{raise}\; V) \;\rightarrow\; (\mathbf{raise}\; V)$	$(\mathbf{raise}_{\,left})$
$(\mathbf{raise}\; V)\, M \;\rightarrow\; (\mathbf{raise}\; V)$	$(\mathbf{raise}_{\,right})$
$(\mathbf{raise}\;(\mathbf{raise}\; V)) \;\rightarrow\; (\mathbf{raise}\; V)$	$(\mathbf{raise}_{\,idem})$
$\mathbf{let}\; y : \neg\alpha \;\mathbf{in}\; V \;\mathbf{handle}\; (y\, x) \Rightarrow N \;\mathbf{end} \;\rightarrow\; V$	$(\mathbf{handle}_{\,simp})$
$\mathbf{let}\; y : \neg\alpha \;\mathbf{in}\; (\mathbf{raise}\; y\, V) \;\mathbf{handle}\; (y\, x) \Rightarrow N \;\mathbf{end} \;\rightarrow\; N[x{:=}V]$	$(\mathbf{handle/raise}_{\,1})$
$\mathbf{let}\; y : \neg\alpha \;\mathbf{in}\; (\mathbf{raise}\; z\, V) \;\mathbf{handle}\; (y\, x) \Rightarrow N \;\mathbf{end} \;\rightarrow\; (\mathbf{raise}\; z\, V) \quad \text{if } z \neq y$	$(\mathbf{handle/raise}_{\,2})$

Table 1. ML-like Reduction Rules

reduction is related to the last three rules: bound variables may become free by reduction.[1] Take for instance Rule ($\mathbf{handle}_{\,simp}$): any free occurrence of y in V is bound in the redex but becomes free in the contractum.

5 Modified Operational Semantics

To circumvent the problem related to the ML-like semantics, we must modify the three last rules of Table 1. This idea gives rise to the modified operational semantics specified by

$$(\lambda x.\,M)\,V \;\to\; M[x:=V] \qquad (\beta_V)$$

$$V_1\,(\texttt{raise}\;V) \;\to\; (\texttt{raise}\;V) \qquad (\texttt{raise}_{\,left})$$

$$(\texttt{raise}\;V)\,M \;\to\; (\texttt{raise}\;V) \qquad (\texttt{raise}_{\,right})$$

$$(\texttt{raise}\;(\texttt{raise}\;V)) \;\to\; (\texttt{raise}\;V) \qquad (\texttt{raise}_{\,idem})$$

$$\texttt{let}\;y:\neg\alpha\;\texttt{in}\;V\;\texttt{handle}\;(y\,x)\Rightarrow N\;\texttt{end} \;\to\; V \quad \text{if } y\notin FV(V) \qquad (\texttt{handle}_{\,simp})$$

$$\begin{array}{l}
\texttt{let}\;y_1:\neg\alpha_1;\;y_2:\neg\alpha_2;\;\ldots;\;y_n:\neg\alpha_n\\
\texttt{in}\;(\texttt{raise}\;y_i\;V)\,\texttt{handle}\;(y_n\,x)\Rightarrow N_n\\
\qquad\qquad\vdots\\
\qquad\quad\mid\;\;(y_2\,x)\Rightarrow N_2\\
\qquad\quad\mid\;\;(y_1\,x)\Rightarrow N_1\\
\texttt{end}\\
\qquad\to\;\begin{array}[t]{l}\texttt{let}\;y_1:\neg\alpha_1;\;y_2:\neg\alpha_2;\;\ldots;\;y_n:\neg\alpha_n\\
\texttt{in}\;N_i[x:=V]\,\texttt{handle}\;(y_n\,x)\Rightarrow N_n\\
\qquad\qquad\vdots\\
\qquad\quad\mid\;\;(y_2\,x)\Rightarrow N_2\\
\qquad\quad\mid\;\;(y_1\,x)\Rightarrow N_1\\
\texttt{end}\end{array}
\end{array} \qquad (\texttt{handle/raise})$$

$$\begin{array}{l}
V\,(\texttt{let}\;y:\neg\alpha\;\texttt{in}\;M\;\texttt{handle}\;(y\,x)\Rightarrow N\;\texttt{end})\\
\qquad\to\;\texttt{let}\;y:\neg\alpha\;\texttt{in}\;V\,M\;\texttt{handle}\;(y\,x)\Rightarrow V\,N\;\texttt{end}
\end{array} \qquad (\texttt{handle}_{\,left})$$

$$\begin{array}{l}
(\texttt{let}\;y:\neg\alpha\;\texttt{in}\;M\;\texttt{handle}\;(y\,x)\Rightarrow N\;\texttt{end})\;O\\
\qquad\to\;\texttt{let}\;y:\neg\alpha\;\texttt{in}\;M\,O\;\texttt{handle}\;(y\,x)\Rightarrow N\,O\;\texttt{end}
\end{array} \qquad (\texttt{handle}_{\,right})$$

$$\begin{array}{l}
(\texttt{raise}\;\texttt{let}\;y:\neg\alpha\;\texttt{in}\;M\;\texttt{handle}\;(y\,x)\Rightarrow N\;\texttt{end})\\
\;\to\;\texttt{let}\;y:\neg\alpha\;\texttt{in}\;(\texttt{raise}\;M)\;\texttt{handle}\;(y\,x)\Rightarrow(\texttt{raise}\;N)\;\texttt{end}
\end{array} \qquad (\texttt{raise/handle})$$

Table 2. Modified Reduction Rules

the reduction rules of Table 2. The three first rules of the modified semantics correspond exactly to the ML-like rules. The fourth rule comes with a proviso that ensures that bound variables may not become free by reduction. Rule (**handle/raise**) is more general that the corresponding rules in the ML-like semantics. In particular, this new rule allows handlers to be used more than once. Finally, the three last rules are necessary to ensure that the execution of programs will not be stuck. These three last rules, which may seem intricate, are nothing but the commuting conversions of disjunction that are used in natural deduction [11]. For instance, Rule (**handle** $_{left}$) corresponds to the following proof reduction:

$$\dfrac{\begin{matrix}\vdots\\ B\to C\end{matrix} \qquad \dfrac{\begin{matrix}[\neg A]\\ \vdots\\ B\end{matrix}\qquad \begin{matrix}A\\ \vdots\\ B\end{matrix}}{B}}{C} \;\;\to\;\; \dfrac{\dfrac{\begin{matrix}\vdots\\ B\to C\end{matrix}\quad \begin{matrix}[\neg A]\\ \vdots\\ B\end{matrix}}{C}\qquad \dfrac{\begin{matrix}\vdots\\ B\to C\end{matrix}\quad \begin{matrix}[A]\\ \vdots\\ B\end{matrix}}{C}}{C}$$

Before investigating further the properties of the modified semantics, we must first

[1] This subject reduction problem is not actually related to the exception handling mechanism of standard ML but rather to the possibility of declaring locally a datatype constructor. Take, for instance the ML expression: `let datatype foo = c in c end`. The New-Jersey SML compiler evaluates this expression to the constructor `c` of type `?.foo`. Here, the question mark suggests that the datatype `foo` has been declared in some unknown module, that is in a lost environment.

determine in what sense the reduction rules of Table 2 specify an operational semantics for $\lambda^{\rightarrow}_{exn}$. The problem is that we have only introduced some notions of reduction without defining any reduction strategy. To settle this, we simply define the result of the evaluation of an expression to be its normal form (if any). Then it remains to prove that this normal form, when it exists, is unique. This is the purpose of the next proposition.

Proposition 5.1 (Church-Rosser Property) *Let $\twoheadrightarrow$ be the reduction relation induced by the notions of reduction of Table 2 (that is the least reflexive, transitive relation containing $\rightarrow$ and compatible with the expression formation rules). If A, B, C are expressions such that $A \twoheadrightarrow B$ and $A \twoheadrightarrow C$ then there exists an expression D such that $B \twoheadrightarrow D$ and $C \twoheadrightarrow D$.*

Proof. The property can be established by using the standard technique due to Tait and Martin-Löf. The different cases are numerous because we are dealing with nine notions of reduction. Nevertheless the proof is not difficult because there are only a few critical pairs. □

To better understand how the modified semantics works, let us see how example 4.1, which gave rise to an uncaught exception, may now be reduced. (We use the compact syntax.)

$$
\begin{array}{ll}
\multicolumn{2}{l}{(\lambda p.\,\langle\, y.(\mathcal{R}\,p\,(\lambda x.\,(\mathcal{R}\,y\,x)))\mid x.x\,\rangle)\,((\lambda x.\,\lambda y.\,\langle\, P.P \mid g.(\mathcal{R}\,g\,x\,y)\,\rangle)\,1\,2)} \\
\rightarrow_{\beta_V} & (\lambda p.\,\langle\, y.(\mathcal{R}\,p\,(\lambda x.\,(\mathcal{R}\,y\,x)))\mid x.x\,\rangle)\,((\lambda y.\,\langle\, P.P \mid g.(\mathcal{R}\,g\,1\,y)\,\rangle)\,2) \\
\rightarrow_{\beta_V} & (\lambda p.\,\langle\, y.(\mathcal{R}\,p\,(\lambda x.\,(\mathcal{R}\,y\,x)))\mid x.x\,\rangle)\,\langle\, P.P \mid g.(\mathcal{R}\,(g\,1\,2))\,\rangle \\
\rightarrow_{h_{left}} & \langle\, P.(\lambda p.\,\langle\, y.(\mathcal{R}\,p\,(\lambda x.\,(\mathcal{R}\,y\,x)))\mid x.x\,\rangle)\,P \\
 & \qquad \mid g.(\lambda p.\,\langle\, y.(\mathcal{R}\,p\,(\lambda x.\,(\mathcal{R}\,y\,x)))\mid x.x\,\rangle)\,(\mathcal{R}\,(g\,1\,2))\,\rangle \\
\rightarrow_{\beta_V} & \langle\, P.\langle\, y.(\mathcal{R}\,P\,(\lambda x.\,(\mathcal{R}\,y\,x)))\mid x.x\,\rangle \\
 & \qquad \mid g.(\lambda p.\,\langle\, y.(\mathcal{R}\,p\,(\lambda x.\,(\mathcal{R}\,y\,x)))\mid x.x\,\rangle)\,(\mathcal{R}\,(g\,1\,2))\,\rangle \\
\rightarrow_{h/r} & \langle\, P.\langle\, y.(\lambda p.\,\langle\, y.(\mathcal{R}\,p\,(\lambda x.\,(\mathcal{R}\,y\,x)))\mid x.x\,\rangle)\,(\mathcal{R}\,((\lambda x.\,(\mathcal{R}\,y\,x))\,1\,2))\mid x.x\,\rangle \\
 & \qquad \mid g.(\lambda p.\,\langle\, y.(\mathcal{R}\,p\,(\lambda x.\,(\mathcal{R}\,y\,x)))\mid x.x\,\rangle)\,(\mathcal{R}\,(g\,1\,2))\,\rangle \\
\rightarrow_{h_{simp}} & \langle\, y.(\lambda p.\,\langle\, y.(\mathcal{R}\,p\,(\lambda x.\,(\mathcal{R}\,y\,x)))\mid x.x\,\rangle)\,(\mathcal{R}\,((\lambda x.\,(\mathcal{R}\,y\,x))\,1\,2))\mid x.x\,\rangle \\
\rightarrow_{\beta_V} & \langle\, y.(\lambda p.\,\langle\, y.(\mathcal{R}\,p\,(\lambda x.\,(\mathcal{R}\,y\,x)))\mid x.x\,\rangle)\,(\mathcal{R}\,((\mathcal{R}\,y\,1)\,2))\mid x.x\,\rangle \\
\rightarrow_{r_{right}} & \langle\, y.(\lambda p.\,\langle\, y.(\mathcal{R}\,p\,(\lambda x.\,(\mathcal{R}\,y\,x)))\mid x.x\,\rangle)\,(\mathcal{R}\,(\mathcal{R}\,(y\,1)))\mid x.x\,\rangle \\
\rightarrow_{r_{idem}} & \langle\, y.(\lambda p.\,\langle\, y.(\mathcal{R}\,p\,(\lambda x.\,(\mathcal{R}\,y\,x)))\mid x.x\,\rangle)\,(\mathcal{R}\,(y\,1))\mid x.x\,\rangle \\
\rightarrow_{r_{left}} & \langle\, y.(\mathcal{R}\,y\,1)\mid x.x\,\rangle \\
\rightarrow_{h/r} & \langle\, y.1 \mid x.x\,\rangle \\
\rightarrow_{h_{simp}} & 1
\end{array}
$$

As expected, $\lambda^{\rightarrow}_{exn}$ with the modified semantics satifies the subject reduction property.

Proposition 5.2 (Subject Reduction Property) *Let Γ be a typing environment, A be an expression, and α be a type such that $\Gamma \vdash A : \alpha$. If $A \rightarrow B$, according to the one-step reduction relation induced by the notions of reduction of Table 2, then $\Gamma \vdash B : \alpha$.*

Proof. As usual, the proof is done by induction on the derivation of $A \rightarrow B$. A substituiton lemma and a few easy lemmas concerning the typing relation are needed. □

The above proposition immediately generalises to the reduction relation $\twoheadrightarrow$. By Proposition 3.5, we know that there cannot exist closed expressions of type $\perp$. This implies that there also cannot exist closed expressions of the form (`raise` M) that are well-typed. Hence, as a corollary of Proposition 5.2, any closed expression that is well-typed cannot be reduced to an expression of the form (**raise** M). Therefore, if one sees $\lambda^{\rightarrow}_{exn}$ as an idealised programming language, we get the interesting property that the evaluation of well-typed programs never gives rise to uncaught exceptions.

6 Relation between the two Operational Semantics

The results so far are encouraging. We have introduced a simple calculus of exception handling whose type system amounts, through the Curry-Howard isomorphism, to classical logic. We have also provided this calculus with notions of reduction that make sense from a proof-theoretical point of view. The resulting system satisfies properties of interest such as Church-Rosser and subject reduction. Nevertheless, we achieved this last result by modifying the ML-like semantics, while our prime motivation was to analyse ML-like exception handling. This raises several questions. How far did we modify the ML-like semantics? Can the new semantics be seen as an extension of the ML-like ones, or did we obtain something completely different?

In this section, we answer the above questions by proving that, as far as non-exceptional values are concerned, the modified semantics is a conservative extention of the ML-like one. This means that whenever a program P evaluates to a non-exceptional value V according to the ML-like semantics, this program P evaluates to the same value V according to the modified semantics, where programs are defined to be closed expressions of ground type.

First we define the notion of non-exceptional value.

Definition 6.1 Non-exceptional values *are defined by the following grammar:*

$$U \ ::= \ c \mid x \mid \lambda x.T$$

Note that, as far as ground types different from $\bot$ are concerned, the notion of non-exceptional value coincides exactly with the one of value. From now on, the uppercase Roman letter U (with possible subscripts) will range over non-exceptional values.

In order to state the property that we intend to prove in this section, we must define some standard reduction strategies. To this end, we introduce the notion of applicative context.

Definition 6.2 Applicative contexts *are defined by the following grammar, where* [ ] *stands for the empty context:*

$$C \ ::= \ [\ ] \mid V\,C \mid C\,T \mid (\texttt{raise}\,C) \mid \texttt{let}\;y:\neg\tau\;\texttt{in}\;C\;\texttt{handle}\;(y\,x)\Rightarrow T\;\texttt{end}$$

We denote by $C[A]$ *the expression obtained by plugging an expression* A *into an applicative context* C.

The notion of applicative context allows the one of standard reduction to be defined.

Definition 6.3 *Let* A *and* B *be expressions. We say that* A reduces standardly in one step to B, *according to the ML-like notions of reduction if and only if there exist expressions* A', B' *and an applicative context* C *such that*

(i) $A \equiv C[A']$,

(ii) $A \equiv C[B']$,

(iii) A', B' *is one of the redex/contractum pairs specified by Table 1,*

where $\equiv$ *stands for the syntactic identity (modulo* α*-conversion).*

The ML-like one-step standard reduction relation from A *to* B *is denoted by* $A \rightarrow_{ML} B$. *The refexive, transitive closure of this relation is written* $\twoheadrightarrow_{ML}$.

The modified one-step standard reduction ($\rightarrow_{exn}$) *and its reflexive, transitive closure* ($\twoheadrightarrow_{exn}$) *are defined similarly.*

In the case of programs, i.e. closed terms of ground types, it can be shown that the standard reduction is a normalizing strategy. We are now ready to state and prove that the modified semantics is conservative over the ML-like one.

Proposition 6.4 (Conservation Property) *Let P be a program and U be a non-exceptional value. If $P \twoheadrightarrow_{ML} U$ then $P \twoheadrightarrow_{exn} U$.*

Proof. The complete proof requires an induction on the number of **handle** constructs that are nested into the applicative contexts. We give here the proof for the basic case and leave the inductive case to the reader.

We proceed by induction on the length of the standard reduction sequence $P \twoheadrightarrow_{ML} U$.

Since $P \twoheadrightarrow_{ML} U$, we know that there exist a redex/contractum pair A, B and an applicative context C such that:

$$P \equiv C[A] \rightarrow_{ML} C[B] \twoheadrightarrow_{ML} U$$

There are seven cases, according to the notion of reduction to which the redex/contractum pair A, B belongs.

In the case of β_V, $\mathtt{raise}_{left}$, $\mathtt{raise}_{right}$, and $\mathtt{raise}_{idem}$ the proof is immediate because these four notions of reduction are kept unchanged in the modified semantics.

In the case of $\mathtt{handle}_{simp}$, we have $A \equiv \langle y.V | x.N \rangle$ and $B \equiv V$. Then, using $\mathtt{handle}_{left}$, $\mathtt{handle}_{right}$, and **raise/handle**, we have:

$$C[\langle y.V \mid x.N \rangle] \twoheadrightarrow_{exn} \langle y.C[V] \mid x.C[N] \rangle.$$

Then, by induction hypothesis,

$$\langle y.C[V] \mid x.C[N] \rangle \twoheadrightarrow_{exn} \langle y.U \mid x.C[N] \rangle.$$

Since U is non-exceptional, $y \notin FV(U)$. Therefore, by modified $\mathtt{handle}_{simp}$, we get

$$\langle y.U \mid x.C[N] \rangle \rightarrow_{exn} U.$$

The proof in the last two cases ($\mathtt{handle/raise}_1$ and $\mathtt{handle/raise}_2$) is similar. □

7 CPS-Interpretation

The modified semantics appears now as a conservative extension of the ML-like ones. Nevertheless, one may still wonder if Table 2 is just an ad-hoc adaptation of the ML-like reduction rules, designed to make Proposition 6.4 hold, or if there is anything canonical in the modified semantics. We already answered partially this question by suggesting that the modified reduction rules correspond to (well-known) proof-theoretic conversions.

In this section, we justify further the modified operational semantics by showing that there is a logical embedding (in the sense of [12]) of $\lambda^{\rightarrow}_{exn}$ into the simply typed λ-calculus. To this end we introduce a *continuation-passing-style* translation of the expressions of $\lambda^{\rightarrow}_{exn}$. This CPS-translation is an adaptation of Plotkin's [19].

Definition 7.1 (CPS-translation) *The* CPS-translation $\overline{M}$ *of an expression M is inductively defined as follows:*

(i) $\overline{c} = \lambda k.\, k\, c$;

(ii) $\overline{x} = \lambda k.\, k\, x$;

(iii) $\overline{y} = \lambda k.\, k\, (\lambda v.\, \lambda k.\, k\, (y\, v))$;

(iv) $\overline{\lambda x.\, M} = \lambda k.\, k\, (\lambda x.\, \overline{M})$;

(v) $\overline{M\,N} \;=\; \lambda k.\overline{M}\,(\lambda m.\overline{N}\,(\lambda n.\,m\,n\,k))$;
(vi) $\overline{(\mathcal{R}\,M)} \;=\; \lambda k.\overline{M}\,(\lambda x.\,x)$;
(vii) $\overline{\langle\, y.M \mid x.N\,\rangle} \;=\; \lambda k.\,(\lambda y.\overline{M}k\,)\,(\lambda x.\overline{N}\,k)$.

This CPS-translation is correct with respect to the modified reduction rules. To prove this, we need one auxiliary definition and three technical lemmas.

Definition 7.2 *The auxiliary function Ψ, sending values to values, is defined as follows:*

(i) $\Psi(c) \;=\; c$;
(ii) $\Psi(x) \;=\; x$;
(iii) $\Psi(y) \;=\; \lambda v.\,\lambda k.\,k\,(y\,v)$;
(iv) $\Psi(\lambda x.\,M) \;=\; \lambda x.\,\overline{M}$;
(v) $\Psi(y\,V) \;=\; y\,\Psi(V)$.

Lemma 7.3 $\overline{V} \twoheadrightarrow_\beta \lambda k.\,k\,\Psi(V)$, *for any value* V.

Proof. By induction on the structure of V. □

Lemma 7.4 $\overline{M[x{:=}V]} \twoheadrightarrow_\beta \overline{M}[x{:=}\Psi(V)]$, *for any expression* M *and any value* V.

Proof. By induction on the structure of M, using Lemma 7.3 for the base case. □

Lemma 7.5 *Let* M *be an expression. If* $k \notin FV(M)$ *then* $\lambda k.\,\overline{M}\,k \rightarrow_\beta \overline{M}$.

Proof. Because $\overline{M}$ is an abstraction for any M. □

We are now in the position of proving that the CPS-translation of Definition 7.1 is compatible with the modified semantics. More precisely, we intend to prove that the translation preserves conversion between terms.

Proposition 7.6 (Correctness of the CPS-Translation) *Let* A, B *be expressions. If* $A \rightarrow B$, *according to the modified semantics, then* $\overline{A} =_\beta \overline{B}$.

Proof. The proof is done by induction on the derivation of $A \rightarrow B$. The inductive steps are straightforward. As for the basic cases, we focus on three of them, leaving the others, which are similar, to the reader.
(**handle**$_{simp}$)

$$\begin{aligned}\overline{\langle\, y.V \mid x.N\,\rangle} &= \lambda k.\,(\lambda y.\,\overline{V}\,k)\,(\lambda x.\,\overline{N}\,k) && \\ &=_\beta \lambda k.\,\overline{V}\,k && y \notin FV(\overline{V}) \\ &=_\beta \overline{V} && \text{by Lemma 7.5}\end{aligned}$$

(**handle/raise**)

$$\begin{aligned}&\overline{\langle\, (y_i).(\mathcal{R}\,y_j\,V) \mid (x_i.N_i)\,\rangle} \\ &=_\beta \lambda k.\,(\lambda y_1.\,\ldots(\lambda y_n.\,(\lambda k.\,\overline{y_j\,V}\,(\lambda x.\,x))\,k)\,(\lambda x_n.\,\overline{N_n}\,k)\cdots)\,(\lambda x_1.\,\overline{N_1}\,k) \\ &=_\beta \lambda k.\,(\lambda y_1.\,\ldots(\lambda y_n.\,\overline{y_j\,V}\,(\lambda x.\,x))\,(\lambda x_n.\,\overline{N_n}\,k)\cdots)\,(\lambda x_1.\,\overline{N_1}\,k) \\ &=_\beta \lambda k.\,(\lambda y_1.\,\ldots(\lambda y_n.\,(\lambda k.\,k\,(y_j\,\Psi(V)))\,(\lambda x.\,x))\,(\lambda x_n.\,\overline{N_n}\,k)\cdots)\,(\lambda x_1.\,\overline{N_1}\,k) \\ &\qquad\qquad\qquad\qquad\qquad\qquad\qquad\qquad\qquad\qquad\qquad\qquad \text{by Lemma 7.3} \\ &=_\beta \lambda k.\,(\lambda y_1.\,\ldots(\lambda y_n.\,y_j\,\Psi(V))\,(\lambda x_n.\,\overline{N_n}\,k)\cdots)\,(\lambda x_1.\,\overline{N_1}\,k) \\ &=_\beta \lambda k.\,(\lambda y_1.\,\ldots(\lambda y_n.\,(\lambda x_j.\,\overline{N_j}\,k)\,\Psi(V))\,(\lambda x_n.\,\overline{N_n}\,k)\cdots)\,(\lambda x_1.\,\overline{N_1}\,k) \\ &=_\beta \lambda k.\,(\lambda y_1.\,\ldots(\lambda y_n.\,\overline{N_j}[x_j{:=}\Psi(V)]\,k)\,(\lambda x_n.\,\overline{N_n}\,k)\cdots)\,(\lambda x_1.\,\overline{N_1}\,k) \\ &=_\beta \lambda k.\,(\lambda y_1.\,\ldots(\lambda y_n.\,\overline{N_j[x_j{:=}V]}\,k)\,(\lambda x_n.\,\overline{N_n}\,k)\cdots)\,(\lambda x_1.\,\overline{N_1}\,k) \quad \text{by Lemma 7.4} \\ &=_\beta \overline{\langle\, (y_i).N_j[x_j{:=}V] \mid (x_i.N_i)\,\rangle}\end{aligned}$$

(handle$_{left}$)

$$\begin{array}{ll}
\overline{V\,\langle y.M \mid x.N\rangle} & \\
= & \lambda k.\overline{V}\,(\lambda m.\,(\lambda k.\,(\lambda y.\overline{M}\,k)\,(\lambda x.\overline{N}\,k))\,(\lambda n.\,m\,n\,k)) \\
=_\beta & \lambda k.\,(\lambda k.\,k\,\Psi(V))\,(\lambda m.\,(\lambda k.\,(\lambda y.\overline{M}\,k)\,(\lambda x.\overline{N}\,k))\,(\lambda n.\,m\,n\,k)) \quad \text{by Lemma 7.3} \\
=_\beta & \lambda k.\,(\lambda k.\,(\lambda y.\overline{M}\,k)\,(\lambda x.\overline{N}\,k))\,(\lambda n.\,\Psi(V)\,n\,k) \\
=_\beta & \lambda k.\,(\lambda y.\overline{M}\,(\lambda n.\,\Psi(V)\,n\,k))\,(\lambda x.\overline{N}\,(\lambda n.\,\Psi(V)\,n\,k)) \\
=_\beta & \lambda k.\,(\lambda y.\,(\lambda k.\,(\lambda k.\,k\,\Psi(V))\,(\lambda m.\overline{M}\,(\lambda n.\,m\,n\,k)))\,k) \\
& \qquad (\lambda x.\,(\lambda k.\,(\lambda k.\,k\,\Psi(V))\,(\lambda m.\overline{N}\,(\lambda n.\,m\,n\,k)))\,k) \\
=_\beta & \lambda k.\,(\lambda y.\,(\lambda k.\overline{V}\,(\lambda m.\overline{M}\,(\lambda n.\,m\,n\,k)))\,k)\,(\lambda x.\,(\lambda k.\overline{V}\,(\lambda m.\overline{N}\,(\lambda n.\,m\,n\,k)))\,k) \\
& \qquad \text{by Lemma 7.3} \\
= & \lambda k.\,\overline{V\,M}\,k[y:=\lambda x.\overline{V\,N}\,k] \\
= & \overline{\langle y.V\,M \mid x.V\,N\rangle}
\end{array}$$

□

Griffin has shown that Plotkin's CPS-translation induces, at the level of the types, a negative translation of classical logic into intuitionistic one [12].

Definition 7.7 *Griffin's* negative translation *is defined as follows:*
$\overline{\alpha} = \neg\neg\alpha^*$, *where:*

(i) $\bot^* = \bot$;

(ii) $a^* = a$, *for a atomic;*

(iii) $(\alpha \to \beta)^* = \alpha^* \to \overline{\beta}$

Our CPS-translation and Griffin's negative translation commute with the typing relation of $\lambda^{\to}_{exn}$ and the one of the simply-typed λ-calculus. This is stated by the next propostion, which shows that there is a logical embedding of $\lambda^{\to}_{exn}$ into the simply-typed λ-calculus.

Proposition 7.8 (Logical Interpretation) *Let* $\vdash_{exn}$ *and* $\vdash_\lambda$ *stand respectively for the typing relation of* $\lambda^{\to}_{exn}$ *and the one of the simply-typed* λ*-calculus. If* $\vdash_{exn} M : \alpha$ *then* $\vdash_\lambda \overline{M} : \overline{\alpha}$, *for any expression* M *and any type* α.

Proof. We prove that, whenever the sequent

$$(x_i : \alpha_i),\ (y_j : \neg\beta_j) \vdash_{exn} M : \alpha$$

is derivable, so is the sequent

$$(x_i : \alpha_i^*),\ (y_j : \neg\beta_j^*) \vdash_\lambda \overline{M} : \overline{\alpha}.$$

We proceed by induction on the derivation of the typing judgement. We give the proof for the **handle** construct and leave the other cases to the reader.

$$\dfrac{\dfrac{\dfrac{\left\{\begin{array}{l} y : \neg\beta^* \vdash_\lambda \overline{M} : \neg\neg\alpha^* \\ k : \neg\alpha^* \vdash_\lambda k : \neg\alpha^* \end{array}\right.}{k : \neg\alpha^*, y : \neg\beta^* \vdash_\lambda \overline{M}\,k : \bot}}{k : \neg\alpha^* \vdash_\lambda \lambda y.\overline{M}\,k : \neg\neg\beta^*} \quad \dfrac{\dfrac{\left\{\begin{array}{l} x : \beta^* \vdash_\lambda \overline{N} : \neg\neg\alpha^* \\ k : \neg\alpha^* \vdash_\lambda k : \neg\alpha^* \end{array}\right.}{k : \neg\alpha^*, x : \beta^* \vdash_\lambda \overline{N}\,k : \bot}}{k : \neg\alpha^* \vdash_\lambda \lambda x.\overline{N}\,k : \neg\beta^*}}{\dfrac{k : \neg\alpha^* \vdash_\lambda (\lambda y.\overline{M}\,k)\,(\lambda x.\overline{N}\,k) : \bot}{\vdash_\lambda \lambda k.\,(\lambda y.\overline{M}\,k)\,(\lambda x.\overline{N}\,k) : \overline{\alpha}}}$$

□

8 Strong Normalisation

By Proposition 6.4, we know that the ML-like and the modified operational semantics are equivalent for programs that yield non exceptional results. This means that whenever a program terminates and yields some value according to the ML-like semantics, it will terminate and yield the same value according to the modified semantics. On the other hand, by Propositions 3.5 and 5.2, we know that the modified semantics ensures that programs may not yield exceptional results (i.e. uncaught exceptions). Therefore, it seems that these three propositions together allows us to conclude that the modified semantics should be preferred to the original one.

This conclusion, however, is premature because Proposition 6.4 concerns only programs yielding non-exceptional results. Indeed a program raising an uncaught exception according to the ML-like semantics could possibly loop for ever according to the modified semantics. We would then have replaced a property that is observable (the production of an uncaught exception) by some other property that is not observable (the non-termination of a program).

In order to eliminate this last possible objection, we establish the strong normalisation of $\lambda_{exn}^{\rightarrow}$. Technically, we show that any infinite sequence of reductions in $\lambda_{exn}^{\rightarrow}$ induces an infinite sequence of β-reductions in the simply-typed λ-calculus. To this end, we need a syntactic translation of the expressions of $\lambda_{exn}^{\rightarrow}$ into simply-typed λ-terms. The CPS-translation of Definition 7.1 is such a syntactic translation. Nevertheless, we cannot use it as such because it does not properly simulate the reduction relation of $\lambda_{exn}^{\rightarrow}$. Therefore, we introduce a modified CPS-translation.

Definition 8.1 *The* modified CPS-translation $\overline{\overline{M}}$ *of an expression* M *is defined as follows:* $\overline{\overline{M}} = \lambda k.\, M : k$*, where*

(i) $V : K \;=\; K\,\Phi(V)$;

(ii) $(V_1\, V_2) : K \;=\; \Phi(V_1)\,\Phi(V_2)\,K$;

(iii) $(V_1\, N) : K \;=\; N : \lambda n.\, \Phi(V_1)\, n\, K$;

(iv) $(M\, V_2) : K \;=\; M : \lambda m.\, m\, \Phi(V_2)\, K$;

(v) $(M\, N) : K \;=\; M : (\lambda m.\, N : (\lambda n.\, m\, n\, K))$;

(vi) $(\mathcal{R}\, M) : K \;=\; M : (\lambda x.\, x)$;

(vii) $(\langle y.\alpha \mid M.x \rangle N) : K \;=\; M : K\,[y{:=}\lambda x.\, (N : K)]$;

(viii) $\Phi(c) \;=\; c$;

(ix) $\Phi(x) \;=\; x$;

(x) $\Phi(y) \;=\; \lambda v.\, \lambda k.\, k\,(y\, v)$;

(xi) $\Phi(\lambda x.\, M) \;=\; \lambda x.\, \overline{\overline{M}}$;

(xii) $\Phi(y\, V) \;=\; y\, \Phi(V)$.

The modified CPS-translation is compatible with the CPS-translation of the previous section in the sense of the following proposition:

Proposition 8.2 *Let* A *be an expression. Then:*

(a) $\Psi(A) \twoheadrightarrow_\beta \Phi(A)$*, whenever* A *is a value,*

(b) $\overline{A}\, K \twoheadrightarrow_\beta A : K$*, for any expression* K*,*

(c) $\overline{A} \twoheadrightarrow_\beta \overline{\overline{A}}$.

Proof. The proof is by induction on the structure of A. Property (c) is the property of interest, while Properties (a) and (b) are needed to make the induction work. □

Proposition 8.2.(c), together with the subject reduction property of the simply-typed λ-calculus, implies that Proposition 7.8 still holds with the modifed CPS-translation.

As expected, the modified CPS-translation allows the notions of reduction of $\lambda^{\rightarrow}_{exn}$ to be simulated by β-reduction. This property is stated by the two next propositions.

Proposition 8.3 *Let R stand for the notion of reduction β_V or* **handle/raise**. *Let A and B be two expressions such that $A \rightarrow_R B$. Then:*

(a) $A : K \xrightarrow{+}_\beta B : K$, *for any expression K,*

(b) $\overline{\overline{A}} \xrightarrow{+}_\beta \overline{\overline{B}}$,

where $\xrightarrow{+}_\beta$ stands for the transitive (but not reflexive) closure of $\rightarrow_\beta$.

Proof. Property (b), which is the property of interest, is a direct consequence of (a). The proof of (a) is akin to the one of proposition 7.6. Lemmas similar to Lemmas 7.3, 7.4, and 7.5 are needed. □

The other notions of reduction of $\lambda^{\rightarrow}_{exn}$ are invariants of the translation.

Proposition 8.4 *Let R stand for one of the notions of reduction* **raise**$_{left}$, **raise**$_{right}$, **raise**$_{idem}$, **handle**$_{simp}$, **handle**$_{left}$, **handle**$_{right}$, *or* **raise/handle**. *Let A and B be two expressions. If $A \rightarrow_R B$ then $\overline{\overline{A}} \equiv \overline{\overline{B}}$.*

Proof. The proof is similar to that of proposition 7.6. □

We may not yet conclude that $\lambda^{\rightarrow}_{exn}$ is strongly normalisable because there is still the possibility of infinite reduction sequences due to the notions of reduction of Proposition 8.4. For these notions of reduction, we must establish strong normalisation independently. To this end, we introduce a norm on the expressions of $\lambda^{\rightarrow}_{exn}$.

Definition 8.5 *The* norm $|A|$ *of an expression A is inductively defined as follows:*

(i) $\lvert c\rvert = 1$;	(viii) $\#c = 1$;
(ii) $\lvert x\rvert = 1$;	(ix) $\#x = 1$;
(iii) $\lvert y\rvert = 1$;	(x) $\#y = 1$;
(iv) $\lvert\lambda x.\, M\rvert = \lvert M\rvert$;	(xi) $\#\lambda x.\, M = \#M$;
(v) $\lvert M\, N\rvert = (\#N \times \lvert M\rvert) + (\#M \times \lvert N\rvert)$;	(xii) $\#M\, N = \#M \times \#N$;
(vi) $\lvert(\mathcal{R}\, M)\rvert = \lvert M\rvert + \#M$;	(xiii) $\#(\mathcal{R}\, M) = \#M$;
(vii) $\lvert\langle y.\alpha \mid M.x\rangle N\rvert = \lvert M\rvert + \lvert N\rvert$;	(xiv) $\#\langle y.\alpha \mid M.x\rangle N = \#M + \#N + 1$.

Proposition 8.6 *Let R stand for one of the notions of reduction* **raise**$_{left}$, **raise**$_{right}$, **raise**$_{idem}$, **handle**$_{simp}$, **handle**$_{left}$, **handle**$_{right}$, *or* **raise/handle**. *Let A and B be two expressions. If $A \rightarrow_R B$ then $|A| > |B|$.*

Proof. The proof is a straightforward induction on the derivation of $A \rightarrow B$. □

We may now state the main result of this section.

Proposition 8.7 (Strong Normalisation) *Any well-typed term of $\lambda^{\rightarrow}_{exn}$ is strongly normalisable.*

Proof. Consequence of Propositions 7.8, 8.2, 8.3, 8.4, and 8.6. □

9 Related Work and Conclusions

Griffin, in 1990, was the first to stress the relation between sequential control and classical logic [12]. His work is based on Felleisen's syntactic theory of sequential control, which provides an idealisation of Scheme `call/cc`.

Around the same time, Murthy studied the computational content of classical proofs [6, 14]. His work is based mainly on negative translations of classical logic, and CPS-transforms.

The work of Griffin was extended by Barbanera and Berardi [2, 3], who noted that Felleinsen's reduction rules are similar to Prawitz's handling of double negation [20]. They use a control operator akin to Felleisen's $\mathcal{C}$ to extract the computational content of classical proofs of Σ^0_1-sentences.

On the proof-theoretic side, Parigot has introduced the $\lambda\mu$-calculus, an algorithmic interpretation of cut elimination in classical natural deduction [16, 17, 18]. From a computer science point of view, the "classical" constructs of the $\lambda\mu$-calculus may be interpreted in terms of labels and jumps [7].

Independently of Parigot, Rehof and Sørensen have developped a calculus (λ_Δ) reminiscent of the $\lambda\mu$-calculus [21]. They use applications of the form (xM) instead of named terms. Consequently they may simulate Parigot's structural reduction with usual substitution and β-reduction.

In yet another direction, there is Girard's work on classical logic, based on the notion of polarity [10]. Girard does not provide his system with expressions encoding proofs. Nevertheless, Murthy has shown how to extract λ-terms from Girard's system, using Felleisen's operator [15].

More recently, Barbanera and Berardi have introduced an intriguing symmetric λ-calculus based on a syntactic identification between a type and its double negation [4].

All the above systems are based on computational interpretations of double negations. In the case of λ_Δ and of the systems based on Felleisen's theory of control, the treatment of double negations, which is explicit, is based on Prawitz-like reduction rules. In the case of Parigot's $\lambda\mu$-calculus, Girard's LC, and Berardi's symmetric calculus, the treatment of double negations is left implicit (it is hidden somehow in the formal system). Nevertheless, we have shown that first-order $\lambda\mu$-claculus is isomorphic to a subtheory of the Felleisen-Griffin system [8].

To the best of our knowledge, $\lambda^{\rightarrow}_{exn}$ is the only "classical" λ-calculus based on a computational interpretation of the elimination of the excluded-middle law, consequently modelling ML-like exception handling. Nevertheless this difference, which is not as deep as it could seem at first sight, vanishes from a denotational point of view. Indeed, Definitions 7.1 and 7.7 do not differ from the ones given by Griffin for Felleisen's calculus. Consequently, it should be possible to simulate $\lambda^{\rightarrow}_{exn}$ in one of the other calculi by using translations akin to the ones used in [8, 21].

References

[1] A. Appel, D.B. MacQueen, R. Milner, and M. Tofte. Unifying exceptions with constructors in standard ML. Technical Report ECS-LFCS-88-55, Laboratory for Foundations of Computer Science, University of Edinburgh, 1988.

[2] F. Barbanera and S. Berardi. Continuations and simple types: a strong normalization result. In *Proceedings of the ACM SIGPLAN Workshop on Continuations*. Report STAN-CS-92-1426, Stanford University, 1992.

[3] F. Barbanera and S. Berardi. Extracting constructive content from classical logic via control-like reductions. In M. Bezem and J.F. Groote, editors, *Proceedings of the International Conference on Typed Lambda Calculi and Applications*, pages 45–59. Lecture Notes in Computer Science, 664, Springer Verlag, 1993.

[4] F. Barbanera and S. Berardi. A symmetric lambda-calculus for "classical" program extraction. In M. Hagiya and J.C. Mitchell, editors. *Proceedings of the International Symposium on Theretical Aspects of Computer Software*, pages 494–515. Lecture Notes in Computer Science, 789, Springer Verlag, 1994.

[5] H.P. Barendregt. *The lambda calculus, its syntax and semantics.* North-Holland, revised edition, 1984.

[6] R. Constable and C. Murthy. Finding computational content in classical proofs. In G. Huet and G. Plotkin, editors, *Logical Frameworks*, pages 341–362. Cambridge University Press, 1991.

[7] Ph. de Groote. A CPS-translation of the $\lambda\mu$-calculus. In S. Tison, editor, *Proceedings of the 19th International Colloquium on Trees in Algebra and Programming (CAAP'94)*, pages 85–99. Lecture Notes in Computer Science, 787, Springer Verlag, 1994.

[8] Ph. de Groote. On the relation between the $\lambda\mu$-calculus and the syntactic theory of sequential control. In *Proceedings of the 5th International Conference on Logic Programming and Automated Reasoning–LPAR'94*, pages 31–43. Lecture Notes in Computer Science, 822, Springer Verlag, 1994.

[9] J.H. Gallier. On the correspondence between proofs and λ-terms. In Ph. de Groote, editor, *Cahiers du Centre de Logique (Université Catholique de Louvain), Volume 8*, pages 55–138. Academia, Louvain-la-Neuve, 1995.

[10] J.-Y. Girard. A new constructive logic: Classical logic. *Mathematical Structures in Computer Science*, 1:255–296, 1991.

[11] J.-Y. Girard, Y. Lafont, and P. Taylor. *Proofs and Types*, volume 7 of *Cambridge Tracts in Theoretical Computer Science.* Cambridge University Press, 1989.

[12] T. G. Griffin. A formulae-as-types notion of control. In *Conference record of the seventeenth annual ACM symposium on Principles of Programming Languages*, pages 47–58, 1990.

[13] R. Milner, M. Tofte, and R. Harper. *The Definition of Standard ML.* The MIT Press, 1990.

[14] C. R. Murthy. An evaluation semantics for classical proofs. In *Proceedings of the sixth annual IEEE symposium on logic in computer science*, pages 96–107, 1991.

[15] C. R. Murthy. A computational analysis of Girard's translation and LC. In *Proceedings of the seventh annual IEEE symposium on logic in computer science*, pages 90–101, 1992.

[16] M. Parigot. $\lambda\mu$-Calculus: an algorithmic interpretation of classical natural deduction. In A. Voronkov, editor, *Proceedings of the International Conference on Logic Programming and Automated Reasoning*, pages 190–201. Lecture Notes in Artificial Intelligence, 624, Springer Verlag, 1992.

[17] M. Parigot. Classical proofs as programs. In G. Gottlod, A. Leitsch, and D. Mundici, editors, *Proceedings of the third Kurt Gödel colloquium - KGC'93*, pages 263–276. Lecture Notes in Computer Science, 713, Springer Verlag, 1993.

[18] M. Parigot. Strong normalization for second order classical natural deduction. In *Proceedings of the eighth annual IEEE symposium on logic in computer science*, pages 39–46, 1993.

[19] G. D. Plotkin. Call-by-name, call-by-value and the λ-calculus. *Theretical Computer Science*, 1:125–159, 1975.

[20] D. Prawitz. *Natural Deduction, A Proof-Theoretical Study.* Almqvist & Wiksell, Stockholm, 1965.

[21] N.J. Rehof and M.H. Sørensen. The λ_Δ-calculus. In M. Hagiya and J.C. Mitchell, editors, *Proceedings of the International Symposium on Theoretical Aspects of Computer Software - TACS'94*, pages 516–542. Lecture Notes in Computer Science, 789, Springer Verlag, 1994.

A Simple Model for Quotient Types

Martin Hofmann*

Department of Computer Science, University of Edinburgh
JCMB, KB, Mayfield Rd, Edinburgh EH9 3JZ, Scotland

Abstract. We give an interpretation of quotient types within in a dependent type theory with an impredicative universe of propositions (Calculus of Constructions). In the model, type dependency arises only at the propositional level, therefore universes and large eliminations cannot be interpreted. In exchange, the model is much simpler and more intuitive than the one proposed by the author in [10]. Moreover, we interpret a choice operator for quotient types that, under certain restrictions, allows one to recover a representative from an equivalence class. Since the model is constructed syntactically, the interpretation function from the syntax with quotient types to the model gives rise to a procedure which eliminates quotient types by replacing propositional equality by equality relations defined by induction on the type structure ("book equalities").

1 Introduction

Intensional type theories like the Calculus of Constructions have been proposed as a framework in which to formalise mathematics and in which to specify and verify computer programs. Mathematical reasoning and to a lesser extent program specification make use of the possibility to redefine equality on a set by quotienting. Examples are the definition of integers as a quotient of pairs of natural numbers or of finite sets as a quotient of lists. In intensional type theories there is no such facility, which is why when formalising constructions involving quotients one is forced to replace equality on each type by an equivalence relation and to prove that all the functions and predicates one defines preserve these relations. This leads to complicated proofs and large bookkeeping requirements.

We claim that this flaw can be remedied. The key idea is to introduce operators for quotienting and other extensional concepts and to give a translation of the type theory including these operators into the pure type theory, under which types get translated into types together with a (partial) equivalence relation. One may then either print out the translation (for example if for conceptual reasons one is interested in a derivation using partial equivalence relations instead of quotient types.) or hide it and only use it in order to decide definitional equality in the theory with the quotient operators.

In [10] we have given such an interpretation[2] for Martin-Löf type theory without impredicative propositions. The interpretation given there is fairly complicated because it accounts for type dependency in full generality. If we confine

* Supported by a EU HCM fellowship, contract number ERBCHBICT930420

[2] The interpretation of quotient types is not explicitly carried out in [10].

the basic source of type dependency to the dependency of a type of proofs on the corresponding proposition, i.e. $x\colon \mathrm{Prop} \vdash \mathrm{Prf}(x)$ **type** then a much simpler interpretation may be given. For many applications, such a more restrained discipline of type dependency is sufficient. See also Section 6.

We translate a type theory with quotient types and a proof-irrelevant universe of propositions into the pure Calculus of Constructions with inductive types. Types get interpreted as types together with partial equivalence relations ("setoids"), terms get interpreted as terms together with proofs that the relations are respected. As opposed to the approach in [10] type dependency is only interpreted at the level of the relations — the underlying types in the interpretation are all fixed, i.e. do not depend on their context. This not only gives the desired interpretation of quotient types, but also leads to a full separation between proofs and programs which may be interesting to study in its own right.

Related work. Quotient types are available in extensional frameworks without explicit proof objects like Nuprl [7] and HOL [9] and also in the internal language of toposes [16]. The rules given there are quite similar to the ones proposed here. For an intensional calculus with explicit proof objects no extension with quotient types exists in the literature to our knowledge. Categorical formulations of extensional quotient types in simple type theory have recently been studied by Jacobs [11] without giving particular instances of models, though.

The method of interpreting type theory in itself so as to obtain additional features has also been used by Martin-Löf in order to account for proof irrelevance and subset types [15]. The use of categorical model theory (Section 3) to describe such translations appears to be new. Also the choice operator we introduce seems original and hinges on the intensional nature of definitional equality.

The model we give bears some resemblance with realizability interpretations of type theory [3] in which types are interpreted as partial equivalence relations on the natural numbers. Again, the main difference is that these models are extensional and therefore semantic equality is undecidable and the choice operator cannot be interpreted. Also these models do not give rise to a translation of type theory into itself, but only to an assignment of untyped algorithms to proofs.

2 Syntax

We consider a source type theory and a target type theory. The target type theory is a version of the Calculus of Constructions with a type of natural numbers and optionally more inductive types. Its (entirely standard) syntax is given in Appendix A. The reader who is not familiar with the Calculus of Constructions and related systems may want to take a look at this Appendix before reading on. Let us here just summarise the form of the judgements of the target type theory and some peculiarities. A context is a well-formed list of variable declarations $\Gamma = x_1\colon\sigma_1, \ldots, x_n\colon\sigma_n$ where due to type dependency the variables x_i may appear free in σ_j if $i < j \leq n$. The judgements are Γ **ctxt** to mean that

Γ is a well-formed context of variable declarations, $\Gamma \vdash \sigma$ **type** to mean that σ is a type in context Γ, $\Gamma \vdash M : \sigma$ to mean that M is a term of type σ in context Γ, $\Gamma = \Delta$ **ctxt** to mean that Γ and Δ are definitionally equal contexts, $\Gamma \vdash \sigma = \tau$ **type** to mean that σ and τ are definitionally equal types in context Γ, and $\Gamma \vdash M = N : \sigma$ to mean that M and N are definitionally equal terms of type σ in context Γ. So we have a typed definitional equality which is given inductively by a system of rules. It is known (see e.g. [6, 1]) that there is a decision procedure for it using untyped conversion, which is often regarded as the very *definition* of definitional equality. Another thing which is slightly unconventional is that if M : Prop we have Prf(M) **type** rather than M **type** as in the PTS-tradition [2]. So an expression is either a term or a type but never both.

The source type theory is meant to be the internal language of the model which we are going to describe. It supports all the type and term formers and rules present in the target type theory, but in addition provides quotient types. These are defined by the following rules.

$$\frac{\Gamma \vdash \sigma \qquad \Gamma\,,\, s, s' {:} \sigma \vdash R[s,s'] : \mathrm{Prop}}{\Gamma \vdash \sigma/R\ \textbf{type}}\ \text{Q-Form} \qquad \frac{\Gamma \vdash M{:}\sigma}{\Gamma \vdash [M]_R{:}\,\sigma/R}\ \text{Q-Intro}$$

$$\frac{\begin{array}{c}\Gamma \vdash \tau\ \textbf{type} \qquad \Gamma\,,\, s{:}\sigma \vdash M[s] : \tau \qquad \Gamma \vdash N : \sigma/R \\ \Gamma\,,\, s, s'{:}\sigma\,,\, p{:}\,\mathrm{Prf}(R[s,s']) \vdash H : \mathrm{Prf}(M[s] \stackrel{L}{=} M[s'])\end{array}}{\Gamma \vdash \mathbf{plug}_R\, N\, \mathbf{in}\, M\, \mathbf{using}\, H : \tau}\ \text{Q-Elim}$$

$$\frac{\Gamma \vdash \mathbf{plug}_R\, [N]_R\, \mathbf{in}\, M\, \mathbf{using}\, H : \tau}{\Gamma \vdash \mathrm{plug}_R\, [N]_R\, \mathrm{in}\, M\, \mathrm{using}\, H = M[N] : \tau}\ \text{Q-Comp}$$

$$\frac{\begin{array}{l}\Gamma \vdash M, N : \sigma \\ \Gamma \vdash H : \mathrm{Prf}(R[M,N])\end{array}}{\Gamma \vdash \mathrm{Qax}_R(H) : \mathrm{Prf}([M]_R \stackrel{L}{=} [N]_R)}\ \text{Q-Ax} \qquad \frac{\begin{array}{l}\Gamma\,,\, x{:}\sigma/R \vdash P[x] : \mathrm{Prop} \\ \Gamma\,,\, s{:}\sigma \vdash H : \mathrm{Prf}(P[[s]_R]) \\ \Gamma \vdash M : \sigma/R\end{array}}{\Gamma \vdash \mathrm{Qind}_R(H, M) : \mathrm{Prf}(P[M])}\ \text{Q-Ind}$$

Notice that plug and Qind are *binders* as indicated by the typing rules. To reduce clutter we do not explicitly indicate the bound variables in these terms. This may be made precise using de Bruijn indices. The symbol $\stackrel{L}{=}$ refers propositional (Leibniz) equality, see 2.1 below. This syntax is a formalisation of usual mathematical practice. Rule Q-Form allows the formation of a type σ/R from a a relation R on some type σ. We do not require R to be an equivalence relation. Using Q-Intro one constructs elements ("classes") of σ/R from "representatives". The rule Q-Elim allows one to construct functions on the quotient type by definition on representatives. The axiom Q-Ax states that the classes of related elements are equal; the axiom Q-Ind states that σ/R consists of "classes" only. One may also consider a dependent version of Q-Elim where an instance of Q-Ax has to be inserted in order to make the proviso on compatibility typecheck. In the present setting such rule is derivable.

For example, we may define a type of integers as Int := Nat $\times$ Nat$/R_{\mathrm{Int}}$ where $R_{\mathrm{Int}}[u, v{:}\,\mathrm{Nat} \times \mathrm{Nat}] = (u.1 + v.2 \stackrel{L}{=} u.2 + v.1)$. The elimination rule now allows to define functions like addition on the integers and the induction principle together with the equations permits to derive properties of these functions from their implementations. In examples we assume that various such functions such as $+$, $|-|$, and $-$ (negation) have been defined.

These rules differ from the ones e.g. in [11] because of the distinction between propositional equality (for equivalence classes) and definitional equality for computations. Up to propositional equality, however, Jacobs rules are derivable from ours and vice versa.

It is easy to see that an interpretation of types as sets, of the type Prop as the set $\{tt, ff\}$ and of quotient types σ/R as the set-theoretic quotient of σ by the equivalence relation generated by R validates all of the above rules. This is not the case for the choice operator we consider. It is governed by a rule of the form

$$\frac{\Gamma \vdash M : \sigma/R \qquad \text{"certain proviso"}}{\Gamma \vdash \mathrm{choice}(M) : \sigma} \quad \text{Q-CHOICE}$$

together with two associated equality rules

$$\frac{\Gamma \vdash M : \sigma/R \qquad \Gamma \vdash \mathrm{choice}(M) : \sigma}{\Gamma \vdash [\mathrm{choice}(M)]_R = M : \sigma/R} \quad \text{Q-CHOICE-AX}$$

$$\frac{\Gamma \vdash \mathrm{choice}([M]_R) : \sigma \quad \Gamma \vdash M : \sigma}{\Gamma \vdash \mathrm{choice}([M]_R) = M : \sigma} \quad \text{Q-CHOICE-COMP}$$

The first rule may be interpreted set-theoretically using a fixed choice of representatives for each equivalence class. The second rule, which states that choice recovers *the* representative used to form an equivalence class, stands in contradiction to this interpretation, and in fact to any extensional understanding of quotient types. It does make sense, however, in an intensional setting in which quotient types are represented by their underlying types and only the propositional equality is redefined. The "certain proviso" roughly states that the free variables in M are not of quotiented type. This is described in Section 5.2 where we also discuss possible applications of the choice operator. It is crucial for the soundness of the choice operator that propositional equality ($\stackrel{L}{=}$) and definitional equality ($=$) are distinct. The former refers to provable extensional equality whereas the latter refers to intensionally equal "algorithms".

In addition to these quotient types the source type theory contains rules which identify all proofs of a proposition (proof irrelevance) and which define bi-implication as the propositional equality on the type of propositions. Moreover, pointwise propositionally equal functions are propositionally equal and in view of proof irrelevance the Σ-type can be used as a subset type former. Thus, the source type theory approximates to a good deal naive set theory.

Definitional equality in the source type theory is defined semantically via the interpretation; we give, however, a few derived equality rules and in particular all the rules in the target type theory continue to hold. This is in line with the intuitive understanding of definitional equality as induced by definitional expansion. We view quotient types and other extensional concepts as being *defined* in terms of interpretations like the one we give. Since the model is built on syntax and (decidable) definitional equality the semantically defined equality is decidable, too.

2.1 Some Abbreviations

For types σ, τ in some context Γ we write $\sigma \to \tau$ for $\Pi x{:}\, \sigma.\tau$ and $\sigma \times \tau$ for $\Sigma x{:}\, \sigma.\tau$. If C is a term, type, or context, we write $C[x := M]$ for the syntactic capture-free substitution of M for x in C. We use logical connectives $\Rightarrow, \wedge, \Leftrightarrow, tt, ff$ for their encodings in the Calculus of Constructions. If P, Q : Prop then their implication $P \Rightarrow Q$ is defined as the proposition $\forall x{:}\, \mathrm{Prf}(P).Q$ and their conjunction $P \wedge Q$ is defined as the proposition $\forall c{:}\, \mathrm{Prop}.(P \Rightarrow Q \Rightarrow c) \Rightarrow c$. Bi-implication $P \Leftrightarrow Q$ is $(P \Rightarrow Q) \wedge (Q \Rightarrow P)$. The true proposition tt is defined as $\forall c{:}\, \mathrm{Prop}.c \Rightarrow c$. The false proposition ff is defined as $\forall c{:}\, \mathrm{Prop}.\mathrm{Prf}(c)$. For $M, N{:}\,\sigma$ we define Leibniz equality $M \stackrel{L}{=} N$ as $\forall P{:}\, \sigma \to \mathrm{Prop}.(P\ M) \Rightarrow (P\ N)$ It is well-known that the proof rules for these connectives in higher-order intuitionistic logic are derivable [6]. Furthermore, we introduce the notation $\Gamma \vdash P$ true to mean that $\Gamma \vdash P{:}\,\mathrm{Prop}$ and that there exists a term $\Gamma \vdash M : \mathrm{Prf}(P)$. If $\Gamma \vdash P : \mathrm{Prop}$ we sometimes simply write P in the running text to mean $\Gamma \vdash P$ true.

Context morphisms. Let Γ and $\Delta = x_1{:}\,\tau_1, \ldots, x_n{:}\,\tau_n$ be contexts. A *syntactic context morphism* from Γ to Δ is an n-tuple of terms $f = (M_1, \ldots, M_n)$ with $\Gamma \vdash M_1 : \tau_1$, $\Gamma \vdash M_2 : \tau_2[x_1 := M_1]$ through to $\Gamma \vdash M_n : \tau_n[x_1 := M_1] \ldots [x_{n-1} := M_{n-1}]$. In this situation we write $f : \Gamma \Rightarrow \Delta$. Definitional equality is extended canonically to syntactic context morphisms, that is we write $f = f' : \Gamma \Rightarrow \Delta$ if f and f' have the same length and definitionally equal components. The syntactic context morphisms include the *identity* from Γ to Γ which consists of the tuple of variables in Γ and they are closed under composition, i.e. if f is a syntactic context morphism from Γ to Δ and g is a syntactic context morphism from Δ to Θ then the parallel substitution of f into g gives a context morphism from Γ to Θ. In this way the contexts form a category, i.e. composition is associative and the identity is neutral. The empty context is a terminal object in this category.

If Γ is a context of length n then if $\gamma = (\gamma_1, \ldots \gamma_n)$ is an n-tuple of variables then $\gamma{:}\,\Gamma$ denotes the context Γ with the variables renamed to $\gamma_1 \ldots \gamma_n$. Now assume $\Gamma, \Delta \vdash \sigma$ for some type expression σ. By the declaration $\tau[\delta{:}\,\Delta] := \sigma$ the meta-variable τ becomes an abbreviation for the expression σ. This notation emphasises that the variables from Δ are free in τ. For example we may define $\mathrm{eqzero}[n{:}\,\mathrm{Nat}] := (n \stackrel{L}{=} 0)$. Explicit variable names in a substitution may now be omitted, for example eqzero[5] denotes the proposition $5 \stackrel{L}{=} 0$. If $f : \Gamma \Rightarrow \Delta$ and $\Delta \vdash \sigma$ **type** then we write $\sigma[f]$ for the parallel substitution of f into σ. We have $\Gamma \vdash \sigma[f]$ **type** and similarly for terms.

3 Categorical Semantics

It turns out that due to type dependency the verification that a certain translation indeed validates all the rules of type theory is quite involved. It has therefore proven useful to insert the intermediate step of abstract categorical models. One then defines a sound interpretation function once and for all and the task of proving that a certain translation is sound can be reduced to the task of verifying that one has an instance of the abstract model. This may be compared to the correspondance between typed λ-calculus and cartesian closed categories as described in [13]. The role of cartesian close categories is played here by the (by now standard) notion of *categories with attributes* introduced by Cartmell [5]. We only sketch this notion of model and the interpretation of syntax therein. The interested reader is referred to [17, 19, 12].

Definition 1 *A* category with attributes *(cwa) is given by the following data and conditions:*

- *A category* $\mathbf{C}$ *with terminal object* 1. *The objects of* $\mathbf{C}$ *are called* contexts *$(A, B, \Gamma, \Delta, \ldots)$, the morphisms are called* context morphisms *$(f, g, h, \ldots)$. The terminal object* $\mathbf{1}$ *is also called the* empty context. *The unique morphism from a context Γ to* $\mathbf{1}$ *is denoted* $!_\Gamma$.
- *A functor* $\mathrm{Fam} : \mathbf{C}^{\mathrm{op}} \to \mathbf{Sets}$. *If $f \in \mathbf{C}(\Gamma, \Delta)$ is a context morphism and $\sigma \in \mathrm{Fam}(\Delta)$ then $\mathrm{Fam}(f)(\sigma) \in \mathrm{Fam}(\Gamma)$ is abbreviated $\sigma\{f\}$. The elements of* $\mathrm{Fam}(\Gamma)$ *are called* types *or* families *$(\sigma, \tau, \rho \ldots)$ in context Γ. The family $\sigma\{f\} \in \mathrm{Fam}(B)$ is called the* substitution *of σ along f.*
- *If σ is a family in context Γ then there is a context $\Gamma \cdot \sigma$, called the* comprehension *of σ, and a context morphism $\mathrm{p}(\sigma) \in \mathbf{C}(\Gamma \cdot \sigma, \Gamma)$ called the* canonical projection *of σ. Moreover, if in addition $f \in \mathbf{C}(B, \Gamma)$ then there is a morphism $\mathrm{q}(f, \sigma) \in \mathbf{C}(B \cdot \sigma\{f\}, \Gamma \cdot \sigma)$ and*

$$\begin{array}{ccc} B\cdot\sigma\{f\} & \xrightarrow{\mathrm{q}(f,\sigma)} & \Gamma\cdot\sigma \\ {\scriptstyle \mathrm{p}(\sigma\{f\})}\downarrow & & \downarrow{\scriptstyle \mathrm{p}(\sigma)} \\ B & \xrightarrow[f]{} & \Gamma \end{array} \qquad (\ddagger\ddagger)$$

is a pullback diagram.
The assignment $\mathrm{q}(-,-)$ is functorial in the sense that $\mathrm{q}(\mathrm{id}_\Gamma, \sigma) = \mathrm{id}_{\Gamma\sigma}$ and for $g \in \mathbf{C}(A, B)$ also $\mathrm{q}(f \circ g, \sigma) = \mathrm{q}(f, \sigma) \circ \mathrm{q}(g, \sigma\{f\})$. Note that these equations "typecheck" by virtue of functoriality of Fam.

An example for a cwa is the term model of some dependently typed calculus constructed as indicated by the names of the various components of a cwa. Fix a cwa for the rest of this section. Let $\sigma \in \mathrm{Fam}(\Gamma)$. A *section* of σ is a morphism $M : \Gamma \to \Gamma\cdot\sigma$ with $\mathrm{p}(\sigma) \circ M = \mathrm{id}_\Gamma$. The sections correspond to terms of σ in the syntax. If furthermore $f : B \to \Gamma$ then by the universal property of pullbacks there exists a unique section $M\{f\}$ of $\sigma\{f\}$ with $\mathrm{q}(f, \sigma) \circ M\{f\} = M \circ f$. This corresponds to substitution in terms. We write $\mathrm{Sect}(\sigma)$ for the set of sections of σ. Substitution along a canonical projection corresponds to weakening. We introduce some abbreviations for this. If $\sigma, \tau \in \mathrm{Fam}(\Gamma)$, $M \in \mathrm{Sect}(\sigma)$, and $f : B \to \Gamma$ we put$\sigma^+ := \sigma\{\mathrm{p}(\tau)\} \in \mathrm{Fam}(\Gamma\cdot\tau)$ and $M^+ := M\{\mathrm{p}(\tau)\} \in \mathrm{Sect}(\sigma^+)$ and $f^+ := \mathrm{q}(f, \tau) \in \mathbf{C}(B\cdot\tau\{f\}\ , \Gamma\cdot\tau)$. This notation being ambiguous (e.g. f^+ can mean both $\mathrm{q}(f, \sigma)$ and $\mathrm{q}(f, \tau)$) we use it only if the meaning is clear from the context.

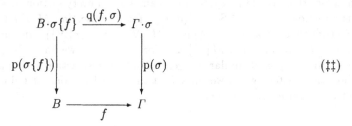

For $\sigma \in \mathrm{Fam}(\Gamma)$ there is a unique section v_σ of $\sigma\{\mathrm{p}(\sigma)\} = \sigma^+$ satisfying $\mathrm{q}(\mathrm{p}(\sigma), \sigma) \circ \mathrm{v}_\sigma = \mathrm{id}_{\Gamma\sigma}$. It interprets the last variable in a nonempty context, i.e. $\Gamma, x{:}\sigma \vdash x : \sigma$. The meanings of the other variables are obtained as weakenings of v. The last variable has in particular the expected property $\mathrm{v}_\sigma\{M\} = M$ for $M \in \mathrm{Sect}(\sigma)$.

Interpretation of the syntax. Provided that suitable interpretations of base types and type constructors are given, an a priori partial interpretation function can be defined on pre-terms, -types, and -contexts in such a way that every context is interpreted as a C-object, every type is interpreted as an element of Fam at the interpretation of its context and finally terms are interpreted as elements of $\mathrm{Sect}(\sigma)$ where σ is the interpretation of its type. Syntactic context morphisms get interpreted as context morphisms. Moreover, the interpretation is such that definitionally equal objects get interpreted as equal objects. The details of this interpretation function are spelled out in [19] for a slightly different but equivalent class of models and in [17] for cwas.

What it means that a cwa is closed under a type former can be almost directly read off from the syntactic rules. For example, closure under dependent products (Π-types) means the following: for every two families $\sigma \in \mathrm{Fam}(\Gamma)$ and $\tau \in \mathrm{Fam}(\Gamma{\cdot}\sigma)$ there is a family $\Pi(\sigma, \tau) \in \mathrm{Fam}(\Gamma)$; for every section $M \in \mathrm{Sect}(\tau)$ there is a section $\lambda_{\sigma,\tau}(M) \in \mathrm{Sect}(\Pi(\sigma, \tau))$; for every section M of $\Pi(\sigma, \tau)$ and $N \in \mathrm{Sect}(\sigma)$ there is a section $\mathrm{App}_{\sigma,\tau}(M,\ N) \in \mathrm{Sect}(\tau\{N\})$; in such a way that $\mathrm{App}_{\sigma,\tau}(\lambda_{\sigma,\tau}(M), N) = M\{N\}$ and for $f : B \to \Gamma$ we have $\Pi(\sigma, \tau)\{f\} = \Pi(\sigma\{f\}, \tau\{\mathrm{q}(f, \sigma)\})$ and similar coherence laws for abstraction and application. We follow this pattern when we show that the model we construct does indeed support the required type-formers.

4 Description of the Model

In the following sections we show how types with partial equivalence relations and suitable morphisms define a category with attributes (4.1,4.2 and Prop. 3). Further down, we show that this cwa supports Π-types (4.3) and a universe of propositions with impredicative quantification, i.e. forms a model of the Calculus of Constructions (4.4). We also show that in the model these propositions satisfy principles of proof irrelevance and extensionality (4.5). In Section 5 we then describe how the model supports the quotient types defined in Section 2.

In the model types will be interpreted as types together with Prop-valued partial equivalence relations. Type dependency is modelled only at the level of the relations, i.e. a family indexed over some type σ is a type τ (not depending on σ) and for each $x{:}\,\sigma$ a partial equivalence relation on τ such that these partial equivalence relations are compatible with the relation on σ.

By analogy to Bishop's definition of sets as assemblies together with an equality relation [4] we call the pairs of types and partial equivalence relations *setoids*.

4.1 Contexts of Setoids

A *context of setoids* is a pair $\Gamma = (\Gamma_{\text{set}}, \Gamma_{\text{rel}})$ where Γ_{set} is a (syntactic) context and Γ_{rel} is a proposition in context $\gamma\colon \Gamma_{\text{set}}, \gamma'\colon \Gamma_{\text{set}}$, i.e. $\gamma\colon \Gamma_{\text{set}}, \gamma'\colon \Gamma_{\text{set}} \vdash \Gamma_{\text{rel}}[\gamma, \gamma'] : \text{Prop}$, in such a way that

$$\gamma, \gamma'\colon \Gamma_{\text{set}}\ ,\ \text{Prf}(\Gamma_{\text{rel}}[\gamma, \gamma']) \vdash \Gamma_{\text{rel}}[\gamma', \gamma]\ \text{true} \qquad \text{SYM}$$
$$\gamma, \gamma', \gamma''\colon \Gamma_{\text{set}}\ ,\ \text{Prf}(\Gamma_{\text{rel}}[\gamma, \gamma'])\ ,\ \text{Prf}(\Gamma_{\text{rel}}[\gamma', \gamma'']) \vdash \Gamma_{\text{rel}}[\gamma, \gamma'']\ \text{true} \qquad \text{TRANS}$$

Here we just write $\text{Prf}(\Gamma_{\text{rel}}[\gamma, \gamma'])$ instead of $p : \text{Prf}(\Gamma_{\text{rel}}[\gamma, \gamma'])$ for some variable p since this variable does not appear in the judgement. We shall adopt this abbreviation in the sequel as well. Two contexts of setoids Γ, Δ are equal if their components are definitionally equal, i.e. if $\vdash \Gamma_{\text{set}} = \Delta_{\text{set}}$ **ctxt** and $\gamma, \gamma'\colon \Gamma_{\text{set}} \vdash \Gamma_{\text{rel}}[\gamma, \gamma'] = \Delta_{\text{rel}}[\gamma, \gamma'] : \text{Prop}$. The implicit witnesses for SYM and TRANS are not compared.

The *empty context of setoids* is given by $\mathbf{1} = (\bullet, tt)$ where SYM and TRANS are validated by the canonical proof of tt.

Morphisms. Let Γ, Δ be contexts of setoids. A *morphism* from Γ to Δ is a (syntactic) context morphism f from Γ_{set} to Δ_{set} with

$$\gamma, \gamma'\colon \Gamma_{\text{set}}\ ,\ \text{Prf}(\Gamma_{\text{rel}}[\gamma, \gamma']) \vdash \Delta_{\text{set}}[f[\gamma], f[\gamma']]\ \text{true} \qquad \text{RESP}$$

Two morphisms are equal if their components are definitionally equal. So a morphism must respect the relations associated to its domain and codomain. Clearly, the composition of two morphisms is a morphism again and the identity is a morphism. Moreover, the unique context morphism into $\mathbf{1}$ trivially satisfies RESP so that we have

Proposition 2 *The contexts of setoids and their morphisms form a category with terminal object.*

Notice that we do not identify "provably equal" morphisms, i.e. f and g with $\gamma\colon \Gamma_{\text{set}}, \text{Prf}(\Gamma_{\text{rel}}[\gamma, \gamma]) \vdash \Delta_{\text{rel}}[f[\gamma], g[\gamma']]$ true. Doing so would render the semantic equality undecidable and therefore the source type theory would be undecidable. The difference between actual semantic equality and provable equality is also important for the choice operator we consider in 5.2 which would be unsound if the two were identified.

4.2 Families of Setoids

Let Γ be a context of setoids. A family of setoids indexed over Γ is a pair $\sigma = (\sigma_{\text{set}}, \sigma_{\text{rel}})$ where σ_{set} is a type *in the empty context* ($\bullet \vdash \sigma$ **type**) and

$$\gamma\colon \Gamma_{\text{set}}\ ,\ s, s'\colon \sigma_{\text{set}} \vdash \sigma_{\text{rel}}[\gamma, s, s'] : \text{Prop}$$

is a family of relations on σ_{set} indexed by Γ_{set} such that

$$\begin{array}{lr}
\gamma{:}\,\Gamma_{\text{set}}\ ,\ s,s'{:}\,\sigma_{\text{set}}\ , & \textsc{Sym}\\
\quad \mathrm{Prf}(\Gamma_{\text{rel}}[\gamma,\gamma])\ ,\ \mathrm{Prf}(\sigma_{\text{rel}}[\gamma,s,s'])\vdash \sigma_{\text{rel}}[\gamma,s',s]\ \text{true} & \\
\gamma{:}\,\Gamma_{\text{set}}\ ,\ s,s',s''{:}\,\sigma_{\text{set}}\ , & \textsc{Trans}\\
\quad \mathrm{Prf}(\Gamma_{\text{rel}}[\gamma,\gamma])\ ,\ \mathrm{Prf}(\sigma_{\text{rel}}[\gamma,s,s'])\ ,\ \mathrm{Prf}(\sigma_{\text{rel}}[\gamma,s',s''])\vdash \sigma_{\text{rel}}[\gamma,s,s'']\ \text{true} & \\
\gamma,\gamma'{:}\,\Gamma_{\text{set}}\ ,\ s,s'{:}\,\sigma_{\text{set}}\ , & \textsc{Comp}\\
\quad \mathrm{Prf}(\Gamma_{\text{rel}}[\gamma,\gamma'])\ ,\ \mathrm{Prf}(\sigma_{\text{rel}}[\gamma,s,s'])\vdash \sigma_{\text{rel}}[\gamma',s,s']\ \text{true} &
\end{array}$$

Two families are equal if their two components $-_{\text{set}}$ and $-_{\text{rel}}$ are definitionally equal. Thus, a family of setoids consists of a Γ_{set}-indexed family of partial equivalence relations on one and the same type σ_{set} with the property that the relations over related elements in Γ_{set} are equivalent (by virtue of Comp and symmetry of Γ_{rel}). It is important that symmetry and transitivity are relativised to "existing" $\gamma{:}\,\Gamma_{\text{set}}$, i.e. those with $\Gamma_{\text{rel}}[\gamma,\gamma]$ since otherwise most of the type formers including quotient types would not go through. Also intuitively this restriction makes sense since σ_{set} and σ_{rel} are meant to be undefined or unconstrained outside the "existing" part of Γ. The set of families of setoids over Γ is denoted by $\mathrm{Fam}(\Gamma)$.

Substitution and comprehension. Let B,Γ be contexts of setoids, $f : B \to \Gamma$ and $\sigma \in \mathrm{Fam}(\Gamma)$. A family $\sigma\{f\} \in \mathrm{Fam}(B)$ is defined by $\sigma\{f\}_{\text{set}} = \sigma_{\text{set}}$ and $\sigma\{f\}_{\text{rel}}[\beta{:}\,B_{\text{set}}\ ,\ s,s'{:}\,\sigma_{\text{set}}] = \sigma_{\text{rel}}[f[\beta],s,s']$. The comprehension of σ is defined by $(\Gamma\cdot\sigma)_{\text{set}} = \Gamma_{\text{set}}, s{:}\,\sigma_{\text{set}}$ and $(\Gamma\cdot\sigma)_{\text{rel}}[(\gamma,s)\ ,\ (\gamma',s')] = \Gamma_{\text{rel}}[\gamma,\gamma'] \wedge \sigma_{\text{rel}}[\gamma,s,s']$.

The morphism $\mathrm{p}(\sigma) : \Gamma\cdot\sigma \to \Gamma$ is defined by $\mathrm{p}(\sigma)[\gamma,s] = \gamma$; the relation is preserved by $\wedge$-elimination. Finally, the morphism $\mathrm{q}(f,\sigma) : B\cdot\sigma\{f\} \to \Gamma\cdot\sigma$ is given by $\mathrm{q}(f,\sigma)[\beta,s] = (f[\beta],s)$. This preserves the relation since f does. It is also readily checked that the square (‡‡) from Def. 1 is a pullback.

Proposition 3 *Contexts of setoids together with their morphisms and families form a cwa—the* setoid model.

Since it is induced by definitional equality in the target type theory, the semantic equality in the setoid model is obviously decidable. We believe that one can obtain a model of extensional type theory (where propositional and definitional equality coincide) by identifying "related" morphisms and contexts and families with equivalent relations. In this model semantic equality would be undecidable and the choice operator (see 5.2) would be unsound.

Sections. Let $\sigma \in \mathrm{Fam}(\Gamma)$. A section of σ is a syntactic context morphism $M : \Gamma_{\text{set}} \Rightarrow \Gamma_{\text{set}}, x{:}\,\sigma_{\text{set}}$ which is the identity on the variables in Γ. Therefore, such a section is entirely determined by its last component. It is appropriate to identify sections with these last components so that henceforth a section of σ will be a term $\Gamma_{\text{set}} \vdash M : \sigma_{\text{set}}$ which respects the relations, i.e.

$$\gamma,\gamma'{:}\,\Gamma_{\text{set}}\ ,\ \mathrm{Prf}(\Gamma_{\text{rel}}[\gamma,\gamma'])\vdash \sigma_{\text{rel}}[\gamma,M[\gamma],M[\gamma']]\ \text{true} \qquad \textsc{Resp}$$

4.3 Dependent Products

Let $\sigma \in \mathrm{Fam}(\Gamma)$ and $\tau \in \mathrm{Fam}(\Gamma \cdot \sigma)$. The dependent product $\Pi(\sigma, \tau)$ is defined by $\Pi(\sigma,\tau)_{\mathrm{set}} = \sigma_{\mathrm{set}} \rightarrow \tau_{\mathrm{set}}$ and $\Pi(\sigma,\tau)_{\mathrm{rel}}[\gamma{:}\, \Gamma_{\mathrm{set}}\ ,\ u, v{:}\, \Pi(\sigma,\tau)_{\mathrm{set}}] = \forall s, s'{:}\, \sigma_{\mathrm{set}}.\sigma_{\mathrm{rel}}[\gamma, s, s'] \Rightarrow \tau_{\mathrm{rel}}[(\gamma, s)\ ,\ u\ s\ ,\ v\ s']$. If $M \in \mathrm{Sect}(\tau)$ then we define $\lambda_{\sigma,\tau}(M)[\gamma{:}\, \Gamma_{\mathrm{set}}] = \lambda s{:}\, \sigma_{\mathrm{set}}.M[\gamma, s]$ and conversely if $M \in \mathrm{Sect}(\Pi(\sigma,\tau))$ and $N \in \mathrm{Sect}(\sigma)$ we define $\mathrm{App}_{\sigma,\tau}(M, N)[\gamma{:}\, \Gamma_{\mathrm{set}}] = (M[\gamma]\ N[\gamma])$ Now by straightforward calculation we obtain:

Proposition 4 *These data endow the setoid model with dependent products.*

In very much the same way we can interpret Σ-types, natural numbers, and various inductive types present in the target type theory. Since the interpretation of these is almost forced we leave it out here.

4.4 Propositions

Our next goal is to identify a type of propositions (and proofs) which allows us to "externalise" the relations associated to each type and to interpret quotient types and extensionality. The semantic version of propositions does not exactly match the syntactic rules, for we make use of the following abstract formulation.

Proposition 5 *Assume some cwa with Π-types. It admits the interpretation of the Calculus of Constructions if the following are given.*

- *distinguished families* $\mathbf{Prop} \in \mathrm{Fam}(\mathbf{1})$ *and* $\mathbf{Prf} \in \mathrm{Fam}(\mathbf{1} \cdot \mathbf{Prop})$
- *for each* $\sigma \in \mathrm{Fam}(\Gamma)$ *and* $S : \Gamma \cdot \sigma \rightarrow \mathbf{1} \cdot \mathbf{Prop}$[3] *a distinguished morphism* $\forall_\sigma(S) : \Gamma \rightarrow \mathbf{1} \cdot \mathbf{Prop}$ *and a distinguished morphism* $\mathrm{ev}_{\sigma,S} : \Gamma \cdot \sigma \cdot \mathbf{Prf}\{\forall_\sigma(S) \circ \mathrm{p}(\sigma)\} \rightarrow \Gamma \cdot \sigma \cdot \mathbf{Prf}\{S\}$ *with* $\mathrm{p}(\mathbf{Prf}\{S\}) \circ \mathrm{ev}_{\sigma,S} = \mathrm{p}(\mathbf{Prf}\{S\})$
- *for each section* $M \in \mathrm{Sect}(\mathbf{Prf}\{S\})$ *a distinguished section* $\lambda_{\sigma,S}(M) \in \mathrm{Sect}(\mathbf{Prf}\{\forall_\sigma(S)\})$ *with* $\mathrm{ev}_{\sigma,S} \circ \lambda_{\sigma,S}(M)\{\mathrm{p}(\sigma)\} = M$

in such a way that all these data are stable under substitution, that is for $f : B \rightarrow \Gamma$, $\sigma \in \mathrm{Fam}(\Gamma)$, $S : \Gamma\sigma \rightarrow \mathbf{1}\mathbf{Prop}$, $M \in \mathrm{Sect}(\mathbf{Prf}\{S\})$ *we have* $\forall_\sigma(S) \circ f = \forall_{\sigma\{f\}}(s \circ f^+)$ *and* $\mathrm{ev}_{\sigma,S} \circ f^{++} = f^{++} \circ \mathrm{ev}_{\sigma\{f\},s \circ f^+}$ *and* $\lambda_{\sigma,S}(M)\{f\} = \lambda_{\sigma\{f\},s \circ f^+}(M\{f^+\})$

Proof (Sketch). One interprets types of the form $\mathrm{Prf}(P)$ as morphisms into $\mathbf{1} \cdot \mathbf{Prop}$. The dependent product $\Pi x{:}\, \sigma.\tau$ gets interpreted by the semantic dependent product if τ is not of the form $\mathrm{Prf}(-)$ and by $\forall$ otherwise. Accordingly for application and abstraction.

In the sequel we denote by "**Prop**" both the family in Fam(1), and its comprehension $\mathbf{1} \cdot \mathbf{Prop}$. To simplify the exposition we assume that the target type theory supports an extensional unit type, i.e. there is a type $\Gamma \vdash 1$ **type** in every context, and a term $\Gamma \vdash \star : 1$ together with the equation $\Gamma \vdash M = \star : 1$ for every $\Gamma \vdash M : 1$. We use the extensional unit type as the underlying type

[3] Upper case is used to avoid confusion with the variable s.

of propositional types. It is possible to get rid of the extensional unit type, by defining families of setoids as being either of the form defined in Section 4.2, or to consist simply of a proposition in context Γ_{set} compatible with Γ_{rel} (in which case it is silently understood that the $-_{\mathrm{set}}$-component is the extensional unit type). Comprehension, substitution, and the type formers have then to be defined by case distinction.

We are ready now ready to define the required ingredients to interpret propositions. The underlying type of **Prop** $\in$ Fam(**1**) in the setoid model is just the syntactic type of propositions.

$$\mathbf{Prop}_{\mathrm{set}} = \mathrm{Prop}$$

In order to identify equivalent propositions we define the relation as bi-implication.

$$\mathbf{Prop}_{\mathrm{rel}}[p, q{:}\,\mathrm{Prop}] = p \Leftrightarrow q$$

This is clearly symmetric and transitive and so a family over **1** has been defined. Note that Comp degenerates to a tautology in the case of families over **1**.

The family **Prf** $\in$ Fam(**1**·**Prop**) is given by

$$\begin{aligned}\mathbf{Prf}_{\mathrm{set}}[p{:}\,\mathrm{Prop}] &= 1\\ \mathbf{Prf}_{\mathrm{rel}}[p{:}\,\mathrm{Prop}\ ,\ x, x'{:}\,1] &= p\end{aligned}$$

where 1 is the extensional unit type. The relation on **Prf** is trivially symmetric and transitive; for Comp we use the relation on **Prop**. More precisely, if $\mathbf{Prop}_{\mathrm{rel}}[p, q]$ and $\mathbf{Prf}_{\mathrm{rel}}[p, x, y]$, i.e. p, then also q, thus $\mathbf{Prf}_{\mathrm{rel}}[q, x, y]$.

Next we define universal quantification. If $\sigma \in \mathrm{Fam}(\Gamma)$ and $S : \Gamma{\cdot}\sigma \rightarrow \mathbf{Prop}$ then we put

$$\forall_\sigma(S)[\gamma{:}\,\Gamma_{\mathrm{set}}] = \forall x{:}\,\sigma_{\mathrm{set}}.\sigma_{\mathrm{rel}}[\gamma, s, s] \Rightarrow S[\gamma, x]$$

For property Resp assume $\gamma, \gamma'{:}\,\Gamma_{\mathrm{set}}$ and $\Gamma_{\mathrm{rel}}[\gamma, \gamma']$. We must show that $\forall_\sigma(S)[\gamma] \Leftrightarrow \forall_\sigma(S)[\gamma']$. So assume $\forall_\sigma(S)[\gamma]$ and $x{:}\,\sigma_{\mathrm{set}}$ with $\sigma_{\mathrm{rel}}[\gamma', x, x]$. Using Sym and Comp we get $\sigma_{\mathrm{rel}}[\gamma, x, x]$, hence $S[\gamma, x]$ by assumption. Property Resp for S using $\Gamma_{\mathrm{rel}}[\gamma, \gamma']$ and $\sigma_{\mathrm{rel}}[\gamma, x, x]$ gives $S[\gamma, x] \Leftrightarrow S[\gamma', x]$ and thus $S[\gamma', x]$ as required. The other direction is symmetric. We see that the relativisation to "existing" $x{:}\,\sigma_{\mathrm{set}}$ is necessary to prove Resp.

For the abstraction let $M \in \mathrm{Sect}(\mathbf{Prf}\{S\})$. We define $\lambda_{\sigma,S}(M)[\gamma{:}\,\Gamma_{\mathrm{set}}] = \star$. To see that this is indeed a section of $\mathbf{Prf}\{\forall_\sigma(S)\}$ assume $\gamma, \gamma'{:}\,\Gamma_{\mathrm{set}}$ and $\Gamma_{\mathrm{rel}}[\gamma, \gamma']$. We must show that

$$\mathbf{Prf}\{\forall_\sigma(S)\}_{\mathrm{rel}}[\gamma, \lambda_{\sigma,S}(M)[\gamma]\ ,\ \lambda_{\sigma,S}(M)[\gamma']]$$

which equals $\forall x{:}\,\sigma_{\mathrm{set}}.\sigma_{\mathrm{rel}}[\gamma, x, x] \Rightarrow S[\gamma, x]$ by definition of $\mathbf{Prf}_{\mathrm{rel}}$ and $\forall$. Now assuming $x{:}\,\sigma_{\mathrm{set}}$ and $\sigma_{\mathrm{rel}}[\gamma, x, x]$ we have that $\Gamma{\cdot}\sigma_{\mathrm{rel}}[(\gamma, x)\ ,\ (\gamma', x)]$, hence $S[\gamma, x]$ by Resp for M.

Finally, the evaluation morphism is given by $\mathrm{ev}_{\sigma,S}[\gamma{:}\,\Gamma_{\mathrm{set}}\ ,\ x{:}\,\sigma_{\mathrm{set}}\ ,\ u{:}\,1] = (\gamma, x, \star)$. We must show that this defines a morphism from $\Gamma{\cdot}\sigma{\cdot}\mathbf{Prf}\{\forall_\sigma(S) \circ \mathrm{p}(\sigma)\}$ to $\Gamma{\cdot}\sigma{\cdot}\mathbf{Prf}\{S\}$. Assume $\gamma, \gamma'{:}\,\Gamma_{\mathrm{set}}$, $x, x'{:}\,\sigma_{\mathrm{set}}$ with $\Gamma_{\mathrm{rel}}[\gamma, \gamma']$ and $\sigma_{\mathrm{rel}}[\gamma, x, x']$;

furthermore assume $\forall x{:}\,\sigma_{\mathrm{set}}.\sigma_{\mathrm{rel}}[\gamma, x, x] \Rightarrow S[\gamma, x]$ (meaning that the two variables of unit type corresponding to $\mathbf{Prf}\{\forall_\sigma(S) \circ \mathrm{p}(\sigma)\}$ are "related" according to the definition of $\mathbf{Prf}_{\mathrm{rel}}$. We must show that $S[\gamma, x]$. But this follows since by SYM and TRANS for Γ we have $\Gamma_{\mathrm{rel}}[\gamma, \gamma]$ and thus $\sigma_{\mathrm{rel}}[\gamma, x, x]$ by SYM and TRANS for σ.

The equations relating evaluation and abstraction follow straightforwardly from the properties of the extensional unit type. Also the stability under substitution of all components follows readily by expanding the definitions.

Proposition 6 *The setoid model admits the interpretation of the Calculus of Constructions.*

4.5 Proof Irrelevance and Extensionality

Next we are going to explore which additional propositions are provable in the setoid model. First, we have a non-provability result:

Proposition 7 (Consistency) *The family* $\mathit{ff} := \mathbf{Prf}\{\forall_{\mathbf{Prop}}(\mathrm{id}_{\mathbf{Prop}})\} \in \mathrm{Fam}(\mathbf{1})$ *has no sections.*

Proof. We have $\mathit{ff}_{\mathrm{set}} = 1$ and $\mathit{ff}_{\mathrm{rel}}[x, y{:}\,1] = \forall p{:}\,\mathrm{Prop}.\mathbf{Prop}_{\mathrm{rel}}[p, p] \Rightarrow p$. A section of ff consists of an element of 1 (necessarily $\star$) such that $\forall p{:}\,\mathrm{Prop}.(p \Leftrightarrow p) \Rightarrow p$ (expanding $\mathbf{Prop}_{\mathrm{rel}}$). But this is not possible by consistency of the syntax.

Next we look at Leibniz equality in the model. If $\sigma \in \mathrm{Fam}(\Gamma)$ and $M, N \in \mathrm{Sect}(\sigma)$ then let $\mathrm{L\_Eq}(M, N) : \Gamma \rightarrow \mathbf{Prop}$ be the denotation of Leibniz equality. By unfolding the definition we find that in the setoid model

$$\begin{aligned}\mathrm{L\_Eq}(M, N)[\gamma{:}\,\Gamma_{\mathrm{set}}] = \forall P{:}\,&\sigma_{\mathrm{set}} \rightarrow \mathrm{Prop}.\\ &(\forall x, x'{:}\,\sigma.\sigma_{\mathrm{rel}}[\gamma, x, x'] \Rightarrow ((P\ \ x) \Leftrightarrow (P\ \ x'))) \Rightarrow\\ &\forall y{:}\,1.(P\ \ M[\gamma]) \Rightarrow (P\ \ N[\gamma])\end{aligned}$$

So (modulo the trivial quantification over 1) M and N are Leibniz equal in the model if they are indistinguishable by observations *which respect the relation on* σ.

Lemma 8 *For* $\sigma \in \mathrm{Fam}(\Gamma)$ *and* $M, N \in \mathrm{Sect}(\sigma)$ *the set* $\mathrm{Sect}(\mathbf{Prf}\{\mathrm{L\_Eq}(M, N)\})$ *is nonempty iff* $\gamma{:}\,\Gamma_{\mathrm{set}}\ ,\ \mathrm{Prf}(\Gamma_{\mathrm{rel}}[\gamma, \gamma]) \vdash \sigma_{\mathrm{rel}}[\gamma, M[\gamma], N[\gamma]]$ true

Proof. Assume a section of $\mathbf{Prf}\{\mathrm{L\_Eq}(M, N)\}$. By definition this means that if $\Gamma_{\mathrm{rel}}[\gamma, \gamma']$ for some $\gamma, \gamma'{:}\,\Gamma_{\mathrm{set}}$ then $M[\gamma]$ and $N[\gamma]$ (not $N[\gamma']$!) are indistinguishable by observations respecting σ_{rel}. Now $\lambda x{:}\,\sigma_{\mathrm{set}}.\sigma_{\mathrm{rel}}[\gamma, M[\gamma], x]$ is such an observation as can be seen using SYM and TRANS for σ and $\Gamma_{\mathrm{rel}}[\gamma, \gamma]$ by virtue of $\Gamma_{\mathrm{rel}}[\gamma, \gamma']$ and SYM, TRANS for Γ. Therefore we can deduce $\sigma_{\mathrm{rel}}[\gamma, M[\gamma], N[\gamma]]$ provided we can show $\sigma_{\mathrm{rel}}[\gamma, M[\gamma], M[\gamma]]$, but this follows from RESP for M. Conversely, if $\sigma[\gamma, M[\gamma], N[\gamma]]$ and $P : \sigma_{\mathrm{set}} \rightarrow \mathrm{Prop}$ respects σ_{rel} then obviously $(P\ \ M[\gamma]) \Rightarrow (P\ \ N[\gamma])$ by definition of "respects".

Thus Leibniz equality "externalises" the relation associated to each family.

Proposition 9 *The following rules can be interpreted in the setoid model.*

$$\frac{\Gamma \vdash A : \text{Prop} \qquad \Gamma \vdash M, N : \text{Prf}(A)}{\Gamma \vdash M = N : \text{Prf}(A)} \ \text{PR-IR}$$

$$\frac{\Gamma \vdash P : \text{Prop} \qquad \Gamma \vdash Q : \text{Prop} \qquad \Gamma \vdash H : \text{Prf}(P \Leftrightarrow Q)}{\Gamma \vdash \text{Bi\_Imp}(P, Q, H) : \text{Prf}(P \stackrel{L}{=} Q)} \ \text{BI-IMP}$$

$$\frac{\Gamma \vdash U, V : \Pi x{:}\sigma.\tau \qquad \Gamma, x{:}\sigma \vdash H : \text{Prf}(U\ x \stackrel{L}{=} V\ x)}{\Gamma \vdash \text{Ext}(H) : U \stackrel{L}{=} V} \ \text{EXT}$$

Proof. We identify syntactic objects with their denotations in the setoid model. Rule PR-IR is immediate because both M and N equal $\star$. For rule BI-IMP assume $P, Q \in \text{Sect}(\textbf{Prop}\{!_\Gamma\})$. The assumption H gives rise to a proof of $P[\gamma] \Leftrightarrow Q[\gamma]$ for every $\gamma : \Gamma_{\text{set}}$ with $\Gamma_{\text{rel}}[\gamma, \gamma]$. But by Lemma 8 this implies that there exists a section of $\textbf{Prf}\{\text{L\_Eq}(P, Q)\}$. The proof for EXT is similar.

Now we shall present a somewhat unexpected feature of Leibniz equality, namely that it behaves like Martin-Löf's identity type in the sense that a Leibniz principle for dependent types can be interpreted from which the definability of Martin-Löf's elimination rule [15] follows using PR-IR (see e.g. [8]).

Proposition 10 *For each $\sigma \in \text{Fam}(\Gamma)$ and $\tau \in \text{Fam}(\Gamma \cdot \sigma)$ and $M, N \in \text{Sect}(\sigma)$ and $P \in \text{L\_Eq}(M, N)$ and $U \in \text{Sect}(\tau\{M\})$ there exists a well-determined section $\text{Subst}_{\sigma,\tau}(P, U) \in \text{Sect}(\tau\{N\})$ in such a way that $\text{Subst}_{\sigma,\tau}(\text{Refl}(M), U) = U$, where $\text{Refl}(M)$ is the canonical section of $\text{L\_Eq}(M, M)$ corresponding to reflexivity. Moreover, this operator Subst is stable under substitution in all its arguments.*

Proof. We define $\text{Subst}_{\sigma,\tau}(P, U)$ simply as U. That this is a section of $\tau\{N\}$ follows from Lemma 8 applied to P and rule COMP for the family τ. The other properties are trivially satisfied.

5 Quotient types

We now turn to the interpretation of quotient types in the model. Recall their defining rules from Section 2 . Let $\sigma \in \text{Fam}(\Gamma)$ and $R : \Gamma \cdot \sigma \cdot \sigma^+ \to \textbf{Prop}$ be a "relation" over σ. Viewed internally R is a term of type $\gamma{:}\, \Gamma_{\text{set}}\ ,\ x, x'{:}\, \sigma_{\text{set}} \vdash$ Prop **type** such that

$$\begin{aligned} &\gamma{:}\, \Gamma_{\text{set}}\ ,\ x, y{:}\, \sigma_{\text{set}}\ ,\ \gamma'{:}\, \Gamma_{\text{set}}\ ,\ x', y'{:}\, \sigma_{\text{set}}\ , \\ &\quad \text{Prf}(\Gamma_{\text{rel}}[\gamma, \gamma'] \wedge \sigma_{\text{rel}}[\gamma, x, x'] \wedge \sigma_{\text{rel}}[\gamma, y, y']) \vdash \\ &\qquad R[x, y] \Leftrightarrow R[x', y'] \text{ true} \end{aligned}$$

We say that R "respects" σ_{rel}. Our aim is to define a new family σ/R on which the relation is given by R. The underlying type remains unchanged by quotienting: $(\sigma/R)_{\text{set}}[\gamma] = \sigma_{\text{set}}[\gamma]$. Now since R is not guaranteed to be symmetric and transitive, we have to take the symmetric and transitive closure of R, moreover we must ensure compatibility with σ_{rel} — the relation already present on σ. It turns out that the right choice for $(\sigma/R)_{\mathrm{rel}}$ is the following higher-order encoding of symmetric, transitive closure.

$$\begin{array}{ll}
(\sigma/R)_{\mathrm{rel}}[\gamma{:}\,\Gamma_{\mathrm{set}}\,,\, s, s'{:}\,\sigma_{\mathrm{set}}] = \forall R'{:}\,\sigma_{\mathrm{set}} \to \sigma_{\mathrm{set}} \to \mathrm{Prop}. & \\
\quad (\forall x, x'{:}\,\sigma_{\mathrm{set}}.(R'\ x\ x') \Rightarrow (R'\ x'\ x)) & (\sigma) \\
\quad \Rightarrow (\forall x, x', x''{:}\,\sigma_{\mathrm{set}}.(R'\ x\ x') \Rightarrow (R'\ x'\ x'') \Rightarrow (R'\ x\ x'')) & (\tau) \\
\quad \Rightarrow (\forall x, x'{:}\,\sigma_{\mathrm{set}}.\sigma_{\mathrm{rel}}[\gamma, x, x'] \Rightarrow (R'\ x\ x')) & (\rho) \\
\quad \Rightarrow (\forall x, x'{:}\,\sigma_{\mathrm{set}}.R[\gamma, x, x'] \Rightarrow \sigma_{\mathrm{rel}}[\gamma, x, x] \Rightarrow \sigma_{\mathrm{rel}}[\gamma, x', x'] \Rightarrow (R'\ x\ x')) & (\kappa) \\
\quad \Rightarrow (R'\ s\ s') &
\end{array}$$

In other words, $(\sigma/R)_{\mathrm{rel}}[\gamma, -, -]$ is the least equivalence relation on σ_{set} which contains $\sigma_{\mathrm{rel}}[\gamma, -, -]$ and $R[\gamma, -, -]$ restricted to the domain of $\sigma_{\mathrm{rel}}[\gamma, -, -]$. We shall call such a relation *suitable*, so $(\sigma/R)_{\mathrm{rel}}$ is the least suitable relation.

5.1 Equivalence Classes

Assume $M \in \mathrm{Sect}(\sigma)$. We want to construct a section $[M]_R$ of σ/R. We put $[M]_R[\gamma] = M[\gamma]$ so $[M]_R$ behaves just like M. We must prove RESP for $[M]_R$. Let $\gamma, \gamma'{:}\,\Gamma_{\mathrm{set}}$ and $\Gamma_{\mathrm{rel}}[\gamma, \gamma']$. Moreover, let $R'{:}\,\sigma_{\mathrm{set}} \to \sigma_{\mathrm{set}} \to \mathrm{Prop}$ be a suitable relation. We must show that $R'[M[\gamma], M[\gamma']]$. Since M is a section we have $\sigma_{\mathrm{rel}}[\gamma, M[\gamma], M[\gamma']]$ so we are done since suitable relations contain $\sigma_{\mathrm{rel}}[\gamma, -, -]$.

For Q-AX assume $M, N \in \mathrm{Sect}(\sigma)$ and $H \in \mathrm{Sect}(\mathrm{Prf}\{R \circ M^{+} \circ N\})$, i.e.

$$\gamma, \gamma'{:}\,\Gamma_{\mathrm{set}}\,,\, \mathrm{Prf}(\Gamma_{\mathrm{rel}}[\gamma, \gamma']) \vdash R[\gamma, M[\gamma], N[\gamma]]\ \mathrm{true}$$

We want to show that $\mathrm{Sect}(\mathrm{Prf}\{\mathrm{L\_Eq}([M]_R, [N]_R)\})$ is nonempty. By Lemma 8 and the definition of $[-]_R$ it suffices to show that

$$\gamma{:}\,\Gamma_{\mathrm{set}}\,,\, \mathrm{Prf}(\Gamma_{\mathrm{rel}}[\gamma, \gamma]) \vdash \sigma/R_{\mathrm{rel}}[\gamma, M[\gamma], N[\gamma]]\ \mathrm{true}$$

In this context we can show $\sigma_{\mathrm{rel}}[\gamma, M[\gamma], M[\gamma]]$ and $\sigma_{\mathrm{rel}}[\gamma, N[\gamma], N[\gamma]]$ using RESP. Thus for every suitable relation $R'{:}\,\sigma_{\mathrm{set}} \to \sigma_{\mathrm{set}} \to \mathrm{Prop}$ we have $(R'\ M[\gamma]\ N[\gamma])$ by (κ) which gives the desired result. Similarly, we can define lifting (Q-ELIM) and induction (Q-IND) so that we conclude

Proposition 11 *The setoid model allows the interpretation of quotient types.*

Effectiveness of quotient types. Here we want to address the question as to whether the converse to Q-AX is also true, i.e. whether we can conclude $R[M, N]$ from $[M]_R \stackrel{L}{=} [N]_R$. In general, this cannot be the case because together with Q-AX this implies that R is an equivalence relation. Quotient types are called *effective* [11, 14] if this condition is already sufficient. It turns out that in the setoid model all quotient types are effective and that this is a purely syntactic consequence of the rules for quotient types and rule BI-IMP.

Proposition 12 *Fix a type theory with quotient types and* BI-IMP. *Let* $\Gamma \vdash \sigma$ *and* $\Gamma\,,\, x, y{:}\,\sigma \vdash R[x, y] : \mathrm{Prop}$ *be an equivalence relation, i.e. in context* Γ *we have* $\forall x{:}\,\sigma.R[x, x]$ true, *and* $\forall x, y{:}\,\sigma.R[x, y] \Rightarrow R[y, x]$ true, *and* $\forall x, y, z{:}\,\sigma.R[x, y] \Rightarrow R[y, z] \Rightarrow R[x, z]$ true. *Then for each* $\Gamma \vdash U, V{:}\,\sigma$ *with* $[U]_R \stackrel{L}{=} [V]_R$ *we have* $R[U, V]$.

Proof. Consider the term $M[y{:}\sigma] = R[U,y]$: Prop. We have Γ , $y,y'{:}\sigma$, $R[y,y'] \vdash M[y] \stackrel{L}{=} M[y']$ by symmetry and transitivity and Bi-Imp. Let H denote the proof of this. Now put $M'[z{:}\sigma/R] = \text{plug}_R\, z \,\text{in}\, M \,\text{using}\, H$: Prop. By Q-Comp $M'[[U]_R]$ equals $R[U,U]$ and $M'[[V]_R]$ equals $R[U,V]$. Now the former is true by reflexivity of R and both are Leibniz equal by assumption on U and V. Therefore $R[U,V]$ is true as well.

5.2 A Choice Operator for Quotient Types

Our goal consists of finding a way of getting hold of a representative for a given element of a quotient type. The idea is that since in the model an element M of a quotient type is nothing but an element of the underlying type, albeit with a weaker Resp requirement, it should under certain circumstances be possible to view M as an element of this underlying type. This is formalised by the rules Q-Choice, Q-Choice-Ax and Q-Choice-Comp from Section 2 which we do not reproduce here for lack of space.

Semantically, we want to interpret the choice operator as the identity, i.e. if $M \in \text{Sect}(\sigma/R)$ then we want to put $\text{choice}(M)[\gamma{:}\,\Gamma_{\text{set}}] = M[\gamma]$. Now from $M \in \text{Sect}(\sigma/R)$ we can deduce that if $\Gamma_{\text{rel}}[\gamma,\gamma']$ for some $\gamma,\gamma'{:}\,\Gamma_{\text{set}}$ then $\sigma/R_{\text{rel}}[\gamma, M[\gamma], M[\gamma']]$. But in order to conclude $\text{choice}(M) = M \in \text{Sect}(\sigma)$ we need to have (the stronger) $\sigma_{\text{rel}}[\gamma, M[\gamma], M[\gamma']]$. One situation in which we can deduce the latter from the former occurs when from $\Gamma_{\text{rel}}[\gamma,\gamma']$ we can conclude that $M[\gamma]$ and $M[\gamma']$ are actually Leibniz equal. Now this situation in turn occurs for example when $\Gamma_{\text{rel}}[\gamma,\gamma']$ entails that γ and γ' are Leibniz equal themselves. This motivates the following definition.

Definition 13 *The set of* non-quotiented *types is defined by the following clauses.*

- Nat *(and other inductive types) and* $\Pi x_1{:}\sigma_1 \ldots \Pi x_n{:}\sigma_n.\text{Prf}(M)$ *for* $n \geq 0$ *are non-quotiented.*
- *If* σ *and* τ *are non-quotiented, so is* $\Sigma x{:}\sigma.\tau$.

A (syntactic) context Γ *is non-quotiented if it is made up out of non-quotiented types only.*

Notice that if $\vdash \Gamma = \Delta$ and Γ is non-quotiented then so is Δ. An easy induction on the definition of "non-quotiented" gives the following semantic counterpart.

Proposition 14 *Let* Γ *be a non-quotiented syntactic context and* $(\Gamma_{\text{set}}, \Gamma_{\text{rel}})$ *be its interpretation in the setoid model.* $\gamma,\gamma'{:}\,\Gamma_{\text{set}}$, $\Gamma_{\text{rel}}[\gamma,\gamma'] \vdash \gamma.i \stackrel{L}{=} \gamma'.i$ true *for all* $i \leq$ *the length of* Γ.

Now we could semantically justify the application of choice in non-quotiented contexts, i.e. the above "certain proviso" would be that Γ is non-quotiented. With such a rule we would, however lose the syntactic weakening and substitution properties which are implicit in e.g. the typing rule for application. For example if $\vdash M : \sigma/R$ we can infer $\vdash \text{choice}(M) : \sigma$ because the empty context is non-quotiented, but we would not have $x{:}\sigma/R \vdash \text{choice}(M) : \sigma$ although x does not occur in M. Therefore, we close the rule up under arbitrary substitution and weakening and thus arrive at the following definitive rule for choice:

$$\frac{\Gamma \vdash M : \sigma/R \qquad \begin{array}{l}\text{There exists a non-quotiented context } \Delta \text{ and} \\ \text{a type } \tau \text{ and a term } N \text{ with } \Delta \vdash N : \tau \text{ and a} \\ \text{syntactic context morphism } f : \Gamma \Rightarrow \Delta \text{ such} \\ \text{that } M \equiv N[f] \text{ and } \sigma/R \equiv \tau[f].\end{array}}{\Gamma \vdash \mathrm{choice}(M) : \sigma} \ \text{Q-CHOICE}$$

Here $\equiv$ means syntactic identity and not just definitional equality so that the side condition is decidable since the possible substitutions f are bounded by the size of M. The condition is e.g. satisfied for $z{:}\,\mathrm{Int} \vdash |z| : \mathrm{Int}$ with $\Delta = \mathrm{Nat}$ and $f[z{:}\,\mathrm{Int}] = |z|$, but it does not hold for $z{:}\,\mathrm{Int} \vdash z + z : \mathrm{Int}$

Proposition 15 *The above rules* Q-CHOICE, Q-CHOICE-COMP, *and* Q-CHOICE-AX *can be soundly interpreted in the setoid model.*

Proof. We interpret choice(M) like M. Suppose that $\Gamma \vdash \mathrm{choice}(M) : \sigma$ by Q-CHOICE then $M \equiv N[f]$ for some (syntactic) substitution $f : \Gamma \Rightarrow \Delta$. We identify these syntactic objects with their interpretations. If $\Gamma_{\mathrm{rel}}[\gamma, \gamma']$ then $f[\gamma]$ and $f[\gamma']$ are actually Leibniz equal by Prop. 14 using the assumption that Δ is non-quotiented. So $M[\gamma]$ and $M[\gamma']$ are Leibniz equal and thus related in σ_{rel}. Thus we can assert $M \in \mathrm{Sect}(\sigma)$. Since both choice and $[-]_R$ are interpreted as the identity the rules Q-CHOICE-COMP and Q-CHOICE-AX are validated.

Discussion. The choice operator is quite unusual and seems paradoxical at first so that a few examples explaining its use are in order. Let $-1 := [(0,1)]_{R_{\mathrm{Int}}}$ and $-1' := [(2,3)]_{R_{\mathrm{Int}}}$ where $1, 2, 3$ are abbreviations for the corresponding numerals of type Nat. Now since $0{+}3 \stackrel{L}{=} 1{+}2$ we have $R_{\mathrm{Int}}[(0,1)\ ,(2,3)]$, and thus $-1 \stackrel{L}{=} -1'$. On the other hand $\mathrm{choice}(-1) = (0,1)$ and $\mathrm{choice}(-1') = (2,3)$. It seems as if we could conclude $(0,1) \stackrel{L}{=} (2,3)$ which would be a contradiction. Now $-1 \stackrel{L}{=} -1'$ means $\forall P{:}\,\mathrm{Int} \rightarrow \mathrm{Prop}.(P\ -1) \Rightarrow (P\ -1')$. In order to get a contradiction from this, we would have to instantiate with $P = \lambda z{:}\,\mathrm{Int}.\mathrm{choice}(-1) \stackrel{L}{=} \mathrm{choice}(z)$. But this expression is not well-typed since $z{:}\,\mathrm{Int} \vdash \mathrm{choice}(z)$ (the context $z{:}\,\mathrm{Int}$ contains a quotient type.) is not. Thus, in some sense the choice operator does not respect Leibniz equality. As any term former it does of course respect definitional equality so in an extensional setting where the two equalities are identified choice would be unsound.

The main usage of the choice operator is that it permits to recover the underlying implementation of a function between quotients internally. For example if we have defined some function $F : \mathrm{Int} \rightarrow \mathrm{Int}$ in non-quotiented context Γ (so in particular F may not just be a variable) then we may form $\Gamma, u{:}\,\mathrm{Nat} \times \mathrm{Nat} \vdash F\ [u]_{R_{\mathrm{Int}}} : \mathrm{Int}$, apply choice and abstract from u to obtain $F' := \lambda u{:}\,\mathrm{Nat} \times \mathrm{Nat}.\mathrm{choice}(F\ [u]_{R_{\mathrm{Int}}}) : (\mathrm{Nat} \times \mathrm{Nat}) \rightarrow (\mathrm{Nat} \times \mathrm{Nat})$ Now from Q-CHOICE-AX we obtain $\Gamma\ , u{:}\,\mathrm{Nat} \times \mathrm{Nat} \vdash F\ [u]_{R_{\mathrm{Int}}} = [F'\ u]_{R_{\mathrm{Int}}} : \mathrm{Int}$ and moreover—since R_{Int} is an equivalence relation— $\Gamma \vdash \forall u, v{:}\,\mathrm{Nat} \times \mathrm{Nat}.R_{\mathrm{Int}}[u, v] \Rightarrow R_{\mathrm{Int}}[F'\ u, F'\ v]$ using the above and Prop. 12.

Yet another application of choice arises from the use of quotient types to model "nonconstructive" types like the real numbers. Assume that we have defined a type of real numbers Real as a quotient of a suitable subset of Nat $\rightarrow$ Nat (thought of as of decimal or continued fraction expansions) the quotienting relation being that the difference is a fundamental sequence defined in the usual ε-δ style. Now since different real numbers can be arbitrarily close together, there can be no non-constant function from the quotient type Real to a ground type like Nat. This has been brought forward as a serious argument against the use of quotienting in a constructive setting, since if one has defined a real number one cannot even put one's hands on its first digit! Using the choice operator one can at least solve this last problem. Assume that we have defined a real number R in the empty context, say e or π. We may then form $\vdash \mathrm{choice}(R) : \mathrm{Nat} \rightarrow \mathrm{Nat}$ and from this extract the desired intensional information like the first digit or other things. However, it is still not possible to write a non-constant *function* from the reals to Nat. Notice that this lack is not particular to the setoid model, but directly inherited from the target type theory which does not provide functions from Nat $\rightarrow$ Nat to Nat which respect the "book"-equality for real numbers. Similarly, in the setoid model we have no nonconstant function from Prop to Nat because this is so in the target type theory.

6 Conclusions

As remarked in the Introduction the model does not provide any type dependency other than the one induced by propositions. In particular we have neither universes nor "large eliminations" [1]. Formally, this can be seen as follows. Assume that the target type theory contains an empty type $\mathbf{0}$ and an operator $?_\sigma$ such that $?_\sigma(M) : \sigma$ if $M : \mathbf{0}$. Then in the setoid model we also have this empty type. Now if we had a universe containing the natural numbers and the empty type then we could define a family of types $n{:}\,\mathrm{Nat} \vdash \sigma[n]$ **type** such that $\sigma[0] = \mathbf{0}$ and $\sigma[\mathrm{Suc}(n)] = \mathrm{Nat}$. Now remember that the $-_{\mathrm{set}}$-component is left unchanged upon substitution. So in view of the first equation σ_{set} there would have to be a function $?_\tau : \sigma_{\mathrm{set}} \rightarrow \tau$ for every type τ which contradicts the second equation if $\tau = \mathbf{0}$.

A major application of such families of types is that they allow to derive Peano's fourth axiom [15, 18]. On the level of propositions we are still able to interpret this axiom, that is we have $0 \stackrel{L}{=} \mathrm{Suc}(0) \Rightarrow \mathit{ff}$, provided this holds in the target type theory. The difference is that $\mathrm{Prf}(\mathit{ff})$ is weaker than the empty type because in the presence of an element of $\mathrm{Prf}(\mathit{ff})$ every proposition is true, but not every type is inhabited.

Universes are also used for modularisation and structuring. E.g. using a universe it is possible to define a type of monoids or groups and to identify the construction of the free group over a monoid as a function. We have investigated an extension to the setoid model which admits a universe restricted in the sense that quotient types and inductive types and elimination of them is only allowed within the universe. This is to be reported elsewhere.

The equality in the setoid model is decidable by performing the interpretation and then using a standard decision procedure for definitional equality. One could therefore devise an implementation of the setoid model which would keep track of the interpretations of all the terms and types in the current environment and replace computation on the external terms with computation on these interpretations. We believe that such an implementation would behave even better than an ordinary implementation like Lego because in the model proofs and programs are entirely separated. At this moment no such implementation exists, but we have defined all the semantic operations in Lego and checked the required semantic equations mechanically. In principle this can be used to machine-check derivations in the source type theory in a variable-free combinatory way. But this is of course far too cumbersome for real applications so that a proper implementation of the translation is called for. However, many applications go through without having the additional *definitional* equalities like PR-IR or Q-CHOICE-COMP, and it is enough to have the corresponding Leibniz equalities. One can then work within a Lego context which provides all the rules we have studied as axioms. In this way we have been able to carry out a number of examples including a proof of Tarski's generalised fixpoint theorem (based on a formalisation by Pollack) and a verification of the alternating bit protocol.

References

1. Thorsten Altenkirch. *Constructions, normalization, and inductive types.* PhD thesis, University of Edinburgh, 1994.
2. Hendrik Barendregt. Functional programming and lambda calculus. Handbook of Theoretical Computer Science, Volume B.
3. Michael Beeson. *Foundations of Constructive Mathematics.* Springer, 1985.
4. Errett Bishop and Douglas Bridges. *Constructive Analysis.* Springer, 1985.
5. J. Cartmell. *Generalized algebraic theories and contextual categories.* PhD thesis, Univ. Oxford, 1978.
6. Thierry Coquand and Gérard Huet. The calculus of constructions. *Information and Computation*, 76:95–120, 1988.
7. Robert Constable et al. *Implementing Mathematics with the Nuprl Development System.* Prentice-Hall, 1986.
8. Herman Geuvers and Benjamin Werner. On the Church-Rosser property for expressive type systems. LICS '94.
9. M. J. C. Gordon and T. F. Melham. *Introduction to HOL.* Cambridge, 1993.
10. Martin Hofmann. Elimination of extensionality for Martin-Löf type theory. LNCS 806.
11. Bart Jacobs. Quotients in Simple Type Theory. submitted.
12. Bart Jacobs. Comprehension categories and the semantics of type theory. *Theoretical Computer Science*, 107:169–207, 1993.
13. J. Lambek and P. Scott. *Introduction to Higher-Order Categorical Logic.* Cambridge, 1985.
14. I. Moerdijk and S. Mac Lane. *Sheaves in Geometry and Logic.* Springer, 1992.
15. B. Nordström, K. Petersson, and J. M. Smith. *Programming in Martin-Löf's Type Theory, An Introduction.* Clarendon Press, Oxford, 1990.
16. Wesley Phoa. An introduction to fibrations, topos theory, the effective topos, and modest sets. Technical Report ECS-LFCS-92-208, LFCS Edinburgh, 1992.

17. Andrew Pitts. Categorical logic. In *Handbook of Theoretical Computer Science*. Elsevier Science, 199? to appear.
18. Jan Smith. The independence of Peano's fourth axiom from Martin-Löf's type theory without universes. *Journal of Symbolic Logic*, 53(3), 1988.
19. Thomas Streicher. *Semantics of Type Theory*. Birkhäuser, 1991.

A Syntax of the target type theory

The sets of *pre-contexts* (Γ), *pre-types* (σ, τ), and *pre-terms* (M, N, L, O, P) are defined by the following grammar. $\Gamma ::= \bullet | \Gamma, x\colon\sigma$ and $\sigma, \tau ::= 1 | \mathrm{Nat} | \Pi x\colon\sigma.\tau | \ \Sigma x\colon\sigma.\tau |$ $\mathrm{Prop} | \mathrm{Prf}(M)$ and $M, N ::= x | \star | 0 | \mathrm{Suc}(M) | \mathrm{R}^{\mathrm{Nat}}_{\sigma}(M, N, O) | (M\ N) | \ (\lambda x\colon\sigma.M | \ (M, N)$ $| M.1 | M.2 | \forall x\colon\sigma.M$ The empty context is valid and if $\Gamma \vdash \sigma_{\mathrm{set}}$ then $\Gamma, x\colon\sigma$ **ctxt** for x fresh.Nat, 1, and Prop are valid types in every valid context. If $\Gamma, x\colon\sigma \vdash \tau$ **type** then $\Gamma \vdash \Pi x\colon\sigma.\tau$ **type** and $\Gamma \vdash \Sigma x\colon\sigma.\tau$ **type**. If $\Gamma \vdash M : \mathrm{Prop}$ then $\Gamma \vdash \mathrm{Prf}(M)$ **type**. The term forming rules are as follows.

$$\frac{\Gamma \vdash \sigma = \tau \ \textbf{type} \qquad \Gamma \vdash M : \sigma}{\Gamma \vdash M : \tau} \qquad \frac{\Gamma, x\colon\sigma, \Delta \ \textbf{ctxt}}{\Gamma, x\colon\sigma, \Delta \vdash x\colon\sigma}$$

$$\frac{\begin{array}{l}\Gamma \vdash M : \sigma \\ \Gamma \vdash N : \tau[M]\end{array}}{\Gamma \vdash (M, N) : \Sigma x\colon\sigma.\tau[x]} \qquad \frac{\Gamma \vdash M : \Sigma x\colon\sigma.\tau[x]}{\Gamma \vdash M.1 : \sigma} \qquad \frac{\Gamma \vdash M : \Sigma x\colon\sigma.\tau[x]}{\Gamma \vdash M.2 : \tau[M.1]}$$

$$\frac{\Gamma, x\colon\sigma \vdash M : \tau[x]}{\Gamma \vdash \lambda x\colon\sigma.M : \Pi x\colon\sigma.\tau[x]} \qquad \frac{\begin{array}{l}\Gamma \vdash M : \Pi x\colon\sigma.\tau[x] \\ \Gamma \vdash N : \sigma\end{array}}{\Gamma \vdash (M\ N) : \tau[N]}$$

$$\frac{\Gamma \ \textbf{ctxt}}{\Gamma \vdash 0 : \mathrm{Nat}} \qquad \frac{\Gamma \vdash M : \mathrm{Nat}}{\Gamma \vdash \mathrm{Suc}(M) : \mathrm{Nat}} \qquad \frac{\begin{array}{l}\Gamma, x\colon\mathrm{Nat} \vdash \sigma[x] \ \textbf{type} \\ \Gamma \vdash M_z : \sigma[0] \quad \Gamma \vdash N\colon\mathrm{Nat} \\ \Gamma, x\colon\mathrm{Nat}, p\colon\sigma[x] \vdash M_s : \sigma[\mathrm{Suc}(x)]\end{array}}{\Gamma \vdash \mathrm{R}^{\mathrm{Nat}}_{\sigma}(M_z, M_s, N) : \sigma[N]}$$

$$\frac{\Gamma \ \textbf{ctxt}}{\Gamma \vdash \star : 1} \qquad \frac{\Gamma, x\colon\sigma \vdash P : \mathrm{Prop}}{\Gamma \vdash \forall x\colon\sigma.P : \mathrm{Prop}}$$

A.1 Equality Rules

Equality is reflexive, symmetric, and transitive and for each context, term, or type forming rule there is a corresponding equality rule. For example

$$\frac{\Gamma \vdash \sigma = \sigma' \ \textbf{type} \qquad \Gamma, x : \sigma \vdash \tau = \tau' \ \textbf{type}}{\Gamma \vdash \Pi x : \sigma.\tau = \Pi x : \sigma'.\tau' \ \textbf{type}}$$

In addition we have the reflection rule for propositions $\Gamma \vdash \mathrm{Prf}(\forall x\colon\sigma.P) = \Pi x\colon\sigma.\mathrm{Prf}(P)$ **type** and the following "computational" equations which are understood to hold under the premise that both left and right hand side have the appropriate type: $\Gamma \vdash \mathrm{App}_{\sigma,\tau}(\lambda x : \sigma.M, N) = M[N] : \tau[N]$, $\Gamma \vdash (M, N).1 = M : \sigma$, $\Gamma \vdash (M, N).2 = N : \tau[M]$, $\Gamma \vdash M = \star : 1$ (rule for the extensional unit type), $\Gamma \vdash \mathrm{R}^{\mathrm{Nat}}_{\sigma[x:\mathrm{Nat}]}(M_z\ ,$ $M_s[x\colon\mathrm{Nat}, p\colon\sigma[x]]\ ,\ 0) = M_z : \sigma[0]$, and $\Gamma \vdash \mathrm{R}^{\mathrm{Nat}}_{\sigma[x:\mathrm{Nat}]}(M_z\ ,\ M_s[x\colon\mathrm{Nat}, p\colon\sigma[x]]\ ,$ $\mathrm{Suc}(N)) = M_s[N, \mathrm{R}_{\sigma}(M_z, M_s, N)] : \sigma[\mathrm{Suc}(N)]$.

Untyped λ-calculus with relative typing *

M. Randall Holmes

Math. Dept., Boise State University, Boise, Idaho, USA 83725

Abstract. A system of untyped λ-calculus with a restriction on function abstraction using relative typing analogous to the restriction on set comprehension found in Quine's set theory "New Foundations" is discussed. The author has shown elsewhere that this system is equiconsistent with Jensen's *NFU* ("New Foundations" with urelements) + Infinity, which is in turn equiconsistent with the simple theory of types with infinity. The definition of the system is given and the construction of a model is described. A semantic motivation for the stratification criterion for function abstraction is given, based on an abstract model of computation. The same line of semantic argument is used to motivate an analogy between the notion of "strongly Cantorian set" found in "New Foundations" and the notion of "data type"; an implementation of absolute types as domains of retractions with strongly Cantorian ranges is described. The implementation of these concepts in a theorem prover developed by the author is sketched.

1 Introduction

We discuss a system of λ-calculus which is, strictly speaking, untyped, but in which abstraction is limited by a relative typing scheme. The criterion used to limit abstraction is essentially the criterion of "stratification" used to limit set comprehension in Quine's set theory "New Foundations", in a version appropriate to a theory of functions. We have shown elsewhere (in [7]) that the system we present is equivalent in consistency strength and expressive power to the theory *NFU* + Infinity ("New Foundations" with urelements plus the Axiom of Infinity) introduced and shown to be consistent relative to *ZFC* by Jensen in [11].

We define the system and describe a model. We then present a semantic motivation for the system in terms of (very abstract) theoretical computer science ideas. We show how to reintroduce a notion of absolute type into this system, via the concept of "strongly Cantorian set" peculiar to Quine-style set theory, and argue for its appropriateness in terms of our semantic motivation. We conclude by sketching some features of an actual implementation of these ideas in a theorem prover.

We introduced systems of synthetic combinatory logic equivalent in strength to "New Foundations" and some of its fragments in [7] and [8]; the description

* **Supported by the U. S. Army Research Office, grant no. DAAH04-94-G-0247**

of the kind of λ-calculus discussed here is implicit in those papers. An extensive discussion of the set theory *NFU* with some discussion of the possible application of related systems of combinatory logic in computer science is found in our [9]; the definition of the λ-calculus we discuss here is briefly stated in that paper. Good references for the kind of set theory alluded to here are Rosser's [12] (old) and Forster's excellent [4] (current).

2 A Simply Typed λ-Calculus

It seems best to us to motivate the untyped system we will present by first introducing a typed version. The type labels in the typed system will be the natural numbers $0, 1, 2 \ldots$. Type $n + 1$ can be understood as the type $n \rightarrow n$ of functions from type n to type n, and each type n is identified with the type $n \times n$, the cartesian product of type n with itself. This very simple type structure is analogous to the type structure of the simple theory of types of Russell and Whitehead, as simplified by Ramsay, in which each type can be thought of as the power set of the preceding type. More complicated function and product types can be coded in this type system by the device of using (possibly iterated) constant functions of an object to represent it in higher types. For example, maps from type $n + 1$ to type n could be encoded as maps of type $n + 2$ all of whose values are constant functions (of type $n + 1$, used to represent their type n values). In this way, all the types of a more usual typed λ-calculus with all cartesian product and arrow types can be identified with subtypes of the more limited supply of types in this system.

Atomic terms are constants π_1, π_2 in each type above 0 , a constant *eq* in each type above 1 and a countable supply of variables of each type. If T and U are terms of type n, (T, U) is a term of type n (this expresses the identification of type n with $n \times n$). If T is a term of type $n+1$ and U is a term of type n, $T(U)$ is a term of type n. If T is a term of type n and x is a variable of type n, then $(\lambda x)(T)$ is a term of type $n + 1$. (These two conditions express the identification of type $n + 1$ with type $n \rightarrow n$.)

Observe that the types of all subterms of a term can be inferred from the type of the larger term. We will take advantage of this and generally not express type superscripts.

The axioms of the theory are those of first-order logic plus the following non-logical axioms:

Proj: $\pi_i(x_1, x_2) = x_i$ for $i = 1, 2$
Prod: $(f, g)(x) = (f(x), g(x))$
Prodλ: $(\lambda x)(T, U) = ((\lambda x)(T), (\lambda x)(U))$
Surj: $(\pi_1(x), \pi_2(x)) = x$
Abstl: $(\lambda x)(T)(x) = T$
Eq: $eq(x, y) =$ if $x = y$ then π_1 else π_2
Nontriv: $\pi_1 \neq \pi_2$

Each of these axioms is "typically ambiguous"; any type which makes sense may be assigned to the two sides of an axiom (and then the types of subterms can be inferred as observed above). The axioms (Prodλ) and (Abst1) are axiom schemes with term parameters represented by T and U.

An axiom of extensionality

Ext: if for all x, $f(x) = g(x)$, then $f = g$

is certainly consistent with this system but will not be adjoined to it for our purposes. A model of this system is easily obtained by taking as type 0 any infinite set X and selecting a surjective map from $X \times X$ onto X to determine the pairing relation, then identifying type 1 with X^X and in general, if the set Y implements type n, using Y^Y to implement type $n+1$. Axiom (Prod) determines the definition of the ordered pair in each type above 0. The axiom (Ext) is satisfied in this model, of course.

3 The Stratified λ-Calculus

Just as in Russell's type theory, the phenomenon of "typical ambiguity" presents itself. Each definable object and theorem has a precise analogue in each higher type (not necessarily in each lower type). Following Quine in his motivation of the set theory "New Foundations", we suggest that the types may be identifiable with one another, in which case the type indices can profitably be omitted from the theory entirely, obtaining a type free theory, but that only those instances of (Abst1) should be retained which involve λ-terms which would well-formed in the typed theory with a suitable assignment of types. Thus, for example, in the untyped λ-calculus each function f has the fixed point $(\lambda x)(f(x(x)))((\lambda x)(f(x(x))))$, but this cannot be typed with integer types as above, and so this λ-term is not associated with an instance of (Abst1) (in fact, it is not even a well-formed term under the rules stated below).

We present the term formation criteria of the type free theory in the form of two mutually recursive definitions.

Definition 1. An "atomic term" of the type free theory is any one of π_1, π_2, eq, or one of a countable supply of variables. A "well-formed term" of the type free theory is either an atomic term, a string of the form (T, U) or $T(U)$, where T and U are well-formed terms, or a string of the form $(\lambda x)(T)$, where T is a well-formed term and x is a variable which does not occur in T with "relative type" (see Definition 2) other than 0. The class of well-formed terms is the smallest class closed under these constructions.

Definition 2. Let T be a well-formed term of the type-free theory. The "relative type" of each occurrence of a subterm of T is an integer chosen as follows: the relative type of T with respect to itself is 0; if an occurrence of the subterm (U, V) has relative type n, the obvious occurrences of U and V are assigned relative type n; if an occurrence of the subterm $U(V)$ has relative type n, the obvious occurrences of U and V are assigned relative types $n+1$ and n, respectively; if an occurrence of the subterm $(\lambda x)(U)$ has relative type n, then the relative types assigned to the obvious occurrences of x and U will be $n-1$ (x will not occur in U with a relative type other than 0 by Definition 1 above).

Note that negative relative types are possible; this exploits Wang's observation in [14] that the simple theory of types or the type theory described above can be extended to allow all negative types, which follows from a simple compactness argument.

We give an example. The term $x(x)$ is clearly well-formed. The relative type of the first occurrence of x in this term is 1; the relative type of the second occurrence is 0. Since the variable x occurs in the term $x(x)$ with a relative type other than 0, the term $(\lambda x)(x(x))$ is *not* well-formed (more generally, the term $(\lambda x)(f(x(x)))$ used in the construction of a fixed point of f is not well-formed).

We call λ-terms which are well-formed in the type-free theory "stratified" λ-terms, and we call this theory the stratified λ-calculus.

The axioms of the stratified λ-calculus are typographically the same as those given for the typed theory above, except, of course, that type indices are not to be supplied.

Theorem 3. *The stratified λ-calculus is consistent (if the usual set theory is consistent).*

Proof. A model for the stratified λ-calculus may be constructed as follows (this construction is given in the author's Ph. D. thesis [6]). Choose an infinite base set X as in the construction of a model of the typed theory, supplied with a bijection with $X \times X$ used for the representation of pairs of elements of X as elements of X. Let $\perp$ denote a fixed element of X (it is necessary for the pair $(\perp, \perp)$ to be represented by $\perp$). We indicate the construction of a cumulative system of types indexed by the ordinals based on X. If the type indexed with α is represented by a set Y, the type indexed by $\alpha+1$ will be represented by the set Y^Y of all functions from Y to Y. The difficulty with this is that Y^Y does not include Y; we resolve this by choosing a subset of type $\alpha+1$ at each stage to be identified with type α (details of this identification follow). Each element of type 0 is identified with its constant function in type 1. When type α has been embedded in type $\alpha+1$, we represent the object of type $\alpha+1$ associated with an object x of type α as x^+. Let f be an element of type $\alpha+1$; we identify it with the function f^+ in type $\alpha+2$ which takes each x^+ (of type $\alpha+1$) to $(f(x))^+$ and each element of type $\alpha+1$ not identified with any element of type α to $\perp$. This is how we embed type $\alpha+1$ in type $\alpha+2$. The cumulative nature of the type system thus defined allows the type indexed by a limit ordinal λ to be defined as the union of all types $\alpha < \lambda$. Note that an object of a limit type λ has already been assigned a value as a function of type $\alpha+1$ for each type $\alpha < \lambda$, and its values as a function in each type are identified by construction, so it has already been in effect assigned a value as a function of type $\lambda+1$ (with value $\perp$ everywhere off its natural domain). Thus, each type λ can be embedded in type $\lambda+1$, which is needed for the process of cumulative embedding of types described above to continue past limit ordinals. Essentially this technique of making function types cumulative was introduced by Dana Scott in connection with domain theory (see [13]). It is straightforward to verify that this cumulatively typed structure satisfies all the axioms of the typed theory, if the pair in higher types is defined in accordance with axiom (Prod).

Now consider this structure in a nonstandard model of set theory with an automorphism j moving a nonstandard ordinal α upward. The model of the stratified λ-calculus we construct will be type α in this structure. Pairing in the model will coincide with pairing in type α. Function application $f(x)$ for f and x objects of type α will be redefined as $j(f)(x)$. The effect of this redefinition is to cause the functions in type $j^{-1}(\alpha)+1$ to be assigned the extensions of the functions in type $\alpha+1$ (which are all the functions from type α into itself); objects of higher type will each be assigned the same extension as some element of this type. The denotation of $(\lambda x)(T)$ will be the unique object of type $j^{-1}(\alpha)+1$ with the correct extension, for each stratified term T (that each such object exists requires verification). It is straightforward to verify that all axioms of stratified λ-calculus hold in this model; proofs in [9] can be adapted to this purpose, for instance (originally derived from the model construction for *NFU* given by Maurice Boffa in [1]). □

We recapitulate an example from above to make it clearer what is happening. The term $x(x)$ presents no difficulty (for x some fixed object); it will be represented by $j(x)(x)$, where j is the external automorphism. However, we cannot expect there to be a function $(\lambda x)(x(x))$ in the model of stratified λ-calculus, because it would be interpreted by the inverse image under j of a map taking each x in type α to $j(x)(x)$, and the latter map cannot be expected to exist (and indeed will not exist) because j is an external automorphism of the nonstandard model of set theory used.

We have demonstrated in [7] that the consistency strength of the stratified λ-calculus described here is exactly that of *NFU* + Infinity; the strength of the latter theory is known to be the same as that of the theory of types with infinity. Stronger extensions of these theories are possible, paralleling the strong extensions possible for the usual set theory.

Note that the model is NOT extensional. The type-free theory with (Ext) has the same strength as Quine's "New Foundations" with full extensionality, whose consistency remains an open question. But the non-extensional version is mathematically entirely adequate for applications. A canonical function with each extension is available in type $j^{-1}(\alpha)+1$ in the model; it happens that there are infinitely many additional objects with each extension in the higher types. A disconcerting feature of "New Foundations" is that the Axiom of Choice can be disproven; this is not the case in *NFU* or in the stratified λ-calculus without extensionality. The model of the stratified λ-calculus constructed above will satisfy AC (suitably expressed) if the underlying set theory satisfies AC.

Axiom (Abst1) can be used to define a notion of β-reduction (applicable only to stratified λ-terms); the fact that the theory with integer types is a subsystem of the usual typed λ-calculus with arrow types and cartesian products, mod an identification of types, allows us to see that this notion of reduction is Church-Rosser and strongly normalizing. A direct proof of this is given in the author's thesis [6]. The fact that the theory with integer types is a subsystem of the usual typed λ-calculus with arrow types and cartesian products also makes it clear that the stratified λ-calculus is very weak, considered as a programming language. It can be strengthened by the addition of a stratified iteration or recursion combinator (acting on Church numerals) (I thank a referee for bringing the need for this comment to my attention).

It is possible to replace the axiom scheme (Abst1) with finitely many instances of the scheme and obtain a system of synthetic combinatory logic, in which the variable binding operator is eliminated. We introduce a new term construction: if T is a term, K$[T]$ is a term (the constant function of T), and if K$[T]$ is assigned relative type n, T is assigned relative type $n-1$. New atomic terms Abst and Λ are introduced. The axioms used to replace (Abst1) are

Const: $\mathrm{K}[x](y) = x$
Abst2: $\mathrm{Abst}(f)(g)(x) = f(\mathrm{K}[x])(g(x))$
Lambda: $\Lambda(x)(y) = x(y); \Lambda(\Lambda(x)) = \Lambda(x); \Lambda(x,y) = (\Lambda(x), \Lambda(y))$
Ext2: if $f(x) = g(x)$ for all x, then $\Lambda(f) = \Lambda(g)$

The interpretation of K$[T]$ is obvious; Abst is a variation of the S combinator which respects stratification; Λ is the canonical retraction $(\lambda f)(\lambda x)(f(x))$ sending each object to a uniquely determined object with the same extension, for which special axioms are needed to preserve the weak extensionality implicit in the fact that substitutions for terms containing variables can be made into λ-terms. This synthetic combinatory logic also has a typed version with integer types as above. The notion of relative type for subterms of terms of the synthetic theory is defined as in Definition 2, except that the clause for λ-terms is replaced by a clause to the effect that if an occurrence of K$[U]$ in a term T is assigned type n relative to T, then the obvious occurrence of U is assigned relative type $n-1$.

In our paper [7], we proved the following abstraction theorem:

Theorem 4. *For each term T of the language of the synthetic theory and variable x which occurs in T with no relative type other than 0, there is a term "$(\lambda x)(T)$" in which the variable x does not occur such that $(\lambda x)(T)(x) = T$ is a theorem of the synthetic theory.*

Note that the term formation rules and axioms of the synthetic theory make no reference to relative type at all; relative type is introduced only in order to state the abstraction theorem above. This is analogous to the situation with the set theories: *NF* or *NFU* can be finitely axiomatized using finitely many instances of stratified comprehension (so not mentioning stratification in the definition of the theory) from which stratified comprehension can be proven as a meta-theorem. Such an axiomatization was first presented by Hailperin in [5]; we give a more accessible one in [9].

4 A Semantic Motivation for the Stratified λ-Calculus

We introduce an (extremely abstract) model of computation in terms of which we will give a semantic motivation for the stratification criterion for functional abstraction (it is worth noting that this is also an indirect semantic motivation for the set theories like *NF*, which are often regarded as motivated purely by a "syntactical trick").

Our abstract machine has an infinite set X of "addresses". There is a bijection pair:$X \times X \rightarrow X$: we use pair(x, y) to represent the pair of addresses (x, y). We refer to functions from X into X as "abstract programs"; the "state" of our machine is specified by a function state$X \rightarrow X^X$; state(x) may be thought of as the program stored in address x.

When x and y are addresses, we use (x, y) and $x(y)$ to abbreviate pair(x, y) and state$(x)(y)$, respectively. We introduce $(\lambda x)(T)$, for each term T and variable x, to represent an address (if there is such an address – an arbitrary one is chosen if there is more than one) such that $(\lambda x)(T)(x) = T$ for each address x. λ-terms can be thought of as program specifications.

We now consider the "program" $\Delta = (\lambda x)(x(x))$ as a sample of the kind of program which is ruled out by the stratification criterion for function abstraction. We suggest that there is an a priori reason to rule out any obligation to provide a program with specification Δ, based essentially on the notion of security of abstract data types. Moreover, this a priori reason, considered in full generality, motivates precisely the stratification criterion for function abstraction.

In our model of programming, programs are stored in and manipulated by reference to addresses. An abstract program is simply a function from X into X; but a program as implemented on our abstract machine has the additional feature of being stored in a particular address. Security of the data type "program" as represented on our machine requires that when we manipulate an address as representing a program, we use no information except information about its extension as a function. But the program Δ tells us to take the program stored in address x and apply it to address x; here we are expected to consider x both as a program and as the address in which that program is stored. A formal way to see that Δ violates the security of the program data type is to consider that a permutation of the assignments of programs to addresses could send the program stored in x to a new address y, and that the new value of $y(y)$ could differ from the old value of $x(x)$, although the new extension of y would be precisely the old extension of x. (A very precise result illuminating the relationship between stratification and permutations in a set-theoretical context is found in Forster's [4] (p. 87); similar reasoning is applied to typed λ-calculus in [3]) A preliminary version of the criterion which excludes program specifications like Δ would assert that if a parameter is used to represent an abstract program we should not have access to information about its identity as an address and vice versa.

The criterion can be sharpened by observing that there are more than two abstractions implemented by addresses on our machine: each address is an address and represents an abstract program via the state function; but one can also interpret the values of the abstract program represented by an address as abstract programs themselves, and thus interpret an address as a function from abstract programs to abstract programs. In this way we obtain a sequence of abstractions which are implemented by addresses on our machine precisely analogous to the sequence of types denoted by integers above; the objects at each level of abstraction are functions from the set of objects of the previous level of abstraction into itself.

The criterion restricting program specifications (λ-terms) is that each parameter in a specification should appear as representing an object at just one level of abstraction in this scheme; and this is precisely the criterion of stratification. To exploit the fact that the same address represents objects at two different levels of abstraction would be a violation of the security of the representations of the various levels of abstraction. The treatment of pairing has not been discussed, but it is precisely analogous to what is done above; it is technically tricky to see that axiom (Prod) and similar axioms are reasonable assumptions (that they are practicable is verified by our model construction above).

The failure of extensionality in the stratified λ-calculus as modelled above does not compromise the data type security argument. If we suppose that multiple addresses on our machine represent the same program, we must conclude that the proper identity criteria for addresses (considered as programs, as they are at all levels of abstraction except level 0) must be extensional; addresses are to be identified if they contain the same abstract program. We nonetheless end up with failure of extensionality when we attempt to interpret programs whose extensions do not respect this relation of identity of extension (which send diverse objects with the same extension to objects of different extensions); such programs must be assigned extensions which do respect the new identity criterion, presumably extensions already implemented by programs respecting the identity criteria, and will correspond to the many non-canonical programs with each extension found in the model described above. An example of a method for making such assignments would be to assign a given program a default value $\perp$ at each "value" (extension) at which it originally returned values with different extensions. The argument sketched in this paragraph is based on Marcel Crabbés proof that the theory *SC* with the axiom scheme of stratified comprehension alone interprets *NFU* (with its apparently stronger form of weak extensionality), given in [2].

As in the set theory "New Foundations" and its variants such as *NFU*, all objects are actually of the same kind, but abstraction (in set theory, set comprehension) is limited by a scheme of relative typing. The idea that all objects are ultimately of the same sort, but that they should be treated in abstractions (program specifications) as belonging to specific roles, is a very reasonable general outlook in computer science, where all objects, of whatever intended type, are in fact implemented as sequences of binary digits, but the type of the intended referent of an object controls the ways in which it can be manipulated in a program.

Of course, it is not necessary to use relative typing schemes in programming; it is not strictly necessary to acknowledge overtly that all objects are of one underlying sort. But it is often tempting, for it is often the case that an operation which is useful on one type is actually useful on objects of many related types or even of all types. A certain amount of polymorphism is implemented in our scheme; a very simple example is the presence of the identity function $(\lambda x)(x)$ which acts on all objects whatsoever. More useful further examples of polymorphic objects definable in stratified λ-calculus are the Church numerals "n" $= (\lambda f)(\lambda x)(f^n(x))$ for each natural number n. The polymorphism provided by this scheme is somewhat limited, however, by comparison with the kind of polymorphism provided in the second-order polymorphic λ-calculus, or, to give a concrete example, in the type system of the computer language *ML*. The motive behind polymorphic type systems is similar to the motive behind the attempt to adopt a relative typing scheme. We will discuss the implementation of data types usual to computer science below; we do not take them to be implemented by the very simple scheme of relative types used in the definition of stratification!

5 Interpreting Data Types in the Stratified λ-Calculus

The scheme of relative typing which underlies the definition of stratification is a very simple type structure, not adequate for the needs of computer science. We present a second notion of absolute type, essentially orthogonal to the notion of relative type used in stratification, which corresponds to a fundamental concept of set theory in the style of "New Foundations" which has no analogue in the usual set theory *ZFC*.

We briefly describe the analogous situation in set theory. Cantor's theorem in the usual set theory establishes that the power set (the collection of all subsets) of a set A must have a larger cardinality than A. If this argument is applied to the universal set V, we obtain "Cantor's paradox"; the power set of the universe, the collection of all sets, is shown to be larger than the universe! Of course, the conclusion to be drawn from this is that there is no such set V in the usual set theory. In *NF* and related set theories, there *is* a universal set, so this argument must break down somehow. The outcome is best understood by referring to the theory of types, where A and its power set are objects of different types which cannot be compared, and the result analogous to Cantor's theorem proves instead that the set of all one-element subsets of A (the set of singletons of elements of A) is smaller in cardinality than the power set of A. The rather odd result in *NF* and the related set theories is that the universe V is larger than the set of all singletons. The map $(\lambda x)(\{x\})$, which would provide a bijection between these two sets, is seen not to exist, which is not especially surprising, since its definition is unstratified.

But Cantor's theorem will hold in its original form for any set A which has the same cardinality as its image under the singleton operation. Such sets are said to be "Cantorian". A set A is said to be "strongly Cantorian" if the restriction of the singleton operation to A defines a function; such a set is certainly Cantorian. The properties "Cantorian" and "strongly Cantorian" are unstratified and do not define sets.

The "class" of strongly Cantorian sets includes all sets of standard finite size. If the assertion that the set of natural numbers is strongly Cantorian (it is provably Cantorian) is adopted as an axiom, then all finite sets are strongly Cantorian as well. This axiom, Rosser's Axiom of Counting, is known to be consistent with *NFU* (and strengthens it essentially); its analogue will hold in our model of stratified λ-calculus iff the automorphism j used in the construction fixes each natural number. The "class" of strongly Cantorian sets is closed under cartesian product, power set, and the construction of function spaces. Strongly Cantorian sets are "small" in relation to the big sets like V; it has been suggested that they are the analogues of the small sets studied in the usual set theory, whose criterion for sethood is "limitation of size". This position is strengthened by the fact that stratification restrictions can be subverted for variables restricted to strongly Cantorian sets; the type of any reference to such a variable can be shifted by replacing references to the object with references to its singleton (we will show how this is done in the context of stratified λ-calculus below). We have discussed in our paper [9] and our unpublished [10] how this analogy can be shown to be precise under suitable assumptions.

We suggest, and will motivate in terms of our abstract model of programming, the proposition that the notion "strongly Cantorian set" is a reasonable analogue in our scheme to the notion of "data type". We see above that the usual types of booleans and natural numbers can be accommodated, and that the class is closed under the most obvious type constructors.

We represent sets in our theory as ranges of retractions (maps t such that $t(t(x)) = t(x)$ for all x). The best analogue of the singleton map in our theory is the operation which takes each x to $K[x] = (\lambda y)(x)$, its constant function. A retraction t represents a strongly Cantorian set iff there is a map K_t such that $K_t(t(x)) = K[t(x)]$ for each x. Notice that the relative type of x can be changed by applying this equation, if we know that x belongs to the range of t (and so is fixed by t); this allows the subversion of the stratification restrictions for variables restricted to strongly Cantorian sets. For example, if t is a retraction whose range is strongly Cantorian, we can define $\Delta_t = (\lambda x)(t(x)(t(x)))$ as $(\lambda x)(K_t^{-1}(K[t(x)])(t(x)))$; the first λ-term requires definition because it is unstratified and so ill-formed; but the trick of replacing the occurrence of $t(x)$ at relative type 1 with $K_t^{-1}(K[t(x)])$ lowers the type of the first occurrence of x to 0, making the term stratified. An occurrence of $t(x)$ can be raised in type by replacing it with $K_t(t(x))(0)$ (the use of 0 is just as a convenient constant). The net effect is that variables with a type retraction t applied to them (effectively, variables restricted to the range of t) do not need to satisfy the stratification restrictions, if t is a retraction with strongly Cantorian domain.

What does the existence of K_t tell us about the range of t in the context of our abstract model of computation? It allows us to correlate $K_t(t(x))$, with $K[t(x)]$; if the occurrence of $t(x)$ is to be understood as taking the role of an address on the right, it must be understood as taking the role of an abstract program on the left. In other words, K_t is a representation as a function of the correlation of objects in the range of t considered as programs and as addresses, which is exactly what the stratification criterion operates to prevent in general. Metaphorically, K_t tells us exactly how the programs in the range of t are stored in the "memory" of our abstract machine; a class of objects which we know how to store in memory seems like a reasonable abstract description of the concept "data type". Concretely, we have seen above that commonly used data types are subsumed under this notion. Type constructors are represented by operations (not always functions) on retractions in a straightforward way. Notice that types themselves, being functions, are first-class objects in this scheme.

We are aware that there is nothing novel about the use of retractions to represent types; this is a very commonly used technical device. The (hoped-for) novelty is all in the use of the notion of "strongly Cantorian set", in the guise "retraction with strongly Cantorian range".

We exhibit the construction of arrow types. Given a retraction s and a retraction t, we can construct a retraction $s \rightarrow t$ whose domain is the set of functions from the domain of s to the domain of t easily: $s \rightarrow t = (\lambda f)(t \circ f \circ s)$ (composition of functions is a stratified construction). An important thing to note is that the relative type of $s \rightarrow t$ is one higher than that of s and t; thus, the arrow type constructor is not a function (the cartesian product constructor does turn out to be a function). This construction is not presented as being original in any way; we are simply pointing out its features in the context of stratified λ-calculus.

We have investigated the representation of more complex type systems, including the second-order polymorphic λ-calculus mentioned above. To achieve polymorphism of this kind, in which objects can be constructed with type parameters, it is necessary to have an actual strongly Cantorian set of types to which we restrict ourselves; the collection of all absolute types (strongly Cantorian sets) is not a set at all, much less a strongly Cantorian set, so cannot be an absolute type. One reason that the set of types in such a scheme must be strongly Cantorian is that the arrow type constructor is type-raising; on a strongly Cantorian set of types, it can be represented by a function, but not on the whole domain of types. It seems more natural in this context to have a hierarchy of ever more inclusive notions of type, analogous to the universes of Martin-Löf type theory (though certainly different in detail).

We think that it is interesting that the notion of "strongly Cantorian set", peculiar to the Quine-style set theories, may turn out to have a useful role in our extension of this set theoretical paradigm to computer science. We also think that it is interesting to see the notion of type break into two orthogonal parts, the relative type scheme restricting the formation of global abstractions and the representation of data types as local domains on which we have additional information which allows us to subvert these global restrictions.

6 The Use of these Concepts in a Theorem Proving Project Sketched

We now discuss the ways in which we have used these concepts in the actual implementation of an equational theorem prover (a technical report on this project should be available from the author fairly shortly). This project was originally not intended to implement stratified λ-calculus; but it turned out that implementing stratification was the easiest way to modify an earlier version of the prover to support higher-order reasoning.

The prover we have implemented avoids the use of bound variables; there are no λ-terms or other variable-binding operations. The situation where stratification is applied is in definitions of functions and operators. The syntax of the input language allows unary prefix and binary infix operators. Each operator is assigned a relative type for each of its arguments (only if these are zero is the operator actually a function). For example, the operator @ (function application) has relative type 1 for its first argument and relative type 0 for its second argument. If the user attempts to define a function Δ by the equation $\Delta @ x = x @ x$, the prover will attempt to type all subterms of the expression and discover that x cannot be consistently assigned the same type everywhere, so the definition will be rejected: if the term $x @ x$ is assigned type 0, we must assign type 1 to the first occurrence of x and type 0 to the second occurrence of x, and when the same variable must be assigned two different types the stratification procedure fails. It is also possible to define new prefixes and infixes which have different relative types from their arguments (and so are not genuine functions); the operators which form singletons and constant functions or the arrow type constructor operating on retractions would be examples. Their definitions also require attention to stratification (the technical term for such operators is "stratified but inhomogeneous").

An amusing fact about the implementation of stratification is that it is a matching operation. A usual application of matching checks whether each variable in one term is structurally correlated with the same subterm of another term, checking for parallelism of structure and consistency of assignments to the same variable everywhere in the term. The stratification checker in the prover assigns an integer to each subterm of a term (this always succeeds), but needs to check further whether the assignments of integers to occurrences of variables are consistent; it shares essential code with the matching functions of the prover. Thomas Forster has made the same observation (oral communication).

If only stratified equational axioms are used, the prover can be understood as working in the typed theory described above, with a system of type inference replacing explicit typing, which allows polymorphic function definitions introducing the defined function in all types simultaneously; this parallels the mathematical practice of those who work in Russell's type theory. However, the absence of explicit typing makes it tempting to introduce unstratified axioms (an example would be the defining equation of any K_t for a retraction t with strongly Cantorian domain). The subversion of the type system by introducing K_t's has been seen above to be useful in allowing more general function abstraction for terms with variables restricted by application of retractions to strongly Cantorian domains; once unstratified equations are introduced, we are working in the stratified λ-calculus (given the absence of bound variables, we may to be said to be working in the equivalent synthetic system).

Automatic features of the prover allow the absolute typing scheme using retractions to be largely inferred rather than explicit as well; occurrences of retractions can be removed where unnecessary (where their presence can be restored by a type inference) automatically and reintroduced equally automatically when needed; they can be treated as if they were type labels in appropriate contexts. Type inference for the usual constructed types can be implemented in a straightforward fashion. Since types in the sense of strongly Cantorian domains are represented by retractions, they are first-class objects in this system, which allows definition of new type constructors and implementation of extended type inference schemes by the user.

7 Acknowledgements

I appreciate the financial support of the U. S. Army Research Office for my automated theorem proving research. I appreciate the comments of the anonymous referees of this paper, whose specific concerns I have addressed in several places. A general concern raised by more than one referee is addressed here: the motive of this paper is to suggest that results originally proved in the completely theoretical area of *NF*-style set theory may have practical applications (or at least, applications closer to practical considerations); in this kind of paper it is not surprising that the main results are proved elsewhere and also that there is perhaps too much philosophical speculation of a vague nature. The aim of bringing these matters to the attention of a more practically minded community is to diminish this vagueness, and I thank the referees for helping me to make this possible.

References

1. Boffa, M. "*ZFJ* and the consistency problem for *NF*", in *Jahrbuch der Kurt Gödel Gesellschaft*, 1988, pp. 102-6.
2. Crabbé, M. On *NFU*. *Notre Dame Journal of Formal Logic*, vol. 33 (1992), pp. 112-119.
3. Forster, T. E. "A semantic characterization of the well-typed formulae of λ-calculus". *Theoretical Computer Science*, vol. 110 (1993), pp. 405-418.
4. Forster, T. E. *Set theory with a universal set*, Oxford logic guides no. 20. Clarendon Press, Oxford, 1992.
5. Hailperin, T. "A set of axioms for logic". *Journal of Symbolic Logic*, vol. 9 (1944), pp. 1-19.
6. Holmes, M. R. "Systems of combinatory logic related to Quine's 'New Foundations'", Ph.D. thesis, State University of New York at Binghamton, 1990.
7. Holmes, M. R. "Systems of combinatory logic related to Quine's 'New Foundations'". *Annals of Pure and Applied Logic*, vol. 53 (1991), pp. 103-133.
8. Holmes, M. R. "Systems of combinatory logic related to predicative and 'mildly impredicative' fragments of Quine's 'New Foundations'". *Annals of Pure and Applied Logic*, vol. 59 (1993), pp. 45-53.

9. Holmes, M. R. "The set-theoretical program of Quine succeeded, but nobody noticed". *Modern Logic*, vol. 4, no. 1 (1994), pp. 1-47.
10. Holmes, M. R. "Strong axioms of infinity in *NFU*", preprint.
11. Jensen, R. B. "On the consistency of a slight (?) modification of Quine's *NF*". *Synthese*, vol. 19 (1969), pp. 250-63.
12. Rosser, J. B. *Logic for Mathematicians*. McGraw-Hill, reprinted (with appendices) by Chelsea, New York, 1978.
13. Scott, Dana. "Continuous lattices", in *Springer Lecture Notes in Mathematics*, no. 274, pp. 97-136.
14. Wang, H. "Negative types". *Mind*, vol. 61 (1952), pp. 366-8.

Final Semantics for untyped λ-calculus*

Furio Honsell[1] and Marina Lenisa[1,2]

[1] Dipartimento di Matematica e Informatica
Università di Udine, Italy.
{honsell,lenisa}@dimi.uniud.it
[2] Dipartimento di Informatica
Università di Pisa, Italy.
lenisa@di.unipi.it

Abstract. *Proof principles* for reasoning about various semantics of untyped λ-calculus are discussed. The semantics are determined operationally by fixing a particular *reduction strategy* on λ-terms and a suitable set of *values*, and by taking the corresponding *observational equivalence* on terms. These principles arise naturally as co-induction principles, when the observational equivalences are shown to be induced by the *unique* mapping into a *final F-coalgebra*, for a suitable functor F. This is achieved either by *induction on computation steps* or exploiting the properties of some, computationally adequate, inverse limit denotational model. The final F-coalgebras cannot be given, in general, the structure of a "denotational" λ-model. Nevertheless the "final semantics" can count as *compositional* in that it induces a *congruence*. We utilize the intuitive categorical setting of *hypersets* and functions. The importance of the principles introduced in this paper lies in the fact that they often allow to factorize the complexity of proofs (of observational equivalence) by "straight" induction on computation steps, which are usually lengthy and error-prone.

Introduction

In this paper we present various proof-principles for reasoning about various semantics of λ-calculus which arise in the literature.

The word *semantics* is often used ambiguously. In fact, it can either refer to an *interpretation function* mapping terms into a space of denotations, or to a *procedure* for determining an equivalence relation, or to an *equivalence relation* on terms *per se* (e.g. the one induced by an interpretation function). We shall be careful in distinguishing between semantics and semantical equivalence. The proof principles that we introduce in this paper will be, therefore, rules for establishing semantical equivalence.

All the semantics that we consider originate operationally by fixing a suitable reduction strategy $\to_\sigma$ on λ-terms (denoted by Λ), or closed λ-terms (denoted

* Work supported by EEC Science contract MASK, HCM contract "Lambda Calcul Typé" and MURST 40% and 60% grants.

by Λ^0). A reduction strategy is a procedure for determining, for each term, a suitable β-redex appearing in it, to be contracted. A (possibly non-deterministic) strategy can be formalized as a relation $\rightarrow_\sigma \subseteq \Lambda \times \Lambda$ ($\Lambda^0 \times \Lambda^0$) such that, if $(M, N) \in \rightarrow_\sigma$ (also written infix as $M \rightarrow_\sigma N$), then N is a possible result of applying the reduction strategy $\rightarrow_\sigma$ to M. Reduction ***paths*** are defined, as usual, by repeatedly applying the reduction strategy (possibly zero times). The set of terms which do not belong to the domain of $\rightarrow_\sigma$ are partitioned into two disjoint sets: the set of ***σ-values***, denoted by Val_σ and the set of ***σ-deadlocks***.

Given a fixed strategy $\rightarrow_\sigma$, we can define the ***evaluation relation*** $\Downarrow_\sigma \subseteq \Lambda \times \Lambda$ ($\Lambda^0 \times \Lambda^0$), such that $M \Downarrow_\sigma N$ holds if and only if there exists a reduction path leading from M to N and N is a σ-value. If there exists N such that $M \Downarrow_\sigma N$, we shall say that the reduction $\rightarrow_\sigma$ ***halts successfully*** (terminates) on N, otherwise we shall write $M \not\Downarrow_\sigma$ and say that the reduction strategy $\rightarrow_\sigma$ ***diverges*** on M or reaches a ***deadlock*** from M. Often, the relation $\Downarrow_\sigma$ can be axiomatized pleasantly using Plotkin's S.O.S. style (see [16]).

Once we have a reduction strategy we may well say that we have an operational semantics, for we can imagine a machine which evaluates terms by implementing the given strategy. A natural semantical equivalence to define on terms is the ***observational equivalence***. This arises if we consider programs as ***black boxes*** and take them to be equivalent if we cannot tell them apart by ***observing*** that for a given ***program context*** the machine halts successfully when one is used as a subprogram but does not halt when the other is used as a subprogram.

Definition 1 (σ-observational Equivalence). Let $\rightarrow_\sigma$ be a reduction strategy and let M, N be programs, i.e. closed λ-terms. The ***observational equivalence*** $\approx_\sigma$ is defined by

$$M \approx_\sigma N \;\text{ iff } \forall C[\,].(C[M], C[N] \in \Lambda^0 \Rightarrow (C[M] \Downarrow_\sigma \Leftrightarrow C[N] \Downarrow_\sigma)).$$

Notice that σ-observational equivalence is a congruence for every σ.

In this paper we consider many different reduction strategies, and corresponding observational equivalences, appearing in the literature. These strategies include those realized by the two most widespread implementations of functional programming languages, i.e.***call-by-value*** and ***lazy***. We shall focus in particular on the former. This was defined by Plotkin in [15] and is the one implemented by Landin's SECD machine or by the FAM. The observational equivalence induced by this strategy has been investigated in [6]. The latter strategy, which is the one used by lazy functional languages, induces an observational equivalence which has been studied in [1, 13]. Other strategies considered in the paper are:
- the "initial" leftmost strategy which stops on ***head normal forms***; its observational equivalence is induced by the canonical D_∞ model of Scott (see [20]);
- the ***β-normalizing*** strategy, i.e. the"complete" leftmost strategy, whose observational equivalence, we conjecture coincides with the equivalence induced by the model defined in [5];
- the "restricted" leftmost strategy which reduces the leftmost β-redex whose operand is in normal form and which can be shown to terminate exactly on

strongly normalizing terms; we conjecture that its observational equivalence is exactly the equivalence induced by the model defined in [10] for the λN^0-calculus;
- a non-deterministic strategy with a non-trivial set of deadlock terms, which induces as observational equivalence the congruence which cannot be realized by a continuously complete C.P.O. model (see [11]).

The importance of establishing observational equivalence of terms is unquestionable, e.g. it is at the core of the software development methodology of "program transformation". However, showing equivalences by induction on computation steps, is a lengthy and error-prone activity. Powerful proof-principles, which allow to factorize this difficult task, are therefore very precious.

In recent years, much attention has been devoted to so-called "coinduction principles", see e.g. [14, 7]. These concepts arose independently in two unrelated areas: that of semantics for concurrent languages [12] and that of non-wellfounded sets [8]. Aczel made the connection and pioneered their categorical account in terms of *final coalgebras* (see [2, 3]). Recently, various authors have deeply investigated *final semantics* (see e.g. [18, 7, 19]). They have developed a general categorical methodology for deriving coinduction principles whenever the semantical equivalence is induced by a final semantics. In this paper we apply this methodology to various semantics of λ-calculus. We proceed as follows. Given a reduction strategy $\rightarrow_\sigma$ and a set Val_σ, we *endow* Λ^0 (possibly in various ways), with the structure of a F-coalgebra, for suitable functors F. These functors are then proved to be "well behaved" in the sense that they admit a final F-coalgebra and preserve "weak kernel pairs" (see Appendix A and [18] for categorical definitions). Hence the *final interpretation given* by the unique morphism into the final F-coalgebra *induces* an equivalence which can be characterized by a coinduction principle, i.e. as the union of all F-bisimulations. Finally we are left with the task of showing that this equivalence coincides with the observational equivalence *determined* by the original reduction strategy.

The final F-coalgebra cannot be given, in general, the usual structure of a "denotational" λ-model. Nevertheless the final semantics can count as *compositional* in that the equivalence induced by it is a *congruence*, w.r.t. the syntactical constructors of the language. Actually, this is precisely what implies, in all the cases discussed here, that the final semantics induces the observational equivalence. Technically, it is quite hard to show that the final semantics induces a congruence. We prove it either by *induction on computation steps* or exploiting some property of the observational equivalence itself, which, in turn, can be proved exploiting the properties of some, computationally adequate, inverse limit denotational model.

We could have considered various categorical settings, such as that of complete metric spaces or complete partial orders. We prefer to utilize that of *hypersets* and functions, because this categorical setting is natural and it allows to keep the naive set-theoretic intuition. Hypersets are *non-wellfounded sets* belonging to a Universe of $ZF_0^-(U)FCU$. This is a Zermelo Frænkel like set-theory with extensionality "up to" *Urelementen* and with the Axiom of Foundation replaced by the "unique" antifoundation axiom FCU, which generalizes the antifoundation axiom X_1 introduced by Forti and Honsell in 1982 (also called AFA by Aczel [2]).

The paper is organized as follows. In Section 1 we present the reduction strategies and the corresponding evaluation relations that we shall deal with. In Section 2, we apply the final methodology. In particular we introduce, for each reduction strategy, a suitable functor and we derive a sound coinduction principle. In Section 3 we focus on the call-by-value observational equivalence and, applying again the final methodology, we derive yet ***another*** coinduction principle. In Section 4 we present an "induction-coinduction" principle for the call-by-value observational equivalence inspired by [14]. Final remarks, comments and conjectures appear in Section 5. In Appendix A we recall some useful definitions.

The authors are grateful to Mariangiola Dezani, Gordon Plotkin, Simona Ronchi Della Rocca, Jan Rutten and Daniele Turi for useful discussions.

1 Operational Semantics

Throughout the paper we use standard λ-calculus concepts and notation as defined in [4].

In this Section we present six operational semantics for λ-calculus. More precisely we give six reduction strategies together with the corresponding evaluation relations. The latter are presented using Plotkin's S.O.S. style, (cfr. [16]).

Definition 2 ($\rightarrow_v$ strategy, $\Downarrow_v$ evaluation). The *lazy call-by-value* strategy $\rightarrow_v \subseteq \Lambda^0 \times \Lambda^0$ reduces the *leftmost* β-redex, not appearing within a λ-abstraction, whose argument is a λ-abstraction. $Val_v = \{\lambda x.M \mid M \in \Lambda\} \cap \Lambda^0$. The evaluation $\Downarrow_v$ is the least binary relation over $\Lambda^0 \times Val_v$ satisfying the following rules:

$$\frac{}{\lambda x.M \Downarrow_v \lambda x.M} \qquad \frac{M \Downarrow_v \lambda x.P \quad N \Downarrow_v Q \quad P[Q/x] \Downarrow_v U}{MN \Downarrow_v U}$$

Definition 3 ($\rightarrow_l$ strategy, $\Downarrow_l$ evaluation). The *lazy call-by-name* strategy $\rightarrow_l \subseteq \Lambda^0 \times \Lambda^0$ reduces the *leftmost* β-redex not appearing within a λ-abstraction. $Val_l = \{\lambda x.M \mid M \in \Lambda\} \cap \Lambda^0$. The evaluation $\Downarrow_l$ is the least binary relation over $\Lambda^0 \times Val_l$ satisfying the following rules:

$$\frac{}{\lambda x.M \Downarrow_l \lambda x.M} \qquad \frac{M \Downarrow_l \lambda x.P \quad P[N/x] \Downarrow_l Q}{MN \Downarrow_l Q}$$

Definition 4 ($\rightarrow_h$ strategy, $\Downarrow_h$ evaluation). The *eager call-by-name* strategy $\rightarrow_h \subseteq \Lambda \times \Lambda$ reduces the *leftmost* β-redex, if the term is not in head normal form. Val_h is the set of λ-terms in head normal form. The evaluation $\Downarrow_h$ is the least binary relation over $\Lambda \times Val_h$ satisfying the following rules:

$$\frac{}{xM_1 \ldots M_n \Downarrow_h xM_1 \ldots M_n}\ n \geq 0 \qquad \frac{M \Downarrow_h N}{\lambda x.M \Downarrow_h \lambda x.N}$$

$$\frac{M[N/x]M_1 \ldots M_n \Downarrow_h P}{(\lambda x.M)NM_1 \ldots M_n \Downarrow_h P}\ n \geq 0$$

Definition 5 ($\rightarrow_n$ strategy, $\Downarrow_n$ evaluation). The *normalizing* strategy $\rightarrow_n \subseteq \Lambda \times \Lambda$ reduces the leftmost β-redex. Val_n is the set of λ-terms in normal form. The evaluation $\Downarrow_n$ is the least binary relation over $\Lambda \times Val_n$ satisfying the following rules:

$$\frac{M_1 \Downarrow_n M_1' \quad \ldots \quad M_n \Downarrow_n M_n'}{xM_1 \ldots M_n \Downarrow_n xM_1' \ldots M_n'}\; n \geq 0 \qquad\qquad \frac{M \Downarrow_n N}{\lambda x.M \Downarrow_n \lambda x.N}$$

$$\frac{M[N/x]M_1 \ldots M_n \Downarrow_n P}{(\lambda x.M)NM_1 \ldots M_n \Downarrow_n P}\; n \geq 0$$

Definition 6 ($\rightarrow_i$ strategy, $\Downarrow_i$ evaluation). The *normalizing call-by-value* strategy $\rightarrow_i \subseteq \Lambda \times \Lambda$ reduces the leftmost β-redex whose argument is a normal form. Val_i is the set of λ-terms in normal form. The evaluation $\Downarrow_i$ is the least binary relation over $\Lambda \times Val_i$ satisfying the following rules:

$$\frac{M_1 \Downarrow_i M_1' \quad \ldots \quad M_n \Downarrow_i M_n'}{xM_1 \ldots M_n \Downarrow_i xM_1' \ldots M_n'}\; n \geq 0 \qquad\qquad \frac{M \Downarrow_i N}{\lambda x.M \Downarrow_i \lambda x.N}$$

$$\frac{N \Downarrow_i P \quad M[P/x]M_1 \ldots M_n \Downarrow_i V}{(\lambda x.M)NM_1 \ldots M_n \Downarrow_i V}\; n \geq 0$$

Definition 7 ($\rightarrow_e$ strategy, $\Downarrow_e$ evaluation). A closed term $\lambda x.M$ is said to be an "eraser" if $x \notin FV(M)$. The non-deterministic strategy $\rightarrow_e \subseteq \Lambda^0 \times \Lambda^0$ rewrites closed λ-terms which are not erasers by reducing any β-redex. $Val_e = \{M \in \Lambda^0 \mid M \text{ is an eraser}\}$. Normal forms which are not erasers are the $\rightarrow_e$-deadlock terms. The evaluation relation $\Downarrow_e$ is the least binary relation over $\Lambda^0 \times Val_e$ satisfying the following rules:

$$\frac{M \in Val_e}{M \Downarrow_e M} \qquad\qquad \frac{C[(\lambda x.M)N] \notin Val_e \quad C[M[N/x]] \Downarrow_e P}{C[(\lambda x.M)N] \Downarrow_e P}$$

2 Final Descriptions of observational Equivalences

In this section we give a first series of "final" accounts of the observational equivalences induced by the evaluation relations defined in the previous section. Each of these accounts gives rise to a particular coinductive characterization of the observational equivalence under consideration.

We work in the category *Class* whose objects are the classes of non-well-founded sets belonging to a Universe of $ZF_0^-(U)FCU$, and the arrows are the functional classes. The theory $ZF_0^-(U)FCU$ is a Zermelo-Frænkel-like set theory with extensionality, "up to" the proper class U of Urelementen (atoms), and with the axiom of Foundation replaced by the anti-foundation axiom FCU of [9] (see Definition 23 in Appendix A). This axiom is the version of X_1 (AFA) "up to" Urelementen, see [8, 2]. The Axiom FCU implies that the Universe is *strongly extensional*, i.e. that sets are unique "up to" bisimulations which preserve atoms (see Definition 23 in Appendix A, and [8, 2, 9] for more details). Alternatively we could have used the category $Class^*$ of [18], the category of $C.P.O.$'s and *strict functions*, or the category of *complete metric spaces* and *non-distance increasing functions*.

Given an evaluation relation $\Downarrow_\sigma$, we will proceed uniformly as follows (see [18]):

1. we endow the set Λ^0 with a structure of F-coalgebra, for a suitable functor $F : Class^* \to Class^*$;
2. we prove that the functor has a final F-coalgebra;
3. we define the interpretation function $\mathcal{M}_\sigma$ as the unique F-coalgebra morphism from the F-coalgebra on Λ^0 into the final F-coalgebra;
4. we prove that the the equivalence induced by $\mathcal{M}_\sigma$ is given by the union of all F-bisimulations on the F-coalgebra on Λ^0;
5. we prove that the equivalence induced by $\mathcal{M}_\sigma$ is $\approx_\sigma$.

The steps 1 and 5 above are motivated and simplified if we introduce and discuss the notion of *applicative equivalence*.

Definition 8. Let $\approx_\sigma^{app} \subseteq \Lambda^0 \times \Lambda^0$, for $\sigma \in \{v, l, h, n, i, e\}$, be the *applicative equivalence* defined by

$$M \approx_\sigma^{app} N \iff \forall P_1, \ldots, P_n \in \Lambda^0. \, (M P_1 \ldots P_n \Downarrow_\sigma \Leftrightarrow N P_1 \ldots P_n \Downarrow_\sigma) \, .$$

In general $\approx_\sigma^{app}$ is not a congruence, but for all $\sigma \in \{v, l, h, n, e\}$, we can prove that $\approx_\sigma^{app}$ coincides with $\approx_\sigma$ and hence it is a congruence.[3] The proofs of these facts will be outlined in the sequel.

The relation $\approx_\sigma^{app}$, for $\sigma \in \{v, l, h, n, i, e\}$ can be characterized coinductively as the greatest fixed point of a suitable monotone operator. Namely:

Definition 9. Let X be a set and $\Psi : \mathcal{P}(X \times X) \to \mathcal{P}(X \times X)$ be an operator. A *Ψ-bisimulation* is a relation $R \subseteq X \times X$ s.t. $R \subseteq \Psi(R)$.

If Ψ is monotone, then the greatest fixed point of Ψ is the greatest Ψ-bisimulation.

Lemma 1. *The applicative equivalence $\approx_\sigma^{app}$, for $\sigma \in \{v, l, h, n, i, e\}$, can be viewed as the greatest fixed point of the monotone operator $\Psi_\sigma : \mathcal{P}(\Lambda^0 \times \Lambda^0) \to \mathcal{P}(\Lambda^0 \times \Lambda^0)$ defined by*

$$\begin{aligned}\Psi_\sigma(R) = \{(M, N) \mid (&M \not\Downarrow_\sigma \ \wedge N \not\Downarrow_\sigma \ \wedge \forall P \in \Lambda^0. \, ((MP, NP) \in R)) \vee \\ (&M \Downarrow_\sigma \ \wedge N \Downarrow_\sigma \ \wedge \forall P \in \Lambda^0. \, ((MP, NP) \in R))\}.\end{aligned}$$

Proof. Let $\nu R.\Psi_\sigma(R)$ be the greatest fixed point of Ψ_σ. It is immediate to show that $\approx_\sigma^{app}$ is a Ψ_σ-bisimulation, hence $\approx_\sigma^{app} \subseteq \nu R.\Psi_\sigma(R)$. In order to show the converse, i.e. $\approx_\sigma^{app} \supseteq \nu R.\Psi_\sigma(R)$, we prove first, by induction on the length of $\overrightarrow{P}$ ($\overrightarrow{P}$ abbreviates $P_1 \ldots P_n$ for $n \geq 0$), that:
$(M, N) \in \nu R.\Psi_\sigma(R) \implies (M\overrightarrow{P}, N\overrightarrow{P}) \in \nu R.\Psi_\sigma(R)$.
Hence, reasoning by contradiction, we get $\approx_\sigma^{app} \supseteq \nu R.\Psi_\sigma(R)$.

□

Notation. Throughout the paper we will denote by $X \to Y$ the class of all functions defined on X taking values in Y; and we will denote by $X + Y$ the "disjoint sum" of X and Y, e.g. $\{v\} \times X \cup \{u\} \times Y$, where v and u are two distinct "fresh" atoms.

[3] We conjecture that the same holds also for $\sigma = i$.

2.1 A final Description of $\approx_v$ in the Hyperset Setting

The set Λ^0 can be endowed with a coalgebra structure appropriate for dealing with $\approx_v$ as follows:

Definition 10. i) Let $F_v : Class \to Class$ be the endofunctor defined by

$$F_v(X) = (\Lambda^0 \to X) + \{*\} ,$$

where $*$ is a generic atom; the definition of F_v on morphisms is canonical.
ii) Let $(\Lambda^0,\ \alpha_v)$ be the F_v-coalgebra defined by

$$\alpha_v(M) = \begin{cases} (u, *) & \text{if } M \not\Downarrow_v \\ (v, \{(N, MN) \mid N \in \Lambda^0\}) & \text{if } M \Downarrow_v . \end{cases}$$

Lemma 2. *The functor F_v has a greatest fixed point $\overline{X_v}$ such that $(\overline{X_v},\ id)$ is a final F_v-coalgebra.*

Proof. One can easily extend the "Special Final Coalgebra Theorem" (see [2] and Corollary 4.23 of [18]) to $ZF_0^-(U)FCU$. The functor F_v can be easily seen to satisfy the appropriate generalizations, to $ZF_0^-(U)FCU$, of the hypotheses of the above theorem: i.e. it is set-continuous, inclusion preserving and uniform on maps.

□

Definition 11. Let $\mathcal{M}_v : \Lambda^0 \to \overline{X_v}$ be the unique F_v-morphism from the F_v-coalgebra $(\Lambda^0,\ \alpha_v)$ to the F_v-coalgebra $(\overline{X_v},\ id)$, i.e.:

$$\mathcal{M}_v(M) = \begin{cases} (u, *) & \text{if } M \not\Downarrow_v \\ (v, \{(N,\ \mathcal{M}_v(MN)) \mid N \in \Lambda^0\}) & \text{if } M \Downarrow_v . \end{cases}$$

The following lemma can be proved straightforwardly.

Lemma 3. *R is a Ψ_v-bisimulation if and only if R is a F_v-bisimulation*

Lemma 4. *Let $M, N \in \Lambda^0$. Then:*

$$\mathcal{M}_v(M) = \mathcal{M}_v(N) \iff M \approx_v^{app} N .$$

Proof. One can easily see that F_v weakly preserves kernel pairs, and hence, by Corollary 3.9 of [18], the equivalence induced by the final morphism is the greatest F_v-bisimulation. Now, the thesis follows immediately from Lemma 3.

□

Lemma 5 (Theorem 33 of [6]). $\approx_v = \approx_v^{app}$

Using the above lemmata we can establish the validity of the following proof principle:

Theorem 6. *Let $M, N \in \Lambda^0$, then the following coinduction principle holds:*

$$\frac{(M,N) \in R \quad R \text{ is a } F_v\text{-bisimulation}}{M \approx_v N}$$

Remark. A more general coinduction principle can be given introducing the notion of Ψ_v-bisimulation up to $\approx_v$ following [12]. Since for any R, S, if $R \subseteq \Psi_v(R_1)$ and $S \subseteq \Psi_v(S_1)$ then $R \circ S \subseteq \Psi_v(R_1 \circ S_1)$, the following principle holds

$$\frac{(M,N) \in R \quad R \subseteq \Psi_v(\approx_v \circ R \circ \approx_v)}{M \approx_v N}$$

2.2 A final Description of $\approx_l$ in the Hyperset Setting

The set Λ^0 can be endowed with a coalgebra structure appropriate for dealing with $\approx_l$ as follows:

Definition 12. i) Let $F_l : Class \to Class$ be the endofunctor defined by

$$F_l(X) = (\Lambda^0 \to X) + \{*\} ,$$

where $*$ is a generic atom; the definition of F_l on morphisms is canonical.
ii) Let $(\Lambda^0, \ \alpha_l)$ be the F_l-coalgebra defined by

$$\alpha_l(M) = \begin{cases} (u, *) & \text{if } M \not\Downarrow_l \\ (v, \{(N, MN) \mid N \in \Lambda^0\}) & \text{if } M \Downarrow_l . \end{cases}$$

Following the same lines of reasoning as in Section 2.1, provided we prove $\approx_l = \approx_l^{app}$, we can eventually show that:

Theorem 7. *Let $M, N \in \Lambda^0$.*
i) Then

$$\mathcal{M}_l(M) = \mathcal{M}_l(N) \iff M \approx_l N .$$

ii) The following coinduction principle holds:

$$\frac{(M,N) \in R \quad R \text{ is a } F_l\text{-bisimulation}}{M \approx_l N}$$

The coincidence of the observational equivalence with the applicative equivalence can be proved in various ways (see e.g. [1]). Here we give a syntactical proof, similar to one of those in [1].

Lemma 8. $\approx_l = \approx_l^{app}$

Proof. Clearly $\approx_l \subseteq \approx_l^{app}$. In order to show the converse, we proceed by induction on computation steps. Suppose by contradiction that there is a context $C[\,]$ such that $C[P] \Downarrow_l$ and $C[P'] \not\Downarrow_l$. Choose a context $C_{min}[\,]$ satisfying the property above such that the length of a path starting from $C_{min}[P]$, converging to a value, is minimal. Since $C_{min}[P] \Downarrow_l$ and $C_{min}[P'] \not\Downarrow_l$, an occurrence of P must necessarily appear as the head of a term in the $\to_l$-reduction path leading from $C_{min}[P]$ to a value. Consider the first time in which an occurrence of P appears in the head in the reduction starting from $C_{min}[P]$. Then we have $C_{min}[P] \to_l^*$

$PC'[P]$ and $C_{min}[P'] \rightarrow_l^* P'C'[P']$, for some context $C'[\,]$. By definition of applicative equivalence, we have $P'C'[P'] \approx_l^{app} PC'[P']$. Now $P \Downarrow_l$, otherwise we have immediately a contradiction. Therefore, suppose $P \Downarrow_l \lambda x.M$, we have
$PC'[P] \rightarrow_l^* (\lambda x.M)C'[P] \rightarrow_l M[C'[P]/x] = C''[P] \Downarrow_l$
$PC'[P'] \rightarrow_l^* (\lambda x.M)C'[P'] \rightarrow_l M[C'[P']/x] = C''[P'] \not\Downarrow_l$,
for some context $C''[\,]$ such that $C''[P]$ converges to a value with a path whose length is strictly less than the length of the converging path of $C_{min}[P]$.

□

2.3 A final Description of $\approx_h$ in the Hyperset Setting

The set Λ^0 can be endowed with a coalgebra structure appropriate for dealing with $\approx_h$ as follows:

Definition 13. i) Let $F_h : Class \rightarrow Class$ be the endofunctor defined by

$$F_h(X) = (\Lambda^0 \rightarrow X) + \{*\}\,,$$

where $*$ is a generic atom; the definition of F_h on morphisms is canonical.
ii) Let $(\Lambda^0,\ \alpha_h)$ be the F_h-coalgebra defined by

$$\alpha_h(M) = \begin{cases} (u, *) & \text{if } M \not\Downarrow_h \\ (v, \{(N, MN) \mid N \in \Lambda^0\}) & \text{if } M \Downarrow_h\,. \end{cases}$$

Following the same lines of reasoning as in Section 2.1, we can show:

Theorem 9. *Let* $M, N \in \Lambda^0$.
i) Then

$$\mathcal{M}_h(M) = \mathcal{M}_h(N) \iff M \approx_h N\,.$$

ii) The following coinduction principle holds:

$$\frac{(M,N) \in R \quad R \textit{ is a } F_h\textit{-bisimulation}}{M \approx_h N}$$

The proofs of the lemmata necessary for showing Theorem 9 are similar to those in Section 2.1, but for:

Lemma 10. $\approx_h = \approx_h^{app}$

A proof of this lemma can be achieved along the lines of the corresponding lemma in Section 2.2, extending the notion of $\approx_h^{app}$ to open terms. It can be also obtained using Wadsworth's extension of Böhm's "separability" Theorem (see [20]).

2.4 A final Description of $\approx_n$ in the Hyperset Setting

The $\approx_n$ equivalence does not equate all $\rightarrow_n$-divergent terms, in fact it is not true that a divergent term applied to any argument always diverges. Hence, in order to define a non well-founded final semantics which induces the $\approx_n$ equivalence, we consider the following functor and coalgebra on λ-terms, which allow us to distinguish correctly between divergent λ-terms whose applicative behavior is different:

Definition 14. i) The endofunctor $F_n : Class \rightarrow Class$ is defined by

$$F_n(X) = (\Lambda^0 \rightarrow X) + (\Lambda^0 \rightarrow X) ,$$

the definition of F_n on morphisms is canonical.
ii) Let $(\Lambda^0,\ \alpha_n)$ be the F_n-coalgebra defined by

$$\alpha_n(M) = \begin{cases} (u, \{(N, MN) \mid N \in \Lambda^0\}) & \text{if } M \not\Downarrow_n \\ (v, \{(N, MN) \mid N \in \Lambda^0\}) & \text{if } M \Downarrow_n . \end{cases}$$

Following the same lines of reasoning as in Section 2.1, we can show:

Theorem 11. *Let $M, N \in \Lambda^0$.*
i) Then

$$\mathcal{M}_n(M) = \mathcal{M}_n(N) \iff M \approx_n N .$$

ii) The following coinduction principle holds:

$$\frac{(M,N) \in R \quad R \text{ is a } F_n\text{-bisimulation}}{M \approx_n N}$$

The proofs of the lemmata necessary for showing Theorem 11 are similar to those used in Section 2.1, but for:

Lemma 12. $\approx_n = \approx_n^{app}$

A proof of this lemma can be achieved along the lines of the corresponding lemma in Section 2.2, extending the notion of $\approx_n^{app}$ to open terms.

2.5 A final Description of $\approx_i$ in the Hyperset Setting

A final semantics inducing the $\approx_i^{app}$ equivalence can be given similarly to the previous cases where all divergent terms are equated, say $\rightarrow_h$. We conjecture the coincidence between $\approx_i$ and $\approx_i^{app}$.

2.6 A final Description of $\approx_e$ in the Hyperset Setting

The equivalence $\approx_e$ does not equates all $\rightarrow_e$-divergent terms. A final semantics for handling this observational equivalence can be given along the lines of that for $\approx_n$ in Section 2.4. The coincidence between $\approx_e$ and $\approx_e^{app}$ can be proved using a model theoretic argument similar to that utilized for $\approx_v$ in [6]. More details will be given in a forthcoming paper.

3 Yet another final Description of $\approx_v$ in the Hyperset setting

In this section we present another final semantics, inducing the $\approx_v$ equivalence, which makes use of a functor different from the one considered in the previous section. From this semantics we derive yet another coinduction principle for establishing $\approx_v$. In particular we prove that $\approx_v^{app}$ can be viewed as the greatest fixed point of the following monotone operator:

Definition 15. Let $\Phi_v : \mathcal{P}(\Lambda^0 \times \Lambda^0) \to \mathcal{P}(\Lambda^0 \times \Lambda^0)$ be the operator defined by $\Phi_v(R) = \{(M,N) \mid (M \not\Downarrow_v \;\wedge N \not\Downarrow_v) \vee (M \Downarrow_v \;\wedge N \Downarrow_v \;\wedge \forall P \exists Q. ((P,Q) \in R \wedge (MP, NQ) \in R) \wedge \forall P \exists Q. ((P,Q) \in R \wedge (NP, MQ) \in R))\}$.

Definition 16. i) Let $G_v : Class \to Class$ be the endofunctor defined by

$$G_v(X) = \mathcal{P}(X \times X) + \{*\} ,$$

where $*$ is a generic atom; the definition on morphisms is canonical.
ii) Let (Λ^0, β_v) be the G_v-coalgebra defined by

$$\beta_v(M) = \begin{cases} (u,*) & \text{if } M \not\Downarrow_v \\ (v, \{(N, MN) \mid N \in \Lambda^0\}) & \text{if } M \Downarrow_v . \end{cases}$$

Lemma 13. *R is a Φ_v-bisimulation if and only if R is a G_v-bisimulation.*

Proof. ($\Rightarrow$) The assertion follows from the definition of G_v-bisimulation.
($\Leftarrow$) By contradiction.

□

The following lemma is proved as Lemma 2.

Lemma 14. *The functor G_v has a greatest fixed point $\overline{Y_v}$ such that $(\overline{Y_v},\ id)$ is a final G_v-coalgebra.*

Definition 17. Let $\mathcal{N}_v : \Lambda^0 \to \overline{Y_v}$ be the unique G_v-morphism from the G_v-coalgebra (Λ^0, β_v) to the G_v-coalgebra $(\overline{Y_v},\ id)$, i.e.:

$$\mathcal{N}_v(M) = \begin{cases} (u,*) & \text{if } M \not\Downarrow_v \\ (v, \{(\mathcal{N}_v(N),\ \mathcal{N}_v(MN)) \mid N \in \Lambda^0\}) & \text{if } M \Downarrow_v . \end{cases}$$

Now we prove that the greatest Φ_v-bisimulation, $\sim_v$, coincides with the applicative equivalence $\approx_v^{app}$, and hence that the equivalence induced by $\mathcal{N}_v$ coincides with $\approx_v$. To this end we introduce the syntactical counterparts of the relations $\equiv_n$ introduced in Section 3.1 of [6].

Definition 18. Let P be the initial solution in the category $C.P.O.^E$ of the domain equation $D = [D \to_\perp D]_\perp$, i.e. $P = lim_{\leftarrow} P_n$. Let Λ_n^0 be the set of all closed λ-terms whose interpretation, up to isomorphism, belongs to P_n.
i) For all n, let $\approx_v^n \subseteq \Lambda_n^0 \times \Lambda_n^0$ be the relation inductively defined as follows
$\approx_v^0 = \Lambda_0^0 \times \Lambda_0^0$

$\approx_v^{n+1} = \{(M,N) \in \Lambda_{n+1}^0 \times \Lambda_{n+1}^0 \mid \forall P, Q \in \Lambda_n^0.(P \approx_v^n Q \Rightarrow MP \approx_v^n NQ)\}$.
ii) For all n, let $\sqsubseteq_v^n \subseteq \Lambda_n^0 \times \Lambda_n^0$ be the relation inductively defined as follows
$\sqsubseteq_v^0 = \Lambda_0^0 \times \Lambda_0^0$
$\sqsubseteq_v^{n+1} = \{(M,N) \in \Lambda_{n+1}^0 \times \Lambda_{n+1}^0 \mid M \in \Lambda_0^0 \vee (M \notin \Lambda_0^0 \wedge N \notin \Lambda_0^0 \wedge \forall P, Q \in \Lambda_n^0.(P \sqsubseteq_v^n Q \Rightarrow MP \sqsubseteq_v^n NQ))\}$.
iii) Since the projections $\pi_n : P \to P_n$ are all λ-definable (see [6]), say by Π_n, we define, for all $M \in \Lambda^0$, M_n to be the term $(\Pi_n M) \in \Lambda_n^0$.

In the sequel we use freely that for all $M, N \in \Lambda^0$, $(M_{n+1}N) =_P (M_{n+1}N_n) =_P (M_{n+1}N_n)_n =_P (MN_n)_n$, where $=_P$ denotes equality in the model P; and also that application is monotone w.r.t. $\sqsubseteq_P$.

The following lemma is instrumental:

Lemma 15. *o) For all $n \in \omega$, $\approx_v^n$ is* P-saturated, *i.e. if $M \approx_v^n N$, $M =_P M'$ and $N =_P N'$, then $M' \approx_v^n N'$. Moreover for all $n \in \omega$, $\sqsubseteq_v^n$ is* P-saturated, *i.e. if $M \sqsubseteq_v^n N$, $M \sqsubseteq_P M'$ and $N \sqsubseteq_P N'$, then $M' \sqsubseteq_v^n N'$.*
i) For all $n \in \omega$, $\approx_v^n = \sqsubseteq_v^n \cap \sqsupseteq_v^n$.
ii) For all $n \in \omega$, $\approx_v^n$ is an equivalence relation.
iii) For all $M, N \in \Lambda^0$, $M \approx_v^{app} N \iff \forall n \in \omega.\ M_n \approx_v^n N_n$.
iv) $\approx_v^{app} \subseteq \sim_v$.
v) For all n: $\sim_v^n = \approx_v^n$, where $\sim_v^n \subseteq \Lambda_n^0 \times \Lambda_n^0$ is inductively defined as follows:
$\sim_v^0 = \Lambda_0^0 \times \Lambda_0^0$
$\sim_v^{n+1} =$
$\{(M,N) \in \Lambda_{n+1}^0 \times \Lambda_{n+1}^0 \mid \forall P \in \Lambda_n^0 \exists Q \in \Lambda_n^0((P,Q) \in \sim_v^n \wedge (MP, NQ) \in \sim_v^n) \wedge \forall Q \in \Lambda_n^0 \exists P \in \Lambda_n^0((Q,P) \in \sim_v^n \wedge (NQ, MP) \in \sim_v^n)\}$.
vi) For all $M, N \in \Lambda^0$, $M \sim_v N \iff \forall n \in \omega.\ M_n \sim_v^n N_n$.

Proof. o) Both assertions are easily proved by induction on n. We show only the first one. The case $n = 0$ is trivial. Suppose $M \approx_v^{n+1} N$, $M =_P M'$ and $N =_P N'$, then if $P \approx_v^n Q$, we have $MP \approx_v^n NQ$, hence by induction hypothesis $M'P \approx_v^n N'Q$.
i) The assertion is proved by induction on n, using the fact that, for all n and m such that $n + m \geq 1$,
$M \approx_v^{n+m} N \wedge P_1 \sqsubseteq_v^{n+m-1} Q_1 \wedge \ldots \wedge P_m \sqsubseteq_v^n Q_m \Rightarrow$
$MP_1 \ldots P_m \sqsubseteq_v^n NQ_1 \ldots Q_m$,
which is easily provable by induction on n, regrouping $n + 1 + m$.
ii) Symmetry is proved straightforwardly by induction on n. Reflexivity follows from the fact that, for all n and m such that $n + m \geq 1$,
$(M \in \Lambda_{n+m}^0 \wedge P_1 \approx_v^{n+m-1} Q_1 \wedge \ldots \wedge P_m \approx_v^n Q_m) \Rightarrow$
$MP_1 \ldots P_m \approx_v^n MQ_1 \ldots MQ_n$,
which is easily provable by induction on n, regrouping $n + 1 + m$.
Finally, transitivity is proved straightforwardly by induction on n, using reflexivity.
iii) Immediate from Theorem 33 of [6].
iv) The assertion is easily proved by coinduction.
v) The assertion is proved by induction on n. If $n = 0$ the thesis is trivially true.

Let $\sim_v^n = \approx_v^n$. Then using the induction hypothesis and reflexivity of $\approx_v^n$, one gets immediately $\approx_v^{n+1} \subseteq \sim_v^{n+1}$. Now suppose by contradiction that $\sim_v^{n+1} \not\subseteq \approx_v^{n+1}$. Then there exist M, N such that $M \sim_v^{n+1} N$ and $M \not\approx_v^{n+1} N$, i.e. there are $P, T \in \Lambda_n^0$ such that $P \approx_v^n T$ and $MP \not\approx_v^n NT$. But there is $Q \in \Lambda_n^0$ such that $P \sim_v^n Q$ and $MP \sim_v^n NQ$. By induction hypothesis, $P \approx_v^n Q$ and $MP \approx_v^n NQ$, hence, by (ii), $NQ \approx_v^n NT$ and $MP \approx_v^n NT$, which is a contradiction.
vi) The implication ($\Leftarrow$) follows immediately from (iii), (iv) and (v). The other implication is proved by induction on n. If $n = 0$ the thesis is trivially true. Let $M_n \sim_v^n N_n$. We will show that $\forall P \in \Lambda_n^0\ \exists Q \in \Lambda_n^0$ such that $P \sim_v^n Q$ and $M_{n+1}P \sim_v^n N_{n+1}Q$. Let $P \in \Lambda_n^0$, then there is Q such that $P \sim_v Q$ and $MP \sim_v NQ$. By induction hypothesis, (v) and (o) $P \approx_v^n P_n \approx_v^n Q_n$, $(MP)_n \approx_v^n (NQ)_n$ and $M_{n+1}P \approx_v^n (NQ)_n$. Now $(NQ)_n \supseteq_v^n N_{n+1}Q_n \approx_v^n N_{n+1}P$, and in particular $M_{n+1}P \supseteq_v^n N_{n+1}Q_n$. Now we show the converse, i.e. $M_{n+1}P \subseteq_v^n N_{n+1}Q_n$. By definition of $\sim_v$, there exists T such that $Q_n \sim_v T$ and $MT \sim_v NQ_n$. By induction hypothesis and (v), $Q_n \approx_v^n T_n$, hence $T_n \approx_v^n P$, and $(MT)_n \approx_v^n (NQ_n)_n \approx_v^n N_{n+1}Q_n$. Hence, $N_{n+1}Q_n \approx_v^n (MT)_n \supseteq_v^n M_{n+1}P$. Summing up we have $N_{n+1}Q_n \approx_v^n M_{n+1}P$; and hence $P \sim_v^n Q_n$ and $M_{n+1}P \sim_v^n N_{n+1}Q_n$.

□

Now we can give:

Theorem 16. *The greatest Φ_v-bisimulation, $\sim_v$, coincides with the applicative equivalence $\approx_v^{app}$.*

Proof. The thesis follows immediately from points (iii), (v) and (vi) of lemma 15.

□

Proceeding as in the previous section we can now prove:

Theorem 17. *Let $M, N \in \Lambda^0$.*
i) Then

$$\mathcal{N}_n(M) = \mathcal{N}_n(N) \iff M \approx_v N \ .$$

ii) The following coinduction principle holds:

$$\frac{(M,N) \in R \quad R \text{ is a } G_v\text{-bisimulation}}{M \approx_v N}$$

4 A syntactical induction-coinduction Principle for $\approx_v$

In this section we prove the soundness of an induction-coinduction principle for establishing $\approx_v$-equivalence. This principle should deserve more investigation; however, it can be viewed as a syntactical version of the semantical induction-coinduction principle appearing in [14]. We use the notation introduced in the previous section.

Definition 19. Let $T : \mathcal{P}(\Lambda^0 \times \Lambda^0) \times \mathcal{P}(\Lambda^0 \times \Lambda^0) \to \mathcal{P}(\Lambda^0 \times \Lambda^0) \times \mathcal{P}(\Lambda^0 \times \Lambda^0)$ be the operator defined by

$T(R^-, R^+) =$
$(\{(M,N) \mid (M \not\Downarrow_v \wedge N \not\Downarrow_v) \vee (M \Downarrow_v \wedge N \Downarrow_v \wedge \forall (P,Q) \in R^+ . \, (MP, NQ) \in R^-)\},$
$\{(M,N) \mid (M \not\Downarrow_v \wedge N \not\Downarrow_v) \vee (M \Downarrow_v \wedge N \Downarrow_v \wedge \forall (P,Q) \in R^- . \, (MP, NQ) \in R^+)\})$.

Definition 20. A relation $R \subseteq \Lambda^0 \times \Lambda^0$ is *λ-inclusive* if for all $M, N \in \Lambda^0$, if for all $n \in \omega$ there exist M', N' such that $(M', N') \in R$, $M' \approx_v M_n$ and $N' \approx_v N_n$, then $(M, N) \in R$.

Theorem 18. *Let R^-, R^+ be two relations on $\Lambda^0 \times \Lambda^0$ such that R^+ is λ-inclusive. Then the following principle holds*

$$\frac{R^- \subseteq \pi_1(T(R^-, R^+)) \quad \pi_1(T(R^+, R^-)) \subseteq R^+}{R^- \subseteq \approx_v \subseteq R^+}$$

Proof. First of all we prove by induction on n that, for all n,

$$\widetilde{R}_n^- \subseteq \approx_v^n \subseteq R^+ ,$$

where $\widetilde{R}_n^- = \{(M, N) \in \Lambda_n^0 \times \Lambda_n^0 \mid \exists (P, Q) \in R^- . (P_n \approx_v M \;\wedge\; Q_n \approx_v N)\}$.

The base case ($n = 0$) follows trivially from the hypotheses of the principle. Suppose that $\widetilde{R}_n^- \subseteq \approx_v^n \subseteq R^+$. Then, since $(\approx_v^{n+1}, \approx_v^{n+1}) = T(\approx_v^n, \approx_v^n) \cap (\Lambda_{n+1}^0 \times \Lambda_{n+1}^0)$ and T is monotone in the first component and antimonotone in the second component, we have:

1. $\approx_v^{n+1} \subseteq \pi_1(T(\approx_v^n, \widetilde{R}_n^-)) \cap (\Lambda_{n+1}^0 \times \Lambda_{n+1}^0) = \pi_1(T(\approx_v^n, R^-)) \subseteq \pi_1(T(R^+, R^-)) \cap (\Lambda_{n+1}^0 \times \Lambda_{n+1}^0) \subseteq R^+$,
   where the equality is established using $(MN_n)_n \approx_v (M_{n+1}N_n)$.
2. $\widetilde{R}_{n+1}^- \subseteq \pi_1(T(\widetilde{R}_n^-, \approx_v^n)) \cap (\Lambda_{n+1}^0 \times \Lambda_{n+1}^0) \subseteq \pi_1(T(\approx_v^n, \approx_v^n)) \cap (\Lambda_{n+1}^0 \times \Lambda_{n+1}^0) = \approx_v^{n+1}$,
   where the first inclusion follows, using $(MN_n)_n \approx_v (M_{n+1}N_n)$, from $R^- \subseteq \pi_1(T(R^-, \approx_v^n))$, which in turn is a consequence of the left hypothesis of the principle and the induction hypothesis.

Now the inclusion $\approx_v \subseteq R^+$ follows immediately from the fact that R^+ is λ-inclusive, while the inclusion $R^- \subseteq \approx_v$ can be directly obtained by contradiction. □

5 Final Remarks

The constructions carried out in this paper raise many open questions and all should deserve more investigations. For lack of space, we can give here only a list of conjectures, claims and concise remarks. We shall elaborate on them in a forthcoming paper.

1. Coinductive characterizations of $\approx_\sigma^{app}$ are useful in factoring out the complexity of establishing observational equivalences between λ-terms. It would be interesting to compare the strength of coinduction principles to that of other tools, e.g. "approximation theorems" such as arise from "computationally adequate" mathematical models (see [20, 6, 11]). Here are some equivalences on which to test the power of coinduction principles:
   - divergent terms are $\approx_\sigma$ for $\sigma \in \{v, l, h, i\}$;
   - *black holes* (i.e. closed λ-terms M s.t. $\forall P.\ MP \rightarrow_\sigma^* M$) are $\approx_\sigma$ for $\sigma \in \{v, l, h, i, n, e\}$;
   - appropriate classes of fixed point operators are $\approx_\sigma$ for each σ;
   - many identities involving fixed points (e.g. the double iteration identity, i.e. $Fix(\lambda x.(Fix(\lambda y.fxy))) \approx_\sigma Fix(\lambda x.fxx)$), hold, for each σ, for appropriate classes of fixed points.
2. We have considered only coinductive characterizations of equivalence relations. We could have discussed, more generally, coinductive principles for establishing partial orders such as ***observational approximation*** and ***applicative approximation***. These are obtained by replacing the bi-implication in the "equitermination" predicate by a simple implication.
3. All the final semantics that we have introduced do not yield "standard" denotational models for λ-calculus. Nevertheless, they can count as ***compositional***, in that they induce observational equivalences, which are congruences w.r.t. the syntactical operators of the language. They can be seen to provide, in effect, alternative presentations of the, obviously ***fully abstract***, term model. To this end it is useful to extend the equivalences considered to open λ-terms:
   Let $\sigma \in \{v, l, h, n, i, e\}$ and let $P, P' \in \Lambda$ be s.t. $FV(P, P') \subseteq \{x_1, \ldots, x_n\}$. We say that $P \approx_\sigma^{app} P'$ if and only if, for all $P_1, \ldots, P_n \in \Lambda^0$, $P[P_i/x_i] \approx_\sigma^{app} P'[P_i/x_i]$.
   Models could have been defined also using the technique of ***processes as terms*** introduced by J.J.M.M.Rutten (see [17, 19]).
4. We conjecture that both the ***syntactical induction-coinduction*** principle and the alternative final description of $\approx_v$ could be defined and shown to hold for all the observational equivalences discussed in the paper. For instance, the definitions and proofs, presented here for $\approx_v$, can be readily adapted to the case of $\approx_e$. The crucial fact is that both equivalences have "computationally adequate" inverse limit models with λ-definable projections. And hence one can obtain $\approx_\sigma$ (for $\sigma \in \{v, e\}$) by defining inductively on approximations a quotient of the interior of the model. This is shown for $\approx_v$ in [6]; the appropriate denotational model for $\approx_e$ is the one discussed in [11].
5. Purely set theoretic models in $ZF^- X_1$, where values are modeled by set-theoretic functions, can be readily obtained for $\approx_v$ and $\approx_l$ if we modify the definition of the functors F_v, F_l, G_v by replacing the disjoint union by the set-theoretic union and the atom $*$ by the empty set.

References

1. S.Abramsky, L.Ong, *Full Abstraction in the Lazy Lambda Calculus,* Information and Computation, 105(2):159–267, 1993.
2. P.Aczel, *Non-wellfounded sets,* Number 14, Lecture Notes CSLI, 1988.
3. P.Aczel, N.Mendler, *A final coalgebra theorem* Category Thepry and Computer Science Proceedings, D.Pitt et al. eds., Springer LNCS n.389:357-365, 1989.
4. H.Barendregt, *The Lambda Calculus, its Syntax and Semantics,* North Holland, Amsterdam, 1984.
5. M.Coppo, M.Dezani-Ciancaglini, M.Zacchi, *Type Theories, Normal Forms and* D_∞*-Lambda-Models,* Information and Computation, 72(2):85–116, 1987.
6. L.Egidi, F.Honsell, S.Ronchi Della Rocca, *Operational, denotational and logical Descriptions: a Case Study,* Fundamenta Informaticae, 16(2):149–169, 1992.
7. M.Fiore, *A Coinduction Principle for Recursive Data Types Based on Bisimulation,* 8th LICS Conference Proceedings, IEEE Computer Society Press:110-119, 1993.
8. M.Forti, F.Honsell, *Set Theory with Free Construction Principles,* Annali Scuola Normale Sup. Pisa, Cl. Sci., (IV), 10:493–522, 1983.
9. M.Forti, F.Honsell, M.Lenisa, *Processes and Hyperuniverses,* MFCS'94 Conference Proceedings, I.Privara et al. eds., Springer LNCS n.841:352-363, 1994.
10. F.Honsell, M.Lenisa, *Some Results on Restricted* λ*-calculi,* MFCS'93 Conference Proceedings, A.Borzyszkowski et al. eds., Springer LNCS n.711:84-104, 1993.
11. F.Honsell, S.Ronchi Della Rocca, *An approximation theorem for topological lambda models and the topological incompleteness of lambda calculus*, J. of Computer and System Sciences (45) 1:49-75, 1992.
12. R.Milner, *Operational and Algebraic Semantics of Concurrent Processes,* Handbook of Theoretical Computer Science, Ch.19, 1990.
13. C.H.L.Ong, *The lazy lambda calculus: an investigation into the foundations of functional programming,* Ph.D. thesis, Imperial College of Science and Technology, University of London, 1988.
14. A.M.Pitts, *Relational Properties of Recursively Defined Domains,* 8th LICS Conference Proceedings, IEEE Computer Society Press:86-97, 1993.
15. G.D.Plotkin, *Call-by-name, Call-by-value and the* λ*-calculus,* Theoretical Computer Science (1):125-159, 1975.
16. S.Ronchi Della Rocca, *International Summer School in Logic for Computer Science,* Chambery 28/6 – 9/7 1993, lecture notes.
17. J.J.M.M.Rutten, *Processes as terms: non-wellfounded models for bisimulation,* Math.Struct.Comp.Sci., 2(3):257–275, 1992.
18. J.J.M.M.Rutten, D.Turi, *On the Foundations of Final Semantics: Non-Standard Sets, Metric Spaces, Partial Orders,* REX Conference Proceedings, J.deBakker et al. eds., Springer LNCS n.666:477-530, 1993.
19. D.Turi, B.Jacobs, *On final Semantics for applicative and non-deterministic languages,* Fifth Biennial Meeting on Category Theory and Computer Science, Amsterdam, 1993.
20. C.P.Wadsworth, *The relation between computational and denotational properties for Scott's* D_∞*-models of the* λ*-calculus,* SIAM J. of Computing, 5(3):488-521 ,1976.

Appendix A

We recall some categorical definitions (for more details see [18]).

Definition 21 (F-coalgebra). Let C be a category and $F : C \to C$ an endofunctor.
i) A *F-coalgebra* is a pair (A, α), where A is an object of C and $\alpha : C \to F(C)$ is a morphism of C.
ii) Let C_F be the category whose objects are F-coalgebras and whose morphisms are F-coalgebra morphisms. A F-coalgebra morphism $f : (A, \alpha) \to (B, \beta)$ is an arrow $f : A \to B$ in the category C such that the following diagram commutes:

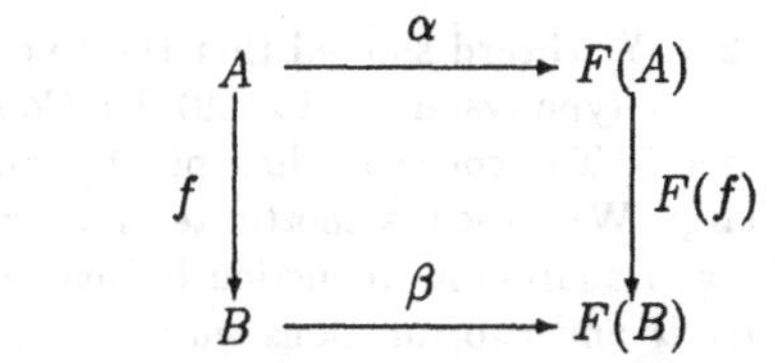

Definition 22 (F-bisimulation). Let C be a category with products. Let $F : C \to C$ be an endofunctor and (A, α) a F-coalgebra. A *F-bisimulation on the F-coalgebra* (A, α) is a relation $R \subseteq A \times A$ such there exists an arrow $\gamma : R \to F(R)$ which makes the following diagram commutes:

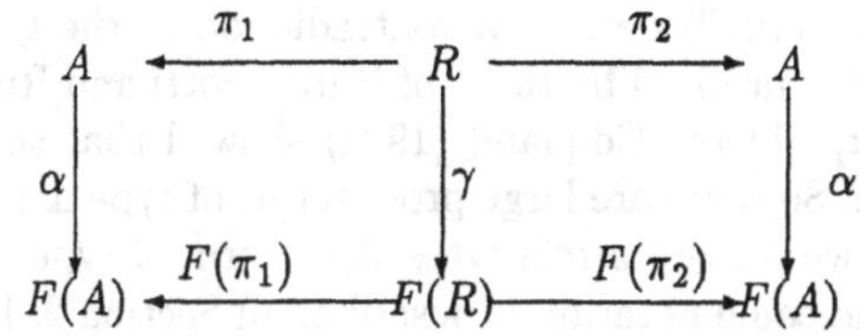

In the following definition we recall the Antifoundation Axiom FCU and the Super Strong Extensionality Axiom up to the set of atoms U ($SSExtU$), which is a consequence of FCU and gives an interesting characterization of the equality between sets (for more details see [8]):

Definition 23. i) Unique Free Construction with respect to a set U of atoms FCU:
Let X be a set such that $X \cap U = \emptyset$. For every function $f : X \to \mathcal{P}(X \cup U)$ there is a unique function $g : X \to V$ verifying

$$g(x) = (f(x) \cap U) \cup \{g(y) \mid y \in f(x) \cap X\}, \quad \forall x \in X.$$

ii) Super Strong Extensionality Axiom up to the set of atoms U $SSExtU$:
Let V be the set theoretic Universe. Two sets X, Y are equal if and only if there exists a relation $R \subseteq V \times V$ such that $(X, Y) \in R$ and $R \subseteq (R)^+$, where $(\ )^+$ is the operator on relations defined by

$$\begin{aligned}(R)^+ = \{(X,Y) \mid\ & X \cap U = Y \cap U \ \text{ and} \\ & \forall W \in (f(X) \setminus U).\exists Z \in (Y \setminus U).\ (W, Z) \in R \ \text{ and} \\ & \forall W \in (Y \setminus U).\exists Z \in (X \setminus U).\ (W, Z) \in R\}.\end{aligned}$$

(Categorically, R is a $\mathcal{P}(\ ) + U$-bisimulation on the coalgebra $(V,\ id)$.)

A Simplification of Girard's Paradox

Antonius J.C. Hurkens

Klaasstokseweg 7, 5443 NS Haps, The Netherlands
e-mail: hurkens@sci.kun.nl

Abstract. In 1972 J.-Y. Girard showed that the Burali-Forti paradox can be formalised in the type system U. In 1991 Th. Coquand formalised another paradox in U^-. The corresponding proof terms (that have no normal form) are large. We present a shorter term of type $\bot$ in the Pure Type System λU^- and analyse its reduction behaviour. The idea is to construct a universe $\mathcal{U}$ and two functions such that a certain equality holds. Using this equality, we prove *and* disprove that a certain object in $\mathcal{U}$ is well-founded.

1 Introduction

Jean-Yves Girard (1972) derived a contradiction in the type system U by formalising a paradox inspired by those of Burali-Forti and Russell. By formalising another paradox, Thierry Coquand (1994) showed that the type system U^- is also inconsistent. So there are large proof terms of type $\bot$ in these type systems.

In Section 3 we present a relatively short term of type $\bot$ in λU^-. This Pure Type System and some notation is described in Section 2. In the last section we show that the β-reduction behaviour of the proof term is very simple.

In the other sections we will see that the proof has the same ingredients as Burali-Forti's paradox: a universe $\mathcal{U}$, a relation $<$ on $\mathcal{U}$, an object Ω in $\mathcal{U}$, and the question whether Ω is well-founded or not.

In Section 4 we describe Burali-Forti's paradox and some simplifications. We analyse the connection between the universe of all ordinals at its power set. In Section 5 we introduce *paradoxical* universes. These are connected to their power set in such a way that we can derive a Burali-Forti like contradiction. This can be formalised in Pure Type Systems. The formalisation can be simplified by considering *powerful* universes. In Section 6 we see how these universes are connected to the power set of their power set.

2 Pure Type Systems

In this section, we describe some Pure Type Systems. For more details, see for example (Barendregt 1992) or (Geuvers 1993).

2.1 The Pure Type Systems λHOL, λU^-, and λU

The typed λ-calculus λHOL (*Higher Order Logic*) is the Pure Type System (with β-conversion) given by the *sorts* $*$, $\square$, and $\triangle$, the *axioms* $* : \square$ and $\square : \triangle$, and the *rules* $(*, *)$, $(\Box, \Box)$, and $(\Box, *)$. It is a consistent system, even if one adds the rule $(\Delta, *)$. Adding the rule $(\Delta, \Box)$, one gets the Pure Type System λU^-. Adding both rules, one gets λU.

2.2 Typing Terms in a Pure Type System

Each term A in a Pure Type System is either a variable x, a sort s, a product $\Pi x : B. C$, an abstraction $\lambda x : B. C$, or an application $(B\ C)$.

By $B[C/x]$ we denote the result of substituting the term C for the free occurrences of the variable x in B (renaming bound variables if necessary). By $=_\beta$ we denote the equivalence relation between terms that is induced by β-reduction: replacing a subterm of the form $(\lambda x : A. B\ C)$ by the term $B[C/x]$. If a term does not contain such a subterm, then the term is called *normal.*

In a Pure Type System, we can derive formal judgements $x_1 : A_1, \cdots, x_n : A_n \vdash B : C$, expressing that B *has type* C in the given context, that is, assuming that for $i = 1, \ldots, n$, variable x_i has type A_i.

We start in the empty context. If, in some context, A has type s for some sort s, then we are allowed to introduce a new variable x of type A.

The context gives the types of some variables.

The axioms give the types of some sorts.

We use the rules (s', s) to type products as follows: if A has type s' and (under the extra assumption $x : A$) B has type s, then (in the original context) $\Pi x : A. B$ also has type s.

If $\Pi x : A. B$ has type s and (under the extra assumption $x : A$) C has type B, then (in the original context) $\lambda x : A. C$ has type $\Pi x : A. B$.

If F has type $\Pi x : A. B$ and C has type A, then $(F\ C)$ has type $B[C/x]$.

Finally, we use β-reduction to change types: if A has type B, $B =_\beta C$, and C has type s, then we may conclude that A has type C.

Note that if a variable, abstraction or application has type A, then A is of type s for some sort.

2.3 Some Useful Properties of λU

Two terms A and B are β-equal if and only if for some C, both A and B reduce to C. If term B has a type D, then this type is unique up to β-equality. Furthermore, if B β-reduces to C, then C is also a term of type D.

We can calculate the *level* of a term (and its subterms) in a given context $x_1 : A_1, \cdots, x_n : A_n$ as follows: The sorts $*$, $\Box$, and Δ have level 2, 3, and 4, respectively. The level of variable x_i is one less then the level of A_i in the context $x_1 : A_1, \cdots, x_{i-1} : A_{i-1}$. The level of a product $\Pi x : B. C$ or an abstraction $\lambda x : B. C$ is the level of C in the extended context $x_1 : A_1, \cdots, x_n : A_n, x : B$. The level of an application $(B\ C)$ is the level of B in the original context.

One can prove that if B has type C in some context, then the level of B is one less than that of C. So each term has level 0, 1, 2, 3, or 4. One can also show that no term in λU contains a subterm of lower level (in the corresponding context). This implies that if we use a rule (s, s) to form a product $\Pi x : B.C$, then $\text{level}(x) < \text{level}(B) = \text{level}(s) - 1 = \text{level}(C)$, so the variable x has no free occurrence in C.

It turns out that each term A of level 1 is ***strongly normalising***: there is no infinite sequence $A \to_\beta A' \to_\beta A'' \to_\beta \cdots$ of β-reduction steps. The terms of higher level are normal, since each abstraction or application has level 0 or 1.

2.4 The Five Levels of Terms in λU

We describe the five levels and introduce some notation to distinguish terms of different levels.

The only term of level 4 is $\triangle$ and the only term of level 3 is $\Box$.

We will call the terms of level 2 *sets* or *universes*. We think of $*$ as the *set* of all propositions. We use calligraphic letters $\mathcal{X}, \ldots$ for set variables.

We will call the terms of level 1 *objects*. Objects $\varphi, \chi, \ldots$ of type $*$ are called *propositions*. We use italic letters $x, \ldots$ for object variables.

Finally, the terms of level 0 are called *proofs* or *proof terms*. We use natural numbers $0, 1, \ldots$ for proof variables. These correspond exactly to the labels of assumptions in a natural deduction in Gentzen's style.

Using the rule $(\Box, \Box)$, we can form the *set* of all functions from a set $\mathcal{S}$ to a set $\mathcal{T}$:

$$(\mathcal{S} \to \mathcal{T}) \equiv \Pi x : \mathcal{S}.\mathcal{T}$$

In particular, the *power set* of $\mathcal{S}$ can be seen as the set of all predicates on $\mathcal{S}$:

$$\wp\mathcal{S} \equiv (\mathcal{S} \to *)$$

Using the rule $(\triangle, \Box)$, which is not allowed in λHOL, we can form a 'polymorphic domain' $\Pi\mathcal{X} : \Box.\mathcal{T}$ (where $\mathcal{X}$ may occur in $\mathcal{T}$). This product of level 2 has no clear set-theoretical interpretation. The products corresponding to the rules $(*, *)$, $(\Box, *)$, and $(\triangle, *)$ are propositions:

$$\begin{aligned} [\varphi \Rightarrow \chi] &\equiv \Pi 0 : \varphi.\chi \\ \forall x : \mathcal{S}.\chi &\equiv \Pi x : \mathcal{S}.\chi \\ \forall \mathcal{X} : \Box.\chi &\equiv \Pi \mathcal{X} : \Box.\chi \end{aligned}$$

Other connectives can be defined as usual. We only need falsehood and negation:

$$\begin{aligned} \bot &\equiv \forall p : *.p \\ \neg\varphi &\equiv [\varphi \Rightarrow \bot] \end{aligned}$$

There are two kinds of abstractions and applications of level 1. We introduce some new notation only for the 'polymorphic' ones:

$$\Lambda\mathcal{X} : \Box.c \equiv \lambda\mathcal{X} : \Box.c \qquad\qquad \{b\ \mathcal{T}\} \equiv (b\ \mathcal{T})$$

Here b and c are objects and $\mathcal{T}$ is a set.

There are three kinds of abstractions and applications of level 0:

$$\begin{array}{rlr} \textsf{suppose } n : \varphi.\, P \equiv \lambda n : \varphi.\, P & \qquad & [P\ Q] \equiv (P\ Q) \\ \textsf{let } x : \mathcal{S}.\, P \equiv \lambda x : \mathcal{S}.\, P & & \langle P\ c\rangle \equiv (P\ c) \\ \textsf{let } \mathcal{X} : \Box.\, P \equiv \lambda \mathcal{X} : \Box.\, P & & \langle P\ \mathcal{T}\rangle \equiv (P\ \mathcal{T}) \end{array}$$

Note that for proofs P and Q, the application $[P\ Q]$ corresponds to *modus ponens* in a natural deduction.

3 A Term of Type $\bot$ in λU^-

We consider the following universe:

$$\mathcal{U} \equiv \Pi\mathcal{X} : \Box.\, ((\wp\wp\mathcal{X} \rightarrow \mathcal{X}) \rightarrow \wp\wp\mathcal{X})$$

For each term t of type $\wp\wp\mathcal{U}$, we define a term of type $\mathcal{U}$:

$$\tau t \equiv \Lambda\mathcal{X} : \Box.\, \lambda f : (\wp\wp\mathcal{X} \rightarrow \mathcal{X}).\, \lambda p : \wp\mathcal{X}.\, (t\ \lambda x : \mathcal{U}.\, (p\ (f\ (\{x\ \mathcal{X}\}\ f))))$$

For each term s of type $\mathcal{U}$, we define a term of type $\wp\wp\mathcal{U}$:

$$\sigma s \equiv (\{s\ \mathcal{U}\}\ \lambda t : \wp\wp\mathcal{U}.\, \tau t)$$

(So we do not consider σ and τ as *terms*.)

We define normal terms of type $\wp\mathcal{U}$ and $\mathcal{U}$, respectively:

$$\begin{aligned} &\Delta \equiv \lambda y : \mathcal{U}.\, \neg\forall p : \wp\mathcal{U}.\, [(\sigma y\ p) \Rightarrow (p\ \tau\sigma y)] \\ &\Omega \equiv \text{the normal form of } \tau\ \lambda p : \wp\mathcal{U}.\, \forall x : \mathcal{U}.\, [(\sigma x\ p) \Rightarrow (p\ x)] \end{aligned}$$

In other words, $\Omega \equiv \Lambda\mathcal{X} : \Box.\, \lambda f : (\wp\wp\mathcal{X} \rightarrow \mathcal{X}).\, \lambda p : \wp\mathcal{X}.\, \forall x : \mathcal{U}.\, [(\sigma x\ \lambda y : \mathcal{U}.\, (p\ (f\ (\{y\ \mathcal{X}\}\ f)))) \Rightarrow (p\ (f\ (\{x\ \mathcal{X}\}\ f)))]$.

We claim that the following is a term of type $\bot$ in λU^-:

$$\begin{gathered} [\textsf{suppose } 0 : \forall p : \wp\mathcal{U}.\, [\forall x : \mathcal{U}.\, [(\sigma x\ p) \Rightarrow (p\ x)] \Rightarrow (p\ \Omega)]. \\ [[\langle 0\ \Delta\rangle\ \textsf{let } x : \mathcal{U}.\, \textsf{suppose } 2 : (\sigma x\ \Delta).\, \textsf{suppose } 3 : \forall p : \wp\mathcal{U}.\, [(\sigma x\ p) \Rightarrow (p\ \tau\sigma x)]. \\ [[\langle 3\ \Delta\rangle\ 2]\ \textsf{let } p : \wp\mathcal{U}.\, \langle 3\ \lambda y : \mathcal{U}.\, (p\ \tau\sigma y)\rangle\ ]]\ \textsf{let } p : \wp\mathcal{U}.\, \langle 0\ \lambda y : \mathcal{U}.\, (p\ \tau\sigma y)\rangle] \\ \textsf{let } p : \wp\mathcal{U}.\, \textsf{suppose } 1 : \forall x : \mathcal{U}.\, [(\sigma x\ p) \Rightarrow (p\ x)].\, [\langle 1\ \Omega\rangle\ \textsf{let } x : \mathcal{U}.\, \langle 1\ \tau\sigma x\rangle]] \end{gathered}$$

Note that each subterm (except for the term itself) is normal. One easily verifies that (in the empty context) there is no normal term of type $\bot$ in λU^-. At the end of this article, we analyse the β-reduction behaviour of this proof term.

The proof is simple in the sense that it contains just 6 applications corresponding to *modus ponens*. In order to get an idea of the influence of abbreviations, one can also calculate the *length*: the total number of applications, abstractions, products, and occurrences of variables and sorts. For example, the terms abbreviated by $\bot$, $\mathcal{U}$, Δ, and Ω have length 3, 15, 241, and 145. The complete proof term has length 2039.

In order to explain the idea of this proof, we first describe the paradox of Burali-Forti.

4 Burali-Forti's Paradox

Cesare Burali-Forti (1897) published a result that lead to the first paradox in naive set theory. He showed that there are different ordinal numbers α and β such that neither $\alpha < \beta$ nor $\beta < \alpha$, which contradicts a result of Georg Cantor (1897). (In fact, Burali-Forti considered *perfectly ordered* classes instead of well-orderings, so one has to adapt his proof in order to get a contradiction.)

A binary relation $\prec$ on a set $\mathcal{X}$ is called a *well-ordering* if it is connected (for all different x and y in $\mathcal{X}$, $x \prec y$ or $y \prec x$) and well-founded (there is no infinite descending sequence $\ldots \prec x_2 \prec x_1 \prec x_0$ in $\mathcal{X}$). Then it is also irreflexive and transitive. Each member x of $\mathcal{X}$ determines an *initial segment* of $(\mathcal{X}, \prec)$: the set $\{y \in \mathcal{X} | y \prec x\}$, ordered by the restriction of $\prec$ to this set.

An *ordinal number* is the *order type* of a well-ordered set. Let α and β be the order types of the well-ordered sets $(\mathcal{X}, \prec)$ and $(\mathcal{Y}, \prec')$. Then $\beta = \alpha$ expresses that $(\mathcal{Y}, \prec')$ is isomorphic to $(\mathcal{X}, \prec)$ and $\beta < \alpha$ expresses that $(\mathcal{Y}, \prec')$ is isomorphic to an initial segment of $(\mathcal{X}, \prec)$. (This is well-defined, since isomorphic well-ordered sets have isomorphic initial segments.) It is equivalent to the existence of a monotone function from $(\mathcal{Y}, \prec')$ to an initial segment of $(\mathcal{X}, \prec)$.

Assuming that the relation $<$ on the collection $\mathcal{NO}$ of all ordinal numbers is connected, Burali-Forti (could have) showed that it is a well-ordering. So it has an order type Ω.

Let α be the order type of a well-ordered set $(\mathcal{X}, \prec)$. Then the function that assigns to each x in $\mathcal{X}$ the order type of the initial segment of $(\mathcal{X}, \prec)$ determined by x, is an isomorphism from $(\mathcal{X}, \prec)$ to the initial segment of $(\mathcal{NO}, <)$ determined by α. This shows that for each ordinal α, $\alpha < \Omega$. In particular, $\Omega < \Omega$. This contradicts the fact that $<$ is a well-ordering.

4.1 Simplifications of Burali-Forti's Paradox

Burali-Forti's paradox can be simplified in such a way that Cantor's result is irrelevant. Girard (1972) considered the universe $\mathcal{UO}$ of all *orderings without torsion*: irreflexive, transitive relations such that different elements determine non-isomorphic initial segments. The definition of $<$ can be extended to $\mathcal{UO}$. Then the following contradictory statements can be proved in system U:

> An ordering without torsion is not isomorphic to any of its initial segments. $(\mathcal{UO}, <)$ is an ordering without torsion. Each ordering without torsion is isomorphic to an initial segment of $(\mathcal{UO}, <)$.

Coquand (1986) formalised a version by considering the universe of order types of transitive, well-founded relations (and using the definition of $<$ in terms of monotone functions). This version is similar to the paradox of Dimitry Mirimanoff (1917):

> A set x is *well-founded* (with respect to the membership relation) if no infinite descending sequence $\ldots \in x_1 \in x_0 \in x$ exists. The collection $\mathcal{WF}$ of all well-founded sets is well-founded, so $\mathcal{WF} \in \mathcal{WF}$. This contradicts the well-foundedness of $\mathcal{WF}$.

A still simpler paradox is that of Bertrand Russell (1903):

> Let Δ be the collection of all sets x such that $x \notin x$. Then the proposition $\Delta \in \Delta$ is equivalent to its negation.

One could try to formalise this paradox in a type system like λU as follows:

Define some universe $\mathcal{U}$, together with a function σ from $\mathcal{U}$ to its power set $\wp\mathcal{U}$ and a function τ in the other direction, such that for each term X of type $\wp\mathcal{U}$, $(\sigma\ (\tau\ X))$ is β-equal to X. For x and y in $\mathcal{U}$, write $y \in x$ instead of $((\sigma\ x)\ y)$. Write $\{x|x \notin x\}$ instead of $\lambda x : \mathcal{U}. \neg\ x \in x$ and let Δ be the term $(\tau\ \{x|x \notin x\})$. Then the term $\Delta \in \Delta$ of type $*$ is β-equal to its negation. So [suppose $0 : \Delta \in \Delta$. [0 0] suppose $0 : \Delta \in \Delta$. [0 0]] is a proof term of type $\perp$.

However, as noted by Coquand (1986), Russell's paradox cannot be formalised in this way since each proposition has a normal form. (Of course, in an inconsistent system each proposition is *provable* equivalent to its negation.)

4.2 From Ordinal Numbers to Collections of Ordinal Numbers and Back

We return to Burali-Forti's paradox and analyse the connection between $\mathcal{NO}$ and its power set.

For each ordinal number α, let $\sigma\alpha$ be the collection of all smaller ordinals. Let X be a collection of ordinals and let τX be the order type of $(X, \prec)$, where $\prec$ is the restriction of $<$ to X. Then, by definition of $<$, for each ordinal β, $\beta < \tau X$ expresses that β is the order type of some initial segment of $(X, \prec)$. Now assume that for each α in X, all smaller ordinals are also in X. Then each initial segment of $(X, \prec)$ is of the form $(\sigma\alpha, \prec')$ for some α in X, where $\prec'$ is the restriction of $<$ to $\sigma\alpha$. Therefore $\sigma\tau X = \{\beta|\beta < \tau X\} = \{\beta|\beta$ is of the form $\tau\sigma\alpha$ for some α in $X\} = \{\tau\sigma\alpha|\alpha$ in $X\}$.

In fact one can show that for each α, $\tau\sigma\alpha = \alpha$, but we will see that we do not need that in order to get a contradiction.

5 Paradoxical Universes

5.1 From a Universe to Its Power Set and Back

Let us call a universe $\mathcal{U}$, together with functions $\sigma : \mathcal{U} \to \wp\mathcal{U}$ and $\tau : \wp\mathcal{U} \to \mathcal{U}$, *paradoxical* if for each X in $\wp\mathcal{U}$, $\sigma\tau X = \{\tau\sigma x|x$ in $X\}$.

Each function $f : \mathcal{S} \to \mathcal{T}$ induces a function $f_\star : \wp\mathcal{S} \to \wp\mathcal{T}$ as follows: for each subset X of $\mathcal{S}$, $f_\star X = \{fx|x$ in $X\}$. Using this notation, we see that $(\mathcal{U}, \sigma, \tau)$ is paradoxical if and only if the composition $\sigma \circ \tau$ is equal to $(\tau \circ \sigma)_\star$. Note that if $(\mathcal{U}, \sigma, \tau)$ is paradoxical, then $(\wp\mathcal{U}, \sigma_\star, \tau_\star)$ is also paradoxical: $\sigma_\star \circ \tau_\star = (\sigma \circ \tau)_\star = (\tau \circ \sigma)_{\star\star} = (\tau_\star \circ \sigma_\star)_\star$. (Here we need extensionality: if two sets have the same elements, then they are equal.)

5.2 Example of a Paradoxical Universe

Let $\mathcal{U}$ be the universe of all *triples* $(\mathcal{A}, \prec, a)$ consisting of a set $\mathcal{A}$, a binary relation $\prec$ on $\mathcal{A}$, and an element a of $\mathcal{A}$. For each triple $(\mathcal{A}, \prec, a)$, let $\sigma(\mathcal{A}, \prec, a)$ be the collection of all triples of the form $(\mathcal{A}, \prec, b)$, where $b \prec a$. So σ is a function from $\mathcal{U}$ to $\wp\mathcal{U}$. It induces a relation $<$ on $\wp\mathcal{U}$ as follows:

For all collections X and Y of triples, $Y < X$ if and only if Y is in $\sigma_\star X$, that is, if Y is of the form $\sigma(\mathcal{A}, \prec, a)$ for some triple $(\mathcal{A}, \prec, a)$ in X.

For each X in $\wp\mathcal{U}$, let τX denote the triple $(\wp\mathcal{U}, <, X)$.

Now $\sigma\tau X = \sigma(\wp\mathcal{U}, <, X) = \{(\wp\mathcal{U}, <, Y) | Y < X\} = \{\tau Y | Y \text{ is in } \sigma_\star X\} = \{\tau\sigma(\mathcal{A}, \prec, a) | (\mathcal{A}, \prec, a) \text{ in } X\}$.

5.3 Contradiction from the Existence of a Paradoxical Universe

Let $(\mathcal{U}, \sigma, \tau)$ be paradoxical. It is possible to derive a contradiction similar to Russell's paradox:

> Let $\approx$ be the least equivalence relation on $\mathcal{U}$ such that for each x in $\mathcal{U}$, $x \approx \tau\sigma x$. Define a relation $\in$ on $\mathcal{U}$ as follows: $y \in x$ if and only if $y \approx z$ for some z in σx. Let $\Delta \equiv \tau\{x | x \notin x\}$. Prove that for each y in $\mathcal{U}$, $y \in \Delta$ if and only if $y \notin y$. Take $y = \Delta$.

We will derive a contradiction in another way.

Elements of $\mathcal{U}$ will be denoted by $x, y, \ldots$ and subsets of $\mathcal{U}$ by $X, Y, \ldots$.

If y is in σx, then we say that y is a *predecessor* of x and we write $y < x$. Since $(\mathcal{U}, \sigma, \tau)$ is paradoxical, the predecessors of $\tau\sigma x$ are the elements of the form $\tau\sigma y$ for some predecessor y of x (take $X = \sigma x = \{y | y < x\}$). So if $y < x$ then $\tau\sigma y < \tau\sigma x$. (We will use the special case $y = \tau\sigma x$.)

There are several ways to define well-foundedness. The following formulation immediately leads to the principle of *proof by transfinite induction* (without using classical logic or the axiom of choice). Furthermore, the only quantifiers and connectives that it uses are 'for all' and 'if ... then'.

We call X *inductive* if the following holds: for each x, if each predecessor of x is in X, then x itself is in X. We say that x is *well-founded* if x is in each inductive X. (One can easily prove that $\{x | x \text{ is well-founded}\}$ is the *least* inductive subset of $\mathcal{U}$, but we do not use this fact.)

Let $\Omega \equiv \tau\{x | x \text{ is well-founded}\}$. Since $(\mathcal{U}, \sigma, \tau)$ is paradoxical, the predecessors of Ω are of the form $\tau\sigma w$ for some well-founded w.

We claim that Ω is well-founded:

Let X be inductive. In order to show that Ω is in X, we only need to show that each predecessor of Ω is in X. Such a predecessor is of the form $\tau\sigma w$ for some well-founded w. We want to show that w belongs to the set $\{y | \tau\sigma y \text{ is in } X\}$. This follows from the fact that this set is inductive:

Let x be such that for each $y < x$, $\tau\sigma y$ is in X. Then $\tau\sigma x$ is in X since X is inductive and each predecessor of $\tau\sigma x$ is in X, since such a predecessor is of the form $\tau\sigma y$ for some $y < x$.

Note that, until now, we only used the fact that for each X, $\sigma\tau X \subseteq \{\tau\sigma x | x$ in $X\}$. Using the other inclusion, we now show that Ω is *not* well-founded:

Suppose that Ω is well-founded. Then $\tau\sigma\Omega$ is of the form $\tau\sigma w$ for some well-founded w, so $\tau\sigma\Omega$ is a predecessor of Ω. On the other hand, $\tau\sigma\Omega \not< \Omega$, since Ω is well-founded and the set $\{y | \tau\sigma y \not< y\}$ is inductive:

Let x be such that for each $y < x$, $\tau\sigma y \not< y$. Then $\tau\sigma x \not< x$. For suppose that $\tau\sigma x < x$. Then $\tau\sigma\tau\sigma x \not< \tau\sigma x$ (take $y = \tau\sigma x$). But $\tau\sigma\tau\sigma x$ is of the form $\tau\sigma y$ for some $y < x$, so $\tau\sigma\tau\sigma x$ is a predecessor of $\tau\sigma x$.

5.4 Formalisation in Pure Type Systems

The preceding derivation of a contradiction from the existence of a paradoxical universe can be formalised in λHOL: we can find a term of type $\bot$ in the context $\mathcal{U} : \Box$, $\sigma : (\mathcal{U} \to \wp\mathcal{U})$, $\tau : (\wp\mathcal{U} \to \mathcal{U})$, $0 : \forall X : \wp\mathcal{U}. (\sigma\ (\tau\ X)) =_{\wp\mathcal{U}} \lambda u : \mathcal{U}. \exists x : \mathcal{U}. ((X\ x) \wedge u =_{\mathcal{U}} (\tau\ (\sigma\ x)))$. Here for each set $\mathcal{A}$, $=_{\mathcal{A}}$ denotes Leibniz equality on $\mathcal{A}$. Instead of $=_{\wp\mathcal{U}}$ one can also take the weaker relation of 'having the same elements'. Since the proof does not use *ex falso sequitur quodlibet* at all, $\bot$ can be replaced by any formula φ.

We need a stronger Pure Type System to prove $\bot$ in the empty context. Let $\mathcal{U}$ be the paradoxical universe given in the example. Using the rule $(\triangle, \Box)$, we formalise the power set $\wp\mathcal{U}$ as the term $\Pi\mathcal{X} : \Box. ((\mathcal{X} \to \wp\mathcal{X}) \to \wp\mathcal{X})$ of type $\Box$. In other words, we read $\Pi u : \mathcal{U}. *$ as abbreviation for $\Pi\mathcal{A} : \Box. \Pi\prec : (\mathcal{A} \to \wp\mathcal{A}). \Pi a : \mathcal{A}. *$. It is not necessary to find a term corresponding to $\mathcal{U}$ itself. For example, $\forall u : \mathcal{U}. (X\ u)$ stands for $\forall\mathcal{A} : \Box. \forall\prec : (\mathcal{A} \to \wp\mathcal{A}). \forall a : \mathcal{A}. ((\{X\ \mathcal{A}\} \prec)\ a)$. Note that the rule $(\triangle, *)$ is needed for the quantification over $\Box$. So this can be done in λU.

One can also formalise the preceding paradox in λU$^-$, using for example the paradoxical universe $\Pi\mathcal{X} : \Box. ((\wp\mathcal{X} \to \mathcal{X}) \to \mathcal{X})$ or the following one:

$$\mathcal{U} \equiv \Pi\mathcal{X} : \Box. ((\wp\mathcal{X} \to \mathcal{X}) \to \wp\mathcal{X})$$

Define a term of type $(\wp\mathcal{U} \to \mathcal{U})$:

$$\tau \equiv \lambda X : \wp\mathcal{U}. \Lambda\mathcal{A} : \Box. \lambda c : (\wp\mathcal{A} \to \mathcal{A}). \lambda a : \mathcal{A}. \varphi$$

Here φ expresses that a is of the form $(c\ (\{x\ \mathcal{A}\}\ c))$ for some $x : \mathcal{U}$ such that $(X\ x)$. (Note that $(\{x\ \mathcal{A}\}\ c) : \wp\mathcal{A}$, so $(c\ (\{x\ \mathcal{A}\}\ c)) : \mathcal{A}$.) This can be done without defining $\exists$, $\wedge$, and $=_{\mathcal{A}}$, as follows:

$$\varphi \equiv \forall P : \wp\mathcal{A}. [\forall x : \mathcal{U}. [(X\ x) \Rightarrow (P\ (c\ (\{x\ \mathcal{A}\}\ c)))] \Rightarrow (P\ a)]$$

Define a term of type $(\mathcal{U} \to \wp\mathcal{U})$:

$$\sigma \equiv \lambda x : \mathcal{U}. (\{x\ \mathcal{U}\}\ \tau)$$

Then one easily verifies that $(\mathcal{U}, \sigma, \tau)$ is paradoxical. In fact, for each X of type $\wp\mathcal{U}$, $(\sigma\ (\tau\ X))$ is β-equal to the term corresponding to "the intersection of all subsets P of $\mathcal{U}$ containing $(\tau\ (\sigma\ x))$ for each x in X". This simplifies the formal proof term, since β-conversion between two propositions φ and χ does not "count" as a proof step: if P is a proof term of type φ, then P also has type χ.

In this way, one finds a term of type $\bot$ in λU^- that uses *modus ponens* 12 times. It is of the form $[P\ Q]$, where P is a normal term of type "Ω is not well-founded" and Q is a normal term of type "Ω is well-founded". The terms $\mathcal{U}$, Ω, "Ω is well-founded", P, and Q have length 11, 163, 285, 1849, and 1405.

6 Powerful Universes

The proof term that we presented earlier, is shorter and has a simpler reduction behaviour. Furthermore, we defined terms τt and σs without using quantifiers or connectives. The main idea of the simplification is to consider the power set of the power set of some universe $\mathcal{U}$. In fact, we already considered $\wp\wp\mathcal{U}$ implicitly: Let for each subset C of $\wp\mathcal{U}$, $\bigcap C$ be the *intersection* of all members Y of C, that is, $\bigcap C \equiv \{y|\text{for each } Y \text{ in } C,\ y \text{ is in } Y\}$. Then, by definition, $\{x|x \text{ is well-founded}\} \equiv \bigcap\{X|X \text{ is inductive}\}$ and for each X in $\wp\mathcal{U}$, $\{\tau\sigma x|x \text{ in } X\} \equiv \bigcap\{Y|\text{for each } x \text{ in } X,\ \tau\sigma x \text{ is in } Y\}$. In the example of a paradoxical universe, we defined $\sigma(\mathcal{A},\prec,a) \equiv \bigcap\{X|\text{for each } b \prec a,\ (\mathcal{A},\prec,b) \text{ is in } X\}$. The relation $\prec$ on $\mathcal{A}$ induces a function $s : \mathcal{A} \to \wp\wp\mathcal{A}$ as follows: $sa = \{B \text{ in } \wp\mathcal{A}|\text{for each } b \prec a,\ b \text{ is in } B\}$. In terms of this function, $\sigma(\mathcal{A},\prec,a) = \bigcap\{X|\{b \text{ in } \mathcal{A}|(\mathcal{A},\prec,b) \text{ is in } X\} \text{ is in } sa\}$. Note that if $\prec$ is (Leibniz) equality on $\mathcal{A}$, then the function s can be defined without using quantifiers or connectives: $sa = \{B \text{ in } \wp\mathcal{A}|a \text{ is in } B\}$.

By using the fact that no set is isomorphic to the power set of its power set, John Reynolds (1984) proved that there is no set-theoretic model of polymorphic (or second-order) typed λ-calculus. By refining this result and using a computer, Coquand (1994) found a formal proof of a contradiction in system U^-. He considered the universe $\mathsf{A}_0 \equiv \Pi\mathcal{X} : \Box.\,((\wp\wp\mathcal{X}\to\mathcal{X})\to\mathcal{X})$ and defined functions $\mathsf{match} : \mathsf{A}_0 \to \wp\wp\mathsf{A}_0$ and $\mathsf{intro} : \wp\wp\mathsf{A}_0 \to \mathsf{A}_0$. Then he showed that these functions constitute an isomorphism with respect to certain partial equivalence relations. In fact, $(\mathsf{A}_0, \mathsf{match}, \mathsf{intro})$ can also be used to formulate a Burali-Forti like paradox: it is an example of a *powerful universe.*

6.1 From a Universe to the Power Set of Its Power Set and Back

Let us call a universe $\mathcal{U}$, together with functions $\sigma : \mathcal{U} \to \wp\wp\mathcal{U}$ and $\tau : \wp\wp\mathcal{U} \to \mathcal{U}$, *powerful* if for each C in $\wp\wp\mathcal{U}$, $\sigma\tau C = \{X|\text{the set } \{y|\tau\sigma y \text{ is in } X\} \text{ is in } C\}$.

Each function $f : \mathcal{S} \to \mathcal{T}$ induces a function $f^\star : \wp\mathcal{T} \to \wp\mathcal{S}$ as follows: for each subset Y of $\mathcal{T}$, $f^\star Y = \{x \text{ in } \mathcal{S}|fx \text{ is in } X\}$. Using this notation, we see that $(\mathcal{U},\sigma,\tau)$ is powerful if and only if the composition $\sigma\circ\tau$ is equal to $(\tau\circ\sigma)^{\star\star}$. Note that if $(\mathcal{U},\sigma,\tau)$ is powerful, then $(\wp\mathcal{U},\tau^\star,\sigma^\star)$ is also powerful: $\tau^\star\circ\sigma^\star = (\sigma\circ\tau)^\star = (\tau\circ\sigma)^{\star\star\star} = (\sigma^\star\circ\tau^\star)^{\star\star}$. (Here we do not need extensionality.)

6.2 Example of a Powerful Universe

Let $\mathcal{U}$ be the universe of all *triples* $(\mathcal{A}, s, a)$ consisting of a set $\mathcal{A}$, a function $s : \mathcal{A} \to \wp\wp\mathcal{A}$, and an element a of $\mathcal{A}$. For each triple $(\mathcal{A}, s, a)$, let $\sigma(\mathcal{A}, s, a)$ be the collection of all subsets X of $\mathcal{U}$ such that $\{b \text{ in } \mathcal{A} | (\mathcal{A}, s, b) \text{ is in } X\}$ is in sa. Since σ is a function from $\mathcal{U}$ to $\wp\wp\mathcal{U}$, σ^{**} is a function from $\wp\wp\mathcal{U}$ to $\wp\wp\wp\wp\mathcal{U}$. For each C in $\wp\wp\mathcal{U}$, let τC denote the triple $(\wp\wp\mathcal{U}, \sigma^{**}, C)$.

In order to verify that $(\mathcal{U}, \sigma, \tau)$ is powerful, let C in $\wp\wp\mathcal{U}$ and X in $\wp\mathcal{U}$. Then the following propositions are equivalent (by definition):

X is in $\sigma\tau C$;
X is in $\sigma(\wp\wp\mathcal{U}, \sigma^{**}, C)$;
$\{b \text{ in } \wp\wp\mathcal{U} | (\wp\wp\mathcal{U}, \sigma^{**}, b) \text{ is in } X\}$ is in $\sigma^{**}C$;
$\sigma^{*}\{b \text{ in } \wp\wp\mathcal{U} | \tau b \text{ is in } X\}$ is in C;
$\{y \text{ in } \mathcal{U} | \tau\sigma y \text{ is in } X\}$ is in C.

6.3 Contradiction from the Existence of a Powerful Universe

Let $(\mathcal{U}, \sigma, \tau)$ be powerful. We will derive a contradiction in a similar way as for paradoxical universes.

Elements of $\mathcal{U}$ will be denoted by $x, y, \ldots$ and subsets of $\mathcal{U}$ by $X, Y, \ldots$. For each x, σx is in $\wp\wp\mathcal{U}$. $(\mathcal{U}, \sigma, \tau)$ is powerful, so:

$$\sigma\tau\sigma x = \{X | \text{the set } \{y | \tau\sigma y \text{ is in } X\} \text{ is in } \sigma x\}$$

We say that y is a *predecessor* of x (and we write $y < x$) if for each X in σx, y is in X (in other words, if y is in $\bigcap \sigma x$). One can easily prove that if $y < x$ then $\tau\sigma y < \tau\sigma x$. We will only do this for the special case $y = \tau\sigma x$. Note that if X is in σx, then each predecessor of x is in X.

X is called *inductive* if the following holds: for each x in $\mathcal{U}$, if X is in σx, then x is in X. We say that x is *well-founded* if x is in each inductive X. (Note that it is *not* clear whether $\{x | x \text{ is well-founded}\}$ is inductive: if one tries to prove this, one would like to use something like: if Y is in σx and $Y \subseteq X$, then X is in σx.)

Let $\Omega \equiv \tau\{X | X \text{ is inductive}\}$. $(\mathcal{U}, \sigma, \tau)$ is powerful, so:

$$\sigma\Omega = \{X | \text{ the set } \{y | \tau\sigma y \text{ is in } X\} \text{ is inductive}\}$$

We claim that Ω is well-founded:

Let X be inductive. In order to prove that Ω is in X, we only need to show that X is in $\sigma\Omega$. In other words, we show that the set $\{y | \tau\sigma y \text{ is in } X\}$ is inductive. So let x be in $\mathcal{U}$. Since X is inductive, we have the following: if X is in $\sigma\tau\sigma x$, then $\tau\sigma x$ is in X. In other words, if the set $\{y | \tau\sigma y \text{ is in } X\}$ is in σx, then x is in $\{y | \tau\sigma y \text{ is in } X\}$. This is exactly what we had to prove.

In order to show that Ω is *not* well-founded, we first prove that the set $\{y | \tau\sigma y \not< y\}$ is inductive:

Let x be such that $\{y | \tau\sigma y \not< y\}$ is in σx. Then $\tau\sigma x \not< x$. For suppose that $\tau\sigma x < x$. In other words, for each X in σx, $\tau\sigma x$ is in X. Applying this to the set $\{y | \tau\sigma y \not< y\}$, which is in σx, we see that $\tau\sigma\tau\sigma x \not< \tau\sigma x$. On the other hand,

$\tau\sigma\tau\sigma x < \tau\sigma x$: Let X be in $\wp\mathcal{U}$. We have to show the following: if X is in $\sigma\tau\sigma x$, then $\tau\sigma\tau\sigma x$ is in X. In other words, if $\{y|\tau\sigma y$ is in $X\}$ is in σx, then $\tau\sigma x$ is in $\{y|\tau\sigma y$ is in $X\}$. This follows from the assumption that $\tau\sigma x < x$, i.e. for each Y in σx, $\tau\sigma x$ is in Y.

Now suppose that Ω is well-founded. Then, since $\{y|\tau\sigma y \not< y\}$ is inductive, $\tau\sigma\Omega \not< \Omega$. On the other hand, $\tau\sigma\Omega < \Omega$: Let X be in $\wp\mathcal{U}$. We have to show: if X is in $\sigma\Omega$, then $\tau\sigma\Omega$ is in X. In other words, if the set $\{y|\tau\sigma y$ is in $X\}$ is inductive, then Ω is in $\{y|\tau\sigma y$ is in $X\}$. This follows from the assumption that Ω is well-founded, i.e. for each inductive Y, Ω is in Y.

7 Reduction Behaviour

Douglas Howe (1987) used a computer to study the reduction behaviour of a massive term corresponding to one particular proof of Girard's paradox. Just like the proofs we gave, it did not use *ex falso sequitur quodlibet*, so $\bot$ can be replaced by a variable of type $*$. Using this, Howe constructed a *looping combinator* (but not a fixed-point combinator). (See (Coquand and Herbelin 1994) and also (Geuvers and Werner 1994).)

We now return to the proof term that we presented in Section 3. It formalises the preceding derivation of a contradiction, using some other powerful universe than the two that we mentioned earlier.

One easily verifies that $(\mathcal{U}, \lambda s : \mathcal{U}.\, \sigma s, \lambda t : \wp\wp\mathcal{U}.\, \tau t)$ is powerful: in fact, for each term t of type $\wp\wp\mathcal{U}$, the term $\sigma\tau t$ β-reduces to $\lambda p : \wp\mathcal{U}.\, (t\ \lambda x : \mathcal{U}.\, (p\ \tau\sigma x))$. One can calculate the normal form of $(\{\tau\sigma\tau\sigma \cdots \tau\sigma x\ \mathcal{X}\}\ f)$, where x, $\mathcal{X}$, and f are variables. It contains nested expressions of the form $\lambda p : \wp\mathcal{X}.\, (\sigma x'\ \lambda x'' : \mathcal{U}.\, (p\ (f\ \cdots)))$.

For each term s of type $\mathcal{U}$, let $\Theta^0 s$ be s and, for each natural number n, let $\Theta^{n+1}s$ be the normal form of $\tau\sigma\Theta^n s$. For each term p of type $\wp\mathcal{U}$, let $\Theta^\star_0 p$ be p and, for each natural number n, let $\Theta^\star_{n+1}p$ be $\lambda y : \mathcal{U}.\, (p\ \Theta^{n+1}y)$. Then, for variables x and p, for each natural number n, the normal form of $(\sigma\Theta^n x\ p)$ is $(\sigma x\ \Theta^\star_n p)$.

The fact that β-reduction of the proof term goes on indefinitely, is caused by steps that correspond to the rule $(*, *)$, that is, replacing a subterm of the form [suppose $n : \varphi.\, P\ Q$] by $P[Q/n]$. One can show that each infinite sequence of β-reduction steps, starting with a term in λU, contains such a step. So we can concentrate on "big steps": steps that correspond to $(*, *)$, followed by a maximal sequence of steps corresponding to other rules.

Let n be a natural number. We first define two propositions:

$$\varphi_n \equiv \forall p : \wp\mathcal{U}.\, [\forall x : \mathcal{U}.\, [(\sigma x\ \Theta^\star_n p) \Rightarrow (p\ \Theta^n x)] \Rightarrow (p\ \Theta^n \Omega)]$$
$$\psi_n \equiv \forall x : \mathcal{U}.\, [(\sigma x\ \Theta^\star_n \Delta) \Rightarrow \neg\forall p : \wp\mathcal{U}.\, [(\sigma x\ \Theta^\star_n p) \Rightarrow (p\ \Theta^{n+1}x)]]$$

So φ_n expresses that $\Theta^n\Omega$ is in each subset X of $\mathcal{U}$ for which $\{y|\Theta^n y$ is in $X\}$ is inductive. Note that φ_0 is "Ω is well-founded" and φ_{n+1} is the normal form of "$\Theta^{n+1}\Omega < \Theta^n\Omega$". The proposition ψ_n expresses that $\{y|\Theta^{n+1}y \not< \Theta^n y\}$ is inductive. We also define five proof terms:

$$R_n \equiv \mathsf{let}\ p : \wp\mathcal{U}.\,\mathsf{suppose}\ 1 : \forall x : \mathcal{U}.\,[(\sigma x\ \Theta^\star_n p) \Rightarrow (p\ \Theta^n x)].\ [\langle 1\ \Omega\rangle\ \mathsf{let}\ x : \mathcal{U}.\,\langle 1\ \tau\sigma x\rangle]$$

$$M_n \equiv \mathsf{let}\ x : \mathcal{U}.\,\mathsf{suppose}\ 2 : (\sigma x\ \Theta^\star_n \Delta).\ \mathsf{suppose}\ 3 : \forall p : \wp\mathcal{U}.\,[(\sigma x\ \Theta^\star_n p) \Rightarrow (p\ \Theta^{n+1} x)].\ [[\langle 3\ \Delta\rangle\ 2]\ \mathsf{let}\ p : \wp\mathcal{U}.\,\langle 3\ \lambda y : \mathcal{U}.\,(p\ \tau\sigma y)\rangle]$$

$$P_n \equiv \mathsf{suppose}\ 4 : \psi_n.\,\mathsf{suppose}\ 0 : \varphi_n.\,[[\langle 0\ \Delta\rangle\ 4]\ \mathsf{let}\ p : \wp\mathcal{U}.\,\langle 0\ \lambda y : \mathcal{U}.\,(p\ \tau\sigma y)\rangle]$$

$$L_n \equiv \mathsf{suppose}\ 0 : \varphi_n.\,[[\langle 0\ \Delta\rangle\ M_n]\ \mathsf{let}\ p : \wp\mathcal{U}.\,\langle 0\ \lambda y : \mathcal{U}.\,(p\ \tau\sigma y)\rangle]$$

$$Q_n \equiv \mathsf{suppose}\ 4 : \psi_n.\,[\langle 4\ \Omega\rangle\ \mathsf{let}\ x : \mathcal{U}.\,\langle 4\ \tau\sigma x\rangle]$$

Then R_n proves φ_n, M_n proves ψ_n, P_n proves $[\psi_n \Rightarrow \neg\varphi_n]$, L_n proves $\neg\varphi_n$, and Q_n proves $[\psi_n \Rightarrow \neg\varphi_{n+1}]$.

Note that $[L_0\ R_0]$ is the proof term that we presented in Section 3.

For each natural number n, $[[P_n\ M_n]\ R_n]$ reduces in one step to $[L_n\ R_n]$. (Variable 4 disappears.) This reduces in a big step to $[[Q_n\ M_n]\ R_{n+1}]$. (Variable 0 disappears and some occurrences of 1 are renamed as 4.) This reduces to $[[P_{n+1}\ M_{n+1}]\ R_{n+1}]$. (Variable 4 disappears and some occurrences of 2 and 3 are renamed as 4 and 0.)

So these proof terms of type $\perp$ in λU^- reduce in three big steps to a similar proof term: only the types of the proof variables change a little bit.

References

Barendregt, H.P.: Typed lambda calculi, in: *Handbook of Logic in Computer Science (Vol. 2)*, S. Abramsky et al. (editors), Clarendon Press, Oxford (1992)

Burali-Forti, C.: Una questione sui numeri transfiniti, *Rendiconti del Circolo Matematico di Palermo* **11** (1897) 154–164

Cantor, G.: Beiträge zur Begründung der transfiniten Mengenlehre, II, *Mathematische Annalen* **49** (1897) 207–246

Coquand, Th.: An analysis of Girard's paradox, in: *Proceedings Symposium on Logic in Computer Science: Cambridge, Massachusetts, June 16–18, 1986*, IEEE Computer Society Press, Washington, D.C. (1986) 227–236

Coquand, Th.: A New Paradox in Type Theory, in: *Logic and philosophy of science in Uppsala: papers from the 9th international congress of logic, methodology and philosophy of science*, D. Prawitz, D. Westerstaahl (editors), Kluwer Academic Publishers, Dordrecht (1994) ?–?

Coquand, Th., Herbelin, H.: *A*-translation and looping combinators in pure type systems, *Journal of Functional Programming* **4** (1994) 77–88

Geuvers, J.H.: *Logics and Type Systems*, Proefschrift, Katholieke Universiteit Nijmegen (1993)

Geuvers, H., Werner, B.: On the Church-Rosser property for Expressive Type Systems and its Consequences for their Metatheoretic Study, in: *Proceedings of the Ninth Annual Symposium on Logic in Computer Science, Paris, France*, IEEE Computer Society Press, Washington, D.C. (1994) 320–329

Girard, J.-Y.: *Interprétation fonctionnelle et élimination des coupures de l'arithmétique d'ordre supérieur*, Thèse de Doctorat d'État, Université Paris VII (1972)

Howe, D.J.: The Computational Behaviour of Girard's Paradox, in: *Proceedings Symposium on Logic in Computer Science: Ithaka, New York, June 22–25, 1987*, IEEE Computer Society Press, Washington, D.C. (1987) 205–214

Mirimanoff, D.: Les antinomies de Russell et de Burali-Forti et le problème fondamental de la théorie des ensembles, *L'Enseignement Mathématique* **19** (1917) 37–52

Reynolds, J.C.: Polymorphism is not Set-Theoretic, in: *Semantics of Data Types*, G. Kahn et al. (editors), Lecture Notes in Computer Science **173**, Springer-Verlag, Berlin Heidelberg (1984) 145–156

Russell, B.: *The Principles of Mathematics*, Cambridge University Press, Cambridge, G.B. (1903)

Basic Properties of Data Types with Inequational Refinements (Extended Abstract)

Hidetaka Kondoh

Advanced Research Laboratory, Hitachi, Ltd.
Hatoyama, Saitama 350-03, Japan
kondoh@harl.hitachi.co.jp

Abstract

In this work we propose a typed functional wide-spectrum (i.e. both programming and specification) language Final, an extension of Cardelli-Wegner's Fun enriched with the fixed-point construction on expressions and inequational refinements (assertions) for types. The inequational assertion has the form $\forall\, x_1 : \sigma_1. \ \ldots\ .\forall\, x_n : \sigma_n . e_1 \leqslant e_2 : \tau$ and its intuitive meaning is that if the expression e_1 terminates, then it gives a value that e_2 does but the computation of e_1 may diverge. Hence it may be seen, in some sense, a weaker form of equations in algebraic specification of abstract data types. The type system has a subtype relation as Fun does, and the subtype relation reflects the strength of assertions. We demonstrate that inequationally refined record types nicely model (the representation type of) abstract data types à la algebraic specification by giving some examples. Then we show that the type system is a conservative extension of the original one, and give a domain-theoretic semantics to its sublanguage without existential types. This semantics assures that we can freely use general recursive functions over inequationally refined types, that is, for any function $f : \tau \to \tau$, its least fixed-point belongs to τ even if τ is a refined type. This shows that our language, Final, is an appropriate base language in designing modular wide-spectrum languages with both general recursions and equationally specified data types which allow us to manipulate implementation modules of abstract data types with general recursion as we do for ordinal data like integers.

1. Introduction

There are essentially two approaches to formal modeling of abstract data types (*ADTs*): one is from logic, using typed λ-calculi; the other is from algebra, using first-order equational logic. But neither succeeds to fully capture our intuitions on abstract data types in computer programming. The motivation of our work originates from the editors' Foreword of the proceedings of International Workshop of Semantics of Data Types [KMP 84]. It says that *"The Symposium was intended to bring these somewhat disparate groups together with a view to promoting a common language ...,"* but unfortunately there have been hardly any efforts to integrate logical and algebraic approaches to abstract data types by now. In this paper, we propose a type system incorporating inequations as *assertions* (we reserve the term *axiom* for meta-theoretical usages) with the 2nd-order calculus Fun of Cardelli and Wegner. It shows that our calculus supports a novel inheritance mechanism based on algebra-like structures of data types and conservatively extends the original calculus. Moreover the subsystem without existential types is sound with respect to the complete partial equivalence relation semantics of data types.

An ADT hides two kinds of information; one is the type of the representation of the data structure to be abstracted by that ADT, which we call the *representation type*; the other is a suite of implementations of *operations associated with* that ADT, which are well-typed with respect to the representation type chosen for that ADT, and we call the particular suite of types of associated operations for an ADT its *implementation type*. Then the inheritances between ADTs are the order relationships based on the relative "richness" of structures, e.g., stacks vs. dequeues, queues vs. dequeues, etc.

The algebraic approach models an ADT as a many-sorted first-order equational theory. Such theory is specified by a set of operator symbols (*signatures*) and equations to define the behavior of the associated operations (denoted by signatures) of the ADT [Ehrig and Mahr 85], but there is still disagreement about whether an ADT should be interpreted as the class of initial algebras or that of all algebras, etc. Furthermore, this approach has had little success in treating higher-order functions and in extending to higher-order logic systems for polymorphism, dependent types, etc.

The logical approach uses record types as the basic tool for modeling ADTs. Recent works on this approach can be classified into two streams; one is the modeling of inheritances in ADTs, introduced in Cardelli's pioneering paper, "A Semantics of Multiple Inheritances" [Cardelli 84]; the other focuses on the formalization of the information hiding mechanism of ADTs by existentially quantified types, a concept originated in "Abstract Data Types Have Existential Type" [Mitchell and Plotkin 85]. We summarize the correspondence between ADTs and existentially quantified record types according to the Cardelli-Mitchell-Plotkin modeling (Mitchell and Plotkin originally have used product types rather than record types, but we use record types as in [Cardelli and Wegner 85]).

abstract data type	existentially quantified record type
implementation type	record type
associated operator symbol	record field label
suite of associated operations	record value
associated operation	value bound to a record field label
inheritance relation	subtype relation on record types

Note that Cardelli's work and most works in his stream are on object-oriented programming rather than on ADTs. But Cardelli's idea to the method inheritances is applicable to inheritances of associated operations of ADTs by explicit parameterizations of methods with respect to objects' internal states, i.e., instance variables.

The essential problem in the above modeling is the ignorance of *algebraic structures* of ADTs in the sense of the algebraic approach. This problem has been pointed out by Reynolds in [Reynolds 83 and 85]. Thus the logical approach is far from satisfactory. In other words, only *anarchic* algebras correspond to record types, while the enriched record types of our system denote *non-anarchic* algebra-like structures.

Please note that results are shown without proofs in this extended abstract due to the limitation of the space.

2. Why Is Refinement Necessary?

In this section we show the problem of the Cardelli-Mitchell-Plotkin modeling of ADTs with a concrete example. To display examples compactly, we informally use Standard ML like syntax [MTH 90] for global definitions.

Example. *Stack* (of natural numbers) has as its equipped operations: *new*, to create a empty stack; *isnew*, to check a stack of its emptiness; *push*, to add some number to a stack; *top*, to see the the top (= lastly pushed) element; and *pop*, to remove the top element from a stack. Suppose we have *List* as a standard type constructor, and *Nat* and *Bool* as base types in Fun, and we select the list of natural numbers as the representation type for *Stack*, i.e.:

type *StackValRep* = *List*[*Nat*];

Then we can define the implementation type:

```
type StackOpImpl = {new: StackValRep,
                    isnew: StackValRep → Bool,
                    push: Nat → StackValRep → StackValRep,
                    top: StackValRep → Nat,
                    pop: StackValRep → StackValRep};
```

Now we can give a suite of implementations of equipped operations of the type *Stack* as a record expression as follows:

```
val aStackOpImpl = {new = nil,
                    isnew = λs: StackValRep.isnull(s),
                    push = λi: Nat.λs: StackValRep.cons(i)(s),
                    pop = λs: StackValRep.tail(s),
                    top = λs: StackValRep.head(s)};
```

This behaves in the *last-in first-out* manner as expected for stacks. By this mechanism, for example,

```
aStackOpImpl.top(
    aStackOpImpl.pop(aStackOpImpl.push(2)(aStackOpImpl.push(1)(aStackOpImpl.new))))
```

yields 1. On the other hand, with the following suite of implementations

```
val anotherStackOpImpl = {new = nil,
                          isnew = λs: StackValRep.isnull(s),
                          push = λi: Nat.λs: StackValRep.cons(i)(s),
                          pop = fix(λp: StackValRep → StackValRep.λs: StackValRe
                                    if length(s) ≤ 1 then nil else cons(head(s))(p
                          top = fix(λt: StackValRep → Nat.λs: StackValRep.
                                    if length(s) ≤ 1 then head(s) else t(tail(s)))}
```

where *length* is the usual length function for lists, the value of the expression

```
anotherStackOpImpl.top(anotherStackOpImpl.pop(
    anotherStackOpImpl.push(2)(anotherStackOpImpl.push(1)(anotherStackOpImpl.new))))
```

is 2, since the *anotherStackOpImpl* acts in the *first-in first-out* fashion. In fact, *anotherStackOpImpl* is an implementation suite adequate for queues rather than for stacks but still has the type *StackOpImpl*.

From this example, we can see that the Cardelli-Mitchell-Plotkin modeling cannot distinguish between behaviors of stacks and of queues, and treats identically stacks and queues having the same type. This limitation of their approach is the problem which this work attempts to solve.

3. The Language Final

In this section, we define the language Final (Fun with inheritances of algebra-like structures), a superset of Fun enriched with inequational refinement types.

The syntax of Final is shown in Fig. 1. The original Fun (with fixed-point construction) is the sublanguage generated by production rules without (∗) marks, and $\mathbf{Exp}_{\mathrm{Fun}} \subset \mathbf{Exp}$ and $\mathbf{Type}_{\mathrm{Fun}} \subset \mathbf{Type}$ denote the set of expressions and that of types, respectively, belonging to Fun.

In the rules, the order of occurrences of assertions in a refined type is insignificant as is the case for the order of field labels/tags in record/variant types, and the change of implementation variables bound by refined-types is also insignificant like in the ususal λ-binding. Before stating syntactical constraints to Final, we need a definition, which is analogous to the notion of *active subexpression* in [Plotkin 77].

Definition 1. Let $e \in \mathbf{Exp}$ and $v \in \mathbf{Var} \cup \mathbf{IVar}$. Then v *occurs formally strictly in* e iff one of the following conditions holds:

(1) $e \equiv x$ and $v \equiv x$;

(2) $e \equiv r$ and $v \equiv r$;

(3) $e \equiv e'.l$ and v occurs formally strictly in e';

(4) $e \equiv e'e''$ and v occurs formally strictly in e';

(5) $e \equiv \lambda x{:}\,\sigma.e'$ and $v \not\equiv x$ and and v occurs formally strictly in e';

(6) $e \equiv \mathbf{case}\ e'\ l_1\ \mathbf{then}\ e_1, \ldots, l_n\ \mathbf{then}\ e_n$ and v occurs formally strictly in e';

(7) $e \equiv \Lambda t.e'$ and v occurs formally strictly in e';

(8) $e \equiv e'[\tau]$ and v occurs formally strictly in e';

(9) $e \equiv \mathbf{pack}_{\exists t<:\sigma.\tau}(\sigma', e')$ and v occurs formally strictly in e';

(10) $e \equiv \mathbf{open}\ e'\ \mathbf{as}\ (t, x)\ \mathbf{in}\ e''$ and either one of the followings holds:

   (a) v occurs formally strictly in e'', or

   (b) v occurs formally strictly in e' and x occurs formally strictly in e''; or

(11) $e \equiv \mathbf{fix}(e')$ and v occurs formally strictly in e'.

Then syntactical constraints to Final are:

(a) each assertion of a refined type must be closed by **forall** quantification except for free occurrences of the implementation variable bound by the refined type containing that assertion; and

(b) the implementation variable bound by a refined type must occur formally strictly in the left-hand expression of each assertion of that refined type.

The first constraint is necessary for giving semantics for the proof theory of Final, while the second one is essential for constructing semantics of Final.

	$e \in$ **Exp**	The set of expressions:
	$e ::= c$	constants,
	$\mid x$	(ordinary) variables,
(*)	$\mid r$	implementation variables,
	$\mid \lambda x{:}\,\sigma.e$	abstractions,
	$\mid e_1 e_2$	applications,
	$\mid \{l_1 = e_1, \ldots, l_n = e_n\}$	record terms ($n \geq 0$),
	$\mid e.l$	field selections,
	$\mid [l = e]$	taggings,
	$\mid$ **case** e **of**	
	l_1 **then** $e_1, \ldots, l_n$ **then** e_n	selections by tag ($n \geq 0$),
	$\mid \Lambda t <: \sigma.e$	type abstractions,
	$\mid e[\tau]$	type applications,
	$\mid \mathbf{pack}_{\exists t <: \sigma.\tau}(\sigma', e)$	packing up implemantations,
	$\mid$ **open** e **as** (t, x) **in** e'	unpacking implemantations,
	$\mid \mathbf{fix}(e)$	recursions by fixed-points.
	$\sigma, \tau \in$ **Type**	The set of types:
	$\sigma ::= \top$	the largest type for unbounded $\forall$ and $\exists$,
	$\mid \iota$	base types,
	$\mid t$	type variables,
	$\mid \sigma_1 \to \sigma_2$	functional types,
	$\mid \{l_1{:}\,\sigma_1, \ldots, l_n{:}\,\sigma_n\}$	record types ($n \geq 0$),
	$\mid [l_1{:}\,\sigma_1, \ldots, l_n{:}\,\sigma_n]$	variant types ($n \geq 0$),
	$\mid \forall t <: \sigma.\tau$	bounded universally quantified types,
	$\mid \exists t <: \sigma.\tau$	bounded existentially quantified types,
(*)	$\mid \{r{:}\,\sigma \mid \phi_1, \ldots, \phi_k\}$	refined types ($k \geq 0$).
(*)	$\phi, \psi \in$ **Assertion**	The set of assertions:
(*)	$\phi ::= e_1 \leqslant e_2 : \sigma$	atomic assertions,
(*)	$\mid$ **forall** $x{:}\,\sigma.\phi$	quantified assertions.

Note: We leave details of the following syntactic categories unspecified:

	$x \in$ **Var**	The set of (ordinary) variables;
(*)	$r \in$ **IVar**	The set of implementation variables;
	$t \in$ **TVar**	The set of type variables;
	$c \in$ **Const**	The set of constant symbols;
	$l \in$ **Label**	The set of record field labels and variant tags;
	$\iota \in$ **BaseType**	The finite set of base types.

Figure 1. Syntax of Final.

$$\{\text{BASE}\}\quad C \triangleright \iota <: \iota$$

$$\{\text{TOP}\}\quad C \triangleright \sigma <: \top$$

$$\{\text{TVAR}\}\quad C[t <: \sigma] \triangleright t <: \sigma$$

$$\{\text{TRANS}\}\quad \frac{C \triangleright \sigma_1 <: \sigma_2 \qquad C \triangleright \sigma_2 <: \sigma_3}{C \triangleright \sigma_1 <: \sigma_3}$$

$$\{\text{ARROW}\}\quad \frac{C \triangleright \sigma' <: \sigma \qquad C \triangleright \tau <: \tau'}{C \triangleright \sigma \to \tau \ <: \ \sigma' \to \tau'}$$

$$\{\text{RECORD}\}\quad \frac{C \triangleright \sigma_1 <: \tau_1 \ \ldots \ C \triangleright \sigma_n <: \tau_n}{C \triangleright \{l_1 : \sigma_1, \ldots, l_n : \sigma_n, \ldots, l_{n+m} : \sigma_{n+m}\} \ <: \ \{l_1 : \tau_1, \ldots, l_n : \tau_n\}}$$

$$\{\text{VARIANT}\}\quad \frac{C \triangleright \sigma_1 <: \tau_1 \ \ldots \ C \triangleright \sigma_n <: \tau_n}{C \triangleright [l_1 : \sigma_1, \ldots, l_n : \sigma_n] \ <: \ [l_1 : \tau_1, \ldots, l_n : \tau_n, \ldots, l_{n+m} : \tau_{n+m}]}$$

$$\{\text{ALL}\}\quad \frac{C[t <: \sigma] \triangleright \tau <: \tau'}{C \triangleright \forall t <: \sigma.\tau \ <: \ \forall t <: \sigma.\tau'} \qquad (t \notin \text{FTV}(C))$$

$$\{\text{EXIST}\}\quad \frac{C[t <: \sigma] \triangleright \tau <: \tau'}{C \triangleright \exists t <: \sigma.\tau \ <: \ \exists t <: \sigma.\tau'} \qquad (t \notin \text{FTV}(C))$$

$$(*)\ \{\text{REFINE}\}\quad \frac{\bigwedge_{i=1}^{k} (\varnothing, \{r : \sigma\}, C \triangleright \phi_i[r_1 := r]) \vdash_{\text{FINAL}} \bigwedge_{j=1}^{l} (\varnothing, \{r : \tau\}, C \triangleright \psi_j[r_2 := r]) \qquad C \triangleright \sigma <: \tau}{C \triangleright \{r_1 : \sigma \mid \phi_1, \ldots, \phi_k\} \ <: \ \{r_2 : \tau \mid \psi_1, \ldots, \psi_l\}}$$

where

- $\bigwedge_{i=1}^{k} (\varnothing, \{r : \sigma\}, C \triangleright \phi_i[r_1 := r]) \vdash_{\text{FINAL}} \bigwedge_{j=1}^{l} (\varnothing, \{r : \tau\}, C \triangleright \psi_j[r_2 := r])$ is a short-hand notation meaning that for each $1 \leq j \leq l$,
  $$\varnothing, \{r : \sigma\}, C \triangleright \phi_1[r_1 := r], \ldots, \varnothing, \{r : \sigma\}, C \triangleright \phi_k[r_1 := r] \vdash_{\text{FINAL}} \varnothing, \{r : \tau\} \triangleright \psi_j[r_2 := r];$$
- r is a fresh implementation variable.

Figure 2. The Subtyping Axioms and the Rules of Final.

The subtype relation on Final and its sublanguage Fun is defined in Fig. 2 where (*) again means that such axiom/rule is specific to Final. A *subtyping judgment* has the form

$$C \triangleright \sigma <: \tau$$

where C is a set of constraints on type variables. Note that our subtype relation <: is not a partial order but a preorder. We write $C \triangleright \sigma \simeq \tau$ if $C \triangleright \sigma <: \tau$ and vice versa.

Now we define the type system, which we call FINAL, of Final. A judgment of the system is either a subtyping judgment or one of the form

$$\Gamma, \Delta, C \triangleright e \leqslant e' : \tau$$

where Γ and Δ are syntactic type assignment for (ordinary) variables and for implementation variables, respectively. The axioms and rules for inference of assertions is shown in Fig. 3. Note that we show the subtyping system and the assertion-inference system separately for the sake of readability, these two systems, however, are *mutually recursive*; it is *not* the case for FUN.

A judgment of the form $\Gamma, \Delta, C \rhd e \leqslant e : \tau$ (note that both sides of the inequation is an identical expression) is called a *typing judgment* and is abbreviated as $\Gamma, \Delta, C \rhd e : \tau$.

We often abbreviate a symmetrical pair of inequations as an equation, e.g.

$$\textbf{forall}\ x_1{:}\,\sigma_1.\cdots.\textbf{forall}\ x_n{:}\,\sigma_n.e = e' : \tau \stackrel{\text{abbrev}}{==}$$
$$\textbf{forall}\ x_1{:}\,\sigma_1.\cdots.\textbf{forall}\ x_n{:}\,\sigma_n.e \leqslant e' : \tau,\ \textbf{forall}\ x_1{:}\,\sigma_1.\cdots.\textbf{forall}\ x_n{:}\,\sigma_n.e' \leqslant e : \tau.$$

Note that each rule, $\langle\beta_{\text{func}}\rangle$, $\langle\beta_{\text{record}}\rangle$, $\langle\beta_{\text{variant}}\rangle$, $\langle\beta_{\forall}\rangle$, $\langle\beta_{\exists}\rangle$ and $\langle\beta_{\text{fix}}\rangle$, whose conclusion is an equational form, actually denotes a pair of rules by this convention.

$$\langle\text{VAR}\rangle \quad \Gamma[x:\sigma], \Delta, C \rhd x \leqslant x : \sigma$$

$$(*)\ \langle\text{IVAR}\rangle \quad \Gamma, \Delta[r:\tau], C \rhd r \leqslant r : \tau$$

$$\langle\text{CONST}\rangle \quad \Gamma, \Delta, C \rhd c_{ij} \leqslant c_{ij} : \iota_i$$

$$(*)\ \langle\text{TRANS}\rangle \quad \frac{\Gamma, \Delta, C \rhd e_1 \leqslant e_2 : \sigma \qquad \Gamma, \Delta, C \rhd e_2 \leqslant e_3 : \sigma}{\Gamma, \Delta, C \rhd e_1 \leqslant e_3 : \sigma}$$

$$\langle\text{WEAK}\rangle \quad \frac{\Gamma, \Delta, C \rhd e_1 \leqslant e_2 : \sigma}{\Gamma[x:\sigma'], \Delta, C \rhd e_1 \leqslant e_2 : \sigma} \qquad (x \notin \text{FV}(e_1) \cup \text{FV}(e_2))$$

$$(*)\ \langle\text{IWEAK}\rangle \quad \frac{\Gamma, \Delta, C \rhd e_1 \leqslant e_2 : \sigma}{\Gamma, \Delta[r:\tau], C \rhd e_1 \leqslant e_2 : \sigma} \qquad (r \notin \text{FV}(e_1) \cup \text{FV}(e_2))$$

$$\langle\text{SUBTYPE}\rangle \quad \frac{\Gamma, \Delta, C \rhd e_1 \leqslant e_2 : \sigma \qquad \sigma <: \sigma'}{\Gamma, \Delta, C \rhd e_1 \leqslant e_2 : \sigma'}$$

$$(*)\ \langle\text{forall-I}\rangle \quad \frac{\Gamma[x:\sigma], \Delta, C \rhd \phi}{\Gamma, \Delta, C \rhd \textbf{forall}\ x{:}\,\sigma.\phi} \qquad (x \notin \text{dom}(\Gamma))$$

$$(*)\ \langle\text{forall-E}\rangle \quad \frac{\Gamma, \Delta, C \rhd \textbf{forall}\ x{:}\,\sigma.\phi \qquad \Gamma, \Delta, C \rhd e : \sigma}{\Gamma, \Delta, C \rhd \phi[x := e]}$$

$$(*)\ \langle\beta_{\text{func}}\rangle \quad \frac{\Gamma[x:\sigma], \Delta, C \rhd e \leqslant e : \sigma' \qquad \Gamma, \Delta, C \rhd e' \leqslant e' : \sigma}{\Gamma, \Delta, C \rhd (\lambda x{:}\,\sigma.e)e' = e[x := e'] : \sigma'}$$

$$\langle\text{ABS}\rangle \quad \frac{\Gamma[x:\sigma], \Delta, C \rhd e \leqslant e' : \sigma'}{\Gamma, \Delta, C \rhd (\lambda x{:}\,\sigma.e) \leqslant (\lambda x{:}\,\sigma.e') : \sigma \to \sigma'}$$

$$\langle\text{APPL}\rangle \quad \frac{\Gamma, \Delta, C \rhd e_1 \leqslant e_1' : \sigma' \to \sigma \qquad \Gamma, \Delta, C \rhd e_2 \leqslant e_2' : \sigma'}{\Gamma, \Delta, C \rhd (e_1 e_2) \leqslant (e_1' e_2') : \sigma}$$

$$(*)\ \langle\beta_{\text{record}}\rangle \quad \frac{\Gamma, \Delta, C \rhd e_1 \leqslant e_1 : \sigma_1 \ \ldots\ \Gamma, \Delta, C \rhd e_n \leqslant e_n : \sigma_n}{\Gamma, \Delta, C \rhd \{l_1 = e_1, \ldots, l_n = e_n\}.l_i = e_i : \sigma_i} \qquad (1 \leq i \leq n)$$

$$\langle\text{RECORD}\rangle \quad \frac{\Gamma, \Delta, C \rhd e_1 \leqslant e_1' : \sigma_1 \ \ldots\ \Gamma, \Delta, C \rhd e_n \leqslant e_n' : \sigma_n}{\Gamma, \Delta, C \rhd \{l_1 = e_1, \ldots, l_n = e_n\} \leqslant \{l_1 = e_1', \ldots, l_n = e_n'\} : \{l_1{:}\,\sigma_1, \ldots, l_n{:}\,\sigma_n\}}$$

$$\langle\text{SELECT}\rangle \quad \frac{\Gamma, \Delta, C \rhd e \leqslant e' : \{l_1{:}\,\sigma_1, \ldots, l_n{:}\,\sigma_n\}}{\Gamma, \Delta, C \rhd e.l_i \leqslant e'.l_i : \sigma_i} \qquad (1 \leq i \leq n)$$

Figure 3. The Axioms and the Rules for Assertions of Final. (to be continued)

$$(*)\ \langle\beta_{\text{variant}}\rangle\ \frac{\Gamma,\Delta,C \rhd e:\sigma_i \quad \Gamma,\Delta,C \rhd e_1 \leqslant e_1:\sigma_1\to\sigma' \ \ldots\ \Gamma,\Delta,C \rhd e_n \leqslant e_n:\sigma_n\to\sigma'}{\Gamma,\Delta,C \rhd \textbf{case}\ [l_i = e]\ \textbf{of}\ l_1\ \textbf{then}\ e_1,\ldots,l_n\ \textbf{then}\ e_n = e_i e:\sigma'} \quad (1\le i\le n)$$

$$\langle\text{VARIANT}\rangle\ \frac{\Gamma,\Delta,C \rhd e \leqslant e':\sigma}{\Gamma,\Delta,C \rhd [l = e] \leqslant [l = e'] : [l{:}\sigma]}$$

$$\langle\text{CASE}\rangle\ \frac{\Gamma,\Delta,C \rhd e \leqslant e' : [l_1{:}\sigma_1,\ldots,l_n{:}\sigma_n] \quad \Gamma,\Delta,C \rhd e_1 \leqslant e_1':\sigma_1\to\sigma' \ \ldots\ \Gamma,\Delta,C \rhd e_n \leqslant e_n':\sigma_n\to\sigma'}{\Gamma,\Delta,C \rhd \textbf{case}\ e\ \textbf{of}\ l_1\ \textbf{then}\ e_1,\ldots,l_n\ \textbf{then}\ e_n \leqslant \textbf{case}\ e'\ \textbf{of}\ l_1\ \textbf{then}\ e_1',\ldots,l_n\ \textbf{then}\ e_n' : \sigma'}$$

$$(*)\ \langle\beta_\forall\rangle\ \frac{\Gamma,\Delta,C[t <: \sigma] \rhd e \leqslant e:\tau \quad C \rhd \sigma' <: \sigma}{\Gamma,\Delta,C \rhd (\Lambda t <: \sigma.e)[\sigma'] = e[t := \sigma'] : \tau[t := \sigma']} \quad (t \notin \text{FTV}(\Gamma)\cup\text{FTV}(C))$$

$$\langle\text{TABS}\rangle\ \frac{\Gamma,\Delta,C[t <: \sigma] \rhd e \leqslant e':\tau}{\Gamma,\Delta,C \rhd \Lambda t <: \sigma.e \leqslant \Lambda t <: \sigma.e' : \forall t <: \sigma.\tau} \quad (t \notin \text{FTV}(\Gamma)\cup\text{FTV}(C))$$

$$\langle\text{TAPPL}\rangle\ \frac{\Gamma,\Delta,C \rhd e \leqslant e' : \forall t <: \sigma.\tau \quad C \rhd \sigma' <: \sigma}{\Gamma,\Delta,C \rhd e[\sigma'] \leqslant e'[\sigma'] : \tau[t := \sigma']}$$

$$(*)\ \langle\beta_\exists\rangle\ \frac{\Gamma,\Delta,C \rhd e \leqslant e : \tau[t := \sigma'] \quad \Gamma[x : \tau[t' := \sigma']],\Delta,C \rhd e' \leqslant e' : \tau' \quad C \rhd \sigma' <: \sigma}{\Gamma,\Delta,C \rhd \textbf{open}\ (\textbf{pack}_{\exists t<:\sigma.\tau}(\sigma',e))\ \textbf{as}\ (t',x)\ \textbf{in}\ e' = e'[t' := \sigma'][x := e] : \tau'}$$

$$\langle\text{PACK}\rangle\ \frac{\Gamma,\Delta,C \rhd e \leqslant e' : \tau[t := \sigma'] \quad C \rhd \sigma' <: \sigma}{\Gamma,\Delta,C \rhd \textbf{pack}_{\exists t<:\sigma.\tau}(\sigma',e) \leqslant \textbf{pack}_{\exists t<:\sigma.\tau}(\sigma',e') : \exists t <: \sigma.\tau}$$

$$\langle\text{OPEN}\rangle\ \frac{\Gamma,\Delta,C \rhd e \leqslant e' : \exists t' <: \sigma.\tau \quad \Gamma[x : \tau[t' := t]],\Delta,C[t <: \sigma'] \rhd e'' \leqslant e''' : \tau'}{\Gamma,\Delta,C \rhd \textbf{open}\ e\ \textbf{as}\ (t,x)\ \textbf{in}\ e'' \leqslant \textbf{open}\ e'\ \textbf{as}\ (t,x)\ \textbf{in}\ e''' : \tau'} \quad \begin{pmatrix} t \notin \text{FTV}(\Gamma)\cup \\ \text{FTV}(C)\cup \\ \text{FTV}(\tau') \end{pmatrix}$$

$$(*)\ \langle\text{ASSERT}\rangle\ \frac{\Gamma,\Delta,C \rhd e \leqslant e : \{r{:}\sigma \mid \phi_1,\ldots,\phi_k\}}{\Gamma,\Delta,C \rhd \phi_i[r := e]} \quad (1\le i\le k)$$

$$(*)\ \langle\text{REFINE}\rangle\ \frac{\Gamma,\Delta,C \rhd e \leqslant e : \{r{:}\sigma \mid \phi_1,\ldots,\phi_k\} \quad \Gamma,\Delta,C \rhd \phi_{k+1}[r := e]}{\Gamma,\Delta,C \rhd e \leqslant e : \{r{:}\sigma \mid \phi_1,\ldots,\phi_{k+1}\}}$$

$$(*)\ \langle\beta_{\text{fix}}\rangle\ \frac{\Gamma,\Delta,C \rhd e \leqslant e : \sigma\to\sigma}{\Gamma,\Delta,C \rhd \textbf{fix}(e) = e(\textbf{fix}(e)) : \sigma}$$

$$\langle\text{FIX}\rangle\ \frac{\Gamma,\Delta,C \rhd e \leqslant e' : \sigma\to\sigma}{\Gamma,\Delta,C \rhd \textbf{fix}(e) \leqslant \textbf{fix}(e') : \sigma}$$

Figure 3. (continued)

The type system, FUN, of Fun is a subsystem of FINAL with typing judgments of the form $\Gamma, \emptyset, C \rhd e : \tau$ (where $\emptyset$ denotes the empty type assignment or the empty constraint) and axioms and rules not marked by $(*)$. The deducibility in FINAL (and in FUN) is shown by $\vdash_{\text{FINAL}}$ (and $\vdash_{\text{FUN}}$, respectively). We usually omit the subscripts and simply write $\vdash$ when there is no danger of confusion.

We end this section by showing how our refined types solve the problem that we have pointed out in Section 2. We can define the implementation type of *stack* in the last section as a refined type. To make examples more readable, we omit implementation variables whose type is a record type in assertions and simply write *l* for *r.l* when there is no danger of confusion.

type *StackOpImpl* = {*aStackOpImpl* : {*new*: *StackValRep*,
isnew: *StackValRep* → *Bool*,
push: *Nat* → *StackValRep* → *StackValRep*,
top: *StackValRep* → *Nat*,
pop: *StackValRep* → *StackValRep*}

(∗push-pop∗) | **forall** *i*: *Nat*.**forall** *s*: *StackValRep*.
pop(*push*(*i*)(*s*)) ⩽ *s* : *StackValRep*,
forall *i*: *Nat*.**forall** *s*: *StackValRep*.*top*(*push*(*i*)(*s*)) ⩽ *i* : *Nat*,
forall *i*: *Nat*.**forall** *s*: *StackValRep*.*isnew*(*push*(*i*)(*s*)) ⩽ *false* : *Bool*,
isnew(*new*) ⩽ *true* : *Bool*};

Now consider another type *StackoidOpImpl*:

type *StackoidOpImpl* = {*aStackoidOpImpl* : {*new*: *StackValRep*,
isnew: *StackValRep* → *Bool*,
push: *Nat* → *StackValRep* → *StackValRep*,
top: *StackValRep* → *Nat*,
pop: *StackValRep* → *StackValRep*}

(∗push²-pop²∗) | **forall** *i*: *Nat*.**forall** *j*: *Nat*.**forall** *s*: *StackValRep*.
pop(*pop*(*push*(*i*)(*push*(*j*)(*s*)))) ⩽ *s* : *StackValRep*,
forall *i*: *Nat*.**forall** *s*: *StackValRep*.*top*(*push*(*i*)(*s*)) ⩽ *i* : *Nat*,
forall *i*: *Nat*.**forall** *s*: *StackValRep*.
isnew(*push*(*i*)(*s*)) ⩽ *false* : *Bool*,
isnew(*new*) ⩽ *true* : *Bool*};

Clearly the assertion (∗push-pop∗) in *StackOpImpl* is stronger than (∗push²-pop²∗) in *StackoidOpImpl*; hence, any stack can be used as a stackoid. Therefore, *StackOpImpl* inherits the structure of *StackoidOpImpl*, and this fact is expressed in FINAL as the subtype relationship: ∅ ▷ *StackOpImpl* <: *StackoidOpImpl*.

Compare the definition of the type *QueueOpImpl* with the same signature as *StackOpImpl*:

type *QueueOpImpl* = {*aQueueOpImpl* : {*new*: *StackValRep*,
isnew: *StackValRep* → *Bool*,
push: *Nat* → *StackValRep* → *StackValRep*,
top: *StackValRep* → *Nat*,
pop: *StackValRep* → *StackValRep*}
| **forall** *i*: *Nat*.**forall** *s*: *StackValRep*.
pop(*push*(*i*)(*s*)) ⩽ **if** *isnew*(*s*) **then** *s*
else *push*(*i*)(*pop*(*s*)) : *StackValRep*,
forall *i*: *Nat*.**forall** *s*: *StackValRep*.
top(*push*(*i*)(*s*)) ⩽ **if** *isnew*(*s*) **then** *i* **else** *top*(*s*) : *Nat*,
forall *i*: *Nat*.**forall** *s*: *StackValRep*.*isnew*(*push*(*i*)(*s*)) ⩽ *false* : *Bool*,
isnew(*new*) ⩽ *true* : *Bool*};

Then clearly $\not\vdash \varnothing \rhd$ *StackOpImpl* <: *QueueOpImpl* and $\not\vdash \varnothing \rhd$ *QueueOpImpl* <: *StackOpImpl*. Moreover, for *aStackOpImpl* and *anotherStackOpImpl* in Section 2, we can show

$$\vdash \varnothing, \varnothing, \varnothing \rhd aStackOpImpl : StackOpImpl$$

and

$$\vdash \varnothing, \varnothing, \varnothing \rhd anotherStackOpImpl : QueueOpImpl$$

as we have pointed out in Section 2 (assuming appropriate assertions on list operations are given).

4. Proof Theoretical Relationships between FUN and FINAL

In this section we investigate the proof theoretical properties of the type system FINAL. Especially we show the system is a conservative extension of FUN.

First, we define classes of types, expressions, bases, sets of constraints, and judgments of Final having correspondences in Fun.

Definition 2.

(1) A type, σ, of Final is said to be *assertion-free* iff $\sigma \in \mathbf{Type}_{\mathrm{Fun}}$.

(2) An expression, e, of Final is said to be *assertion-free* iff $e \in \mathbf{Exp}_{\mathrm{Fun}}$.

(3) A basis, Γ, is *assertion-free* iff Γ assigns an assertion-free type to each variable.

(4) A set of constraints, C, is *assertion-free* iff the upper bound type of each constraint in C is assertion-free.

(5) A judgment, Σ, of FINAL is *assertion-free* iff Σ has either one of the following forms:

 (a) Σ is a subtyping judgment, $C \rhd \sigma <: \tau$, where σ and τ are assertion-free; or

 (b) Σ is a typing judgment, $\Gamma, \varnothing, C \rhd e : \sigma$, where Γ, C, e and σ are all assertion-free.

Next we define a function which removes all assertions from types.

Definition 3. The function $(\cdot)^* : \mathbf{Type} \to \mathbf{Type}_{\mathrm{Fun}}$ is defined such that

(1) $\top^* = \top$,

(2) $t^* = t$,

(3) $\iota^* = \iota$,

(4) $(\sigma \to \tau)^* = \sigma^* \to \tau^*$,

(5) $\{l_1{:}\,\sigma_1, \ldots, l_n{:}\,\sigma_n\}^* = \{l_1{:}\,\sigma_1{}^*, \ldots, l_n{:}\,\sigma_n{}^*\}$,

(6) $[l_1{:}\,\sigma_1, \ldots, l_n{:}\,\sigma_n]^* = [l_1{:}\,\sigma_1{}^*, \ldots, l_n{:}\,\sigma_n{}^*]$,

(7) $(\forall t <: \sigma.\tau)^* = \forall t <: \sigma^*.\tau^*$,

(8) $(\exists t <: \sigma.\tau)^* = \exists t <: \sigma^*.\tau^*$,

(9) $\{r{:}\,\tau \mid \phi_1, \ldots, \phi_k\}^* = \tau^*$.

This $(\cdot)^*$ is extended on **Exp**, **Assertion**, bases, constraints, and judgments of FINAL in the obvious way. Then we can show the following property for assertion-free judgments.

Theorem 4. *If an assertion-free judgment Σ is provable in* FINAL, *then there is a proof of Σ comprising only assertion-free judgments.* ∎

As stated before, any assertion-free judgment Σ of FINAL is also a well-formed judgment of FUN.

Corollary 5 Conservative Extension Theorem. *The theory* FINAL *is a conservative extension of* FUN. *That is, for any assertion-free judgment* Σ *of* FINAL,

$$\vdash_{\text{FINAL}} \Sigma \iff \vdash_{\text{FUN}} \Sigma. \blacksquare$$

The type system of Final is clearly undecidable since it has a power to specify a kind of partial correctness of functional programs. But the system restores decidability by forgetting all assertions, hence our system FINAL can be viewed as a type system for specification/verification while FUN is its decidable subsystem for compile-time type-checking. Then the following theorem states that "*all correct programs pass the compiler.*"

Theorem 6. *For any* $\Gamma, \Delta, C\ \sigma$, *and any* e *without implementation variables,*

$$\vdash_{\text{FINAL}} \Gamma, \Delta, C \triangleright e : \sigma \implies \vdash_{\text{FUN}} \Gamma^*, \varnothing, C^* \triangleright e^* : \sigma^*. \blacksquare$$

5. Semantics of Algebraic Types and Algebraic Inheritances

In this section, we give a denotational semantics of Final and show that the theory FINAL in Section 4 is sound with respect to this semantics. First we give a semantics for expressions using the type-free interpretation of expressions. The semantic domain $\mathbf{D}$ for the interpretation is the complete partially ordered set (*cpo*) satisfying the following domain equation using the inverse-limit construction. For details on this construction and on cpos, we follow [Plotkin 83] and [Barendregt 81].

$$v \in \mathbf{D} \cong \mathbf{A}_0 \oplus \cdots \mathbf{A}_n \oplus \mathbf{F} \oplus \mathbf{R} \oplus \mathbf{U} \oplus \mathbf{W}$$

where

- $\mathbf{A}_i$ is the cpo of values of the base type ι_i $(1 \le i \le n)$;
- $f \in \mathbf{F} = [\mathbf{D} \to \mathbf{D}]$ is for function values;
- $q \in \mathbf{R} = [\mathbf{Label}_\perp \to_\perp \mathbf{D}]$ is for record values;
- $u \in \mathbf{U} = [\mathbf{Label}_\perp \times \mathbf{D}]$ is for variant (tagged union) values;
- $\mathbf{W} \stackrel{\text{def}}{=} \{?\}_\perp$

  where

  ? is the value modeling run-time type errors
  and we write its image as *wrong*, i.e. $\mathit{wrong} \stackrel{\text{def}}{=} \mathit{in}_{\mathbf{W}}(?)$;
- $\oplus$ indicates the coalesced sum construction of cpos;
- $\to$ is the domain constructor of function space.
- $\to_\perp$ is the domain constructor of strict function space.

We also need a few auxiliary domains for environments:

$\epsilon \in \mathbf{Env} = \mathbf{EEnv} \times \mathbf{IEnv}$	the domain of environments;
$\zeta \in \mathbf{EEnv} = \mathbf{Var}_\perp \to_\perp \mathbf{D}$	the domain of valuations for ordinary variables;
$\xi \in \mathbf{IEnv} = \mathbf{IVar}_\perp \to_\perp \mathbf{D}$	the domain of valuations for implementation variables.

We interpret each expression of Final via its erasure; in other words, we give Final a type-free interpretation. The semantic equations for expressions are shown in Fig. 4 (here we assume a semantic function $\mathcal{K}_i$ for each base type ι_i for the interpretation of its constants) where $\mathit{in}_{\mathbf{X}}$, $\mathit{out}_{\mathbf{X}}$ and $\mathit{is}_{\mathbf{X}}$ are usual primitive operations for sum domains.

$$\mathcal{E} : \mathbf{Exp} \to \mathbf{Env} \to \mathbf{D}$$

$$
\begin{aligned}
\mathcal{E}[\![x]\!]\varepsilon &= \textit{let } \langle\zeta,\xi\rangle = \varepsilon \textit{ in } \zeta[\![x]\!] \textit{ end};\\
\mathcal{E}[\![r]\!]\varepsilon &= \textit{let } \langle\zeta,\xi\rangle = \varepsilon \textit{ in } \xi[\![r]\!] \textit{ end};\\
\mathcal{E}[\![c_{ij}]\!]\varepsilon &= \textit{in}_{\mathbf{A}_i}(\mathcal{K}_i[\![c_{ij}]\!]);\\
\mathcal{E}[\![\lambda x{:}\sigma.e]\!]\varepsilon &= \textit{let } \langle\zeta,\xi\rangle = \varepsilon \textit{ in } \textit{in}_{\mathbf{F}}(\lambda v \in \mathbf{D}.\mathcal{E}[\![e]\!]\langle\zeta[x \mapsto v],\xi\rangle) \textit{ end};\\
\mathcal{E}[\![ee']\!]\varepsilon &= \textit{if } \textit{is}_{\mathbf{F}}(\mathcal{E}[\![e]\!]\varepsilon) \textit{ then } \textit{out}_{\mathbf{F}}(\mathcal{E}[\![e]\!]\varepsilon)(\mathcal{E}[\![e']\!]\varepsilon) \textit{ else wrong};\\
\mathcal{E}[\![\{l_1 = e_1,\ldots,l_n = e_n\}]\!]\varepsilon &= \textit{in}_{\mathbf{R}}(\lambda l \in \mathbf{Label}_\perp.\textit{if } l = l_1 \textit{ then } \mathcal{E}[\![e_1]\!]\varepsilon\\
&\qquad \textit{elseif} \ldots\\
&\qquad \textit{elseif } l = l_n \textit{ then } \mathcal{E}[\![e_n]\!]\varepsilon\\
&\qquad \textit{else wrong});\\
\mathcal{E}[\![e.l]\!]\varepsilon &= \textit{if } \textit{is}_{\mathbf{R}}(\mathcal{E}[\![e]\!]\varepsilon) \textit{ then } \textit{out}_{\mathbf{R}}(\mathcal{E}[\![e]\!]\varepsilon)(l) \textit{ else wrong};\\
\mathcal{E}[\![[l = e]]\!]\varepsilon &= \textit{in}_{\mathbf{U}}\langle l, \mathcal{E}[\![e]\!]\varepsilon\rangle;\\
\mathcal{E}[\![\mathbf{case}\ e\ \mathbf{of}\ l_1\ \mathbf{then}\ e_1,\ldots,l_n\ \mathbf{then}\ e_n]\!]\varepsilon &= \textit{if } \textit{is}_{\mathbf{U}}(\mathcal{E}[\![e]\!]\varepsilon) \textit{ then}\\
&\quad \textit{let } \langle l,v\rangle = \textit{out}_{\mathbf{U}}(\mathcal{E}[\![e]\!]\varepsilon) \textit{ in}\\
&\qquad \textit{if } l = l_1 \textit{ then}\\
&\qquad\quad \textit{if } \textit{is}_{\mathbf{F}}(\mathcal{E}[\![e_1]\!]\varepsilon) \textit{ then } \textit{out}_{\mathbf{F}}(\mathcal{E}[\![e_1]\!]\varepsilon)(v) \textit{ else wrong}\\
&\qquad \textit{elseif} \ldots\\
&\qquad \textit{elseif } l = l_n \textit{ then}\\
&\qquad\quad \textit{if } \textit{is}_{\mathbf{F}}(\mathcal{E}[\![e_n]\!]\varepsilon) \textit{ then } \textit{out}_{\mathbf{F}}(\mathcal{E}[\![e_n]\!]\varepsilon)(v) \textit{ else wrong}\\
&\qquad \textit{else wrong}\\
&\quad \textit{end}\\
&\quad \textit{else wrong};\\
\mathcal{E}[\![\Lambda t.e]\!]\varepsilon &= \mathcal{E}[\![e]\!]\varepsilon;\\
\mathcal{E}[\![e[\sigma]]\!]\varepsilon &= \mathcal{E}[\![e]\!]\varepsilon;\\
\mathcal{E}[\![\mathbf{fix}(e)]\!]\varepsilon &= \textit{if } \textit{is}_{\mathbf{F}}(\mathcal{E}[\![e]\!]\varepsilon) \textit{ then let } f = \textit{out}_{\mathbf{F}}(\mathcal{E}[\![e]\!]\varepsilon) \textit{ in } \bigsqcup_n f^n(\perp_{\mathbf{D}}) \textit{ end}\\
&\quad \textit{else wrong}.
\end{aligned}
$$

Figure 4. The Semantic Equations for Expressions of Final.

Next we give a semantics for types based on a kind of partial equivalence relation models.

Definition 7. Let X be a set.

(1) A *partial equivalence relation* (*per* for short) on X is a symmetric and transitive binary relation on X.

(2) Let P be a per on X. Then define the *domain* of P, $|P|$, by:

$$|P| \stackrel{\text{def}}{=} \{v \in X \mid \langle v,v\rangle \in P\}.$$

(3) Let P and Q be pers on X. Then define the function space per, $P \to Q$, by:

$$\langle f,g\rangle \in P \to Q \quad \stackrel{\text{def}}{\iff} \quad \forall v,v' \in X.[\langle v,v'\rangle \in P \Rightarrow \langle f(v),g(v')\rangle \in Q].$$

(4) Let P and Q be pers on X. Then define the product per, $P \times Q$, by:

$$\langle\langle v,w\rangle,\langle v',w'\rangle\rangle \in P \times Q \quad \stackrel{\text{def}}{\iff} \quad \langle v,v'\rangle \in P \text{ and } \langle w,w'\rangle \in Q.$$

(5) Let P be a per on X and $x \in |P|$. Then define

$$[x]_P \stackrel{\text{def}}{=} \{y \in X \mid \langle x, y\rangle \in P\}.$$

(6) Let P be a per on X and $S \subseteq X$. Then define the *restriction of P on S*, $P\lceil S$, by:

$$P\lceil S \stackrel{\text{def}}{=} \{\langle u, v\rangle \in P \mid u \in S \text{ and } v \in S\}.$$

In order to interpret types as pers on **D**, we require the domain of each per corresponding to a type to be a sub-cpo of **D**.

Definition 8.

(1) Let P be a per on the cpo **D**. Then P is *complete* iff P satisfies both of the following conditions:
   (a) $\langle \bot_{\mathbf{D}}, \bot_{\mathbf{D}}\rangle \in P$; and
   (b) P is closed under lubs of ω-chains, i.e.,

$$\forall i \in \omega. \langle v_i, w_i\rangle \in R \implies \langle \bigsqcup_{i\in\omega} v_i, \bigsqcup_{i\in\omega} w_i\rangle \in P.$$

(2) **CPER** denotes the collection of complete pers (*cpers* for short) on **D**.

It is easily shown that $\langle \mathbf{CPER}, \subseteq\rangle$ is a complete lattice and has greatest lower bounds as intersections. Note that least upper bounds in **CPER** are not simple unions in general.

The semantic equations for types are shown in Fig. 5, where $\eta \in \mathbf{TEnv} = \mathbf{TVar}_\bot \to_\bot \mathbf{CPER}$ is a valuation for type variables.

$$\mathcal{T} : \mathbf{Type} \to \mathbf{TEnv} \to \mathbf{CPER}$$

$$\mathcal{T}[\![\top]\!]\eta = (\mathbf{D} - \{\mathit{wrong}\}) \times (\mathbf{D} - \{\mathit{wrong}\});$$
$$\mathcal{T}[\![\iota_i]\!]\eta = \mathbf{A}_i \times \mathbf{A}_i;$$
$$\mathcal{T}[\![\sigma_1 \to \sigma_2]\!]\eta = \mathcal{T}[\![\sigma_1]\!]\eta \to \mathcal{T}[\![\sigma_2]\!]\eta;$$
$$\mathcal{T}[\![\{l_1{:}\,\sigma_1, \ldots, l_n{:}\,\sigma_n\}]\!]\eta = \bigcap_{i=1}^{n} \{\langle q, q'\rangle \mid q, q' \in \mathbf{R} \text{ and } \langle q(l_i), q'(l_i)\rangle \in \mathcal{T}[\![\sigma_i]\!]\eta\};$$
$$\mathcal{T}[\![[l_1{:}\,\sigma_1, \ldots, l_n{:}\,\sigma_n]]\!]\eta = \bigcup_{i=1}^{n} \{\langle\langle l_i, v\rangle, \langle l_i, v'\rangle\rangle \mid \langle l_i, v\rangle, \langle l_i, v'\rangle \in \mathbf{U} \text{ and } \langle v, v'\rangle \in \mathcal{T}[\![\sigma_i]\!]\eta\};$$
$$\mathcal{T}[\![\forall t <: \sigma.\tau]\!]\eta = \bigcap\{\mathcal{T}[\![\tau]\!]\eta[t \mapsto R] \mid R \in \mathbf{CPER} \text{ and } R \subseteq \mathcal{T}[\![\sigma]\!]\eta\};$$
$$\mathcal{T}[\![\{r{:}\,\tau \mid \phi_1, \ldots, \phi_k\}]\!]\eta = \mathit{let}\ R = \mathcal{T}[\![\tau]\!]\eta$$
$$\mathit{and}\ S = \{v \in \mathbf{D} \mid \bigwedge_{j=1}^{k} \mathcal{A}[\![\phi_j]\!]\langle \bot_{\mathbf{EEnv}}, [r \mapsto v]\rangle\eta\}$$
$$\mathit{in}\ R\lceil S\ \mathit{end}.$$

Figure 5. The Semantic Equations for Types of Final.

We interpret each assertion as an element of non-pointed **T**, since an assertion must be always either *true* or *false* even if evaluation of some of the expressions contained in it would not terminate.

$$\mathcal{A} : \textbf{Assertion} \rightarrow \textbf{Env} \rightarrow \textbf{TEnv} \rightarrow \textbf{T}$$

$$\mathcal{A}[\![e_1 \leqslant e_2 : \sigma]\!]\varepsilon\eta = (\mathcal{E}[\![e_1]\!]\varepsilon \sqsubseteq \mathcal{E}[\![e_2]\!]\varepsilon) \wedge (\mathcal{E}[\![e_1]\!]\varepsilon \in \mathcal{T}[\![\sigma]\!]\eta) \wedge (\mathcal{E}[\![e_2]\!]\varepsilon \in \mathcal{T}[\![\sigma]\!]\eta);$$
$$\mathcal{A}[\![\textbf{forall } x{:}\,\sigma.\phi]\!]\varepsilon\eta = \textit{let } \langle\zeta,\xi\rangle = \varepsilon \textit{ in } \forall v \in |\mathcal{T}[\![\sigma]\!]\eta|.\mathcal{A}[\![\phi]\!](\zeta[x \mapsto v],\xi)\eta \textit{ end}.$$

Figure 6. The Semantic Equations for Assertions of Final.

The semantic function $\mathcal{T}$ is well-defined by the syntactical constraint (a) to Final which we imposed in Section 3.

Proposition 9.

(1) *Let* $P \in$ **CPER** *and* $S \subseteq$ **D** *be pointed and closed under lubs of* ω*-chains. Then* $P\lceil S \in$ **CPER**.

(2) *Let* $P, Q \in$ **CPER**. *Then* $P \times Q, P \rightarrow Q \in$ **CPER**.

(2) *Let* $\{P_i\}_i \subseteq$ **CPER** *be a family of cpers. Then* $\bigcap_i P_i \in$ **CPER**. ∎

Theorem 10 Well-definedness of $\mathcal{T}$. *$\mathcal{T}$ is well-defined. That is, for each* $\tau \in$ **Type** *and for any* $\eta \in$ **TEnv**, $\mathcal{T}[\![\tau]\!]\eta \in$ **CPER**. ∎

Intuitively speaking, our notion of type is a collection of values satisfying *at least* some particular properties (the set of operation actable to the value, constraints on the value which can be specified by a set of inequations).

Hence it is natural to request that, if each value of an ω-chain satisfies such properties, then the supreme of the chain must also satisfy those properties. This corresponds to the completeness condition requested to our pers (the pointedness is necessary since we want to have fix on all types).

On the other hand, when $v_1 \sqsubseteq v_2$, v_1 has less information than v_2 does, so v_1 may not satisfy some of the properties that v_2 does. This is the reason why we have not requested the downward closedness like in ideals [Cardelli 84] nor the closedness under approximations like in the class of pers used in [Cardone 91] and [Amadio and Cardelli 94].

One drawback of our semantics is that the computation of a functional application cannot be performed within a type in general. To be more concrete, let f be a function from type σ to τ and a be a value of σ, then $f(a)$ must be calculated using bases $\langle e_i\rangle_{i\in\omega}$ of a. The point is that some of these bases may not belong to the sub-cpo (the domain of the cper) corresponding to the type of a, σ, hence we must perform this calculation in the whole domain **D**. This is the cost we have paid for our more expressive type system.

We now turn to the soundness of our type theory FINAL with respect to this semantics.

Definition 11.

(1) A type environment η is said to *respect a set of constraints* C (notation: $\eta \models C$) iff $\eta[\![t]\!]\eta \subseteq \mathcal{T}[\![\sigma]\!]\eta$ for any constraint $[t <: \sigma] \in C$.

(2) An environment $\varepsilon = \langle\zeta,\mu\rangle$ is said to *respect bases* Γ, Δ (notation: $\varepsilon \models \Gamma, \Delta$) iff it satisfies both of the following two conditions:

(a) $\zeta \models \Gamma$, i.e., for any variable $x \in \mathrm{dom}(\Gamma)$ and any type environment η, $\zeta[\![x]\!] \in |\mathcal{T}[\![\Gamma(x)]\!]\eta|$; and

(b) $\xi \models \Delta$, i.e., for any implementation variable $r \in \mathrm{dom}(\Delta)$ and any type environment
$\xi[\![r]\!] \in |\mathcal{T}[\![\Delta(r)]\!]\eta|$.

(3) Let Σ be a judgment of FINAL. Then Σ is *satisfied* under an environment $\varepsilon = \langle \zeta, \mu \rangle$ and a type environment η (notation: $\varepsilon, \eta \models \Sigma$) iff either one of the following cases holds:

(a) when $\Sigma \equiv C \rhd \sigma <: \tau$,

$$\eta \models C \quad \Longrightarrow \quad \mathcal{T}[\![\sigma]\!]\eta \subseteq \mathcal{T}[\![\tau]\!]\eta;$$

(b) when $\Sigma \equiv \Gamma, \Delta \rhd \phi$,

$$\varepsilon \models \Gamma, \Delta \text{ and } \eta \models C \quad \Longrightarrow \quad \mathcal{A}[\![\phi]\!]\varepsilon\eta = \mathit{true}.$$

especially, when Σ is a typing judgment, $\Gamma, \Delta, C \rhd e : \sigma$,

$$\varepsilon \models \Gamma, \Delta \mathcal{E}[\![e]\!]\varepsilon \in |\mathcal{T}[\![\sigma]\!]\eta|;$$

(4) Let Σ be a judgment of FINAL. Then Σ is *valid* (notation: $\models \Sigma$) iff $\varepsilon \models \Sigma$ for any environment $\varepsilon \in \mathbf{Env}$.

Theorem 12 Soundness Theorem. *The theory* FINAL *is sound with respect to this semantics; i.e., for any judgment* Σ *of* FINAL,

$$\vdash \Sigma \quad \Longrightarrow \quad \models \Sigma. \ \blacksquare$$

Versions of this theorem have been presented in forms for special cases (cf. [Cardelli 84]):

Corollary 13 Semantical Soundness Theorem. *If an expression is syntactically typable, then it does not cause any run-time type error. That is,*

$$\vdash \Gamma, \Delta, C \rhd e : \sigma \quad \Longrightarrow \quad \forall \varepsilon \models \Gamma, \Delta \text{ and } \eta \models C.[\mathcal{E}[\![e]\!]\varepsilon \in |\mathcal{T}[\![\sigma]\!]\eta|].$$

In other words,

$$\vdash \Gamma, \Delta, C \rhd e : \sigma \quad \Longrightarrow \quad \forall \varepsilon \models \Gamma, \Delta \text{ and } \eta \models C.[\mathcal{E}[\![e]\!]\varepsilon \neq \mathit{wrong}]. \ \blacksquare$$

Corollary 14 Semantical Subtyping Theorem. *Let* σ *and* τ *be types of* Final. *Then*

$$C \rhd \sigma <: \tau \quad \Longrightarrow \quad \forall \eta \models C.[\mathcal{T}[\![\sigma]\!]\eta \subseteq \mathcal{T}[\![\tau]\!]\eta]. \ \blacksquare$$

6. Directions of Future Research and Related Works

What we have shown in this paper is that the type system with inequational assertions is a natural extension of a polymorphic typed λ-calculus with record types and the complete partial equivalence relation model is rich enough to interpret such types.

Our system can be called a type system combining programming types (usual types of Fun) and specification (inequational assertions as *partial correctness* requirements; *cf*. an equational assertion may roughly correspond to *total correctness*, since it means that the computation of the left-hand side *must terminate* and the result must be just the same value of the right-hand side while an inequation means the computation of the left-hand side can be diverging) used to write specifications for *verification* as well as *executable* programs, hence our Final is a good candidate for foundations of type systems of *functional wide-spectrum languages* such as Extended ML [Sannella and Tarlecki 89] based on functional *programming* language Standard ML. But Standard ML and its wide-spectrumized Extended ML are *two level*, the lower level is for ordinary values such as integers, lists, ... etc., i.e. functional programming, and the higher one is for modules, i.e. functorial programming. And these two levels are separated in these languages. On the other hand, Fun and our Final are one-level and modules are represented as record values. This uniform feature allows programmers to handle modules more freely and modules can be constructed using general recursion, which is not available in ML-style functorial programming.

The sentence "Final is an appropriate base language in designing modular wide-spectrum languages with both general recursions and equationally specified data types" in the abstract may need some explanation, since Final allows only inequations as assertions. The problem caused by equations is that data types with equational assertions may be empty, and this emptiness prohibit general recursions on such types. To assure pointedness for equationally asserted types, each equation must have formally strict occurrence of the implementation variable in both sides. This syntactical condition on equations is too restrictive and exclude most interesting data types such as *Stack*, *Queue* in Section 3 and so on. In other words, if we want to use general recursion in module handling, then we must abandon equations and adopt inequations instead for specifications of modules.

As foundations of wide-spectrum languages, the Martin-Löf type system [Martin-Löf 84] is also such a system [NPS 90]. But it has two drawbacks: the first is that it does not support fixed-points and limits only total functions losing some computable total functions. The second is rather pragmatic problem, writing a specification of an ADT with his type system using equality types for assertions, means that the execution of the extracted program for the ADT contains a construction of the proof of "this ADT is correctly implemented," which is intuitively irrelevant for the execution of the *intended* program. The Göteborg group has introduced their Subset Theory [NPS 90] to remedy this inefficiency, but it does not preserve De Bruijn-Curry-Howard correspondence which is the main merit of the original Martin-Löf system.

The present work is just the first step toward the goal of incorporating algebraic structures with logical types with domain-theoretical foundations. The most interesting issue, both from theoretical and from practical viewpoint is to find sufficient syntactical constraints for inequations so that denotations of refined types are closed under approximation. If this is possible, we can introduce recursive types to our language, and this allows us to define familiar recursive types like *list*, *tree*, ... etc. If we strengthen the syntactical constraint (b) in Section 3 as:

(b') the implementation variable bound by a refined type must occur formally strictly in the left-hand expression of each assertion of that refined type *and must not occur in the right-hand expression*;

then it is easily shown that the domain of the per denoted by refined types is not only closed under approximation but also *downward closed* (i.e. an ideal in the sense of the ideal model [Cardelli 84]). This constraint still allows us to specify *Stack* in Section 3, but is too restrictive, since this excludes refined types such as *Queue* in the same section. So we must find more modest constraints for assertions.

Another interesting issue is to give a domain-theoretic semantics of the type system containing both existential types and inequational refinements. Until now, the only known model of existential types is based on intervals of ideals [Martini 88] and Cardone has pointed out that the profinite pers (cpers closed under approximations) cannot be used to model such types [Cardone 91]. Recently, Abadi and Cardelli claims that such pers can interpret bounded existential types [Abadi and Cardelli 94]. It is not clear their approach is adaptable to inequationally refined types. As we have shown that inequational refinements and existential types does *not* cause any syntactical problem, since Final is a conservative extension of Fun. And inequational refinements and existential types allow us to give the full modeling of ADTs. For example, we can define the abstract data type of the group-like structure as:

$$\begin{aligned} \textbf{type}\ \mathit{Group} = \exists G.\{\mathit{aGroupOp} : \{&(\_\cdot\_) : G \to G \to G,\\ &(\_)^{-1} : G \to G,\\ &e : G\}\\ \mid\ &\textbf{forall}\ x, y, z : G.x \cdot (y \cdot z) = (x \cdot y) \cdot z : G,\\ &\textbf{forall}\ x : G.x \cdot (x)^{-1} = e : G,\\ &\textbf{forall}\ x : G.e \cdot x \leqslant x : G\}. \end{aligned}$$

So the domain-theoretic semantics of full-scale Final (without the well-known encoding of $\exists$ by $\forall$ and $\rightarrow$ in the second-order systems [PAC 94]) is a very interesting issue.

Relating our results to T-algebras is another interesting theme. For strict and continuous $f \in \mathbf{D} \rightarrow \mathbf{D}$ and continuous $g \in \mathbf{D} \rightarrow \mathbf{D}$, predicates of the form $P(v) \equiv f(v) \sqsubseteq g(v)$ are ω-inductive, and such predicates has a strong connection with T-algebras (cf. [Plotkin 83], Chapter 5, Theorem 4). [Lehmann and Smyth 81] used T-algebras to interpret types with operations in domain theory. But as Pierce has pointed out in [Pierce 91, p. 41], *"this construction works only for algebras without equations. The framework has apparently never been extended to include algebras with equations."*

Acknowledgements

The author wish to express his deepest thanks to Professor Henk Barendregt for his invaluable advice on an earlier version of this paper and for his warm encouragement. He also sincerely thanks to anonymous referees of this conference, TLCA'95, who pointed out many mistakes in the draft of this paper and gave the author many critical but constructive comments and valuable suggestions.

References

[Abadi and Cardelli 94] Abadi, M. and L. Cardelli: A Semantics of Object Types, *9th IEEE Conf. of Logic in Computer Science*, 332–341 (1994).

[Barendregt 81] Barendregt, H. P.: *The Lambda Calculus: Its Syntax and Semantics*, North-Holland, Amsterdam (1981).

[Cardelli 84] Cardelli, L.: A Semantics of Multiple Inheritances, in [KMP 84], 51–67; a revised version appeared in *Inform. Comput.* **76**, 138–164 (1988).

[Cardelli and Wegner 85] Cardelli, L. and P. Wegner: On Understanding Types, Data Abstraction, and Polymorphism, *ACM Comput. Surv.* **17** (1985).

[Cardone 91] Cardone, F.: Recursive Types for Fun, *Theoret. Comput. Sci.* **83**, 29–56 (1991).

[Ehrig and Mahr 85] Ehrig, H. and B. Mahr: *Fundamentals of Algebraic Specification 1*, Springer-Verlag, Berlin (1985).

[KMP 84] Kahn, G., D. B. MacQueen, and G. Plotkin (eds.): *Semantics of Data Types*, Proceedings of International Symposium, Sophia-Antipolis, June 1984, Lecture Notes in Computer Science **173**, Springer-Verlag, Berlin (1984).

[Lehmann and Smyth 81] Lehmann, D. J. and M. B. Smyth: Algebraic Specification of Data Types: A Synthetic Approach, *Math. Syst. Theory* **14**, 97–139 (1981).

[Martini 88] Martini, S.: Bounded Quantification Have Interval Models, *1988 ACM Conf. on LISP and Functional Programming*, 164–173 (1988).

[Martin-Löf 84] Martin-Löf, P.: *Intuitionistic Type Theory*, Bibliopolis, Napoli (1984).

[Mitchell and Plotkin 85] Mitchell, J. C. and G. D. Plotkin: Abstract Types Have Existential Type, *12th ACM Symp. on Principles of Programming Languages*, 37–51; a revised version appeared in *ACM Trans. Prog. Lang. Syst.* **10**, 470-502 (1988).

[MTH 90] Milner, R. et al.: *The Definition of Standard ML*, MIT Press, Cambridge MA (1990).

[NPS 90] Nordström, B. et al.: *Programming in Martin-Löf's Type Theory*, Clarendon Press, Oxford (1990).

[Pierce 91] Pierce, B. C.: *Basic Category Theory for Computer Scientists*, MIT Press, Cambridge MA (1991).

[Plotkin 77] Plotkin, G. D.: LCF Considered as a Programming Language, *Theoret. Comput. Sci.* **5**, 223–255 (1977).

[Plotkin 83] Plotkin, G.D.: *Domains*, Advanced Postgraduate Course Notes, Department of Computer Science, University of Edinburgh (1983).

[PAC 94] Plotkin, G.D., M. Abadi and L. Cardelli: Subtyping and Parametricity, *9th IEEE Conf. of Logic in Computer Science*, 310-319 (1994).

[Reynolds 83] Reynolds, J. C.: Types, Abstraction and Parametric Polymorphism, *Information Processing 83* (R. E. A. Mason ed.), 513–523, North-Holland, Amsterdam (1983).

[Reynolds 85] Reynolds, J. C.: Three Approaches to Type Structures, *TAPSOFT-CAAP '85* (H. Ehrig et al. eds.), Lecture Notes in Computer Science **185**, 97–138, Springer-Verlag, Berlin (1985).

[Sannella and Tarlecki 89] Sannella D. and A. Tarlecki: *Toward Formal Development of ML Programs: Foundations and Methodology — Preliminary Version*, Technical Report ECS-LFCS-89-71, Laboratory for Foundations of Computer Science, Department of Computer Science, University of Edinburgh (1989).

Decidable Properties of Intersection Type Systems

Toshihiko Kurata Masako Takahashi

Department of mathematical and computing sciences
Tokyo institute of technology
kurata@is.titech.ac.jp, masako@is.titech.ac.jp

Abstract

We give positive answers to the following decision problems for the intersection type system $\lambda\wedge$ and its variations; (1) the type checking problem of normalizing terms for $\lambda\wedge$, (2) the inhabitation problem for the system $\lambda\wedge$ without ($\wedge$I)-rule, and (3) the same problem for a typed counterpart of $\lambda\wedge$ (or the intersection type system *à la* Church). Our result (1) contrasts with the well-known negative answer to the type checking problem (of all terms) for $\lambda\wedge$, while (2) and (3) contrast with Urzyczyn's negative answer to the inhabitation problem for $\lambda\wedge$.

1 Introduction

The intersection type system $\lambda\wedge$ (see e.g. [BCD 83], [CC 90]) is an extension of the simple type assignment system $\lambda_{\rightarrow}$, and is introduced to overcome some defects of $\lambda_{\rightarrow}$. Indeed, in the system $\lambda\wedge$ types are invariant under $=_\beta$ and to important terms such as fixed-point combinators meaningful types are assigned. Moreover the notions of strong normalizability, normalizability, and solvability are neatly characterized in the system $\lambda\wedge$.

However as far as decision problems are concerned, what have been known in literature are negative; the type checking problem for $\lambda\wedge$ is undecidable (by Scott's theorem, cf. [Bar 92]), and Urzyczyn's recent result shows that the inhabitation problem for $\lambda\wedge$ is undecidable [Urz 94]. On the other hand, for the simple system $\lambda_{\rightarrow}$ it is well-known that these problems are decidable. Our motivation of the present work is to distinguish rules and other features of $\lambda\wedge$ which make the system undecidable.

First, for $\lambda\wedge$ we prove that the type checking problem of normalizing terms is decidable; that is, the set $\{(\Gamma, A) \mid \Gamma \vdash_{\lambda\wedge} M : A\}$ is decidable when M is normalizing. Next we consider the system $\lambda\wedge$ without $\wedge$-introduction rule, and prove that the inhabitation problem for it is decidable. We also study a variation of the system $\lambda\wedge$ which might be considered as a typed counterpart of $\lambda\wedge$ (or the intersection type system *à la* Church). For the system, among others we prove that the inhabitation problem is decidable.

The paper is organized as follows. In section 2, we prove some fundamental properties of intersection types. In particular, we show that any two types have supremum as well as infimum in the set of intersection types (with respect to the subtype relation $\leq$).

In section 3, we define a typed counterpart $\lambda\wedge'$ of $\lambda\wedge$, and prove relationship between the systems $\lambda\wedge'$, $\lambda\wedge'$ without $\wedge$-introduction rule, and $\lambda\wedge$ without $\wedge$-introduction rule. Also an approximation theorem for $\lambda\wedge'$ is proved.

In section 4, we prove characterization theorems of judgements, one for $\lambda\wedge$ and the other for $\lambda\wedge'$. Then from the theorems we obtain positive answers to type checking problems; the problem of normalizing terms for the system $\lambda\wedge$, and that of β-nf's for the system $\lambda\wedge'$.

In section 5, we study inhabitation problems. From the results in section 3, we see that our inhabitation problems mentioned above can be reduced to the emptiness problem of the set $\{M \text{ in } \beta\text{-nf} \mid \Gamma \vdash_{\lambda\wedge'} M : A\}$. Then based on the characterization theorem for $\lambda\wedge'$ in section 4 the latter problem is shown to be reducible to the emptiness problem of context-free grammars, which is well-known to be decidable.

Some proofs of theorems in sections 3 and 5 are given in Appendix.

2 Intersection types

The set $T_\wedge$ of *types* in intersection type systems consists of atomic types including ω, arrow types $(A \to B)$, and intersection types $(A \wedge B)$ where A, B are types. As usual we will omit some parentheses; for example, we write $A_1 \to A_2 \to \cdots \to A_n$ for $(A_1 \to (A_2 \to \cdots \to (A_{n-1} \to A_n) \cdots))$. *Subtype relation* $\leq$ between types is defined by the following axioms and rules:

$$\begin{array}{ll}
A \leq A \wedge A, & (A \to B) \wedge (A \to C) \leq A \to B \wedge C, \\
A_1 \wedge A_2 \leq A_i \ (i = 1, 2), & A \leq B, B \leq C \Longrightarrow A \leq C, \\
A \leq \omega, & A \leq B, C \leq D \Longrightarrow A \wedge C \leq B \wedge D, \\
\omega \leq \omega \to \omega, & A \leq B, C \leq D \Longrightarrow B \to C \leq A \to D.
\end{array}$$

Note that $\leq$ is a pre-order satisfying $(A_1 \to B_1) \wedge (A_2 \to B_2) \leq A_1 \wedge A_2 \to B_1 \wedge B_2$, and the equivalence relation $=$ induced by $\leq$ (i.e., $A = B \Longleftrightarrow_{def} A \leq B$ and $B \leq A$) satisfies: $A = A \wedge A$, $A \wedge (B \wedge C) = (A \wedge B) \wedge C$, $A \wedge B = B \wedge A$, $(A \to B) \wedge (A \to C) = A \to B \wedge C$, and $A \to B = \omega$ if and only if $B = \omega$. Note also that $A_1 \wedge A_2$ is an infimum of A_1 and A_2. In this paper, for the sake of simplicity, we identify (in notation $\equiv$) conjunctions of types $A_1, A_2, \ldots, A_m$ in any order and any association, and write it as $\bigwedge_{i=1}^m A_i$. We also use notations $\bigwedge_{i \in I} A_i$ or $\bigwedge\{A_i \mid i \in I\}$ for any finite set I. (In particular, when $I = \emptyset$, they stand for ω.) We use $A, B, \ldots$ as metavariables ranging over types, $a, b, \ldots$ for atomic types different from ω, and $I, J, \ldots$ for finite sets (of indices).

The set NT (of *normal types*) and its subset $\mathsf{NT}^\to$ (of *arrow normal types*) are defined by simultaneous recursion, as follows:

1. If $A_1, A_2, \ldots, A_n \in \mathsf{NT}$ $(n \geq 0)$, then $A_1 \to \cdots \to A_n \to a \in \mathsf{NT}^\to$.
2. If $A_1, A_2, \ldots, A_n \in \mathsf{NT}^\to$ $(n \geq 0)$, then $\bigwedge_{i=1}^n A_i \in \mathsf{NT}$. (In particular, $\omega \in \mathsf{NT}$.)

Normal types satisfy the following properties (cf. [Hin 82]).

Lemma 2.1 If $A_i, B_j \in \mathsf{NT}^\to$ $(i \in I, j \in J)$, then

$\bigwedge_{i\in I} A_i \le \bigwedge_{j\in J} B_j \iff \forall j\, \exists i\ (A_i \le B_j)$. □

Lemma 2.2 If $A, B \in \mathsf{NT}^{\to}$, then

$$A \le B \iff \text{for some } p \ge 0,\ a,\ A_i\text{'s and } B_i\text{'s}, \begin{cases} A \equiv A_1 \to \cdots \to A_p \to a, & B \equiv B_1 \to \cdots \to B_p \to a \\ \text{and } B_i \le A_i\ (i = 1, ..., p). \end{cases}$$ □

In next definition and lemma, we introduce the notion of normal type A^* of A, which is shown to be the unique representative of the equivalence class containing A.

Definition 2.3 A *normal type* A^* of A is defined recursively, as follows:

1. $a^* \equiv a$.
2. $(\bigwedge_{i\in I} A_i)^* \equiv \bigwedge_{i\in J} A_i^*$ where J is a minimum subset of I such that $\bigwedge_{i\in J} A_i^* = \bigwedge_{i\in I} A_i^*$.
3. $(A \to B)^* \equiv \begin{cases} \bigwedge_{i\in I}(A^* \to B_i) & \text{if } B^* \equiv \bigwedge_{i\in I} B_i \not\equiv \omega, B_i \in \mathsf{NT}^{\to}\ (i \in I), \\ \omega & \text{if } B^* \equiv \omega. \end{cases}$ □

Note that the condition $\bigwedge_{j\in J} B_j = \bigwedge_{j\in J'} B_j$ is decidable by Lemmas 2.1 and 2.2, which enables us to construct A^* from A effectively.

Lemma 2.4 (1) $A^* \in \mathsf{NT}$; and $A = A^*$.
(2) If $A^* \equiv \bigwedge_{i\in I} A_i$ and $A_i \in \mathsf{NT}^{\to}$ $(i \in I)$, then $\forall i, i' \in I\ (A_i \le A_{i'} \Longrightarrow i = i')$.
(3) $A = B \iff A^* \equiv B^*$.
Proof. (1) and (2) are by induction on types. (3) $A = B$ if and only if $A^* = B^*$ by (1). So it suffices to show that $A^* = B^*$ implies $A^* \equiv B^*$. We will prove it by induction on the number of occurrences of atomic types in A^* and B^*.

Suppose $A^* = B^*$. If $A^*, B^* \in \mathsf{NT}^{\to}$, then they are of the form $A^* \equiv A_1^* \to \cdots \to A_n^* \to a, B^* \equiv B_1^* \to \cdots \to B_n^* \to a$ where $A_i^* = B_i^*$ $(i = 1, ..., m)$ (cf. Lemma 2.2), which implies $A_i^* \equiv B_i^*$ $(i = 1, ..., m)$ by induction hypothesis; hence $A^* \equiv B^*$. For the general case, assume $A^* \equiv \bigwedge_{i\in I} A_i^*$ and $B^* \equiv \bigwedge_{j\in J} B_j^*$ where $A_i^*, B_j^* \in \mathsf{NT}^{\to}$ $(i \in I, j \in J)$. Then by Lemma 2.1 $\forall i \in I\ \exists j \in J\ \exists i' \in I\ (A_{i'}^* \le B_j^* \le A_i^*)$; hence by (2) we get $i' = i$ and so $A_i^* = B_j^*$. This implies $A_i^* \equiv B_j^*$ by induction hypothesis. Thus we get $\{A_i^* \mid i \in I\} \subseteq \{B_j^* \mid j \in J\}$. Likewise for the converse; hence $A^* \equiv \bigwedge_{i\in I} A_i^* \equiv \bigwedge_{j\in J} B_j^* \equiv B^*$. □

Corollary 2.5 The binary relation $A \le B$ is decidable.
Proof. Check whether $A^* \le B^*$ or not, by using Lemmas 2.1 and 2.2. □

The properties of normal types described in Lemmas 2.1 and 2.2 can be generalized to types, as follows.

Lemma 2.6 Suppose $\bigwedge_{i\in I}(A_i \to A_i') \le B$. Then
(1) $B^* \equiv \bigwedge_{j\in J}(C_j \to C_j')$ for some J, C_j's and C_j''s. In particular, if $B \in \mathsf{NT}^{\to}$ then $B \equiv C \to C'$ for some C and C'.
(2) If $B \equiv C \to C' \in \mathsf{NT}^{\to}$, then $\exists i \in I\ (C \le A_i$ and $A_i' \le C')$.
(3) If $B \equiv C \to C' \not\equiv \omega$, then $\exists I'\ (\subseteq I, \ne \emptyset)\ (C \le \bigwedge_{i\in I'} A_i$ and $\bigwedge_{i\in I'} A_i' \le C')$. In particular, if $A \to A' \le C \to C' \not\equiv \omega$, then $C \le A$ and $A' \le C'$.

Proof. (1) follows immediately from the special case with $B \in \mathsf{NT}^{\rightarrow}$. To prove the special case, suppose $A_i'^* \equiv \bigwedge_{k\in K_i} A'_{i,k}$ with $A'_{i,k} \in \mathsf{NT}^{\rightarrow}$ $(i \in I, k \in K_i)$. Then since $\bigwedge_{i\in I}\bigwedge_{k\in K_i}(A_i^* \rightarrow A'_{i,k}) = \bigwedge_{i\in I}(A_i \rightarrow A_i') \le B$ and $B \in \mathsf{NT}^{\rightarrow}$, we know from Lemmas 2.1 and 2.2 that B is of the form $B \equiv C \rightarrow C'$, and moreover $C \le A_i$ and $A_i' \le A'_{i,k} \le C'$ for some $i \in I$ and $k \in K_i$. This proves (1) and (2).

To see (3), suppose $C'^* \equiv \bigwedge_{j\in J} C_j'$ where $C_j' \in \mathsf{NT}^{\rightarrow}$ $(j \in J)$. Then our assumption implies $\bigwedge_{i\in I}(A_i \rightarrow A_i') \le \bigwedge_{j\in J}(C^* \rightarrow C_j') \le C^* \rightarrow C_j'$ for each j. Then by applying (2) we get $\forall j \in J, \exists i_j \in I\ (C^* \le A_{i_j}$ and $A'_{i_j} \le C_j')$. Let $I' = \{i_j | j \in J\}$. Then $I' \ne \emptyset$ since $J \ne \emptyset$, and it satisfies $C \le \bigwedge_{j\in J} A_{i_j} \equiv \bigwedge_{i\in I'} A_i$ and $\bigwedge_{i\in I'} A_i' \equiv \bigwedge_{j\in J} A'_{i_j} \le \bigwedge_{j\in J} C_j' \le C'$. □

Lemma 2.7 Suppose $A \le C \rightarrow C' \ne \omega$. Then $A^* \equiv \bigwedge_{i\in I}(A_i \rightarrow A_i') \wedge A'$, $C \le \bigwedge_{i\in I} A_i$ and $\bigwedge_{i\in I} A_i' \le C'$ for some $I \ne \emptyset$, A_i's, A_i''s and A'.
Proof. Let $A^* \equiv \bigwedge_{i\in I} B_i$ and $C'^* \equiv \bigwedge_{j\in J} C_j'$ where $B_i, C_j' \in \mathsf{NT}^{\rightarrow}$ $(i \in I, j \in J)$. Then by assumption $\bigwedge_{i\in I} B_i \le (C \rightarrow C')^* \equiv \bigwedge_{j\in J}(C^* \rightarrow C_j')$; hence by Lemma 2.1 $\forall j \in J, \exists i_j \in I\ (B_{i_j} \le C^* \rightarrow C_j')$. Since $B_{i_j} \in \mathsf{NT}^{\rightarrow}$, this together with Lemma 2.2 implies $B_{i_j} \equiv A_{i_j} \rightarrow A'_{i_j}, C^* \le A_{i_j}$ and $A'_{i_j} \le C_j'$ $(j \in J)$ for some A_{i_j} and A'_{i_j}. Then as before for $I' = \{i_j | j \in J\}$ $(\ne \emptyset)$, we have $C \le \bigwedge_{i\in I'} A_i$ and $\bigwedge_{i\in I'} A_i' \le \bigwedge_{j\in J} C_j' \le C'$, and moreover $A^* \equiv \bigwedge_{i\in I} B_i \equiv \bigwedge_{i\in I'}(A_i \rightarrow A_i') \wedge A'$ where $A' \equiv \bigwedge_{i\in I-I'} B_i$. □

We can extend these results to types with two or more consecutive arrows. For example, we have the following generalization of Lemma 2.7 and its converse.

Corollary 2.8 Suppose $C \ne \omega$. Then

$$A \le C_1 \rightarrow \cdots \rightarrow C_q \rightarrow C \iff \text{for some } I \ne \emptyset,\ A_{i,j}\text{'s},\ A_i\text{'s and } A',\ \begin{cases} A^* \equiv \bigwedge_{i\in I}(A_{i,1} \rightarrow \cdots \rightarrow A_{i,q} \rightarrow A_i) \wedge A', \\ C_j \le \bigwedge_{i\in I} A_{i,j}\ (j = 1, \ldots, q), \text{ and } \bigwedge_{i\in I} A_i \le C. \end{cases}$$

Proof. (Only if part) By induction on q, using Lemma 2.7. (If part) $A = A^* \le \bigwedge_{i\in I} A_{i,1} \rightarrow \cdots \rightarrow \bigwedge_{i\in I} A_{i,q} \rightarrow \bigwedge_{i\in I} A_i \le C_1 \rightarrow \cdots \rightarrow C_q \rightarrow C$. □

In $\mathrm{T}_\wedge$, types have infima by definition. We will show that they also have suprema (with respect to the subtype relation $\le$). For $A, B \in \mathsf{NT}^{\rightarrow}$, define

$$A \vee B \equiv \begin{cases} A_1 \wedge B_1 \rightarrow \cdots \rightarrow A_n \wedge B_n \rightarrow a \\ \quad \text{if } A \equiv A_1 \rightarrow \cdots \rightarrow A_n \rightarrow a, B \equiv B_1 \rightarrow \cdots \rightarrow B_n \rightarrow a \\ \quad \text{for some } n \ge 0,\ A_i\text{'s},\ B_i\text{'s and } a, \\ \omega \quad \text{otherewise.} \end{cases}$$

We extend the definition to arbitrary $A, B \in \mathrm{T}_\wedge$ by $A \vee B \equiv \bigwedge\{A_i \vee B_j \mid i \in I, j \in J\}$ where $A^* \equiv \bigwedge_{i\in I} A_i$ and $B^* \equiv \bigwedge_{j\in J} B_j$ with $A_i, B_j \in \mathsf{NT}^{\rightarrow}$ $(i \in I, j \in J)$.

Theorem 2.9 For any $A, B \in \mathrm{T}_\wedge$, $A \vee B$ is a supremum of A and B; i.e., $\forall C \in \mathrm{T}_\wedge\ (A \le C \text{ and } B \le C \iff A \vee B \le C)$.
Proof. (If part) is clear from the definition. (Only if part) When $A, B \in \mathsf{NT}^{\rightarrow}$, it is immediate from Lemmas 2.1 and 2.2. In other cases, suppose $A^* \equiv \bigwedge_{i\in I} A_i$,

$B^* \equiv \bigwedge_{j\in J} B_j$ and $C^* \equiv \bigwedge_{k\in K} C_k$ with $A_i, B_j, C_k \in \mathsf{NT}^{\rightarrow}$ $(i \in I, j \in J, k \in K)$. Then by using Lemma 2.1 and the case above for $\mathsf{NT}^{\rightarrow}$ we have

$$\begin{aligned} A \le C \text{ and } B \le C &\Longrightarrow \forall k\, (\exists i\, (A_i \le C_k) \text{ and } \exists j\, (B_j \le C_k)) \\ &\Longrightarrow \forall k\, \exists i\, \exists j\, (A_i \vee B_j \le C_k) \\ &\Longrightarrow \bigwedge\{A_i \vee B_j \mid i \in I, j \in J\} \le \bigwedge_{k\in K} C_k \\ &\Longrightarrow A \vee B \le C. \quad \square \end{aligned}$$

We can easily see that distributive laws and the equality $(A \to B) \vee (C \to D) = A \wedge C \to B \vee D$ hold.

3 Intersection type systems $\lambda\wedge$ and $\lambda\wedge'$

In the standard intersection type system $\lambda\wedge$ (cf. [BCD 83], [CC 90], [Bar 92]), *type assignments* or *judgements* $\Gamma \vdash M : A$ are derived from *axioms* (var) and (ω) by applying *inference rules* ($\to$ I), ($\to$ E), ($\wedge$I) and ($\le$) below. Here the *statement* $M : A$ consists of the *subject* M which is a (type-free) λ-term and the *predicate* A which is a type (in $\mathrm{T}_\wedge$). The *basis* Γ is a finite set of statements whose subjects are term variables distinct each other.

$$\text{(var)} \quad \Gamma \vdash x : A \quad (x : A \in \Gamma) \qquad\qquad (\omega) \quad \Gamma \vdash M : \omega$$

$$(\to \mathrm{I}) \quad \frac{\Gamma, x : B \vdash M : A}{\Gamma' \vdash \lambda x.M : B \to A} \qquad\qquad (\to \mathrm{E}) \quad \frac{\Gamma \vdash M : A \to B \quad \Gamma \vdash N : A}{\Gamma \vdash MN : B}$$

$$(\wedge \mathrm{I}) \quad \frac{\Gamma \vdash M : A \quad \Gamma \vdash M : B}{\Gamma \vdash M : A \wedge B} \qquad\qquad (\le) \quad \frac{\Gamma \vdash M : A \quad A \le B}{\Gamma \vdash M : B}$$

In ($\to$ I)-rule, Γ' is any basis including Γ (so that 'weakening' is derived).

In this paper we consider the system $\lambda\wedge$ and its variations including what one might call a typed counterpart of $\lambda\wedge$, or the intersection type system *à la* Church. In the latter system, which we denote by $\lambda\wedge'$, terms are not type-free but typed; that is, those in which types are assigned to bound variables. Judgements in $\lambda\wedge'$ are derived by the same axioms and rules as above, except that ($\to$ I) is now replaced by the following.

$$(\to \mathrm{I})' \quad \frac{\Gamma, x : B \vdash M : A}{\Gamma' \vdash (\lambda x : B.M) : B \to A}$$

We note that the system $\lambda\wedge'$ is a subsystem of the programming language 'Forsythe' introduced by Reynolds [Rey 88]. In next lemma we show that ($\wedge$I)'-rule is redundant in the system $\lambda\wedge'$, and hence $\lambda\wedge'$ is comparable with the system $\lambda\wedge$ without ($\wedge$I)-rule.[1]

Lemma 3.1 (1) $\Gamma \vdash_{\lambda\wedge'} M : A \Longleftrightarrow \Gamma \vdash_{\lambda\wedge'-(\wedge\mathrm{I})'} M : A$.
(2) $\exists N(\Gamma \vdash_{\lambda\wedge'-(\wedge\mathrm{I})'} N : A$ and $|N| \equiv M)^2 \Longleftrightarrow \Gamma \vdash_{\lambda\wedge-(\wedge\mathrm{I})} M : A$.
Proof. By induction on terms ($\Longrightarrow$ of (1)) and on derivations. $\square$

[1] We write $\lambda\wedge$-($\wedge$I) for the system $\lambda\wedge$ without ($\wedge$I)-rule, and similarly for $\lambda\wedge'$-($\wedge$I)'.
[2] By $|N|$ we mean the type erasure of a typed term N.

The lemma shows that the system $\lambda\wedge'$ is considerably weaker than the system $\lambda\wedge$, though it is much stronger than the simple type system; for example, $\lambda x : A.xx$ is not typable in $\lambda_{\rightarrow}$, but $\vdash_{\lambda\wedge'} \lambda x : A.xx : A \rightarrow B$ when $A \equiv B \wedge (B \rightarrow B)$. More generally, we can prove that for any type-free λ-term M in β-nf, there exists a typed term N such that $M \equiv |N|$ and $\Gamma \vdash_{\lambda\wedge'} N : A$ for some Γ and ω-free type A. It can also be verified that β-reduction and η-reduction preserve types in $\lambda\wedge'$, although neither β-conversion nor η-conversion does. Indeed, for $\Gamma = \{x : a \wedge b\}$ and $M \equiv (\lambda y : a.y)x \rightarrow_\beta x$ we have $\Gamma \vdash_{\lambda\wedge'} M : a$, but $\Gamma \not\vdash_{\lambda\wedge'} M : a \wedge b$, while for $\Delta = \{x : a \rightarrow a\}$ and $N \equiv \lambda y : a \wedge b.xy \rightarrow_\eta x$ we have $\Delta \vdash_{\lambda\wedge'} N : a \wedge b \rightarrow a$ but $\Delta \not\vdash_{\lambda\wedge'} N : a \rightarrow a$.

When we define the notion of finite approximation ($\sqsubseteq$ in notation) as usual (see e.g., [DM 86]), we can prove the following (one-sided) approximation theorem for the system $\lambda\wedge'$.

Theorem 3.2 $\Gamma \vdash_{\lambda\wedge'} M : A \Longrightarrow \exists P \sqsubseteq M\ (\Gamma \vdash_{\lambda\wedge'} P : A)$.
Proof. See Appendix A. □

The converse however does not hold (cf. the counterexample above for β-conversion).

4 Type checking problems for $\lambda\wedge$ and $\lambda\wedge'$

In this section, for each of the systems $\lambda\wedge$ and $\lambda\wedge'$ we provc a characterization theorem of judgements $\Gamma \vdash M : A$ where M is in β-nf, and based on the theorem we give positive answers to the type checking problem of normalizing terms for $\lambda\wedge$ and that of β-nf's for $\lambda\wedge'$.

Lemma 4.1 (Generation lemma for $\lambda\wedge$)
(1) If $A \neq \omega$, then $\Gamma \vdash_{\lambda\wedge} x : A \Longleftrightarrow \exists B\ (x : B \in \Gamma$ and $B \leq A)$.
(2) $\Gamma \vdash_{\lambda\wedge} MN : A \Longleftrightarrow \exists B\ (\Gamma \vdash_{\lambda\wedge} M : B \rightarrow A$ and $\Gamma \vdash_{\lambda\wedge} N : B)$.
(3) If $A \neq \omega$, then $\Gamma \vdash_{\lambda\wedge} \lambda x.M : A \iff$ for some m, A_i's and A_i''s $[A^* \equiv \bigwedge_{i=1}^{m}(A_i \rightarrow A_i')$ and $(\Gamma \setminus x), x : A_i \vdash_{\lambda\wedge} M : A_i'\ (i = 1, ..., m)\ ]$.[3] □

Theorem 4.2 If $A \neq \omega$, then

$$\Gamma \vdash_{\lambda\wedge} \lambda x_1 \cdots x_p.xM_1M_2 \cdots M_q : A \Longleftrightarrow$$
for some $m, A_{i,j}$'s and A_i's,
$$\begin{cases} A^* \equiv \bigwedge_{i=1}^{m}(A_{i,1} \rightarrow \cdots \rightarrow A_{i,p} \rightarrow A_i), \\ \text{for each } i = 1, ..., m, \\ \text{for some } C, n \geq 1, C_{k,j}\text{'s}, C_k\text{'s}, C', \text{and } \Gamma' \\ \begin{cases} x : C \in \Gamma',\ \Gamma' = (\Gamma \setminus x_1, ..., x_p) \cup \{x_1 : A_{i,1}, ..., x_p : A_{i,p}\}, \\ C^* \equiv \bigwedge_{k=1}^{n}(C_{k,1} \rightarrow \cdots \rightarrow C_{k,q} \rightarrow C_k) \wedge C', \\ \Gamma' \vdash_{\lambda\wedge} M_j : \bigwedge_{k=1}^{n} C_{k,j}\ (j = 1, ..., q), \bigwedge_{k=1}^{n} C_k \leq A_i. \end{cases} \end{cases}$$

Proof. From Lemma 4.1 (1), (2) and Corollary 2.8, we have

$$\Gamma \vdash_{\lambda\wedge} xM_1M_2 \cdots M_q : A \Longleftrightarrow$$
for some C, $n \geq 1$, $C_{i,j}$'s, C_i's and C',

[3] By $\Gamma \setminus x_1, ..., x_n$ we mean the basis consisting of statements in Γ whose subjects are different from $x_1, ..., x_n$.

$$\left\{\begin{array}{l} x : C \in \Gamma,\ C^* \equiv \bigwedge_{i=1}^{n}(C_{i,1} \to \cdots \to C_{i,q} \to C_i) \wedge C', \\ \Gamma \vdash_{\lambda\wedge} M_j : \bigwedge_{i=1}^{n} C_{i,j}\ (j = 1, ..., q), \bigwedge_{i=1}^{n} C_i \leq A. \end{array}\right.$$

Then by applying Lemma 4.1 (3) we get the theorem. □

Theorem 4.3 For a basis Γ, a term M in β-nf and a type A, whether $\Gamma \vdash_{\lambda\wedge} M : A$ or not is decidable.

Proof. The proof is by induction on M. First, if $A = \omega$, then the judgement $\Gamma \vdash_{\lambda\wedge} M : A$ always holds. Otherwise, suppose $M \equiv \lambda x_1 \cdots x_p.xM_1M_2 \cdots M_q$. Then by Theorem 4.2, $\Gamma \vdash_{\lambda\wedge} M : A$ holds true if and only if there exist m (≥ 1), $A_{i,j}$'s, A_i's, C_i's, n_i's (≥ 1), $C_{i,k,j}$'s, $C_{i,k}$'s and C'_i's such that

1. $A^* \equiv \bigwedge_{i=1}^{m}(A_{i,1} \to \cdots \to A_{i,p} \to A_i)$,
2. $(x : C_i) \in \Gamma'_i$ $(i = 1, ..., m)$,
3. $C_i^* \equiv \bigwedge_{k=1}^{n_i}(C_{i,k,1} \to \cdots \to C_{i,k,q} \to C_{i,k}) \wedge C'_i$ $(i = 1, ..., m)$,
4. $\Gamma'_i \vdash_{\lambda\wedge} M_j : \bigwedge_{k=1}^{n_i} C_{i,k,j}$ $(i = 1, ..., m,\ j = 1, ..., q)$,
5. $\bigwedge_{k=1}^{n_i} C_{i,k} \leq A_i$ $(i = 1, ..., m)$

where $\Gamma'_i = (\Gamma \setminus x_1, ..., x_p) \cup \{x_1 : A_{i,1}, ..., x_p : A_{i,p}\}$. Note that there are only finitely many combinations of $m, A_{i,j}$'s, A_i's, C_i's, n_i's, $C_{i,k,j}$'s, $C_{i,k}$'s, C'_i's and Γ'_i's satisfying conditions 1, 2 and 3, and moreover that they can be obtained effectively from Γ, M, and A. For each of these combinations it is possible to check whether it satisfies the conditions 4 and 5 by induction hypothesis and Corollary 2.5. □

Corollary 4.4 When M is normalizing, $\Gamma \vdash_{\lambda\wedge} M : A$ or not is decidable.

Proof. Immediate from Theorem 4.3 because β-nf's of normalizing terms can effectively be obtained, and types are invariant under $=_\beta$ in $\lambda\wedge$. □

Similar discussion as lemma 4.1 through theorem 4.3 can be carried out to the system $\lambda\wedge'$, yielding the following.

Lemma 4.5 (Generation lemma for $\lambda\wedge'$)
(1) If $A \neq \omega$, then $\Gamma \vdash_{\lambda\wedge'} x : A \Longleftrightarrow \exists B\, (x : B \in \Gamma$ and $B \leq A)$.
(2) $\Gamma \vdash_{\lambda\wedge'} MN : A \Longleftrightarrow \exists B\, (\Gamma \vdash_{\lambda\wedge'} M : B \to A$ and $\Gamma \vdash_{\lambda\wedge'} N : B)$.
(3) If $A \neq \omega$, then $\Gamma \vdash_{\lambda\wedge'} (\lambda x : B.M) : A \Longleftrightarrow$ for some m, A_i's and A'_i's [$A^* \equiv \bigwedge_{i=1}^{m}(A_i \to A'_i)$, $(\Gamma \setminus x), x : B \vdash_{\lambda\wedge'} M : \bigwedge_{i=1}^{m} A'_i$, and $\bigvee_{i=1}^{m} A_i \leq B$.] □

Theorem 4.6 If $A \neq \omega$, then

$$\Gamma \vdash_{\lambda\wedge'} (\lambda x_1 : B_1 \cdots x_p : B_p.xM_1M_2 \cdots M_q) : A \Longleftrightarrow$$

for some $m, A_{i,j}$'s, A_i's, C, $n \geq 1$, $C_{i,j}$'s, C_i's, C' and Γ'

$$\left\{\begin{array}{l} A^* \equiv \bigwedge_{i=1}^{m}(A_{i,1} \to \cdots \to A_{i,p} \to A_i), \\ (x : C) \in \Gamma',\ \Gamma' = (\Gamma \setminus x_1, ..., x_p) \cup \{x_1 : B_1, ..., x_p : B_p\}, \\ C^* \equiv \bigwedge_{i=1}^{n}(C_{i,1} \to \cdots \to C_{i,q} \to C_i) \wedge C', \\ \Gamma' \vdash_{\lambda\wedge'} M_j : \bigwedge_{i=1}^{n} C_{i,j}\ (j = 1, ..., q), \\ \bigvee_{i=1}^{m} A_{i,k} \leq B_k\ (k = 1, ..., p),\ \ \bigwedge_{i=1}^{n} C_i \leq \bigwedge_{i=1}^{m} A_i. \end{array}\right.$$ □

Theorem 4.7 When M is in β-nf, $\Gamma \vdash_{\lambda\wedge'} M : A$ or not is decidable. □

5 Inhabitation problems for $\lambda\wedge-(\wedge\mathrm{I})$ and $\lambda\wedge'$

In this section, we prove that the inhabitation problems for $\lambda\wedge-(\wedge\mathrm{I})$ and $\lambda\wedge'$ are decidable. First we note that the problems can be reduced to the emptiness problem of the set $\{M \text{ in } \beta\text{-nf} \mid \Gamma \vdash_{\lambda\wedge'} M : A\}$.

Theorem 5.1 The followings are equivalent:

1. $\exists M(\Gamma \vdash_{\lambda\wedge-(\wedge\mathrm{I})} M : A)$.
2. $\exists M(\Gamma \vdash_{\lambda\wedge'} M : A)$.
3. $\exists M$ in β-nf $(\Gamma \vdash_{\lambda\wedge'} M : A)$.

Proof. $1 \Longleftrightarrow 2$ is immediate from Lemma 3.1. $2 \Longrightarrow 3$ is from Theorem 3.2 and the fact that $\Gamma \vdash_{\lambda\wedge'} P : A$ with P in $\beta\bot$-nf implies $\Gamma \vdash_{\lambda\wedge'} N : A$ for the λ-term N in β-nf which is obtained from P by replacing $\bot$ with a variable. $3 \Longrightarrow 2$ is obvious. $\square$

Next we observe that the emptiness of $\{M \text{ in } \beta\text{-nf} \mid \Gamma \vdash_{\lambda\wedge'} M : A\}$ can be reduced to the same problem for the following subsystem $\lambda\wedge''$ of $\lambda\wedge'$: The set of term variables in $\lambda\wedge''$ is restricted to the set $\{x_A \mid A \in \mathrm{T}_\wedge\}$ which is in one-to-one correspondence with the set $\mathrm{T}_\wedge$ of types, and each type A is assigned to x_A but no other variables. In other words, bases of the system $\lambda\wedge''$ are finite subsets of $\{x_A : A \mid A \in \mathrm{T}_\wedge\}$, and abstraction terms in $\lambda\wedge''$ are of the form $\lambda x_A : A.M$. When Γ is such a basis and M is a term in $\lambda\wedge''$, then we write $\Gamma \vdash_{\lambda\wedge''} M : A$ for $\Gamma \vdash_{\lambda\wedge'} M : A$. The following lemma guarantees the reducibility of the emptiness problem of the set $\{M \text{ in } \beta\text{-nf} \mid \Gamma \vdash_{\lambda\wedge'} M : A\}$ to the same problem for $\lambda\wedge''$.

Lemma 5.2 Let Γ be a basis of the system $\lambda\wedge'$ and A be a type. Then

$$\exists M \text{ in } \beta\text{-nf } (\Gamma \vdash_{\lambda\wedge'} M : A) \Longleftrightarrow \exists M \text{ in } \beta\text{-nf } (\varphi(\Gamma) \vdash_{\lambda\wedge''} M : A)$$

where $\varphi(\Gamma) = \{x_B : B \mid (y : B) \in \Gamma \text{ for some } y\}$. $\square$

For the sake of simplicity, in the system $\lambda\wedge''$ we consider the set of types $\mathrm{T}_\wedge$ modulo $=$. Also for simplicity, we will write $\{x_{A_1}, x_{A_2}, ..., x_{A_n}\}$ for the basis $\{x_{A_1} : A_1, x_{A_2} : A_2, ..., x_{A_n} : A_n\}$, and $\lambda x_A.M$ for the term $\lambda x_A : A.M$.

Note that in $\lambda\wedge''$, since bound variables indicate their types, α-conversion does *not* preserve types. For example, $\vdash_{\lambda\wedge''} \lambda x_A.x_A : A \to A$, but $\not\vdash_{\lambda\wedge''} \lambda x_\omega.x_\omega : A \to A$ unless $A = \omega$. Therefore in $\lambda\wedge''$ we do not identify α-convertible terms.

From Theorem 4.6, one can easily obtain the following characterization of judgements in $\lambda\wedge''$.

Corollary 5.3 If $A \neq \omega$, then

$$\Gamma \vdash_{\lambda\wedge''} \lambda x_{B_1} \cdots x_{B_p}.x_C M_1 M_2 \cdots M_q : A \Longleftrightarrow$$
for some m, $A_{i,j}$'s, A_i's, $n \geq 1$, $C_{i,j}$'s, C_i's and C',
$$\begin{cases} A^* \equiv \bigwedge_{i=1}^m (A_{i,1} \to \cdots \to A_{i,p} \to A_i), \\ x_C \in \Gamma \cup \{x_{B_1}, ..., x_{B_p}\}, \\ C^* \equiv \bigwedge_{i=1}^n (C_{i,1} \to \cdots \to C_{i,q} \to C_i) \wedge C', \\ \Gamma, x_{B_1}, ..., x_{B_p} \vdash_{\lambda\wedge''} M_j : \bigwedge_{i=1}^n C_{i,j} \ (j = 1, ..., q), \\ \bigvee_{i=1}^m A_{i,k} \leq B_k \ (k = 1, ..., p), \ \bigwedge_{i=1}^n C_i \leq \bigwedge_{i=1}^m A_i. \end{cases}$$
$\square$

For a basis Γ of $\lambda\wedge''$ and a type A, we write

$$\mathbf{K}(\Gamma, A) =_{def} \{M \in \Lambda''_{\mathrm{nf}} \mid \Gamma \vdash_{\lambda\wedge''} M : A\}$$

where Λ''_{nf} stands for the set of β-nf's in the system $\lambda\wedge''$. Our goal is to get information for deciding whether $\mathbf{K}(\Gamma, A) = \emptyset$ or not. Let $S = \{(\Gamma, A) \mid \Gamma$ is a basis of $\lambda\wedge''$, A is a type$\}$, and $\mathbf{K}$ be the (infinite) sequence $\langle \mathbf{K}(s)\rangle_{s\in S}$ of sets $\mathbf{K}(s)$ ($\subseteq \Lambda''_{\mathrm{nf}}$). Then the content of Corollary 5.3 can be stated as $\mathbf{K}(s) = \mathbf{\Phi}_s(\mathbf{K})$ for $s = (\Gamma, A) \in S$ with $A \neq \omega$ where $\mathbf{\Phi}_s$ is the mapping from sequences $\mathbf{X} = \langle \mathbf{X}(s)\rangle_{s\in S}$ of subsets of Λ''_{nf} to subsets of Λ''_{nf} defined by

$$\begin{aligned}\mathbf{\Phi}_s(\mathbf{X}) =_{def} \{\lambda x_{B_1} \cdots x_{B_p}.x_C M_1 M_2 \cdots M_q \mid\ & \\ & A^* \equiv \textstyle\bigwedge_{i=1}^m (A_{i,1} \to \cdots \to A_{i,p} \to A_i), \\ & x_C \in \Gamma \cup \{x_{B_1}, \dots, x_{B_p}\}, \\ & C^* \equiv \textstyle\bigwedge_{i=1}^n (C_{i,1} \to \cdots \to C_{i,q} \to C_i) \wedge C', \\ & M_j \in \mathbf{X}(\Gamma\cup\{x_{B_1}, \dots, x_{B_p}\}, \textstyle\bigwedge_{i=1}^n C_{i,j})\ (j = 1, \dots, q), \\ & \textstyle\bigvee_{i=1}^m A_{i,k} \leq B_k\ \ (k = 1, \dots, p),\ \bigwedge_{i=1}^n C_i \leq \bigwedge_{i=1}^m A_i \\ & \text{for some } m, A_{i,k}\text{'s}, A_i\text{'s}, n \geq 1, C_{i,j}\text{'s}, C_i\text{'s and } C'\}.\end{aligned}$$

We extend the definition of $\mathbf{\Phi}_s$ by $\mathbf{\Phi}_s(\mathbf{X}) =_{def} \Lambda''_{\mathrm{nf}}$ for $s = (\Gamma, \omega) \in S$ so that $\mathbf{K}(s) = \mathbf{\Phi}_s(\mathbf{K})$ holds for each $s \in S$. Then we know that the sequence $\mathbf{K} = \langle \mathbf{K}(s)\rangle_{s\in S}$ is the unique solution of the simultaneous equation $\mathbf{X} = \mathbf{\Phi}(\mathbf{X})$ where $\mathbf{\Phi}(\mathbf{X})$ stands for the sequence $\langle \mathbf{\Phi}_s(\mathbf{X})\rangle_{s\in S}$ (cf. [TAH 94] Proposition 1.1). Moreover, when we write $\mathbf{K}_n(s) = \{M \in \mathbf{K}(s) \mid \mathrm{height}(M) \leq n\}$[4] ($n \geq 0$, $s \in S$) and $\mathbf{K}_n = \langle \mathbf{K}_n(s)\rangle_{s\in S}$, we have $\mathbf{K}_0 = \boldsymbol{\emptyset}$ (the sequence $\langle\emptyset\rangle_{s\in S}$ of empty set), $\mathbf{K}_{n+1} = \mathbf{K}_n \cup \mathbf{\Phi}(\mathbf{K}_n)$, and $\mathbf{K} = \bigcup_{n=0}^{\infty} \mathbf{K}_n$ (i.e., $\mathbf{K}(s) = \bigcup_{n=0}^{\infty} \mathbf{K}_n(s)$ for each $s \in S$).

We are to decide the emptiness of components $\mathbf{K}(s)$ of $\mathbf{K}$ based on the description $\mathbf{K} = \mathbf{\Phi}(\mathbf{K})$, but to do so directly seems difficult. So we consider a slightly different mapping $\mathbf{\Psi}$, and reduce our problem to the emptiness problem of the (unique) fixed point of $\mathbf{\Psi}$.

For each sequence $\mathbf{X} = \langle \mathbf{X}(s)\rangle_{s\in S}$ of subsets of Λ''_{nf}, let

$$\begin{aligned}\mathbf{\Psi}_s(\mathbf{X}) =_{def} \{\lambda x_{B_1} \cdots x_{B_p}.x_C M_1 M_2 \cdots M_q \mid\ & \\ & A^* \equiv \textstyle\bigwedge_{i=1}^m (A_{i,1} \to \cdots \to A_{i,p} \to A_i), \\ & x_C \in \Gamma \cup \{x_{B_1}, \dots, x_{B_p}\}, \\ & C^* \equiv \textstyle\bigwedge_{i=1}^n (C_{i,1} \to \cdots \to C_{i,q} \to C_i) \wedge C', \\ & M_j \in \mathbf{X}(\Gamma\cup\{x_{B_1}, \dots, x_{B_p}\}, \textstyle\bigwedge_{i=1}^n C_{i,j})\ (j = 1, \dots, q), \\ & \textstyle\bigvee_{i=1}^m A_{i,k} = B_k\ \ (k = 1, \dots, p),\ \bigwedge_{i=1}^n C_i \leq \bigwedge_{i=1}^m A_i \\ & \text{for some } m, A_{i,k}\text{'s}, A_i\text{'s}, n \geq 1, C_{i,j}\text{'s}, C_i\text{'s and } C'\} \\ & \qquad \text{if } s = (\Gamma, A) \in S \text{ and } A \neq \omega; \\ \mathbf{\Psi}_s(\mathbf{X}) =_{def} \{x_\omega\} \qquad & \qquad \text{if } s = (\Gamma, \omega) \in S.\end{aligned}$$

The definition of $\mathbf{\Psi}_s$ for $s = (\Gamma, A)$ with $A \neq \omega$ is same as $\mathbf{\Phi}_s$ except the condition for bound variables x_{B_k}; the types B_k of bound variables vary under the condition $\bigvee_{i=1}^m A_{i,k} \leq B_k$ in $\mathbf{\Phi}_s$, while $\bigvee_{i=1}^m A_{i,k} = B_k$ in $\mathbf{\Psi}_s$.

[4]height(M) means the height of Böhm tree of M.

As in the case of $\mathbf{\Phi}$, we see that the mapping $\mathbf{\Psi}$ has a unique fixed point, which we write $\mathbf{L} = \langle \mathbf{L}(s)\rangle_{s\in S}$. Then as before we have $\mathbf{L} = \bigcup_{n=0}^{\infty} \mathbf{L}_n$ where $\mathbf{L}_0 = \emptyset$ and $\mathbf{L}_{n+1} = \mathbf{L}_n \cup \mathbf{\Psi}(\mathbf{L}_n)$ $(n = 0, 1, ...)$. Between the fixed points $\mathbf{K}$ of $\mathbf{\Phi}$ and $\mathbf{L}$ of $\mathbf{\Psi}$, we can observe the following relations.

Lemma 5.4 $\mathbf{L} \subseteq \mathbf{K}$ (i.e., $\mathbf{L}(s) \subseteq \mathbf{K}(s)$ for each $s \in S$).
Proof. By definition of $\mathbf{\Psi}$, clearly $\mathbf{\Psi}(\mathbf{X}) \subseteq \mathbf{\Phi}(\mathbf{X})$ for each sequence $\mathbf{X}$. On the other hand, since $\mathbf{\Phi}$ is a monotone mapping (i.e., $\mathbf{\Phi}(\mathbf{X}) \subseteq \mathbf{\Phi}(\mathbf{X}')$ if $\mathbf{X} \subseteq \mathbf{X}'$), if $\mathbf{L}_n \subseteq \mathbf{K}_n$ then $\mathbf{L}_{n+1} = \mathbf{L}_n \cup \mathbf{\Psi}(\mathbf{L}_n) \subseteq \mathbf{L}_n \cup \mathbf{\Phi}(\mathbf{L}_n) \subseteq \mathbf{K}_n \cup \mathbf{\Phi}(\mathbf{K}_n) = \mathbf{K}_{n+1}$. Thus we know by induction that $\mathbf{L}_n \subseteq \mathbf{K}_n$ for each n. □

Lemma 5.5 For each $s \in S$, if $\mathbf{K}(s) \neq \emptyset$ then $\mathbf{L}(s) \neq \emptyset$.
Proof. We can verify a stronger statement; from any $M \in \mathbf{K}(s)$ a term $N \in \mathbf{L}(s)$ can be constructed by changing bound variables x_B in M to some $x_{B'}$ with $B' \leq B$ and by replacing some subterms with x_ω. The details of the proof are omitted. □

Corollary 5.6 $\forall s \in S$ $(\mathbf{K}(s) = \emptyset \Longleftrightarrow \mathbf{L}(s) = \emptyset)$. □

Our problems are now reduced to the emptiness problem of $\mathbf{L}(\Gamma, A)$ for given $(\Gamma, A) \in S$. To solve the latter problem, we will use a result in formal language theory; we show that $\mathbf{L}(\Gamma, A)$ is a context-free language (over an alphabet consisting of variables, $\lambda, ., (,)$), and that a context-free grammar generating $\mathbf{L}(\Gamma, A)$ can be constructed from Γ and A. Then we can apply a well-known algorithm in formal language theory to see $\mathbf{L}(\Gamma, A) = \emptyset$ or not.

For the sake of simplicity, in the sequel we will write Γ for $\mathrm{Pred}(\Gamma) =_{def} \{B \mid x_B \in \Gamma\}$. Note that in the system $\lambda\wedge''$ the mapping Pred is a one-to-one correspondence between bases and finite sets of types.

Definition 5.7 Let $s = (\Gamma, A) \in S$. We define three sets T_s ($\subseteq \{x_B \mid B \in \mathrm{T}_\wedge\}$[5] $\cup \{\lambda, ., (,)\}$) of terminal symbols, N_s ($\subseteq S$) of nonterminal symbols, and R_s ($\subseteq \{\varphi \rightsquigarrow \varphi_1\varphi_2\cdots\varphi_m \mid \varphi \in N_s, m \geq 1, \varphi_1, \ldots, \varphi_m \in T_s \cup N_s\}$) of production rules, recursively:

1. $\{x_C \mid C \in \Gamma\} \cup \{\lambda, ., (,)\} \subseteq T_s$ and $s \in N_s$.
2. If $s' = (\Gamma', \omega) \in N_s$, then $x_\omega \in T_s$ and $(s' \rightsquigarrow x_\omega) \in R_s$.
3. If $s' = (\Gamma', A') \in N_s$, $A' \neq \omega$, and

$$C \in \Gamma' \cup \{B_1, \ldots, B_p\},$$
$$C^* \equiv \textstyle\bigwedge_{i=1}^{n}(C_{i,1} \to \cdots \to C_{i,q} \to C_i) \wedge C',$$
$$s_j = (\Gamma' \cup \{B_1, \ldots, B_p\}, \textstyle\bigwedge_{i=1}^{n} C_{i,j}) \quad (j = 1, \ldots, q),$$
$$A'^* \equiv \textstyle\bigwedge_{i=1}^{m}(A'_{i,1} \to \cdots \to A'_{i,p} \to A'_i),$$
$$\textstyle\bigwedge_{i=1}^{n} C_i \leq \bigwedge_{i=1}^{m} A'_i, \quad \bigvee_{i=1}^{m} A'_{i,k} = B_k \quad (k = 1, \ldots, p),$$

then

$$x_{B_1}, \ldots, x_{B_p} \in T_s,\ s_1, \ldots, s_q \in N_s, \text{ and}$$
$$(s' \rightsquigarrow \lambda x_{B_1} \cdots x_{B_p}.x_C s_1 s_2 \cdots s_q) \in R_s. \quad \Box$$

[5] Recall that in $\lambda\wedge''$ we consider types modulo $=$.

Roughly speaking, N_s is the set of s''s in S such that $\mathbf{X}(s')$ appears in the 'unfolding' of $\mathbf{X}(s) = \mathbf{\Psi}_s(\mathbf{X})$, and R_s consists of production rules of the form $s' \leadsto \lambda x_{B_1} \cdots x_{B_p}.x_C s_1 s_2 \cdots s_q$ where $s' \in N_s$, $\lambda x_{B_1} \cdots x_{B_p}.x_C M_1 M_2 \cdots M_q \in \mathbf{\Psi}_{s'}(\mathbf{X})$, and $M_j \in \mathbf{X}(s_j)$ $(j = 1, \ldots, q)$.

If the sets T_s, N_s and R_s are shown to be finite, then $G_s = (T_s, N_s, R_s, s)$ is a context-free grammar, and from the construction it clearly generates the set $\mathbf{L}(s)$.

In order to see the finiteness of sets T_s, N_s and R_s, it suffices to prove the finiteness of N_s. This is because for each $s' \in N_s$ only a finite number of production rules with s' in the lefthand sides are introduced in R_s, and the set T_s of terminal symbols in these rules is finite.

Theorem 5.8 The set N_s is finite for each $s \in S$.
Proof. See Appendix B. □

Theorem 5.9 For any Γ and A, whether $\mathbf{L}(\Gamma, A) = \emptyset$ or not is decidable.
Proof. Let $s = (\Gamma, A)$. First, following Definition 5.7 we enumerate members of sets T_s, N_s and R_s. At some point a same member of N_s appears twice by Theorem 5.8. Then we know all the elements of T_s, N_s and R_s, and the set $\mathbf{L}(s)$ can be generated by the context-free grammar $G_s = (T_s, N_s, R_s, s)$. It is well-known (cf. [HU 79]) that the context-free language $\mathbf{L}(s)$ generated by G_s is empty if and only if $\{M \in \mathbf{L}(s) \mid \text{height}(M) \leq n+1\} = \emptyset$ where n is the number of elements of N_s. Thus the emptiness of $\mathbf{L}(s)$ is decidable. □

Corollary 5.10 The inhabitation problems for $\lambda\wedge$–($\wedge$I) and for $\lambda\wedge'$ are decidable.
Proof. By Theorem 5.1, Lemma 5.2, Corollary 5.6 and Theorem 5.9. □

Appendix A. (Proof of Theorem 3.2)

In order to prove Theorem 3.2, we define *depth* and *length* of types, as follows: If $A^* \equiv \bigwedge_{i=1}^{m}(A_{i,1} \to \cdots \to A_{i,p_i} \to a_i) \not\equiv \omega$, then $\text{dp}(A) = \max\{\text{dp}(A_{i,j}) \mid i = 1, \ldots, m,\ j = 1, \ldots, p_i\} + 1$ and $\text{lg}(A) = \max\{p_i \mid i = 1, \ldots, m\}$. If $A = \omega$, we define $\text{dp}(A) = \text{lg}(A) = 0$. We will write $\text{ord}(A)$ for $(\text{dp}(A), \text{lg}(A))$, and $\leq$ for the lexicographic order on pairs of natural numbers. Note that $\text{ord}(A) = 0 \iff A = \omega$.

Lemma A.1 (1) $\text{ord}(A_i) \leq \text{ord}(\bigwedge_{i \in I} A_i)$ if $A_i \in \mathsf{NT}^{\to}$ $(i \in I)$.
(2) $\text{ord}(A_i) < \text{ord}(A_1 \to A_2)$ $(i = 1, 2)$. □

When γ is a derivation in the system $\lambda\wedge'$ of the form

$$\frac{\begin{array}{cc}\vdots & \vdots \\ \Gamma \vdash \lambda x : A.M : C \to D & \Gamma \vdash N : C\end{array}}{\Gamma \vdash (\lambda x : A.M)N : D}\ (\to \text{E})$$

we say γ is a *cut* represented by the term $(\lambda x : A.M)N$, and its *order* ($\text{ord}(\gamma)$ in notation) is defined by $\text{ord}(C \to D)$.

Lemma A.2 For any derivation π of $\Gamma \vdash_{\lambda\wedge'} M : A$, there exists a term N and a derivation π' of $\Gamma \vdash_{\lambda\wedge'} N : A$ such that $M \twoheadrightarrow_\beta N$ and any cut in π' is of order $(0,0)$.

Proof. Suppose π contains some cuts γ with $\mathrm{ord}(\gamma) \neq (0,0)$. Then among the cuts in π of maximum order, say θ, let

$$\dfrac{\begin{array}{c}\vdots \\ \lambda x : A.M : C \to D\end{array} \quad \begin{array}{c}\vdots\ \delta \\ N : C\end{array}}{(\lambda x : A.M)N : D}\ (\to \mathrm{E}) \\ \vdots\ \zeta$$

be the one which is represented by a shortest term, and call it γ. Without loss of generality, we may assume that the major premiss $\lambda x : A.M : C \to D$ of γ is derived from subderivations of the form

$$\dfrac{\begin{array}{c} x : A \\ \vdots\ \delta_i \\ M : B_i \end{array}}{\lambda x : A.M : A \to B_i}\ (\to \mathrm{I})' \qquad (i = 1, \ldots, m)$$

and some instances of (ω)-rule $\lambda x : A.M : \omega$ by means of $(\wedge\mathrm{I})$- and $(\leq)$-rules. In general, if a statement $M : E$ is derived from $M : E_i$ $(i = 1, \ldots, m)$ by applying only $(\wedge\mathrm{I})$- and $(\leq)$-rules, then $\bigwedge_{i=1}^m E_i \leq E$ holds. Therefore we have $\bigwedge_{i=1}^m (A \to B_i) \leq C \to D \neq \omega$ since $\mathrm{ord}(C \to D) = \mathrm{ord}(\gamma) \neq (0,0)$. Then by Lemma 2.6 (3) we know that there exists a nonempty subset I of $\{1, \ldots, m\}$ such that $C \leq A$ and $\bigwedge_{i \in I} B_i \leq D$. In this case, we replace the cut γ with the following derivation.

$$\dfrac{\begin{array}{c} \dfrac{\begin{array}{c}\vdots\ \delta \\ N : C\end{array} \quad C \leq A}{N : A}\ (\leq) \\ \vdots\ \delta_i' \\ M[x := N] : B_i \\ \vdots\ (\wedge\mathrm{I}) \\ M[x := N] : \bigwedge_{i \in I} B_i \end{array} \qquad \bigwedge_{i \in I} B_i \leq D}{M[x := N] : D}\ (\leq) \\ \vdots\ \zeta'$$

Note that this replacement may create new cuts in δ_i' and/or ζ', and moreover some of them may be of order θ or more. If so, we can replace such cuts once again with derivations which contain only cuts of order less than θ. Indeed, suppose there is a new cut γ' in δ_i' such that $\mathrm{ord}(\gamma') \geq \theta$. Then γ' must be of the form

$$\dfrac{\begin{array}{c} \dfrac{\begin{array}{c}\vdots\ \delta \\ N : C\end{array} \quad C \leq A}{N : A}\ (\leq) \\ \vdots\ (\wedge\mathrm{I}), (\leq) \\ N : D \to E \end{array} \qquad \begin{array}{c}\vdots\ \varepsilon \\ L : D\end{array}}{NL : E}\ (\to \mathrm{E})$$

where N is an abstraction. Then we have $C \leq D \to E \not\equiv \omega$, and hence by Lemma 2.7 we can find $m, n \geq 1$, E_i's, C_j's and C'_j's such that $C^* \equiv \bigwedge_{j=1}^{n}(C_j \to C'_j)$, $E^* \equiv \bigwedge_{i=1}^{m} E_i$, and $\forall i \in \{1, \ldots, m\}\ \exists j_i \in \{1, \ldots, n\}\ (D \leq C_{j_i}$ and $C'_{j_i} \leq E_i)$. In this case, we can replace γ' with the following derivation.

$$\dfrac{\dfrac{\dfrac{\dfrac{\begin{matrix}\vdots\ \delta \\ N : C\end{matrix} \quad C \leq C_{j_i} \to C'_{j_i}}{N : C_{j_i} \to C'_{j_i}}\,(\leq) \quad \dfrac{\begin{matrix}\vdots\ \varepsilon \\ L : D\end{matrix} \quad D \leq C_{j_i}}{L : C_{j_i}}\,(\leq)}{NL : C'_{j_i}}\,(\to \mathrm{E}) \quad C'_{j_i} \leq E_i}{\begin{matrix} NL : E_i \\ \vdots\ (\wedge \mathrm{I}) \\ NL : \bigwedge_{i=1}^{m} E_i \end{matrix}}\,(\leq) \quad \bigwedge_{i=1}^{m} E_i \leq E}{NL : E}\,(\leq)$$

Note that this replacement does not create any cut of order θ or more. By a similar technique, we can also eliminate cuts in ζ' of order θ or more.

By applying the process repeatedly, we can obtain a term N and a derivation π' which satisfy the lemma. □

Theorem A.3 (Theorem 3.2) $\Gamma \vdash_{\lambda\wedge'} M : A \Longrightarrow \exists P \sqsubseteq M\ (\Gamma \vdash_{\lambda\wedge'} P : A)$.

Proof. Suppose N and π' are obtained from a derivation π of $\Gamma \vdash_{\lambda\wedge'} M : A$ by applying Lemma A.2. In the derivation π', each statement occurence whose subject is a β-redex is introduced by (ω)-rule or by a cut of order $(0,0)$. Then by replacing these statements by $\bot : \omega$, we get a derivation of $\Gamma \vdash_{\lambda\wedge'} P : A$ where P is a finite approximation of M. □

Appendix B. (Proof of Theorem 5.8)

In order to prove the finiteness of N_s, we introduce some auxiliary notions.

Definition B.1 Suppose $A^* \equiv \bigwedge_{i=1}^{m}(A_{i,1} \to \cdots \to A_{i,p_i} \to a_i)$. Then we define two subsets $\mathrm{dom}_\wedge(A)$ and $\mathrm{dom}^\vee(A)$ of $\mathrm{T}_\wedge$, as follows:

$$\mathrm{dom}_\wedge(A) = \{\textstyle\bigwedge_{i \in I} A_{i,j} \mid \emptyset \neq I \subseteq \{1, \ldots, m\},\ 1 \leq j \leq p_i\ (i \in I)\ \},$$
$$\mathrm{dom}^\vee(A) = \{\textstyle\bigvee_{i=1}^{m} A_{i,j} \mid 1 \leq j \leq p_i\ (i = 1, \ldots, m)\ \}. \qquad \square$$

Note that both $\mathrm{dom}_\wedge(A)$ and $\mathrm{dom}^\vee(A)$ are finite for each type A. For a set T of types we write $\mathrm{dom}_\wedge(T)$ and $\mathrm{dom}^\vee(T)$ for $\bigcup_{A \in T} \mathrm{dom}_\wedge(A)$ and $\bigcup_{A \in T} \mathrm{dom}^\vee(A)$, respectively.

Definition B.2 For each $s \in S$, $\mathrm{dom}(s)$ $(\subseteq \mathrm{T}_\wedge)$ is defined recursively, as follows:

1. $\Gamma \cup \{A\} \subseteq \mathrm{dom}(s)$.
2. $\mathrm{dom}_\wedge(\mathrm{dom}(s) \cup \mathrm{dom}^\vee(\mathrm{dom}(s))) \subseteq \mathrm{dom}(s)$. □

In order to see the finiteness of N_s, it suffices to show that

$$N_s \subseteq \mathcal{P}(\mathrm{dom}(s) \cup \mathrm{dom}^\vee(\mathrm{dom}(s))) \times \mathrm{dom}(s),$$

and that dom(s) is finite. First we prove the inclusion relation.

Lemma B.3 For each $s \in S$, $N_s \subseteq \mathcal{P}(\text{dom}(s) \cup \text{dom}^{\vee}(\text{dom}(s))) \times \text{dom}(s)$.
Proof. By induction on the structure of N_s. Let $\Delta = \text{dom}(s) \cup \text{dom}^{\vee}(\text{dom}(s))$.
(case 1) If $s = (\Gamma, A)$, then $\Gamma \cup \{A\} \subseteq \text{dom}(s) \subseteq \Delta$ by definition of dom(s). Thus s belongs to $\mathcal{P}(\Delta) \times \text{dom}(s)$.
(case 2) Suppose $s' = (\Gamma', A') \in N_s$ with $A' \neq \omega$, and $s'' = (\Gamma'', A'')$ is added to N_s in the recursion step 3 of Definition 5.7 (i.e., $\mathbf{X}(\Gamma'', A'')$ appears in the definition of $\Psi_{s'}(\mathbf{X})$). Then $\Gamma'' = \Gamma' \cup \{B_1, \ldots, B_p\}$ where $A'^* = \bigwedge_{i=1}^{m}(A'_{i,1} \to \cdots \to A'_{i,p} \to A')$, $B_k = \bigvee_{i=1}^{m} A'_{i,k}$ $(\in \text{dom}^{\vee}(A'))$ $(k = 1, \ldots, p)$, and $A'' = \bigwedge_{i=1}^{n} C_{i,j}$ $(\in \text{dom}_{\wedge}(C))$ where $C \in \Gamma''$, $C^* \equiv \bigwedge_{i=1}^{n}(C_{i,1} \to \cdots \to C_{i,q} \to C_i) \wedge C'$ and $1 \leq j \leq q$. Then for each k, since $B_k \in \text{dom}^{\vee}(A')$ and $A' \in \text{dom}(s)$ (by induction hypothesis), by definition of Δ we have $B_k \in \Delta$. This means that $\Gamma'' = \Gamma' \cup \{B_1, \ldots, B_p\} \subseteq \Delta$ because $\Gamma' \subseteq \Delta$ by induction hypothesis. On the other hand, since $C \in \Gamma'' \subseteq \Delta$, we have $\text{dom}_{\wedge}(C) \subseteq \text{dom}_{\wedge}(\Delta) \subseteq \text{dom}(s)$ by definition of dom(s) and Δ. This proves $A'' \in \text{dom}_{\wedge}(C) \subseteq \text{dom}(s)$. Hence $(\Gamma'', A'') \in \mathcal{P}(\Delta) \times \text{dom}(s)$. □

Next, to see the finiteness of dom(s), we define the degree of types, as follows:

$$\deg(A) = \begin{cases} \max\{\deg(A_{i,1}) + \cdots + \deg(A_{i,p_i}) + p_i \mid i = 1, \ldots, m\} \\ \qquad \text{if } A^* \equiv \bigwedge_{i=1}^{m}(A_{i,1} \to \cdots \to A_{i,p_i} \to a_i) \not\equiv \omega, \\ 0 \qquad \text{if } A^* \equiv \omega. \end{cases}$$

In the following three lemmas, we assume that $A^* \equiv \bigwedge_{i=1}^{m}(A_{i,1} \to \cdots \to A_{i,p_i} \to a_i)$.

Lemma B.4 $\deg(A_{i,j}) < \deg(A)$.
Proof. $\deg(A_{i,j}) < \deg(A_{i,1} \to \cdots \to A_{i,p_i} \to a_i) \leq \deg(A^*) = \deg(A)$. □

Lemma B.5 If $B \in \text{dom}_{\wedge}(A)$, then $\deg(B) < \deg(A)$.
Proof. Since $B^* \equiv \bigwedge_{i \in I} A_{i,j}$ for some I and j, $\deg(B) \leq \max\{\deg(A_{i,j}) \mid 1 \leq i \leq m, 1 \leq j \leq p_i\} < \deg(A)$ by Lemma B.4. □

Lemma B.6 If $C \in \text{dom}_{\wedge}(\text{dom}^{\vee}(A))$, then $\deg(C) < \deg(A)$.
Proof. Let us call each $A_{i,j}$ a *component* of A^*, and similarly for other normal types. Suppose $C \in \text{dom}_{\wedge}(B)$ and $B \in \text{dom}^{\vee}(A)$. Then by definition of $\text{dom}_{\wedge}(B)$, C is an intersection of components of B^*. On the other hand, since $B = \bigvee_{i=1}^{m} A_{i,j}$ for some j $(\leq \min\{p_1, \ldots, p_m\})$, by Definition 2.9 each component of B^* is an intersection of components of components $A_{i,j}$ of A^*. Hence C is also an intersection of components of components of A^*. Therefore by Lemma B.4 we get $\deg(C) < \deg(A)$. □

Corollary B.7 The set dom(s) is finite for each $s \in S$.
Proof. By definition, each element of dom(s) is obtained as B_i for some $i \geq 0$ where $B_0 \in \Gamma \cup \{A\}$, and $B_i \in \text{dom}_{\wedge}(B_{i-1}) \cup \text{dom}_{\wedge}(\text{dom}^{\vee}(B_{i-1}))$ $(i = 1, 2, \ldots)$. Then it is clear that possible B_0's are finite, and for each B_{i-1} the set of B_i's satisfying above condition is also finite. Moreover by Lemmas B.5 and B.6 $\deg(B_i) < \deg(B_{i-1})$, which implies that there is no infinite sequence

$B_0, B_1, B_2, \ldots$. Then by König's lemma we know that the set dom(s) is finite. □

Theorem B.8 (Theorem 5.8) The set N_s is finite for each $s \in S$.
Proof. By Lemma B.3 and Corollary B.7. □

Acknowledgements

The authors are very grateful to Mariangiola Dezani-Ciancaglini for helpful discussions and comments about the subject of this paper.

References

[Bar 92] H.P. Barendregt, Lambda calculi with types, in: *Handbook of Logic in Computer Science vol.II*, ed. S. Abramsky et al., Oxford University Press, 1992.

[BCD 83] H.P. Barendregt, M. Coppo and M. Dezani-Ciancaglini, A filter lambda model and the completeness of type assignment, *J. Symbolic Logic* 48 (1983), pp.931-940.

[CC 90] F. Cardonne and M. Coppo, Two extensions of Curry's type inference system, in: *Logic and Computer Science* ed. P. Odifreddi, Academin Press, pp.19-75, 1990.

[DM 86] M. Dezani-Ciancaglini and I. Margaria, A characterisation of F-complete type assignments, *Theoretical Computer Science* 45 (1986), pp. 121-157.

[Hin 82] J.R. Hindley, The simple semantics for Coppo-Dezani-Sallé types, *Lecture Notes in Computer Science* 137 (1982), pp.212-226.

[HU 79] J.E. Hopcroft and J.D. Ullman, *Introduction to Automata Theory, Languages and Computation*. Addison-Wesley, 1979.

[Rey 88] J.C. Reynolds, Preliminary design of the Programming Language Forsythe, Report CMU-CS-88-159, Carnegie-Mellon University, 1988.

[TAH 94] M. Takahashi, Y. Akama and S. Hirokawa, Normal proofs and their grammar, Lecture Notes in Computer Science 789 (1994), pp.465-493.

[Urz 94] P. Urzyczyn, The emptiness problem for intersection types, *Proceedings of Logic in Computer Science*, IEEE, 1994.

Termination Proof of Term Rewriting System with the Multiset Path Ordering. A Complete Development in the System Coq

François LECLERC

CRIN-CNRS & INRIA-Lorraine
BP 239, 54506 Vandœuvre-lès-Nancy, France
E-mail: Francois.Leclerc@loria.fr

Abstract. We propose a constructive termination proof in the Calculus of Constructions of any finite term rewriting systems whose rules can be oriented by the multiset path ordering. We propose a new proof which consists in an embedding of the rewrite relation into the standard ordering over natural numbers. Then, we show how to derive automatically constructive well-foundedness proofs of rewrite relations. This work has been completely formalised in the Coq system. Such a mechanization is not useless since there was some nontrivial mistakes in the previous proofs of Cichon and Weiermann. Furthermore this kind of development reflects the ability to formalise in Coq parts of nontrivial mathematics.

Introduction

Termination property of term rewriting systems (TRS) is undecidable. Fortunately proof technics can be used to prove termination of a wide range of TRS. A common one consists in embedding the rewrite relation in a well founded ordering which is compatible with the term structure. Thus, it ensures that the normal form of a term can be reached in finitely many steps whatever rewrite strategy is taken. Orderings like the multiset path ordering (mpo) or the lexicographic path ordering (lpo) are good candidates for such proofs.

But we would like to say more in this case about the rewriting process. How much time does it require to be completed, and does the ability of a particular method for proving termination impose limits on the computation time? For instance, lpo can be used to yield a termination proof for the non primitive recursive Ackermann's function, while mpo fails. More precisely, the question is to provide a complexity characterization of the class of the TRS reducible under a given ordering. Derivation length is the simplest estimation for the complexity measure of a function computed by a TRS. Complexity is then given by a function D_R : *term* $\rightarrow$ *nat* which with each term t of size $|t|$ associates the maximal length of all the rewriting sequences starting from t. C. Lautemann [Lau88] answered these questions for termination proofs with polynomial interpretations. A. Cichon [Cic90], D. Hofbauer [Hof90] and A. Weiermann [Wei93] showed that there is a primitive recursive function which bounds D_R for any finite TRS reducible under mpo. The direct theoretic consequence is that any function computed by a finite TRS reducible under mpo is primitive recursive. This result is constructive. The bounding function can be systematically computed for any finite TRS, from its constants and from a precedence (an ordering on the signature) provided that the mpo induced by this precedence achieves in orienting the rules. The proofs of this bounding theorem proposed in [Cic90] and [Wei93] are actually termination proofs for rewriting systems and do not require the well-foundedness property of mpo. In other words they provide a method for translating a termination proof of a TRS with mpo, relying on the impredicative proof of Kruskal's theorem into a constructive one. We propose another proof of this theorem based on an embedding of the rewrite relation into the standard strict ordering over natural numbers instead of monotonic interpretations.

This work was first motivated by two pragmatical reasons. The first one was to formalise, correct, and mechanically check the proof of the bounding theorem since there are errors in its previous proofs. We propose here the first mechanically certified proof of it. Moreover, our proof based on the embedding of the rewrite relation into the standard ordering over natural numbers is valid also for sequences of open terms. The second one was to study the expressivity of the Calculus of Constructions with Inductive Types which is still a prototype. But in addition our method might be used to mechanize constructive well-foundedness proofs of rewrite relations. Actually, there is a need for automatic proof search in the Calculus of Constructions, especially for termination proofs. For instance, when developing programs from pure proofs, we need to justify the recursive structure of programs which is strongly related to the structure of the proofs.

The main result is a proof, that is mechanically certified. In what follows when we write Lemma or Theorem, it means the statement has been mechanically proved using Coq and is a part of the proof of the main result. In the first part of the paper we introduce term rewriting systems theory. In the second part we focus on the specification of mpo and prove that it is a simplification ordering. The embedding of terms in the natural numbers and its monotonicity properties is given in the third part, and we dedicate the fourth one to the proof of the complexity theorem. In the fifth part, we derive a valid termination program for TRS from the bounding theorem proof. Finally, we show how to build automatically well-foundedness proofs of rewrite relations reducible under mpo and we discuss about the computational content of the proof. This paper as well as the proof development is self contained: there is no axiom. All the proofs have been developped in the current version Coq v5.8.

Notations

We choose to present the paper like a fully annoted Coq script. Secondary lemmas and proofs are generally omitted. We give here a very short introduction to some of the characteristic features of Coq. The reader may consult [DFH+91, PM93] for more details.

- $[x : A]$ stands for the usual typed lambda-abstraction λx^A.
- $(x : A)B$ stands for the dependent product $\Pi x^A.B$.
- The arrow symbol associates to the right and the application to the left.
- $\sim$ stands for the negation.

- **Prop and Set**: The two sorts *Set* and *Prop* of Coq distinguish between proofs having a computational content and proofs having only a logical meaning. An object whose type is *Prop* is called a proposition and an object whose type is *Set* is called a specification.
- **Inductive definitions**: the calculus was extended with primitive inductive definitions. In particular the induction principle on natural numbers is now provable in the system. An inductive definition is given by the signature of its constructors. Both inductive sets (inductive definitions of sort *Set*) and inductive predicates (inductive definitions of sort *Prop*) can be defined. However there is a restriction: in the inductive definition of X, X has to occur at strictly positive positions in the type of its constructors.
- **Primitive recursive definitions**: in Coq only primitive recursive schemes are allowed to define funtions. The syntax of Coq is not so easy to read, therefore we use equational definitions.

1 Term Rewriting System

We present an axiomatisation of term rewriting theory in Coq. For this purpose various rewriting notions are introduced such as terms, substitutions, rewrite rules and rewrite relations.

1.1 Quasiterms and Terms

Definition 1 Terms. The algebra of first order terms $\mathcal{T}(\mathcal{F},\mathcal{X})$ built on a finite signature $\mathcal{F}$ and a countable set $\mathcal{X}$ of variables is inductively defined as follows:

1. constants and variables are terms
2. if $f \in F$ is a functional symbol equipped with an arity n and if $t_1,\ldots,t_n$ are terms, then $f(t_1,\ldots,t_n)$ is a term.

Terms are not directly expressible in Coq as an inductive set, since in the second case the constructor would be typed $fun \rightarrow$ (*term list*) $\rightarrow$ *term* where *term* occurs inside the inductive type *List*. This kind of mutually recursive definition is not allowed in the Coq v5.8 system. First order terms coding in Coq has already been devised by Joseph Rouyer [Rou92, RL92] for his work about unification. Because it seems well suited for our current purpose, we choose to reuse his axiomatisation as a Coq proof library.

First, since variables and symbols of the signature form countable sets we do identify them with *nat*. Then we define *quasiterm* as an inductive set with four constructors corresponding respectively to variables, constants, rooted terms and argument lists of rooted terms.

```
Definition var,fun:Set = nat.
Inductive Set quasiterm =
    V:var->quasiterm
  | C:fun->quasiterm
  | Root:fun->quasiterm->quasiterm
  | ConsArg:quasiterm->quasiterm->quasiterm.
```

Terms form a specific subset of quasiterms characterized by a predicate. For this purpose we define an arity function, the length of a quasiterm, and the two control predicates *Simple* and *l_term*. In this development we are dealing only with finite signatures. Therefore, the signature is entirely determined by a given finite list of arities *arity_list*, such that: if $f \in fun = nat$, then the arity of f is the $f + 1^{th}$ element of *arity_list*.

```
Variable arity_list:nat_list.
Definition arity:fun->nat = (list_access arity_list).
```

Length: quasiterm→nat ≡
Length(V(x)) = Length(C(c)) = Length(Root(f,t)) = 1
Length(ConsArg(t_1,t_2)) = Length(t_1) + Length(t_2)

Quasiterms whose length is equal to 1 are called *Simple quasiterms*. We restrict quasiterms to *l_terms*.

```
Inductive Definition l_term:quasiterm->Prop =
 l_term_V:(x:var)(l_term (V x))
|l_term_C:(c:fun)(<nat>(arity c)=0)->(l_term (C c))
|l_term_R:(f:fun)(t:quasiterm)(l_term t)->(<nat>(arity f)=(Length t))
      ->(l_term (Root f t))
|l_term_CA:(t1,t2:quasiterm)(l_term t1)->(l_term t2)->(Simple t1)
       ->(l_term (ConsArg t1 t2)).
```

Finally we call *term* any *Simple quasiterm* which is a *l_term*.

```
Inductive Definition term [t:quasiterm]:Prop =
        term_init:(l_term t)->(Simple t)->(term t).
```

Joseph Rouyer [Rou92] showed that equality in the set of quasiterms is decidable and that both predicates *l_term* and *term* are also decidable. Such properties are essential, since the elimination in a proof corresponds to a case reasoning, and in a context of proof as program paradigm to an "if then else" construction.

In the sequel, all the notions are defined for quasiterms. As a counterpart, we have to ensure when necessary that they are compatible with the term structure. For instance, we shall prove that a quasiterm which is a term remains a term after applying a substitution on it.

1.2 Occurrence and Inclusion of Variables - Ground Quasiterm

The decidable relation expressing that a variable occurs in a term is introduced as an inductive predicate:

```
Inductive Definition is_in [x:var]:quasiterm->Prop =
  is_inV:(is_in x (V x))
| is_inR:(f:fun)(t:quasiterm)(is_in x t)->(is_in x (Root f t))
| is_inCA_l:(t1,t2:quasiterm)(is_in x t1)->(is_in x (ConsArg t1 t2))
| is_inCA_r:(t1,t2:quasiterm)(is_in x t2)->(is_in x (ConsArg t1 t2)).
```

An embedding relation over sets of variables occurring in two terms and ground quasiterms are defined as:

```
Definition infv = [t,s:quasiterm](y:var)(is_in y t)->(is_in y s).
Definition gd_quasiterm = [t:quasiterm](x:var)~(is_in x t).
```

1.3 Quasisubstitution

A quasisubstitution is a function $\sigma : var \rightarrow quasiterm$. We extend quasisubstitutions to quasiterms to get substitutions $Subst : quasisubst \rightarrow quasiterm \rightarrow quasiterm$.

```
Definition quasisubst = var->quasiterm.
```

Subst: quasisubst→quasiterm→quasiterm ≡
Subst(σ,V(x)) = σ(x)
Subst(σ,C(c)) = C(c)
Subst(σ,Root(f,t)) = Root(f,Subst(σ,t))
Subst(σ,ConsArg(t_1,t_2)) = ConsArg(Subst(σ,t_1),Subst(σ,t_2))

We restrict quasisubstitutions to those with only terms as images to get the preservation lemma for substitutions.:

```
Definition termsubst = [s:quasisubst](x:var)(term (s x)).
Lemma term_subst: (s:quasisubst)(termsubst s)->(t:quasiterm)(term t)
                 ->(term (Subst s t))).
```

1.4 Finite Term Rewriting System

Definition 2 finite term rewriting system. A finite term rewriting system $\mathcal{R}$ over $\mathcal{T}(\mathcal{F},\mathcal{X})$ is a finite set of ordered couples $\langle l_i, r_i\rangle \in \mathcal{T}(\mathcal{F},\mathcal{X}) \times \mathcal{T}(\mathcal{F},\mathcal{X})$ such that $Var(r_i) \subset Var(l_i)$, $\forall i\ 1 \leq i \leq n$.

Lists of couples of quasiterm are rewrite rules. This type is called *qrule_list*.

```
Inductive Set qrule_list =
nilqr:qrule_list | consqr:quasiterm->quasiterm->qrule_list->qrule_list.
```

We note *is_qr* the predicate that says that a quasirule belongs to a list of quasirules. We add controls to restrict the form of rewrite rules. That is, left-hand and right-hand sides of a rule are terms and variables of the right-hand side are variables of the left hand-side. The decidability of the following predicate comes from the decidability of *term* and *infv*.

```
Inductive Definition termrule_list:qrule_list->Prop =
  niltr:(termrule_list nilqr)
| constr:(R:qrule_list)(termrule_list R)->(l,r:quasiterm)(term l)->(term r)
       ->(infv r l)->(termrule_list (consqr l r R)).
```

Definition 3 rewrite relation. The rewrite relation $\rightarrow_{\mathcal{R}}$ is the smallest binary relation on $\mathcal{T}(\mathcal{F},\mathcal{X}) \times \mathcal{T}(\mathcal{F},\mathcal{X})$ satisfying:

1. if $\langle l_i, r_i\rangle \in \mathcal{R}$ and if $\sigma \in \mathcal{X} \rightarrow \mathcal{T}(\mathcal{F},\mathcal{X})$, then $\sigma(l_i) \rightarrow_{\mathcal{R}} \sigma(r_i)$
2. if $s \rightarrow_{\mathcal{R}} t$, then $f(\ldots,s,\ldots) \rightarrow_{\mathcal{R}} f(\ldots,t,\ldots)$.

We define the rewrite relation on quasiterms as the smallest relation satisfying:

```
Inductive Definition TRR [R:qrule_list]:quasiterm->quasiterm->Prop =
  TRR_instance:(l,r:quasiterm)(is_qr l r R)->(s:quasisubst)(termsubst s)
            ->(TRR R (Subst s l) (Subst s r))
| TRR_Root:(t,s:quasiterm)(TRR R t s)->(f:fun)(TRR R (Root f t) (Root f s))
| TRR_CA_l:(t,s,u:quasiterm)(TRR R t s)->(TRR R (ConsArg t u) (ConsArg s u))
| TRR_CA_r:(t,s,u:quasiterm)(TRR R t s)->(TRR R (ConsArg u t) (ConsArg u s)).
```

The aim of this paper is to prove a theorem on the derivation bound of rewriting systems. Therefore, we introduce the transitive closure of a rewrite relation with a counting parameter *TRRplus*, and the derivability relation *DRR*.

```
Inductive Definition TRRplus [R:qrule_list]:quasiterm->quasiterm->nat->Prop =
  TRRplus0:(t,s:quasiterm)(TRR R t s)->(TRRplus R t s 0)
| TRRplusS:(t,s,u:quasiterm)(TRR R t s)->(n:nat)(TRRplus R s u n)
        ->(TRRplus R t u (S n)).
```

```
Inductive Definition DRR [R:qrule_list]:quasiterm->quasiterm->Prop =
  DRR0:(t,s:quasiterm)(TRR R t s)->(DRR R t s)
| DRRS:(t,s,u:quasiterm)(DRR R t u)->(DRR R u s)->(DRR R t s).
```

TRR, *TRRplus* and *DRR* are defined on quasiterms. We may check that when quasirules are term_ rules any rewriting sequence starting from a term does yield another term, which means that these relations are compatible with the term structure:

2 Multiset Path Ordering

The multiset path ordering is induced by a precedence, that is an ordering on the symbols of the signature. We assume that this ordering is total. In the following definition = stands for the permutative congruence of terms induced by the precedence, and $\succ\!\!\succ_{mpo}^{mult}$ stands for the multiset extension of $\succ_{mpo}$.

Definition 4 multiset path ordering. The multiset path ordering on $\mathcal{T}(\mathcal{F},\mathcal{X})$ is inductively defined as follows:
$t = f(t_1,\ldots,t_n) \succ_{mpo} g(s_1,\ldots,s_m) = s$ if and only if

1. $f \prec g$ and $t_k \succeq_{mpo} s$, for a $k \in \{1,\ldots,n\}$, or
2. $f \succ g$ and $t \succ_{mpo} s_k$, for all $k \in \{1,\ldots,m\}$, or
3. $f = g$ and $\{t_1,\ldots,t_n\} \succ\!\!\succ_{mpo}^{mult} \{s_1,\ldots,s_m\}$.

The multiset path ordering is a simplification ordering [Der82], which is transitive and irreflexive, closed both under context application and under substitution and contains the subterm ordering. The subterm ordering also contains homeomorphic embedding. From the tree theorem of Kruskal, the homeomorphic embedding is a well-quasi-ordering on the set of terms over a finite vocabulary, and so is well-founded. Therefore the subterm condition suffices for well-foundedness of mpo. But this well-foundedness proof relies on Kruskal's tree theorem proof by Nash Williams which is very impredicative [Gal91].

We describe in the following parts the encoding of mpo induced by a finite total precedence and prove its main properties, excepted the well-foundedness. All along the paper *gt* and *le* stand respectively for the usual orderings > and ≤ on natural numbers.

2.1 Total Precedence

A precedence $\prec$ is a total ordering on a finite signature $\mathcal{F}$. One can enumerate $\mathcal{F}$ in precedence increasing order: $\mathcal{F} = \{f_1, \ldots, f_p\}$ and $f_i \prec f_j$ if $i < j$. One defines a function *prec* : *fun* $\rightarrow$ *nat* built on a precedences list, exactly like for functional arities. The function *prec* associates with each functional symbol a rank in the precedence. Totality of *prec* on $\mathcal{F}$ comes from the decidability of the natural ordering over integers.

```
Variable prec_list:nat_list.
Definition prec:fun->nat = (list_access prec_list).
```

2.2 Argument of a Quasiterm - Deletion - Permutative Congruence

A quasiterm u is an argument of a quasiterm t if it is an immediate subquasiterm of t. We call *arg* this relation.

```
Inductive Definition arg [u:quasiterm]:quasiterm->Prop =
  argf1:(f:fun)(arg u (Root f u))
| argf2:(f:fun)(t:quasiterm)~(Simple t)->(arg u t)->(arg u (Root f t))
| argC1:(t:quasiterm)(arg u (ConsArg u t))
| argC2:(t:quasiterm)(arg u (ConsArg t u))
| argC3:(t,s:quasiterm)~(Simple s)->(arg u s)->(arg u (ConsArg t s)).
```

We prove that *arg* is a decidable relation stable by substitution. We also define a function which deletes the first occurrence of a quasiterm in a quasiterm built with *ConsArg*.

delete: quasiterm$\rightarrow$quasiterm$\rightarrow$quasiterm $\equiv$
Simple(t) $\Rightarrow$ delete(k,t) = t
k=t_1 $\Rightarrow$ delete(k,ConsArg(t_1,t_2)) = t_2
k$\neq t_1$ $\wedge$ k=t_2 $\Rightarrow$ delete(k,ConsArg(t_1,t_2)) = t_1
k$\neq$ t_1 $\wedge$ k$\neq$ t_2 $\Rightarrow$ delete(k,ConsArg(t_1,t_2)) = ConsArg(t_1,delete(k,t_2))

Definition 5 permutative congruence. Terms of equal arities $t = f(t_1, \ldots, t_n)$ and $g(s_1, \ldots, s_n) = s$ are congruent, we note $t \sim s$, if they have identical precedence and there is a permutation π such that: $t_i \sim s_{\pi(i)}$ for all $i \in \{1, \ldots, n\}$.

We prove that this inductive relation called *perm* is decidable, reflexive, symmetric, transitive over l-terms and contains the equality relation on terms. Finally we show that it is compatible with substitution.

2.3 Encoding the Multiset Path Ordering

We have all the elements to define the multiset path ordering. Whereas the definition of the mpo on terms requires two levels (terms and multiset of terms) we are going to use a chararacteristic of quasiterms which basically makes no difference between quasiterms and list of quasiterms (operation *ConsArg*) to give a definition of the mpo on quasiterms on a unique level. The definition is divided in two parts. The first one describes the clauses for simple quasiterms, the second one is dedicated to the clauses where multisets occur.

```
Inductive Definition mpo:quasiterm->quasiterm->Prop =
```

1. mpo for simple quasiterms:

```
 mpofv1:(f:fun)(x:var)(mpo (Root f (V x)) (V x))
|mpofv2:(f:fun)(t:quasiterm)(x:var)(mpo t (V x))->(mpo (Root f t) (V x))

|mpocc:(c,e:fun)(gt (prec c) (prec e))->(mpo (C c) (C e))
|mpofc1:(f:fun)(c:fun)(t:quasiterm)(gt (prec c) (prec f))
      ->(perm prec t (C c))->(mpo (Root f t) (C c))
|mpofc2:(f:fun)(t:quasiterm)(c:fun)(gt (prec c) (prec f))->(mpo t (C c))
      ->(mpo (Root f t) (C c))
|mpofc3:(f:fun)(t:quasiterm)(c:fun)(le (prec c) (prec f))
      ->(mpo (Root f t) (C c))

|mpocf:(c:fun)(f:fun)(t:quasiterm)(gt (prec c) (prec f))->(mpo (C c) t)
     ->(mpo (C c) (Root f t))
|mpoff1:(f,g:fun)(t,s:quasiterm)(gt (prec g) (prec f))
      ->(perm prec t (Root g s))->(mpo (Root f t) (Root g s))
|mpoff2:(f,g:fun)(t,s:quasiterm)(gt (prec g) (prec f))->(mpo t (Root g s))
      ->(mpo (Root f t) (Root g s))
|mpoff3:(f,g:fun)(t,s:quasiterm)(<nat>(prec f)=(prec g))->(mpo t s)
      ->(mpo (Root f t) (Root g s))
|mpoff4:(f,g:fun)(t,s:quasiterm)(gt (prec f) (prec g))->(mpo (Root f t) s)
      ->(mpo (Root f t) (Root g s))
```

2. Multiset extension of mpo:

```
|mpoCv1:(t:quasiterm)(x:var)(mpo (ConsArg (V x) t) (V x))
|mpoCv2:(t:quasiterm)(x:var)(mpo (ConsArg t (V x)) (V x))
|mpoCv3:(t1,t2:quasiterm)(x:var)(mpo t1 (V x))->(mpo (ConsArg t1 t2) (V x))
|mpoCv4:(t1,t2:quasiterm)(x:var)(mpo t2 (V x))->(mpo (ConsArg t1 t2) (V x))

|mpoCc1:(t,s:quasiterm)(c:fun)(perm prec t (C c))->(mpo (ConsArg t s) (C c))
|mpoCc2:(t,s:quasiterm)(c:fun)(perm prec s (C c))->(mpo (ConsArg t s) (C c))
|mpoCc3:(t1,t2:quasiterm)(c:fun)(mpo t1 (C c))->(mpo (ConsArg t1 t2) (C c))
|mpoCc4:(t1,t2:quasiterm)(c:fun)(mpo t2 (C c))->(mpo (ConsArg t1 t2) (C c))

|mpoCf1:(t1,t2,u:quasiterm)(f:fun)(perm prec t1 (Root f u))
      ->(mpo (ConsArg t1 t2) (Root f u))
|mpoCf2:(t1,t2,u:quasiterm)(f:fun)(perm prec t2 (Root f u))
      ->(mpo (ConsArg t1 t2) (Root f u))
|mpoCf3:(t1,t2,u:quasiterm)(f:fun)(mpo t1 (Root f u))
      ->(mpo (ConsArg t1 t2) (Root f u))
|mpoCf4:(t1,t2,u:quasiterm)(f:fun)(mpo t2 (Root f u))
      ->(mpo (ConsArg t1 t2) (Root f u))

|mpocC:(c:fun)(t1,t2:quasiterm)(mpo (C c) t1)->(mpo (C c) t2)
     ->(mpo (C c) (ConsArg t1 t2))
|mpofC:(f:fun)(t,s1,s2:quasiterm)(mpo (Root f t) s1)->(mpo (Root f t) s2)
     ->(mpo (Root f t) (ConsArg s1 s2))
```

```
|mpoCC1:(t1,t2,s1,s2:quasiterm)(k,l:quasiterm)(Simple k)->(Simple l)
      ->(arg k (ConsArg t1 t2))->(arg l (ConsArg s1 s2))->(perm prec k l)
      ->(mpo (delete k (ConsArg t1 t2)) (delete l (ConsArg s1 s2)))
      ->(mpo (ConsArg t1 t2) (ConsArg s1 s2))

|mpoCC2:(t1,t2,s1,s2:quasiterm)
        ((k,l:quasiterm)(Simple k)->(Simple l)->(arg k (ConsArg t1 t2))->
         (arg l (ConsArg s1 s2))->~(perm prec k l)
        )->(mpo (ConsArg t1 t2) s1)->(mpo (ConsArg t1 t2) s2)
      ->(mpo (ConsArg t1 t2) (ConsArg s1 s2)).
```

We prove in the following that *mpo* is a simplification ordering on *l_terms*, that is an irreflexive and transitive relation, compatible with permutative equivalence, closed both under context application and under substitution and containing the subterm ordering.

```
Lemma mpo_irr:(t,s:quasiterm)(l_term t)->(l_term s)->(perm t s)->~(mpo t s).
```

We prove simultaneously subterm and monotonicity properties of *mpo* for l-terms.

```
Lemma mpo_monot:
(s,t:quasiterm)(l_term t)->(l_term s)->((mpo s t)\/(perm s t))
->(((k:quasiterm)(arg k t)->(mpo s k)) /\
   ((f:fun)(mpo (Root f s) t)) /\
   ((u:quasiterm)(l_term (ConsArg s u))->(mpo (ConsArg u s) t)) /\
   ((u:quasiterm)(l_term (ConsArg u s))->(mpo (ConsArg s u) t)) /\
   ((k:quasiterm)(Simple k)->(arg k t)->(mpo s (delete k t)))
  )
/\ (k:quasiterm)(Simple k)->(arg k s)
   ->((mpo (delete k s) t)\/(perm (delete k s) t))->(mpo s t).
```

The proof of the monotonicity lemma is not very hard, but is rather involved. Indeed it is about forty pages. With the monotonicity lemma and with an auxiliary lemma we are able to derive the transitivity proof of *mpo*, which is known not to be easy. The difficulty comes from the multiset case. To deal with this problem we need to introduce the notion of inclusion of a quasiterm in another one, and prove the following lemma:

$$t = \{t_1, \ldots, t_n\} \succ\!\!\succ_{mpo}^{mult} \{s_1, \ldots, s_m\} = s \;\Rightarrow\; \exists i \in 1 \ldots n,$$

sucht that one of the following statement is fulfilled:

1. $t - t_i \succeq_{mpo} s$ or,
2. $\exists K = \{k_1, \ldots, k_j\}.\ K \subset s \wedge (\forall l \in 1 \ldots j.\ t_i \succ_{mpo} k_l) \wedge (t - t_i \succeq_{mpo} s - K)$ or,
3. $t_i \succ_{mpo} s$

```
Lemma mpo_trans: (t,u,v:quasiterm)(l_term t)->(l_term u)->(l_term v)->
                 (mpo t u)->(mpo u v)->(mpo t v).
```

Notice that when restricted to closed terms, *mpo* can be slightly simplified [Les90]. The multiset path ordering on closed terms is total, when the precedence is total. Therefore, in order to compare two multisets, it is sufficient to compare their maximal element. With such a definition, the transitivity becomes very easy to prove. The transitivity lemma is then used to prove the stability of mpo by substitution:

```
Lemma mpo_subst: (s:quasisubst)(termsubst s)->(u,v:quasiterm)(l_term u)
             ->(l_term v)->(mpo u v)->(mpo (Subst s u) (Subst s v)).
```

mpo is a partial relation over quasiterms. For instance, variables are not comparable. But it becomes total when restricted to ground quasiterms. We use the monotonicity lemma for proving the following lemma:

```
Lemma mpo_gd_total: (t,s:quasiterm)(l_term t)->(l_term s)->(gd_quasiterm t)
               ->(gd_quasiterm s)->{(mpo t s)}+{(perm t s)}+{(mpo s t)}
```

Simplification orderings contain the subterm relation, and are well-founded orders by Kruskal's theorem. M. Rathjen and A. Weiermann [RA93] characterized the proof-theoretic strength of Kruskal's theorem and provided a constructive proof. But, their analysis requires powerful systems to denote very large ordinals. Moreover, the main idea of this paper is to yield termination proofs for rewriting systems without taking into account the well-foundedness property of *mpo*.

2.4 TRS Reducible under the Multiset Path Ordering

A finite term rewriting system is reducible under *mpo* if the rewrite relation is embedded into *mpo*. That is, for each rule $\langle l_i, r_i\rangle \in \mathcal{T}(\mathcal{F},\mathcal{X}) \times \mathcal{T}(\mathcal{F},\mathcal{X})$ $l_i \succ_{mpo} r_i$. This notion is expressed by the following decidable predicate:

```
Inductive Definition mpo_reducible:qrule_list->Prop =
  rednil:(mpo_reducilble nilqr)
| redconsqr:(R:qrule_list)(l,r:quasiterm)(mpo l r)->(mpo_reducilble R)
        ->(mpo_reducible (consqr l r R).
```

Moreover, proving that the set of rewrite rules can be oriented with *mpo* is sufficient to make the rewrite relation embeddable into mpo.

```
Lemma mpo_red_TRR: (R:qrule_lists)(termrule_list R)->(mpo_reducible R)
              ->(t,s:quasiterm)(TRR R t s)->(mpo t s).
```

3 Embedding the Rewrite Relation into a Standard Well-founded Ordering

A. Cichon, D. and A. Weiermann use monotonic interpretations in order to prove that any rewriting sequence of closed terms does terminate. The main problem is to choose a good interpretation τ. For instance, polynomials interpretations are not powerful enough to prove the termination of all system reducible under *mpo* [CL92]. Cichon's idea is to interpret each functional symbol as a function belonging to a variant of the Grzegorzcyk hierarchy of number theoretic functions. Instead of interpretations, we propose an embedding of the rewrite relations into some well-founded ordering, namely a mapping τ from terms (possibly free terms) into the natural numbers, such that $\tau(t) > \tau(s)$ whenever t rewrites to s.

Theorem 6. *Given a well-founded ordered set* $(\mathcal{W},\succ)$, *a mapping* $\tau{:}\mathcal{T}(\mathcal{F},\mathcal{X}) \to \mathcal{W}$, *and a term rewriting system* R *over* $\mathcal{T}(\mathcal{F},\mathcal{X})$, R *is terminating if*

$$t \to_{\mathcal{R}} s \;\Rightarrow\; \tau(t) \;>\; \tau(s),\; \forall t,s \in \mathcal{T}(\mathcal{F},\mathcal{X})$$

3.1 A Variant of the Grzegorzcyk-Hierarchy

Definition 7 Gzregorzcyk-Hierarchy. Let $d > 1$ be an integer, the finite levels of the hierarchy $\{F_k : Nat \to Nat\}$, $k \in Nat$ is defined by induction on k:

$F_0(x) = d^{x+1}$,
$F_{k+1}(x) = F_k^{d(1+x)}(x)$, where F^i stands for i-th iterate of F.

We can define the hierachy as a 3-ary function
F': nat→nat→nat→nat ≡
F'(d,0,x) = d^{x+1}
F'(d,k+1,x) = Iter(λx.F'(d,k,x),d(x+1)-1)
where Iter stands for the iterate of function. But, as a convenient notation, we define a 4-ary function
F: nat→nat→nat→nat→nat ≡
F(d,k,x,i) = Iter(λx.F'(d,k,x),i)

We can notice that F is not itself primitive recursive. But for any fixed k, F(d,k) is primitive recursive. In fact F enumerates all the primitive recursive functions.

Lemma 8. *F is monotonic for each argument.*

3.2 Quasi-Embedding

Given $\mathcal{F}= \{f_1, \ldots, f_n\}$ and $d > 1$.

Definition 9 Embedding. We define a mapping τ of terms of $\mathcal{T}(\mathcal{F}, \mathcal{X})$ into the natural numbers as folllows:

$$\tau(f_k(t_1, \ldots, t_n)) = F_{k+1}(d^{\tau(t_1)} + \cdots + d^{\tau(t_n)}) \textit{ and}$$
$$\tau(f_k) = F_{k+1}^d(d), \textit{ when } f_k \textit{ is a constant symbol},$$
$$\tau(x) = 2, \textit{ when } x \textit{ is a variable}$$

where k denotes the rank of f in the precedence.

The use of the Grzegorzcyk hierarchy is related to the ordinal analysis of termination orderings [DO88]. In [Cic90] Adam Cichon gives an ordinal notation system, an embedding of a rewriting relation into this initial segment of ordinals and an ordinal bound for the order type of *mpo*. He shows that the slow growing hierarchy of number theoretic functions indexed by the initial segment, provides a good measure for the size of such orderings. This means that if the index of function f bounds the index of another function g, then f bounds g.

We call qE: quasiterm→nat the quasi-embedding of quasiterms in nat. We shall prove in the remaining of the paper that qE is a good measure for quasiterms in the following sense: rewriting makes the embedding decrease whenever the rewrite rules are oriented with *mpo*.

Lemma 10. *Our embedding of quasiterms enjoys subterm and monotonicity properties and is compatible with permutative equivalence.*

3.3 Natural Upper Bounds for the Embedding of Quasiterms

We define the rank of a quasiterm as follows:
rank: quasiterm→nat ≡
rank(V(x)) = rank(C(c))) = 1
rank(Root(f,t))= rank(t)+1
rank(ConsArg(t_1,t_2)) = rank(t_1) + rank(t_2)

Let $precF = max\{k,\ f_k \in F\}$ and $d > 1$. We show that one can build a function, namely $\lambda t.F_p(d+rank(t))$ where $p = precF + 2$, that bounds the embedding of any term.

Lemma 11. $\forall t \in \mathcal{T}(\mathcal{F},\mathcal{X}),\ \tau(t) < F_p(d+rank(t))$*, where* $p = precF + 2$

3.4 Two Mains Lemmas

The first main lemma is the key of the whole complexity proof. We first state it informally:

Lemma 12 First main lemma. *Let* $t = f_i(t_1,\ldots,t_n) \in \mathcal{T}(\mathcal{F},\mathcal{X})$*, let* σ *be a substitution, and* τ *the embedding of terms into the natural numbers.*

Assume that forall t_k, $k \in 1\ldots n$*, forall* u *such that* $rank(u) \leq d$*:*

$$t_k \succ_{mpo} u \Rightarrow \tau(\sigma t_k) > \tau(\sigma u)$$

Put $\tau_{min} = min\{\tau(\sigma t_i), i \in 1\ldots n\}$.

Then, forall $s \in \mathcal{T}(\mathcal{F},\mathcal{X})$ *such that* $rank(s) < d$*:*

$$t \succ_{mpo} s \Rightarrow \tau(\sigma s) < F_{i+1}(M)$$

where $M = d^{\tau(\sigma t_1)} + \ldots + d^{\tau(\sigma t_n)} - d^{\tau_{min}} + d^{\tau_{min}-1}.rank(s)$.

Proof: by induction on s. The proof is rather technical. So, we do not sketch it here. The difficult point arises when s is a Rooted quasiterm with a head symbol equipped with a precedence equal to the precedence of f.

Notice that $F_{i+1}(d^{\tau(\sigma t_1)} + \ldots + \ldots d^{\tau(\sigma t_n)} - d^{\tau_{min}} + d^{\tau_{min}-1}.rank(s))$ is strictly smaller than $\tau(\sigma(f_i(t_1,\ldots,\ldots t_n)))$ if $rank(s)$ is strictly smaller than d.

Lemma 13 Second main lemma. *Let* t, $s \in \mathcal{T}(\mathcal{F},\mathcal{X})$*, such that* $rank(s) < d$*. Let* σ *be a substitution, and* τ *the embedding of terms, then:*

$$t \succ_{mpo} s \Rightarrow \tau(\sigma t) > \tau(\sigma s).$$

3.5 One Rewrite Step Makes the Embedding Decrease

We first state the theorem informally:

Theorem 14. *Assume that $\mathcal{R}$ is a finite set of rewrite rules over $\mathcal{T}(\mathcal{F},\mathcal{X})$ reducible under $\succ_{mpo}$. Let τ be the embedding of terms. Then, for all terms t and $s \in \mathcal{T}(\mathcal{F},\mathcal{X})$,*

$$t \rightarrow_{\mathcal{R}} s \;\Rightarrow\; \tau(t) \;>\; \tau(s).$$

We now formally state this theorem in Coq. Let *rkR* be the maximal rank of right-hand sides of the rewrite rules, that is:

```
Definition precF:[lp:nat_list](max_nat_list lp).
Definition rkR:qrule_list->nat = [R:qrule_list](max_qrule_list R).
```

Put K greater than the maximum of 1, *rkR* and *precF*:

```
Definition K:qrule_list->nat = (S (max_nat (S O) (max_nat (rkR R) precF))).

Theorem Rewrite_qE: (R:qrule_list)(mpo_reducible R)
   ->(t,s:quasiterm)(l_term t)->(TRR R t s)->(gt (qE (K R) t) (qE (K R) s)).
```

3.6 Bound on Derivation Length

Theorem 15 Cichon-Hofbauer-Weiermann. *Let $\mathcal{R}$ be a finite term rewriting system over $\mathcal{T}(\mathcal{F},\mathcal{X})$ such that $\rightarrow_{\mathcal{R}}$ is contained in $\succ_{mpo}$. Let precF and K, the constants of $\mathcal{R}$.*

$$\textit{If } t_1 \rightarrow_{\mathcal{R}} \ldots \rightarrow_{\mathcal{R}} t_n, \textit{ then}$$

$$n \le F_{precF+2}(K + rank(t_1)).$$

forall $t_1, \ldots, t_n \in \mathcal{T}(\mathcal{F},\mathcal{X})$.

Formally:

```
Theorem Bound_Theorem: (R:qrule_list)(termrule_lists R)->(mpo_reducible R)
                    ->(t,s:quasiterm)(l_term t)->(n:nat)(TRRplus R t s n)
              ->(le (S n) (F (K R) (S (S precF)) (plus (K R) (rank t)) O)).
```

This result provides a complexity characterisation for the finite TRS whose rules can be oriented with the multiset path ordering. It states that for such systems the height of the computation tree of any term is finitely bounded. Furthermore, the bound depends on the size of the starting term and some constants that can be statically determined from the precedence and from the rules of the system. By abstracting on terms the bound is given by a function $\lambda t.F_p(K + rank(t))$ belonging to a finite level of the Grzegorzcyk-Hierarchy which is primitive recursive. It means that any function computed by such a TRS is primitive recursive. It means also that any algorithm expressed with mpo can be reformulated with a primitive recursive scheme.

4 Extracted Program from the Bounding Theorem

It is well known that the proof as programs paradigm allows us to derive certified programs from constructive proofs. Actually, from an intuitionistic proof of $\forall x^A.P(x) \Rightarrow \exists y^B.Q(x,y)$, the system Coq does extract a functional program $f : A \rightarrow B$. And the theory guarantees that $\forall x^A.P(x) \Rightarrow Q(x,(f(x)))$ holds.

The bounding theorem specifies a partial program P which, with a term rewriting system R and with a given signature la and a given precedence lp as inputs, associates either an upper bound for its derivation length, or an exception if the rules of R can not be oriented with *mpo*. We define the program specification inductively as follows:

```
Inductive Definition Termination [R:qrule_list][la;lp:nat_list] : Set =
 Success:(termrule_list R)->(mpo_reducible la lp R)->
         (D:quasiterm->nat)
         ((t,s:quasiterm)(term la t)->(n:nat)(TRRplus R t s n)->(le n (D t)))
       ->(Termination R la lp)

| Rulefailure:~(termrule_list R)->(Termination la lp R)

| Precfailure:~(mpo_reducible la lp R)->(Termination la lp R).
```

The termination program is derived from the proof of this specification:

```
Theorem Prog: (R:qrule_list)(arity_list,prec_list:nat_list)
              (Termination arity_list prec_list R).
```

5 Automatizing Well-foundedness Proofs of Rewrite Relations

The main application of this work is to derive a generic well-foundedness proof for any term rewriting system whose rules can be oriented with the multiset path ordering.

First, we can extend **Theorem 14** to the derivability relation in such a way:

Theorem 16. *Assume that $\mathcal{R}$ is a finite set of rewrite rules over $\mathcal{T}(\mathcal{F},\mathcal{X})$ reducible under $\succ_{mpo}$. Let $\rightarrow_{\mathcal{R}}$ be the induced rewrite relation, and $\rightarrow_{\mathcal{R}^+}$ be the transitive and irreflexive closure of $\rightarrow_{\mathcal{R}}$. Let τ be our interpretation of terms. Then, for all terms t and $s \in \mathcal{T}(\mathcal{F})$,*

$$t \rightarrow_{\mathcal{R}^+} s \Rightarrow \tau(t) > \tau(s).$$

Corollary 17 rec_τ. *Let τ be the embedding of terms into the natural numbers, and P any predicate over terms:*

$$(\forall t.(\forall s.\tau(t) > \tau(s) \Rightarrow P(s)) \Rightarrow P(t)) \Rightarrow \forall t.P(t)$$

Then, with **Theorem 14** and the previous induction scheme, one builds a constructive well-foundedness proof of $\rightarrow_{\mathcal{R}^+}$

Theorem 18 *$rec_{\rightarrow_{\mathcal{R}^+}}$. Assume that $\rightarrow_{\mathcal{R}}$ is contained in $\succ_{mpo}$, let P be any predicate over terms:*

$$(\forall t.(\forall s.t \rightarrow_{\mathcal{R}^+} s \Rightarrow P(s)) \Rightarrow P(t)) \Rightarrow \forall t.P(t)$$

The latter induction principle states the well-foundedness of $\rightarrow_{\mathcal{R}^+}$. It is a termination proof for the function which computes the normal form of a term by succesive rewriting steps. The structure of such a program, corresponding to the direct proof of the induction principle, will be realised without any fix-point. In a Ml-like language it looks like:

```
let W F x = indrec (τ(x)+1) x
    where rec indrec = function
     O → (function x → error)
    | (S p) → (function x → (F x (function y → indrec y p)));;
```

Given any finite vocabulary and any set of rewrite rules, the proof only depends on the choice of a good precedence. We have to bear in mind that if it exists, the precedence needs to be total on the vocabulary and that the induced multiset path ordering has to achieve in orienting the rewrite rules. However, rewriting tools like the rewriting laboratory REVE are designed for such tasks and might be coupled to this search.

Conclusion

In this paper we investigated the termination proof of term rewriting systems with the multiset path ordering in the Calculus of Constructions. We proved that any term rewrite system whose rules can be oriented with the multiset path ordering induced by a given precedence terminates. We proved also a bounding theorem for such systems. Namely, there is a primitive recursive bound for the derivation lengths. We applied these previous results to derive a generic constructive proof of the well-foundness of rewrite relations over free terms. The underlying theory has been formalised in the Calculus of Constructions and all proofs have been certified by the machine. The price to pay for such a garantee is not at all negligible (the Coq script is about 250 pages). In particular the definition of the multiset path ordering over quasiterms and the proof that it is a simplification ordering were very difficult to design and achieve. Furthermore, as previously mentioned, our proof is widely inspired by the works of [Cic90] and [Wei93] where some nontrivial errors occur. A future extension of this work might be to apply these results to the Calculus of Constructions itself. For instance, a new approach described in [Par93] consists in synthesizing an automatic proof from a program, of which it could be extracted. Such a method needs to generate termination proofs of programs. If one allows general recursion schemes for programs, one has to find an adequate ordering and prove its well-foundedness. Such proofs are obviously left to the user. We think that the use of simplification orderings like *mpo* and our proof method could bring some helpful automation for such a task.

Acknowledgements : I would like to thank Adam Cichon for discussions on topics related to this paper. Thanks also to Christine Paulin and to the referees for their helpful comments.

References

[Cic90] E. A. Cichon. Bounds on derivation lenghts from termination proofs. Technical Report CSD-TR-622, Royal Holloway and Bedford New College, 1990.

[CL92] E. A. Cichon and P. Lescanne. Polynomial interpretations and the complexity of algorithms. In D. Kapur, editor, *Proceedings 11th International Conference on Automated Deduction, Saratoga Springs (N.Y., USA)*, volume 607 of *Lecture Notes in Computer Science*, pages 139–147. Springer-Verlag, June 1992.

[Der82] N. Dershowitz. Orderings for term-rewriting systems. *Theoretical Computer Science*, 17:279–301, 1982.

[DFH+91] G. Dowek, A. Felty, H. Herbelin, G. Huet, C. Paulin-Mohring, and B. Werner. The Coq Proof Assistant. User's guide, INRIA-CNRS-ENS, 1991.

[DO88] N. Dershowitz and M. Okada. Proof-theoretic techniques and the theory of rewriting. In *Proceedings 3rd IEEE Symposium on Logic in Computer Science, Edinburgh (UK)*, pages 104–11. IEEE, 1988.

[Gal91] J. Gallier. What's so special about Kruskal's theorem and the ordinal Γ_0? A survey of some results in proof theory. *Annals of Pure and Applied Logic*, 53(3):199–261, September 1991.

[Hof90] D. Hofbauer. Termination proofs by multiset path orderings imply primitive recursive derivation lenghts. In Hélène Kirchner and W. Wechler, editors, *Proceedings 2nd International Conference on Algebraic and Logic Programming, Nancy (France)*, volume 463 of *Lecture Notes in Computer Science*, pages 347–358, 1990.

[Lau88] C. Lautemann. A note on polynomial interpretation. *Bulletin of European Association for Theoretical Computer Science*, 1(36):129–131, October 1988.

[Les90] P. Lescanne. On the recursive decomposition ordering with lexicographical status and other related orderings. *Journal of Automated Reasoning*, 6:39–49, 1990.

[Par93] Catherine Parent. Developing certified programs in the system coq-the program tactic. RR 93-29, ENS, October 1993.

[PM93] C. Paulin-Mohring. Inductive Definitions in the System Çoq - Rules and Properties. In *Proceedings of the conference Typed Lambda Calculus and Applications*, Lecture Notes in Computer Science. Springer-Verlag, 1993.

[RA93] Rathjen.M and Weiermann A. Proof-theoretic investigations on kruskal's theorem. *Annals of pure and applied logic*, 60:49–88, 1993.

[RL92] Joseph Rouyer and Pierre Lescanne. Verification and programming of first-order unification in the calculus of constructions with inductive types, November 1992.

[Rou92] Joseph Rouyer. Développement de l'algorithme d'unification dans le calcul des constructions avec types inductifs. RR 1795, INRIA, November 1992.

[Wei93] Andreas Weiermann. Bounds for derivation lengths from termination proofs with rpo and rlpo. Private communication, 1993.

Typed λ-calculi with explicit substitutions may not terminate

Paul-André Mellies *

Ecole Normale Supérieure, 45 rue d'Ulm, 75005 Paris, France
INRIA Rocquencourt, Domaine de Voluceau, 78153 Le Chesnay Cedex, France
FWI, De Boelelaan 1081a, 1081 HV Amsterdam, Nederland
mellies@cs.vu.nl

Abstract. We present a simply typed λ-term whose computation in the λσ-calculus does not always terminate.

1 The λσ-calculus, introduction

Any effective implementation of the λ-calculus requires some control on the substitution to benefit from graph sharing [1] and avoid immediate size explosion. The original λ-calculus cannot describe these controls an easy way. The λσ-calculus was introduced in [2] as a *bridge between the classical λ-calculus and its concrete implementations*. Substitutions become explicit, they can be delayed and stored. The calculus provides a pleasant setting to study substitutions and check implementations.

The syntax of the λσ-calculus contains two classes of objects: terms and substitutions. Terms are written in the De Brujn notation [3].

$$\begin{array}{ll} \textbf{Terms} & a ::= \mathbf{1} | ab | \lambda a | a[s] \\ \textbf{Substitutions} & s ::= id | \uparrow | a \cdot s | s \circ t \end{array}$$

The rule *Beta* is equivalent to the usual β-rule of the λ-calculus. The other rules, called σ-rules, expose how substitutions are pushed inside the terms and performed.

$$\begin{array}{ll} Beta & (\lambda a)b \to a[b \cdot id] \\ \\ App & (ab)[s] \to a[s]b[s] \\ Abs & (\lambda a)[s] \to \lambda(a[\mathbf{1}.(s \circ \uparrow)]) \\ Clos & a[s][t] \to a[s \circ t] \\ Map & (a \cdot s) \circ t \to a[t] \cdot (s \circ t) \\ Ass & (s_1 \circ s_2) \circ s_3 \to s_1 \circ (s_2 \circ s_3) \\ \\ VarId & \mathbf{1}[id] \to \mathbf{1} \\ VarCons & \mathbf{1}[a.s] \to a \\ IdL & id \circ s \to s \\ ShiftId & \uparrow \circ\, id \to \uparrow \\ ShiftCons & \uparrow \circ\, (a \cdot s) \to s \end{array}$$

* This work was partly supported by the Esprit BRA CONFER.

When carried out inside the λ-calculus, any reduction of a typed λ-term M reaches its normal form. Some $\lambda\sigma$-reductions can mimic the λ-reductions and terminate too. Others can be more subtle and compute M in a non-standard way. However, does any $\lambda\sigma$-computation of a typed term normalise it? The question was much debated and investigated with hopes for a positive answer. The major clue was the strong normalisation of the σ-rules which was proved effective in [4] and then [5][6] on any $\lambda\sigma$-term. It makes a non terminating $\lambda\sigma$-computation continually create and reduce new *Beta*-redexes, which seems to contradict the typed structure of the term.

However, we present here a closed and simply typed λ-term whose computation in the $\lambda\sigma$-calculus does not always terminate. The $\lambda\sigma$-reductions are thus not strictly bound to the λ-reductions, which is a surprise.

2 Basic intuitions

Let M be the simply typed λ-term $\lambda v.(\lambda x.(\lambda y.y)((\lambda z.z)x))((\lambda w.w)v)$. Like any typed term its $\lambda\sigma$-computation may normalise it. Next section, we show that it may also not terminate.

Building such a non terminating strategy on M requires precision. The σ-rules enjoy strong normalisation on any $\lambda\sigma$-term. The *Beta*-rule mimics the β-rule whose computation on any well typed λ-term strongly terminates. This shows that non termination must come from thin interactions between the *Beta* and σ-rules. Let $(\lambda a)b$ be a λ-term and s a substitution on top of it. We study next two natural strategies to reduce the root *Beta*-redex and begin the propagation of s.

One standard strategy begins to reduce the *Beta*-redex

$$((\lambda a)b)[s] \rightarrow (a[b \cdot id])[s] \ \ \mathit{Beta}$$

and then propagate the two substitutions s and $(b \cdot id)$ inside a using σ-rules. If carried on, the σ-computation terminates on a λ-term c.

Another natural strategy begins with the two σ-rules *App* and *Lambda* in order to propagate s through the *Beta*-redex. We call s and s' the two copies of s by *App*.

$$\begin{array}{ll} ((\lambda a)b)\ [s] & \\ \rightarrow ((\lambda a)[s])\ b[s'] & \mathit{App} \\ \rightarrow (\lambda(a[\mathbf{1} \cdot s \circ \uparrow]))\ b[s'] & \mathit{Lambda} \end{array}$$

It then computes the root *Beta*-redex:

$$\rightarrow a[\mathbf{1} \cdot s \circ \uparrow][b[s'] \cdot id] \ \ \mathit{Beta}$$

The two substitutions $(\mathbf{1} \cdot (s \circ \uparrow))$ and $(b[s'] \cdot id)$ are then propagated inside a using σ-rules. If carried on the process terminates again on the same λ-term c.

The property of strong normalisation seems natural in both computations. However, remark that the second strategy duplicates the substitution s with the rule *App*. The duplications by *App* are safe to strong termination when carried out within the scope of the σ-rules. Intuitively, the duplicated substitutions then are kept disjoint during σ-reductions and cannot interact. We show next how introducing *Beta*-redexes may combine two disjoint substitutions and provide a potential non terminating strategy to the calculus.

The combining strategy begins with the two σ-rules *App* and *Lambda* which propagate s through the *Beta*-redex:

$$\begin{array}{ll} ((\lambda a)b)[s] & \\ \to ((\lambda a)[s])b[s'] & App \\ \to (\lambda a[\mathbf{1} \cdot s \circ \uparrow])b[s'] & Lambda \end{array}$$

We call $s_1 = s$. The situation is clear. The two substitutions $\mathbf{1} \cdot (s_1 \circ \uparrow)$ and s' stand over the two disjoint terms: a and b. The *Beta*-redex mixes them:

$$\to a[\mathbf{1} \cdot s_1 \circ \uparrow][b[s'] \cdot id] \ Beta$$

The substitution $\mathbf{1} \cdot (s_1 \circ \uparrow)$ still acts on a whereas $(b[s'] \cdot id)$ and hence s' may be propagated through a and also $s_1 \circ \uparrow$. The propagation begins with some σ-rules:

$$\begin{array}{lll} \to a[(\mathbf{1} \cdot s_1 \circ \uparrow) \circ (b[s'] \cdot id)] & Clos & \\ \to a[\mathbf{1}[b[s'] \cdot id] \cdot (s_1 \circ \uparrow) \circ (b[s'] \cdot id)] & Map & \\ \to a[b[s'] \cdot (s_1 \circ \uparrow) \circ (b[s'] \cdot id)] & VarCons & \\ \to a[b[s'] \cdot s_1 \circ \underbrace{(\uparrow \circ (b[s'] \cdot id))}_{s_2}] & Ass & (\star) \end{array}$$

The rule *Map* duplicates $(b[s'] \cdot id)$ and divides its propagation in two distinct works. The first one is essential. It is devoted to substitute $b[s']$ in a via the substitution of $\mathbf{1}$. The second one is superfluous. It intends to substitute $b[s']$ inside $s_1 \circ \uparrow$ although no variable in s is bound to b: s_2 is therefore vacuous. Applying *ShiftCons* at that point would clarify the situation to $a[b[s'] \cdot (s_1 \circ id)]$ which roughly corresponds to a term obtained from $((\lambda a)b)[s]$ with the first strategy:

$$\begin{array}{ll} ((\lambda a)b)[s] & \\ \to (a[b \cdot id])[s] & Beta \\ \to a[(b \cdot id) \circ s] & Clos \\ \to a[b[s'] \cdot (id \circ s_1)] & Map \\ \to a[b[s'] \cdot s_1] & IdL \end{array}$$

Suppose that s_1 is $((\lambda a)b) \cdot id$. The substitution s_1 in $(\star)$ may then capture the useless s_2 with σ-rules, and duplicate it:

$$
\begin{array}{ll}
s_1 \circ s_2 = ((\lambda a)b) \cdot id) \circ s_2 & \\
\to ((\lambda a)b)[s_2] \cdot (id \circ s_2) & Map \\
\to^2 ((\lambda a)[s_2])(b[s_2]) \cdot s_2 & App + IdL \\
\to (\lambda(a[1 \cdot s_2 \circ \uparrow]))(b[s_2]) \cdot s_2 & Lambda \\
\to a[1 \cdot s_2 \circ \uparrow][b[s_2] \cdot id] \cdot s_2 & Beta \\
\to a[(1 \cdot s_2 \circ \uparrow) \circ (b[s_2] \cdot id)] \cdot s_2 & Clos \\
\to a[1[b[s_2] \cdot id] \cdot (s_2 \circ \uparrow) \circ (b[s_2] \cdot id)] \cdot s_2 & Map \\
\to a[b[s_2] \cdot (s_2 \circ \uparrow) \circ (b[s_2] \cdot id)] \cdot s_2 & VarCons \\
\to a[b[s_2] \cdot s_2 \circ \underbrace{(\uparrow \circ (b[s_2] \cdot id))}_{s_3}] \cdot s_2 & Ass
\end{array}
$$

Let t be any substitution. Call $\mathbf{rec}(t) = \uparrow \circ (b[t] \cdot id)$.

The substitution we obtain from $s_1 \circ s_2$ contains the substitution $s_2 \circ s_3 = s_2 \circ \mathbf{rec}(s_2)$ as a subterm. More generally, $s_1 = (\lambda a)b \cdot id$ behaves like a duplicator: any substitution $s_1 \circ t$ may be computed to a substitution containing $t \circ \mathbf{rec}(t)$. If the substitution $s_2 = \mathbf{rec}(s_1)$ behaves like a duplicator too then $s_2 \circ s_3$ may be reduced to a substitution containing $s_3 \circ \mathbf{rec}(s_3)$.

This sounds like the beginning of an infinite iteration. Let us call $(s_n)_{n>0}$ the sequence defined by s_1 and $s_{n+1} = \mathbf{rec}(s_n)$ and suppose that $(s_k \circ t)$ may be reduced for any k to a substitution which contains $t \circ \mathbf{rec}(t)$. The substitution $s_k \circ s_{k+1}$ may be computed to a substitution containing $s_{k+1} \circ \mathbf{rec}(s_{k+1}) = s_{k+1} \circ s_{k+2}$. The process may therefore be iterated for ever and provide a non terminating computation of $((\lambda a)b)[s]$.

3 The counter-example

3.1 The proof

Let us introduce the sequence $(s_i)_{i>0}$ of substitutions:

Definition

- $s_1 = (\lambda 1)1 \cdot id$
- $\mathbf{rec}(t) = \uparrow \circ (1[t] \cdot id)$
- $s_{n+1} = \mathbf{rec}(s_n)$
- $\mathbf{C}_x(y) = \uparrow \circ (1[y] \cdot x)$
- $\mathbf{D}_x(y) = 1[1[x] \cdot y] \cdot x$

The further lemma describes how s_1 duplicates a substitution t and nests its two copies.

Lemma 1 Duplication Step. $s_1 \circ t \to^+ \mathbf{D}_t(t \circ \mathbf{rec}(t))$

Proof:

$((\lambda \mathbf{1})\mathbf{1} \cdot id) \circ t$
$\rightarrow ((\lambda \mathbf{1})\mathbf{1})[t] \cdot id \circ t$ — *Map*
$\rightarrow^2 (\lambda \mathbf{1})[t]\mathbf{1}[t] \cdot t$ — *App + IdL*
$\rightarrow (\lambda \mathbf{1}[\mathbf{1} \cdot t \circ \uparrow])\mathbf{1}[t] \cdot t$ — *Abs*
$\rightarrow (\mathbf{1}[\mathbf{1} \cdot t \circ \uparrow][\mathbf{1}[t] \cdot id] \cdot t$ — *Beta*
$\rightarrow \mathbf{1}[(\mathbf{1} \cdot t \circ \uparrow) \circ (\mathbf{1}[t] \cdot id)] \cdot t$ — *Clos*
$\rightarrow \mathbf{1}[\mathbf{1}[\mathbf{1}[t] \cdot id] \cdot (t \circ \uparrow) \circ (\mathbf{1}[t] \cdot id)] \cdot t$ — *Map*
$\rightarrow^2 \mathbf{1}[\mathbf{1}[t] \cdot t \circ (\uparrow \circ (\mathbf{1}[t] \cdot id))] \cdot t$ — *VarCons + Ass*
$= \mathbf{1}[\mathbf{1}[t] \cdot t \circ \mathbf{rec}(t)] \cdot t$

□

The further lemma explains how s_n captures any substitution t step by step.

Lemma 2 Capture Step. $\mathbf{rec}(s) \circ t \rightarrow^+ \mathbf{C}_t(s \circ t)$

Proof:

$(\uparrow \circ (\mathbf{1}[s] \cdot id)) \circ t$
$\rightarrow \uparrow \circ ((\mathbf{1}[s] \cdot id) \circ t)$ — *Ass*
$\rightarrow \uparrow \circ (\mathbf{1}[s][t] \cdot (id \circ t))$ — *Map*
$\rightarrow^2 \uparrow \circ (\mathbf{1}[s \circ t] \cdot t)$ — *Clos + IdL*

□

We use our two lemmas on $s_n \circ s_{n+1}$:

$$s_n \circ s_{n+1} \;=\; \mathbf{rec}(\mathbf{rec}(...\mathbf{rec}(s_1))) \circ s_{n+1} \qquad \text{written with } (n-1)\ \mathbf{rec}.$$

It may be reduced with a capture step:

$$\rightarrow^+ \mathbf{C}_{s_{n-1}}(\mathbf{rec}(\mathbf{rec}(...\mathbf{rec}(s_1))) \circ s_{n+1}) \qquad \text{with } (n-2)\ \mathbf{rec}.$$

...with $(n-2)$ capture steps more:

$$\rightarrow^+ \mathbf{C}_{s_{n-1}}(\mathbf{C}_{s_{n+1}}(..:\mathbf{C}_{s_{n-1}}(s_1 \circ s_{n-1}))) \qquad \text{with } (n-1)\ \mathbf{C}_{s_{n+1}}(.).$$

...and the duplication step:

$$\rightarrow^+ \mathbf{C}_{s_{n+1}}(\mathbf{C}_{s_{n-1}}(...\mathbf{C}_{s_{n-1}}(\mathbf{D}_{s_{n-1}}(s_{n+1} \circ \mathbf{rec}(s_{n+1})))))$$

$$= \mathbf{C}_{s_{n-1}}(\mathbf{C}_{s_{n+1}}(...\mathbf{C}_{s_{n-1}}(\mathbf{D}_{s_{n+1}}(s_{n+1} \circ s_{n+2}))))$$

We obtain a substitution with $(s_{n+1} \circ s_{n+2})$ inside. It proves that the $\lambda\sigma$-computation of $(s_n \circ s_{n+1})$ may keep on incrementing k on $(s_{n+k} \circ s_{n+k+1})$ and never terminate.

We give below an explicit report of the process. Let us write C^n any function C applied n times:

Proposition

a. $s_{k+1} \circ s_{n+1} \to^+ \mathbf{C}_{s_{n+1}}(s_k \circ s_{n+1})$
b. $s_n \circ s_{n+1} \to^* \mathbf{C}^{n-1}_{s_{n+1}}(s_1 \circ s_{n+1})$
c. $s_1 \circ s_{n+1} \to^+ \mathbf{D}_{s_{n+1}}(s_{n+1} \circ s_{n+2})$
d. $s_n \circ s_{n+1} \to^+ \mathbf{C}^{n-1}_{s_{n+1}}(\mathbf{D}_{s_{n+1}}(s_{n+1} \circ s_{n+2}))$
e. $s_1 \circ s_1 \to^+ \mathbf{D}_{s_1}(s_1 \circ s_2)$

Corollary *The $\lambda\sigma$-computation of $(s_1 \circ s_1)$ may not terminate.*

3.2 The term

Let M be the closed and simply typed λ-term:

$$\lambda v.(\lambda x.(\lambda y.y)((\lambda z.z)x))((\lambda w.w)v)$$

It is translated in the De Brujn notation as:

$$\lambda(\ (\lambda(\lambda 1)((\lambda 1)1))\ ((\lambda 1)1)\ )$$

We show next that the $\lambda\sigma$-computation of M may not terminate. Yet, many $\lambda\sigma$-reductions compute M to its normal form. For instance:

$$\begin{array}{ll}
\lambda(\ (\lambda(\lambda 1)((\lambda 1)1))\ ((\lambda 1)1)\ ) & \\
\to^2 \lambda(\ (\lambda(\lambda 1)(1[1\cdot id]))\ (1[1\cdot id])\ ) & Beta + Beta \\
\to^2 \lambda(\ (\lambda((\lambda 1)1))\ 1\ ) & VarCons + VarCons \\
\to \lambda(\ (\lambda(1[1\cdot id]))\ 1\ ) & Beta \\
\to \lambda(\ (\lambda 1)\ 1\ ) & Varcons \\
\to \lambda(\ 1[1\cdot id]\ ) & Beta \\
\to \lambda 1 & Varcons
\end{array}$$

Proposition $\lambda((\lambda(\lambda 1)((\lambda 1)1))((\lambda 1)1)) \to^* \lambda(1[s_1 \circ s_1])$.

Proof:

$$\begin{array}{ll}
\lambda((\lambda(\lambda 1)((\lambda 1)1))((\lambda 1)1)) & \\
\to \lambda((\lambda(1[(\lambda 1)1\cdot id]))((\lambda 1)1)) & Beta \\
\to \lambda(1[s_1][(\lambda 1)1\cdot id] & Beta \\
\to \lambda(1[s_1 \circ s_1]) & Clos
\end{array}$$

□

Theorem *The $\lambda\sigma$-computation of M may not terminate.*

One should remark that the two rules $VarId$ and IdL are used for clarity's sake. Six rules only are required for the example: *Beta*, *App*, *Abs*, *Clos*, *Map* and *Ass*.

One can also check that similarly a non terminating $\lambda\sigma$-computation may occur on $\lambda v.(\lambda x.(\lambda y.A)((\lambda z.B)C))((\lambda w.D)E)$ with λ-terms A,B,C,D,E.

4 Conclusion

We give an example of a simply typed term whose computation in the $\lambda\sigma$-calculus does not always terminate. To our knowledge, the example cannot be avoided in any system with explicit substitution and composition.

The $\lambda\sigma$-calculus was designed to describe the actual implementations, not to strongly normalise any typed term. The discovery that some gap exists between the two things is an important result of the theory. It shows that a natural implementation may have unexpected behaviours, which justifies the interest for explicit substitutions.

New techniques should be investigated to avoid the cycling interactions between the *Beta*-rule and the σ-rules. Calculi without composition strongly normalise on typed terms, see [7], but more power on substitutions is often required, at least for confluence, see [8]. We believe that designing a calculus with composition of substitutions, confluence on open terms and strong termination on typed terms is the right theoretical and technical goal.

References

1. C.P. Wadsworth. *Semantics and Pragmatics of the Lambda Calculus*. PhD thesis, Oxford Universtity, 1971.
2. M. Abadi L. Cardelli P.-L. Curien J-J. Lévy. Explicit substitutions. *Journal of Functionnal Programming*, 1(4):375–416, 1991.
3. N. De Bruijn. Lambda-calculus notation with nameless dummies, a tool for automatic formula manipulation. *Indag. Mat.*, 34:381–392, 1972.
4. T. Hardin A. Laville. Proof of termination of the rewriting system subst on ccl. *Theoretical Computer Science*, 46:305–312, 1986.
5. P.-L. Curien T. Hardin A. Ríos. Strong normalization of substitutions. *Lecture Notes in Computer Science*, 629:209–217, 1992.
6. H. Zantema. Termination of term rewriting by interpretation. *Lecture Notes in Computer Science*, 656, 1993.
7. P. Lescanne J. Rouyer-Degli. The calculus of explicit substitutions $\lambda\upsilon$. *Submitted to the Journal of Functionnal Programming*, 1993.
8. T. Hardin, J.-J. Lévy, A Confluent Calculus of Substitutions, *France-Japan Artificial Intelligence and Computer Science Symposium, Izu*, 1989.

On Equivalence Classes of Interpolation Equations

Vincent Padovani

Université PARIS VII-C.N.R.S
U.R.A. 753
Equipe de Logique Mathématique
2 Place Jussieu - Case 7012
75251 PARIS CEDEX 05 - (FRANCE)
padovani@logique.jussieu.fr

Abstract. An Interpolation Equation is an equation of the form $[(x)c_1 \ldots c_n = b]$[1], where $c_1 \ldots c_n$, b are simply typed terms containing no instantiable variable. A natural equivalence relation between two interpolation equations is the equality of their sets of solutions. We prove in this paper that given a typed variable x and a simply typed term b, the quotient by this relation of the set of all interpolation equations of the form $[(x)w_1 \ldots w_p = b]$ contains only a finite number of classes, and relate this result to the general study of Higher Order Matching.

1 Introduction

Interpolation Equations are particular instances of the *Higher Order Matching* problem, which is the problem of determining, given two simply typed terms a and b, whether there exists a substitution σ such that $\sigma(a)$ and b normalize to the same term, or equivalently, the problem of solving the equation $a =_\beta b$ (written $[a = b]$) where b contains no instantiable variables.

The decidability of Higher Order Matching is still open. The Third Order Matching problem, or particular case of instantiable variables being of order at most three, has been proven decidable by Gilles Dowek in [4].

An interpolation equation is a matching problem of the form $[(x)c_1 \ldots c_n = b]$, where $c_1 \ldots c_n$, b are normal terms containing no instantiable variable, and b is of atomic type. The set of solutions of this equation is defined as the set of all terms t such that $(t)c_1 \ldots c_n$ is well typed, and normalizes to b. A natural equivalence relation between two interpolation equations is the equality of their sets of solutions (in general, infinite). Write $\sim$ the relation thus defined. We prove in this paper the two following results:

1. Given a typed variable x and a term b, the quotient by the relation $\sim$ of the set of all interpolation equations of the form $[(x)w_1 \ldots w_p = b]$ contains only a *finite* number of classes.

[1] we write $(u)v$ the application of u to v, and $(x)v_1 \ldots v_n$ for $(\ldots((x)v_1)v_2 \ldots v_{n-1})v_n$

2. The decidability of the following problem implies the decidability of Higher Order Matching:

   "Given two finite sets of interpolation equations Φ and Ψ, determine whether there exists a term t such that for each $E \in \Phi$, t is a solution of E, for each $F \in \Psi$, t is not a solution of F."

We have proven in [5] and [6]: the decidability of this latter problem in two particular cases:

1. the case where all equations are at most fourth order. As a consequence, we get the decidability of Fourth Order Matching.
2. the case where all right members of the equations considered are first-order constants. As a consequence, we get the decidability of Atomic Matching (the problem of solving a finite set of equations whose right members are all first-order constants).

2 Terms

We assume that the reader is familiar with the notions of λ-term, β and η-reduction and type systems. These notions will not be redefined, and the reader is invited to refer to [3] or [1] for an introduction to these notions.

We first inductively define a set of types (starting from a finite set of type variables, the set of atomic types, and using the symbol $\rightarrow$ as a binary connective). Considering three kinds of typed terms variables - constant, local and instantiable - we build the set of Simply Typed Terms, following a given set of rules. The rules for λ-abstraction (third and fourth rules) are used in a special way, according to the following requirement: local variables are the only kind of variables that may be bound in the terms produced by these rules. In other words, a constant symbol or an instantiable variable, appearing in a considered term, is always free in this term.

2.1 Types

We first consider a language consisting of: a *finite* set of constants $\mathcal{O}$, and a binary connective $\rightarrow$. The set $\mathcal{T}$ of all formulas of this language is inductively defined as follows:

0) $\mathcal{O} \subset \mathcal{T}$.
1) $A, B \in \mathcal{T} \Rightarrow (A \rightarrow B) \in \mathcal{T}$.

We write $A_1 \ldots A_k \rightarrow A$ for $(A_1 \rightarrow (\ldots A_k \rightarrow A)\ldots)$. We call *order* of a formula the integer computed as follows:

0) $\text{Ord}\ (\circ) = 1 \quad \text{for } \circ \in \mathcal{O}$,
1) $\text{Ord}\ (A_1 \ldots A_k \rightarrow \circ) = \sup\ (\text{Ord}\ (A_1), \ldots, \text{Ord}\ (A_k)) + 1 \quad \text{for } \circ \in \mathcal{O}$.

2.2 Typed Variables

Given an infinite, countable set of variables $\mathcal{X} = \{x, y \ldots\}$, we consider an application from $\mathcal{X}$ to the set of formulas such that each element of $\mathcal{T}$ has an infinite number of antecedents. For each formula A in $\mathcal{T}$, we call set of *variables of type* A the (infinite) subset of $\mathcal{X}$ of all antecedents of A. We call *typed variables* all pairs of the form (x, B), written $x : B$, where B is the type of x.

From now on, we will deal with three particular sets of typed variables $\mathcal{C}$, $\mathcal{L}$ and $\mathcal{I}$, called respectively set of *constants*, set of *local variables*, and set of *instantiable variables*, with the following properties:

$\mathcal{C}$, $\mathcal{L}$ and $\mathcal{I}$ are mutually disjoint;
$\mathcal{C}$ is finite;
$\mathcal{L}$ and $\mathcal{I}$ both contain an infinite number of variables of each type.

2.3 Simply Typed Terms

A *context* Γ is defined as a finite subset of the union of $\mathcal{C}$, $\mathcal{L}$ and $\mathcal{I}$. Assuming that a typed variable $x : A$ is not already in Γ, we write $\Gamma, x : A$ for $\Gamma \cup \{x : A\}$.

Given a formula A, a context Γ, and a term t of pure λ-calculus (written with elements of $\mathcal{X}$ as variables), we define the notion "t is a simply typed (typable) term of type A, in the context Γ", written $\Gamma \vdash t : A$, by means of the following rules:

1) $x : A \vdash x : A$ for $x : A \in \mathcal{C}, \mathcal{L}$ or $\mathcal{I}$,
2) if $\Gamma \vdash u : A \to B$ and $\Gamma' \vdash v : A$ then $(\Gamma \cup \Gamma') \vdash (u)v : B$,
3) if $\Gamma, x : A \vdash u : B$ and $x : A \in \mathcal{L}$, then $\Gamma \vdash \lambda x.u : A \to B$,
4) if $\Gamma \vdash u : B$, $x : A \in \mathcal{L}$ and $x : A \notin \Gamma$, then $\Gamma \vdash \lambda x.u : A \to B$.

Since, for any variable x, there exists a unique type A such that $(x : A)$, an immediate induction on the length of proofs leads to the following result:

Proposition 1. *If a term t of pure λ-calculus is simply typable (in the sense defined above) then there exists a unique context Γ and a unique type A such that $\Gamma \vdash t : A$. If $\Gamma = \{x_1 : A_1, \ldots, x_k : A_k\}$ then $\{x_1, \ldots, x_k\}$ is the set of all variables free in t.*

Definition 2. For any simply typable term t, the set Γ and the formula A such that $\Gamma \vdash t : A$ will be called the *context of* t and the *type of* t respectively. We define the *order of* t as the order of its type.

Remark. The definition presented above is slightly different from the usual definition of simply typed terms (which can be found for instance in [4]), generally presented as follows:

i) $\Gamma, x : A \vdash^* x : A$ for $\Gamma, x : A$ included in the union of $\mathcal{C}$, $\mathcal{L}$ and $\mathcal{I}$,
ii) if $\Gamma \vdash^* u : A \rightarrow B$ and $\Gamma \vdash^* v : A$ then $\Gamma \vdash^* (u)v : B$,
iii) if $\Gamma,\ y : A \vdash^* u : B$ and $y : A \in \mathcal{L}$, then $\Gamma \vdash^* \lambda y.u : A \rightarrow B$,

If $\Gamma \vdash^* t : A$, then for any $\Gamma^* \supset \Gamma$, $\Gamma^* \vdash^* t : A$, therefore the notion of "context of a typed term $t : A$" is not well defined in this system.

However, if $\Gamma \vdash t : A$, then $\Gamma \vdash^* t : A$ and conversely, if $\Gamma^* \vdash^* t : A$, then there exists a unique context $\Gamma \subset \Gamma^*$ such that $\Gamma \vdash t : A$. Thus, the well-known results of strong normalization of all typable terms and the stability of their typing under β-reduction hold for our presentation.

3 Reduction on Terms

We assume that the set of local variables $\mathcal{L}$ is split into two infinite subsets $\mathcal{A}$ and $\mathcal{P}$, each of these sets containing an infinite number of variables of each type. Elements of $\mathcal{A}$ will be called *active* variables, elements of $\mathcal{P}$ *passive* variables. In the following, these two kinds of variables will allow us to discern immediately in a given term the variables which cannot take part in the process of reduction of this term.

3.1 α-Equivalence, $\mathcal{S}$-Terms, Terms

We write $\equiv$ the α-equivalence on terms of λ-calculus. Renamings of bound variables in Simply Typed Terms are assumed to respect the kind (active or passive) and the type of the variables renamed.

Example 1. the variable $x \in \mathcal{A}$ of type B in $\lambda x.x : B \rightarrow B$ may only be renamed by an active variable of the same type. For any active variable $y : B$, $\lambda x.x : B \rightarrow B$ and $\lambda y.y : B \rightarrow B$ are α-equivalent. For $l \in \mathcal{P}$, $\lambda x.x$ and $\lambda l.l$ are not α-equivalent. For $z : D$ with $B \neq D$, $\lambda x.x : B \rightarrow B$ and $\lambda z.z : D \rightarrow D$ are not α-equivalent.

We let $\mathcal{S}$ be the set of Simply Typed Terms, and define $\overline{\mathcal{S}}$ as the quotient of this set by the α-equivalence $(\mathcal{S}/\equiv)$. By convention, elements of $\overline{\mathcal{S}}$ and $\mathcal{S}$ will be called *terms* and $\mathcal{S}$*-terms* respectively. Greek letters shall be used to denote arbitrary $\mathcal{S}$-terms. An $\mathcal{S}$-term τ of the α-class (the term) t will be called *a representative* of t.

3.2 β-reduction

The definition of β-reduction used in this section is borrowed from [3]. The β-reduction on terms is the least binary relation β reflexive, transitive, and including the relation β_0 defined by the following rules:

0) if t is an element of $\mathcal{C}$, $\mathcal{L}$ or $\mathcal{I}$ then $t\,\beta_0\,t'$ is false for all t'.
1) if $t = \lambda z\, u$, then $t\,\beta_0\,t'$ if and only if $t' = \lambda z\, u'$ with $u\,\beta_0\,u'$.
2) if $t = (u)v$, then $t\,\beta_0\,t'$ if and only if: either $t' = (u')v$, with $u\,\beta_0\,u'$,
or $t' = (u)v'$, with $v\,\beta_0\,v'$,
or $u = \lambda x\, w$, $t' = w[u/x]$.

We let $\beta^* \subset \beta$ be the least binary relation reflexive, transitive, and including the relation β_0^* defined by:

i) β_0^* satisfies conditions (0) and (1).
ii) if $t = (u)v$, then $t\,\beta_0^*\,t'$ if and only if: either $t' = (u')v$, with $u\,\beta_0^*\,u'$,
or $t' = (u)v'$, with $v\,\beta_0^*\,v'$,
or $u = \lambda x\; w$, $x \in \mathcal{A}$, $t' = w[u/x]$.

3.3 β-Normal, η-Long Forms

Let $t = \lambda x_1 \ldots x_m.(x)u_1 \ldots u_p : A_1 \ldots A_n \to o$ (where $m \leq n$ and o is an atomic type i.e $o \in \mathcal{O}$) be a β-normal term. A *β-normal η-long form* of t is defined as a term of same type of the form

$$t' = \lambda x_1 \ldots x_m x_{m+1} \ldots x_n.(x)u'_1 \ldots u'_p x'_{m+1} \ldots x'_n$$

where u'_i is a β-normal η-long form of u_i, and x'_i is a β-normal η-long form of x_i. From now on, all normal terms will be supposed to be in β-normal η-long form.

Remark. By definition of α-equivalence, every β-normal term has only a finite number of η-long forms. Furthermore, if $t, v_1, \ldots, v_n$ are β-normal terms such that $(t)v_1 \ldots v_n$ is well-typed and first-order, $(t)v_1 \ldots v_n$ normalizes to b iff there exists t^*, $v_1^*, \ldots v_n^*$, b^*, η-long forms of $t, v_1, \ldots, v_n, b$ respectively, such that $(t^*)v_1^* \ldots v_n^*$ normalizes to b^*. Therefore, we may restrict whitout loss of generality the set of normal terms to the set of β-normal η-long forms.

3.4 Restriction of the Set of Terms

For $t \in \overline{\mathcal{S}}$, the notation $t = \lambda y_1 \ldots y_n.u$ supposes: for every representative ν of u, $\lambda y_1 \ldots y_n.\nu$ is a representative of t; the variables $y_1, \ldots, y_n$ are distinct; the term u is first-order. For $\mathcal{Y} = (y_1, \ldots, y_n)$, we write $\lambda\mathcal{Y}.u$ for $\lambda y_1 \ldots y_n.u$.

In the remaining, we will focus on a particular subset of $\overline{\mathcal{S}}$, the set $\overline{\mathcal{S}}_0$ defined by the following rules:

0) for every $x : o$ in $\mathcal{C}, \mathcal{A}, \mathcal{P}$ or $\mathcal{I}$ with o atomic, $x \in \overline{\mathcal{S}}_0$.
1) let $u : o \in \overline{\mathcal{S}}_0$ with o atomic. For every sequence of active variables $(y_1 : A_1, \ldots, y_n : A_n)$, $\lambda y_1 \ldots y_n.u : A_1 \ldots A_n \to o \in \overline{\mathcal{S}}_0$.
2) let $v_1 : A_1, \ldots, v_n : A_n \in \overline{\mathcal{S}}_0$. For every $x : A_1 \ldots A_n \to o$ in $\mathcal{A}$ or $\mathcal{I}$, $(x)v_1 \ldots v_n : o \in \overline{\mathcal{S}}_0$.

3) let $u_1 : o_1, \ldots, u_n : o_n \in \overline{\mathcal{S}}_0$ with $o_1, \ldots, o_n$ atomic. Let $\mathcal{K}_1, \ldots, \mathcal{K}_n$ be finite sequences of passive variables. Let $A_1, \ldots, A_n$ be the types of $\lambda\mathcal{K}_1.u_1, \ldots, \lambda\mathcal{K}_n.u_n$.
For every $K : A_1 \ldots A_n \to o$ in $\mathcal{C}$ or $\mathcal{P}$, $(K)\lambda\mathcal{K}_1.u_1 \ldots \lambda\mathcal{K}_n.u_n \in \overline{\mathcal{S}}_0$.
4) let $w_0 : A_1 \ldots A_n \to o$, $w_1 : A_1 \ldots w_n : A_n \in \overline{\mathcal{S}}_0$.
$(w_0)w_1 \ldots w_n : o \in \overline{\mathcal{S}}_0$.

Note that all normal terms in $\overline{\mathcal{S}}_0$ are in β-normal η-long form and conversely, for any $\mathcal{S}$-term τ in β-normal η-long form, there exists a unique renaming ρ of bound variables in τ (which may require to change the kind of the variables renamed) such that the class of $\rho(\tau)$ is an element of $\overline{\mathcal{S}}_0$.

Thus, we may assume without loss of generality that every term on β-normal η-long form is an element of $\overline{\mathcal{S}}_0$.

Remark. A term in $\overline{\mathcal{S}}_0$ of non-atomic type may only be obtained by application of rule 1, that is to say, if $t : B_1 \ldots B_n \to o \in \overline{\mathcal{S}}_0$, then there exists an $\mathcal{A}$-sequence $(x_1 : B_1, \ldots, x_n : B_n)$ and a first order $u : o \in \overline{\mathcal{S}}_0$ such that $t = \lambda x_1 \ldots x_n.u$.

Remark. The β^*-reduction on α-equivalence classes forbids the reduction of a redex of the form $(\lambda l.u)v$ where l is a passive variable. However, the following lemma proves that this relation is sufficient to reduce (in the usual sense) all non-normal elements of $\overline{\mathcal{S}}_0$.

Proposition 3. *Let $u{:}A$, $v{:}B \in \overline{\mathcal{S}}_0$. For any $x{:}B \in \mathcal{A}$ or $\mathcal{I}$, $u[v/x]{:}A \in \overline{\mathcal{S}}_0$.*

Proof. Straightforward induction on the number of rules used in the proof of $u : A \in \overline{\mathcal{S}}_0$. Note that the conclusion does not hold if we allow x to be in $\mathcal{C}$ or $\mathcal{P}$.

Lemma 4. *Let $t : A \in \overline{\mathcal{S}}_0$. If $t\,\beta_0\,t'$ then $t\,\beta_0^*\,t'$ and $t' : A \in \overline{\mathcal{S}}_0$.*

Proof. Induction on the number of rules used in the proof of $t : A \in \overline{\mathcal{S}}_0$. The cases of rules 1, 2 and 3 are immediate, so we only treat in details the case of rule 4, $t = (\lambda x \lambda y_1 \ldots y_n.u)v\, v_1 \ldots v_n : o$ with $t' = (\lambda y_1 \ldots y_n.u[v/x])v_1 \ldots v_n$ (we may assume that $y_1, \ldots, y_n$ are not free in v). Since $\lambda x \lambda y_1 \ldots y_n.u : A \to B$ is of higher order, $x, y_1, \ldots, y_n$ are necessarily active variables hence $t\,\beta_0^*\,t'$. By the preceding proposition, $\lambda y_1 \ldots y_n.u[v/x] : B \in \overline{\mathcal{S}}_0$. By rule 4, $t' : o \in \overline{\mathcal{S}}_0$. □

Lemma 5. *The Church-Rosser property holds for the β-reduction on simply typed terms (and in particular, for the β^*-reduction on $\overline{\mathcal{S}}_0$)*

Proof. See for instance [2]. □

Through sections 4 and 5, the set of terms will be restricted to $\overline{\mathcal{S}}_0$. We will call β-reduction the β^*-reduction on terms and β_0 the relation β_0^*. We will write $\simeq$ the β^*-equivalence.

4 Pattern Matching

Definition 6. A *matching problem* is by definition a finite set of equations of the form $[a = b]$, where a, b are normal terms and b contains no instantiable variable. A *solution* of a matching problem Ψ is a finite substitution σ on the set of instantiable variables free in Ψ to the set of normal terms, such that for each equation $[a = b] \in \Psi$, $\sigma(a)$ normalizes to b. We call *order of* Ψ the maximal order of an instantiable variable in this problem.

Note that we can assume without loss of generality that Ψ consists of a single equation of first order members: from $\{[a_1 = b_1], \ldots, [a_m = b_m]\}$, we construct the matching equation $[(K)a_1 \ldots a_m = (K)b_1 \ldots b_m]$, where K is a new constant of adequate type. Obviously, Ψ and this equation have the same set of solutions.

Definition 7. An *interpolation equation* E of *arguments* $(c_1, \ldots, c_n)$, of *result* b is by definition a matching equation of the form $[(x)c_1 \ldots c_n = b : \mathrm{o}]$ where x is instantiable, b is of atomic type, $c_1, \ldots, c_n$ (and by definition, b) contain no instantiable variable. A *solution* of E is a normal term t such that $(t)c_1 \ldots c_n$ is well-typed and normalizes to b.

Two interpolation equations E and E' will be called *equivalent*, written $E \sim E'$, if and only if they have the same set of solutions.

Definition 8. We call *interpolation problem* any finite set of interpolation equations. A *dual problem* (Φ, Ψ) is by definition a pair of interpolation problems whose equations contain the same instantiable variable. A *solution* of (Φ, Ψ) is a normal term t such that for each $E \in \Phi$, t is a solution of E, and for each $F \in \Psi$, t is a not a solution of F.

4.1 Accessible Contexts

From now on, we allow the constants appearing in $\overline{\mathcal{S}}_0$-terms to be either in $\mathcal{C}$, or in a new separated set of constants, the set *Nil*: this set contains, for each atomic type o, a new element nil_{o} of type o. As seen in the following, we do not need to explicitly differentiate these constants, i.e we will write *nil* all elements of *Nil*. For any set of variables $\mathcal{Z} = \{z_1 : A_1, \ldots, z_n : A_n\}$, and for any term u, we will write $u\,[\mathcal{Z} \leftarrow Nil]$ the term $u\,[\lambda\mathcal{X}_1.nil : A_1/z_1 \ldots \lambda\mathcal{X}_n.nil : A_n/z_n]$ (where all elements of $\mathcal{X}_1, \ldots, \mathcal{X}_n$ are active variables of expected type).

Lemma 9. *Let* $u \in \overline{\mathcal{S}}_0$. *For any set of variables* $\mathcal{Z}$, *there exists a term* $v \in \overline{\mathcal{S}}_0$ *such that* $u' = u[\mathcal{Z} \leftarrow Nil]\,\beta\, v$.

Proof. Straightforward induction on the number of rules in the proof of $u \in \overline{\mathcal{S}}_0$. The only non-trivial case is $u = (K)\lambda\mathcal{K}_1.u_1 \ldots \mathcal{K}_p.u_p$ with $K \in \mathcal{Z}$, $\underline{K} \in \mathcal{P}$. In this case, $u' = (\lambda\mathcal{X}.nil)\lambda\mathcal{K}.u'_1 \ldots \lambda\mathcal{K}_p.u'_p \notin \overline{\mathcal{S}}_0$. As $\mathcal{X} \subset \mathcal{A}$, $u'\,\beta\, nil \in \overline{\mathcal{S}}_0$. □

Lemma 10. *Let Δ be any context disjoint from Nil. Call set of* accessible contexts *the set of all subsets of* $(\Delta \cup Nil)$.

i) Let b be a normal term such that $\Delta \vdash b : o$ with o atomic. Let E be any interpolation equation of result b. There exists an equation E' equivalent to E, whose arguments are of accessible context.

ii) Let $b_1, \ldots, b_m$ be normal terms of same atomic type o such that $\Delta \vdash b_1, \ldots, b_m$. Let $\Phi = \{[a_1 = b_1], \ldots, [a_m = b_m]\}$ be any interpolation problem. Let Ψ be any interpolation problem such that the result of each element of Ψ contains no free element of Nil. If (Φ, Ψ) has a solution, then (Φ, Ψ) has a solution of accessible context.

Proof. We write $\Gamma \vdash c_1, \ldots, c_n$ the relation "$c_1, \ldots, c_n$ are normal terms elements of $\overline{S}_0$, and the union of their contexts is included in Γ". Remark that $u \, \beta_0 \, v$ implies $u[w/x] \, \beta_0 \, v[w/x]$. If x is not free in v, then $u[w/x] \, \beta_0 \, v$. Hence,

i) suppose E is of the form $[(x)c_1 \ldots c_n = b]$, with $\Gamma, \Delta \vdash c_1, \ldots, c_n$. For each i, let $c'_i = c_i[\Gamma \leftarrow Nil]$. For every t, $(t)c_1 \ldots c_n \, \beta \, b$ if and only if $(t)c'_1 \ldots c'_n \, \beta \, b$, that is to say, E and $E' = [(x)c'_1 \ldots c'_n = b]$ are equivalent.

ii) Let t be any solution of (Φ, Ψ). Suppose $\Gamma, \Delta_0 \vdash t : A$, where $\Delta_0 \subset \Delta$ and Γ contains no element of Δ. Let $t' = t[\Gamma \leftarrow Nil]$. Let $E = [(x)c_1 \ldots c_n = b_j]$ be any element of Φ. Since no element of Γ is free in b_j, t' is still a solution of E. Let $[(x)d_1 \ldots d_n = e]$ be any element of Ψ. Since e contains no element of *Nil*, t' is still not a solution of F. In other words, t' is still a solution of (Φ, Ψ). □

5 Equivalence Classes of Equations

The aim of this section is to prove the following result :

> Let b be a normal term such that $\mathcal{C} \vdash b : o$ with o atomic. Let A be any type. The quotient by $\sim$ of the set of all interpolation equations of the form $[(x)w_1 \ldots w_n = b]$ where x is of type A contains only a finite number of classes.

As a corollary of this result, we will prove that the decidability of Dual Interpolation implies the decidability of Pattern Matching.

5.1 Characterization Theorem

We give in this section a necessary and sufficient condition on t, t' in $[(x)t = b]$, $[(x)t' = b]$, so that these two equations are equivalent.

Definition 11. We assume the existence of a computable function *Rep* that given a term w, returns a representative ε of w such that for any z, z is not simultaneously free and bounded in ε, and "λz" appear at most once in ε.

Definition 12. For any normal term b such that $\mathcal{C} \vdash b : o$ with o atomic, we write $RSub(b)$ the set of α-equivalence classes of all first order subterms of $Rep(b)$. Remark that as b is normal, first order and of constant context, all free variables in the elements of $RSub(b)$ are in the union of $\mathcal{C}$ and $\mathcal{P}$.

Definition 13. Let S be a finite set of terms, and let t, t' be two terms of same type $A = A_1 \dots A_n \to o$. We will say that t and t' are *parallel on* S (or S-*parallel*) if and only if $\forall v_1 : A_1, \dots, \forall v_n : A_n, \forall s \in S, (t)v_1 \dots v_n \,\beta\, s \Leftrightarrow (t')v_1 \dots v_n \,\beta\, s$.

Remark that S-parallelism is an equivalence relation. In the particular case of $n = 0$, t and t' are S-parallel if and only if either $t \notin S$ and $t' \notin S$, or t and t' are both in S and in that case, are equal terms.

Proposition 14. *Let b be a normal term such that $\mathcal{C} \vdash b : o$ with o atomic. let $S = RSub(b)$. Let $\diamond$ be any atomic type.*

Let $(w : B_1 \dots B_p \to \diamond)$, $(e_1 : B_1), \dots, (e_p : B_m)$ be arbitrary terms. Let $t : A$ and $t' : A$ be two S-parallel terms of constant context. Let $z : A$ be a fresh active variable. Then $(w[t/z])e_1 \dots e_p$ and $(w[t'/z])e_1 \dots e_p$ are S-parallel terms.

Proof. We fix t and t', and prove the result by induction on P, and for each P, by induction on N, where P is the sum of the length of all normalizations of $(w[t/z])e_1 \dots e_p$, N is the number of rules used in the proof of $w \in \overline{S}_0$. Since all terms in $\overline{S}_0$ are strongly normalizing, this induction is well founded. We consider the last rule used in the proof of $w \in \overline{S}_0$.

If it is rule 0, $w = x$ is a first order variable. Either x and z are distinct and $w[t/z] = w[t'/z] = x$, or $x = z$ and $w[t/z] = t$, S-parallel to $w[t'/z] = t'$.

If it is rule 1, w is of the form $\lambda x_1 \dots x_p.v$ with $p \neq 0$. We assume that z and $x_1, \dots, x_p$ are distinct. Since z is not free in $e_1 \dots e_p$, $(w[t/z])e_1 \dots e_p$ β-reduces to $v[t/z][e_1/x_1 \dots e_p/x_p] = v[t/z][\mathbf{e}/\mathbf{x}] = v[\mathbf{e}/\mathbf{x}][t/z]$. The sum of the length of all normalizations of this latter term is at most $(P-p)$ hence, by induction hypothesis, $(w[t/z])e_1 \dots e_n \,\beta\, v[\mathbf{e}/\mathbf{x}][t/z]$ and $v[\mathbf{e}/\mathbf{x}][t'/z] \simeq (w[t'/z])e_1 \dots e_p$ (recall that the symbol $\simeq$ stands for the β-equivalence) are S-parallel terms.

If it is rule 2, w is of the form $(x)v_1 \dots v_m$, with $m \neq 0$ and $x \in \mathcal{A}$. Note that, as x is an active variable, if x is not equal to z then $w[t/z]$ and $w[t'/z]$ cannot normalize in S. Suppose w of the form $(z)v_1 \dots v_m$ and t of the form $\lambda y_1 \dots y_m.t_0$. Then $w[t/z] = (t)v_1[t/z] \dots v_m[t/z] \,\beta\, t_0[v_1[t/z]/y_1 \dots v_m[t/z]/y_m]$ $= t_0[v_1/y_1 \dots v_m/y_m][t/z] = t_0[\mathbf{v}/\mathbf{y}][t/z]$. The sum of the length of all normalizations of this latter term is at most $(P-m)$ hence, by induction hypothesis, $(t)v_1[t/z] \dots v_m[t/z] \simeq t_0[\mathbf{v}/\mathbf{y}][t/z]$ and $t_0[\mathbf{v}/\mathbf{y}][t'/z] \simeq (t)v_1[t'/z] \dots v_m[t'/z]$ are S-parallel terms. Furthermore, as t and t' are S-parallel, $(t)v_1[t'/z] \dots v_m[t'/z]$ and $(t')v_1[t'/z] \dots v_m[t'/z] = w[t'/z]$ are S-parallel terms.

If it is rule 3, w is of the form $(K)\lambda\mathcal{K}_1.u_1 \dots \lambda\mathcal{K}_m.u_m$, with $K \in \mathcal{C}$ or $\mathcal{P}$. Suppose for instance that $w[t/z]$ normalizes to $s \in S$. In this case, we may assume that $\mathcal{K}_1, \dots, \mathcal{K}_m$ are such that $s = (K)\lambda\mathcal{K}_1.s_1 \dots \lambda\mathcal{K}_m.s_m$ with $s_1, \dots, s_m \in S$.

Then $w[t/z] = (K)\lambda\mathcal{K}_1.u_1[t/z]\ldots\lambda\mathcal{K}_m.u_m[t/z] \simeq (K)\lambda\mathcal{K}_1.s_1\ldots\lambda\mathcal{K}_m.s_m$. By induction hypothesis on N, for each j, $u_j[t/z] \simeq u_j[t'/z] \simeq s_j$ therefore $w[t/z] \simeq (K)\lambda\mathcal{K}_1.s_1\ldots\lambda\mathcal{K}_m.s_m \simeq w[t'/z]$. The converse hypothesis ($w[t'/z] \simeq s' \in S$) leads to a similar conclusion.

If it is rule 4, w is of the of the form $(\lambda x_1...x_n.u)v_1...v_n$ with $n \neq 0$. We assume that $x_1,...,x_n$, z are distinct. Then $w[t/z]\ \beta\ u[t/z][v_1[t/z]/x_1...v_n[t/z]/x_n] = u[v_1/x_1\ldots v_n/x_n][t/z] = u[\mathbf{v}/\mathbf{x}][t/z]$. The sum of the length of all normalizations of this latter term is at most $(P-n)$ hence, by induction hypothesis, $w[t/z] \simeq u[\mathbf{v}/\mathbf{x}][t/z]$ and $u[\mathbf{v}/\mathbf{x}][t'/z] \simeq w[t'/z]$ are S-parallel terms. □

Theorem 15. *Let b be a normal term such that $C \vdash b : o$ with o atomic. Let t, t' be two terms of constant context. The equations $[(x)t = b]$ and $[(x)t' = b]$ are equivalent if and only if t and t' are parallel on RSub(b).*

Proof. Suppose t and t' $RSub(b)$-parallel. By the proposition 14, for every u, $u[t/z] \simeq b \Leftrightarrow u[t'/z] \simeq b$, thus $(\lambda z.u)t \simeq b \Leftrightarrow (\lambda z.u)t' \simeq b$ i.e. $[(x)t = b]$ and $[(x)t' = b]$ are equivalent. Conversely, suppose for instance that for $v_1\ldots v_n$, $(t)v_1\ldots v_n \simeq s \in RSub(b)$ and $(t')v_1\ldots v_n \simeq s' \neq s$. Let $\chi_1\ldots\chi_p$ be the sequence of symbols ("λ", "(", ")" or a typed variable) equal to $\varepsilon = Rep(b)$. Let j, k be such that $\chi_j\ldots\chi_{j+k} = \varepsilon_0$, representative of s. Replace this subsequence in ε by a representative of $(z)v_1\ldots v_n$. Call b^* the α-class of the sequence of symbols thus defined. As t is of constant context, no bounded variable in ε is free in t hence, there exists a representative of $b^*[t/z]$ of the form $\chi_1\ldots\chi_{j-1}\,\kappa\,\chi_{j+k+1}\ldots\chi_p$ where κ is a representative of $(t)v_1\ldots v_n$. The normal form of $b^*[t/z]$ has a representative of the form $\chi_1\ldots\chi_{j-1}\,\varepsilon_0\,\chi_{j+k+1}\ldots\chi_p$ i.e. is equal to b. As t' is of constant context, no bounded variable in ε is free in t' hence, there exists a representative of $b^*[t'/z]$ of the form $\chi_1\ldots\chi_{j-1}\,\kappa'\,\chi_{j+k+1}\ldots\chi_p$ where κ' is a representative of $(t')v_1\ldots v_n$. The normal form of $b^*[t'/z]$ has a representative of the form $\chi_1\ldots\chi_{j-1}\,\varepsilon'_0\,\chi_{j+k+1}\ldots\chi_p$ where ε'_0 is a representative of $s' \neq s$ i.e. is distinct from b. Thus, $\lambda z.b^*$ is a solution of $[(x)t = b]$ and $[(x)t' \neq b]$, i.e. these equations are not equivalent. □

5.2 Specifying the Context of Solutions

We may add some new equations to a dual problem (Φ, Ψ) in order to forbid a particular set of variables from appearing in every solution of accessible context of the new problem. Consider, for instance, $E = [(x)A = A]$, where A is a first order constant. This interpolation equation has only two solutions, $\lambda y.y$ and $\lambda y.A$. Let $F = [(x)B = B]$, with $B \neq A$. The only solution of $\{E, F\}$, $\lambda y.y$, does not contain A. The following proposition generalizes this simple example.

Definition 16. We will say that two ordered sets of variables $\mathcal{Z}$, $\mathcal{Z}'$ are in one to one corespondance if and only if they are of the form $\mathcal{Z} = (z_1 : A_1, \ldots, z_n : A_n)$ and $\mathcal{Z}' = (z'_1 : A_1, \ldots, z'_n : A_n)$. In this case, we write $[\mathcal{Z}'/\mathcal{Z}]$ the substitution $[z'_1/z_1\ldots z'_n/z_n]$.

Lemma 17. *Let s be a normal term such that $C_0, \mathcal{K} \vdash s : \diamond$ with $C_0 \subset \mathcal{C}$, $\mathcal{K} \subset \mathcal{P}$ and $\diamond$ atomic. Let $\mathcal{K}^*$ be a new subset of $\mathcal{P}$ in one to one correspondance with $\mathcal{K}$. Let $E = [(x)v_1 \ldots v_n = s]$ be an interpolation equation. Let $E^* = E[\mathcal{K}^*/\mathcal{K}] = [(x)v_1^* \ldots v_n^* = s^*]$. Then,*

i) $\forall w$, w is a solution of $\Phi = \{E, E^*\} \Leftrightarrow w[\mathcal{K}\mathcal{K}^* \leftarrow Nil]$ is a solution of Φ;

ii) for any t of constant context, t is a solution of $E \Leftrightarrow t$ is a solution of E^*.

Proof. i) As s and s^* contain no element of *Nil*, if $w[\mathcal{K}\mathcal{K}^* \leftarrow Nil]$ is a solution of Φ then w is a solution of Φ. Conversely, let w be any solution of Φ. Suppose for instance that $w[\mathcal{K} \leftarrow Nil]$ is not a solution of Φ. Let $\mathcal{Z}$ be a new subset of $\mathcal{P}$ in one to one correspondance with $\mathcal{K}$. Let $w_0 = w[\mathcal{Z}/\mathcal{K}]$. let s_0, s_0^* be the normal forms of $(w_0)v_1 \ldots v_n$ and $(w_0)v_1^* \ldots v_n^*$ respectively. Then at least one element of $\mathcal{Z}$ is free in s_0 or s_0^*. Otherwise,
$(w[\mathcal{K} \leftarrow Nil])v_1 \ldots v_n = (w_0)v_1 \ldots v_n[\mathcal{Z} \leftarrow Nil] \;\beta\; s_0[\mathcal{Z} \leftarrow Nil] = s_0$,
$s_0 = s_0[\mathcal{K}/\mathcal{Z}] \simeq (w_0)v_1 \ldots v_n[\mathcal{K}/\mathcal{Z}] = (w)v_1 \ldots v_n \;\beta\; s$, and
$(w[\mathcal{K} \leftarrow Nil])v_1^* \ldots v_n^* = (w_0)v_1^* \ldots v_n^*[\mathcal{Z} \leftarrow Nil] \;\beta\; s_0^*[\mathcal{Z} \leftarrow Nil] = s_0^*$,
$s_0^* = s_0^*[\mathcal{K}/\mathcal{Z}] \simeq (w_0)v_1^* \ldots v_n^*[\mathcal{K}/\mathcal{Z}] = (w)v_1^* \ldots v_n^* \;\beta\; s^*$, a contradiction.
Since $(w[\mathcal{Z}/\mathcal{K}])v_1 \ldots v_n \simeq s_0 \Leftrightarrow (w[\mathcal{Z}/\mathcal{K}])v_1^* \ldots v_n^* \simeq s_0[\mathcal{K}^*/\mathcal{K}]$, $s_0^* = s_0[\mathcal{K}^*/\mathcal{K}]$. Hence for any $z \in \mathcal{Z}$, z is free in s_0 iff z is free in s_0^*. As $s^* = s_0^*[\mathcal{K}/\mathcal{Z}]$, we conclude that s^* contains a $\mathcal{K}$-occurrence, a contradiction. The proof of "$w[\mathcal{K}^* \leftarrow Nil]$ is a solution of Φ" is symmetrical.
(ii) Indeed, $(t)v_1 \ldots v_n \;\beta\; s \Leftrightarrow ((t)v_1 \ldots v_n)[\mathcal{K}^*/\mathcal{K}] = (t)v_1^* \ldots v_n^* \;\beta\; s[\mathcal{K}^*/\mathcal{K}] = s^*$. □

5.3 Finiteness Lemma

Preliminaries

Definition 18. For any context Δ disjoint from *Nil*, and for any type A, we write *Terms*(A, Δ, Nil) the (in general infinite) subset of $\overline{\mathcal{S}}_0$ of all normal terms of type A, of context included in the union of Δ and *Nil*.

For any normal term s such that $\Delta \vdash s : \diamond$ where $\diamond$ is atomic, we write *Equ*(A, s) the set of all interpolation equations of the form $[(x)c_1 \ldots c_n = s]$, where x is an instantiable variable of type A and for $A = A_1 \ldots A_n \rightarrow \diamond$, each c_i is an element of *Terms*(A_i, Δ, Nil).

Lemma 19. *Let Δ be any context disjoint from Nil. Let s be any normal term such that $\Delta \vdash s : \diamond$, and let A be any type. The cardinal of (Equ(A, s)/ $\sim$) is equal to the cardinal of the quotient by $\sim$ of the set of all interpolation equations of the form $[(x)w_1 \ldots w_n = s]$, where x is of type A.*

Proof. Clear, by lemma 10. For any interpolation equation $E = [(x)w_1 ... w_n = s]$, there exists in *Equ*(A, s) an equation equivalent to E.

Proposition 20. *For $E = [(x)c_1 \dots c_n = b]$, $F = [(x)d_1 \dots d_n = b]$, E and F are equivalent if and only if for each i, $[(z)c_i = b]$ and $[(z)d_i = b]$ are equivalent.*

Proof. By induction on n. Suppose E and F equivalent, $[(z)c = b]$ and $[(z)d = b]$ equivalent. For $\lambda y y_1 \dots y_n.u = \lambda y\mathcal{Y}.u$, assume that $y, y_1, \dots, y_n$ are not free in $c, d, c_1 \dots c_n, d_1 \dots d_n$. Then $(\lambda y\mathcal{Y}.u)c\, c_1 \dots c_n\, \beta\, b \Leftrightarrow (\lambda\mathcal{Y}.u[c/y])c_1 \dots c_n\, \beta\, b \Leftrightarrow$ $(\lambda\mathcal{Y}.u[c/y])d_1 \dots d_n\, \beta\, b$ (as E and F are equivalent) $\Leftrightarrow (\lambda y.u[d_1/y_1 \dots d_n/y_n])c\, \beta\, b$ $\Leftrightarrow (\lambda y.u[d_1/y_1 \dots d_n/y_n])d\, \beta\, b$ (as $[(z)c = b]$ and $[(z)d = b]$ are equivalent) $\Leftrightarrow$ $(t)d\, d_1 \dots d_n\, \beta\, b$ □

Key Lemma From now on, we fix an enumeration of the set of all terms, the set of all variables, vand the set of all interpolation equations.

Lemma 21. *Let N be an arbitrary order.*

1) Let b be any normal term such that $\mathcal{C} \vdash b : \circ$ with $\circ$ atomic. For any type $A = A_1 \dots A_n \to \circ$ of order at most N, the quotient by the relation $\sim$ of the set

$$\mathrm{Equ}(A, b) = \{\, [(x)c_1 \dots c_n = b] \mid x : A \text{ and } \forall i\, c_i \in \mathit{Terms}(A_i, \mathcal{C}, \mathit{Nil})\}$$

contains a finite number of classes

2) There exists a function Ω_N satisfying the two following properties

> *i) For any type of A of order at most N, and for any normal $s : \diamond$ of context included in the union of $\mathcal{C}$ and $\mathcal{P}$ with $\diamond$ atomic, $\Omega_N(A, s)$ contains a unique representative of each class in $(\mathrm{Equ}(A, s)/\sim)$.*
>
> *ii) If Dual Interpolation of order $(N - 1)$ is decidable, then Ω_N is computable.*

Proof. By induction on N. The case $N = 1$ is immediate, since for any $x : \circ \in \mathcal{I}$, the set $\mathit{Equ}(\circ, b)$ contains a unique equation of instantiable variable x, $[x = b]$. Suppose $N > 1$.

1) Let $S = \mathit{RSub}(b)$. Let $\mathcal{K}$ be the set of all passive variables free in the elements of S. Let $\mathcal{K}^*$ be a new subset of $\mathcal{P}$ in one to one correspondance with $\mathcal{K}$. For any type D of order at most $(N - 1)$, we define the finite set of dual problems $\mathit{Car}(D, b)$ as follows:

> Let z be the first instantiable variable of type D. For each $s \in S$, for each $E \in \Omega_{N-1}(D, s)$, let $E^* = E[\mathcal{K}^*/\mathcal{K}]$. Define $P = \mathit{Car}(D, b)$ as the finite set of all dual problems (Φ, Ψ) of instantiable variable z satisfying
>
> - $\Phi \cup \Psi = \{E \mid \exists s \in S, E \in \Omega_{N-1}(D, s)\} \cup \{E^* \mid \exists s \in S, E \in \Omega_{N-1}(D, s)\}$,
>
> - for any $s \in S$, for any $E \in \Omega_{N-1}(D, s)$, $\{E, E^*\} \subset \Phi$ or $\{E, E^*\} \subset \Psi$.

Remark that for any $s \in S$ and for $\mathcal{K}_0$ defined as a new set of constants in one to one correspondance with $\mathcal{K}$, $s_0 = s[\mathcal{K}_0/\mathcal{K}]$ is of constant context, $(\mathit{Equ}(D, s)/\sim)$

and $(Equ(D, s_0)/\sim)$ have same cardinal. Therefore, by induction hypothesis, for each s in S, $(Equ(D, s)/\sim)$ and thereby $\Omega_{N-1}(D, s)$ are finite sets. Hence, $Car(D, b)$ is a finite set.

All elements of $Car(D, b)$ are then dual problems of order at most $(N-1)$. We let $SDual(D, b)$ be the least finite set containing, for each element (Φ, Ψ) in $Car(D, b)$ *which has a solution*[2], the first solution of this problem of minimal context. Then,

• Let (Φ, Ψ) be any element of $Car(D, b)$. By lemma 10 (ii), all solutions of (Φ, Ψ) of minimal context are of context included in the union of $\mathcal{C}$, $\mathcal{K}$, $\mathcal{K}^*$ and Nil. By lemma 17 (i) and by definition of Car, there is no solution of (Φ, Ψ) which is at once of minimal context and containing a free element of $(\mathcal{K} \cup \mathcal{K}^*)$. Therefore, all elements of $SDual(D, b)$ are of context included in the union of $\mathcal{C}$ and Nil.

• If (Φ, Ψ), (Φ', Ψ') are two distinct elements of $Car(D, b)$, then there exists at least one pair E, E^* in Φ which is in Ψ', or (if Φ is empty) at least one pair in Φ' which is in Ψ. By lemma 17 (ii), for any term c of constant context, c is a solution of E if and only if c is a solution of E^*, hence c cannot be at once a solution of (Φ, Ψ) and (Φ', Ψ'). Therefore, for any term c of constant context, of type D, there exists, a unique dual problem in $Car(D, b)$ of which c is a solution.

• For any $s \in RSub(b)$, for any $F = [(z)w_1 \ldots w_n = s]$ in $Equ(D, s)$, there exists in $\Omega_{N-1}(D, s)$ an equation $[(z)v_1 \ldots v_n = s]$ equivalent to F. Therefore, for any terms t, t' of constant context and of type D,

t, t' are solutions of the same problem $(\Phi, \Psi) \in Car(D, b)$,

$\Leftrightarrow$ for every $s \in S$, for every $F \in Equ(D, s)$,
t is a solution of F if and only if t' is a solution of F,

$\Leftrightarrow$ t and t' are S-parallel terms (by definition of parallelism),

$\Leftrightarrow$ $[(x)t = b]$ and $[(x)t' = b]$ are equivalent (by theorem 15).

Hence, the finite set $\{[(x)t = b] \mid t \in SDual(D, b)\}$ contains a unique representative of each class in $(Equ(D \to o, b)/\sim)$ i.e. this latter set contains a finite number of classes. By proposition 20, for $A = D_1 \ldots D_n \to o$ the finite set $\{[(x)t_1 \ldots t_n = b] \mid \forall i, t_i \in SDual(D_i, b)\}$ contains for each class in $(Equ(A, b)/\sim)$, a unique representative of this class i.e. this latter set contains only a finite number of classes.

2) i) We may extend the function Ω_{N-1} to Ω_N by the following definition:

> Let s be a normal term such that $\mathcal{C}_0, \mathcal{K} \vdash s : \diamond$ with $\mathcal{C}_0 \subset \mathcal{C}$, $\mathcal{K} \subset \mathcal{P}$ and $\diamond$ atomic. Let $\mathcal{K}_0$ be a new set of constants in one to one correspondance with $\mathcal{K}$. For any type $A = D_1 \ldots D_n \to \diamond$ of order N,
>
> let $\Omega_N(A, s) = \{[(x)t_1 \ldots t_n = s] \mid \forall i, t_i[\mathcal{K}_0/\mathcal{K}] \in SDual(D_i, s[\mathcal{K}_0/\mathcal{K}])\}$

[2] As we don't know whether Dual Interpolation of order $N-1$ is decidable, the function $SDual$ may be not computable

ii) The functions *RSub* and Ω_1 are computable. For any $1 < P < N$, if Ω_P is computable, then the restriction of *Car* to types of order at most P is computable. The decidability of Dual Interpolation of order P implies that the function *SDual* restricted to types of order P is computable and thereby, implies that the function Ω_{P+1} is computable. □

6 Main Results

Theorem 22. *Let $b : o$ be a normal term of atomic type. Let $A = A_1 \ldots A_n \to o$ be an arbitrary type. The quotient by the relation $\sim$ of set of interpolations equations*

$$EQU(A,b) = \{\, [(x)c_1 \ldots c_n = b] \mid x : A,\ c_1 : A_1,\ \ldots ,c_n : A_n\}$$

contains only a finite number of classes.

Proof. We may assume that b is of constant context, by substituting new constants for all free variables in this term. We may also assume that every β-normal term is on η-long form. At last, we may assume that every term on β-normal η-long form (in particular, b) is an element of $\overline{S}_0$ (by adjusting the kind of bound variables in all terms considered) and that the β-reduction is restricted to β^*. The conclusion follows then from the preceding lemma and lemma 19. □

Theorem 23. *Let N be an arbitrary order. The decidability of Dual Interpolation of order N implies the decidability of Pattern Matching of order N.*

Proof. Indeed, by lemma 21, the decidability of Dual Interpolation of order N implies that the function Ω_{N+1} is a computable function. Let $z_1 : A_1 \ldots z_n : A_n$ be instantiable variables of order at most N. Let $A = A_1 \ldots A_n \to o$. For any normal term b such that $C \vdash b : o$, let Σ be the finite set containing, for each $[(x)t_1 \ldots t_n = b]$ in $\Omega_{N+1}(A,b)$, the substitution $[t_1/z_1 \ldots t_n/z_n]$. (since A is of order at most $N+1$, by hypothesis, the set Σ is computable). Let $F = [u[z_1 \ldots z_n] = b]$ be a matching equation. Let $[t_1^*/z_1 \ldots t_n^*/z_n]$ be an arbitrary solution of F. Then $\lambda y_1 \ldots y_n.u[y_1/z_1 \ldots y_n/z_n]$ $(y_1 \ldots y_n \in \mathcal{A})$ is a solution of $[(x)t_1^* \ldots t_n^* = b]$; there exists in $\Omega_{N+1}(A,b)$ an equation $[(x)t_1 \ldots t_n = b]$ equivalent to $[(x)t_1^* \ldots t_n^* = b]$; $u[t_1/z_1 \ldots t_n/z_n]$ still normalizes to b.

In other words, the set Σ contains a solution of F. □

7 Conclusion

So far, the results presented in this paper leave open the issue of the decidability of Pattern Matching. Since we do not consider the problem of solving simultaneously equations *and* inequations between simply typed terms (i.e. we do not consider inequations of the form $[a \neq b]$, where b contains no instantiable variables), Pattern Matching could be decidable without Dual Interpolation being decidable for all orders. The methods used in [5] and [6] in order to prove the decidability of Fourth Order Matching and Atomic Matching are quite different, and both rely on properties specific to these particular cases.

References

1. Barendregt, H.: The Lambda Calculus, its Syntax and Semantics. North Holland (1981), (1984)
2. Hindley J.R., Seldin, J.P.: Introduction to Combinators and λ-Calculus. Cambridge University Press, Oxford (1986)
3. Krivine J.L.: Lambda Calculus, Types and Models. Ellis Horwood series in computer and their applications (1993) 1–66
4. Dowek G.: Third Order Matching is Decidable. Proceedings of Logic in Computer Science, Annals of Pure and Applied Logic (1993)
5. Padovani V.: Fourth Order Dual Interpolation is Decidable. Manuscript (1994)
6. Padovani V.: Atomic Matching is Decidable. Manuscript (1994)

Strict Functionals for Termination Proofs

Jaco van de Pol and Helmut Schwichtenberg [1]
Mathematisches Institut, Universität München
jaco@phil.ruu.nl schwicht@rz.mathematik.uni-muenchen.de

A semantical method to prove termination of higher order rewrite systems (HRS) is presented. Its main tool is the notion of a strict functional, which is a variant of Gandy's notion of a hereditarily monotonic functional [1]. The main advantage of the method is that it makes it possible to transfer ones intuitions about why an HRS should be terminating into a proof: one has to find a "strict" interpretation of the constants involved in such a way that the left hand side of any rewrite rule gets a bigger value than the right hand side. The applicability of the method is demonstrated in three examples.

- An HRS involving map and append.
- The usual rules for higher order primitive recursion in Gödel's T.
- Derivation terms for natural deduction systems. We prove termination of the rules for β–conversion and permutative conversion for logical rules including introduction and elimination rules for the existential quantifier. This has already been proved by Prawitz in [5]; however, our proof seems to be more perspicuous.

Technically we build on [7]. There a notion of a strict functional and simultaneously of a strict greater–than relation $>_{\mathrm{str}}$ between monotonic functionals is introduced. The main result then is the following. Let M be a term in β normal form and $\Box \in \mathrm{FV}(M)$. Then for any strict environment U and all monotonic $\mathbf{f}$ and $\mathbf{g}$, one has $\mathbf{f} >_{\mathrm{mon}} \mathbf{g} \implies [\![M]\!]_{U[\Box\mapsto\mathbf{f}]} >_{\mathrm{str}} [\![M]\!]_{U[\Box\mapsto\mathbf{g}]}$. From this van de Pol derives the technique described above for proving termination of higher order term rewrite systems, generalizing a similar approach for first order rewrite systems (cf. [3, p. 367]). Interesting applications are given in [7].

Here a slight change in the definition of strictness is exploited (against the original conference paper; cf. [7, Footnote p. 316]). This makes it possible to deal with rewrite rules involving types of level > 2 too, and in particular with proof theoretic applications. In order to do this some theory of strict functionals is developed. We also add product types, which are necessary to treat e.g. the existential quantifier.

[1] Both authors are partially supported by the Science Twinning Contract SC1*–CT91–0724 of the European Community.

1. Monotonicity and Strictness

Let ρ, σ, τ denote simple types over some base types ι (containing at least o), composed with $\to$ and $\times$. For simplicity we consider the sets $\mathcal{T}_\rho$ of all functionals of type ρ over some ground domains $\mathcal{T}_\iota$. The ground domains are provided with some partial order $>_\iota$.

Definition. *For any type ρ we define the set $\mathcal{M}_\rho \subseteq \mathcal{T}_\rho$ of monotonic functionals of type ρ and simultaneously a relation $\geq$ on $\mathcal{T}_\rho$.*

(i) (a) $\mathcal{M}_\iota = \mathcal{T}_\iota$.

(b) $\mathbf{f} \in \mathcal{M}_{\sigma\to\tau} \iff$ *for all* $\mathbf{x}, \mathbf{y} \in \mathcal{M}_\sigma$, $\mathbf{f}(\mathbf{x}) \in \mathcal{M}_\tau$ *and if* $\mathbf{x} \geq \mathbf{y}$ *then* $\mathbf{f}(\mathbf{x}) \geq \mathbf{f}(\mathbf{y})$.

(c) $\mathcal{M}_{\sigma\times\tau} = \mathcal{M}_\sigma \times \mathcal{M}_\tau$.

(ii) (a) $n \geq_\iota m \iff n >_\iota m$ *or* $n = m$.

(b) $\mathbf{f} \geq_{\sigma\to\tau} \mathbf{g} \iff$ *for all* $\mathbf{x} \in \mathcal{M}_\sigma, \mathbf{f}(\mathbf{x}) \geq_\tau \mathbf{g}(\mathbf{x})$.

(c) $\langle \mathbf{a}, \mathbf{b}\rangle \geq_{\sigma\times\tau} \langle \mathbf{c}, \mathbf{d}\rangle \iff \mathbf{a} \geq_\sigma \mathbf{c}$ *and* $\mathbf{b} \geq_\tau \mathbf{d}$.

We will use the following notation: $\vec{\sigma} \to \rho$ denotes the type $\sigma_1 \to \cdots \sigma_n \to \rho$. Let $\mathbf{x}(0)$ denote the left component of the pair $\mathbf{x}$ and $\mathbf{x}(1)$ its right component. This allows us to write projections and applications in a uniform way. Furthermore, simply typed terms M, N are introduced as usual: Typed variables x, y, z, application MN, abstraction $\lambda x\, M$, pairing $\langle M, N\rangle$ and projections $\pi_i(M)$ for $i = 0, 1$. Projections are also written $M0$ and $M1$. We use standard notions of free and bound variables, substitution, interpretation of M in a domain under environment U (denoted by $[\![M]\!]_U$). Using the new notation for projections, the previous definition can be written very compactly as:

Definition. *For any type ρ we define the set $\mathcal{M}_\rho \subseteq \mathcal{T}_\rho$ of monotonic functionals of type ρ and simultaneously a relation $\geq$ on $\mathcal{T}_\rho$.*

(i) $\mathbf{f} \in \mathcal{M} \iff$ *for all* $\vec{\mathbf{x}}, \vec{\mathbf{y}} \in \mathcal{M} \cup \{0, 1\}$, *if* $\vec{\mathbf{x}} \geq \vec{\mathbf{y}}$, *then* $\mathbf{f}(\vec{\mathbf{x}}) \geq \mathbf{f}(\vec{\mathbf{y}})$.

(ii) $\mathbf{f} \geq \mathbf{g} \iff$ *for all* $\vec{\mathbf{x}} \in \mathcal{M} \cup \{0, 1\}$, $\mathbf{f}(\vec{\mathbf{x}}) \geq \mathbf{g}(\vec{\mathbf{x}})$.

Here $\vec{\mathbf{x}}$ and $\vec{\mathbf{y}}$ only range over vectors for which $\mathbf{f}(\vec{\mathbf{x}})$ and $\mathbf{f}(\vec{\mathbf{y}})$ are of base type. $\vec{\mathbf{x}} \geq \vec{\mathbf{y}}$ means: For all i such that $\mathbf{x}_i \in \mathcal{M}$, $\mathbf{x}_i \geq \mathbf{y}_i$.

Lemma 1. *For any term M of the simply typed λ–calculus we have*

(i) $[\![M]\!]_U \in \mathcal{M}$ *for any monotonic environment U.*

(ii) $U \geq V \implies [\![M]\!]_U \geq [\![M]\!]_V$ *for monotonic environments U, V.*

Proof by simultaneous induction on M (standard). □

Definition. $\mathbf{f} >_{\text{mon}} \mathbf{g} \iff$ *for all* $\vec{\mathbf{x}} \in \mathcal{M} \cup \{0,1\}$, $\mathbf{f}(\vec{\mathbf{x}}) > \mathbf{g}(\vec{\mathbf{x}})$.

Remark. Gandy's definition of hereditarily monotonic functionals from [1] has the following form. For any type ρ he defines the set $\mathcal{G}_\rho \subseteq \mathcal{T}_\rho$ of hereditarily monotonic functionals of type ρ and simultaneously a relation $>_{\text{Gandy}}$ on $\mathcal{T}_\rho$ by

(i) $\mathbf{f} \in \mathcal{G} \iff$ for all $\vec{\mathbf{x}}, \vec{\mathbf{y}} \in \mathcal{G}$, if $\vec{\mathbf{x}} >_{\text{Gandy}} \vec{\mathbf{y}}$, then $\mathbf{f}(\vec{\mathbf{x}}) > \mathbf{f}(\vec{\mathbf{y}})$. Here $\vec{\mathbf{x}} >_{\text{Gandy}} \vec{\mathbf{y}}$ means that at least once we have $\mathbf{x}_i >_{\text{Gandy}} \mathbf{y}_i$ and otherwise $\mathbf{x}_j = \mathbf{y}_j$.

(ii) $\mathbf{f} >_{\text{Gandy}} \mathbf{g} \iff$ for all $\vec{\mathbf{x}} \in \mathcal{G}$, $\mathbf{f}(\vec{\mathbf{x}}) > \mathbf{g}(\vec{\mathbf{x}})$.

This definition is not well suited for termination proofs. Consider e.g. the term $xz(\lambda y\, 0)$, where x is interpreted by $\mathbf{x} \in \mathcal{G}$. Then also in the case $[\![M]\!] >_{\text{Gandy}} [\![N]\!]$ one cannot conclude $[\![xM(\lambda y\, 0)]\!] > [\![xN(\lambda y\, 0)]\!]$, since $[\![\lambda y\, 0]\!] \notin \mathcal{G}$. Hence Gandy in [1] had to restrict himself to λ–I–terms. As an alternative it is tempting to replace "for all $\vec{\mathbf{x}}, \vec{\mathbf{y}} \in \mathcal{G}$" in (i) by "for all $\vec{\mathbf{x}}, \vec{\mathbf{y}} \in \mathcal{M}$". Furthermore it turns out to be useful to add $\mathbf{f} \in \mathcal{M}$ to the right hand side of (i) and also $\mathbf{f} \geq \mathbf{g}$ to the right hand side of (ii). On pairs, the order $>_{\text{Gandy}}$ is defined pointwise in [1]. We propose a change to obtain a more well suited order for termination proofs. If in a pair $\langle M, N\rangle$, M rewrites to M', with $[\![M]\!] > [\![M']\!]$, one wants to conclude that the corresponding interpretation gets smaller. These considerations motivate the following definition:

Definition. *For any type ρ we define the set $\mathcal{S}_\rho \subseteq \mathcal{M}_\rho$ of strict functionals of type ρ and simultaneously a relation $>_{\text{str}}$ on $\mathcal{T}_\rho$.*

(i) $\mathbf{f} \in \mathcal{S} \iff \mathbf{f} \in \mathcal{M}$, *and for all* $\vec{\mathbf{x}}, \vec{\mathbf{y}} \in \mathcal{M} \cup \{0,1\}$, *if* $\vec{\mathbf{x}} >_{\text{str}} \vec{\mathbf{y}}$, *then* $\mathbf{f}(\vec{\mathbf{x}}) > \mathbf{f}(\vec{\mathbf{y}})$. *Here* $\vec{\mathbf{x}} >_{\text{str}} \vec{\mathbf{y}}$ *means that at least once we have* $\mathbf{x}_i >_{\text{str}} \mathbf{y}_i$ *and otherwise* $\mathbf{x}_j \geq \mathbf{y}_j$.

(ii) $\mathbf{f} >_{\text{str}} \mathbf{g} \iff \mathbf{f} \geq \mathbf{g}$ *and*

(a) (base type) $\mathbf{f} > \mathbf{g}$; *or*

(b) (arrow type) for all $\mathbf{x} \in \mathcal{S}$, $\mathbf{f}(\mathbf{x}) >_{\text{str}} \mathbf{g}(\mathbf{x})$; *or*

(c) (product type) $\mathbf{f}(0) >_{\text{str}} \mathbf{g}(0)$ *or* $\mathbf{f}(1) >_{\text{str}} \mathbf{g}(1)$.

Remark. In [7] a very similar modification of Gandy's definition is used. In a preliminary version, the requirement $\mathbf{f} \geq \mathbf{g}$ in (ii) was missing. For the examples considered in [7], which only concern rewrite rules for constants of level ≤ 2, this makes no difference. However, if one considers higher order rewrite rules like those for the primitive recursion operators in Gödel's T, then it is necessary to be able to infer $\mathbf{f} \geq \mathbf{g}$ from $\mathbf{f} >_{\text{str}} \mathbf{g}$. This property is not satisfied without this requirement. (For the proof consider two functionals $\mathbf{f}, \mathbf{g}$ of level 2 satisfying for all $\mathbf{x} \in \mathcal{S}$ the inequality $\mathbf{f}(\mathbf{x}) > \mathbf{g}(\mathbf{x})$. Now modify these functionals on the non–strict, but monotonic functions, e.g. by giving $\mathbf{f}$ on $[\![\lambda z\, 0]\!]$ the value 0 and $\mathbf{g}$ on $[\![\lambda z\, 0]\!]$ the value 1.)

From the definition it is clear that from $\mathbf{f} \in \mathcal{S}$ and $\mathbf{x} \in \mathcal{M} \cup \{0,1\}$ we can conclude $\mathbf{f}(\mathbf{x}) \in \mathcal{S}$. Furthermore from $\mathcal{S} \subseteq \mathcal{M}$ we get immediately $\mathbf{f} >_{\text{mon}} \mathbf{g} \implies \mathbf{f} >_{\text{str}} \mathbf{g}$.

Theorem. *Let M be a term in β normal form and $\Box \in \mathrm{FV}(M)$. Then for any strict environment U and all $\mathbf{f}, \mathbf{g} \in \mathcal{M}$*

$$\mathbf{f} >_{\mathrm{mon}} \mathbf{g} \implies [\![M]\!]_{U[\Box\mapsto\mathbf{f}]} >_{\mathrm{str}} [\![M]\!]_{U[\Box\mapsto\mathbf{g}]}.$$

Proof by induction on M. Let M be in long normal form. Let $\mathbf{f}, \mathbf{g} \in \mathcal{M}$ with $\mathbf{f} >_{\mathrm{mon}} \mathbf{g}$ be given. Then $\geq$ holds by Lemma 1(ii).

Case $\lambda\vec{x}.\Box\vec{M}$. Let $\vec{\mathbf{x}} \in \mathcal{S}$ and $V := U[\vec{x} \mapsto \vec{\mathbf{x}}]$. From $\mathbf{f} >_{\mathrm{mon}} \mathbf{g}$ we get

$$\mathbf{f}([\![\vec{M}]\!]_{V[\Box\mapsto\mathbf{f}]}) >_{\mathrm{mon}} \mathbf{g}([\![\vec{M}]\!]_{V[\Box\mapsto\mathbf{f}]})$$

and therefore also $>_{\mathrm{str}}$. Furthermore from $\mathbf{f} >_{\mathrm{mon}} \mathbf{g}$ we obtain $\mathbf{f} \geq \mathbf{g}$, hence $[\![\vec{M}]\!]_{V[\Box\mapsto\mathbf{f}]} \geq [\![\vec{M}]\!]_{V[\Box\mapsto\mathbf{g}]}$. Now $\mathbf{g}([\![\vec{M}]\!]_{V[\Box\mapsto\mathbf{f}]}) \geq \mathbf{g}([\![\vec{M}]\!]_{V[\Box\mapsto\mathbf{g}]})$ follows because $\mathbf{g} \in \mathcal{M}$.

Case $\lambda\vec{x}.y\vec{M}$ with $y \neq \Box$. Let $\vec{\mathbf{x}} \in \mathcal{S}$ and $V := U[\vec{x} \mapsto \vec{\mathbf{x}}]$. For any i with $\Box \in \mathrm{FV}(M_i)$ we have $[\![M_i]\!]_{V[\Box\mapsto\mathbf{f}]} >_{\mathrm{str}} [\![M_i]\!]_{V[\Box\mapsto\mathbf{g}]}$ by IH, hence $[\![\vec{M}]\!]_{V[\Box\mapsto\mathbf{f}]} >_{\mathrm{str}} [\![\vec{M}]\!]_{V[\Box\mapsto\mathbf{g}]}$. Since $V(y) \in \mathcal{S}$, we obtain $[\![y\vec{M}]\!]_{V[\Box\mapsto\mathbf{f}]} > [\![y\vec{M}]\!]_{V[\Box\mapsto\mathbf{g}]}$.

Case $\lambda\vec{x}.\langle M_0, M_1\rangle$. Then $\Box \in \mathrm{FV}(M_i)$ for some $i \in \{0,1\}$. Let $\vec{\mathbf{x}} \in \mathcal{S}$ and $V := U[\vec{x} \mapsto \vec{\mathbf{x}}]$. By IH $[\![M_i]\!]_{V[\Box\mapsto\mathbf{f}]} >_{\mathrm{str}} [\![M_i]\!]_{V[\Box\mapsto\mathbf{g}]}$ for this i. □

This theorem shows that the strict functionals form an interesting class. In the rest of this section we will explore the strict functionals and in the next section it will be shown how to use them in termination proofs. The first question is of course, whether there exist such functionals at all. To construct strict functionals, we surely need them on the base types. Hence we assume that for any tuple $\iota_1, \dots, \iota_n, \iota$ of base types we are given a strict function $+$ of type $\iota_1 \to \dots \to \iota_n \to \iota$ (written in infix notation, or as prefix $\sum$; we will write 0^ι for $+^\iota$.) Using this $+$, we simultaneously define special functionals $\mathbf{S}^\sigma$ (a *strict* functional of type σ for any σ) and $\mathbf{M}_\sigma$ (a *measure* functional of type $\sigma \to o$), where o is one of the base types. In this definition, $\vec{\mathbf{S}}^{\vec{\rho}}$ denotes $\mathbf{S}^{\rho_1}, \cdots, \mathbf{S}^{\rho_n}$, and $\vec{\mathbf{M}}(\vec{\mathbf{f}})$ is to be read as $\mathbf{M}(\mathbf{f}_{i_1}), \cdots, \mathbf{M}(\mathbf{f}_{i_k})$, where $\mathbf{f}_{i_1}, \cdots, \mathbf{f}_{i_k}$ are the proper arguments among $\vec{\mathbf{f}}$, i.e. not the 0 and 1 used for projections. These shortcuts will be used frequently. In the last equation $\mathbf{S}^\sigma(\vec{\mathbf{f}})$ is to be of base type.

Definition.

$$\begin{aligned}
\mathbf{M}_{\vec{\sigma}\to\iota}(\mathbf{f}) &:= +^{\iota\to o}(\mathbf{f}(\vec{\mathbf{S}}^{\vec{\sigma}}))\\
\mathbf{M}_{\vec{\sigma}\to\rho\times\tau}(\mathbf{f}) &:= \mathbf{M}_\rho(\mathbf{f}(\vec{\mathbf{S}}^{\vec{\sigma}},0)) + \mathbf{M}_\tau(\mathbf{f}(\vec{\mathbf{S}}^{\vec{\sigma}},1))\\
\mathbf{S}^\sigma(\vec{\mathbf{f}}) &:= \sum \vec{\mathbf{M}}(\vec{\mathbf{f}})
\end{aligned}$$

In examples, we assume that the $+^{\vec{\iota}\to\iota}$ are chosen in such a way that $0^{\iota_1} + \dots + 0^{\iota_n} = 0^\iota$ holds for any combination of base types and $+^{\iota\to\iota}$ is the identity. For instance, we may take 0 in $\mathbf{N}$ with usual order and addition, or else take the empty list in $\mathbf{N}^*$ and let $+$ be concatenation and $>$ be the comparison of lengths. Under these assumptions $\mathbf{M}_\sigma = \mathbf{S}^{\sigma\to o}$. By induction on the types one can see immediately that $\mathbf{M}(\mathbf{S}) = 0$.

Here are some examples:

$$\begin{aligned}
\mathbf{S}^\iota &= 0^\iota,\\
\mathbf{S}^{\iota\to\iota}(\mathbf{x}) &= \mathbf{x},\\
\mathbf{S}^{(\iota\to\iota)\to\iota}(\mathbf{f}) &= \mathbf{f}(0),\\
\mathbf{S}^{((\iota\to\iota)\to\iota\to\iota)\to(\iota\to\iota)\to\iota\to\iota}(\mathbf{F},\mathbf{f},\mathbf{x}) &= \mathbf{F}(\mathbf{S}^{\iota\to\iota},0) + \mathbf{f}(0) + \mathbf{x},\\
\mathbf{S}^{\iota\times\iota\to\iota}\langle\mathbf{x},\mathbf{y}\rangle &= \mathbf{x}+\mathbf{y},\\
\mathbf{S}^{\iota\times\iota\to\iota\times\iota}\langle\mathbf{x},\mathbf{y}\rangle &= \langle\mathbf{x}+\mathbf{y},\mathbf{x}+\mathbf{y}\rangle,\\
\mathbf{S}^{\sigma\times\tau} &= \langle\mathbf{S}^\sigma,\mathbf{S}^\tau\rangle.
\end{aligned}$$

Lemma 2. *For any type ρ, both $\mathbf{M}_\rho$ and $\mathbf{S}^\rho$ are strict.*

Proof by simultaneous induction on the type ρ. If ρ is a base type, then $\mathbf{S}^\iota = 0^\iota$ and $\mathbf{M}_\iota(n) = +^{\iota\to o}(n)$. Strictness is clear. So let ρ be some compound type.

Let $\vec{\mathbf{x}},\vec{\mathbf{y}} \in \mathcal{M}\cup\{0,1\}$ be given, with $\vec{\mathbf{x}} \geq \vec{\mathbf{y}}$ and $\mathbf{S}^\rho(\vec{\mathbf{x}})$ of base type. The $\vec{\mathbf{M}}$ in $\vec{\mathbf{M}}(\vec{\mathbf{x}})$ all have type smaller than ρ, so they are strict by IH, hence monotonic. This yields that $\vec{\mathbf{M}}(\vec{\mathbf{x}}) \geq \vec{\mathbf{M}}(\vec{\mathbf{y}})$, so also $\sum\vec{\mathbf{M}}(\vec{\mathbf{x}}) \geq \sum\vec{\mathbf{M}}(\vec{\mathbf{y}})$. This proves monotonicity of $\mathbf{S}^\rho$. Next, assume that $\vec{\mathbf{x}} >_{\mathrm{str}} \vec{\mathbf{y}}$ holds. Then $\geq$ holds, and for some i, $\mathbf{x}_i >_{\mathrm{str}} \mathbf{y}_i$. By IH $\mathbf{M}(\mathbf{x}_i) > \mathbf{M}(\mathbf{y}_i)$, so $\sum\vec{\mathbf{M}}(\vec{\mathbf{x}}) > \sum\vec{\mathbf{M}}(\vec{\mathbf{y}})$. This proves that $\mathbf{S}^\rho$ is strict.

Next we prove strictness of $\mathbf{M}_\rho$. Let $\rho = \vec{\rho}\to\tau$, with τ not an arrow type. Let $\mathbf{f},\mathbf{g} \in \mathcal{M}_\rho$ be given. Note that the $\vec{\mathbf{S}}^{\vec{\rho}}$ are strict by IH, hence they are monotonic too. So if $\mathbf{f} \geq \mathbf{g}$, then $\mathbf{f}(\vec{\mathbf{S}}^{\vec{\rho}}) \geq \mathbf{g}(\vec{\mathbf{S}}^{\vec{\rho}})$. Moreover, if $\mathbf{f} >_{\mathrm{str}} \mathbf{g}$ then $\mathbf{f}(\vec{\mathbf{S}}^{\vec{\rho}}) >_{\mathrm{str}} \mathbf{g}(\vec{\mathbf{S}}^{\vec{\rho}})$. In case τ is a base type, this proves both monotonicity and strictness of $\mathbf{M}_\rho$. Otherwise, $\tau = \tau_0\times\tau_1$, and we use that $\mathbf{M}_{\tau_0}$ and $\mathbf{M}_{\tau_1}$ are strict by IH, and hence monotonic too. The monotonicity of $\mathbf{M}_\rho$ then follows from monotonicity of the projections and $+$. For strictness, note that either $\mathbf{f}(\vec{\mathbf{S}},0) >_{\mathrm{str}} \mathbf{g}(\vec{\mathbf{S}},0)$ or $\mathbf{f}(\vec{\mathbf{S}},1) >_{\mathrm{str}} \mathbf{g}(\vec{\mathbf{S}},1)$. For the other component $\geq$ holds. Now strictness of $\mathbf{M}_\rho$ follows from strictness of the $\mathbf{M}_{\tau_i}$ and of $+^{o\to o\to o}$. □

The success of the method, to be developed in Section 2, depends on finding strict functionals. By now, we have only seen the $\mathbf{S}$ functionals as examples. The following lemma enables us to find a lot more strict functionals:

Lemma 3. *For any strict functional $\mathbf{G}$ and monotonic functional $\mathbf{H}$, the functional $\mathbf{F}$ defined by $\mathbf{F}(\vec{\mathbf{x}}) := \mathbf{G}(\vec{\mathbf{x}}) + \mathbf{H}(\vec{\mathbf{x}})$, is strict.*

Proof. Let $\vec{\mathbf{x}} >_{\mathrm{str}} \vec{\mathbf{y}}$ for some monotonic $\vec{\mathbf{x}}$ and $\vec{\mathbf{y}}$. Then $\mathbf{G}(\vec{\mathbf{x}}) > \mathbf{G}(\vec{\mathbf{y}})$ (by strictness of $\mathbf{G}$). By the definition of $>_{\mathrm{str}}$, we obtain $\vec{\mathbf{x}} \geq \vec{\mathbf{y}}$, hence by monotonicity of $\mathbf{H}$, $\mathbf{H}(\vec{\mathbf{x}}) \geq \mathbf{H}(\vec{\mathbf{y}})$. This yields $\mathbf{F}(\vec{\mathbf{x}}) > \mathbf{F}(\vec{\mathbf{y}})$. □

Note that this result doesn't hold if one drops the requirement $\mathbf{f} \geq \mathbf{g}$ in the definition of $\mathbf{f} >_{\mathrm{str}} \mathbf{g}$. So this addition is motivated by the fact that it enables us to find more strict functionals easily. We proceed with showing that one cannot get smaller strict functionals.

Lemma 4. *Consider the special case that the only ground domain is* N *with usual ordering and addition. Then for any* $\mathbf{f} \in \mathcal{S}_\sigma$, $\mathbf{f} \geq \mathbf{S}^\sigma$.

Proof. We use an operation $\mathbf{L}_\sigma$ (*lower* by 1) on functionals, defined by induction on the type. $\mathbf{L}_\sigma$ takes two arguments, a functional $\mathbf{f}$ of type σ and a sequence $\vec{\mathbf{a}}$ in $\mathcal{M} \cup \{0,1\}$, such that $\mathbf{f}(\vec{\mathbf{a}})$ is of base type. The result of $\mathbf{L}_\sigma(\mathbf{f}, \vec{\mathbf{a}})$ will be of type σ. We will write $\mathbf{L}_{\vec{\mathbf{a}}}(\mathbf{f})$ for $\mathbf{L}_\sigma(\mathbf{f}, \vec{\mathbf{a}})$.

$$\begin{aligned}
\mathbf{L}_\epsilon(n) &:= \begin{cases} 0 & \text{if } n = 0 \\ n-1 & \text{otherwise} \end{cases} \\
\mathbf{L}_{\langle \mathbf{a}, \vec{\mathbf{a}} \rangle}(\mathbf{f}, \mathbf{x}) &:= \mathbf{L}_{\vec{\mathbf{a}}}(\mathbf{f}(\mathbf{x})) \\
\mathbf{L}_{\langle 0, \vec{\mathbf{a}} \rangle}(\langle \mathbf{x}, \mathbf{y} \rangle) &:= \langle \mathbf{L}_{\vec{\mathbf{a}}}(\mathbf{x}), \mathbf{y} \rangle \\
\mathbf{L}_{\langle 1, \vec{\mathbf{a}} \rangle}(\langle \mathbf{x}, \mathbf{y} \rangle) &:= \langle \mathbf{x}, \mathbf{L}_{\vec{\mathbf{a}}}(\mathbf{y}) \rangle.
\end{aligned}$$

Note that the $\vec{\mathbf{a}}$ is only used to know which of the components of a product to lower. With induction on the types, it is easy to see that for any $\vec{\mathbf{a}}$ and monotonic $\mathbf{x}$,

(i) $\mathbf{L}_{\vec{\mathbf{a}}}$ is monotonic, and

(ii) $\mathbf{M}(\mathbf{L}_{\vec{\mathbf{a}}}(\mathbf{x})) = \mathbf{L}_\epsilon(\mathbf{M}(\mathbf{x}))$.

We now prove the lemma by a main induction on σ. For the base type, we have to show that $m \geq 0$ for $m \in$ N, which clearly holds. If $\sigma = \rho \times \tau$, observe that by IH, for any strict pair $\langle \mathbf{x}, \mathbf{y} \rangle$, $\langle \mathbf{x}, \mathbf{y} \rangle \geq \langle \mathbf{S}^\rho, \mathbf{S}^\tau \rangle$, and that the latter equals $\mathbf{S}^\sigma$. If $\sigma = \rho \to \tau$, we have to prove that for monotonic $\mathbf{x}$, $\mathbf{f}(\mathbf{x}) \geq \mathbf{S}^\sigma(\mathbf{x})$. This is proved with induction on $\mathbf{M}(\mathbf{x})$.

If $\mathbf{M}(\mathbf{x}) = 0$, we use that $\mathbf{f}(\mathbf{x})$ is strict, hence $\mathbf{f}(\mathbf{x}) \geq \mathbf{S}^\tau$ (main IH). Now for monotonic $\vec{\mathbf{x}}$ we obtain $\mathbf{f}(\mathbf{x}, \vec{\mathbf{x}}) \geq \mathbf{S}^\tau(\vec{\mathbf{x}}) = \mathbf{M}(\mathbf{x}) + \mathbf{S}^\tau(\vec{\mathbf{x}}) = \mathbf{S}^\sigma(\mathbf{x}, \vec{\mathbf{x}})$.

If $\mathbf{M}(\mathbf{x}) = n+1$, we can find $\vec{\mathbf{a}}$ with elements among $\mathbf{S}$ and 0 and 1, such that $\mathbf{x}(\vec{\mathbf{a}}) \geq 1$. Define $\mathbf{y} := \mathbf{L}_{\vec{\mathbf{a}}}(\mathbf{x})$. By (i) above, $\mathbf{y}$ is monotonic. We first show, that $\mathbf{x} >_{\mathtt{str}} \mathbf{y}$. It suffices to show that $\mathbf{x}(\vec{\mathbf{z}}) > \mathbf{y}(\vec{\mathbf{z}})$, where $\vec{\mathbf{z}}$ is obtained from $\vec{\mathbf{a}}$ by replacing the real arguments by arbitrary strict functionals. (i.e. the 0 and 1s for projections are not replaced.) By IH, we have that $\vec{\mathbf{z}} \geq \vec{\mathbf{S}}$, hence $\mathbf{x}(\vec{\mathbf{z}}) \geq \mathbf{x}(\vec{\mathbf{a}}) \geq 1$. Hence $\mathbf{y}(\vec{\mathbf{z}}) = \mathbf{x}(\vec{\mathbf{z}}) - 1$.

Now we show that for monotonic $\vec{\mathbf{x}}$, $\mathbf{f}(\mathbf{x}, \vec{\mathbf{x}}) \geq \mathbf{S}^\sigma(\mathbf{x}, \vec{\mathbf{x}})$. Note that $\mathbf{f}(\mathbf{x}, \vec{\mathbf{x}}) > \mathbf{f}(\mathbf{y}, \vec{\mathbf{x}})$, because $\mathbf{f}$ is strict. By (ii) above $\mathbf{M}(\mathbf{y}) = n$. Hence we can apply the inner IH, and obtain $\mathbf{f}(\mathbf{y}, \vec{\mathbf{x}}) \geq \mathbf{S}^\sigma(\mathbf{y}, \vec{\mathbf{x}}) = n + \mathbf{S}(\vec{\mathbf{x}})$, hence $\mathbf{f}(\mathbf{x}, \vec{\mathbf{x}}) \geq n + 1 + \mathbf{S}(\vec{\mathbf{x}}) = \mathbf{S}^\sigma(\mathbf{x}, \vec{\mathbf{x}})$. □

So we have found out that $\mathbf{S}(\vec{\mathbf{x}}) + \mathbf{H}(\vec{\mathbf{x}})$ is strict in $\vec{\mathbf{x}}$ for monotonic $\mathbf{H}$ and that $\mathbf{S}$ is a minimal strict one. One might wonder if all strict functionals have the form $\mathbf{S}$ + monotonic. However, this is not the case. Consider $\mathbf{F}(\mathbf{f}) := \mathbf{f}(1)$, of type $(o \to o) \to o$. This is clearly strict. But the difference between $\mathbf{F}$ and $\mathbf{S}$ is not monotonic: Put $\mathbf{f}(n) := \max(1, n)$ and $\mathbf{g}(n) := n$. Then $\mathbf{f}$ and $\mathbf{g}$ are both monotonic, and $\mathbf{f} \geq \mathbf{g}$. But $\mathbf{g}(1) - \mathbf{g}(0) > \mathbf{f}(1) - \mathbf{f}(0)$.

2. Termination

To be able to apply the theorem above to prove termination we of course need to know that $>_{\text{str}}$ is a well-founded partial ordering on any $\mathcal{T}_\rho$. This can be proved if we assume that for the base types ι we are given domains $\mathcal{T}_\iota$ together with well-founded (partial) orderings $>_\iota$.

Proposition. *$>_{\text{str}}$ is well-founded on any $\mathcal{T}_\rho$.*

Proof. Let $(\mathbf{x}_i)_{i\in\mathbf{N}}$ of type ρ be given. Consider $(\mathbf{M}_\rho(\mathbf{x}_i))_{i\in\mathbf{N}}$. □

Following [7] we define a higher order term rewrite system (HRS) to be given by rules $L \mapsto R$ with closed terms L, R of the same type ρ. Then $M_1 \to M_2$ (M_1 rewrites to M_2) is defined to mean that we have a β–normal term M with $\Box \in \mathrm{FV}(M)$ such that for some rule $L \mapsto R$

$$M_1 = M[L/\Box]{\downarrow_\beta} \qquad \text{and} \qquad M_2 = M[R/\Box]{\downarrow_\beta}.$$

Here $N{\downarrow_\beta}$ denotes the β–normal form of the term N; β is defined as usual for arrow types, and for product types by the two rules $\pi_i\langle M_0, M_1\rangle \mapsto M_i$.

Note that we only require from L, R that they are closed terms of the same type. Closedness is not a restriction, but it avoids substitutions in the definition of a rewrite step. If L and R are not closed, one can simulate the step $M[L^\sigma] \to M[R^\sigma]$ by $M[(\lambda\vec{x}.L)\vec{x}^\sigma] \to M[(\lambda\vec{x}.R)\vec{x}^\sigma]$, where $\vec{x}$ is the list of variables occurring in l or r. Hence this notion of an HRS is quite liberal (and e.g. strictly includes the one given by Nipkow in [4]). The reason for this liberality is of course that termination results get stronger that way. See [8] for a comparison with other higher-order rewrite formats.

Example. Consider the rule $\lambda x.x + x \mapsto \lambda x\, x$. Then

$$\lambda u, v.c(\lambda w.wu + wu)(v + v) \to \lambda u, v.c(\lambda w.wu)v$$

using the term $M := \lambda u, v.c(\lambda w.\Box(wu))(\Box v)$.

Now we obtain as in [7] the following method to prove termination of higher order rewrite systems.

(1) For the base types ι choose domains $\mathcal{T}_\iota$ together with well-founded (partial) orders $>_\iota$. Furthermore find for any tuple $\iota_1, \dots, \iota_n, \iota$ of base types a strict function $+$ of type $\iota_1 \to \dots \to \iota_n \to \iota$.

(2) Find an appropriate strict interpretation of the constants.

(3) For any rule $L \mapsto R$ of the higher order rewrite system show that

$$[\![L]\!] >_{\text{mon}} [\![R]\!].$$

Theorem. *Any HRS satisfying (1)–(3) is terminating.*

Proof. Assume that we have $(M_i)_{i\in\mathbb{N}}$ such that $M_i \to M_{i+1}$ for all $i \in \mathbb{N}$. Let U be a strict interpretation. Then we obtain

$$\begin{array}{rcll} [\![M_i]\!]_U & = & [\![M[L/\Box]]\!]_U & \\ & = & [\![M]\!]_{U[\Box\mapsto[\![L]\!]]} & \\ & >_{\mathrm{str}} & [\![M]\!]_{U[\Box\mapsto[\![R]\!]]} & \text{since } [\![L]\!] >_{\mathrm{mon}} [\![R]\!] \\ & = & [\![M[R/\Box]]\!]_U & \\ & = & [\![M_{i+1}]\!]_U. & \end{array}$$

This contradicts the well-foundedness of $>_{\mathrm{str}}$. □

In Section 3, termination of Gödel's T is proved using this method. Section 4 contains a termination proof for the proper reductions and permutative conversions on derivation terms of first order logic. We first treat a well known small example, to illustrate the use of the proposed strategy to prove termination of HRSs.

Consider terms built up from the constants

$$\begin{array}{llll} \mathrm{nil} & : o & \mathrm{append} & : o \to o \to o \\ \mathrm{cons} & : o \to o \to o & \mathrm{map} & : (o \to o) \to o \to o. \end{array}$$

The types are chosen such that e.g. $\mathrm{map}(\lambda x\, \mathrm{append}(x,x), \ell)$ is well typed. Terms of type o represent finite lists of lists. The functions map and append are defined via the following rewrite rules (for readability, we drop the initial λs):

$$\begin{array}{rcll} \mathrm{append}(\mathrm{nil},\ell) & \mapsto & \ell & \text{(i)} \\ \mathrm{append}(\mathrm{cons}(k,\ell),m) & \mapsto & \mathrm{cons}(k,\mathrm{append}(\ell,m)) & \text{(ii)} \\ \mathrm{map}(f,\mathrm{nil}) & \mapsto & \mathrm{nil} & \text{(iii)} \\ \mathrm{map}(f,\mathrm{cons}(k,\ell)) & \mapsto & \mathrm{cons}(f(k),\mathrm{map}(f,\ell)) & \text{(iv)} \\ \mathrm{append}(\mathrm{append}(k,\ell),m) & \mapsto & \mathrm{append}(k,\mathrm{append}(\ell,m)) & \text{(v)} \\ \mathrm{map}(f,\mathrm{append}(\ell,k)) & \mapsto & \mathrm{append}(\mathrm{map}(f,\ell),\mathrm{map}(f,k)) & \text{(vi)} \end{array}$$

To prove termination, we have to satisfy (1), (2) and (3) above. For the ground domain, we choose $\mathbb{N}$, with the usual order and addition. The interpretation of the constants is specified in the following way:

$$\begin{array}{rlrl} [\![\mathrm{nil}]\!] & := 1 & & \\ [\![\mathrm{cons}]\!](m,n) & := m+n+1 & [\![\mathrm{map}]\!](f,n) & := \sum_{i=0}^{n} f(i) + 3n + 1 \\ [\![\mathrm{append}]\!](m,n) & := 2m+n+2 & & \end{array}$$

The interpretations of nil, cons and append are obviously strict. Strictness of $[\![\mathrm{map}]\!]$ follows e.g. by Lemma 3, if we write its definition as

$$(f(0)+n) + \Big(\sum_{i=1}^{n} f(i) + 2n + 1\Big).$$

Hence (1) and (2) are fulfilled. We still have to check (3). In the sequel k, ℓ, m, f are arbitrary values for the corresponding variables. Note that f ranges over monotonic functionals. For rule (v) e.g. the check boils down to the true inequality $2 \cdot (2\ell + k + 2) + m + 2 > 2\ell + (2k + m + 2) + 2$. We don't present all calculations here, but let us yet verify the most difficult one, rule (vi):

$$\begin{aligned}
& [\![\mathrm{map}(f, \mathrm{append}(\ell, k))]\!] \\
= \; & \sum_{i=0}^{2\ell+k+2} f(i) + 3 \cdot (2\ell + k + 2) + 1 \\
= \; & \sum_{i=0}^{\ell} f(i) + \sum_{i=\ell+1}^{2\ell+1} f(i) + \sum_{i=2\ell+2}^{2\ell+k+2} f(i) + 6\ell + 3k + 7 \\
> \; & \sum_{i=0}^{\ell} f(i) + \sum_{i=0}^{\ell} f(i) + \sum_{i=0}^{k} f(i) + 6\ell + 3k + 5 \quad \text{because } f \text{ is monotonic} \\
= \; & 2 \cdot (\sum_{i=0}^{\ell} f(i) + 3\ell + 1) + (\sum_{i=0}^{k} f(i) + 3k + 1) + 2 \\
= \; & [\![\mathrm{append}(\mathrm{map}(f, \ell), \mathrm{map}(f, k))]\!]
\end{aligned}$$

For all rules, this relation between left- and right hand side hold. Therefore the HRS under consideration is terminating.

3. Example: Higher order primitive recursion

We now apply this method to prove termination for the canonical rules associated with higher order primitive recursion from Gödel's T. These are based on constants Rec of type $\rho \to (o \to \rho \to \rho) \to o \to \rho$, for any type ρ.

$$\begin{aligned}
\mathrm{Rec}(g, h, 0) &\mapsto g, \\
\mathrm{Rec}(g, h, s(x)) &\mapsto h(x, \mathrm{Rec}(g, h, x)).
\end{aligned}$$

As ground domain we choose N with the usual addition + and the usual ordering >. Then (1) is clearly satisfied. For (2) we choose a strict interpretation of the constants Rec, as follows.

$$\begin{aligned}
[\![\mathrm{Rec}]\!](\mathbf{g}, \mathbf{h}, 0, \vec{\mathbf{x}}) &= \mathbf{g}(\vec{\mathbf{x}}) + \mathbf{S}(\mathbf{g}, \mathbf{h}, \vec{\mathbf{x}}) + 1, \\
[\![\mathrm{Rec}]\!](\mathbf{g}, \mathbf{h}, n+1, \vec{\mathbf{x}}) &= \mathbf{h}(n, [\![\mathrm{Rec}]\!](\mathbf{g}, \mathbf{h}, n), \vec{\mathbf{x}}) + [\![\mathrm{Rec}]\!](\mathbf{g}, \mathbf{h}, n, \vec{\mathbf{x}}) + 1.
\end{aligned}$$

The strictness of $[\![\mathrm{Rec}]\!]$ can be seen as follows.

First we show that $[\![\mathrm{Rec}]\!](\mathbf{g}, \mathbf{h}, n)$ for $\mathbf{g}, \mathbf{h} \in \mathcal{M}$ and any n is monotonic, by induction on n. *Case* 0. $[\![\mathrm{Rec}]\!](\mathbf{g}, \mathbf{h}, 0)$ is monotonic, since $\mathbf{g}$ is. *Case* $n+1$. $[\![\mathrm{Rec}]\!](\mathbf{g}, \mathbf{h}, n+1)$ is monotonic, since $[\![\mathrm{Rec}]\!](\mathbf{g}, \mathbf{h}, n)$ and $\mathbf{h}$ are monotonic.

Hence we get $[\![\mathrm{Rec}]\!] \in \mathcal{M}$ as follows. Let $\mathbf{g}, \mathbf{h}, n, \vec{\mathbf{x}} \in \mathcal{M}$. It suffices to show that by decreasing these arguments in $\mathcal{M}$ in the sense of $\geq$ the value $[\![\mathrm{Rec}]\!](\mathbf{g}, \mathbf{h}, n, \vec{\mathbf{x}})$ will get at most smaller. This clearly holds if n is decreased. For the other possibilities we fix n. In the case $n = 0$ the claim is obvious, in case $n + 1$ we need the monotonicity of $[\![\mathrm{Rec}]\!](\mathbf{g}, \mathbf{h}, n)$.

Now we can show that $[\![\mathrm{Rec}]\!]$ is strict. $[\![\mathrm{Rec}]\!] \in \mathcal{M}$ has already been proved. Let $\mathbf{g}, \mathbf{h}, n, \vec{\mathbf{x}} \in \mathcal{M}$. It remains to show that by decreasing exactly one of these arguments in $\mathcal{M}$ in the sense of $>_{\mathsf{str}}$ the value $[\![\mathrm{Rec}]\!](\mathbf{g}, \mathbf{h}, n, \vec{\mathbf{x}})$ gets strictly smaller. This again clearly holds if n is decreased. For the other possibilities we fix n and use Lemma 3:

First note that $[\![\mathrm{Rec}]\!](\mathbf{g}, \mathbf{h}, n, \vec{\mathbf{x}}) = \mathbf{S}(\mathbf{g}, \mathbf{h}, \vec{\mathbf{x}}) + \mathbf{H}(\mathbf{g}, \mathbf{h}, n, \vec{\mathbf{x}})$, where $\mathbf{H}$ is defined by

$$\begin{aligned} \mathbf{H}(\mathbf{g}, \mathbf{h}, 0, \vec{\mathbf{x}}) &= \mathbf{g}(\vec{\mathbf{x}}) + 1, \\ \mathbf{H}(\mathbf{g}, \mathbf{h}, n+1, \vec{\mathbf{x}}) &= \mathbf{h}(n, \mathbf{S}(\mathbf{g}, \mathbf{h}) \oplus \mathbf{H}(\mathbf{g}, \mathbf{h}, n), \vec{\mathbf{x}}) + \mathbf{H}(\mathbf{g}, \mathbf{h}, n, \vec{\mathbf{x}}) + 1; \end{aligned}$$

here we have written $\mathbf{x} \oplus \mathbf{y}$ for the functional which takes the value $\mathbf{x}(\vec{\mathbf{z}}) + \mathbf{y}(\vec{\mathbf{z}})$ on $\vec{\mathbf{z}}$. This can be proved easily by induction on n. Since $\mathbf{H} \in \mathcal{M}$ can be proved just as we proved $[\![\mathrm{Rec}]\!] \in \mathcal{M}$ above, it follows from Lemma 3 that $[\![\mathrm{Rec}]\!](\mathbf{g}, \mathbf{h}, n, \vec{\mathbf{x}})$ is strict for fixed n.

For the proof of (3) let us first consider the rule $\mathrm{Rec}(g, h, 0) \mapsto g$. We have to show that for monotonic $\mathbf{g}, \mathbf{h}, \vec{\mathbf{x}}$ we have

$$[\![\mathrm{Rec}]\!](\mathbf{g}, \mathbf{h}, 0, \vec{\mathbf{x}}) > \mathbf{g}(\vec{\mathbf{x}}).$$

This holds because of the summand 1 in the first defining equation for $[\![\mathrm{Rec}]\!]$. For the rule $\mathrm{Rec}(g, h, s(x)) \mapsto h(x, \mathrm{Rec}(g, h, x))$ we have to show that for monotonic $\mathbf{g}, \mathbf{h}, \vec{\mathbf{x}}$ we have

$$[\![\mathrm{Rec}]\!](\mathbf{g}, \mathbf{h}, n+1, \vec{\mathbf{x}}) > \mathbf{h}(n, [\![\mathrm{Rec}]\!](\mathbf{g}, \mathbf{h}, n), \vec{\mathbf{x}}).$$

This clearly holds because of the summand 1 in the second equation of the definition for $[\![\mathrm{Rec}]\!]$.

4. Example: Permutative Conversions

The next example comes from proof theory in the style of Prawitz. In [5] several reductions are given, to bring proofs into a certain normal form. These are divided in *proper reductions* and *permutative conversions.* Strong normalization is then proved via a refined notion of strong computability, *strong validity.* In [1] also examples taken from proof theory occur. There a normalization proof is given via hereditarily monotonic functionals, but the permutative conversions are not dealt with. We also refer to [2] for another adaptation of Gandy's approach, which can be extended to the full calculus including permutative conversions (See [2, Exc. 2.C.10]). Instead of bounding reduction lengths by functionals, Girard uses the length of a specific reduction path, given by a weak normalization theorem for the full calculus.

We present a termination proof for the whole calculus, including the permutative conversions. However, for simplicity we don't include disjunction. We first reduce the calculus with derivation terms to an HRS. Termination of this HRS is proved using the method of Section 2. The translation to an HRS is such that termination of the derivation terms immediately follows.

Definition. *Derivation terms are defined simultaneously with the set of free assumption variables (FA) occurring in them. We use A, B, C for formulae; d, e, f for derivation terms; r, s for object terms; x, y for object variables and u, v for assumption variables; i ranges over 0, 1.*

u^A	$(\lambda x\, d^A)^{\forall x A}$, provided $x \notin \mathrm{FV}(B)$ for any $u^B \in \mathrm{FA}(d)$
$(\lambda u^A\, d^B)^{A \to B}$	$(d^{\forall x A(x)} r)^{A(r)}$
$(d^{A \to B} e^A)^B$	$\langle r, d^{A(r)} \rangle^{\exists x A(x)}$
$\langle d^A, e^B \rangle^{A \wedge B}$	$(\varepsilon x u^A . d^{\exists x A} e^B)^B$, provided $x \notin \mathrm{FV}(B)$ and $x \notin \mathrm{FV}(C)$ for any $v^C \in \mathrm{FA}(e) \setminus \{u\}$.
$\pi_i(d^{A_0 \wedge A_1})^{A_i}$	

We define $\mathrm{FA}(\varepsilon x u.de) := \mathrm{FA}(d) \cup (\mathrm{FA}(e) \setminus \{u\})$. *In the other cases the set of free assumption variables is defined as usual.*

The following conversion rules are taken from [5]. The first four are the *proper reductions*, the last four are called *permutative conversions*. Again i ranges over 0, 1.

$$
\begin{array}{rlrl}
(\lambda u\, d)e & \mapsto d[u := e] & (\varepsilon x u.de)f & \mapsto \varepsilon x u.d\,(ef) \\
\pi_i\langle d_0, d_1\rangle & \mapsto d_i & \pi_i(\varepsilon x u.de) & \mapsto \varepsilon x u.d\,\pi_i(e) \\
(\lambda x\, d)r & \mapsto d[x := r] & (\varepsilon x u.de)r & \mapsto \varepsilon x u.d\,(er) \\
\varepsilon x u.(\langle r, d\rangle e) & \mapsto e[x, u := r, d] & \varepsilon x u.(\varepsilon y v.de)f & \mapsto \varepsilon x u.d\,\varepsilon y v.ef
\end{array}
$$

To translate this calculus into an HRS, we first have to transform formulae into types. This is done by removing the dependencies on object terms, also called *collapsing*. This technique is also used in [6, p. 560]. Collapsing A will be denoted by A^*. In the following definition, P is a predicate symbol.

$$
\begin{array}{rlrl}
P(\vec{t})^* & = o & & \\
(A \to B)^* & = A^* \to B^* & (\exists x A)^* & = o \times A^* \\
(A \wedge B)^* & = A^* \times B^* & (\forall x A)^* & = o \to A^*
\end{array}
$$

Clearly, A^* is a type for any formula A. The difference between implication and quantification disappears. Existential quantifiers and conjunctions are translated into product types.

The derivation terms are translated too. We introduce a new constant $\exists^-$ to model the ε-construct. In the definition of a rewrite step, β-normalization is performed implicitly. To avoid these implicit steps, we introduce another constant I, to block the β-redexes. So for any type σ (and τ) we have the following constants, which make the signature of the HRS we are constructing:

$$
I_\sigma \;: \sigma \to \sigma \qquad \exists^-_{\sigma,\tau} \;: o \times \sigma \to (o \to \sigma \to \tau) \to \tau
$$

To describe the translation precisely, we extend the collapse function on derivation terms:

$$
\begin{aligned}
(u^A)^* &= u^{A^*} \\
(\lambda u^A\, d)^* &= \lambda u^{A^*}\, d^* \\
\langle d, e\rangle^* &= \langle d^*, e^*\rangle \\
(\lambda x\, d)^* &= \lambda x^o\, d^* \\
\langle r, d\rangle^* &= \langle r, d^*\rangle
\end{aligned}
\qquad
\begin{aligned}
(d^{A\to B} e)^* &= I_{A^*\to B^*}(d^*, e^*) \\
\pi_i(d^{A\wedge B})^* &= I_{A^*\wedge B^*}(d^*, i) \\
(d^{\forall x A} r)^* &= I_{o\to A^*}(d^*, r) \\
(\varepsilon x u^A.d e^B)^* &= \exists^-_{A^*,B^*}(d^*, \lambda x^o u^{A^*}.e^*)
\end{aligned}
$$

Clearly $(d^A)^*$ gives a term of type A^* for any derivation term d. Due to the blocking I, d^* cannot contain subterms MN, with M an assumption variable, an abstraction or a pair. So d^* is in β-normal form, even after substituting β-normal terms for free assumption variables. Furthermore, it is easy to see that $(d[u := e])^* = d^*[u := e^*]$.

Finally, we present the rewrite rules of the HRS. These are all well typed instances of the following schemata; i ranges over 0, 1.

$$
\begin{aligned}
I_\sigma(x) &\mapsto x & \text{(i)} \\
\exists^-_{\sigma,\tau}(\langle r, d\rangle, e) &\mapsto e(r, d) & \text{(ii)} \\
I_{\sigma\to\tau}(\exists^-_{\rho,\sigma\to\tau}(d, e), f) &\mapsto \exists^-_{\rho,\tau}(d, \lambda x^o\, u^\rho.I_{\sigma\to\tau}(e(x, u), f)) & \text{(iii)} \\
\pi_i(I_{\sigma_0\times\sigma_1}(\exists^-_{\rho,\sigma_0\times\sigma_1}(d, e))) &\mapsto \exists^-_{\rho,\sigma_i}(d, \lambda x^o\, u^\rho.\pi_i(I_{\sigma_0\times\sigma_1}(e(x, u)))) & \text{(iv)} \\
\exists^-_{\sigma,\tau}(\exists^-_{\rho,o\times\sigma}(d, e), f) &\mapsto \exists^-_{\rho,\tau}(d, \lambda x^o\, u^\rho.\exists^-_{\sigma,\tau}(e(x, u), f)) & \text{(v)}
\end{aligned}
$$

It is not difficult to check that if $d \to e$ for derivation terms d and e, then also $d^* \to e^*$ with the rules just described. The first rule deals with proper reductions for $\to$, $\wedge$ and $\forall$; the second with the proper $\exists$-reduction. The third takes care of permutative conversions with $\to$ and $\forall$, the fourth with $\wedge$ and the last rule deals with the permutative conversion for $\exists$. We give as an example the proper $\to$-reduction. Consider the rewrite step $(\lambda u\, d)e \to d[u := e]$. The first derivation term translates to $I(\lambda u\, d^*, e^*)$. Now rule (i) is applicable. Literal replacement yields $(\lambda u\, d^*)e^*$, which has to be rewritten to β-normal form, due to the definition of a rewrite step. This normal form is $d^*[u := e^*]$, which is exactly the translation of the second derivation term.

Next we prove termination of the HRS, by carrying out the strategy of Section 2. As domain we (again) choose N (with standard order and addition). The interpretation of I is defined by

$$[\![I]\!](\mathbf{f}, \vec{\mathbf{z}}) := \mathbf{S}(\mathbf{f}, \vec{\mathbf{z}}) + \mathbf{f}(\vec{\mathbf{z}}) + 1.$$

This is strict by Lemma 3, and clearly $[\![I]\!](\mathbf{x}) >_{\text{mon}} \mathbf{x}$ for any monotonic $\mathbf{x}$. This already proves termination of the proper reduction rules for $\to$, $\wedge$ and $\forall$ and in particular of the simply typed lambda calculus with products. (Note however, that we used the unique β-normal form of simply typed terms. In fact, weak normalization suffices at meta-level.)

Due to the presence of the permutative conversions, it is more difficult to find a well-suited interpretation of $\exists^-$. We first need auxiliary functionals $\mathbf{A}_\sigma$ of type $\sigma \to \sigma$, which calculate the price of repeated $\to$ and $\times$-eliminations. Here the value of the blocking constant has to be taken into account. This leads to the following definition:

$$\begin{aligned}
\mathbf{A}_o(n) &:= n+1, \\
\mathbf{A}_{\sigma\to\tau}(\mathbf{f},\mathbf{x}) &:= \mathbf{A}_\tau([\![I]\!](\mathbf{f},\mathbf{x})), \\
\mathbf{A}_{\rho_0\times\rho_1}(\mathbf{f},i) &:= \mathbf{A}_{\rho_i}([\![I]\!](\mathbf{f},i)), \text{ for } i=0,1.
\end{aligned}$$

With induction on the type and using strictness of $[\![I]\!]$, one easily checks that $\mathbf{A}$ is strict. Also $\mathbf{A}(\mathbf{x}) >_{\mathrm{mon}} \mathbf{x}$ can be proved with induction. Let $\mathbf{A}^n(\mathbf{x})$ denote the n–fold application of $\mathbf{A}$ on $\mathbf{x}$. We write $\mathbf{x} \oplus \mathbf{y}$ for the functional which takes the value $\mathbf{x}(\vec{\mathbf{z}}) + \mathbf{y}(\vec{\mathbf{z}})$ on $\vec{\mathbf{z}}$. Now we can define

$$[\![\exists^-_{\sigma,\tau}]\!](\mathbf{d},\mathbf{e}) = \mathbf{A}_\tau^{2^{\mathbf{S}(\mathbf{d})}}(\mathbf{e}(\pi_0(\mathbf{d}),\mathbf{S}^\sigma \oplus \pi_1(\mathbf{d}))).$$

Let us first explain the intuition behind this interpretation. Due to the β-rule for $\exists^-$, we need a subterm $\mathbf{e}(\pi_0(\mathbf{d}),\pi_1(\mathbf{d}))$. The summand $\mathbf{S}\oplus$ is added to achieve strictness in $\mathbf{e}$. With a permutative conversion, the second argument of the $\exists^-$ gets bigger. After an application of rule (iii), the argument f appears inside the $\exists^-$. Note however, that the type of the involved $\exists^-$ goes down. So the value of an $\exists^-$ of higher type has to count for the value of f, which is still raised by the value of the blocking I. This explains the occurrence of $\mathbf{A}$ (which is defined by induction on the types). The same intuition applies to rule (iv). The last permutative conversion is still more involved. Here the type doesn't go down. The only thing which goes down is the left argument of the $\exists^-$–symbols involved. So the value of $\exists^-$ has to weigh its first argument rather high, to compensate for the increasing second argument. This explains the $2^{\mathbf{S}(\mathbf{d})}$ in the previous definition.

Monotonicity of $[\![\exists^-]\!]$ follows from monotonicity of $\mathbf{A}$. Next strictness is proved. Let $\mathbf{e},\mathbf{f},\mathbf{x},\mathbf{y}$ be monotonic. If $\mathbf{x} >_{\mathrm{str}} \mathbf{y}$, then by monotonicity of $\mathbf{e}$, $\mathbf{e}(\pi_0(\mathbf{x}),\mathbf{S}\oplus\pi_1(\mathbf{x})) \geq \mathbf{e}(\pi_0(\mathbf{y}),\mathbf{S}\oplus\pi_1(\mathbf{y}))$. Furthermore $2^{\mathbf{S}(\mathbf{x})} > 2^{\mathbf{S}(\mathbf{y})}$. Because $\mathbf{A}(\mathbf{x}) >_{\mathrm{mon}} \mathbf{x}$ for all $\mathbf{x}$, it follows that $[\![\exists^-]\!](\mathbf{x},\mathbf{e}) >_{\mathrm{mon}} [\![\exists^-]\!](\mathbf{y},\mathbf{e})$. This proves strictness in the first argument. Next, assume that $\mathbf{e} >_{\mathrm{str}} \mathbf{f}$. Note that both $\pi_0(\mathbf{x})$ and $\mathbf{S}\oplus\pi_1(\mathbf{x})$ are strict (the first is of base type, the second by Lemma 3). Hence $\mathbf{e}(\pi_0(\mathbf{x}),\mathbf{S}\oplus\pi_1(\mathbf{x})) >_{\mathrm{str}} \mathbf{f}(\pi_0(\mathbf{x}),\mathbf{S}\oplus\pi_1(\mathbf{x}))$. Now $[\![\exists^-]\!](\mathbf{x},\mathbf{e}) >_{\mathrm{mon}} [\![\exists^-]\!](\mathbf{x},\mathbf{f})$ follows from strictness of $\mathbf{A}$. This proves strictness in the second argument. Strictness in the next arguments directly follows from strictness of $\mathbf{A}$.

Now we verify condition (3) from Section 2 for the last three rules. First we show this for the proper $\exists^-$–rule. Let $\mathbf{r}$, $\mathbf{d}$ and $\mathbf{e}$ be monotonic. Then, using $\mathbf{A}(\mathbf{x}) >_{\mathrm{mon}} \mathbf{x}$ for monotonic $\mathbf{x}$, we get:

$$[\![\exists^-]\!](\langle\mathbf{r},\mathbf{d}\rangle,\mathbf{e}) >_{\mathrm{mon}} \mathbf{e}(\mathbf{r},\mathbf{S}\oplus\mathbf{d}) \geq \mathbf{e}(\mathbf{r},\mathbf{d}).$$

Hence, in any monotonic environment $[\![\exists^-\langle r,d\rangle e]\!] >_{\mathrm{mon}} [\![erd]\!]$.

Next we verify the same relation for rules (iii) and (iv), permutative conversions for $\to$, $\forall$ and $\wedge$. These two rules can be written as:

$$I(\exists^-(d,e),f) \mapsto \exists^-(d,\lambda x\, u.I(e(x,u),f)),$$

where f is a term or 0 or 1 for the projections.

Let $\mathbf{d}, \mathbf{e}, \mathbf{f}, \vec{\mathbf{z}}$ be monotonic. Put $\mathbf{a} := [\![\exists^-]\!](\mathbf{d}, \mathbf{e})$; $\mathbf{b}(\mathbf{x}, \mathbf{u}) := [\![I]\!](\mathbf{e}(\mathbf{x}, \mathbf{u}), \mathbf{f})$ and $\mathbf{c} := \mathbf{e}(\pi_0(\mathbf{d}), \mathbf{S} \oplus \pi_1(\mathbf{d}))$. Note that $\mathbf{a} \geq \mathbf{A}(\mathbf{c}) >_{\text{mon}} \mathbf{c}$. We have to show that $[\![I]\!](\mathbf{a}, \mathbf{f}, \vec{\mathbf{z}}) > [\![\exists^-]\!](\mathbf{d}, \mathbf{b}, \vec{\mathbf{z}})$.

$$\begin{aligned}
[\![I]\!](\mathbf{a}, \mathbf{f}, \vec{\mathbf{z}}) &= \mathbf{S}(\mathbf{a}, \mathbf{f}, \vec{\mathbf{z}}) + \mathbf{a}(\mathbf{f}, \vec{\mathbf{z}}) + 1 \\
&> [\![\exists^-]\!](\mathbf{d}, \mathbf{e}, \mathbf{f}, \vec{\mathbf{z}}) \\
&= \mathbf{A}^{2^{\mathbf{S}(\mathbf{d})}}(\mathbf{c})(\mathbf{f}, \vec{\mathbf{z}}). \\
[\![\exists^-]\!](\mathbf{d}, \mathbf{b}, \vec{\mathbf{z}}) &= \mathbf{A}^{2^{\mathbf{S}(\mathbf{d})}}(\mathbf{b}(\pi_0(\mathbf{d}), \mathbf{S} \oplus \pi_1(\mathbf{d})))(\vec{\mathbf{z}}) \\
&= \mathbf{A}^{2^{\mathbf{S}(\mathbf{d})}}([\![I]\!](\mathbf{c}, \mathbf{f}))(\vec{\mathbf{z}}).
\end{aligned}$$

So it suffices to prove that $\mathbf{A}^{n+1}(\mathbf{c})(\mathbf{f}) \geq \mathbf{A}^{n+1}([\![I]\!](\mathbf{c}, \mathbf{f}))$. This is proved by induction on n. If $n = 0$, both terms are equal by definition of $\mathbf{A}$. The successor case uses that $[\![I]\!](\mathbf{x}) >_{\text{mon}} \mathbf{x}$, for all monotonic $\mathbf{x}$:

$$\begin{aligned}
\mathbf{A}^{n+2}(\mathbf{c})(\mathbf{f}) &= \mathbf{A}(\mathbf{A}^{n+1}(\mathbf{c}), \mathbf{f}) \\
&= \mathbf{A}([\![I]\!](\mathbf{A}^{n+1}(\mathbf{c}), \mathbf{f})) \quad \text{by definition of } \mathbf{A} \\
&> \mathbf{A}(\mathbf{A}^{n+1}(\mathbf{c})(\mathbf{f})) \\
&\geq \mathbf{A}(\mathbf{A}^{n+1}([\![I]\!](\mathbf{c}, \mathbf{f}))) \quad \text{by IH} \\
&= \mathbf{A}^{n+2}([\![I]\!](\mathbf{c}, \mathbf{f})).
\end{aligned}$$

Finally, we have to prove condition (3) for the $\exists^-\exists^-$ permutative conversion,

$$\exists^-_{\sigma,\tau}(\exists^-_{\rho,o\times\sigma}(d, e), f) \mapsto \exists^-_{\rho,\tau}(d, \lambda x^o u^\rho . \exists^-_{\sigma,\tau}(e(x, u), f)).$$

Let $\mathbf{d}, \mathbf{e}, \mathbf{f}$ be monotonic. Put $\mathbf{a} := [\![\exists^-]\!](\mathbf{d}, \mathbf{e})$; $\mathbf{b}(\mathbf{x}, \mathbf{u}) := [\![\exists^-]\!](\mathbf{e}(\mathbf{x}, \mathbf{u}), \mathbf{f})$ and $\mathbf{c} := \mathbf{e}(\pi_0(\mathbf{d}), \mathbf{S} \oplus \pi_1(\mathbf{d}))$. We have to show that $[\![\exists^-]\!](\mathbf{a}, \mathbf{f}) >_{\text{mon}} [\![\exists^-]\!](\mathbf{d}, \mathbf{b})$. Again we have $\mathbf{a} \geq \mathbf{A}(\mathbf{c}) >_{\text{mon}} \mathbf{c}$, so $\mathbf{S}(\mathbf{a}) > \mathbf{S}(\mathbf{c})$. From the left hand side of the rule it is clear that $\mathbf{a}$ is of product type. Hence, $\mathbf{S}(\mathbf{a}) = \mathbf{a}(0) + \mathbf{S}(\mathbf{a}(1))$. Because $\mathbf{S}(\mathbf{a}(1)) > 0$, we obtain $\mathbf{S}(\mathbf{a}) > \mathbf{a}(0) = \mathbf{A}^{2^{\mathbf{S}(\mathbf{d})}}(\mathbf{c})(0) \geq 2^{\mathbf{S}(\mathbf{d})} \geq \mathbf{S}(\mathbf{d}) + 1$. Hence

$$2^{\mathbf{S}(\mathbf{a})} \geq 2^{\max\{\mathbf{S}(\mathbf{d})+1, \mathbf{S}(\mathbf{c})\}+1} \geq 1 + 2^{\mathbf{S}(\mathbf{d})} + 2^{\mathbf{S}(\mathbf{c})}.$$

Now we can compute:

$$\begin{aligned}
[\![\exists^-]\!](\mathbf{a}, \mathbf{f}) &= \mathbf{A}^{2^{\mathbf{S}(\mathbf{a})}}(\mathbf{f}(\pi_0(\mathbf{a}), \mathbf{S} \oplus \pi_1(\mathbf{a}))) \\
&>_{\text{mon}} \mathbf{A}^{2^{\mathbf{S}(\mathbf{d})}}(\mathbf{A}^{2^{\mathbf{S}(\mathbf{c})}}(\mathbf{f}(\pi_0(\mathbf{a}), \mathbf{S} \oplus \pi_1(\mathbf{a})))) \\
&\geq \mathbf{A}^{2^{\mathbf{S}(\mathbf{d})}}(\mathbf{A}^{2^{\mathbf{S}(\mathbf{c})}}(\mathbf{f}(\pi_0(\mathbf{c}), \mathbf{S} \oplus \pi_1(\mathbf{c})))) \\
&= \mathbf{A}^{2^{\mathbf{S}(\mathbf{d})}}([\![\exists^-]\!](\mathbf{c}, \mathbf{f})) \\
&= \mathbf{A}^{2^{\mathbf{S}(\mathbf{d})}}(\mathbf{b}(\pi_0(\mathbf{d}), \mathbf{S} \oplus \pi_1(\mathbf{d}))) \\
&= [\![\exists^-]\!](\mathbf{d}, \mathbf{b}).
\end{aligned}$$

We have shown that for all rules, the left hand side is greater than the right hand side. Hence the HRS is terminating. This directly implies termination for the calculus with derivation terms presented at the beginning of this section.

References

[1] Robin O. Gandy. Proofs of strong normalization. In J.P. Seldin and J.R. Hindley, editors, *To H.B. Curry: Essays on Combinatory Logic, Lambda Calculus and Formalism*, pages 457–477. Academic Press, 1980.

[2] Jean-Yves Girard. *Proof Theory and Logical Complexity*. Bibliopolis, Napoli, 1987.

[3] Gerard Huet and Derek Oppen. Equations and rewrite rules — a survey. In *Formal Language Theory — Perspectives and Open Problems*, pages 349–405. Academic Press, 1980.

[4] Tobias Nipkow. Orthogonal higher-order rewrite systems are confluent. In M. Bezem and J.F. Groote, editors, *Typed Lambda Calculi and Applications*, volume 664 of *Lecture Notes in Computer Science*, pages 306–317, Berlin, 1993. Springer.

[5] Dag Prawitz. Ideas and results in proof theory. In J.E. Fenstad, editor, *Proceedings of the Second Scandinavian Logic Symposium*, pages 235–307. North-Holland, Amsterdam, 1971.

[6] Anne S. Troelstra and Dirk van Dalen. *Constructivism in Mathematics. An Introduction*, volume 121, 123 of *Studies in Logic and the Foundations of Mathematics*. North-Holland, Amsterdam, 1988.

[7] Jaco van de Pol. Termination proofs for higher-order rewrite systems. In J. Heering, K. Meinke, B. Möller, and T. Nipkow, editors, *Higher-Order Algebra, Logic and Term Rewriting (HOA '93)*, volume 816 of *Lecture Notes in Computer Science*, pages 305–325, Berlin, 1994. Springer.

[8] Vincent van Oostrom. *Confluence for Abstract and Higher-Order Rewriting*. PhD thesis, Vrije Universiteit, Amsterdam, 1994.

A Verified Typechecker *

Robert Pollack

Dept. of Computing Science
Chalmers Univ. of Technology and Univ. of Göteborg
S-412 96 Göteborg SWEDEN
pollack@cs.chalmers.se

1 Introduction

In [MP93] we describe the early stages of a formal development of the theory of Pure Type Systems (PTS), expressed in the Extended Calculus of Constructions (ECC) with inductive types [Luo94] and checked by the LEGO proof development system [LP92, JP93, JP94]. We gave two long-term motivations for that work: to construct a verified type checking program for some class of PTS, and to be a realistic example of formal mathematics. As for the latter goal, our current formal theory of PTS has well over 3000 definitions and lemmas, and continues to grow. In this paper I describe partial attainment of the former goal: a verified type checking program for a class of PTS of practical interest, but one not yet efficient enough to actually execute in interesting cases.

Techniques for type checking the Calculus of Constructions (CC) and its extensions were well known [Hue89], but there were some difficulties for other PTS. In [vBJMP94] we clarify the problem of type checking all PTS, giving satisfactory solutions for functional PTS and for semifull PTS, and some rather complicated theorems about the general case. While [vBJMP94] was written in informal language, many of its results were formalized and checked in LEGO. However, the "satisfactory solution" to type checking given in [vBJMP94] is expressed as a syntax-directed formal system (an inductively defined relation), not as an executable algorithm (a function), the idea being that the type checking function is obtained by just following the rules of the formal system, which, being syntax directed, leaves no choices to be made.

A verified typechecker for CC is described in [DB93]. Working in the Boyer-Moore logic, the authors prove only the soundness of their typechecker, not its completeness (compare with our corollary 7), and even so, do not prove all the lemmas used in the soundness result. In contrast, our development is based on a completely formalized theory of PTS. [DB93] inspired me to formally verify a type checking function.

Given the work in [vBJMP94], there are still two problems to be solved to have a verified type checking program for a class of PTS; (a) the algorithm

* This work was supported by the ESPRIT BRA on TYPES and by the British SERC, and was done at the University of Edinburgh, and at Chalmers University of Technology and University of Göteborg

for deciding PTS judgements from those of the syntax-directed system, and (b) termination of the process of applying the rules of the syntax-directed system. The current paper fills these gaps, although for simplicity I will restrict to *semi-full* and *functional* PTS (these terms will be defined below, but see [vBJMP94] for a more detailed explanation), which includes the Calculus of Constructions (CC) and λP, the language of the Edinburgh Logical Framework.

Since many PTS have normalization theorems that we cannot, or do not wish to prove in ECC, I give a partial correctness proof for a typechecker that uses a partially correct normalization program.

The type system ECC [Luo94] is not a PTS because of its cumulativity of universes. However, it is very similar to a PTS, and most of what is discussed in this paper can be done uniformly for a more general class I call *Cumulative Type Systems* (CTS), which includes ECC. See [Pol94] for details of this development, expressed in formal LEGO notation.

Acknowledgement I thank Bert Jutting and James McKinna, my co-workers on much of the material underlying this paper, and dedicate this paper to Bert Jutting, who wanted [vBJMP94] to contain explicit algorithms.

2 Retrospective on Previous Work

In this paper I will use informal notation, and only briefly review Pure Type Systems, the type checking problem, and the syntax-directed system for semi-full PTS. This material is detailed in [vBJMP94, Pol94].

2.1 Pure Type Systems

A PTS is a tuple $(\mathcal{V}, \mathcal{S}, \mathsf{ax}, \mathsf{rl})$ where $\mathcal{V}$ is an infinite set of *variables*, ranged over by x, y; $\mathcal{S}$ is a set of *sorts*, ranged over by s, t; and $\mathsf{ax} \subseteq \mathcal{S} \times \mathcal{S}$ and $\mathsf{rl} \subseteq \mathcal{S} \times \mathcal{S} \times \mathcal{S}$ are relations that parameterize the typing judgement.

The terms are given by the grammar

atoms	$\alpha ::= x \mid s$	*variable, sort*
terms	$M ::= \alpha \mid \lambda x{:}M.M \mid \Pi x{:}M.M \mid M\ M$	*atom, lambda, pi, application*

As usual, $\lambda x{:}A.B$ and $\Pi x{:}A.B$, bind x in B but not in A. M, N, A, B, C, D, E, a, b range over terms. We write $A \twoheadrightarrow B$ and $A \simeq B$ for beta-reduction and beta-conversion respectively.

Contexts, ranged over by Γ, Δ, are lists of variable-term pairs, written as

$$\textbf{context} \quad \Gamma ::= \bullet \mid \Gamma[x{:}A] \qquad \textit{empty, non-empty}$$

The typing judgement of PTS has shape $\Gamma \vdash M : A$, and is defined inductively by the rules in table 1. We say M is a *PTS-term* (resp. *PTS-type*) iff $\exists \Gamma, A \,.\, \Gamma \vdash M : A$ (resp. $\exists \Gamma, A \,.\, \Gamma \vdash A : M$), and M is a *PTS-object* iff it is a PTS-term or a PTS-type.

The basic theory of PTS is now well known [Bar91, Ber90, GN91, Bar92, vBJ93], and I will only mention a few definitions and lemmas used in this paper.

Ax	$\bullet \vdash s_1 : s_2$	$\mathsf{ax}(s_1{:}s_2)$
Start	$\dfrac{\Gamma \vdash A : s}{\Gamma[x{:}A] \vdash x : A}$	$x \notin \Gamma$
vWeak	$\dfrac{\Gamma \vdash y : C \qquad \Gamma \vdash A : s}{\Gamma[x{:}A] \vdash y : C}$	$x \notin \Gamma$
sWeak	$\dfrac{\Gamma \vdash s : C \qquad \Gamma \vdash A : s}{\Gamma[x{:}A] \vdash s : C}$	$x \notin \Gamma$
Pi	$\dfrac{\Gamma \vdash A : s_1 \qquad \Gamma[x{:}A] \vdash B : s_2}{\Gamma \vdash \Pi x{:}A.B : s_3}$	$\mathsf{rl}(s_1, s_2, s_3)$
Lda	$\dfrac{\Gamma[x{:}A] \vdash M : B \qquad \Gamma \vdash \Pi x{:}A.B : s}{\Gamma \vdash \lambda x{:}A.M : \Pi x{:}A.B}$	
App	$\dfrac{\Gamma \vdash M : \Pi x{:}A.B \qquad \Gamma \vdash N : A}{\Gamma \vdash M\,N : [N/x]B}$	
Conv	$\dfrac{\Gamma \vdash M : A \qquad \Gamma \vdash B : s}{\Gamma \vdash M : B}$	$A \simeq B$

Table 1. The Informal Typing Rules of PTS

Generation lemmas are the *inversion* of the inductive definition of $\vdash$; they say that each shape of judgement can only be constructed in certain ways. For example, any derivation of a judgement $\Gamma \vdash \Pi x{:}A.M : B$ must end with an instance of the Pi rule followed by zero or more instances of the Conv rule, so we can read off from the premises of the Pi rule some conditions that must hold for this to be the case.

Correctness of types Every type is itself well typed: for all Γ, M ,A

$$\Gamma \vdash M : A \quad \Rightarrow \quad \exists s \,.\, A = s \text{ or } \Gamma \vdash A : s.$$

If $\exists t \,.\, \mathsf{ax}(s{:}t)$ then s is called a *typedsort*; if $\neg\exists t \,.\, \mathsf{ax}(s{:}t)$, s is called a *topsort*. The left disjunct in the conclusion of this lemma, $A = s$, is necessary in case A is a topsort. Topsort was defined originally by Berardi [Ber90], who made some simple but pretty observations about it. These notions are used in the next section.

Closure under reduction (*subject reduction* and *predicate reduction*). Judgements are preserved by reduction: for all Γ, M, A, Γ', M', A'

$$\Gamma \vdash M : A,\ \Gamma \twoheadrightarrow \Gamma',\ M \twoheadrightarrow M',\ A \twoheadrightarrow A' \quad \Rightarrow \quad \Gamma' \vdash M' : A'.$$

Functional PTS A PTS is *functional* iff

- $\mathsf{ax}(s{:}t)$ and $\mathsf{ax}(s{:}u)$ implies $t = u$, and
- $\mathsf{rl}(s_1, s_2, t)$ and $\mathsf{rl}(s_1, s_2, u)$ implies $t = u$.

That is, a PTS is functional iff ax and rl are the graphs of partial functions. Functional PTS have *uniqueness of types* up to conversion: for all Γ, M, A, B

$$\Gamma \vdash M : A, \ \Gamma \vdash M : B \quad \Rightarrow \quad A \simeq B.$$

Any PTS with this type uniqueness property also has the *subject expansion* property: for all Γ, M, A, N, B

$$\Gamma \vdash M : A, \ N \twoheadrightarrow M, \ \Gamma \vdash N : B \quad \Rightarrow \quad \Gamma \vdash N : A.$$

While subject reduction says that terms don't lose types under reduction, subject expansion says that terms don't gain types under reduction.

2.2 Type Checking and Type Synthesis

To *decide* a proposition, P, is to prove P or $\neg P$; and we write $\mathsf{decidable}(P)$ for P or $\neg P$. The Type Checking (TC) problem, for Γ, M and A, is to decide $\Gamma \vdash M : A$. The Type Synthesis (TS) problem for Γ and M is to decide $\exists A \,.\, \Gamma \vdash M : A$.

We will use TS to solve TC for a class of PTS (section 3). Our strategy for solving TS for this class of PTS (section 4) is to find an inductive definition of a relation that is equivalent in some way to PTS, but deterministic in the sense that, given Γ and M, there will be no choice of which rule is the root of any derivation over Γ and M, or of what type is derived by that rule. Such a rule application will have premises which need to be satisfied, and the subjects of these subgoals should be determined by the subject, Γ and M, of the previous goal. Further, all the side conditions of the rules should be solvable. This is what I am calling, informally, a *syntax directed* definition.

2.3 Semifull PTS

A PTS is called *full* iff for all s_1, s_2 there exists s_3 with $\mathsf{rl}(s_1, s_2, s_3)$. In full PTS the right premise of the LDA-rule can be simplified. The purpose of that premise, $\Gamma \vdash \Pi x{:}A.B : s$, is to assure type correctness. But we know from the left premise that $\Gamma \vdash A : s_A$ for some s_A (by inversion); and that either B is a sort, or B has a type which is a sort (by type correctness). As long as $\Gamma[x{:}A] \vdash B : s_B$ for some s_B, we can conclude that for full PTS there exists s with $\mathsf{rl}(s_A, s_B, s)$, so $\Pi x{:}A.B$ is well typed. This suggests replacing the right premise of the LDA-rule by the requirement that B is not a topsort, or, making a positive statement, if B is a sort, then B is a typedsort. We can generalize this idea somewhat beyond full PTS.

Definition 1 Semifull. A PTS is *semi-full* iff for all s_1

$$(\exists s_2, s_3 \,.\, \mathsf{rl}(s_1, s_2, s_3)) \quad \Rightarrow \quad \forall s_2 \,\exists s_3 \,.\, \mathsf{rl}(s_1, s_2, s_3).$$

SDSF-AX	$\bullet \vdash_{sdsf} s_1 : s_2$	$\mathsf{ax}(s_1{:}s_2)$
SDSF-STRT	$\dfrac{\Gamma \vdash_{sdsf} A : X}{\Gamma[x{:}A] \vdash_{sdsf} x : A}$	$X \twoheadrightarrow s,\ x \notin \Gamma$
SDSF-vWK	$\dfrac{\Gamma \vdash_{sdsf} y : C \qquad \Gamma \vdash_{sdsf} A : X}{\Gamma[x{:}A] \vdash_{sdsf} y : C}$	$X \twoheadrightarrow s,\ x \notin \Gamma$
SDSF-sWK	$\dfrac{\Gamma \vdash_{sdsf} s : C \qquad \Gamma \vdash_{sdsf} A : X}{\Gamma[x{:}A] \vdash_{sdsf} s : C}$	$X \twoheadrightarrow s',\ x \notin \Gamma$
SDSF-PI	$\dfrac{\Gamma \vdash_{sdsf} A : X \qquad \Gamma[x{:}A] \vdash_{sdsf} B : Y}{\Gamma \vdash_{sdsf} \Pi x{:}A.B : s_3}$	$\mathsf{rl}(s_1, s_2, s_3)$ $X \twoheadrightarrow s_1,\ Y \twoheadrightarrow s_2$
SDSF-LDA	$\dfrac{\Gamma \vdash_{sdsf} A : X \qquad \Gamma[x{:}A] \vdash_{sdsf} M : B}{\Gamma \vdash_{sdsf} \lambda x{:}A.M : \Pi x{:}A.B}$	$X \twoheadrightarrow s_1,\ \mathsf{rl}(s_1, s_2, s_3)$ $B \in \mathcal{S} \Rightarrow$ typedsort B
SDSF-APP	$\dfrac{\Gamma \vdash_{sdsf} M : X \qquad \Gamma \vdash_{sdsf} N : Y}{\Gamma \vdash_{sdsf} M\,N : [N/x]B}$	$X \twoheadrightarrow \Pi x{:}A.B,\ Y \simeq A$

Table 2. Syntax-directed semi-full PTS

While the Pure Calculus of Constructions, CC, and various extensions with type universes are full, the Edinburgh Logical Framework, λP, is only semi-full. To the best of my knowledge, this definition first appeared in [Pol92] where I used it to give a syntax directed presentation of a class of type theories including CC and λP. That paper, in improved form, is published as a section in [vBJMP94]. The notion is also used in [Geu93], where it is credited to [vBJMP94].

For our purposes, the interesting fact about semifull PTS is that, because of the possibility to simplify the LDA rule mentioned above, we can give an alternative presentation of such systems: relation sdsf (syntax directed semifull) is defined inductively by the rules of table 2. Notice that sdsf has no rule corresponding to the conversion rule, CONV of PTS; this is why it is syntax directed. sdsf is arrived at by permuting the conversion rule of PTS downward through all premises of all the other rules; thus sdsf has judgements differing from PTS at most by a final use of the conversion rule (see lemma 2). However, this permutation cannot be completely carried out, so some reduction and conversion side conditions get left behind on some rules. Since sdsf no longer has a conversion rule, where PTS required that a term *be* a sort or a Pi, sdsf can only require that term to *reduce to* a sort or a Pi. SDSF-LDA also has the side condition discussed above that is the residual of the right premiss of rule LDA.

The relationship between PTS and sdsf is formalized by the following lemma, proved in detail in [vBJMP94, Pol94]:

Lemma 2 adequacy and faithfulness.

Adequacy $\forall\Gamma, M, A\,[\Gamma \vdash M : A \Rightarrow \exists E \,.\, \Gamma \vdash_{sdsf} M : E \text{ and } E \simeq A]$
Faithfullness *For semifull PTS,* $\forall\Gamma, M, A\,[\Gamma \vdash_{sdsf} M : A \Rightarrow \Gamma \vdash M : A]$

2.4 Efficiency of sdsf

The derivations of sdsf and PTS are very inefficient, as the number of rule applications is exponential in the size of the conclusion. All leaves have an empty context and all branches must build their own context with the start and weakening rules. Since some rules mention the same context in both premises, it must be checked in branches leading to both premises. In [Pol94], I formalize a presentation of PTS that avoids this blow-up by checking validity of the context incrementally, but this presentation is more difficult to work with than the one given in table 1, and I have not yet carried its development through to a type checking algorithm.

3 Type Checking Using Type Synthesis

In [vBJMP94] we went to a lot of trouble to make sdsf deterministic, and in section 4 we will see that the effort has payed off; type synthesis is computable for sdsf under some assumptions. The question for this section is how to typecheck PTS assuming we already have a type synthesis algorithm for sdsf.

3.1 Characterizing Semifull PTS: an abstract typechecker

Lemma 2 does not yet explicitly characterize semifull PTS in terms of sdsf, but we have:

***Lemma 3 Characterization of* PTS.** *For functional, semifull PTS,* $\forall\Gamma, M, A$

$$\begin{gathered}\Gamma \vdash M : A \\ \Leftrightarrow \\ \exists E \,.\, \Gamma \vdash_{sdsf} M : E \text{ and } (A \in \mathcal{S} \text{ or } \exists D \,.\, \Gamma \vdash_{sdsf} A : D) \text{ and } E \simeq A\end{gathered} \tag{1}$$

Proof.

$\Rightarrow$ By lemma 2 we have E with $\Gamma \vdash_{sdsf} M : E \simeq A$. Also by type correctness of PTS, for some sort s, $A = s$ or $\Gamma \vdash A : s$. In the first case we are done. In the second case $\exists D \,.\, \Gamma \vdash_{sdsf} A : D$ by lemma 2.

$\Leftarrow$ By lemma 2 we have $\Gamma \vdash M : E \simeq A$. If A is a sort, then $E \twoheadrightarrow A$, and we are done by predicate reduction of PTS. Otherwise $\exists D \,.\, \Gamma \vdash_{sdsf} A : D$, hence $\Gamma \vdash A : D$. If we knew D reduced to some sort, s, we would be done, for then $\Gamma \vdash A : s$ by predicate reduction, and $\Gamma \vdash M : A$ by rule Conv (table 1). Thus the following claim finishes this lemma:

Claim. $\forall\Gamma, M, E, A, D \,.\, (\Gamma \vdash M : E \simeq A \text{ and } \Gamma \vdash A : D) \Rightarrow \exists s \,.\, D \twoheadrightarrow s$

Proof of claim. By type correctness, for some t either $E = t$ or $\Gamma \vdash E : t$. In the first case $A \twoheadrightarrow t$, so $\Gamma \vdash t : D$ by subject reduction, and D reduces to some sort by the generation lemma. In the second case, let X be a common reduct of E and A. Then $\Gamma \vdash X : t$ and $\Gamma \vdash X : D$ by subject reduction, so $D \twoheadrightarrow t$ by type unicity of functional PTS. (Using the Typing Lemma from [vBJ93] in place of type unicity of functional PTS, this claim, and the whole lemma, are seen to hold for all PTS. This more general proof is not yet checked in LEGO.) □

Lemma 3 is suggestive of an algorithm for type checking PTS given type synthesis for **sdsf**: to decide $\Gamma \vdash M : A$, compute an **sdsf**-type, E, for (Γ, M), check if A itself is a correct **sdsf**-type, and see if E converts with A. In this algorithmic reading, the order of the disjuncts is important in the RHS of equivalence (1); the test for $E \simeq A$ is written last, as it must be deferred until we know that both E and A are PTS-objects, as only then can we expect E and A to be normalizing, hence this conversion test to be decidable.

The lemma is still unsatisfactory for type checking; the quantifier $\exists E$ in the RHS not only allows M to fail to have an **sdsf**-type in Γ, (in which case M fails to have a PTS type in Γ) but also requires us to search through all **sdsf** types of M in Γ. If the PTS is not functional the rules SDSF-AX and SDSF-PI can produce different types for the same subject. I will assume the PTS is functional, as this covers almost every case of practical interest including the three type systems of LEGO. With this proviso, it is easy to prove

Lemma 4* sdsf *has unique types. *For functional PTS, for all Γ, M, A*

$$\Gamma \vdash_{sdsf} M : A \text{ and } \Gamma \vdash_{sdsf} M : B \quad \Rightarrow \quad A \simeq B$$

Thus any **sdsf**-type, E, for (Γ, M) will do in the algorithmic reading of equivalence (1). If $E \not\simeq A$ then A is not a PTS-type for (Γ, M), and there is no use looking further.

Remark. This is the essential use of functionality in type synthesis and type checking for semi-full PTS; the other uses in this paper could be replaced by more difficult argument, but to have a suitably deterministic syntax-directed system for non-functional systems requires the technique of *sort variables*, *schematic terms*, and *constraints*, which is not discussed in this paper (see [HP91, vBJMP94]). Functionality plays a more important role in type checking for PTS that are *not* semi-full, as in this case the failure of subject expansion for non-functional PTS becomes problematic, and we must normalize some terms to be sure we have all of their types. In fact the general normalizing PTS with certain conditions decidable (see section 4.1) does have decidable type checking [vBJ93, vBJMP94], but this is much more complicated to prove.

Remark (Principal Types). When this approach is developed for Cumulative Type Systems such as ECC (see [Pol94]), something more interesting happens. ECC, although a functional CTS, does not have unique types up to conversion, but it does have a notion of *principal type* (defined by Luo [Luo94]), which characterizes all the types of a term. In this setting lemma 4 is replaced by a lemma saying that sdsf computes the principal type of a term. ECC, with principal types, is well behaved; for an (informal) development of type checking for a system similar to ECC, but not having principal types, see [HP91].

Remark. Lemma 4 is surprisingly weak. sdsf is called "syntax-directed", but it has two sources of non-determinism which explain why sdsf-types are unique only up to conversion.

1. I am being informal about variable names, but in detail there is a need to choose fresh variables in the rules SDSF-PI and SDSF-LDA, so that sdsf types are unique only up to alpha-conversion.
2. In the rule SDSFAPP the side condition that the type of M reduces to $\Pi x{:}A.B$ allows non-deterministic reduction as long as it stops at a Pi. Many of the other rules have non-deterministic reduction to a sort, but a sort is a normal form, so this causes no multiplicity of types. A Pi is not necessarily a normal form, but is a weak-head normal form, and we should, for moral purity, replace $\twoheadrightarrow$ with weak-head reduction in this rule[2]. Instead, for our present purposes, we will accept that sdsf types are unique only up to beta-conversion. We are still free to compute this side condition using weak-head reduction when we construct an algorithm in section 4.

Guided by equivalence (1), and keeping in mind the need to use uniqueness of types for functional PTS, we are almost ready to use a program for sdsf-TS to decide PTS judgements. First we need to consider decidability of some side conditions.

3.2 Decidability of Side Conditions

Let us assume that PTS-terms are normalizing. Then by correctness of types, PTS-types are normalizing, and conversion of PTS-objects is decidable. Furthermore it is decidable for a PTS-object whether it reduces to a sort, and whether it reduces to a Pi.

For the moment, we will also assume that TS for sdsf is decidable, i.e., defining

$$\mathsf{decide\text{-}sdsf}(\Gamma, M) \;\triangleq\; \mathsf{decidable}(\exists A \,.\, \Gamma \vdash_{sdsf} M : A)$$

we assume $\forall \Gamma, M \,.\, \mathsf{decide\text{-}sdsf}(\Gamma, M)$. In section 4 we will discharge the second of these assumptions, and in section 5 we will consider discharging the first.

3.3 A Type Checking Algorithm

With these assumptions, TC for PTS is decidable.

[2] James McKinna [McK94] has formalized a theory of weak-head reduction and weak-head normal forms in our setting that is adequate for this example. It is used to handle similar issues in formalizing [vBJMP94].

Lemma 5 TC is decidable. *For functional, semifull PTS, assuming every PTS-term has a normal form, and that* **sdsf-TS** *is decidable*

$$\forall \Gamma, M, A \,.\, \mathsf{decidable}(\Gamma \vdash M : A).$$

Proof. With our assumptions, all the questions on the RHS of equivalence (1) are decidable: if $\Gamma \vdash_{sdsf} M : E$ for some E, either A is a sort or A has an sdsf-type, and $E \simeq A$, then $\Gamma \vdash M : A$. This positive outcome is the easy part of the proof; more tedious is showing that if one of the conditions fails then $\Gamma \vdash M : A$ is *not* derivable.

By assumption $\mathsf{decidable}(\exists E \,.\, \Gamma \vdash_{sdsf} M : E)$. In case of the right disjunct, choosing the negative outcome for the lemma, we want to prove

$$(\neg \exists E \,.\, \Gamma \vdash_{sdsf} M : E) \quad \Rightarrow \quad \neg \Gamma \vdash M : A$$

which follows by contraposition from lemma 2. Thus we may assume the left disjunct, $\Gamma \vdash_{sdsf} M : E$ for some E. Hence $\Gamma \vdash M : E$, and by lemma 4,

$$\Gamma \vdash M : A \quad \Leftrightarrow \quad (A \in \mathcal{S} \text{ or } \exists D \,.\, \Gamma \vdash_{sdsf} A : D) \text{ and } E \simeq A.$$

Now it's clear how to finish, as under our assumptions it is decidable whether A is a sort, and whether A has a sdsf-type, and, if either of these are true, it is decidable whether $E \simeq A$. □

4 Type Synthesis

Again assuming that PTS-objects are normalizing, we will show that sdsf-TS is decidable, thus discharging the second assumption made in the proof of lemma 5.

4.1 Decidability of Side Conditions

Recall that by lemma 2 every sdsf-object is also a PTS-object, so the comments about decidability of conversion and reduction of section 3.2 apply here as well.

Decidable properties of **ax** *and* **rl**. Three more restrictions on the PTS are needed. Among other things, an sdsf-TS algorithm will decide $\exists X \,.\, \bullet \vdash_{sdsf} s_1 : X$ for arbitrary s_1. Since the only possible derivation of $\bullet \vdash_{sdsf} s_1 : X$ has shape:

$$\frac{\mathsf{ax}(s_1{:}s_2)}{\bullet \vdash_{sdsf} s_1 : s_2}\ \textsc{sdsfAx}.$$

we have

$$\exists s_2 \,.\, \mathsf{ax}(s_1{:}s_2) \quad \Leftrightarrow \quad \exists X \,.\, \bullet \vdash_{sdsf} s_1 : X.$$

That is, an sdsf-TS algorithm decides the property *typedsort*. As ax is an arbitrarily given relation, we cannot hope to decide typedsort in general (because of its existential quantifier), even if ax is decidable; so in order to prove that sdsf-TS is decidable we must assume that typedsort is decidable.

Similarly, if $\mathsf{ax}(t_1{:}s_1)$ and $\mathsf{ax}(t_2{:}s_2)$, we have

$$\exists u \,.\, \mathsf{rl}(s_1, s_2, u) \;\Leftrightarrow\; \exists X \,.\, \bullet \vdash_{sdsf} \Pi v{:}t_1.t_2 : X$$

i.e. sdsf-TS decides $\exists u.\mathsf{rl}(s_1, s_2, u)$; and

$$\exists u_2, u_3 \,.\, \mathsf{rl}(s_1, u_2, u_3) \;\Leftrightarrow\; \exists X \,.\, \bullet \vdash_{sdsf} \lambda x{:}t_1.t_2 : X$$

i.e. sdsf-TS decides $\exists u_2, u_3.\mathsf{rl}(s_1, u_2, u_3)$. Thus we must also assume the properties

$$\begin{aligned} \textit{ruledsort}\ s_1 &= \exists s_2, s_3 \,.\, \mathsf{rl}(s_1, s_2, s_3) \\ \textit{ruledsorts}\ s_1\ s_2 &= \exists s_3 \,.\, \mathsf{rl}(s_1, s_2, s_3) \end{aligned}$$

are decidable.

4.2 A Type Synthesis Algorithm

Recall the definition of **decide-sdsf** from section 3.2.

Lemma 6 TS is decidable. *For semifull, functional PTS, assuming every PTS-term has a normal form, and that typedsort, ruledsort and ruledsorts are decidable:*

$$\forall \Gamma, M \,.\, \textit{decide-sdsf}(\Gamma, M)$$

Proof. The proof follows the syntax-directed rules of sdsf. Although there is much detailed argument to be done, the major question is whether this procedure terminates. For this purpose we use a well-founded induction measure "the sum of the length of M and lengths of the terms appearing in Γ". The lngth of a term is required to have two properties:

- a term has positive lngth (even atomic terms),
- a term has lngth strictly greater than that of any proper subterm.

The Lngth of a context is defined by:

$$\begin{aligned} \mathsf{Lngth}(\bullet) &= 0 \\ \mathsf{Lngth}(\Gamma[x{:}A]) &= \mathsf{Lngth}(\Gamma) + \mathsf{lngth}(A) \end{aligned}$$

Now the induction measure is defined[3]:

$$\mathsf{LNGTH}(\Gamma, M) = \mathsf{Lngth}(\Gamma) + \mathsf{lngth}(M)$$

By induction on the measure $\mathsf{LNGTH}(\Gamma, M)$ it suffices to show

$$\begin{aligned} &\forall M, \Gamma \,. \\ &\quad [\forall m, \gamma \,.\, \mathsf{LNGTH}(\gamma, m) < \mathsf{LNGTH}(\Gamma, M) \;\Rightarrow\; \textsf{decide-sdsf}(\gamma, m)\,] \;\Rightarrow \\ &\quad \textsf{decide-sdsf}(\Gamma, M). \end{aligned} \qquad (2)$$

[3] Compare this with the lexicographic induction on $\mathsf{Lngth}(\Gamma)$ followed by the structure of M used informally in [Luo94] Definition 5.12.

The phrase

$$\forall m, \gamma \,.\, \mathsf{LNGTH}(\gamma, m) < \mathsf{LNGTH}(\Gamma, M) \;\Rightarrow\; \textsf{decide-sdsf}(\gamma, m)$$

is the well-founded induction hypothesis.

We show (2) by cases on the shape of M (i.e. by structural induction on M, not using the structural induction hypotheses). For each shape of term M, use the appropriate rule(s) of sdsf to compute its type. We do two cases.

(**M is a sort: $M = s$**) SDSF-AX and SDSF-SWK are the only rules constructing an sdsf-type for a sort. If $\Gamma = \bullet$ then only SDSF-AX can apply. By assumption *typedsort* is decidable: $\textsf{decidable}(\exists t \,.\, \textsf{ax}(s{:}t))$. In case of the left disjunct use SDSF-AX to return a proof of $\bullet \vdash_{sdsf} s : t$; in case of the right disjunct, no rule can apply, so return a proof of $\neg\exists A \,.\, \bullet \vdash_{sdsf} s : A$.

Now we may assume Γ is not $\bullet$, so $\Gamma = \Delta[q, A]$. Only rule SDSF-SWK can apply. Fail (i.e. return a proof of $\neg\exists A \,.\, \Delta[q, A] \vdash_{sdsf} s : A$) if $q \in \Delta$ because SDSF-SWK cannot apply, so assume $q \notin \Delta$. Addressing the left premise, by induction hypothesis on (Δ, s), $\textsf{decide-sdsf}(\Delta, s)$. Fail in case of the right disjunct, otherwise $\Delta \vdash_{sdsf} s : C$ for some C. Now address the right premiss; by induction hypothesis on (Δ, A)[4], $\textsf{decide-sdsf}(\Delta, A)$. Fail in case of the right disjunct, otherwise $\Delta \vdash_{sdsf} A : X$ for some X. X is a PTS-type, so it is decidable if X reduces to a sort; if not then fail, if so use SDSF-SWK to return $\Gamma \vdash_{sdsf} s : C$.

(**M is an application: $M = N\,L$**) The only rule that can apply is SDSF-APP. $\textsf{decide-sdsf}(\Gamma, N)$ by induction hypothesis on (Γ, N); fail in case of the right disjunct, otherwise $\Gamma \vdash_{sdsf} N : X$ for some X. Similarly for the right premise: fail if it is not derivable, or $\Gamma \vdash_{sdsf} L : Y$ for some Y. Fail if X does not reduce to some pi, otherwise have $X \twoheadrightarrow \Pi x{:}A.B$ for some x, A and B. We have $\Gamma \vdash N : X$, so by predicate reduction $\Gamma \vdash N : \Pi x{:}A.B$, by type correctness $\Gamma \vdash \Pi x{:}A.B : Z$, and by the generation lemma for pi, A is a PTS-term. Thus we can decide if $Y \simeq A$; fail if not, and use SDSFAPP to return $\Gamma \vdash_{sdsf} N\,L : [L/v]B$ if so. □

PTS-TC. Putting lemmas 5 and 6 together, we have

Corollary 7 TC is decidable. *For semifull, functional PTS, assuming every PTS-term has a normal form, and that typedsort, ruledsort and ruledsorts are decidable:*

$$\forall \Gamma, M, A \,.\, \textit{decidable}(\Gamma \vdash M : A)$$

Remark. We showed that decidability of *typedsort* is a necessary condition for decidability of TS. As TC differs from TS in having no existential quantifier in the statement of the problem, one might think that decidability of PTS-TC requires only that ax be decidable, not that *typedsort* be decidable. However this is not the case, because the derivation skeleton

$$\dfrac{\dfrac{\textsf{ax}(s{:}t)}{\bullet \vdash s : t}\;\textsc{Axiom}}{[x{:}s] \vdash x : s}\;\textsc{Start}$$

[4] here we use that $\textsf{lngth}(s)$ is positive, so $\mathsf{LNGTH}(\Delta, A) < \mathsf{LNGTH}(\Delta[q, A], s)$.

shows that

$$[x{:}s] \vdash x : s \;\Leftrightarrow\; \exists t \,.\, \mathsf{ax}(s{:}t).$$

Similarly, one can see that decidability of PTS-TC implies *ruledsort* and *ruledsorts* are decidable.

5 Executable Typecheckers?

In lemma 6 and corollary 7 we have decision procedures for a class of PTS. Is this a reasonable class, and can these procedures actually be executed? All the assumptions except semifullness and functionality are necessary conditions for decidability of TS and TC[5]. However, it is desirable to program sound, if not necessarily complete, type checking algorithms for PTS where these conditions are not provably satisfied, or not satisfied at all. Since non-normalizability of well-typed terms is the most interesting example of this problem, I will consider partial correctness of TS and TC given a partially correct normalization program, and then, briefly, consider questions of efficiency.

5.1 Partial Correctness

λP meets the assumptions of corollary 7, and I expect to be able to formally prove that λP is normalizing, and produce a normalization algorithm for it. CC meets the assumptions of corollary 7, but it is a real challenge to formally prove normalization of CC in ECC. It is very probably not possible to prove that ECC is normalizing within ECC. Further, we might be interested in typechecking a system like λ^* [Bar91] which gives types to non-normalizing terms, but otherwise meets all the assumptions of corollary 7. It is possible to write a program in ECC that, for any n, computes n steps of reduction on any lambda term (I leave open how to count the steps). This is enough to program sound, but possibly incomplete, TS and TC programs for any PTS meeting all the assumptions of corollary 7 except, possibly, normalization. Let me temporarily defer the objection that this is a cheap specification, met even by the program that always fails.

Representing partial functions. To represent partiality in ECC, I will use the notion of an *option type*. For example, in SML there is the type

```
datatype 'a option = SOME of 'a | NONE;
```

There are two ways to produce an element of `'a option`, one of them, `SOME`, requires evidence, i.e. an object of type `'a`, while the other, `NONE`, requires nothing. An object of type `'a option` can be destructed to see if it contains evidence or not. For our purpose of representing partiality, `NONE` is better named `MAYBE`, because it contains no information at all, not the information that some computation actually fails to terminate.

[5] In [Pol94] I show that the approach of this paper can be extended to include ECC; in [vBJMP94] we show how to remove the requirement for semifullness, and (less satisfactorily) the requirement for functionality.

Partial Normalization Functions. Inductively define a predicate $\mathsf{optNormalizing}(M)$ with the constructors

$$\frac{\mathsf{normalizing}(M)}{\mathsf{optNormalizing}(M)}\ \text{ONSOME} \qquad \frac{}{\mathsf{optNormalizing}(M)}\ \text{ONMAYBE}.$$

Any proof of

$$\forall M\ .\ \ldots\ \Rightarrow\ \mathsf{optNormalizing}(M) \tag{3}$$

is a sound normalization program: given M (and possibly some other data) it returns either a normal form of M, or no information. Let $\rightrightarrows$ be some unspecified effective reduction strategy, and n some unspecified number. I have in mind an implementation of specification (3) that computes up to n steps of $\rightrightarrows$ starting from M, and returns ONSOME if a normal form is reached, and ONMAYBE otherwise.

If we are satisfied to typecheck PTS that we believe are normalizing, then the reduction strategy used in a partial normalization program is not of (theoretical) importance. However, if we want to typecheck non-normalizing PTS, we should use a reduction strategy, $\rightrightarrows$, that is *cofinal* in the sense that if $A \twoheadrightarrow B$ then $\exists C\ .\ A \rightrightarrows^{*} C$ and $B \twoheadrightarrow C$. An example of such a strategy is *complete development*, that contracts all the redexes in a term at once. This relation, which we also use for our proof of Church-Rosser, is already formalized in LEGO.

Remark. The normalization function required by corollary 7 is specified by

$$\forall M\ .\ \mathsf{PTS\text{-}term}(M) \Rightarrow \mathsf{normalizing}(M). \tag{4}$$

Consider the use of the premise $\mathsf{PTS\text{-}term}(M)$ in specification (4). For the supplier of a program meeting this specification, this premise provides a "recursor" necessary for normalizing M, as general recursion is not available in ECC. Since our plan for implementing partial normalization is just to compute n steps of some unspecified reduction strategy, we don't need this premise. However, for the user of the normalization program, i.e. a TS or TC program, this premise is hygienic: it prevents such a user from committing to normalization of a term until it is known that this term is really well typed, thus preventing unnecessary incompleteness. Even for a non-normalizing PTS such as λ^{*}, it is hard to find non-normalizing typable terms, so we might well decide to keep this premise in a specification of partial normalization, even though it is not used in the computation of partial normalization.

Partial TS and TC. Inductively define a relation $\mathsf{optTypChk}(\Gamma, M, A)$ with constructors

$$\frac{\Gamma \vdash M : A}{\mathsf{optTypChk}(\Gamma, M, A)}\ \text{OTCSOME} \qquad \frac{}{\mathsf{optTypChk}(\Gamma, M, A)}\ \text{OTCMAYBE}.$$

Any proof of

$$\forall \Gamma, M, A\ .\ \ldots\ \Rightarrow\ \mathsf{optTypChk}(\Gamma, M, A) \tag{5}$$

is a sound typechecker: given Γ, M, A (and possibly some other data) it returns either a derivation of $\Gamma \vdash M : A$, or no information. In particular, modifying the specification of corollary 7 we have:

$$\begin{aligned} \forall \Gamma, M, A \,.\quad & \ldots \\ & (\forall X \,.\, \mathsf{PTS\text{-}term}(X) \Rightarrow \mathsf{optNormalizing}(X)) \Rightarrow \\ & \mathsf{optTypChk}(\Gamma, M, A). \end{aligned} \tag{6}$$

A proof of specification (6) is at hand: the proof of corollary 7 given above will do, except that we call the partial normalization program instead of a total normalization program, and fail if the normalization program fails. Also we can omit the reasoning justifying the negative disjuncts of that lemma, as we have now allowed ourselves to return "no information" without any evidence at all.

Remark. We could throw away less information than I have suggested by having a third constructor for optTypChk

$$\frac{\neg \Gamma \vdash M : A}{\mathsf{optTypChk}(\Gamma, M, A)}\ \text{OTCNONE}.$$

With this definition, the proof of specification (6) can retain the information on failure that the proof of corollary 7 already contains.

Remark. For brevity I have shown the partial versions of only the normalizing and TypChk relations. In fact, partiality propagates throughout a proof of specification (6), and we need partial versions of several other relations, such as whether a term reduces to a sort, and whether two terms convert.

Completeness. Although we can see that the only cause of failure of our partial typechecker is failure of the given partial normalization program, specification (6) is very weak, as it allows returning OTCMAYBE without any justification. In order to be more precise, we can fix a reduction strategy, and index all our relations with the number of steps of reduction they may use. Then we can express that "if a judgement is derivable there is some n such that the typechecker succeeds on that judgement when allowed at most n steps of reduction". This works even for non-normalizing PTS if we use a cofinal reduction strategy. Some technical details of such an approach are worked out in section 7 of [vBJMP94].

5.2 Efficiency

A proof of corollary 7 or specification (6) is a (partially) correct typechecking program for some type theories we are interested in. Can we expect to *actually* run such a type checker?

One reason why we cannot actually run our typechecker is the size of sdsf-derivation trees; this is discussed in section 2.4. More efficient formulations of sdsf are known; they are harder to reason about, but we have no choice if we hope to execute a verified typechecker.

Another reason we cannot run our partially correct typechecker on the currently distributed LEGO is that LEGO is *very* slow at computing in its object languages. Sorting short lists has been known to take hours; an enterprising user [Bai93] actually burned 56 hours on a big workstation factoring a small polynomial. One reason for this is that LEGO, built to be an interactive proofchecker, does not use internal representation selected for fast computation, but for simplicity and a clear correspondence with the user's concrete representation. However, there is some recent work on the problem of "intensional representations" with efficient computation, e.g. [NW93], and there is no reason why a proofchecker cannot be much better than LEGO in this regard.

Finally, and most difficult in the long term, is the problem of efficiently executing the computational content of constructive proofs. There is now much literature about program extraction from constructive proofs, and some type theory implementations, such as Nuprl [Con86] and Coq [DFH+93] have an extraction mechanism, although LEGO does not. Whether such approaches can produce feasible programs from a proof of something like corollary 7 remains to be seen.

References

[Bai93] Anthony Bailey. Representing algebra in LEGO. Master's thesis, University of Edinburgh, 1993.

[Bar91] Henk Barendregt. Introduction to Generalised Type Sytems. *J. Functional Programming*, 1(2):125–154, April 1991.

[Bar92] Henk Barendregt. Lambda calculi with types. In Abramsky, Gabbai, and Maibaum, editors, *Handbook of Logic in Computer Science*, volume II. Oxford University Press, 1992.

[Ber90] Stefano Berardi. *Type Dependence and Constructive Mathematics*. PhD thesis, Dipartimento di Informatica, Torino, Italy, 1990.

[Con86] Robert L. Constable, *et. al. Implementing Mathematics with the Nuprl Proof Development System*. Prentice–Hall, Englewood Cliffs, NJ, 1986.

[DB93] Gilles Dowek and Robert Boyer. Towards checking proof checkers. In Herman Geuvers, editor, *Informal Proceedings of the Nijmegen Workshop on Types for Proofs and Programs*, May 1993.

[DFH+93] Dowek, Felty, Herbelin, Huet, Murthy, Parent, Paulin-Mohring, and Werner. The Coq proof assistant user's guide, version 5.8. Technical report, INRIA-Rocquencourt, February 1993.

[Geu93] Herman Geuvers. *Logics and Type Systems*. PhD thesis, Department of Mathematics and Computer Science, University of Nijmegen, 1993.

[GN91] Herman Geuvers and Mark-Jan Nederhof. A modular proof of strong normalization for the calculus of constructions. *Journal of Functional Programming*, 1(2):155–189, April 1991.

[HP91] Robert Harper and Robert Pollack. Type checking with universes. *Theoretical Computer Science*, 89:107–136, 1991.

[Hue89] Gérard Huet. The constructive engine. In R. Narasimhan, editor, *A Perspective in Theoretical Computer Science.* World Scientific Publishing, 1989. Commemorative Volume for Gift Siromoney.

[JP93] Claire Jones and Randy Pollack. Incremental changes in LEGO: 1993. Available by anonymous ftp with LEGO distribution, May 1993.

[JP94] Claire Jones and Randy Pollack. Incremental changes in LEGO: 1994. Available by anonymous ftp with LEGO distribution, May 1994.

[LP92] Zhaohui Luo and Robert Pollack. LEGO proof development system: User's manual. Technical Report ECS-LFCS-92-211, LFCS, Computer Science Dept., University of Edinburgh, The King's Buildings, Edinburgh EH9 3JZ, May 1992. Updated version. Available by anonymous ftp with LEGO distribution.

[Luo94] Z. Luo. *Computation and Reasoning: A Type Theory for Computer Science.* International Series of Monographs on Computer Science. Oxford University Press, 1994.

[McK94] James McKinna. Typed λ-calculus formalized: Church-Rosser and standardisation theorems. In preparation, 1994.

[MP93] James McKinna and Robert Pollack. Pure Type Sytems formalized. In M.Bezem and J.F.Groote, editors, *Proceedings of the International Conference on Typed Lambda Calculi and Applications, TLCA'93*, pages 289–305. Springer-Verlag, LNCS 664, March 1993.

[NW93] Gopalan Nadathur and Debra Sue Wilson. A notation for lambda terms I: A generalization of environments. Technical Report Technical Report CS-1993-22, Duke University, 1993.

[Pol92] R. Pollack. Typechecking in Pure Type Sytems. In *Informal Proceedings of the 1992 Workshop on Types for Proofs and Programs, Båstad, Sweden*, pages 271–288, June 1992. Available by ftp.

[Pol94] Robert Pollack. *The Theory of LEGO: A Proof Checker for the Extended Calculus of Constructions.* PhD thesis, University of Edinburgh, 1994. Available by anonymous ftp from `ftp.cs.chalmers.se` in directory `pub/users/pollack`.

[vBJ93] L.S. van Benthem Jutting. Typing in Pure Type Sytems. *Information and Computation*, 105(1):30–41, July 1993.

[vBJMP94] L.S. van Benthem Jutting, James McKinna, and Robert Pollack. Checking algorithms for Pure Type Systems. In Henk Barendregt and Tobias Nipkow, editors, *Types for Proofs and Programs: International Workshop TYPES'93, Nijmegen, May 1993*, volume 806 of *LNCS*, pages 19–61. Springer-Verlag, 1994.

Categorical semantics of the call-by-value λ-calculus

A. Pravato*‡ S. Ronchi della Rocca*‡ L. Roversi†‡

Abstract

The denotational semantics of the call-by-value λ-calculus in a categorical setting is given. Furthermore, a particular model based on coherence domains is studied.

1 Introduction

The call-by-value λ-calculus is a restriction of the classical λ-calculus, based on the notion of *value*. A value is a term which is either a variable or an abstraction. The call-by-value λ-calculus is obtained from the classical one by restricting the evaluation rule (the β-rule) to redexes whose operand is a value. The call-by-value λ-calculus was introduced by Plotkin [14] in order to define a paradigmatic language for modeling two important features both present in the implementation of many real programming languages: the call-by-value and the lazy evaluation. An evaluation is call-by-value if it evaluates parameters before they have been passed. It is lazy if it evaluates function bodies only when parameters are supplied. These features were implemented in the SECD machine, defined by Landin [13] for computing λ-terms. Here we will deal with the semantics of the "pure" (i.e., without constants) call-by-value λ-calculus $(\lambda\beta_v)$. Following the work of Plotkin, an operational semantics can be defined for it, inducing the following equivalence: given two terms M and N,

$$M \sim_v N \Leftrightarrow (\forall \text{ closing context } C[\,].\ C[M] \text{ reduces to a value } \Leftrightarrow C[N] \text{ reduces to a value }).$$

This definition of operational semantics corresponds to the Leibnitz principle for programs. Namely, a program (closed term) is characterized by its observational behaviour, and so two subprograms (terms) will be equivalent if they can be replaced each other in the same program without changing its behaviour. In a

*Addr.: Università degli studi di Torino, Dipartimento di Informatica, C.so Svizzera 185 - 10149 TORINO. E-mail: {pravato,ronchi}@di.unito.it

†Addr.: Università degli studi di Pisa, Dipartimento di Informatica, C.so Italia 40 - 56125 PISA. E-mail: rover@di.unipi.it

‡Work partially supported by the HCM project CHRX-CT92-0046 "Typed λ-calculus"

language without constants, like $\lambda\beta_v$, the simplest observational property is the termination one.

A definition of a model for $\lambda\beta_v$ was given in [4], following the Hindley-Longo approach for defining a model for λ-calculus [8]. Moreover, in [4], the denotational semantics of $\lambda\beta_v$ was studied over Scott domains. Namely, the model arising from the initial solution of the domain equation: $D = [D \rightarrow_\perp D]_\perp$, where $[D \rightarrow_\perp D]_\perp$ denotes the lifted space of strict continuous functions, has been investigated. In particular, it was proved that the model is correct but not complete w.r.t. the operational semantics, i.e., denotational equivalence implies the operational one but not vice-versa. However, the correctness is enough for proving interesting operational properties of the languages, like extensionality on values, for example. However, in [4], a fully abstract model, i.e. both correct and complete, has been built by a collapse of the preceding model, based on a notion of bisimulation.

It would be interesting to give also a categorical characterization of a $\lambda\beta_v$ model. Remember that models of λ-calculus have a very nice categorical description: they are all, and only, the *reflexive* objects of a *cartesian closed category* with enough points, where an object A is reflexive if, and only if, A^A is a retract of it (Notation: $A \triangleright A^A$). A categorical characterization of $\lambda\beta_v$-models cannot be obtained by modifying or restricting the previous one: in fact, looking at the model $[D \rightarrow_\perp D]_\perp$, the category of Scott domains and strict continuous functions is not cartesian closed.

In this paper we give a categorical definition of a $\lambda\beta_v$-model. We start from looking at a different setting for studying the semantics of $\lambda\beta_v$, namely the *coherence domains*, defined by Girard [6], [5]. The natural counterpart in coherence domains of the Scott domain $[D \rightarrow_\perp D]_\perp$ is $!(D \multimap D)$, where $\multimap$ is the linear implication. This correspondence was first stated by Girard, and it was used in [7] for building an optimal reduction machine for β-reduction, translating $\lambda\beta_v$ into a variation of proof-nets. It turns out that a suitable class of categories for interpreting $\lambda\beta_v$ is a restriction of the one defined in [2] for interpreting the multiplicative fragment of linear logic. This class of categories is general enough for grasping models built in very different settings. In fact, we prove that every space D such that either D is a Scott domain and $D \triangleright [D \rightarrow_\perp D]_\perp$ or D is a coherence domain and $D \triangleright !(D \multimap D)$, induces a categorical model of $\lambda\beta_v$. Unfortunately, this is not a complete characterization of the $\lambda\beta_v$-models. Indeed, in [9], a model for $\lambda\beta_v$ was studied by introducing a particular kind of coherence domains: the "pointed" one. It is easy to see that this model is not an instance of our categorical definition.

Moreover, we study the coherence model $\mathcal{M}_!$ based on the initial solution of the domain equation $D = !(D \multimap D)$. For studying the theory induced by this model we adapt the technical tools introduced in [10] for reasoning about the interpretations of λ-terms in models of λ-calculus built on qualitative domains. It is interesting to compare the $\lambda\beta_v$-theory of $\mathcal{M}_!$ with the $\lambda\beta_v$-theories of the other two models previously cited. All models are correct but not complete w.r.t. the operational semantics. The two coherence models, $\mathcal{M}_!$ and that one studied in [9], induce different theories. Namely, in the model in [9] the extensionality on values holds, while it does not hold in $\mathcal{M}_!$. However, $\mathcal{M}_!$ equates every two β-convertible closed and normalizing terms of λI-calculus, while the model in [9] does not. In the model studied in [4], based on Scott domains, both these two equalities holds.

In the paper we will assume a basic knowledge of both category theory and coherence domains.

2 The call-by-value λ-calculus

The call-by-value λ-calculus ($\lambda\beta_v$) is defined by the pair (Λ, *Val*), where Λ is the set of terms of pure λ-calculus over a set of variables *Var*, and *Val* is the *set of values*, defined as $Val = Var \cup \{\lambda x.M \mid M \in \Lambda\}$. Free and bound variables are defined as usual. For every term M, $\mathcal{FV}(M)$ denotes the set of free variables in M. $\Lambda^o \subset \Lambda$ is the set of all closed terms. Terms in Λ are considered modulo α-equivalence, i.e, up to the name of bound variables.

Definition 1 *i) The* call-by-value reduction *is:*

$$(\beta_v) \qquad (\lambda x.M)N \rightarrow_v [N/x]M \text{ if } N \in Val,$$

where $[N/x]M$ denotes the substitution of N for every free occurrence of x in M, avoiding the capture of free variables in N. The contextual, reflexive and transitive closure of $\rightarrow_v$ is $\Longrightarrow_v$. The symmetric closure of $\Longrightarrow_v$ is $=_v$.

ii) A term M is valuable *iff it β_v-reduces to a value. Notation: $M \Longrightarrow_v Val$.*

(β_v) leads to the definition of the *input-output relation* $M \Downarrow_v V$, where $M \in \Lambda^o$ and $V \in Val$. The judgments $M \Downarrow_v V$ are derivable by the rules

$$\frac{}{\lambda x.M \Downarrow_v \lambda x.M} \qquad \frac{M \Downarrow_v \lambda x.M' \quad N \Downarrow_v V' \quad [V'/x]M' \Downarrow_v V}{MN \Downarrow_v V}$$

We have that $M \Downarrow_v V$ implies $M \Longrightarrow_v V$. The other implication does not hold.

Let $M \Downarrow_v$ denotes $\exists V.M \Downarrow_v V$. The relation $\Downarrow_v$ induces an *operational equivalence* among terms based on the *observational behaviour* w.r.t. the *termination property*: $(M \sim_v N) \Leftrightarrow (\forall C[\,].\, C[M], C[N] \in \Lambda^o \Rightarrow (C[M] \Downarrow_v \Leftrightarrow C[N] \Downarrow_v))$.

3 Syntactical model for $\lambda\beta_v$

A model for the call-by-value λ-calculus must provide a semantic account of valuable terms. This can be achieved by defining a subset of the interpretation domain, the set of semantic values, where valuable terms are interpreted. Environments must map variables to semantic values since variables are values.

A general definition of a model for $\lambda\beta_v$, following the Hindley-Longo approach for defining a λ-calculus model [8], has been given in [4]. We recall here this definition for sake of completeness.

Definition 2 *A* syntactical model for the call-by-value λ-calculus *is a structure* $\mathcal{M} = \langle S, V, \bullet, \mathcal{V}\rangle$, *where* $V \subset S$ *is the set of semantic values,* $\bullet : S \times S \to S$ *and* $\mathcal{V} : \Lambda \times (Var \to V) \to S$ *satisfies the following conditions:*

1. $\mathcal{V}[x]\rho = \rho(x)$,
2. $\mathcal{V}[M\ N]\rho = \mathcal{V}[M]\rho \bullet \mathcal{V}[N]\rho$,
3. *if* $d \in V$ *then* $\mathcal{V}[\lambda x.M]\rho \bullet d = \mathcal{V}[M]\rho_x^d$,
4. *if* $\forall x \in \mathcal{FV}(M).\, \rho(x) = \rho'(x)$ *then* $\mathcal{V}[M]\rho = \mathcal{V}[M]\rho'$,
5. *if* $y \notin \mathcal{FV}(M)$ *then* $\mathcal{V}[\lambda x.M]\rho = \mathcal{V}[\lambda y.[y/x]M]\rho$,
6. *if* $\forall d \in V.\, \mathcal{V}[M]\rho_x^d = \mathcal{V}[N]\rho_x^d$ *then* $\mathcal{V}[\lambda x.M]\rho = \mathcal{V}[\lambda x.N]\rho$,
7. *if* $M \Longrightarrow_v Val$ *then* $\forall \rho.\, \mathcal{V}[M]\rho \in V$,

where ρ_x^d *behaves as* ρ *on every* $y \neq x$, *while* $\rho(x) = d$.

4 A categorical interpretation of $\lambda\beta_v$

In this section we define the properties a category must enjoy for interpreting $\lambda\beta_v$. The class of categories we are going to define is a restriction of the class given in [2], for modeling the multiplicative and exponential fragment of intuitionistic linear logic. The relation between the semantics of $\lambda\beta_v$ and (this fragment of) linear logic will be clear in a few. Indeed, we shall prove that a natural setting for interpreting $\lambda\beta_v$ is the category of coherence domains.

Definition 3 *A* call-by-value linear category **Cbv** *is a category such that:*

- **Cbv** *is monoidal symmetric with respect to the bifunctor* $\otimes$. *The unity of* $\otimes$ *is* $\mathbb{1}$.
- **Cbv** *is closed with respect to the bifunctor* $\multimap$, *i.e. for all* $A, B, C \in Obj_{\mathbf{Cbv}}$ *there exists a natural isomorphism*

$$\Lambda_{A,B,C} : Hom_{\mathbf{Cbv}}(A \otimes B, C) \to Hom_{\mathbf{Cbv}}(A, B \multimap C).$$

- *In* **Cbv** *there is a comonad* $(!, \delta :! \to !!, out :! \to ID_{\mathbf{Cbv}})$ *such that:*
  - *the functor* ! *is monoidal symmetric and the maps out and* δ *are monoidal natural transformations, where* $ID_{\mathbf{Cbv}}$ *is the identity endofunctor over* **Cbv**;
  - *if, by an abuse of notation,* $\mathbb{1}$ *is also the obvious constant functor on the category* **Cbv**, *then there exist natural transformations* $E :! \to \mathbb{1}$ *and* $Dup :! \to ! \otimes !$ *such that, for all* $A \in Obj_{\mathbf{Cbv}}$, $(!A, Dup_A, E_A)$ *is a comonoid;*

- *for all $A, B \in Obj_{\mathbf{Cbv}}$, $m_{A,B} : !A \otimes !B \to !(A \otimes B)$ and $m_{1\!\!1} : 1\!\!1 \to !1\!\!1$ are the morphisms making ! a monoidal functor and δ and out monoidal natural transformations;*
- *δ is an element both of $\otimes$-**coalg**$_{\mathbf{Cbv}}((!A, D_A), (!!A, D_{!A}))$ and of $1\!\!1$-**coalg**$_{\mathbf{Cbv}}((!A, E_A), (!!A, E_{!A}))$.*

- *There is an object $\mathcal{D} \in Obj_{\mathbf{Cbv}}$ such that $\mathcal{D} \triangleright !(\mathcal{D} \multimap \mathcal{D})$ under $F : \mathcal{D} \to !(\mathcal{D} \multimap \mathcal{D})$ and $G : !(\mathcal{D} \multimap \mathcal{D}) \to \mathcal{D}$.*

We call $\mathcal{D}$ the model *object of* **Cbv**.

In the following, for a better reading, we shall drop subscripts and superscripts on the morphisms of **Cbv**, when they will be clear from the context.

Notation Let $A, B, C, \ldots$ range over $Obj_{\mathbf{Cbv}}$.

- We name $ev_{B,C}$ the evaluation morphism such that, for all $f : A \otimes B \to C$, the following diagram commutes:

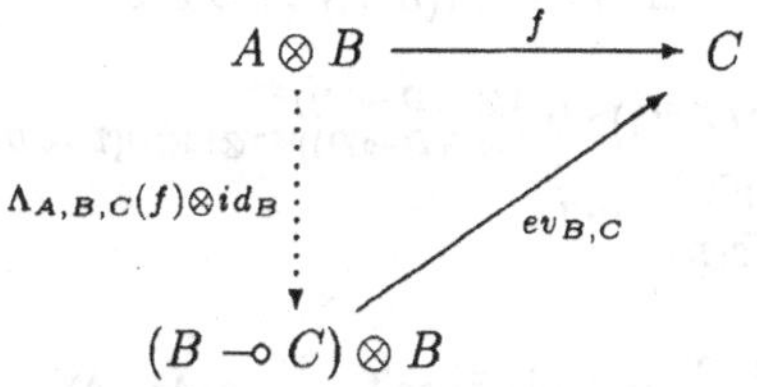

- Let A be either a morphism or an object of the category. By $A^{\otimes n}$ we denote the tensor product $A \otimes \cdots \otimes A$, n times.

In the next definition, we introduce some morphisms useful for defining the interpretation of a term in a simple and concise way, while preserving the correctness.

Definition 4 • *For all $A_1, \ldots, A_n \in Obj_{\mathbf{Cbv}}$ and for every permutation σ of the sequence $1, \ldots, n$, we call $Exc^{A_1 \otimes \cdots \otimes A_n}_{A_{\sigma(1)} \otimes \cdots \otimes A_{\sigma(n)}}$ the natural isomorphism between $A_1 \otimes \cdots \otimes A_n$ and $A_{\sigma(1)} \otimes \cdots \otimes A_{\sigma(n)}$. The isomorphism exists because of the definition of* **Cbv**.

- *Let $H^{\gamma}_{\gamma'} : \gamma \to \gamma'$ be an abbreviation for $Exc^{\eta}_{\gamma'} \circ (Dup_{1\!\!1} \otimes Dup_{A_1} \otimes \cdots \otimes Dup_{A_n})$, where $\gamma = !1\!\!1 \otimes !A_1 \otimes \cdots \otimes !A_n$ and $\eta = !1\!\!1 \otimes !1\!\!1 \otimes !A_1 \otimes !A_1 \otimes \cdots \otimes !A_n \otimes !A_n$ and $n \geq 0$.*

- *Let assume to have $j \in \{1, \ldots, n\}$. For every $1 \leq i \neq j \leq n$, let $A_i \in Obj_{\mathbf{Cbv}}$ have the form $!B_i$ and $A_j \in Obj_{\mathbf{Cbv}}$. Then $\pi^j_{A_1 \otimes \cdots \otimes A_n} : A_1 \otimes \cdots \otimes A_n \to A_j$ is the morphism $iso_{A_j} \circ (E_{B_1} \otimes \cdots \otimes E_{B_{j-1}} \otimes id_{A_j} \otimes E_{B_{j+1}} \otimes \cdots \otimes E_{B_n})$, where iso_A is the natural isomorphism between $1\!\!1 \otimes \cdots \otimes 1\!\!1 \otimes A \otimes 1\!\!1 \otimes \cdots \otimes 1\!\!1$ and A.*

- *Let $m_{A_1,\ldots,A_n}$ be the obvious generalization of $m_{A,B}$:*
  - $m^n_A : !!\mathbb{1} \otimes (!A)^{\otimes n} \to !(!\mathbb{1} \otimes A^{\otimes n})$ *is defined as* $m^n_A = m_{!\mathbb{1},A,\ldots,A}$,
  - $\mathsf{m}^n : (\mathbb{1}^{\otimes n}) \to !(\mathbb{1}^{\otimes n})$ *is defined as* $\mathsf{m}^n = m_{\mathbb{1},\ldots,\mathbb{1}} \circ m^{\otimes n}_{\mathbb{1}}$.

4.1 The interpretation function

Following [1] and [11], given a **Cbv** category with a model object $\mathcal{D}$, we will interpret a term M, with free variables $\{x_1, \ldots, x_n\}$, as a morphism from $!\mathbb{1} \otimes \mathcal{D}^{\otimes n}$ to $\mathcal{D}$. $!\mathbb{1}$ is needed for interpreting closed terms.

Definition 5 *Let $M \in \Lambda$ such that $\mathcal{FV}(M) \subseteq \{x_1, \ldots, x_n\}$. Let $\mathcal{C}(\mathcal{D})$ be a* **Cbv** *category with $\mathcal{D}$ as model object. The* interpretation function $[\![.]\!]^{\mathcal{C}(\mathcal{D})}$ *such that $[\![x_1, \ldots, x_n \vdash M]\!]^{\mathcal{C}(\mathcal{D})} \in Hom(!\mathbb{1} \otimes \mathcal{D}^{\otimes n}, \mathcal{D})$ is defined by induction on M as follows:*

1. $[\![x_1, \ldots, x_n \vdash x_i]\!]^{\mathcal{C}(\mathcal{D})} = G \circ \pi^{i+1}_{!\mathbb{1}\otimes(!(\mathcal{D}\multimap\mathcal{D}))^{\otimes n}} \circ (id_{!\mathbb{1}} \otimes F^{\otimes n})$,

2. $[\![x_1, \ldots, x_n \vdash MN]\!]^{\mathcal{C}(\mathcal{D})} = ev_{\mathcal{D},\mathcal{D}} \circ ((out_{\mathcal{D}\multimap\mathcal{D}} \circ F \circ h) \otimes g) \circ r$,
   where $r =$
   $((id_{!\mathbb{1}} \otimes G^{\otimes n}) \otimes (id_{!\mathbb{1}} \otimes G^{\otimes n})) \circ H^{!\mathbb{1}\otimes(!(\mathcal{D}\multimap\mathcal{D}))^{\otimes n}}_{!\mathbb{1}\otimes(!(\mathcal{D}\multimap\mathcal{D}))^{\otimes n}\otimes!\mathbb{1}\otimes(!(\mathcal{D}\multimap\mathcal{D}))^{\otimes n}} \circ (id_{!\mathbb{1}} \otimes F^{\otimes n})$,
   $h = [\![x_1, \ldots, x_n \vdash M]\!]^{\mathcal{C}(\mathcal{D})}$ *and*
   $g = [\![x_1, \ldots, x_n \vdash N]\!]^{\mathcal{C}(\mathcal{D})}$,

3. $[\![x_1, \ldots, x_n \vdash \lambda x.M]\!]^{\mathcal{C}(\mathcal{D})} = G \circ !f^{\approx} \circ m^n_{!(\mathcal{D}\multimap\mathcal{D})} \circ (\delta_{\mathbb{1}} \otimes \delta^{\otimes n}_{(\mathcal{D}\multimap\mathcal{D})}) \circ (id_{!\mathbb{1}} \otimes F^{\otimes n})$,
   where
   $f = \Lambda_{!\mathbb{1}\otimes\mathcal{D}^{\otimes n},\mathcal{D},\mathcal{D}}([\![x_1, \ldots, x_n, x \vdash M]\!]^{\mathcal{C}(\mathcal{D})}) \in Hom(!\mathbb{1} \otimes \mathcal{D}^{\otimes n}, \mathcal{D} \multimap \mathcal{D})$,
   $f^{\approx} = f \circ (id_{!\mathbb{1}} \otimes G^{\otimes n}) \in Hom(!\mathbb{1} \otimes (!(\mathcal{D} \multimap \mathcal{D}))^{\otimes n}, \mathcal{D} \multimap \mathcal{D})$,
   and, thus, $!f^{\approx} \in Hom(!(!\mathbb{1} \otimes !(\mathcal{D} \multimap \mathcal{D})^{\otimes n}), !(\mathcal{D} \multimap \mathcal{D}))$.

4.2 The categorical model

In this subsection we will prove that a model object of a **Cbv** category is a syntactical model of $\lambda\beta_v$, as defined in Definition 2.

First of all, we must introduce in the categorical setting all the necessary notions for defining a syntactical model.

Definition 6 *Let $\mathcal{C}(\mathcal{D})$ denote a* **Cbv** *category with a model object $\mathcal{D}$. The* categorical model $\mathcal{M}^{\mathcal{C}(\mathcal{D})}$ *is the structure $\langle S, V, \bullet, \mathcal{V}\rangle$, where:*

- $S = Hom(\mathbb{1}, \mathcal{D})$. *Let notice that $Hom(\mathbb{1}, \mathcal{D}) \approx Hom(\mathbb{1}^{\otimes n}, \mathcal{D})$ for all $n \geq 1$. Therefore, we consider S modulo isomorphisms.*
- $V = \{f \mid f \in Hom(\mathbb{1}^{\otimes n}, \mathcal{D})$ *and* $\exists h \in Hom(\mathbb{1}^{\otimes n}, \mathcal{D} \multimap \mathcal{D}).f = G \circ !h \circ \mathsf{m}^n\}$,
- $f \bullet g = ev_{\mathcal{D},\mathcal{D}} \circ ((out_{\mathcal{D}\multimap\mathcal{D}} \circ F \circ f) \otimes g)$, *for every pair of morphisms $f, g \in Hom(\mathbb{1}, \mathcal{D})$,*

- $\mathcal{V}[M]\rho = [\![x_1,\ldots,x_n \vdash M]\!]^{\mathcal{C}(\mathcal{D})} \circ (m_{\mathbb{1}} \otimes \rho(x_1) \otimes \cdots \otimes \rho(x_n))$, *where* $\mathcal{FV}(M) = \{x_1,\ldots,x_n\}$. *We call every* $\rho(x_i)$ environment component. *Since* ρ *maps variables to values, every environment component* $\rho(x_i)$ *is of the form* $Go!h_i \circ m_{\mathbb{1}}$, *for some* $h_i (1 \leq i \leq n)$.

  In this way if $x \notin \{x_1,\ldots,x_n\}$, *then* ρ^d_x *is* $(m_{\mathbb{1}} \otimes \rho(x_1) \otimes \cdots \otimes \rho(x_n)) \otimes d$, *where* $d \in V$.

Looking at Definition 6 from a set theoretical perspective, the morphisms in S select the elements of $\mathcal{D}$. The set V of semantic values contains morphisms obtained through the ! operator. Moreover, an environment component is a morphism of S which is a semantic value.

Theorem 1 *Let* $\mathcal{C}(\mathcal{D})$ *be a* **Cbv** *category with a model object* $\mathcal{D}$. *Its categorical model* $\mathcal{M}^{\mathcal{C}(\mathcal{D})}$ *is a syntactical model for the call-by-value* λ*-calculus.*

Proof. We shall prove that $\mathcal{M}^{\mathcal{C}(\mathcal{D})}$ satisfies Definition 2. To prove condition 1 we use the commuting diagram between the natural transformation E (look at the definition of π, Definition 4) and a morphism of the form $!h$. Condition 2 comes both from the definition of $\bullet$ and from the requirement that δ belongs both to $\otimes$-**coalg**$_{\mathbf{Cbv}}((!A, D_A), (!!A, D_{!A}))$ and $\mathbb{1}$-**coalg**$_{\mathbf{Cbv}}((!A, E_A), (!!A, E_{!A}))$. Condition 3 is proved as follows. Let $\mathcal{FV}(\lambda x.M) = \{x_1,\ldots,x_n\}$. In what follows, for simplicity, we use the more compact notation: $\forall i \in \{1,\ldots,n\}.\rho(x_i) = \rho_i$. Let also $d \in V$. Step by step, we have that:

$$\begin{aligned}
&\mathcal{V}[\lambda x.M]_\rho \bullet d = \\
&\quad = ev \circ ((out \circ F \circ [\![x_1,\ldots,x_n \vdash \lambda x.M]\!]^{\mathcal{C}(\mathcal{D})} \circ (m_{\mathbb{1}} \otimes \rho_1 \otimes \cdots \otimes \rho_n)) \otimes d) \\
&\quad = ev \circ ((\Lambda_{!\mathbb{1}\otimes\mathcal{D}^{\otimes n},\mathcal{D},\mathcal{D}}([\![x_1,\ldots,x_n,x \vdash M]\!]^{\mathcal{C}(\mathcal{D})}) \circ (m_{\mathbb{1}} \otimes \rho_1 \otimes \cdots \otimes \rho_n)) \otimes d) \\
&\quad = ev \circ (\Lambda_{!\mathbb{1}\otimes\mathcal{D}^{\otimes n},\mathcal{D},\mathcal{D}}([\![x_1,\ldots,x_n,x \vdash M]\!]^{\mathcal{C}(\mathcal{D})}) \otimes id) \circ ((m_{\mathbb{1}} \otimes \rho_1 \otimes \cdots \otimes \rho_n) \otimes d) \\
&\quad = [\![x_1,\ldots,x_n,x \vdash M]\!]^{\mathcal{C}(\mathcal{D})} \circ (m_{\mathbb{1}} \otimes \rho_1 \otimes \cdots \otimes \rho_n \otimes d) = \mathcal{V}[M]\rho^d_x,
\end{aligned}$$

where the second step is mainly due to the commuting diagram in Figure 1. Conditions 4, 5 and 6 are trivially satisfied. To show condition 7, remember that a term N is valuable if, and only if, N β_v-reduces either to a variable or to a term $\lambda x.M$, for some M. In both cases every morphism h obtained in the interpretation, when composed with an environment, becomes a value. In the first case because of the definition of environment. In the second case, assuming $\rho_i = Go!h_i \circ m_{\mathbb{1}} (1 \leq i \leq n)$:

$$\begin{aligned}
&\mathcal{V}[\lambda x.M]_\rho = \\
&\quad = Go!f^{\approx} \circ m^n_{!(\mathcal{D}\multimap\mathcal{D})} \circ (\delta_{\mathbb{1}} \otimes \delta^{\otimes n}_{(\mathcal{D}\multimap\mathcal{D})}) \circ (id_{!\mathbb{1}} \otimes F^{\otimes n}) \circ (m_{\mathbb{1}} \otimes \rho_1 \otimes \cdots \otimes \rho_n) \\
&\quad = Go!f^{\approx} \circ m^n_{!(\mathcal{D}\multimap\mathcal{D})} \circ ((\delta_{\mathbb{1}} \circ m_{\mathbb{1}}) \otimes (\delta_{(\mathcal{D}\multimap\mathcal{D})} d h_1 \circ m_{\mathbb{1}}) \otimes \cdots \otimes (\delta_{(\mathcal{D}\multimap\mathcal{D})} d h_n \circ m_{\mathbb{1}})) \\
&\quad = Go!f^{\approx} \circ m^n_{!(\mathcal{D}\multimap\mathcal{D})} \circ ((\delta_{\mathbb{1}} \circ m_{\mathbb{1}}) \otimes (!!h_1 \circ \delta_{\mathbb{1}} \circ m_{\mathbb{1}}) \otimes \cdots \otimes (!!h_n \circ \delta_{\mathbb{1}} \circ m_{\mathbb{1}})) \\
&\quad = Go!f^{\approx} \circ m^n_{!(\mathcal{D}\multimap\mathcal{D})} \circ ((!m_{\mathbb{1}} \circ m_{\mathbb{1}}) \otimes (!!h_1 \circ !m_{\mathbb{1}} \circ m_{\mathbb{1}}) \otimes \cdots \otimes (!!h_n \circ !m_{\mathbb{1}} \circ m_{\mathbb{1}})) \\
&\quad = Go!f^{\approx} \circ m^n_{!(\mathcal{D}\multimap\mathcal{D})} \circ (!m_{\mathbb{1}} \otimes !(!h_1 \circ m_{\mathbb{1}}) \otimes \cdots \otimes !(!h_n \circ m_{\mathbb{1}})) \circ m^{\otimes n}_{\mathbb{1}}
\end{aligned}$$

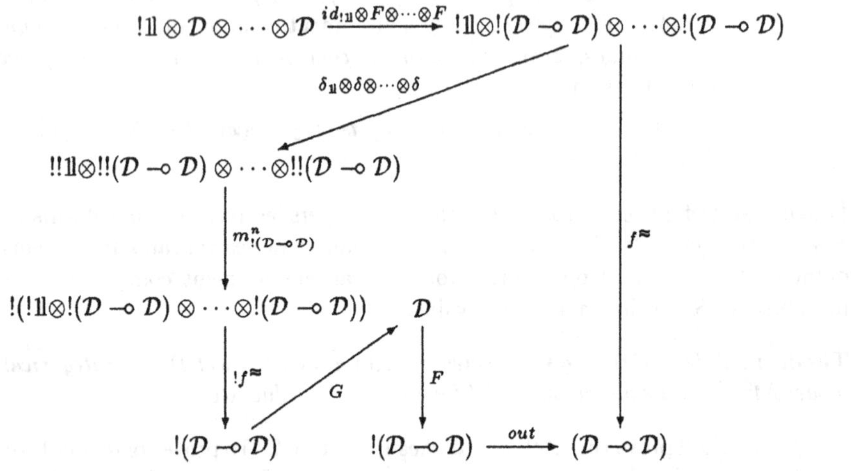

Figure 1: Commuting diagram for the application

$$= G \circ !f^{\approx} \circ !(\mathrm{m}_{\mathbb{1}} \otimes (!h_1 \circ \mathrm{m}_{\mathbb{1}}) \otimes \cdots \otimes (!h_n \circ \mathrm{m}_{\mathbb{1}})) \circ m_{\mathbb{1},\ldots,\mathbb{1}} \circ \mathrm{m}_{\mathbb{1}}^{\otimes n}$$
$$= G \circ !(f^{\approx} \circ (\mathrm{m}_{\mathbb{1}} \otimes (!h_1 \circ \mathrm{m}_{\mathbb{1}}) \otimes \cdots \otimes (!h_n \circ \mathrm{m}_{\mathbb{1}}))) \circ \mathrm{m}^n.$$

□

5 Two instances of Cbv

The definition of a categorical model allows to characterize some models of $\lambda\beta_v$ in different settings. The next theorem will deal with Scott domains. It proves that every model of $\lambda\beta_v$ belonging to the class defined in [3] is a categorical model in the previous sense.

Theorem 2 *Let $\mathcal{D}$ be a* Scott domain *such that $\mathcal{D} \rhd [\mathcal{D} \rightarrow_{\perp} \mathcal{D}]_{\perp}$. Then $\mathcal{D}$ gives a categorical model for $\lambda\beta_v$.*

Proof. Let us consider the category of Scott domains, and take:

- $\otimes$ as the functor for the *smash product*,
- $\multimap$ as the functor for the *strict continuous functions*,
- ! as the *lifting* functor $\perp$, whose canonical immersion and projection pair realizes the comonad,

for making $\mathcal{C}(\mathcal{D})$ a **Cbv** category with model object $\mathcal{D}$. The proof follows from Theorem 1. □

In [4] the initial solution to $D = [D \to_\perp D]_\perp$ in the category of Scott domain is extensively studied.

5.1 Coherence semantics of $\lambda\beta_v$

In this subsection we study a *coherence semantics* for $\lambda\beta_v$, namely, we use a category with *coherence domains* as objects and *linear functions* as morphisms.

We recall that a coherence domain A is defined by a set of atoms $|A|$ on which a symmetric and transitive *compatibility relation* is defined. The elements of A are all, and only, sets containing compatible atoms, ordered by the set inclusion relation. Therefore, $\emptyset \in A$. If A and B are two coherence domains, a function $f : A \to B$ is *linear* iff it is both stable and $f(\bigcup \mathcal{A}) = \bigcup\{f(a) \mid a \in \mathcal{A}\}$ for every $\mathcal{A} \subseteq A$. Details are in [12].

Notation Let A be a coherence domain. Atoms of A will be ranged over by $a, a_1, \ldots$. 1 denotes the unique atom of $\mathbb{1}$, namely $\mathbb{1} = \{\emptyset, \{1\}\}$. $c(a, a')$ means that a and a' are compatible. Let $\otimes$, $!$, and $\multimap$ be the functors over coherence domains such that:

- $|A \otimes B| = \{[a, b] \mid a \in |A| \text{ and } b \in |B|\}$, where $c([a, b], [a', b'])$ iff both $c(a, a')$ and $c(b, b')$.
- $|!A| = \{d \mid d \text{ is a finite element of } A\}$, where, if $d, d' \in |!A|$, then $c(d, d')$ iff $d \cup d' \in A$.
- $|A \multimap B| = \{(a, b) \mid a \in |A| \text{ and } b \in |B|\}$, where $c((a, b), (a', b'))$ iff $c(a, a')$ implies both $c(b, b')$ and if $b = b'$, then $a = a'$.
  The elements of $A \multimap B$ are linear traces of linear functions from A to B. If $f : A \to B$ is a linear function, then its *linear trace* is denoted by $ltr(f)$ and it is used as follows: $f(\{a_i \mid i \in I\}) = \{b_i \mid (a_i, b_i) \in ltr(f)\}$.

Theorem 3 *Let $\mathcal{D}$ be a* Coherence domain *such that $\mathcal{D} \rhd !(\mathcal{D} \multimap \mathcal{D})$. Then $\mathcal{D}$ gives a categorical model for $\lambda\beta_v$.*

Proof. We shall prove that the category of Coherence domains with $\mathcal{D}$ as model object is a **Cbv** category. Then, the result follows from Theorem 1. Let $\otimes$, $!$, and $\multimap$ be as in the previous notation. Let use the following definitions:

- If f is a linear function from A to B, then

$$ltr(!f) = \{(\emptyset, \emptyset)\} \cup \{(\{a_{i_1}, \ldots, a_{i_k}\}, \{b_{i_1}, \ldots, b_{i_k}\}) \mid (a_{i_j}, b_{i_j}) \in ltr(f) \text{ and } \{a_{i_1}, \ldots, a_{i_k}\}, \{b_{i_1}, \ldots, b_{i_k}\} \text{ are sets of compatible elements}$$

- The linear traces of the natural transformations for the comonad are

$$ltr(\delta_A) = \{(\bigcup \Omega, \Omega) \mid \Omega \in |!!A|\}, \qquad ltr(out_A) = \{(\{\alpha\}, \alpha) \mid \alpha \in |A|\}.$$

- The linear traces of the morphisms making ! monoidal are

$$ltr(m_{\mathbb{1}}) = \{(1,\emptyset),(1,\{1\})\}$$

$$\begin{aligned} ltr(m_{A_1,\ldots,A_n}) = \; & \{([\overbrace{\emptyset,\ldots,\emptyset}^{n}],\emptyset)\}\cup \\ & \{([\{a_1\},\ldots,\{a_n\}],\{[a_1,\ldots,a_n]\}) \mid \{a_i\} \in |!A_i| \text{ and} \\ & \quad \{[a_1,\ldots,a_n]\} \in |!(A_1 \otimes \ldots \otimes A_n)|\}. \end{aligned}$$

Remember that m_A^n and m^n are defined starting from $m_{A_1,\ldots,A_n}$.

- The linear traces of the morphisms giving the comonoid are

$$ltr(E_A) = \{(\emptyset,1)\}, \qquad ltr(Dup_A) = \{(\alpha,[\alpha_1,\alpha_2]) \mid \alpha_1 \cup \alpha_2 = \alpha \in |!A|\}.$$

□

5.1.1 The least solution of the domain equation $D = !(D \multimap D)$

By Theorem 3, it follows that we can build a canonical model $\mathcal{M}_! = \langle S, V, \bullet, \mathcal{V}\rangle$, for $\lambda\beta_v$, induced by the initial solution of the domain equation $D = !(D \multimap D)$ in the category of coherence domains and linear functions. Such solution is $\mathcal{D} = \lim_{n\to\infty} D_n$, where $|D_0| = \emptyset$, $D_{n+1} = !(D_n \multimap D_n)$ and $F : \mathcal{D} \to !(\mathcal{D} \multimap \mathcal{D})$ and $G : !(\mathcal{D} \multimap \mathcal{D}) \to \mathcal{D}$ are the induced isomorphisms.

Let us consider $\mathcal{M}_!$. The elements h of S are morphisms whose linear traces have components of the shape $(1,d)$, where $d \in |\mathcal{D}|$. Therefore, every h allows to pick up elements of $\mathcal{D}$ in the following way: $h(\{1\}) = \{d \mid (1,d) \in ltr(h)\}$. For this reason, we feel free to replace the morphisms of S by the elements of $\mathcal{D}$ they represent. In the same way, we can consider that the set of semantic values V is directly the set of elements of $\mathcal{D}$ selected by the morphisms in V. Let see the form of such semantic values, remembering that every value is of the shape $G \circ !h \circ \mathsf{m}^n$, for some h. An example follows. Let $h : \mathbb{1} \to (\mathcal{D} \multimap \mathcal{D})$ be such that $ltr(h) = \{(1,(d_1,d_2)),(1,(e_1,e_2))\}$. Then $ltr(!h) = \{(\emptyset,\emptyset),(\{1\},\{(d_1,d_2)\}),(\{1\},\{(e_1,e_2)\}),(\{1\},\{(d_1,d_2),(e_1,e_2)\})\}$, and $ltr(!h \circ m_{\mathbb{1}}) = \{(1,\emptyset),(1,\{(d_1,d_2)\}),(1,\{(e_1,e_2)\}),(1,\{(d_1,d_2),(e_1,e_2)\})\}$. The element of $\mathcal{D}$ selected by $G \circ !h \circ m_{\mathbb{1}}$ is $\{\emptyset,\{(d_1,d_2)\},\{(e_1,e_2)\},\{(d_1,d_2),(e_1,e_2)\}\}$, up to the isomorphism G.

Proposition 1 *A semantic value is, up to G, an element $\Delta \in !(\mathcal{D} \multimap \mathcal{D})$ such that if $d \in \Delta$, then $\mathcal{P}(d) \subseteq \Delta$, where $\mathcal{P}(d)$ is the power set of d.*

Proof. From the definition of !, given in the proof of Theorem 3, and the above observation. □

As consequence, we will take $\mathcal{V}[M]\rho = ([\![x_1,\ldots,x_n \vdash M]\!]^{\mathcal{C}(\mathcal{D})} \circ ((m_{\mathbb{1}} \otimes \rho(x_1) \otimes \cdots \otimes \rho(x_n)))(\{[1,1,\ldots,1]\}))$, where $\mathcal{FV}(M) \subseteq \{x_1,\ldots,x_n\}$. Moreover, every $\rho(x_i)$ abbreviates $\rho(x_i)(\{1\})(1 \le i \le n)$.

Notice that the model $\mathcal{M}_!$ contains values such that: (i) they have the same functional behaviour, (ii) they are comparable w.r.t. the set inclusion relation, and (iii) they are different. For example, let consider $\{\emptyset, \{(a,b)\}, \{(c,d)\}\}$ and $\{\emptyset, \{(a,b)\}, \{(c,d)\}, \{(a,b),(c,d)\}\}$.

It is particularly difficult to reason about the interpretation of call-by-value λ-terms in a given coherence domain. It is, therefore, desirable to build a formal system for correctly reasoning about denotation of terms. In [10], a formal system for dealing with interpretations of λ-terms in models built from qualitative domains was defined. We shall adapt that technique to the model $\mathcal{M}_!$.

We define a type assignment system, where types are names for the atoms of $\mathcal{D}$, i.e., types are finite linear traces of linear functions from $\mathcal{D}$ to $\mathcal{D}$. The type system proves judgments $\Gamma \vdash M : \alpha$, where M is a term, α is a type, and Γ is a basis. Every basis is a partial function from variables to types. The intended meaning of every judgment is that $\Gamma \vdash M : \alpha$ iff the atom of $\mathcal{D}$ with name α belongs to the interpretation of M in a suitable environment connected to Γ.

Definition 7 *i) The set T of types is the smallest set containing:*

- *the costant ν,*
- *for all $\sigma, \tau \in T$ the* arrow type *$\sigma \to \tau$, and*
- *for all $\sigma, \tau \in T$ such that* $\mathtt{comp}(\sigma,\tau)$ *the* intersection type *$\sigma \wedge \tau$,*

where $\mathtt{comp}(\sigma,\tau)$ *holds iff the atoms represented by σ and τ are compatible. This predicate can be formalized as in [10].*

ii) $\approx$ is the least equivalence relation on T such that $\sigma \wedge \nu \approx \sigma$, $(\sigma \wedge \tau) \approx (\tau \wedge \sigma)$, $(\sigma \wedge \tau) \wedge \mu \approx \sigma \wedge (\tau \wedge \mu)$, for all $\sigma, \tau, \mu \in T$.

The relation between types (modulo $\approx$) and linear traces are expressed by the following isomorphism:

Definition 8 *The isomorphism $* : T/_{\approx} \longrightarrow |\mathcal{D}|$ is so defined:*

- $\nu^* = \emptyset$,
- $(\bigwedge_{i \in I}(\alpha_i \to \beta_i))^* = \{(\alpha_i^*, \beta_i^*) \mid i \in I\}$.

Proposition 2 *i) $*$ is injective and surjective.*

ii) $\sigma \in T/_{\approx} \Longrightarrow \sigma^* \in |\mathcal{D}|$.

iii) $d \in |\mathcal{D}| \Longrightarrow \exists \sigma \in T/_{\approx}$ s.t. $\sigma^* = d$.

□

From now on we will consider types modulo $\approx$.

Definition 9 *i) A* basis Γ *is a partial function with finite domain from variables to types. A basis* Γ *such that* $\Gamma(x_i) = \alpha_i \ (i=1,\ldots,n)$ *will be denoted by* $x_1:\alpha_1,\ldots,x_n:\alpha_n$. $\Gamma_1 \wedge \Gamma_2$ *is the union of the basis* Γ_1 *with the basis* Γ_2 *where, if* $x \in dom(\Gamma_1) \cap dom(\Gamma_2)$ *then* $(\Gamma_1 \wedge \Gamma_2)(x) = \Gamma_1(x) \wedge \Gamma_2(x)$. Γ_ν *denotes a basis such that* $\forall x \in dom(\Gamma_\nu).\Gamma_\nu(x) = \nu$. `Comp` *checks the compatibility between the basis to which it is applied, i.e.* $\texttt{Comp}(\Gamma_1, \Gamma_2)$ *holds iff* $\forall x \in dom(\Gamma_1) \cap dom(\Gamma_2).\texttt{comp}(\Gamma_1(x), \Gamma_2(x))$.

ii) The call-by-value coherence type assignment system $\vdash$ *proves judgments* $\Gamma \vdash M : \alpha$, *where* Γ *is a basis,* M *is a term and* α *is a type. The rules of the system are the following:*

$$\frac{}{x:\alpha \vdash x:\alpha}(var) \qquad\qquad \frac{}{\Gamma_\nu \vdash \lambda x.M:\nu}(\nu)$$

$$\frac{\Gamma \vdash M:\sigma \qquad x \notin dom(\Gamma)}{\Gamma, x:\nu \vdash M:\sigma}(weak)$$

$$\frac{(\Gamma_i, x:\alpha_i \vdash M:\beta_i)_{i\in I} \qquad \forall h,k \in I.\texttt{Comp}(\Gamma_h, \Gamma_k)}{\bigwedge_{i\in I}\Gamma_i \vdash \lambda x.M:\bigwedge_{i\in I}\alpha_i \to \beta_i}(\to I)$$

$$\frac{\Gamma_1 \vdash M:\alpha \to \beta \qquad \Gamma_2 \vdash N:\alpha \qquad \texttt{Comp}(\Gamma_1,\Gamma_2)}{\Gamma_1 \wedge \Gamma_2 \vdash MN:\beta}(\to E)$$

Theorem 4 (Soundness) *Let* Γ *be a basis and* M *be a term.*

$$\textit{If} \quad \Gamma \vdash M:\alpha \quad \textit{then} \quad \alpha^* \in \mathcal{V}[M]\rho_\Gamma,$$

where $\rho_\Gamma(x) = \mathcal{P}((\Gamma(x))^*)$, *if* $x \in dom(\Gamma)$, $\rho_\Gamma(x) = \{\emptyset\}$ *otherwise.*

Proof. By induction on derivation. □

Theorem 5 (Completeness) *Let* M *be a* λ*-term,* $x_1,\ldots,x_n$ *a sequence of all free variables of* M *and* ρ *an n-environment. Let assume* $d \in \mathcal{V}[M]\rho$. *There exist a type* σ *and a basis* Γ, *such that* $\Gamma \vdash M:\sigma$, *where both* $\sigma^* = d$ *and, if* $\Gamma(x_i) = \alpha$, *then* $\alpha^* \in \rho(x_i)$.

Proof. By induction on the structure of M. □

The following theorem relates valuable terms with values in $\mathcal{D}$.

Theorem 6 *Let* M *be a* λ*-term.*

$$M \textit{ is valuable} \iff \Gamma_\nu \vdash M:\nu.$$

Proof. ($\Longrightarrow$) Easy.
($\Longleftarrow$) To prove this implication we need a notion of approximation. Following [4] we define an extension of the λ-calculus, and we shall prove that the interpretation of a term M of $\lambda\beta_v$ is the union of the interpretations of a set of normal forms of the extended language, which we shall call the *approximants* of M.

Definition 10 *i) The set of terms Λ^Ω is inductively defined as Λ, starting from $Var \cup \{\Omega\}$. The set of values of Λ^Ω is $V_\Omega = Var \cup \{\lambda x.M \mid M \in \Lambda^\Omega\}$ The reduction rules are (β_v) and the following ones:*

$$(\Omega)\ \Omega M \to_\Omega \Omega,$$
$$(\Omega)\ M\Omega \to_\Omega \Omega.$$

ii) The set of approximants of a $\lambda\beta_v$ term M is:

$$\mathcal{A}(M) = \{N \mid \ N \in \Lambda^\Omega \text{ and } N \text{ is a } \beta_v\text{-}\Omega\text{-normal form obtained from some } M' =_{\beta_v} M, \text{ by substituting some of its subterms with } \Omega\}$$

Ω represents the undefined non valuable term. The typing rules are naturally extended to Λ^Ω. It immediate to verify that Ω has no typings, while $\lambda x.\Omega$ has the unique typing $\Gamma_\nu \vdash \lambda x.\Omega : \nu$.

Proposition 3 *Let Γ be a basis and $M \in \Lambda$.*

$$\Gamma \vdash M : \alpha \Longleftrightarrow \exists N \in \mathcal{A}(M).\ \ \Gamma \vdash N : \alpha.$$

Proof. (Hints) The proof of the *only if* part is based on a procedure normalizing derivations of $\vdash$. Let a *possible cut* be a subderivation ending by $(\multimap E)$ such that: (i) its major premise ends by $(\multimap I)$, which is followed by a (possibly empty) sequence of (*weak*), (ii) if its minor premise is $\Delta \vdash N : \nu$, then Δ is Γ_ν, and (iii) it contains at least one cut as subderivation. Let a *cut* be a *possible cut* such that: (i) the subject of the minor premise is a value, and (ii) it does not contain any other cut. Let the *degree of a cut* be the number of symbols in the type of the major premise of $(\multimap E)$. Let the *measure of a derivation* be the 3-tuple $\langle$*number of possible cuts, maximum degree of a cut, number of cuts of maximum degree*$\rangle$. A derivation with measure $\langle 0,0,0\rangle$ is *normal.* An effective procedure which eliminates a cut of maximum degree at each step can be designed. The elimination of a cut in a derivation D proving $\Gamma \vdash M : \alpha$ corresponds to the reduction of one of the innermost β_v-redexes of M. Every cut elimination transforms D into D' proving $\Gamma \vdash M' : \alpha$, where $M \Longrightarrow_v M'$. If D' is normal and M' contains redexes, they can only occur in subterms $\lambda x.N$ which are subjects of a rule (ν) of D'. The replacement of every one of such subterms by $\lambda x.\Omega$ gives the needed approximant.

To prove the *if* part, let assume $N \in \mathcal{A}(M)$. There are $n \geq 0$ terms $N_1, \ldots, N_n$ such that, by replacing the i-th occurrence of Ω in N by N_i, a term $M' =_v M$ is obtained. Since $\Gamma \vdash N : \alpha$ implies $\Gamma \vdash M' : \alpha$, by Theorems 4 and 5 we have $\Gamma \vdash M : \alpha$. □

Finally, we can complete the proof of the ($\Leftarrow$) side of Theorem 6, proceeding by absurdum. Let M be a non-valuable term. Every $N \in \mathcal{A}(M)$ is either of the form $xN_1 \ldots N_n$, with $N_i \neq \Omega$, or Ω $(1 \leq i \leq n)$. In both cases M does not have type ν from the basis Γ_ν. □

As consequence, $\mathcal{M}_!$ characterizes the valuable terms of Λ^Ω.

Theorem 7 *Let* $M \in \Lambda$.

i) If M *is valuable, then* $\forall \rho.\mathcal{V}[M]\rho$ *is a value.*

ii) If M *is not valuable, then* $\exists \rho.\mathcal{V}[M]\rho$ *is not a value.*

Proof. *i)* From Theorem 1.
ii) From Theorem 6, ($\Leftarrow$). □

So $\mathcal{M}_!$ is correct w.r.t. the operational equivalence:

Theorem 8 (Correctness) *For all terms* M, N,

$$(\mathcal{M}_! \models M = N) \Longrightarrow (M \sim_v N).$$

□

The interpretation of the terms is not closed under the η_v-equivalence, which is the equivalence relation induced among terms by restricting the classical η-rule to values in the following way:

$$(\eta_v) \qquad \lambda x.Mx \to_{\eta_v} M \quad \text{if } M \text{ is a value and } x \notin \mathcal{FV}(M).$$

Let see an example, using the type assignment system. If we start from the basis $\Gamma \equiv y : (\alpha \to \beta) \wedge (\mu \to \gamma)$, then we can derive $y : (\alpha \to \beta) \wedge (\mu \to \gamma)$. Clearly, this typing is not derivable for $\lambda x.yx$ from the same basis.

In [4], it was proved that the operational equivalence $\sim_v$ is closed under the η_v-equality. This immediately implies that $\mathcal{M}_!$ *is not fully-abstract*, namely:

Theorem 9 (Incompleteness) $(M \sim_v N) \not\Rightarrow (\mathcal{M}_! \models M = N)$, *for all* M *and* N.

Remember that the λI-calculus is the λ-calculus where every $\lambda x.M$ is legal iff $x \in \mathcal{FV}(M)$. $\mathcal{M}_!$ does not separate two different closed β_v-normal forms belonging to the λI-calculus, if they reduce to the same β-normal form.

Theorem 10 *For all closed* M *and* N *belonging to the* λI*-calculus, if both* M *and* N β*-reduce to the same* β*-normal form then* $\mathcal{M}_! \models M = N$.

Proof. (Hint) Derivations of $\vdash$ enjoy a generalization of the normalization property proved in Proposition 3. Namely, let define a *cut* as a possible cut in the proof of Proposition 3, while dropping point (iii). Now, the elimination of a cut is exactly the reduction of a β*-redex* whose argument is well typed. This implies that: (i) the current term has not unsolvable subterms, (ii) if a subterm is erased by such β-reduction, it is a value. Point (ii) is a consequence of the following two facts: if $M \Longrightarrow_v M'$ and $x \notin \mathcal{FV}(M)$, then $\lambda x.M$ can only be assigned either the type ν or a type of the form $\nu \rightarrow \sigma$, for some σ, and non-valuable terms cannot have typing ν from the basis Γ_ν, by Theorem 6. The points (i) and (ii) are both verified by the normalizing terms of the λI-calculus.

The statement does not hold for open terms. In fact, $(\lambda xy.xy)(zw) =_\beta \lambda y.zwy$, but $(\lambda xy.xy)(zw)$ and $\lambda y.zwy$ are not equal in $\mathcal{M}_!$ since $\Gamma_\nu \vdash \lambda y.zwy : \nu$ is a derivable judgment, while $\Gamma_\nu \vdash (\lambda xy.xy)(zw) : \nu$ is not. □

The property proved in the previous theorem is quite natural. In all programming languages, a different operational behaviour between call-by-name and call-by-value evaluation can be found exactly in presence both of diverging computations and the ability of erasing arguments. The theory induced on Λ by $\mathcal{M}_!$ differs from the theory induced by the model over coherence domains, given in [9]. In [9] the η_v-equivalence holds since the model is extensional, but Theorem 10 does not. In the theory induced by Scott domain semantics of [4] both properties hold.

Acknowledgments We would like to thank Jean Yves Girard, Yves Lafont, Furio Honsell and Eugenio Moggi for very stimulating and useful discussions on the topic of this paper.

References

[1] A. Asperti and G. Longo. *Categories, Types and Structures.* Foundations of Computing. The MIT Press, 1991.

[2] N. Benton, G. Bierman, V. de Paiva, and M. Hyland. Term assignment for intuitionistic linear logic. Technical Report 262, Computer Laboratory, University of Cambridge, August 1990.

[3] M. Dezani-Ciancaglini, F. Honsell, and S. Ronchi della Rocca. Models for theories for functions strictly depending on all their arguments (abstract). *The Journal of Symbolic Logic*, 51(3):845 – 846, 1986.

[4] L. Egidi, F. Honsell, and S. Ronchi della Rocca. Operational , denotational and logical descriptions: a case study. *Fundamenta Informaticae*, 16(2):149 – 169, February 1992.

[5] J.Y. Girard. The system F of variable types, fifteen years later. *Theoretical Computer Science*, 45:159 – 192, 1986.

[6] J.Y. Girard, Y. Lafont, and P. Taylor. *Proofs and Types.* Cambridge University Press, 1989.

[7] G. Gonthier, M. Abadi, and J.J. Lévy. The geometry of optimal lambda reductions. In *Proc. 19th Symp. on Principles of Programming Languages (POPL 92)*, 1992.

[8] R. Hindley and G. Longo. *To H.B. Curry: Essay on Combinatory Logic, Lambda Calculus and Formalisms*, chapter Lambda-calculus models. J.R.Hindley, J.P. Seldin editors, Academic Press, 1980.

[9] F. Honsell and M. Lenisa. Some results on the full-abstraction problem for restricted λ-calculi. In *LNCS 711: Proceedings of 18th. Symposium on MFCS'93 (Poland)*, pages 84 – 104. Springer-Verlag, 1993.

[10] F. Honsell and S. Ronchi della Rocca. *Reasoning about interpretations in qualitative lambda models*, pages 505–522. M. Broy and C.B. Jones editors, North-Holland, 1990.

[11] C.P.J. Koymans. Models of the lambda calculus. *Information and Control*, 52(3):306 – 332, 1982.

[12] Y. Lafont. Introduction to linear logic. Lecture notes for the Summer School on Constructive logic and Category theory (Isle of Thorns), August 1988.

[13] P.J. Landin. The mechanical evaluation of expressions. *Computer Journal*, 6(4), 1964.

[14] G. Plotkin. Call-by-name, call-by-value and the λ-calculus. *Theoretical Computer Science*, 1:125 – 159, 1975.

A Fully Abstract Translation between a λ-Calculus with Reference Types and Standard ML

Eike Ritter*[1] and Andrew M. Pitts**[2]

[1] Oxford University Computing Laboratory, Wolfson Building, Parks Road, Oxford OX1 3QD, UK

[2] Cambridge University Computer Laboratory, Pembroke Street, Cambridge CB2 3QG, UK

Abstract. This paper describes a syntactic translation for a substantial fragment of the core Standard ML language into a typed λ-calculus with recursive types and imperative features in the form of reference types. The translation compiles SML's use of declarations and pattern matching into λ-terms, and transforms the use of environments in the operational semantics into a simpler relation of evaluation to canonical form. The translation is shown to be 'fully abstract', in the sense that it both preserves and reflects observational equivalence (also commonly called contextual equivalence). A co-inductively defined notion of applicative bisimilarity for lambda calculi with state is developed to establish this result.

1 Introduction

To apply techniques from operational and denotational semantics to higher order functional programming languages it is often convenient to define a translation of the programming language into a typed λ-calculus with appropriate additional features and study the typed λ-calculus instead. The transfer of semantic results about the typed λ-calculus back to ones about the original programming language relies on properties of the translation which can be rather hard to establish and which seem to be addressed rather rarely in the literature. Here we study such a translation for a substantial fragment of the core Standard ML language [MTH90] containing recursively defined datatypes and imperative features in the form of reference types. The translation compiles SML's use of declarations and pattern matching into λ-terms, and transforms the use of environments in the operational semantics into a simpler relation of evaluation to canonical form. We show that this translation is 'fully abstract', in the sense that it both preserves and reflects observational equivalence (also commonly called contextual equivalence).

* Research partially supported by EU ESPRIT project CLICS-II.

** Research partially supported by UK EPSRC project GR/G53279.

Our motivation for establishing such a result stems from ongoing work on logics for verifying properties of SML programs (in particular, equivalence of programs). It is more convenient to develop such logics on top of a suitable typed lambda calculus rather than the programming language itself. However, one then needs results such as the one established here to relate equivalence of SML programs to equivalence of the corresponding lambda calculus expressions. Preservation and reflection of observational equivalence along the translation is certainly to be expected since apart from the use of patterns and environments, the two languages are quite similar. However the proof of this property is not straightforward, because it is hard to reason directly about observational equivalence for programming languages such as SML that involve dynamic creation of local storage for higher order functions.

Riecke [Rie93] investigates similar types of problem, namely how to show which translations between call-by-value and call-by-name PCF are fully abstract. He defines a fully abstract model for the two languages to examine the translations. He transforms the question of preserving observational equivalence into the question whether a term in one language and its translation have meanings in the fully abstract models that are related by a logical relation. This method works only if fully abstract domain-theoretic models of the languages are easily available. This is true for the simple language PCF (albeit after the addition of a parallel-or function), but definitely not for SML. Indeed the presence of dynamically created, mutable storage locations for recursively defined data involving higher order functions makes the construction of adequate, let alone fully abstract, denotational models somewhat tedious and hard to use in practice.

In this paper we establish the full abstraction property of our translation using methods based on operational semantics rather than denotational semantics. This entails working directly with the notion of observational equivalence of language expressions, rather than with equality of their denotations. However, the quantification over program contexts in the definition of observational equivalence makes it rather unamenable to the standard techniques associated with operational semantics (such as various forms of structural induction). For the simple language PCF, Milner's context lemma [Mil77] overcomes this problem by showing that one can restrict to a very simple collection of applicative contexts when establishing instances of PCF observational equivalence. Inspired by Milner's result, Mason and Talcott [MT91, **ciu** Theorem] prove a much harder context lemma for an untyped λ-calculus with mutable local storage locations. Although the restriction on contexts in their result is not as great as one might hope for, nevertheless they show that a substantial number of properties of observational equivalence result from it. However, Milner's result can be usefully generalized in a different direction. It is equivalent to a characterization of observational equivalence in PCF in terms of a co-inductively defined notion of applicative bisimilarity. Such notions were transferred from the study of concurrent processes to untyped λ-calculus by Abramsky [AO93] and have been applied to various kinds of lambda calculi, type theory and pure functional programming by Howe [How89], Ong [Ong93] and Gordon [Gor93]. Here we introduce a notion of applicative bisimilarity for functional languages with state and show that Howe's method in [How89] can be adapted to prove that it is a congruence. The difficulty that has to be overcome to do this for languages with state concerns the more complicated form of the operational semantics compared with that for pure functional languages. We were guided to our definition by analogy with the co-inductive characterization of equality in recursively defined domains given in [Pit94], applied to the domains needed for a denotational semantics of the kind of typed λ-calculus considered here.

As a consequence of its congruence property, the relation of applicative bisimilarity implies that of observational equivalence. Unlike the situation for non-imperative (deterministic) functional languages, the two notions do not coincide. This is due to the subtle nature of observational equivalence in the presence of dynamically created local state: see [PS93], for example. Nevertheless the notion of applicative bisimilarity we give has sufficient properties (all relatively easy to establish, apart from the congruence property) to yield the full abstraction properties of our translation of SML into a typed λ-calculus.

The paper is structured as follows. In section 2 we define the typed λ-calculus we consider, the corresponding fragment of Standard ML, and translations between the calculi in each direction. The next section defines the notion of applicative bisimulation used to show the full abstractness of the translations, and proves that it is a congruence relation. Afterwards we apply this result to show that both translations preserve and reflect observational equivalence and are in fact mutually inverse to each other up to observational equivalence.

2 The Calculi

First we define a λ-calculus with reference types, λML, and then the appropriate fragment of Standard ML.

2.1 The λ-Calculus λML

We consider a typed λ-calculus with the following features: higher-order, recursively defined functions; (local, monomorphic) references to mutable locations; and globally defined, mutually recursive datatypes. In fact we define a family of languages, parameterized by a function $decl : TypConst \to Typ$ used to give the (recursive) definition $\kappa = decl(\kappa)$ of some *type constants* κ in the finite set *TypConst*, and where *Typ*, the collection of *types*, is defined by the grammar

$$\sigma ::= \kappa \mid \texttt{unit} \mid \texttt{bool} \mid \texttt{zs} \times \sigma \mid \sigma \Rightarrow \sigma \mid \sigma \ \texttt{ref} \,,$$

where $\kappa \in TypConst$ and σ `ref` is the type expression for reference types. The type constants κ act like mutually recursively defined datatypes that are global and static in the sense that they are fixed once and for all when the language is defined. There are no construction mechanisms for building new recursive datatypes from old ones in the language itself.

The expressions of the calculus are given as follows, where x is taken from an infinite set of identifiers and a from an infinite set of address names:

$$e ::= x \mid \mathtt{in}^{\kappa}(e) \mid \mathtt{out}(e) \mid () \mid e := e \mid \mathtt{true} \mid \mathtt{false} \mid \mathtt{if}\ e\ \mathtt{then}\ e\ \mathtt{else}\ e$$
$$\mid e = e \mid (e, e) \mid \mathtt{fst}(e) \mid \mathtt{snd}(e) \mid \lambda x^{\sigma}.e \mid ee$$
$$\mid \mathtt{rec}\ f(x) =_{\sigma} e\ \mathtt{in}\ f(e) \mid \mathtt{ref}(e) \mid !e \mid a$$

The expression $\mathtt{rec}\ f(x) =_{\sigma} e'\ \mathtt{in}\ f(e)$ defines f as a recursive function and applies it to e, and $!e$ accesses the value stored in the address to which e evaluates. The notion of bound and free variables are defined as usual. An expression is *closed* if it has no free identifiers. The *canonical expressions* are the closed expressions generated by the grammar:

$$c ::= \mathtt{in}^{\kappa}(c) \mid () \mid \mathtt{true} \mid \mathtt{false} \mid (c, c) \mid \lambda x^{\sigma}.e \mid a$$

Remark We do not have any sum types or alternate clauses in the definition of recursive datatypes. Those constructions are used in modelling standard datatypes like lists. This omission simplifies the presentation—see Section 2.3 for the reason, but the same ideas that are described in this paper yield the full abstraction of the translations for the extended calculus as well.

A *type assignment* Γ is a finite function from identifiers to types. A typing assertion takes the form $\Gamma \vdash e : \sigma$ where Γ is a type assignment, e an expression, σ a type and the identifiers of e are contained in the domain of Γ. The rules are listed in the appendix. Note that for every expression there is at most one type σ such that $\Gamma \vdash e : \sigma$.

A *state* s is a finite function from addresses to canonical expressions such that all addresses contained in the canonical expressions are contained in the domain of s. We write $s + \{a \mapsto c\}$ for the state obtained from s by extending its domain by $a \notin dom(s)$ and mapping a to c. We write $s \cup \{a \mapsto c\}$ for the state obtained from s by mapping $a \in dom(s)$ to c and otherwise acting like s.

The evaluation relation takes the form $\langle s, e\rangle \Downarrow \langle s', c\rangle$ where s and s' are states such that $dom(s) \subseteq dom(s')$, e is any expression and c is canonical. The valid instances of the relation are inductively defined by a set of rules listed in the appendix. They use the sequentiality convention employed in the definition of SML [MTH90], which states that a rule of the form

$$\frac{e_1 \Downarrow c_1 \qquad e_2 \Downarrow c_2 \qquad \cdots \qquad e_n \Downarrow c_n}{e \Downarrow c}$$

is an abbreviation for

$$\frac{\langle s_1, e_1\rangle \Downarrow \langle s_2, c_1\rangle \qquad \langle s_2, e_2\rangle \Downarrow \langle s_3, c_2\rangle \qquad \cdots \qquad \langle s_n, e_n\rangle \Downarrow \langle s_{n+1}, c_n\rangle}{\langle s_1, e\rangle \Downarrow \langle s_{n+1}, c\rangle}$$

Note that, as in SML, the evaluation strategy is call-by-value.

2.2 The fragment of SML

The fragment of Core SML [MTH90] that we consider here is monomorphic, has no exception mechanism, and only has globally declared datatypes. The pattern matching and the local value declarations, which are the distinctive features of this fragment compared to the λ-calculus λML, are handled exactly as in the full Core SML. The commands for manipulating the state are the same in the fragment and in λML. The fragment we consider has products instead of pattern rows. This is a technical simplification, and it is easy to see how the argument advanced in this paper can be extended to cover expression rows.

The pattern matching and local declarations require separate syntactic categories. Values are the results of evaluating expressions, and only they are bound in local declarations and matched against a pattern. This reflects the call-by-value nature of evaluation in Standard ML. We have the following syntactic categories, called *phrases*, in the fragment:

- Expressions *exp*
- Matches *match*, which are lists of match rules.
- Match rules *mrule*, which associate an expression to a pattern. The expression is evaluated if the given value matches the pattern.
- Declarations *dec*, which associate values to variables.
- Value bindings *valbind*, which perform the binding in declarations.
- Patterns *pat*, which are matched against a value.

The grammar for raw SML phrases is as follows:

$$
\begin{aligned}
exp &::= (exp, exp) \mid \mathtt{true} \mid \mathtt{false} \mid () \mid var \mid \mathtt{ref}\, exp \mid \mathtt{in}^{\kappa}(exp) \mid\, !exp \\
&\quad \mid \mathtt{let}\; dec\; \mathtt{in}\; exp\; \mathtt{end} \mid exp\, exp \mid exp\; \mathtt{id}\; exp \mid exp\colon ty \mid \mathtt{fn}\colon ty\;\, match \\
match &::= mrule \;\langle \mid match\rangle \\
mrule &::= pat \Rightarrow exp \\
dec &::= val\; valbind \mid \mathtt{local}\; dec\; \mathtt{in}\; dec\; \mathtt{end} \mid dec\; \langle ; \rangle\; dec \\
valbind &::= pat = exp\; \langle \mathtt{and}\; valbind\rangle \mid \mathtt{rec}\; valbind \\
pat &::= \_ \mid \mathtt{true} \mid \mathtt{false} \mid () \mid \mathtt{ref}\; pat \mid \mathtt{in}^{\kappa}(pat) \\
&\quad\; var \mid (pat, \ldots, pat) \mid pat\colon ty \mid var\; \langle ty\rangle\; \mathtt{as}\; pat
\end{aligned}
$$

Now we turn to the static semantics, i.e. the typing assignments. The types are the same as in the λ-calculus. The absence of polymorphism in the SML fragment makes it possible to simplify the judgements considerably. Because the judgements are a restriction of those given in the definition of Standard ML [MTH90], we omit the details. An expression is called *closed* if $\emptyset \vdash exp\colon \sigma$, where $\emptyset$ denotes the empty context.

The canonical expressions of the λ-calculus form a subset of all expressions. This is necessary because the evaluation of an application works by first reducing the argument to a canonical expression and then substituting this expression textually into the body of the function. The pattern matching in SML renders a similar definition of the evaluation relation impossible. A function application is evaluated not by substitution, but by matching the argument against the pattern the function specifies and evaluating the selected expression. Hence the result of evaluating a function is a closure, i.e. it contains both the body of the function and the values associated to all free variables in the function. As a consequence of these differences we call the result of an evaluation of a phrase a *value* and not a canonical expression.

The values are given by the grammar:

$$v ::= a \mid \texttt{true} \mid \texttt{false} \mid () \mid (v_1, v_2) \mid \texttt{in}^{\kappa}(v) \mid (match, E, E')$$

where E and E' are environments, i.e. finite functions from variables to values. We will write $(x_i := v_i)$ for such an environment. In the closure $(match, E, E')$ the environment E' contains the values for the variables that are used in recursive function calls and E contains the value for all other variables. We have the following kinds of judgements:

- $s, E \vdash exp \Downarrow v, s'$: The expression M evaluates to the value v in the environment E.
- $s, E, v \vdash match \Downarrow v', s'$: The match *match* is tested against the value v, yielding v' as a result.
- $s, E, v \vdash pat \Downarrow E', s'$: The pattern *pat* is matched against value v and added into the environment.
- $s, E \vdash dec \Downarrow E', s'$: The declaration *dec* is evaluated, resulting in a new declaration E'.
- $s, E \vdash valbind \Downarrow E', s'$: The binding in *valbind* is added to E.

Here s and s' are the states before and after evaluation, where a *state* is a finite function from addresses to values. They are just as in [MTH90] (and employ the sequentiality convention mentioned above). In full SML, evaluation of pattern matches can raise exceptions. Here we just assume a restriction on expressions so that every match succeeds exactly for one pattern.

2.3 The translations between the languages

The translation from the λ-calculus into SML has to express all selectors like `fst` and `snd` by pattern matching. The translation in the other direction is essentially a compilation of pattern matching into the λ-calculus. Both translations are compositional: the translation of a phrase is defined in terms of the translations of its subphrases. The translation of λ-calculus expressions into SML expressions, denoted by $(-)^M$, is given in the appendix and is quite straightforward. The translation in the other direction, denoted by $[\![-]\!]$, is given in the appendix and is more involved. The translations $[\![exp]\!]$ and $[\![match]\!]$ of expressions and matches are λML expressions. The translations $[\![valbind]\!]$ and $[\![dec]\!]$ of value bindings and declarations are *substitutions*, σ, which are finite functions from identifiers to canonical λML expressions, indicated by $[c_i/x_i]$. The translation of a pattern is a tuple $(cond, \sigma)$. The first component *cond* is a boolean expression in λML that characterizes the condition a value has to satisfy to be successfully matched against the pattern. The second component is a substitution that describes how the variables declared by the pattern *pat* are obtained from the value matched against the pattern. The variable *arg* occurring in σ represents this value. The clause for matches $pat_1 \;\texttt{=>}\; exp_1 | \cdots | pat_n \;\texttt{=>}\; exp_n$ describes a λ-term that tests the conditions $cond_1$ to $cond_n$ to select the succeeding pattern pat_i and returns the expressions exp_i with the substitution σ_i applied. A value binding *val pat* = *exp* is translated into a substitution that binds the variables declared in the substitution originating from the translation of the pattern *pat* to the values obtained by performing the match.

As an example, consider the SML-function

$$\texttt{fn} : \texttt{bool} \times \texttt{bool}.(\texttt{true}, x) \;\texttt{=>}\; x \mid (\texttt{false}, y) \;\texttt{=>}\; \texttt{not}(y)\,.$$

The translation of the first pattern is the tuple $(\texttt{fst}(arg), [\texttt{snd}(arg)/x)]$, and the translation of the second pattern is $(\texttt{not}(\texttt{fst}(arg)), [\texttt{snd}(arg)/y])$. Hence the translation of this function into λML is the function

$$\lambda arg\colon \texttt{bool} \times \texttt{bool}.\texttt{if}\ \texttt{fst}(arg)\ \texttt{then}\ \texttt{snd}(arg)\ \texttt{else}\ \texttt{not}(\texttt{snd}(arg))\,.$$

We extend the translation $[\![-]\!]$ simultaneously to values and environments as follows: for a value v, $[\![v]\!]$ is a canonical λML expression defined by induction on the structure of v (details omitted); for an environment $E = (x_i := v_i)$, $[\![E]\!]$ is defined to be the substitution $[[\![v_i]\!]/x_i]$. Finally SML states $s = \{a_i \mapsto v_i\}$ are translated into λML states by defining $[\![s]\!] = \{a_i \mapsto [\![v_i]\!]\}$, and similarly for the action of $(-)^M$ on λML states.

Both translations preserve strong typing, i.e. when an expression e is well-formed in context Γ, then its translation into the other language is well-formed as well. Moreover, both translation are *sound*, i.e. they preserve the validity of evaluation judgments. This statement becomes meaningful only after the addition of judgements about the evaluation of substitutions into the λML-calculus. The SML-judgements for the evaluation of patterns, declarations and value bindings correspond to judgement of this form in the λML-calculus. These properties can be shown by an induction over the derivation of the judgements.

The translation from SML into λML justifies why we omitted sum types and alternates in the definition of recursive datatypes. The construction of a λ-term that describes the selection of a succeeding match becomes significantly more complicated if these constructions are added. The proof that the translation between the calculi with these additional features are fully abstract follows the same line as the one for the calculus used in this paper. To simplify the exposition, we decided therefore not to consider such an extended calculus here.

3 Applicative bisimulation

We aim to show that the two translations both preserve and reflect the evaluation behaviour of expressions in all program contexts. We will do this by showing that $(-)^M$ and $[\![-]\!]$ are mutually inverse[3] up to *observational equivalence*, written

[3] Modulo a restriction on patterns in the SML expressions arising from the fact that for simplicity we are avoiding exception handling: see Section 4.

$\approx_{obs}$. By definition, $e_1 \approx_{obs} e_2$ holds in λML if for all closing boolean contexts $C[-]$ and states s, $\exists s_1.\langle s, C[e_1]\rangle \Downarrow \langle s_1, \mathtt{true}\rangle$ iff $\exists s_2.\langle s, C[e_2]\rangle \Downarrow \langle s_2, \mathtt{true}\rangle$. The definition of $\approx_{obs}$ for SML is similar. The quantification over all contexts makes it very difficult indeed to establish the required properties of the translations directly from the definition of $\approx_{obs}$. Instead, we proceed indirectly by giving co-inductively defined notions of program equivalence for λML and SML, called *applicative bisimilarity* and written $\approx_{app}$. Applicative bisimilarity implies observational equivalence and it is much easier to establish the (equational) properties of $\approx_{app}$ needed to show that the translations between λML and Standard ML are mutually inverse. Because we consider recursive types and higher-order references we cannot define the applicative bisimulation by an induction over the type structure.

The difficult part in extending existing definitions of applicative bisimilarity for pure functional languages [AO93, How89, Gor93] to the languages studied here is the handling of the states involved in the evaluation relation. We were guided to the definition given below by analogy with the co-inductive characterization of equality in recursively defined domains given in [Pit94], applied to the domains needed for a denotational semantics of the λ-calculus λML. We explain the definition first for λML and adapt it afterwards to the fragment of SML.

3.1 Simulation for the λ-calculus λML

The relation of applicative similarity is defined as the greatest fixed point of a certain monotone operator $R \mapsto \overline{R}$ on relations between closed expressions. We start by giving the definition of $\overline{R}$ on canonical expressions. Afterwards we extend the definition of $\overline{R}$ to all closed expressions. Because it is not a priori clear that the extension of $\overline{R}$ to arbitrary expressions coincides with $\overline{R}$ on canonical expressions, we use a different symbol $\langle R\rangle$ for the relation $\overline{R}$ on canonical expressions at the moment.

Definition 1 *Given any family of relations $R = (R_\sigma \subseteq Exp_\sigma \times Exp_\sigma \mid \sigma \in Types)$ between closed λML expressions, let the family of relations $\langle R\rangle_\sigma$ between canonical expressions of type σ be inductively generated by the following rules:*

- $c \,\langle R\rangle_\sigma\, c'$ *if* $c \equiv c'$ *and* σ *is either* `unit`, `bool` *or a reference type.*
- $(c_1, c_2)\langle R\rangle_{\sigma_1\times\sigma_2}(c_1', c_2')$ *if* $c_1\langle R\rangle_{\sigma_1}c_1'$ *and* $c_2\langle R\rangle_{\sigma_2}c_2'$.
- $\lambda x{:}\,\sigma.M \;\langle R\rangle_{\sigma\Rightarrow\sigma'}\; \lambda x{:}\,\sigma.M'$ *if for all canonical expressions* c *of type* σ, *we have* $M[c/x]\; R_{\sigma'}\; M'[c/x]$.
- $\mathsf{in}^\kappa(c)\;\langle R\rangle_\kappa\;\mathsf{in}^\kappa(c')$ *if* $c\,\langle R\rangle_\sigma\, c'$ *and* $decl(\kappa) = \sigma$.

For any two states s_0 and s_1, we will write $s_0\langle R\rangle s_1$ to mean that s_0 and s_1 have the same domain and for all addresses a in that domain $s_0(a)\;\langle R\rangle_\sigma\; s_1(a)$ (where σ `ref` is the type of a).

We extend the definition of $\langle R\rangle_\sigma$ to a relation between all closed expressions of type σ by requiring that whenever the first expression reduces to a canonical expression, the second does so, and that the resulting states respect the relation $\langle R\rangle_\sigma$. This extension is defined as follows:

Definition 2 *With R as in Definition 1, define another family $\overline{R}$ of relations between closed λML expressions of equal type as follows:*

$$M \overline{R_\sigma} M' \text{ iff } \forall s_0, s_1, c \langle s_0, M \rangle \Downarrow \langle s_1, c \rangle \Rightarrow \exists s_1', c'. \langle s_0, M' \rangle \Downarrow \langle s_1, c' \rangle \text{ and } c \langle R \rangle_\sigma c' \text{ and } s_1 \langle R \rangle s_1' .$$

The operator $R \mapsto \overline{R}$ is monotone and hence we can define a relation $\sqsubseteq$, called *applicative similarity*, as the greatest fixed point of this operator. As usual, $\sqsubseteq$ is in fact greatest amongst the post-fixed points, i.e. those R satisfying $R \subset \overline{R}$; such R are called applicative simulations. We extend the relation $\sqsubseteq$ to a relation between open expressions of the same type by defining $M \sqsubseteq M'$ to hold iff $M[c_i/x_i] \sqsubseteq M'[c_i/x_i]$ holds whenever c_i are canonical expressions and $M[c_i/x_i]$ and $M'[c_i/x_i]$ are closed. One can show easily that this extended relation $\sqsubseteq$ is a partial order and that for all canonical expressions c and c', $c\langle\sqsubseteq\rangle c'$ holds iff $c \sqsubseteq c'$ is satisfied. Applicative *bisimilarity*, written $\approx_{app}$, is defined as the symmetrization of $\sqsubseteq$: $M \approx_{app} M'$ iff $M \sqsubseteq M'$ and $M' \sqsubseteq M$.

The most important property of the relation $\sqsubseteq$ is that it is a pre-congruence. By this we mean that for all contexts $C[-]$, we have $C[M] \sqsubseteq C[N]$ whenever $M \sqsubseteq N$. The clause for λ-abstracton in Definition 1 causes a direct proof of this property by induction over the structure of terms to fail. Indeed it is rather difficult to establish this result. Howe [How89] describes a method for showing pre-congruence that has become well established: for example, it is used for a λ-calculus with non-determinism in [Ong93] and a λ-calculus with recursive datatypes and input/output in [Gor93]. One proceeds by defining a relation $\sqsubseteq^*$ that is easily seen to be a pre-congruence and includes $\sqsubseteq$; the inverse inclusion is then shown by a co-inductive argument. Here we have to adapt this method to cope with the presence of states in the operational semantics. We will omit the detailed proofs and sketch only the structure of the argument.

Definition 3 *The relation $\sqsubseteq^*$ between two well-formed expressions of the same type is defined by induction over the structure of expressions as follows:*

- *$x \sqsubseteq^* M$ whenever $x \sqsubseteq M$.*
- *For every operator τ of arity n, $\tau(M_1, \ldots, M_n) \sqsubseteq^* N$ whenever there exist $M_1', \ldots, M_n'$ such that $M_i \sqsubseteq^* M_i'$ for all $1 \leq i \leq n$ and $\tau(M_1', \ldots, M_n') \sqsubseteq N$. (Similarly for variable binding operators.)*

First, one establishes some properties of the relation $\sqsubseteq^*$.

Lemma 4 (i) *If $M \sqsubseteq^* M'$ and $M' \sqsubseteq M''$, then $M \sqsubseteq^* M''$.*
(ii) *$M \sqsubseteq^* M$.*
(iii) *$M \sqsubseteq M'$ implies $M \sqsubseteq^* M'$.*
(iv) *$\sqsubseteq^*$ is a congruence relation.*
(v) *If $M \sqsubseteq^* M'$ and $c \sqsubseteq^* c'$, then $M[c/x] \sqsubseteq^* M'[c'/x]$.*

The crucial part of the congruence proof is the following proposition, which is proved by induction over the derivation of $\langle s_0, M \rangle \Downarrow \langle s_1, c \rangle$.

Proposition 5 *Let M be any closed expression of type σ. Then whenever $\langle s_0, M\rangle \Downarrow \langle s_1, c\rangle$ then for all closed expressions M' of type σ and states s_0' such that $M \sqsubseteq^* M'$, $s_0 \sqsubseteq^* s_0'$ and the addresses of M' are in the domain of s_0', there exists a canonical expression c' and a state s_1' such that $\langle s_0', M'\rangle \Downarrow \langle s_1', c'\rangle$ and $c \sqsubseteq^* c'$ and $s_1 \sqsubseteq^* s_1'$.*

This is sufficient to show

Proposition 6 *$M \sqsubseteq^* M'$ iff $M \sqsubseteq M'$.*

Proof. It follows from Proposition 5 that $\sqsubseteq^*$ restricted to closed expressions is an applicative simulation, i.e. is a post-fixed point of the operation $R \mapsto \overline{R}$. Since the relation $\sqsubseteq$ is the greatest such, we have that $M \sqsubseteq^* M'$ implies $M \sqsubseteq M'$. The reverse implication is just Lemma 4(iii).

The pre-congruence property of applicative similarity immediately yields

Corollary 7 *$M \approx_{app} M'$ implies $M \approx_{obs} M'$.*

However, $\approx_{app}$ does not coincide with $\approx_{obs}$, unlike the case for deterministic, pure functional languages. Partly this is due to the fact that in Definition 2 related states are required to have equal domains. As a consequence if $M \approx_{app} M'$ holds then evaluation of M and M' creates the same number of fresh addresses for local references, whereas such a property of address creation need not hold when $M \approx_{obs} M'$. (For example, $(\lambda x^{\texttt{bool ref}}.())\texttt{ref(true)}$ is observationally equivalent to $()$, but is not applicatively bisimilar to it.) This particular defect is remedied by the more refined notions of applicative equivalence considered in [PS93, PPS94]. We do not need such refinements to establish the full abstraction result of this paper. However, at the moment there is no known notion of applicative equivalence that coincides with observational equivalence for languages with dynamically created local state.

3.2 Simulation for SML

Now we define the applicative bisimulation for the fragment of SML. The idea behind the definition is the same, but the technical details are more complex due to the various syntactic categories of the language. Because the expression $M[c/x]$ in the λ-calculus λML corresponds to the pair $(E + (x := v), exp)$ in SML, we have to consider (environment, expression)-pairs. For this purpose we define a *generalized expression* to be either (E, exp), $(E + v, match)$ or (E, dec), where exp is an expression, v a value, $match$ a match, and dec a declaration. Such a generalized expression is called *closed* if $\emptyset \vdash E{:}\,\Gamma$ and exp, $match$ or dec are well-formed in context Γ. We define the applicative simulation for these generalized expressions.

Definition 8 *Given any type-indexed family R of relations between closed generalized expressions of equal type, a family of relations $\langle R\rangle_\sigma$ between values of each type σ is inductively defined by the following rules:*

- $v_1 \langle R \rangle_\sigma v_2$ *if* $v_1 \equiv v_2$ *and* σ *is* **bool**, **unit**, *or a reference type.*
- $(v_1, v_2)\langle R \rangle_{\sigma_1 \times \sigma_2}(v_1', v_2')$ *if* $v_1 \langle R \rangle_{\sigma_1} v_1'$ *and* $v_2 \langle R \rangle_{\sigma_2} v_2'$.
- $\mathtt{in}^\kappa(v_1)\langle R \rangle_\kappa \mathtt{in}^\kappa(v_2)$ *if* $v_1 \langle R \rangle_\sigma v_2$, *where* $decl(\kappa) = \sigma$.
- $(match_1, E_1, E_1')\langle R \rangle_{\sigma \Rightarrow \sigma'}(match_2, E_2, E_2')$ *if for all values* v *of type* σ, *we have* $(E_1, E_1' + v, match_1) \; R \; (E_2, E_2' + v, match_2)$.

The relation $\langle R \rangle$ is extended to states as for λML: for any two states s_0 and s_1, we write $s_0 \langle R \rangle s_1$ if s_0 and s_1 have equal domains and are related argument-wise by $\langle R \rangle$. We also extend $\langle R \rangle$ to a relation between environments, $E_1 \langle R \rangle E_2$, in exactly the same way. Using these definitions, the extension of $\langle R \rangle$ to all closed generalized expressions is defined as follows:

Definition 9 *Let* R *be as in Definition 8. We define another such family of relations* $\overline{R}$ *as follows:*

(i) $(E_1, exp_1) \, \overline{R} \, (E_2, exp_2)$ *iff* $\forall s_0, s_1, v_1$
$s_0, E_1 \vdash exp_1 \Downarrow v_1, s_1$ *implies* $\exists v_2, s_2$ *such that*
$s_0, E_2 \vdash exp_2 \Downarrow v_2, s_2$ *and* $s_1 \langle R \rangle s_2$ *and* $v_1 \; \langle R \rangle_\sigma \; v_2$ *(where* σ *is the type of* v_1, v_2*).*
(ii) $(E_1 + v_1, match_1) \, \overline{R} \, (E_2 + v_2, match_2)$ *iff* $\forall s_0, s_1, v_1$
$s_0, E_1, v_1 \vdash match_1 \Downarrow v_1, s_1$ *implies* $\exists v_2, s_2$ *such that*
$s_0, E_2, v_2 \vdash match_2 \Downarrow v_2, s_2$ *and* $s_1 \langle R \rangle s_2$ *and* $v_1 \; \langle R \rangle_\sigma \; v_2$ *(where* σ *is the type of* v_1, v_2*).*
(iii) $(E_1, dec_1) \, \overline{R} \, (E_2, dec_2)$ *iff* $\forall s_0, s_1, E_1'$
$s_0, E_1 \vdash dec_1 \Downarrow E_1', s_1$ *implies* $\exists E_2', s_2$ *such that*
$s_0, E_2 \vdash dec_2 \Downarrow E_2', s_2$ *and* $E_1' \langle R \rangle E_2'$ *and* $s_1 \langle R \rangle s_2$.

Let $\sqsubseteq$ be the greatest fixed point of the monotone operator $R \mapsto \overline{R}$. Then $\sqsubseteq$ is extended to open generalized expressions by defining $(E_1, exp_1) \sqsubseteq (E_2, exp_2)$ to hold iff for all values v_i such that $(E_1 + (x_i := v_i), exp_1)$ and $(E_2 + (x_i := v_i), exp_2)$ are closed, we have $(E_1 + (x_i := v_i), exp_1) \sqsubseteq (E_2 + (x_i := v_i), exp_2)$. The relation of applicative bisimilarity for SML is the symmetrization of this relation: $(E_1, exp_1) \approx_{app} (E_2, exp_2)$ iff $(E_1, exp_1) \sqsubseteq (E_2, exp_2)$ and $(E_2, exp_2) \sqsubseteq (E_1, exp_1)$.

A relation $\sqsubseteq^*$ is defined from $\sqsubseteq$ in much the same way as for the λ-calculus λML and the analogues of Lemma 4 and Proposition 5 can be established. Thus $\sqsubseteq$ coincides with $\sqsubseteq^*$ and the latter is a pre-congruence. As in the λ-calculus, the congruence property of the relation $\sqsubseteq$ yields immediately that SML applicative bisimilarity implies observational equivalence.

Theorem 10 *For all closed SML expressions* exp_1 *and* exp_2 *(of equal type),* $(\emptyset, exp_1) \approx_{app} (\emptyset, exp_2)$ *implies* $exp_1 \approx_{obs} exp_2$.

4 The Equivalence

We are now in a position to establish the full abstraction properties of the translations between SML and the λ-calculus λML. We first show that if we translate an expression to the other calculus and back, we get a result that is applicatively bisimilar to the expression we started with. Because we are not considering exception handling in this paper, for this property to hold for the SML expressions we have to restrict to ones involving matches that succeed for exactly one pattern. For example, consider the function $exp = \texttt{fn true => true}$; we have $(\llbracket exp \rrbracket)^M = \texttt{fn}\ arg \texttt{ => true}$, and hence $(exp)\texttt{false}$ raises an exception in SML, whereas $(\llbracket exp \rrbracket)^M\texttt{false}$ evaluates to $\texttt{true}$.

Theorem 11 *For any λML expression e, we have $\llbracket (e)^M \rrbracket \approx_{app} e$. For any generalized SML expression (E, exp) with exp satisfying the restriction on matches mentioned above, we have $(\llbracket E, exp \rrbracket)^M \approx_{app} (E, M)$, where $\llbracket E, exp \rrbracket$ denotes the λML expression obtained by applying the substitution $\llbracket E \rrbracket$ to the expression $\llbracket exp \rrbracket$.*

Proof. **(sketch)** The first statement is shown by induction over the structure of e. For the second statement, we show by induction over the structure of exp, that whenever all values v in E satisfy $(\llbracket v \rrbracket)^M \approx_{app} v$, then $(\llbracket E, exp \rrbracket)^M \approx_{app} (E, exp)$.

The induction steps make use of certain identities that hold up to applicative bisimilarity, which are easily established from the co-inductive definition of $\sqsubseteq$. For example, consider the λML expression $\texttt{fst}(e)$. The translation into SML and back yields the term $(\lambda arg^{\sigma_1 \times \sigma_2}.\texttt{fst}(arg))\llbracket (e)^M \rrbracket$. By induction hypothesis $\llbracket (e)^M \rrbracket \approx_{app} e$ holds. Hence by the congruence property of $\approx_{app}$ established in Section 3.1, we have $\llbracket (\texttt{fst}(e))^M \rrbracket \approx_{app} (\lambda arg^{\sigma_1 \times \sigma_2}.\texttt{fst}(arg))e$. Then $\llbracket (\texttt{fst}(e))^M \rrbracket \approx_{app} \texttt{fst}(e)$ follows from the identity $(\lambda arg^{\sigma_1 \times \sigma_2}.\texttt{fst}(arg))e \approx_{app} \texttt{fst}(e)$, which is easily established from the fact that the preorder $\sqsubseteq$ inducing $\approx_{app}$ is a fixed point (indeed the greatest one) of the monotone operator $R \mapsto \overline{R}$ of Definition 2.

The main result of the paper now follows from the fact that applicative bisimilarity implies observational equivalence, together with the above theorem and the general properties of the translations of compositionality and soundness with respect to evaluation.

Theorem 12 *The translations $\llbracket - \rrbracket$ and $(-)^M$ between the SML fragment and the λ-calculus λML are fully abstract, in the sense that they preserve and reflect observational equivalence:*

$$e_1 \approx_{obs} e_2 \text{ iff } (e_1)^M \approx_{obs} (e_2)^M$$

$$exp_1 \approx_{obs} exp_2 \text{ iff } \llbracket exp_1 \rrbracket \approx_{obs} \llbracket exp_2 \rrbracket$$

(provided exp_1, exp_2 satisfy the restriction on matches mentioned above).

Proof. Combining Corollary 7 with Theorems 10 and 11, we have that $\llbracket (e)^M \rrbracket \approx_{obs} e$ and $(\llbracket exp \rrbracket)^M \approx_{obs} exp$. From this it follows that it is sufficient to just prove that each translation preserves observational equivalence.

So suppose that $e_1 \approx_{obs} e_2$ in λML and that in SML for some boolean context $C[-]$ we have $s_0, \emptyset \vdash C[(e_1)^M] \Downarrow \texttt{true}, s_1$. Then by the soundness property of

$[\![-]\!]$, in λML we have $\langle [\![s_0]\!], [\![\emptyset, C[(e_1)^M]]\!]\rangle \Downarrow \langle [\![s_1]\!], [\![\mathtt{true}]\!]\rangle$. Now $[\![\mathtt{true}]\!] = \mathtt{true}$; and by compositionality of $[\![-]\!]$ and the fact that $\approx_{app}$ is a congruence, we have

$$[\![\emptyset, C[(e_1)^M]]\!] \equiv [\![C]\!][[\![(e_1)^M]\!]] \approx_{app} [\![C]\!][e_1] \approx_{app} [\![C]\!][e_2] \,.$$

Hence (by Proposition 5) for some s_2' $\langle [\![s_0]\!], [\![C]\!][e_2]\rangle \Downarrow \langle s_2', \mathtt{true}\rangle$. Then by the soundness property of $(-)^M$, in SML we have $([\![s_0]\!])^M, \emptyset \vdash ([\![C]\!][e_2])^M \Downarrow (\mathtt{true})^M, (s_2')^M$. Since $(\mathtt{true})^M = \mathtt{true}$, $([\![s_0]\!])^M \approx_{app} s_0$ and $([\![C]\!][e_2])^M \equiv ([\![C]\!])^M[(e_2)^M] \approx_{app} C[(e_2)^M]$, it follows that there is some s_2 with $s_0, \emptyset \vdash C[(e_2)^M] \Downarrow \mathtt{true}, s_2$. Thus by definition of observational equivalence we have that $(e_1)^M \approx_{obs} (e_2)^M$ in SML when $e_1 \approx_{obs} e_2$ in λML. The argument that $[\![-]\!]$ preserves observational equivalence is similar.

5 Conclusions

This paper describes a translation from a fragment of Standard ML to a λ-calculus with reference types, the main feature of which is a compilation of SML's use of pattern matching into λ-terms. We proved that the translation is fully abstract, i.e. that it preserves and reflects observational equivalence. The proof of this property is surprisingly difficult, because it is hard to reason directly about observational equivalence for programming languages such as SML that involve dynamic creation of local storage for higher order functions. A co-inductively defined notion of applicative bisimilarity is used here to obtain the result.

The notion of applicative similarity developed as a means to an end in this paper seems interesting in its own right. It implies observational equivalence, but not vice versa. A closely related notion, which validates more observational equivalences, is developed in [PPS94]. It remains an open problem to find a co-inductive characterization of observational equivalence for languages like SML that combine higher order functions and local state. Nevertheless, we believe that the existing notions of applicative bisimilarity, or refinements of them, may provide simpler methods for verifying program properties for languages like SML compared with denotational methods, or with reasoning directly about observational equivalence (as in [MT91, MT92], for example).

The fragment of Standard ML we have considered is monomorphic in order to avoid the known difficulties with mixing ML polymorphism with reference types—difficulties that are largely irrelevant to the concerns of this paper. For simplicity, we also excluded any exception-handling mechanism and alternate clauses in the definition of recursive datatypes from the fragment. We do not envisage any problem in extending the definition of applicative bisimilarity to cope with these features, although we have not considered this yet. Apart from anything else, such an extension would enable Standard ML's method for evaluating incomplete patterns to be treated.

References

[AO93] S. Abramsky and C.-H. L. Ong. Full abstraction in the lazy lambda calculus. *Information and Computation*, 105:159–267, 1993.

[Gor93] A. Gordon. *Functional Programming and Input/Output.* PhD thesis, University of Cambridge, 1993. Also available as Technical Report No. 285.

[How89] D. J. Howe. Equality in lazy computation systems. In *Proc. 4th Annual Symp. Logic in Computer Science*, pages 198–203. IEEE Computer Society Press, 1989.

[MT91] I. A. Mason and C. L. Talcott. Equivalence in functional languages with effects. *Journal of Functional Programming*, 1:287–327, 1991.

[MT92] I. A. Mason and C. L. Talcott. References, local variables and operational reasoning. In *Proc. 7th Annual Symp. Logic in Computer Science*, pages 186–197. IEEE Computer Society Press, 1992.

[Mil77] R. Milner. Fully abstract models of typed λ-calculi. *Theoretical Computer Science*, 4:1–22, 1977.

[MTH90] R. Milner, M. Tofte, and R. Harper. *The Definition of Standard ML.* MIT Press, Cambridge, MA, 1990.

[Ong93] C.-H. L. Ong. Non-determinism in a functional setting–extended abstract. In *Proc. 8th Annual Symp. Logic in Computer Science, Montréal, Canada*, pages 275–286. IEEE Computer Society Press, 1993.

[PPS94] V. C. V. de Paiva, A. M. Pitts and I. D. B. Stark. A Monadic ML. In preparation.

[Pit94] A. M. Pitts. A co-induction principle for recursively defined domains. *Theoretical Computer Science*, 124:195–219, 1994.

[PS93] A. M. Pitts and I. D. B. Stark. Observable properties of higher order functions that dynamically create local names, or: What's *new*? In *Proc. Int. Symp. on Math. Foundations of Computer Science*, pages 122–141. Lecture Notes in Computer Science No. 711, Berlin, 1993.

[Rie93] J.G. Riecke. Fully abstract translations between functional languages. *Mathematical Structures in Computer Science*, 3:387–415, 1993.

A Appendix

First, we define the translations between λML and the fragment of SML. Second, we list the judgements defining well-typed terms in λML as well as the rules for evaluation to canonical form.

A.1 The translations

The translation $(-)^M$, which maps a λML-type and a λML-term into an SML-type and an SML-expression respectively, is defined as follows:

Types

$$(\kappa_i)^M = \text{type } \kappa_i \text{ in } \mathtt{datatype}\ \kappa_1 = \mathtt{in}^{\kappa_1}\ \mathtt{of}\ \sigma_1 \cdots \text{ and } \kappa_n = \mathtt{in}^{\kappa_n}\ \mathtt{of}\ \sigma_n$$

$$(\mathtt{unit})^M = \mathtt{unit} \qquad (\mathtt{bool})^M = \mathtt{bool}$$
$$(\sigma_1 \times \sigma_2)^M = (\sigma_1)^M \times (\sigma_2)^M \qquad (\sigma_1 \to \sigma_2)^M = (\sigma_1)^M \to (\sigma_2)^M$$
$$(\sigma\ \mathtt{ref})^M = (\sigma)^M\ \mathtt{ref}$$

Expressions

$$
\begin{aligned}
(x)^M &= x & (\mathtt{in}^\kappa(e))^M &= \mathtt{in}^\kappa((e)^M)\\
(())^M &= () & (e_1 := e_2)^M &= (e_1)^M := (e_2)^M\\
(\mathtt{true})^M &= \mathtt{true} & (\mathtt{false})^M &= \mathtt{false}\\
(e_1 = e_2)^M &= (e_1)^M = (e_2)^M & ((e_1, e_2))^M &= ((e_1)^M, (e_2)^M)\\
(\lambda x^\sigma.e)^M &= \mathtt{fn}\ x{:}\,(\sigma)^M \Rightarrow (e)^M & (e_1 e_2)^M &= (e_1)^M (e_2)^M\\
(\mathtt{ref}(e))^M &= \mathtt{ref}((e)^M) & (!e)^M &= !(e)^M
\end{aligned}
$$

$$
\begin{aligned}
(\mathtt{rec}\ f(x) =^\sigma e_1\ \mathtt{in}\ f(e_2))^M &= \mathbf{let\ val\ rec}\ f = \mathbf{fn}\ x{:}\,\sigma \Rightarrow (e_1)^M\ \mathbf{in}\ f((e_2)^M)\ \mathbf{end}\\
(\mathsf{out}(e))^M &= \mathbf{let\ val\ in}^\kappa(x) = (e)^M\ \mathbf{in}\ x\ \mathbf{end}\\
(\mathtt{if}\ e_1\ \mathtt{then}\ e_2\ \mathtt{else}\ e_3)^M &= (\mathbf{fn\ true}{:}\,\mathbf{bool} \Rightarrow (e_2)^M \mid \mathbf{false}{:}\,\mathbf{bool} \Rightarrow (e_3)^M)(e_1)^M\\
(\mathsf{Fst}(e))^M &= \mathbf{let\ val}(x, \_) = (e)^M\ \mathbf{in}\ x\ \mathbf{end}\\
(\mathsf{Snd}(e))^M &= \mathbf{let\ val}(\_, x) = (e)^M\ \mathbf{in}\ x\ \mathbf{end}
\end{aligned}
$$

The translation from SML into λML is given next. The λ-expression $\pi_i(e)$ is an abbreviation for $\mathtt{fst}^i(\mathtt{snd}(e))$.

$$
\begin{aligned}
[\![(exp_1, \ldots, exp_n)]\!] &= ([\![exp_1]\!], \ldots, [\![exp_n]\!]) & [\![\mathtt{true}]\!] &= \mathtt{true}\\
[\![\mathtt{false}]\!] &= \mathtt{false} & [\![()]\!] &= ()\\
[\![\mathtt{in}^\kappa(exp)]\!] &= \mathtt{in}^\kappa([\![exp]\!]) & [\![var]\!] &= var\\
[\![!exp]\!] &= ![\![exp]\!]
\end{aligned}
$$

$$
\frac{[\![dec]\!] = (t_1/x_1, \ldots, t_n/x_n)}{[\![\mathtt{let}\ dec\ \mathtt{in}\ exp\ \mathtt{end}]\!] = (\lambda x_n^{\sigma_1}. \cdots .(\lambda x_n.[\![exp]\!])[\![t_n]\!] \cdots)[\![t_1]\!]}
$$

$$
\begin{aligned}
[\![exp\ exp]\!] &= [\![exp]\!][\![exp]\!]\\
[\![\mathtt{fn}{:}\,ty\ match]\!] &= \lambda arg{:}\,[\![ty]\!].[\![match]\!]
\end{aligned}
$$

$$
\frac{[\![mrule]\!] = (cond, exp)}{[\![mrule\ \langle \mid match\rangle]\!] = \langle \mathtt{if}\ cond\ \mathtt{then}\rangle\ exp\ \langle \mathtt{else}\ [\![match]\!]\rangle}
$$

$$
\frac{[\![pat]\!] = (cond, \sigma)}{[\![pat \Rightarrow exp]\!] = (cond, [\![exp]\!]\sigma)}
$$

$$
\begin{aligned}
[\![val\ valbind]\!] &= [\![valbind]\!]\\
[\![\mathtt{local}\ dec_1\ \mathtt{in}\ dec_2\ \mathtt{end}]\!] &= [\![dec_2]\!][[\![dec_1]\!]]\\
[\![dec_1; dec_2]\!] &= [\![dec_1]\!] + [\![dec_2]\!][[\![dec_1]\!]]
\end{aligned}
$$

$$
\frac{[\![pat]\!] = (cond, \sigma)}{[\![pat = exp]\!] = \sigma[[\![exp]\!]/arg]}
$$

$$
\frac{[\![pat = exp]\!] = \sigma \qquad [\![valbind]\!] = \sigma'}{[\![pat = exp\ \mathtt{and}\ valbind]\!] = \sigma \oplus \sigma'}
$$

$$
\frac{[\![valbind]\!] = [\lambda x_i{:}\,\sigma_i.t_i/f_i]}{[\![\mathtt{rec}\ valbind]\!] = [\lambda x_i{:}\,\sigma_i.\pi_i\mathtt{rec}\ f(arg) = (t_i[\pi_i(f)/f_i][\pi_i(x)/x_i])\ \mathtt{in}\ f(x_i)/f_i]}
$$

$$[\![\_]\!] = (\mathtt{true}, \{\}) \qquad [\![\mathtt{true}]\!] = (arg, \{\})$$
$$[\![\mathtt{false}]\!] = (\mathtt{not}(arg), \{\}) \qquad [\![()]\!] = (\mathtt{true}, ()/arg)$$
$$[\![var]\!] = (\mathtt{true}, [var/arg]) \qquad [\![pat{:}\, ty]\!] = [\![pat]\!]$$

$$\frac{[\![pat_i]\!] = (cond_i, E_i)}{[\![pat_1, \ldots, pat_n]\!] = (\bigwedge cond_i[\pi_i(arg)/arg], \oplus E_i[\pi_i(arg)/arg])}$$

$$\frac{[\![pat]\!] = (cond, E)}{[\![\mathtt{in}^{\kappa}(pat)]\!] = (cond, E)[\mathtt{out}(arg)/arg]}$$

$$\frac{[\![pat]\!] = (cond, E)}{[\![\mathtt{ref}(pat)]\!] = (cond, E)[!(arg)/arg]}$$

$$\frac{[\![pat]\!] = (cond, E)}{[\![var\langle{:}\, ty\rangle\ \mathbf{as}\ pat]\!] = (cond, E \oplus var/arg)}$$

A.2 Rules for λML

The rules for well-typed terms in λML are as follows:

$$\frac{}{\Gamma \vdash x{:}\, \Gamma(x)} \qquad \frac{\Gamma \vdash e{:}\, \sigma}{\Gamma \vdash \mathtt{in}^{\kappa}(e){:}\, \kappa}\ (decl(\kappa) = \sigma)$$

$$\frac{\Gamma \vdash e{:}\, \kappa}{\Gamma \vdash \mathtt{out}(e){:}\, \sigma}\ (\sigma = decl(\kappa)) \qquad \frac{}{\Gamma \vdash ()\,{:}\, \mathtt{unit}}$$

$$\frac{\Gamma \vdash r{:}\, \sigma\ \mathtt{ref}}{\Gamma \vdash (r := e){:}\, \mathtt{unit}} \qquad \frac{}{\Gamma \vdash \mathtt{true}{:}\, \mathtt{bool}}$$

$$\frac{}{\Gamma \vdash \mathtt{false}{:}\, \mathtt{bool}} \qquad \frac{\Gamma \vdash b{:}\, \mathtt{bool} \quad \Gamma \vdash e{:}\, \sigma \quad \Gamma \vdash e'{:}\, \sigma}{\Gamma \vdash (\mathbf{if}\ b\ \mathbf{then}\ e\ \mathbf{else}\ e'){:}\, \sigma}$$

$$\frac{\Gamma \vdash r{:}\, \sigma\ \mathtt{ref} \quad \Gamma \vdash r'{:}\, \sigma\ \mathtt{ref}}{\Gamma \vdash (r = r'){:}\, \mathtt{bool}} \qquad \frac{\Gamma \vdash e{:}\, \sigma' \quad \Gamma \vdash e'{:}\, \sigma'}{\Gamma \vdash (e, e'){:}\, \sigma \times \sigma'}$$

$$\frac{\Gamma \vdash e{:}\, \sigma \times \sigma'}{\Gamma \vdash \mathtt{fst}(e){:}\, \sigma} \qquad \frac{\Gamma \vdash e{:}\, \sigma \times \sigma'}{\Gamma \vdash \mathtt{snd}(e){:}\, \sigma'}$$

$$\frac{\Gamma, x{:}\, \sigma \vdash e{:}\, \sigma'}{\Gamma \vdash \lambda x^{\sigma}.e{:}\, \sigma \to \sigma'} \qquad \frac{\Gamma \vdash e{:}\, \sigma \to \sigma' \quad \Gamma \vdash e'{:}\, \sigma}{\Gamma \vdash ee'{:}\, \sigma'}$$

$$\frac{\Gamma, f{:}\, \sigma \to \sigma', x{:}\, \sigma \vdash e'{:}\, \sigma' \quad \Gamma \vdash e{:}\, \sigma}{\Gamma \vdash \mathbf{rec}\ f(x) =_{\sigma} e'\ \mathbf{in}\ f(e){:}\, \sigma'} \qquad \frac{\Gamma \vdash e{:}\, \sigma}{\Gamma \vdash \mathtt{ref}(e){:}\, \sigma\ \mathtt{ref}}$$

$$\frac{\Gamma \vdash r{:}\, \sigma\ \mathtt{ref}}{\Gamma \vdash !r{:}\, \sigma} \qquad \frac{}{\Gamma \vdash a^{\sigma}{:}\, \sigma\ \mathtt{ref}}$$

The rules for evaluation to canonical expression are as follows:

$$\frac{e \Downarrow c}{\mathtt{in}^{\kappa}(e) \Downarrow \mathtt{in}^{\kappa}(c)} \qquad \frac{e \Downarrow \mathtt{in}^{\kappa}(c)}{\mathtt{out}(e) \Downarrow c}$$

$$\frac{}{() \Downarrow ()} \qquad \frac{\langle s_1, r\rangle \Downarrow \langle s_2, a\rangle \qquad \langle s_2, e\rangle \Downarrow \langle s_3, c\rangle}{\langle s_1, r := e\rangle \Downarrow \langle s_3 \cup \{a \mapsto c\}, ()\rangle}$$

$$\frac{}{\mathtt{true} \Downarrow \mathtt{true}} \qquad \frac{}{\mathtt{false} \Downarrow \mathtt{false}}$$

$$\frac{b \Downarrow \mathtt{true} \qquad e \Downarrow c}{\mathtt{if}\ b\ \mathtt{then}\ e\ \mathtt{else}\ e' \Downarrow c} \qquad \frac{\langle s_1, e\rangle \Downarrow \langle s_2, a\rangle}{\langle s_1, !e\rangle \Downarrow \langle s_2, c\rangle}\ (c = s_2(a))$$

$$\frac{b \Downarrow \mathtt{false} \qquad e' \Downarrow c}{\mathtt{if}\ b\ \mathtt{then}\ e\ \mathtt{else}\ e' \Downarrow c} \qquad \frac{\langle s_1, e\rangle \Downarrow \langle s_2, c\rangle}{\langle s_1, \mathtt{ref}(e)\rangle \Downarrow \langle s_2 \oplus \{a \mapsto c\}, a\rangle}\ (a \notin dom(s_2))$$

$$\frac{r \Downarrow a \qquad r' \Downarrow a'}{r = r' \Downarrow \mathtt{false}}\ (a \neq a') \qquad \frac{e \Downarrow c \qquad e' \Downarrow c'}{(e, e') \Downarrow (c, c')}$$

$$\frac{e \Downarrow (c, c')}{\mathtt{fst}(e) \Downarrow c} \qquad \frac{e \Downarrow (c, c')}{\mathtt{snd}(e) \Downarrow c'}$$

$$\frac{}{\lambda x^{\sigma}.e \Downarrow \lambda x^{\sigma}.e} \qquad \frac{f \Downarrow \lambda x^{\sigma}.e' \qquad e \Downarrow c \qquad e'[c/x] \Downarrow c'}{fe \Downarrow c'}$$

$$\frac{}{a \Downarrow a} \qquad \frac{e \Downarrow c \qquad e'[\lambda x^{\sigma}.\mathtt{rec}\ f(x) =_{\sigma} e'\ \mathtt{in}\ f(x)/f, c/x] \Downarrow c'}{\mathtt{rec}\ f(x) =_{\sigma} e'\ \mathtt{in}\ f(e) \Downarrow c'}$$

$$\frac{r \Downarrow a \qquad r' \Downarrow a}{r = r' \Downarrow \mathtt{true}}$$

Categorical completeness results for the simply-typed lambda-calculus

Alex K. Simpson

LFCS, Department of Computer Science, University of Edinburgh,
JCMB, The King's Buildings, Edinburgh, EH9 3JZ
Email: Alex.Simpson@dcs.ed.ac.uk

Abstract. We investigate, in a categorical setting, some completeness properties of beta-eta conversion between closed terms of the simply-typed lambda calculus. A cartesian-closed category is said to be *complete* if, for any two unconvertible terms, there is some interpretation of the calculus in the category that distinguishes them. It is said to have a *complete interpretation* if there is some interpretation that equates only interconvertible terms. We give simple necessary and sufficient conditions on the category for each of the two forms of completeness to hold. The classic completeness results of, e.g., Friedman and Plotkin are immediate consequences. As another application, we derive a syntactic theorem of Statman characterizing beta-eta conversion as a maximum consistent congruence relation satisfying a property known as typical ambiguity.

1 Introduction

In 1970 Friedman proved that beta-eta conversion is complete for deriving all equalities between the (simply-typed) lambda-definable functionals in the category **Set** [5]. (Incidentally, this result was independently discovered by Plotkin [10], published in [11].) However, in computer science one is often interested in interpretations in other cartesian closed categories (such as the category of complete partial orders and continuous functions). It is natural to ask whether similar completeness results also hold in such cases. For the category of complete partial orders, Plotkin was able to extend Friedman's argument and show that completeness does indeed still hold (see [9, Theorem 5.2.28]). More recently, Berger and Schwichtenberg used different techniques to show that completeness holds relative to any model capable of faithfully representing certain basic operations on syntax [3].

In this paper we investigate such completeness questions in a categorical setting. As is well known, cartesian-closed categories (CCCs) provide a general notion of model for the simply-typed lambda calculus. We ask under what conditions on a CCC, $\mathcal{C}$, does beta-eta conversion derive all equalities between terms which are true in $\mathcal{C}$. Actually, this question is not yet well defined, as different interpretations of base types in $\mathcal{C}$ might induce different equalities. Thus there are two natural strengths of completeness. The weaker form holds when beta-eta conversion derives all those equalities between terms which are true under all interpretations in $\mathcal{C}$. The stronger form holds when there is a single interpretation that equates only terms that are beta-eta convertible. In this paper we give necessary and sufficient conditions on $\mathcal{C}$ for each of the forms of completeness to hold (Theorems 1 and 2). The conditions turn out to be simple ones that are easily checked in particular cases. Moreover, they show the failure of completeness to be the exception rather than the rule.

As an application, we use Theorem 1 to obtain Statman's [16] characterization of beta-eta convertibility as a maximally consistent congruence relation satisfying typical ambiguity (Theorem 3). Indeed, as will be seen, our work is closely related to, and also heavily dependent upon, some fundamental syntactic work of Statman. We shall discuss this dependency further in Section 7.

2 Preliminaries

In order to have a tight connection between the lambda-calculus and cartesian-closed categories we work with a calculus with finite product types. We use $\alpha, \beta, \ldots$ to range over a non-empty set of base types, X, containing a distinguished base type, 0. We use $\sigma, \tau, \ldots$ to range over types which comprise: base types, function types $\sigma \to \tau$, (binary) product types $\sigma \times \tau$, and a unit type $\mathbf{1}$. We work with explicitly typed variables $x^\sigma, y^\tau, \ldots$ although we often omit type labels for convenience. We use $U, V, \ldots$ to range over open terms which are given by the grammar:

$$U ::= x^\sigma \mid \lambda x^\sigma . U \mid U(V) \mid \langle U, V\rangle \mid \pi_1(U) \mid \pi_2(U) \mid *$$

(where $\langle U, V\rangle$ and $\pi_i(U)$ are pairing and projection for product types and $*$ is the canonical element of $\mathbf{1}$) subject to the usual typing constraints. Each term has a unique type and we write U^σ to mean that the type of U is σ. We use $L, M, N, \ldots$ to range over *closed* terms. We write Λ_X for the set of *closed* terms. We write $\Lambda_X^{\rightarrow}$ for those terms in Λ_X that are terms of the usual pure functionally typed lambda-calculus (i.e. those terms all of whose subterms have types built from X using $\to$). We adopt standard conventions such as associating $\to$ to the right and application to the left. We also use evident notation for products of arbitrary finite arity, their tuples and projections.

We assume that the reader is acquainted with the rules for beta-eta convertibility, $=_{\beta\eta}$, between terms of identical type (see, e.g, [1, 4, 7]). Two classes of terms, the *neutral terms* and the *long-$\beta\eta$ normal forms*, are defined by mutual induction. A term is *neutral* if it has one of the following forms: x^σ; or $U(V)$ where U is neutral and V is in long-$\beta\eta$ normal form; or $\pi_i(U)$ where U is neutral. A term is in *long-$\beta\eta$ normal form* if it has one of the following forms: U^α where U is neutral (note the restriction to a base type); or $\lambda x^\sigma . U$ where U is in long-$\beta\eta$ normal form; or $\langle U, V\rangle$ where U and V are both in long-$\beta\eta$ normal form; or $*$. The important fact about long-$\beta\eta$ normal forms is that, for every term U, there is a unique long-$\beta\eta$ normal form, $\beta\eta(U)$, such that $U =_{\beta\eta} \beta\eta(U)$ (see [1, 4, 7]). By this characterization it is clear that $=_{\beta\eta}$ between terms in Λ_X is conservative over the usual beta-eta convertibility between terms in $\Lambda_X^{\rightarrow}$.

Let $\mathcal{C}$ be a cartesian-closed category with distinguished: terminal object, $\mathbf{1}$; binary products, $A \times B$; and exponentials, B^A. (We do not assume that $\mathcal{C}$ has all finite limits.) An interpretation of the calculus in $\mathcal{C}$ is determined by a function $[\![\cdot]\!]$ from X to objects of $\mathcal{C}$. This extends (using the CCC structure of $\mathcal{C}$) to interpret arbitrary types σ as objects $[\![\sigma]\!]$ of $\mathcal{C}$. Then a closed term M^σ is interpreted as a morphism $[\![M]\!] \in \mathcal{C}(\mathbf{1}, [\![\sigma]\!])$. (The interpretation is defined using a more general interpretation of open terms, U^σ, as morphisms from objects interpreting the context of free variables in U to $[\![\sigma]\!]$.) We write $\Lambda_X \to \mathcal{C}$ for the class of all interpretations of the calculus in $\mathcal{C}$. The soundness of beta-eta conversion in CCCs says that $M =_{\beta\eta} N$ implies that, for all $[\![\cdot]\!] : \Lambda_X \to \mathcal{C}$, it holds that $[\![M]\!] = [\![N]\!]$. We shall be interested in when the converse implication holds, and related questions.

Before considering such completeness questions we consider the categorical formulation of what an interpretation of the lambda-calculus in $\mathcal{C}$ is (see [8]). This formulation is in terms of cartesian-closed functors (CC-functors), which are those functors between CCCs that preserve the cartesian-closed structure "on the nose".[1] Let $\mathcal{F}_X$ be the free cartesian-closed category generated by the set of objects X. To give a concrete description, $\mathcal{F}_X$ is the category whose objects are types and whose morphisms from σ to τ are the closed long-$\beta\eta$ normal forms of type $\sigma \to \tau$. The identities and composition are obtained as the long-$\beta\eta$ normal forms of the evident lambda-terms. The freeness of $\mathcal{F}_X$ means that any function $[\![\cdot]\!]$ from X to objects of $\mathcal{C}$ extends to a unique CC-functor, F, from $\mathcal{F}_X$ to $\mathcal{C}$, where "extends" means that $F(\alpha) = [\![\alpha]\!]$. Further, if we write $[\![\cdot]\!]$ for the interpretation of the lambda-calculus induced by the function on X, it holds that, for all M^σ, $[\![M]\!] = F(\beta\eta(\lambda x^1 .\, M)) \in \mathcal{C}(\mathbf{1}, [\![\sigma]\!])$ and, for all long-$\beta\eta$ normal forms $M^{\sigma\to\tau}$ that $F(M) \in \mathcal{C}([\![\sigma]\!], [\![\tau]\!])$ is the evident exponential transpose of $[\![M]\!] \in \mathcal{C}(\mathbf{1}, [\![\tau]\!]^{[\![\sigma]\!]})$. Thus interpretations of the lambda-calculus in $\mathcal{C}$ are essentially equivalent to CC-functors from $\mathcal{F}_X$ to $\mathcal{C}$.

3 The Completeness Theorems

We now define the two forms of completeness we shall be investigating. First the weaker notion, which is the direct converse to the soundness statement above. We say that $\mathcal{C}$ is *complete (for $=_{\beta\eta}$)*[2] if, for all M^σ, N^σ,

$$M =_{\beta\eta} N \text{ iff for all } [\![\cdot]\!] : \Lambda_X \to \mathcal{C},\ \ [\![M]\!] = [\![N]\!].$$

This concept has a natural categorical formulation. Recall that a class of functors from a category $\mathcal{A}$ to a category $\mathcal{B}$ is *collectively faithful* if, for all $A \xrightarrow{f} B$ and $A \xrightarrow{g} B$ in $\mathcal{A}$, whenever it holds that $F(f) = F(g)$ for all functors F in the

[1] The whole discussion here could easily be generalized to deal with functors preserving the structure up to isomorphism. Such functors are categorically more natural, but for our purposes the simpler "on the nose" functors suffice.

[2] It would perhaps be preferable to say that $=_{\beta\eta}$ is complete for $\mathcal{C}$, however this is not so easily shortened.

class, then $f = g$. Thus, using the equivalence between interpretations and CC-functors, $\mathcal{C}$ is complete if and only if the class of CC-functors from $\mathcal{F}_X$ to $\mathcal{C}$ is collectively faithful.

For the stronger notion we require completeness relative to a single interpretation rather than the class of all interpretations. We say that an interpretation $[\![\cdot]\!] : \Lambda_X \to \mathcal{C}$ is *complete (for* $=_{\beta\eta}$*)* if, for all M^σ, N^σ,

$$M =_{\beta\eta} N \text{ iff } [\![M]\!] = [\![N]\!].$$

We say that $\mathcal{C}$ *has a complete interpretation (for* $=_{\beta\eta}$*)* if there exists a complete interpretation $[\![\cdot]\!] : \Lambda_X \to \mathcal{C}$. Again these concepts have natural categorical reformulations. An interpretation is complete if and only if the corresponding CC-functor from $\mathcal{F}_X$ to $\mathcal{C}$ is faithful. Similarly, $\mathcal{C}$ has a complete interpretation if and only if there exists a faithful CC-functor from $\mathcal{F}_X$ to $\mathcal{C}$.

In this paper we characterize the conditions under which $\mathcal{C}$ is complete (Theorem 1) and under which $\mathcal{C}$ has a complete interpretation (Theorem 2). It is also interesting to consider the question of characterizing when a given interpretation $[\![\cdot]\!] : \Lambda_X \to \mathcal{C}$ is complete. This problem is of a different nature as it no longer concerns a property intrinsic to the category $\mathcal{C}$. In the case that $X = \{0\}$, such a characterization (essentially due to Statman) will be obtained in Section 4 (Corollary 4). We do not have such a result for arbitrary X. Some of the problems in obtaining one will be considered in Section 6.

Before giving the characterizations, we consider some motivating examples. First, the category **Set** has a complete interpretation. Indeed any interpretation mapping each base type to an infinite set is complete. This result is proved explicitly in [4], but it is closely related to Friedman's famous completeness theorem [5]. (There is a detailed discussion of the differences in [4].) It is clear then that **Set** is complete, as in general the existence of a complete interpretation implies completeness. The converse is not true. An example that is complete but which has no complete interpretation is the category of finite sets, **FinSet**. The completeness of **FinSet** is proved explicitly in [13], but it is closely related to Theorem 2 of [15] (a result originally due to Plotkin [10]), which is basically a finite model property for beta-eta conversion. The non-existence of a complete interpretation in **FinSet** was essentially observed by Friedman [5]. The reason is simply that there exist types with an infinite number of equivalence classes of closed terms modulo $=_{\beta\eta}$, for example $(0 \to 0) \to 0 \to 0$. Lastly, there do indeed exist cartesian-closed categories that are not complete. Recall that a *preorder* is a category with at most one morphism in each hom-set. It is obvious that any cartesian-closed preorder (for example, any Heyting algebra) is not complete.

The first characterization says that the preorder observation above is the only obstacle to completeness.

Theorem 1 *$\mathcal{C}$ is complete if and only if it is not a preorder.*

So, perhaps surprisingly, completeness turns out to be merely a question of the non-triviality of the hom-sets of $\mathcal{C}$.

We have seen that completeness is determined by the simple cardinality condition that there exists a hom-set with cardinality ≥ 2. Given that the counterexample to a complete interpretation in **FinSet** is also via a cardinality argument, one might wonder whether $\mathcal{C}$ has a complete interpretation if and only if it has an infinite hom-set. This, however, is not the case. For a counterexample take the full subcategory of the co-Kleisli category of the $\omega \times -$ comonad on **Set** determined by those objects that are the image of finite sets under the inclusion from **Set** to the co-Kleisli category. We call this category $\mathbf{FinSet}_{\omega\times -}$. (More concretely, $\mathbf{FinSet}_{\omega\times -}$ has finite sets for objects, and the morphisms from X to Y are those functions from $\omega \times X$ to $\omega \times Y$ that preserve the first component of pairs.) Theorem 2 below gives an elementary way of checking that there is indeed no complete interpretation in $\mathbf{FinSet}_{\omega\times -}$. A more abstract reason for this failure is that any CC-functor from $\mathcal{F}_X$ to $\mathbf{FinSet}_{\omega\times -}$ necessarily factors through the inclusion from **FinSet**. This can be proved using the universal property of the co-Kleisli category as a polynomial category (see [8]) together with the initiality of $\mathcal{F}_X$. We omit the argument.

Nevertheless, a closely related condition does succeed in characterizing the existence of a complete interpretation. We say that an endomorphism $A \xrightarrow{a} A$ is *non-repeating* if all its iterates are distinct (i.e. if $a^h = a^k$ implies $h = k$).

Theorem 2 *$\mathcal{C}$ has a complete interpretation if and only if it contains a non-repeating endomorphism.*

Note that it is not apparent from the definitions of the two forms of completeness that they are independent of the choice of X. Theorems 1 and 2 show this to be the case.

4 Proofs of Theorems 1 and 2

We shall prove Theorem 2 first and then derive Theorem 1 as a consequence. Throughout the proofs we move freely between categorical formulations in terms of CC-functors and syntactic formulations in terms of interpretations. We also move freely between the interpretations of terms as morphisms in $\mathcal{C}(\mathbf{1}, [\![\tau]\!]^{[\![\sigma]\!]})$ and their exponential transposes as morphisms in $\mathcal{C}([\![\sigma]\!], [\![\tau]\!])$.

For the left-to-right implication of Theorem 2, suppose that $[\![\cdot]\!]$ is a complete interpretation. Not surprisingly, a non-repeating endomorphism is given by a successor function on the interpretation of the Church numerals. Specifically, the endomorphism is:[3]

$$[\![(0 \to 0) \to 0 \to 0]\!] \xrightarrow{[\![\lambda x^{(0\to 0)\to 0\to 0}.\,\lambda y^{0\to 0}.\,\lambda z^{0}.(x(y)(y(z)))]\!]} [\![(0 \to 0) \to 0 \to 0]\!]$$

(making use of an exponential transpose as discussed above). It is non-repeating because if its n-th iterate is composed with

$$\mathbf{1} \xrightarrow{[\![\lambda y^{0\to 0}.\,\lambda z^{0}.z]\!]} [\![(0 \to 0) \to 0 \to 0]\!]$$

[3] This and other "diagrams" were prepared using Paul Taylor's Latex diagram macros package.

then one obtains:

$$1 \xrightarrow{[\![\lambda y^{0\to 0}.\,\lambda z^{0}.y^{n}(z)]\!]} [\![(0\to 0)\to 0\to 0]\!],$$

and, by completeness, it is clear that the latter differs for distinct values of n. Incidentally, here we have shown that it is a necessary condition for $[\![\cdot]\!]$ to be complete that the above endomorphism is non-repeating. In Section 6 we show that this is not in general a sufficient condition, even for interpretations of $\Lambda_{\{0\}}$.

For the converse implication, given a non-repeating endomorphism in $\mathcal{C}$, we must construct a faithful CC-functor from $\mathcal{F}_X$ to $\mathcal{C}$.

Proposition 1 *There is a faithful CC-functor from $\mathcal{F}_X$ to $\mathcal{F}_{\{0\}}$.*

Proof. The CC-functor is that determined by the unique function from X to $\{0\}$. This maps any X-type, σ, to the $\{0\}$-type, $\overline{\sigma}$, obtained by replacing every base type α with 0. For any $M^\sigma \in \Lambda_X$ define $\overline{M}$ to be the $\Lambda_{\{0\}}$-term obtained by replacing every variable x^σ in M with $x^{\overline{\sigma}}$. Clearly $\overline{M}$ has type $\overline{\sigma}$. For faithfulness it is enough to show that, for any two distinct long-$\beta\eta$ normal forms $M^\sigma, N^\sigma \in \Lambda_X$, it holds that $\overline{M}$ and $\overline{N}$ are distinct long-$\beta\eta$ normal forms in $\Lambda_{\{0\}}$. This is done by a straightforward induction on the structure of long-$\beta\eta$ normal forms. ⊠

Thus it remains to find a faithful CC-functor from $\mathcal{F}_{\{0\}}$ to $\mathcal{C}$. For this we appeal to a deep syntactic result about the (pure functional) simply-typed lambda-calculus due to Statman [15, Theorem 3]. Define $\top$ to be the type $(0\to 0\to 0)\to 0\to 0$.

Proposition 2 (Statman) *For all $M^\sigma, N^\sigma \in \Lambda^{\to}_{\{0\}}$, it holds that $M =_{\beta\eta} N$ if and only if, for all $L^{\sigma\to\top}$, $L(M) =_{\beta\eta} L(N)$.*

A detailed proof can be found in [12]. Incidentally, in [14, Proposition 1], Statman shows that, for each σ, there exists $L^{\sigma\to\top}$, such that $M =_{\beta\eta} N$ if and only if $L(M) =_{\beta\eta} L(N)$, but we do not need this stronger result here.

Proposition 3 *For all $M^\sigma, N^\sigma \in \Lambda_{\{0\}}$, it holds that $M =_{\beta\eta} N$ if and only if, for all $L^{\sigma\to\top}$, $L(M) =_{\beta\eta} L(N)$.*

Proof. Left-to-right is trivial. For the converse suppose that $M^\sigma \neq_{\beta\eta} N^\sigma$. It is easily shown that σ is isomorphic (in $\mathcal{F}_{\{0\}}$) to a finite product $\sigma_1 \times \ldots \times \sigma_n$ (where $n \geq 0$) of types σ_i built from 0 using $\to$. We write σ' for this product type and $I^{\sigma\to\sigma'}$ for the lambda-term giving (one half of) the isomorphism. Clearly $I(M) \neq_{\beta\eta} I(N)$, so $\langle \pi_1(I(M)),\ldots,\pi_n(I(M))\rangle \neq_{\beta\eta} \langle \pi_1(I(N)),\ldots,\pi_n(I(N))\rangle$. Therefore there is some i for which $\pi_i(I(M)) \neq_{\beta\eta} \pi_i(I(N))$. So $\beta\eta(\pi_i(I(M))) \neq \beta\eta(\pi_i(I(N)))$. But these terms are both normal forms of type σ_i, and hence they are terms of $\Lambda^{\to}_{\{0\}}$ (because all subterms of a normal form have subtypes of its type). So, by Proposition 2, there exists $L^{\sigma_i\to\top}$ such that $L(\pi_i(I(M))) \neq_{\beta\eta} L(\pi_i(I(N)))$. But then $\lambda x^\sigma. L(\pi_i(I(x)))$ is the term of type $\sigma\to\top$ that we are trying to find. ⊠

Corollary 4 *An interpretation, $[\![\cdot]\!] : \Lambda_{\{0\}} \to \mathcal{C}$, is complete if and only if for all $M^{\top}$, $N^{\top}$ it holds that $[\![M]\!] = [\![N]\!]$ implies $M =_{\beta\eta} N$.*

Proof. Left-to-right is trivial. For the converse, suppose that for all $M^{\top}$, $N^{\top}$ it holds that $[\![M]\!] = [\![N]\!]$ implies $M =_{\beta\eta} N$. Suppose that $[\![M^{\sigma}]\!] = [\![N^{\sigma}]\!]$. By the "compositionality" of $[\![\cdot]\!]$ we have, for all $L^{\sigma\to\top}$, that $[\![L(M)]\!] = [\![L(N)]\!]$. Whence, by the assumption, for all $L^{\sigma\to\top}$, we have $L(M) =_{\beta\eta} L(N)$. So, by Proposition 3, $M =_{\beta\eta} N$. Thus $[\![\cdot]\!]$ is indeed complete. ⊠

The corollary gives a necessary and sufficient condition for an interpretation of $\Lambda_{\{0\}}$ in $\mathcal{C}$ to be complete. We use this to obtain a useful sufficient condition. A *very weak natural number object* in $\mathcal{C}$ is an object B together with morphisms:

$$\mathbf{1} \xrightarrow{\ \overline{0}\ } B \xrightarrow{\ s\ } B \underset{\times}{\overset{+}{\leftleftarrows}} B \times B$$

such that, for all m, n, it holds that $\overline{m+n} = + \circ \langle \overline{m}, \overline{n} \rangle$ and $\overline{m \times n} = \times \circ \langle \overline{m}, \overline{n} \rangle$, where we write $\overline{n}$ for the "numeral" morphism $s^n \circ \overline{0}$.[4] A very weak natural number object is said to be *faithful* if all the numerals are distinct (i.e. if $\overline{m} = \overline{n}$ implies $m = n$).

The next lemma generalizes the completeness theorem that appears in Berger and Schwichtenberg [3] (although they work in a non-categorical setting).

Lemma 5 *An interpretation, $[\![\cdot]\!] : \Lambda_{\{0\}} \to \mathcal{C}$, is complete if $[\![0]\!]$ is a faithful very weak natural number object.*

Proof. Let B be $[\![0]\!]$. Let ϕ be the binary function on natural numbers defined by $\phi(m,n) = (m+n)^2 + m + 1$. By simple composition using the very weak natural number morphisms, there is a morphism $B \times B \xrightarrow{\overline{\phi}} B$ such that $\overline{\phi} \circ \langle \overline{m}, \overline{n} \rangle = \overline{\phi(m,n)}$. We shall use this to show that the condition of Corollary 4 is satisfied.

First, it is routine to check that the closed long-$\beta\eta$ normal forms of type $\top$ have the form $\lambda p^{0\to0\to0}.\, \lambda l^0.\, t$ where t is given by the grammar:

$$t ::= l \mid p(t_1)(t_2).$$

Now we define inductively a numerically valued function, $(\cdot)^*$, on the set of such t by:

$$l^* = 0,$$
$$(p(t_1)(t_2))^* = \phi(t_1^*, t_2^*).$$

It is easily seen that $t_1^* = t_2^*$ implies t_1 and t_2 are identical (as ϕ is an injective function from $\mathbf{N} \times \mathbf{N}$ to $\mathbf{N}^+$).

[4] Note that there is no requirement that $+$ and $\times$ satisfy any of the usual algebraic identities.

Now, for any t, we have a morphism $[\![\lambda p.\,\lambda l.\,t]\!] \in \mathcal{C}(\mathbf{1}, \top)$ and we note the evident corresponding $B^{(B\times B)} \times B \xrightarrow{\tilde{t}} B$. We also note the exponential transpose $\mathbf{1} \xrightarrow{\tilde{\phi}} B^{(B\times B)}$ of $\overline{\phi}$. It is easily checked that the composite:

$$\mathbf{1} \xrightarrow{\langle\tilde{\phi},\overline{0}\rangle} B^{(B\times B)} \times B \xrightarrow{\tilde{t}} B$$

is equal to $\overline{t^*}$. So if $[\![\lambda p.\,\lambda l.\,t_1]\!] = [\![\lambda p.\,\lambda l.\,t_2]\!]$ then $\overline{t_1^*} = \overline{t_2^*}$ and thus t_1 and t_2 are identical (as the very weak natural number object is faithful).

To complete the proof, suppose that $[\![M^\top]\!] = [\![N^\top]\!]$. Suppose that $\beta\eta(M) = \lambda p.\,\lambda l.\,t_1$ and $\beta\eta(N) = \lambda p.\,\lambda l.\,t_2$. Then, by the above, t_1 and t_2 are identical so $M =_{\beta\eta} N$. Thus the condition of Corollary 4 is indeed satisfied. ⊠

Note that the condition of the lemma is not necessary for $[\![\cdot]\!]$ to be complete. It fails, for example, for the evidently complete "identity" interpretation of $\Lambda_{\{0\}}$ in $\mathcal{F}_{\{0\}}$ where there is no morphism from $\mathbf{1}$ to $[\![0]\!]$ (and the only endomorphism on $[\![0]\!]$ is the identity).

Let $A \xrightarrow{a} A$ be a non-repeating endomorphism in $\mathcal{C}$. Let B be the object $(A^A)^{(A^A)}$. We use the internal lambda-calculus of $\mathcal{C}$ to define arrows:

$$\begin{aligned}
\overline{0} &= \mathbf{1} \xrightarrow{\lambda f^{A^A}.\lambda a^A.a} B \\
s &= B \xrightarrow{b \;\mapsto\; \lambda f^{A^A}.\lambda a^A.b(f)(f(a))} B \\
+ &= B\times B \xrightarrow{\langle b,b'\rangle \;\mapsto\; \lambda f^{A^A}.\lambda a^A.b(f)(b'(f)(a))} B \\
\times &= B\times B \xrightarrow{\langle b,b'\rangle \;\mapsto\; \lambda f^{A^A}.\lambda a^A.b(b'(f))(a)} B
\end{aligned}$$

making use of standard encodings of successor, addition and multiplication on Church numerals. It is clear that these morphisms show B to be a very weak natural number object. To see that it is faithful note that, by exponential transpose, each $\overline{n}$ gives a morphism $A^A \xrightarrow{\tilde{n}} A^A$ and a gives a morphism $\mathbf{1} \xrightarrow{\tilde{a}} A^A$. It is easily seen that the exponential transpose of the composite $\tilde{n} \circ \tilde{a}$ is $A \xrightarrow{a^n} A$. Thus the numerals must all be distinct as otherwise would contradict a being a non-repeating endomorphism.

It now follows from Lemma 5 that the interpretation $[\![\cdot]\!] : \Lambda_{\{0\}} \to \mathcal{C}$ determined by setting $[\![0]\!] = B$ is complete. Together with Proposition 1, this completes the proof of Theorem 2.

We now turn to Theorem 1. The left-to-right implication is trivial. For the converse, suppose that $\mathcal{C}$ is not a preorder. We shall show that there is a faithful CC-functor, F, from $\mathcal{F}_X$ to $\mathcal{C}^\omega$ (the countably infinite power of $\mathcal{C}$), which is indeed a CCC. Given such an F, a collectively faithful set of CC-functors from $\mathcal{F}_X$ to $\mathcal{C}$ is $\{\pi_i \circ F \mid i \in \omega\}$ where π_i is is the i-th projection from $\mathcal{C}^\omega$ to $\mathcal{C}$ (it is easily checked that the projections are CC-functors), from which it is clear that the class of all CC-functors is collectively faithful.

To obtain F we use Theorem 2, by which it suffices to find a non-repeating endomorphism in $\mathcal{C}^\omega$. As $\mathcal{C}$ is not a preorder, suppose that f and g are two distinct morphisms in $\mathcal{C}(A,B)$. For $n \geq 1$ define $B_n = B^{B^n}$ where B^n is the n-fold product of B with itself. For $i \in \{0,\ldots,n-1\}$ define:

$$\begin{aligned} \overline{i_n} &= 1 \xrightarrow{\lambda c^{B^n}.\pi_i(c)} B_n \\ s_n &= B_n \xrightarrow{d \;\mapsto\; \lambda c^{B^n}.d(\langle \pi_2(c),\ldots,\pi_n(c),\pi_1(c)\rangle)} B_n. \end{aligned}$$

Clearly $s_n \circ \overline{i_n} = \overline{j_n}$ where j is $i+1$ modulo n. We now show that $\overline{0_n},\ldots,\overline{(n-1)_n}$ are all distinct. Let $B^n \xrightarrow{\tilde{i_n}} B$ be the exponential transpose of i_n. It is clear that the composite:

$$A \xrightarrow{\langle \overbrace{f,\ldots,f}^{j},g,\ldots,g\rangle} B^n \xrightarrow{\tilde{i_n}} B$$

is equal to f if $i \leq j$ and is equal to g otherwise. This shows that $j > i$ implies $\overline{i_n} \neq \overline{j_n}$ (as $f \neq g$), so $\overline{0_n},\ldots,\overline{(n-1)_n}$ are indeed all distinct. It is now clear that

$$(B_1, B_2, \ldots) \xrightarrow{(s_1, s_2, \ldots)} (B_1, B_2, \ldots)$$

is a non-repeating endomorphism in $\mathcal{C}^\omega$, as required.

5 Typical Ambiguity

In this section we apply Theorem 1 to obtain a syntactic characterization of $=_{\beta\eta}$ as, in a sense to be defined below, a maximally consistent congruence relation satisfying typical ambiguity (Theorem 3). For the calculus $\Lambda_{\{0\}}^{\rightarrow}$, this result is originally due to Statman [16]. Although the theorem for Λ_X is easily derived from Statman's result for $\Lambda_{\{0\}}^{\rightarrow}$, it is an interesting application of our completeness results to obtain it instead as a consequence of Theorem 1. As a matter of fact, we shall also see that one can turn the tables and derive Theorem 1 from Theorem 3. Thus, in some sense, Theorem 1 is a semantic counterpart to the syntactic Theorem 3.

First we introduce the necessary notation to state Theorem 3. Given a function ϕ from X to types, we write $\sigma[\phi]$ for the type obtained by simultaneously replacing each occurrence of a base type α in σ with $\phi(\alpha)$. Similarly, we write $M[\phi]$ for the term obtained by replacing all variables x^σ in M with $x^{\sigma[\phi]}$. If M has type σ then $M[\phi]$ has type $\sigma[\phi]$. Such a substitution of types clearly corresponds to a CC-functor from $\mathcal{F}_X$ to itself.

Let $\sim$ be a well-typed equivalence relation on Λ_X (i.e. one for which $M \sim N$ implies M and N are of identical type) such that $M =_{\beta\eta} N$ implies $M \sim N$. We say that $\sim$ is a *congruence* if $M^\sigma \sim N^\sigma$ implies that, for all $L^{\sigma\rightarrow\tau}$, $L(M) \sim L(N)$ (the other properties of a congruence relation follows from this because $\sim$ contains $=_{\beta\eta}$). We say that $\sim$ is *consistent* if, for some σ, there exist two terms, M^σ and N^σ, such that $M \not\sim N$. We say that $\sim$ satisfies *typical ambiguity* if, for all type-valued functions, ϕ, on X, it holds that $M \sim N$ implies $M[\phi] \sim N[\phi]$.

Theorem 3 *If $\sim$ is a consistent congruence relation containing $=_{\beta\eta}$ and $\sim$ satisfies typical ambiguity then $M \sim N$ if and only if $M =_{\beta\eta} N$.*

To prove the theorem, suppose that $\sim$ satisfies the assumptions. We construct a category $\mathcal{F}_X/\!\sim$ as follows. The objects of $\mathcal{F}_X/\!\sim$ are types. The morphisms from σ to τ are the equivalence classes of the set of closed terms of type $\sigma \to \tau$ modulo $\sim$, and we write $[M]$ for the equivalence class of M. The identities and composition are evident. It is easily checked that $\mathcal{F}_X/\!\sim$ is a CCC, using the fact that $\sim$ extends $=_{\beta\eta}$ and the congruence property of $\sim$. Further, by the consistency property, $\mathcal{F}_X/\!\sim$ is not a preorder.

Lemma 6 *Given any $[\![\cdot]\!] : \Lambda_X \to \mathcal{F}_X/\!\sim$, define ϕ from X to types by $\phi(\alpha) = [\![\alpha]\!]$. Then $[\![M]\!] = [\lambda x^1 . M[\phi]]$ in $\mathcal{F}_X/\!\sim (1, \sigma[\phi])$.*

This is proved by induction on the structure of M. The induction, which involves going through interpretations of open terms, is routine.

Now suppose that $M \sim N$. Let $[\![\cdot]\!]$ be any interpretation in $\mathcal{F}_X/\!\sim$, and define ϕ as above. By typical ambiguity, $M[\phi] \sim N[\phi]$. Whence, by the congruence property, $\lambda x^1 . M[\phi] \sim \lambda x^1 . N[\phi]$. So it follows from the lemma that $[\![M]\!] = [\![N]\!]$. We have shown that, for any $[\![\cdot]\!]$, we have that $[\![M]\!] = [\![N]\!]$. Thus Theorem 1 implies that $M =_{\beta\eta} N$. This proves Theorem 3.

As commented above, one can also derive Theorem 1 from Theorem 3. To this end, suppose that $\mathcal{C}$ is not a preorder. Define a well-typed equivalence relation, $\sim$, by:

$$M \sim N \text{ iff for all } [\![\cdot]\!] : \Lambda_X \to \mathcal{C}, \quad [\![M]\!] = [\![N]\!].$$

By the soundness of $=_{\beta\eta}$, we have that $\sim$ contains $=_{\beta\eta}$. Theorem 1 says that $M =_{\beta\eta} N$ if and only if $M \sim N$. To show this we need only verify that $\sim$ satisfies the conditions of Theorem 3. The congruence property is straightforward (it holds because of the "compositionality" of $[\![\cdot]\!] : \Lambda_X \to \mathcal{C}$). Consistency follows from $\mathcal{C}$ not being a preorder, as it is easy to find an interpretation such that $[\![\lambda x^0 . \lambda y^0 . x]\!] \neq [\![\lambda x^0 . \lambda y^0 . y]\!]$. It remains to show typical ambiguity. First we note the lemma below, which is proved by a straightforward induction on the structure of M (again involving interpretations of open terms).

Lemma 7 *Given any ϕ from X to types and interpretation $[\![\cdot]\!] : \Lambda_X \to \mathcal{C}$, let $[\![\cdot]\!]'$ be the interpretation determined by $[\![\alpha]\!]' = [\![\phi(\alpha)]\!]$. Then $[\![M[\phi]]\!] = [\![M]\!]'$.*

Suppose that $M \sim N$. Let $[\![\cdot]\!]$ be any interpretation. By the lemma, we have that $[\![M[\phi]]\!] = [\![M]\!]'$ and $[\![N[\phi]]\!] = [\![N]\!]'$. Now $M \sim N$, so by the definition of $\sim$ we have that $[\![M]\!]' = [\![N]\!]'$. Therefore $[\![M[\phi]]\!] = [\![N[\phi]]\!]$. So $M[\phi] \sim N[\phi]$, and $\sim$ does indeed satisfy typical ambiguity.

The derivation of Theorem 1 from Theorem 3, gives a proof of Theorem 1 not involving Theorem 2. However, Statman's proof of Theorem 3 (for $\Lambda^{\rightarrow}_{\{0\}}$) also relies on the reduction of $=_{\beta\eta}$ to the single type $\top$ (Proposition 2), on which our proof of Theorem 2 was based. It is an interesting fact that an alternative direct proof of Theorem 3 is possible using a typed version of the Böhm-out technique [2, Ch. 10]. The details are beyond the scope of this paper.

6 Complete Interpretations

In this section we consider the problem of obtaining a characterization of when a given interpretation is complete. Corollary 4 already characterizes when an interpretation $[\![\cdot]\!] : \Lambda_{\{0\}} \to \mathcal{C}$ is complete. We consider whether this characterization can be improved in a natural way. We also consider whether it generalizes to interpretations of Λ_X for an arbitrary X. Although the results we obtain are negative, they do illustrate well some of the more delicate aspects of the completeness questions.

One natural question is whether Corollary 4 can be improved by simplifying the type of M and N from $\top$ to $(0 \to 0) \to 0 \to 0$. Below, we use logical relations to construct a model answering this questions in the negative. This negative answer justifies the comment made at the end of our proof of the left-to-right implication of Theorem 2. In general it is an insufficient condition for an interpretation $[\![\cdot]\!] : \Lambda_{\{0\}} \to \mathcal{C}$ to be complete that the interpretation of the successor function on Church numerals be a non-repeating endomorphism.

The category $\mathbf{R_3}$ is defined as follows. Its objects A are pairs $(|A|, R_A)$ where $|A|$ is a set and R_A is a ternary relation on $|A|$ such that $R_A(a, a, a)$ for all $a \in |A|$. The morphisms from A to B are those functions $f : |A| \to |B|$ such that, for all $a_1, a_2, a_3 \in |A|$, it holds that $R_A(a_1, a_2, a_3)$ implies $R_B(f(a_1), f(a_2), f(a_3))$. This category is cartesian closed with: $|\mathbf{1}| = \{\emptyset\}$ where $R_1(\emptyset, \emptyset, \emptyset)$ holds; and $|A \times B| = |A| \times |B|$ with $R_{A\times B}(\langle a_1, b_1\rangle, \langle a_2, b_2\rangle, \langle a_3, b_3\rangle)$ if and only if $R_A(a_1, a_2, a_3)$ and $R_B(b_1, b_2, b_3)$; and $|B^A| = \mathbf{R_3}(A, B)$ with $R_{B^A}(f_1, f_2, f_3)$ if and only if, for all $a_1, a_2, a_3 \in |A|$, it holds that $R_A(a_1, a_2, a_3)$ implies $R_B(f_1(a_1), f_2(a_2), f_3(a_3))$. The details are easily checked. Let A be the object of $\mathbf{R_3}$ defined by $|A| = \omega$ and $R_A(l, m, n)$ if and only if either $l = m = n$ or $l + 1 = m = n - 1$.

Define an interpretation $[\![\cdot]\!] : \Lambda_{\{0\}} \to \mathbf{R_3}$ by setting $[\![0]\!] = A$. We claim that, for all $M^{(0\to0)\to0\to0}, N^{(0\to0)\to0\to0}$, it holds that $[\![M]\!] = [\![N]\!]$ implies $M =_{\beta\eta} N$. Note that the closed long-$\beta\eta$ normal forms of $(0 \to 0) \to 0 \to 0$ have the form $\lambda s.\, \lambda z.\, s^n(z)$ for $n \geq 0$. As the function $n \mapsto n + 1$ is in $\mathbf{R_3}(A, A)$, and hence in $|A^A|$, it is easily seen that any two distinct long-$\beta\eta$ normal forms of $(0 \to 0) \to 0 \to 0$ get interpreted as different functionals in $[\![(0 \to 0) \to 0 \to 0]\!]$. The claim follows.

Despite completeness for the type $(0 \to 0) \to 0 \to 0$, it turns out that $[\![\cdot]\!]$ is not complete. By Corollary 4, we know that the incompleteness must already arise for terms of type $\top$.

Lemma 8 *A function $f : A \times A \to A$ is in $\mathbf{R_3}(A \times A, A)$ if and only if, for some $k \geq 0$, it holds that f is one of: $\langle m, n\rangle \mapsto k$; or $\langle m, n\rangle \mapsto m + k$; or $\langle m, n\rangle \mapsto n + k$.*

Proof. The right-to-left implication is easily checked. For the converse, suppose that $f \in \mathbf{R_3}(A \times A, A)$. Set $k = f(0, 0)$. We have that: $R_{A\times A}(\langle 0, 0\rangle, \langle 0, 1\rangle, \langle 0, 2\rangle)$, and $R_{A\times A}(\langle 0, 0\rangle, \langle 1, 0\rangle, \langle 2, 0\rangle)$, and $R_{A\times A}(\langle 0, 0\rangle, \langle 1, 1\rangle, \langle 2, 2\rangle)$. So there are apparently five choices for the four values $(f(0, 0), f(0, 1), f(1, 0), f(1, 1))$ namely: *(i)* (k, k, k, k); *(ii)* $(k, k, k, k + 1)$; *(iii)* $(k, k, k + 1, k + 1)$; *(iv)* $(k, k + 1, k, k + 1)$; *(v)* $(k, k + 1, k + 1, k + 1)$.

However, *(ii)* and *(v)* are impossible. We show this for *(v)*. Clearly *(v)* requires that $f(0,2) = k+2$ and, because $R_{A\times A}(\langle 1,0\rangle, \langle 1,1\rangle, \langle 1,2\rangle)$, that $f(1,2) = k+1$. But then, as $R_{A\times A}(\langle 0,2\rangle, \langle 1,2\rangle, \langle 2,2\rangle)$, there is no possible value for $f(2,2)$.

We claim that for the other cases: *(i)* determines f to be $\langle m,n\rangle \mapsto k$; *(iii)* determines f to be $\langle m,n\rangle \mapsto m+k$; and *(iv)* determines f to be $\langle m,n\rangle \mapsto n+k$. We show this for *(iv)*. Clearly *(iv)* determines that $f(2,0) = k$ and that $f(2,1) = k+1$. Now a simple inductive argument shows, for all m, that $f(m,0) = k$ and $f(m,1) = k+1$. But then it is clear that $f(m,2) = k+2$, and another inductive argument shows that indeed $f(m,n) = n+k$. ⊠

It is now straightforward to show that, for example, the two distinct long-$\beta\eta$ normal forms, $\lambda p.\, \lambda l.\, p(p(l)(l))(l)$ and $\lambda p.\, \lambda l.\, p(p(l)(p(l)(l)))(l)$, of type T, are interpreted as the same functional in $[\![\mathsf{T}]\!]$ (as are any two "trees" such that both leftmost branches have the same length, h say, and both rightmost branches have length k say). Thus we have shown that completeness cannot be reduced to completeness for the single type $(0 \to 0) \to 0 \to 0$.

Another direction in which one might hope to improve Corollary 4 would be to characterize the complete interpretations of Λ_X for arbitrary X. One would prefer a characterization that is both simple and useful (like Corollary 4), but unfortunately we do not have one. Here we content ourselves with showing that a most naïve attempt at a generalization of Corollary 4 fails. Specifically, define T_α to be the type $(\alpha \to \alpha \to \alpha) \to \alpha \to \alpha$. We show that it is not necessarily the case that $[\![\cdot]\!] : \Lambda_X \to \mathcal{C}$ is complete when, for all $\alpha \in X$, for all M^{T_α}, N^{T_α}, it holds that $[\![M]\!] = [\![N]\!]$ implies $M =_{\beta\eta} N$. For a counterexample take $X = \{0, 0'\}$ and $\mathcal{C} = \mathbf{Set} \times \mathbf{Set}$. Define $[\![0]\!] = (\omega, \emptyset)$ and $[\![0']\!] = (\emptyset, \omega)$. By the completeness of Λ_X in **Set** we have that, for all $\alpha \in X$, for all M^{T_α}, N^{T_α}, it holds that $[\![M]\!] = [\![N]\!]$ implies $M =_{\beta\eta} N$. However, one sees that $[\![0 \to 0']\!]$ is interpreted as $(\emptyset, \mathbf{1})$ and so, for example, the two distinct terms (modulo $=_{\beta\eta}$) of $(0 \to 0') \to (0 \to 0') \to (0 \to 0')$ are interpreted as the same (unique) point of $[\![(0 \to 0') \to (0 \to 0') \to (0 \to 0')]\!]$. It follows that $[\![\cdot]\!]$ is not complete. We leave the finding of a useful characterization of complete interpretations of Λ_X as an open question. A related question is to find the simplest set of types to which $=_{\beta\eta}$ can be reduced in the manner of Proposition 2.

7 Discussion

It is clear that the work presented in this paper is heavily dependent on old results of Statman. In particular we use Theorem 3 of [15] (our Proposition 2) in a critical way, and our Theorem 2 is not too difficult a consequence of it. Further, we saw in Section 5 that Theorem 1 could also be derived as a fairly straightforward consequence of Statman's typical ambiguity theorem. However, although our main results follow without too much effort from Statman's work, the elegance and generality of our theorems makes them compelling semantic alternatives to Statman's syntactic results. We also hope that the present paper will have the effect of drawing attention to Statman's results, whose implications deserve to be better known.

Two departures from Statman's work are that we work with a calculus with unit and product types and that we allow more than one base type. The former difference is overcome using the characterization of $=_{\beta\eta}$ in terms of long $\beta\eta$-normal forms, which until quite recently was a field of active research (see, e.g., [1, 4, 7]). The latter difference turns out to be irrelevant in the case of Theorems 1 and 2 (as is shown by Proposition 1). In Section 6 we saw that this difference is non-trivial for the question of characterizing when an interpretation is complete.

It is interesting to compare our work with Statman's own semantic application of his syntactic results. In [17] he states his important *1-Section Theorem* giving necessary and sufficient conditions for an interpretation of $\Lambda_{\{0\}}^{\rightarrow}$ in a Henkin model to be complete. (See [12] for a detailed discussion and proof of the theorem.) The 1-Section Theorem is closely related to our Corollary 4, but it goes further, reducing completeness at the second-order type $\top$ to a property of elements of first-order types in a countable direct-product of the model. However, in doing so, the 1-Section Theorem makes essential use of the "well-pointedness" of Henkin models. There is a natural analogue of the 1-Section Theorem for well-pointed cartesian-closed categories, but not for general cartesian-closed categories. In this paper we have preferred not to consider results that apply only to well-pointed categories. After all, one of the benefits of the categorical setting is that non-well-pointed structures (such as closed-term categories) are handled alongside (the more set-theoretic) well-pointed structures in a uniform semantic framework. Note that our derivation of Theorem 3 from Theorem 1 made essential use of the applicability of our results to non-well-pointed categories.

One question is whether the results can be generalized to give completeness results for Λ_X augmented with typed constants. Categorically, one then considers CC-functors from the free cartesian closed category generated by a graph. Čubrić used Friedman's techniques to show that there is a faithful CC-functor from any such free CCC to **Set** [4]. Unfortunately, our proofs do not extend in this way, as Proposition 2 fails once constants are added to the syntax.

Another interesting question is whether the purely categorical formulations of Theorems 1 and 2 extend to other kinds of categories with structure. It seems likely that both results will generalize to bicartesian closed categories. The main obstacle in proving such a generalization is to get a good handle on equality in the internal language. It is already difficult to generalize long-$\beta\eta$ normal forms (although see [6] for progress on this question), let alone the deep syntactic results of Statman. On the other hand, for recursion theoretic reasons, it is clear that our results do not generalize to cartesian-closed categories with a natural numbers object.

Acknowledgements

I thank Aurelio Carboni, Eugenio Moggi and Pino Rosolini for useful feedback when I presented this work in Genoa. This research was carried out under an EPSRC postdoctoral fellowship.

References

1. Y. Akama. On Mints' reduction for ccc-calculus. In M. Bezem and J. F. Groote, editors, *Typed Lambda Calculi and Applications, Proceedings of TLCA '93*. LNCS 664, Springer Verlag, 1993.
2. H. P. Barendregt. *The Lambda Calculus, its Syntax and Semantics*. North Holland, Amsterdam, 1984. Second edition.
3. U. Berger and H. Schwichtenberg. An inverse of the evaluation functional for typed λ-calculus. In *Proceedings of 6th Annual Symposium on Logic in Computer Science*, pages 203 - 211, 1991.
4. D. Čubrić. Embedding of a free cartesian closed category into the category of sets. *Journal of Pure and Applied Algebra*, to appear, 1995.
5. H. Friedman. Equality between functionals. In R. Parikh, editor, *Logic Colloquium*, Springer-Verlag, New York, 1975.
6. N. Ghani. $\beta\eta$-equality for coproducts. *This volume*, 1995.
7. C. B. Jay and N. Ghani. The virtues of eta-expansion. *Journal of Functional Programming*, to appear, 1995.
8. J. Lambek and P. J. Scott. *Introduction to Higher Order Categorical Logic*. Number 7 in Cambridge studies in advanced mathematics. Cambridge University Press, 1986.
9. E. Moggi. *The Partial Lambda-Calculus*. Ph.D. thesis, Department of Computer Science, University of Edinburgh, 1988. Available as LFCS report no. ECS-LFCS-88-63.
10. G. D. Plotkin. Lambda-definability and logical relations. Technical Report SAI-RM-4, School of Artificial Intelligence, University of Edinburgh, 1973.
11. G. D. Plotkin. Lambda-definability in the full type hierarchy. In J. P. Seldin and J. R. Hindley, editors, *To H. B. Curry: Essays on Combinatory Logic, Lambda Calculus and Formalism*. Academic Press, New York, 1980.
12. J. G. Riecke. Statman's 1-Section Theorem. *Information and Computation*, to appear, 1995.
13. S. Soloviev. The category of finite sets and CCCs. *Journal of Soviet Mathematics*, 22:1387–1400, 1983.
14. R. Statman. On the existence of closed terms in the typed λ-calculus I. In J. P. Seldin and J. R. Hindley, editors, *To H. B. Curry: Essays on Combinatory Logic, Lambda Calculus and Formalism*. Academic Press, New York, 1980.
15. R. Statman. Completeness, invariance and λ-definability. *Journal of Symbolic Logic*, 47:17–26, 1982.
16. R. Statman. λ-definable functionals and $\beta\eta$ conversion. *Archiv fur Math. Logik und Grund.*, 23:21–26, 1983.
17. R. Statman. Equality between functionals revisited. In L. A. Harrington *et al*, editors, *Harvey Friedman's Research on the Foundations of Mathematics*. Elsevier Science Publishers, 1985.

Third-Order Matching in the Presence of Type Constructors

(Extended Abstract)

Jan Springintveld[1]

Department of Philosophy, Utrecht University

P.O. Box 80126, 3508 TC Utrecht, The Netherlands

Abstract

We show that it is decidable whether a third-order matching problem in $\lambda\underline{\omega}$ (an extension of the simply typed lambda calculus with type constructors) has a solution or not. We present an algorithm which, given such a problem, returns a solution for this problem if the problem has a solution and returns *fail* otherwise.

1 Introduction

It is well-known that type theory is a good basis for the implementation of proof checkers. The man-machine interaction of proof checking can be considerably improved if some kind of matching algorithm can be implemented for the terms of the underlying type theory. For if one wants to prove $\phi(t)$ for a certain formula ϕ and term t, and one already has a proof H of $\forall x\phi(x)$, one would like it to be sufficient to indicate that H should be used without having to mention t. The proof checker should be able to match $\phi(x)$ with $\phi(t)$. Here t can be an object (natural number, boolean, ...) or a function, a functional, etc. To deal with this, one needs an algorithm for *higher-order matching.* So we ask: can we find such an algorithm? In other words: is the higher-order matching problem decidable for the underlying type theory? The starting point of this paper is the Calculus of Constructions (CoC), a type theory which features polymorphism, dependent types and type constructors (see [2]). Unfortunately, while second-order matching is decidable in CoC [4], third-order matching is undecidable [3]. It is important to understand why third-order and hence higher-order matching is undecidable in CoC: is it a specific feature (polymorphism, dependent types, type constructors) that is responsible for the undecidability or is the undecidability caused by the interaction of these features? It is precisely for this kind of questions that Barendregt's lambda cube [1] of subsystems of CoC was developed. Starting from the simply typed lambda calculus the cube is erected by three systems, each supporting one of the abovementioned features: λP supports dependent types, $\lambda 2$ supports polymorphism and $\lambda\underline{\omega}$ supports type constructors. The other systems in the cube combine these features in all possible combinations. It is proved in [3] that in λP third-order matching is undecidable. In $\lambda 2$, higher-order matching is undecidable ([7]); fourth-order matching in $\lambda 2$ has been shown decidable in [14]. It is (to our knowledge) an open question whether matching of finite order $n \geq 4$ is decidable in $\lambda 2$. In $\lambda\omega$, the combination of $\lambda 2$ and $\lambda\underline{\omega}$, fourth-order matching is undecidable [3]. It is (again to our knowledge) open whether third-order matching is decidable in $\lambda\omega$.

In this paper, we show that third-order matching is decidable in $\lambda\underline{\omega}$. In other words, we show that the mere presence of type constructors is not sufficient to make third-order matching undecidable. At first sight, this is not surprising, since $\lambda\underline{\omega}$ is a weak extension of the simply typed lambda calculus and in [5] it is proved that third-order matching is decidable in the simply typed lambda calculus (this result was recently extended to fourth-order matching [12, 13]). But it becomes more surprising when one realizes that the type

[1]This work is supported by the Netherlands Computer Science Research Foundation (SION) with financial support of the Netherlands Organisation for Scientific Research (NWO).

structure of $\lambda\underline{\omega}$ is rich enough to encode, in types of order 3, the second-order unification problem, which is undecidable by [9]. The main argument in the proof of the decidability of third-order matching is that second-order unification can be avoided by defining and solving matching problems in a suitable order.

This paper is organized as follows. In Section 2, we present the two systems with which we shall be concerned in this paper: $\lambda\tau$ (the simply typed lambda calculus with one base type O) and $\lambda\underline{\omega}$. In Section 3, we present terminology concerning matching problems and solutions. We also give an example which illustrates the difficulties that are involved in the proof of our main result. In Section 4, we prove the decidability of third-order matching. Almost all proofs of lemmas are omitted; these can be found in the full version, [15].

Acknowledgements. I would like to thank Gilles Dowek for his hospitality during my stay at INRIA Rocquencourt and for the fruitful discussions we had there. I also thank Marc Bezem, Jan Friso Groote, Jaco van de Pol and Alex Sellink for critical remarks and helpful suggestions.

2 The systems $\lambda\tau$ and $\lambda\underline{\omega}$

In this section we introduce two typed lambda calculi: $\lambda\tau$, the simply typed lambda calculus with one base type O, and $\lambda\underline{\omega}$, the simply typed lambda calculus with type variables and type constructors (we consider this system with the Conversion Rule for $\beta\eta$-conversion). For more information on these systems the reader is referred to [1] or [8].

Definition 2.1 *(Terms and reductions). Pseudo-terms* are given by the following abstract syntax: $\mathcal{T} ::= \mathcal{C} \mid \mathcal{V} \mid \mathcal{T}\mathcal{T} \mid \lambda\mathcal{V}{:}\mathcal{T}.\mathcal{T} \mid \mathcal{T}{\rightarrow}\mathcal{T}$. Here $\mathcal{C}$ is an infinite set of constants and $\mathcal{V}$ is an infinite set of variables; x, y, y', y_1, ... range over $\mathcal{V}$. Among the constants, three elements are singled out: O, $*$ and $\Box$. Roman letters range over $\mathcal{T}$. We define $\mathcal{K}_\Box ::= * \mid \mathcal{K}_\Box{\rightarrow}\mathcal{K}_\Box$ and $\mathcal{K}_* ::= O \mid \mathcal{K}_*{\rightarrow}\mathcal{K}_*$. We apply the usual conventions concerning brackets; so ABC means $(AB)C$ and $A{\rightarrow}B{\rightarrow}C$ means $A{\rightarrow}(B{\rightarrow}C)$. The set of free variables of A is defined as usual and denoted by $FV(A)$. Also the substitution of A for x in B (denoted by $B[x := A]$) and the relations $\rightarrow_\beta$, $\twoheadrightarrow_\beta$, $\rightarrow_\eta$, $\twoheadrightarrow_\eta$, $\twoheadrightarrow_{\beta\eta}$, $=_\beta$ and $=_{\beta\eta}$ are defined on pseudo-terms as usual. Syntactic equality (modulo α-conversion) is denoted by $\equiv$.

Definition 2.2 *(Contexts and Judgements).* In this paper Q, Q_1, ... range over $\{\exists, \forall\}$. $Qx : B$ is called a *(quantified) declaration.* A *(quantified) pseudo-context* is a finite ordered sequence of quantified declarations $Q_i x_i : C_i$, where the x_i are pairwise distinct. Pseudo-contexts are denoted by capital Greek letters Δ, Γ, Γ_0,... The empty context is denoted by $\langle\rangle$. If $\exists x : C$ occurs in Γ, then x is said to be *existential in* Γ. If $\forall x : C$ occurs in Γ, then x is said to be *universal in* Γ. If every declaration in Γ is of the form $\exists x : C$, then Γ is an *existential context.* The intuition behind the quantification of variables is that universal variables are considered to be constant, in the sense that solutions to matching problems are not allowed to substitute terms for them; substitutions are only allowed to substitute terms for existential variables.

If $\Gamma \equiv \langle Q_1 x_1 : A_1, \ldots, Q_n x_n : A_n \rangle$, then $dom(\Gamma) = \{x_1, \ldots, x_n\}$ and $\Gamma, Qx : B$ denotes $\langle Q_1 x_1 : A_1, \ldots, Q_n x_n : A_n, Qx : B\rangle$. (In general, we denote the concatenation of Γ and Δ by Γ, Δ.) If the declaration $Qx : C$ occurs in Γ, then $\Gamma(x)$ denotes C. We write $\Gamma \subseteq \Delta$ (resp. $\Gamma \subseteq_\Box \Delta$) if each $Qx{:}A$ in Γ (resp. each $Qx{:}A$ in Γ such that $A \in \mathcal{K}_\Box$) also occurs in Δ.

A *judgement* is of the form $\Gamma \vdash A : B$, where Γ is a quantified pseudo-context and A and B are pseudo-terms. When we want to indicate that a judgement is derived in a system λo, we write $\Gamma \vdash_{\lambda o} A : B$. If $\Gamma \vdash_{\lambda o} A : B$, then we call A *closed in* Γ if all free variables x in A are universal in Γ and moreover $\Gamma(x)$ is closed in Γ_x. We use the abbreviation $\Gamma \vdash A : B : C$ for $\Gamma \vdash A : B$ and $\Gamma \vdash B : C$. A pseudo-term A is called *legal* when there exist a pseudo-context

Axiom	$\langle\rangle \vdash c : s$		if $c : s \in Ax_{\lambda\circ}$
Start	$\dfrac{\Gamma \vdash B : s}{\Gamma, Qx{:}B \vdash x{:}B}$		if $x \notin \Gamma$
Weakening	$\dfrac{\Gamma \vdash A : B \quad \Gamma \vdash B' : s}{\Gamma, Qx{:}B' \vdash A : B}$		if $x \notin \Gamma$
Product	$\dfrac{\Gamma \vdash A_1 : s \quad \Gamma \vdash A_2 : s}{\Gamma \vdash A_1 {\to} A_2 : s}$		
Application	$\dfrac{\Gamma \vdash A_1 : B_1 {\to} B_2 \quad \Gamma \vdash A_2 : B_1}{\Gamma \vdash A_1 A_2 : B_2}$		
Abstraction	$\dfrac{\Gamma, Qx{:}A_1 \vdash A_2 : B_2 \quad \Gamma \vdash A_1 {\to} B_2 : s}{\Gamma \vdash \lambda x{:}A_1.A_2 : A_1 {\to} B_2}$		
Conversion	$\dfrac{\Gamma \vdash A : B' \quad \Gamma \vdash B : s}{\Gamma \vdash A : B}$		if $B' =_{\beta\eta} B$

Table 1: The rules

Γ and a pseudo-term B such that $\Gamma \vdash A : B$ or $\Gamma \vdash B : A$. A pseudo-context Γ is called legal when there exist pseudo-terms A and B such that $\Gamma \vdash A : B$.

Definition 2.3 *(The systems).* The systems $\lambda\tau$ and $\lambda\underline{\omega}$ are defined using the rules in Table 1. Here s ranges over $Sort_{\lambda\circ}$, where $Sort_{\lambda\circ} \subseteq C$ is a set of *sorts*. Furthermore $Ax_{\lambda\circ}$ is a set of statements of the from $c : s$, where $c \in C$. The systems $\lambda\tau$ and $\lambda\underline{\omega}$ can be obtained from the rules by taking $Sort_{\lambda\tau} = \{*\}$, $Sort_{\lambda\underline{\omega}} = \{*, \Box\}$, $Ax_{\lambda\tau} = \{O : *\}$ and $Ax_{\lambda\underline{\omega}} = \{* : \Box\}$. In this paper we let $\lambda\circ$ range over $\lambda\tau$ and $\lambda\underline{\omega}$ and (except in Table 1) s over $\{*, \Box\}$. In $\lambda\tau$, the Conversion Rule is redundant.

A typical judgement in $\lambda\underline{\omega}$ is:

$$\forall A : *, \exists Y : *{\to}* \vdash_{\lambda\underline{\omega}} \lambda x{:}Y((\lambda\beta{:}*.\beta{\to}\beta)A).x : (Y((\lambda\beta{:}*.\beta{\to}\beta)A)){\to}(Y(A{\to}A)).$$

The systems $\lambda\tau$ and $\lambda\underline{\omega}$ enjoy many well-known properties such as Strong Normalization, Confluence, Subject Reduction and Unicity of Types (all w.r.t. $\beta\eta$-reduction). See [1] or [8]. Next, we give some basic facts concerning the terms in our systems.

Lemma 2.4.

1. $\Gamma \vdash_{\lambda\underline{\omega}} A : \Box \Leftrightarrow$ *(Γ legal and $A \in \mathcal{K}_\Box$) and* $\Gamma \vdash_{\lambda\tau} A : s \Leftrightarrow$ *(Γ legal and $s \equiv *$ and $A \in \mathcal{K}_*$).*

2. *If $\Gamma \vdash_{\lambda\circ} A : s$, then for all $x \in dom(\Gamma)$ such that $\Gamma_x \vdash_{\lambda\circ} \Gamma(x) : *$ we have: $x \notin FV(A)$. So A contains no object variables.*

3. *If $\Gamma \vdash_{\lambda\circ} A : *$ and Γ, A are normal, then A is called a* type *and A is of one of the three following forms: (i) O; (ii) $xA_1 \ldots A_n$, for some $n \geq 0$, where (for $0 \leq i \leq n$) A_i is normal and $\Gamma \vdash_{\lambda\circ} A_i : B_i : \Box$, for some term B_i; (iii) $A_1{\to}A_2$, where (for $i = 1, 2$) A_i is normal and $\Gamma \vdash_{\lambda\circ} A_i : *$. In the first two cases, A is called an* atomic type*. In the third case, A is called an* arrow type.

4. *If* $\Gamma \vdash_{\lambda\text{o}} A : B : *$ *and* Γ, A *are normal, then* A *is of one of the two following forms: (i)* $xA_1 \ldots A_n$*, for some* $n \geq 0$*, where (for* $1 \leq i \leq n$*)* A_i *is normal and* $\Gamma \vdash_{\lambda\text{o}} A_i : B_i : *$ *for some term* B_i*; (ii)* $\lambda x{:}A_1.A_2$*, where* A_1 *is as described in (3),* A_2 *is normal and* $\Gamma, x : A_1 \vdash_{\lambda\text{o}} A_2 : B_2 : *$*, for some term* B_2*.*

5. *Suppose* $\Gamma \vdash_{\lambda\text{o}} A : s$*, where* A *is normal. Using the brackets convention we can write* A *uniquely as* $A_1 \rightarrow \cdots \rightarrow A_n \rightarrow B$*, with* $n \geq 0$ *and* B *atomic. Unless stated otherwise, we assume that such terms* A *are written in this way.*

Unless stated otherwise, we assume that terms are in η-long-β-normal form (LNF). This notion is defined, e.g., in [6]. Lemma 2.4 remains true when we replace 'normal' by 'in LNF'. We need a slightly non-standard notion of head normal form for types. First we say what a *domain* is. If A has a subterm of the form $\lambda x{:}A_1.A_2$, then A_1 is called a *domain in A*.

Definition 2.5. (In this definition terms are not assumed to be in LNF.) Let $\Gamma \vdash_{\lambda\text{o}} A : s$. Then A is in *head-normal form (HNF)* if $A \equiv O$, $A \equiv *$, $A \equiv xA_1 \ldots A_n$ or $A \equiv A_1 \rightarrow A_2$ with A_1, A_2 in HNF. A term A is in *D-HNF* if every domain in A is in HNF. Note that a term in LNF is also in (D)-HNF.

3 Matching problems

We present the notions of a *matching problem* and a *solution* for such a problem along the lines of [4], [6]. We give an example which is characteristic of matching problems and solutions in $\lambda\underline{\omega}$.

Definition 3.1. Suppose $\Gamma \vdash_{\lambda\text{o}} A : s$, where A is in HNF. We define $ord_\Gamma(A)$, the *order of A in* Γ, as follows. $ord_\Gamma(A)$ equals 2 if $A \equiv *$; 1 if $A \equiv O$ or $A \equiv xA_1 \ldots A_n$ and x is universal in Γ; ∞ if $A \equiv xA_1 \ldots A_n$ and x is existential in Γ; $\max(\{1 + ord_\Gamma(A_1), ord_\Gamma(A_2)\})$ if $A \equiv A_1 \rightarrow A_2$. The definition by cases is O.K. by Lemma 2.4. By convention, $\max(\{n, \infty\}) = \infty$ and $n + \infty = \infty$. When Γ is clear from the context, we simply speak about the order of A. Note that if A is closed in Γ, then $ord_\Gamma(A)$ is finite.

Definition 3.2.

1. A *substitution* is a finite set of triples $\langle x_i\,;\,\gamma_i\,;\,M_i\rangle$, such that the x_i are pairwise distinct, γ_i is an existential context and $dom(\gamma_i)$ consists of fresh variables, possibly occurring in M_i. We let σ, σ', τ, ... range over substitutions.

2. If $\langle x\,;\,\gamma\,;\,M\rangle \in \sigma$, then we say that σ *binds* x. Put $dom(\sigma) = \{x \mid x \text{ bound by } \sigma\}$. M is called a *substitution term*, γ a *substitution context*. To indicate that the substitution context is an "auxiliary" context, we denote it by a small Greek letter.

3. A substitution σ is extended to a function on pseudo-terms as follows: $\sigma(c) = c$ (for $c \in \mathcal{C}$); $\sigma(x) = M$ if $\langle x\,;\,\gamma\,;\,M\rangle \in \sigma$ and x otherwise; $\sigma(A_1A_2) = \sigma(A_1)\sigma(A_2)$; $\sigma(\lambda x{:}A_1.A_2) = \lambda x{:}\sigma(A_1).\sigma(A_2)$; $\sigma(A_1 \rightarrow A_2) = \sigma(A_1) \rightarrow \sigma(A_2)$.

4. A substitution σ is extended to a function on pseudo-contexts as follows: $\sigma(\langle\rangle) = \langle\rangle$; $\sigma(\Gamma, Qx : C) = \sigma(\Gamma), \gamma$ if $Q = \exists$ and $\langle x\,;\,\gamma\,;\,M\rangle \in \sigma$; otherwise $\sigma(\Gamma, Qx : C) = \sigma(\Gamma), Qx : \sigma(C)$.

5. Let Γ be a legal context in λo. Then we call σ *well-typed in* Γ when the following three requirements are satisfied: (i) σ binds no variables that are universal in Γ; (ii) $\sigma(\Gamma)$ is legal in λo; (iii) for all existential variables x in Γ that are bound by σ we have that $\sigma(\Gamma_x), \gamma \vdash_{\lambda\text{o}} M : \sigma(\Gamma(x))$, where $\langle x\,;\,\gamma\,;\,M\rangle$ is the unique triple in σ that binds x. Note that the empty substitution, denoted by $\emptyset$, is well-typed in any legal context. In general we have that if $\Gamma \vdash_{\lambda\text{o}} A : B$ and σ is well-typed in Γ, then $\sigma(\Gamma) \vdash_{\lambda\text{o}} \sigma(A) : \sigma(B)$.

6. Let σ and τ be substitutions. We define $\sigma \circ \tau = \{\langle x\,;\, \sigma(\gamma)\,;\, \sigma(t)\rangle \mid \langle x\,;\, \gamma\,;\, t\rangle \in \tau\} \cup \{\langle x\,;\, \gamma\,;\, t\rangle \in \sigma \mid x \text{ not bound by } \tau\}$. One can prove that if Γ is legal, τ well-typed in Γ and σ well-typed in $\tau(\Gamma)$, then $\sigma \circ \tau$ is well-typed in Γ and $(\sigma \circ \tau)(\Gamma) = \sigma(\tau(\Gamma))$.

7. Let σ be well-typed in some context Γ which is legal in $\lambda\circ$ and suppose $dom(\sigma) \subseteq dom(\Gamma)$. We can write σ uniquely as $\sigma = \sigma_\Box \cup \sigma_*$, where σ_s is the set of triples $\langle x, \gamma, M\rangle$ such that $\Gamma_x \vdash_{\lambda\circ} \Gamma(x) : s$. For a fixed sort s, σ is said to be an s-substitution if for every triple $\langle x\,;\, \gamma\,;\, M\rangle$ we have $\Gamma_x \vdash_{\lambda\circ} \Gamma(x) : s$. Note that $\sigma_\Box$ and σ_* depend on Γ. In general, $\sigma_\Box$ is not well-typed in Γ (substitution terms may depend on variables declared in substitution contexts in σ_*) but by arranging things in a suitable way we can assume that $\sigma_\Box$ *is* well-typed in Γ.

Definition 3.3.

1. A *matching problem in* $\lambda\circ$ is a triple $\langle\Gamma\,;\, A\,;\, B\rangle$, where Γ is a quantified context such that (i) there exists a term C such that $\Gamma \vdash_{\lambda\circ} A : C$ and $\Gamma \vdash_{\lambda\circ} B : C$; (ii) B is closed in Γ. Note that C is closed in Γ (recall that we assume that terms and types are in LNF!). If $\Gamma \vdash_{\lambda\circ} C : *$, then we say that $\langle\Gamma\,;\, A\,;\, B\rangle$ is a matching problem *for objects*; if $\Gamma \vdash_{\lambda\circ} C : \Box$, then we say that $\langle\Gamma\,;\, A\,;\, B\rangle$ is a matching problem *for types*. A matching problem $\langle\Gamma\,;\, A\,;\, B\rangle$ is *of order n* if the types of the existential variables in Γ have order at most n in Γ. *In this paper, we assume that the type of* every *variable in* Γ *is of finite order. In the full version [15] we show that this can be assumed without loss of generality.*

2. A *solution* for a matching problem $\langle\Gamma\,;\, A\,;\, B\rangle$ in $\lambda\circ$ is a substitution σ, well-typed in Γ, such that $\sigma(A) =_{\beta\eta} B$. σ is called a solution for a collection of matching problems $\{P_i \mid i \in I\}$ if σ is a solution for every P_i. By a standard argument, a set of matching problems $\{\langle\Gamma\,;\, A_i\,;\, B_i\rangle \mid 1 \le i \le n\}$ (for some $n \in \mathbb{N}$) can be encoded as a single matching problem. We call the matching problems in such a set Γ-*compatible*.

Now we give an example of a third-order matching problem for objects in $\lambda\underline{\omega}$. It illustrates the difficulties which have to be overcome in order to prove the decidability of third-order matching in $\lambda\underline{\omega}$.

Example 3.4. We define $\langle\Gamma\,;\, t_1\,;\, t_2\rangle$, a third-order matching problem for objects in $\lambda\underline{\omega}$. Take

- $\Gamma \equiv \langle \forall B : *, \forall X : * \to *, \exists A_1 : *, \exists A_2 : *, \forall a_1 : (XB), \forall a_2 : (XB),$
  $\forall g : (XB) \to (XB) \to (XB) \to B, \exists y_1 : (XA_1), \exists y_2 : (XA_1),$
  $\exists f : ((XA_2) \to (XA_2) \to (XA_2)) \to (XA_1) \to (XA_1) \to (XB)\rangle$
- $t_1 \equiv$
  $g(f(\lambda x_1{:}(XA_2).\lambda x_2{:}(XA_2).x_1)y_1y_2)$
  $(f(\lambda x_1{:}(XA_2).\lambda x_2{:}(XA_2).x_2)y_1y_2)$
  $(f(\lambda x_1{:}(XA_2).\lambda x_2{:}(XA_2).x_1)y_2y_1)$
- $t_2 \equiv ga_1a_2a_2$.

The reader is invited to check that this problem has a unique solution:

$$\sigma = \{\langle A_1\,;\, \langle\rangle\,;\, B\rangle, \langle A_2\,;\, \langle\rangle\,;\, B\rangle, \langle y_1\,;\, \langle\rangle\,;\, a_1\rangle, \langle y_2\,;\, \langle\rangle\,;\, a_2\rangle, \langle f\,;\, \langle\rangle\,;\, \lambda z_1{:}(XB) \to (XB) \to (XB).\lambda z_2{:}(XB).\lambda z_3{:}(XB).z_1z_2z_3\rangle\}.$$

Note that this matching problem has as only solution a substitution which *unifies* the terms XA_1 and XA_2. For the proof of decidability of third-order matching in $\lambda\underline{\omega}$ it is essential that full (third-order) unification of types can be avoided. For it is in general undecidable whether a third-order unification problem for types has a solution or not (see [9], [3]). The idea is to avoid full unification of types by defining and solving matching problems for types in some specific order.

Let us take a closer look at how unification can be avoided in our example. The variable f has three occurrences in t_1 and the term that is substituted for f each time takes a term whose type is initially $(XA_2)\to(XA_2)\to(XA_2)$ and which after application to two arguments (y_1, y_2 of type XA_1) should yield a term of type XB. So XA_2 has to be matched with XB. This constitutes a matching problem because XB is closed. Thereafter we know that the type of the two arguments (i.e. XA_1) has to be matched with XB. This again constitutes a matching problem. Note that if we would first try to match XA_1 with XA_2 we would be faced with a unification problem. Our algorithm is a generalization of this idea. The order is implemented as follows. First we define (and solve) matching problems that arise when we try to find terms that have to be substituted for those existential variables that have type $S_1\to\cdots\to S_n\to S$, *where* S *is closed.* (In this case we start with f.) Then we apply the solutions for these matching problems to the types of the other existential variables (in this case y_1 and y_2), in the hope that the respective types (in this case XA_1) become closed (in this case the solution changes XA_1 to XB). If this hope is fulfilled then we define (and solve) the matching problems that arise when we try to find terms that have to be substituted for these variables w.r.t. their (new) types. For completeness we show that if the initial problem has a solution then the matching problems have a solution *and* all existential variables are treated (in case substituting a term for such a variable is essential to obtain a solution).

4 Decidability of third-order matching

As explained in the previous section, the proof of decidability of third-order matching for objects hinges on the possibility to avoid having to solve (third-order) unification problems for types. The strategy is to decompose these unification problems into third-order matching problems for types. This of course only makes sense if it is indeed decidable whether a third-order matching problem for types in $\lambda\underline{\omega}$ has a solution or not. Without going into any detail, we state that this is the case.

Proposition 4.1. *It is decidable whether a third-order matching problem for types in* $\lambda\underline{\omega}$ *has a solution.*

Proof. Third-order matching problems for types in $\lambda\underline{\omega}$ can be encoded as second-order matching problems in $\lambda\tau$. By [10], it is decidable whether a second-order matching problem in $\lambda\tau$ has a solution. ⊠

In the remainder of this paper we will show that it is decidable whether a third-order matching problem for objects in $\lambda\underline{\omega}$ has a solution. First we define a translation that maps such a problem $P = \langle\Gamma\,;\,A\,;\,B\rangle$ to a third-order matching problem $|P|$ in $\lambda\tau$ and solutions σ for P to solutions $|\sigma|$ for $|P|$. Then we will prove that a substitution σ is a solution for P iff σ is well-typed in Γ and $|\sigma|$ is a solution for $|P|$. This divides the task of finding solutions for P in two parts: find solutions τ for $|P|$ and see if we can "lift" such solutions to substitutions τ' that are well-typed in Γ and such that $|\tau'| = \tau$. Dowek [5] has shown that to find solutions for $|P|$ it does no harm to restrict one's attention to a search space whose cardinality is bounded by a function value depending only on the size of $|P|$. Given such a solution τ, we will try to lift τ in two stages. First we decorate τ in a straightforward way: given an existential variable x of type $S_1\to\cdots\to S_n\to S$ in Γ and a triple $\langle x\,;\,\gamma\,;\,t\rangle$ in τ, where $t \equiv \lambda x_1{:}|S_1|\ldots\lambda x_n{:}|S_n|.yt_1\ldots t_m$, we decorate t to $\lambda x_1{:}S_1\ldots\lambda x_n{:}S_n.yt'_1\ldots t'_m$, where t'_j is the decorated version of t_j. This procedure need not yield terms that are well-typed in $\lambda\underline{\omega}$. In order to change these terms to well-typed ones, we define (starting from Γ and τ) a third-order matching problem *Match* for types in $\lambda\underline{\omega}$ such that if this problem has a solution ρ then the composition of ρ with the decorated substitution is a substitution θ that is well-typed in Γ and such that $|\theta| = \tau$ (hence $|\theta|$ is a solution for $|P|$ and θ is a solution for P).

4.1 Flattening types

We define a map, $|\cdot|$, that replaces all atomic subtypes by O. This map is extended to contexts and substitutions. We show that it preserves judgements, order, $\beta\eta$-reduction, the property of being a matching problem of finite order and the property of being a solution for such a problem. When we say below that a term is in (D-) HNF, this term is not assumed to be normal or in LNF.

Definition 4.2.

1. Suppose $\Gamma \vdash_{\lambda\underline{\omega}} A : s$, where A is in HNF. We define $|A|_{\mathrm{T}}$ by induction on the structure of A. $|*|_{\mathrm{T}} = O$; $|xA_1 \cdots A_n|_{\mathrm{T}} = O$; $|A_1{\rightarrow}A_2|_{\mathrm{T}} = |A_1|_{\mathrm{T}}{\rightarrow}|A_2|_{\mathrm{T}}$. By inspecting Definition 2.5, one easily verifies that the case distinction is O.K.
2. Let Γ be legal in $\lambda\underline{\omega}$ (and in HNF) and suppose that for every domain D in A we have $\Gamma \vdash_{\lambda\underline{\omega}} D : s$ (and D is in HNF). We define $|A|$ by induction on the structure of A. $|c| = c$ (for $c \in C$); $|x| = x$; $|A_1A_2| = |A_1||A_2|$; $|\lambda x{:}A_1.A_2| = \lambda x{:}|A_1|_{\mathrm{T}}.|A_2|$; $|A_1{\rightarrow}A_2| = |A_1|{\rightarrow}|A_2|$.
3. We extend the definition to legal contexts in HNF. $|\langle\rangle| = \langle\rangle$; $|\Gamma, Qx : A| = |\Gamma|, Qx : |A|_{\mathrm{T}}$ if $\Gamma \vdash_{\lambda\underline{\omega}} A : *$; $|\Gamma, Qx : A| = |\Gamma|$ if $\Gamma \vdash_{\lambda\underline{\omega}} A : \Box$.
4. Let Γ be legal in $\lambda\underline{\omega}$, σ well-typed in Γ and $dom(\sigma) \subseteq dom(\Gamma)$. Then $|\sigma| = \{\langle x\,;\,|\gamma|\,;\,|S|\rangle \mid \langle x\,;\,\gamma\,;\,S\rangle \in \sigma_*\}$. (We need to have $dom(\sigma) \subseteq dom(\Gamma)$, because otherwise σ_* is not defined.)

Lemma 4.3. *Suppose* $\Gamma \vdash_{\lambda\underline{\omega}} A : B$, *where* Γ *and* A *are in* (D-) HNF. *The term* $|A|$ *is in* D-HNF *and if* A *is in* LNF, *then* $|A|$ *is in* LNF.

Lemma 4.4.

1. *Suppose that* $\Gamma \vdash_{\lambda\underline{\omega}} A : s$ *and that* Γ *and* A *are in* HNF. *Then* $|\Gamma| \vdash_{\lambda\tau} |A|_{\mathrm{T}} : *$. *If* $s \equiv *$ *and* $ord_\Gamma(A)$ *is finite then* $ord_\Gamma(A) = ord_{|\Gamma|}(|A|_{\mathrm{T}})$.
2. *Suppose that* $\Gamma \vdash_{\lambda\underline{\omega}} A : B : *$ *and* Γ, A *and* B *are in* (D-) HNF. *Then* $|\Gamma| \vdash_{\lambda\tau} |A| : |B|_{\mathrm{T}} : *$.

Corollary 4.5. *Let* $P = \langle\Gamma\,;\,A\,;\,B\rangle$ *be a matching problem of order* n *for objects in* $\lambda\underline{\omega}$. *Then* $\langle|\Gamma|\,;\,|A|\,;\,|B|\rangle$ *is a matching problem of order* n *in* $\lambda\tau$. *We denote it by* $|P|$.

Proof. By Lemma 4.3 and Lemma 4.4. ⊠

Lemma 4.6. *Assume that all terms in the premises of (1) and (2) are in* (D-) HNF.

1. *Suppose* $\Gamma, \exists x : C, \Delta \vdash_{\lambda\underline{\omega}} A : *$. *Suppose* $\Gamma \vdash_{\lambda\underline{\omega}} D : C$ *and* $ord_\Gamma(A)$ *is finite. Then* $A[x := D]$ *is in* HNF *and* $|A[x := D]|_{\mathrm{T}} \equiv |A|_{\mathrm{T}}$.
2. *Suppose* $\Gamma, Qx : C, \Delta \vdash_{\lambda\underline{\omega}} A : B : *$ *and* $\Gamma \vdash_{\lambda\underline{\omega}} D : C : *$. *Then* $A[x := D]$ *is in* D-HNF *and* $|A[x := D]| \equiv |A|[x := |D|]$.

Proof. (1), (2): induction on the structure of A. The intuition for (1) is as follows. A is composed of arrows and atomic types $yA_1 \ldots A_n$. Since the order of A is finite, y is not existential in Γ, so is not x. So $(yA_1 \ldots A_n)[x := D] \equiv y(A_1[x := D]) \ldots (A_n[x := D])$. ⊠

Lemma 4.7. *Suppose* $\Gamma \vdash_{\lambda\underline{\omega}} A : B : *$, *where* Γ *and* A *are in* (D-) HNF. *Then*
$(A \twoheadrightarrow_{\beta\eta} A') \Rightarrow |A| \twoheadrightarrow_{\beta\eta} |A'|$.

Proof. By induction on the generation of $\twoheadrightarrow_{\beta\eta}$, using Lemma 4.6. ⊠

Proposition 4.8. *Let* $P_1 = \langle \Gamma \,;\, A \,;\, B \rangle$ *be a matching problem of order n for objects in* $\lambda\underline{\omega}$, σ *a solution for* P_1 *with* $dom(\sigma) \subseteq dom(\Gamma)$. *Then* $|\sigma|$ *is a solution for* $P_2 = \langle |\Gamma| \,;\, |A| \,;\, |B| \rangle$.

Proof. This follows essentially from the fact that $|\cdot|$ preserves substitution (Lemma 4.6) and $\beta\eta$-reduction (Lemma 4.7). ⊠

The following lemma is a key step in the main proof. It allows us to cut up the problem of finding solutions for third-order matching problems in $\lambda\underline{\omega}$ into two relatively easy subproblems.

Lemma 4.9. *Let* $P = \langle \Gamma \,;\, A \,;\, B \rangle$ *be a third-order matching problem for objects in* $\lambda\underline{\omega}$. *Let* σ *be a substitution, well-typed in* Γ, $dom(\sigma) \subseteq dom(\Gamma)$. *Suppose that* $|\sigma|$ *is a solution for* $|P|$. *Then* σ *is a solution for* P.

4.2 Decompositions and standard solutions

In [5], Dowek defines for each third-order matching problem P in $\lambda\tau$ and each solution σ for P, a set $\Phi(P, \sigma)$ of third-order matching problems in $\lambda\tau$. He uses this set to analyse the effect of σ on P in order to strip σ of superfluous parts. We will also use this set to analyse the effect of σ on P, but with the aim of lifting σ to a substitution that is well-typed in $\lambda\underline{\omega}$.

From this point on we use the following convention. If $\Gamma \vdash_{\lambda o} A : B$ and σ is well-typed in Γ, then we write $\sigma(A)$ for $\mathrm{Inf}_{\sigma(\Gamma)}(\sigma(A))$.

Definition 4.10. Let $P = \langle \Gamma \,;\, A \,;\, B \rangle$ be a third-order matching problem in $\lambda\tau$ and let σ be a solution for P. By induction on the length of A we define $\Phi(P, \sigma)$, a tree in which some of the nodes are labeled with triples (these triples can be proved to be third-order matching problems).

- $A \equiv xA_1 \ldots A_n$, where x is universal in Γ. Because σ is a solution for P, $B \equiv xB_1 \ldots B_n$ and σ is a solution for $\langle \Gamma \,;\, A_i \,;\, B_i \rangle$, for every $1 \le i \le n$. We let $\Phi(P, \sigma)$ consist of an unlabeled root and subtrees $\Phi(\langle \Gamma \,;\, A_1 \,;\, B_1 \rangle, \sigma)$, ..., $\Phi(\langle \Gamma \,;\, A_n \,;\, B_n \rangle, \sigma)$ (ordered from left to right).
- $A \equiv \lambda x{:}A_1.A_2$. Again because σ is a solution for P, $B \equiv \lambda x{:}A_1.B_2$ and σ is a solution for $\langle \Gamma, \forall x{:}A_1 \,;\, A_2 \,;\, B_2 \rangle$. Put $\Phi(P, \sigma) = \Phi(\langle \Gamma, \forall x{:}A_1 \,;\, A_2 \,;\, B_2 \rangle, \sigma)$.
- $A \equiv xA_1 \ldots A_n$, where x is existential in Γ. Suppose $\exists x : S_1 \to \cdots \to S_n \to S$ occurs in Γ. For all i, $1 \le i \le n$, write $A_i \equiv \lambda y_1{:}R_1 \ldots \lambda y_m{:}R_m.A_i'$ and consider the normal form of $\sigma(x)\sigma(A_1) \ldots \sigma(A_{i-1}) z_i \sigma(A_{i+1}) \ldots \sigma(A_n)$, where z_i is a fresh variable of type S_i. If z_i occurs in this normal form, put $C_i = \sigma(A_i)$, $C_i' = \sigma(A_i')$ and $H_i = \Phi(\langle \Gamma, \forall y_1 : R_1, \ldots, \forall y_m : R_m \,;\, A_i' \,;\, C_i' \rangle, \sigma)$. (One can prove that H_i is defined.) Otherwise put $C_i := z_i$ and H_i a tree consisting of an unlabeled node (we call z_i a *dummy variable*). Let $\{z_{i_1}, \ldots, z_{i_k}\}$ be the set of new variables thus obtained. Now the tree $\Phi(P, \sigma)$ consists of a root labeled with $\langle \Gamma, \forall z_{i_1} : S_{i_1}, \ldots, \forall z_{i_k} : S_{i_k} \,;\, xC_1 \ldots C_n \,;\, B \rangle$ and subtrees $H_1, \ldots, H_n$ (ordered from left to right).

Compare the form of the triples in $\Phi(P,\sigma)$ with the $\forall\exists\forall$ format in [11]. Note that each label of a node in $\Phi(P,\sigma)$ is of the form $\langle\Delta\,;\, xC_1\dots C_n\,;\, D\rangle$. The phrase *a triple P' in* $\Phi(P,\sigma)$ will mean a node p' in $\Phi(P,\sigma)$ labeled with P'. The *depth* of P' in $\Phi(P,\sigma)$ is the number of labeled nodes in the path from the root to p', not counting p'. The depth of $\Phi(P,\sigma)$ is the maximal depth of a triple in $\Phi(P,\sigma)$. Let $Q = \langle\Delta\,;\, xC_1\dots C_n\,;\, D\rangle$ and $Q' = \langle\Delta'\,;\, x'C'_1\dots C'_n\,;\, D'\rangle$ be triples in $\Phi(P,\sigma)$. Then we say that *the head x is below the head x'* if Q is below Q'. The tree structure of $\Phi(P,\sigma)$ is only occasionally used; mostly we identify $\Phi(P,\sigma)$ with the set of its labels.

Lemma 4.11. *Let $P = \langle\Gamma\,;\, A\,;\, B\rangle$ be a third-order matching problem for objects in $\lambda\tau$ and let σ be a solution for P. Then $\Phi(P,\sigma)$ is a collection of third-order matching problems for objects in $\lambda\tau$. Moreover, σ is a solution for $\Phi(P,\sigma)$ and if τ is a solution for every matching problem in $\Phi(P,\sigma)$, then τ is a solution for P.*

Proof. cf. [5]. ⊠

Below we develop some terminology to relate matching problems in $\Phi(|\langle\Gamma\,;\, A\,;\, B\rangle|,\tau)$ to the original problem $\langle\Gamma\,;\, A\,;\, B\rangle$ (more specifically: to Γ).

Definition 4.12. Suppose $\Gamma\vdash_{\lambda_0} A : C : *$. By induction on the length of A we define a tree, *comp*(Γ, A) (*comp* for *companion*). The nodes of *comp*(Γ, A) are labeled with contexts. Write $A \equiv \lambda x_1{:}S_1\dots\lambda x_n{:}S_n.xA_1\dots A_m$. Then *comp*$(\Gamma, A)$ consists of a root labeled with $\Gamma' \equiv \Gamma, \forall x_1 : S_1,\dots,\forall x_n : S_n$ and subtrees *comp*(Γ', A_1), ..., *comp*(Γ', A_m) (ordered from left to right).

Let $P = \langle\Gamma\,;\, A\,;\, B\rangle$ be a third-order matching problem for objects in $\lambda\underline{\omega}$. Write T_1 for the underlying tree of *comp*(Γ, A). When we write T_2 for the underlying tree of *comp*$(|\Gamma|, |A|)$ then we see that T_2 equals T_1. Let τ be a solution for $|P|$. Let T_3 be the underlying tree of $\Phi(|P|,\tau)$. Then it is easy to see that T_3 can be viewed as the result of replacing some subtrees in T_2 by leaves. Thus every path starting from the root in T_3 corresponds to a path starting from the root of T_2 and hence corresponds to a path starting from the root of T_1. This gives a correspondence between nodes in T_3 and nodes in T_1. Now for each triple $\langle\Delta\,;\, C\,;\, D\rangle$ in $\Phi(|P|,\tau)$ we define the *$\lambda\underline{\omega}$-companion to Δ* as the label of the corresponding node in T_1.

As mentioned before, Dowek uses the sets defined above to transform solutions into more efficient solutions. By slightly changing this transformation, these efficient solutions may be assumed to have a certain desirable "standard" form.

Definition 4.13. Suppose $\Gamma\vdash_{\lambda\tau} t : T$. Let Z be a set of variables in Γ. Then t is called *standard in Γ and Z* if for every $x\in FV(t)$ such that $x\notin Z$ we have that either x is universal in Γ, or x is existential in Γ, $\Gamma(x)$ is atomic and x has a unique occurrence in t.

We extend the definition to substitutions. Let σ be well-typed in Γ and $dom(\sigma)\subseteq dom(\Gamma)$. Then we call σ *standard in* Γ if, first, for all triples $\langle x\,;\,\gamma\,;\,\lambda x_1{:}S_1.\dots\lambda x_n{:}S_n.yt_1\dots t_m\rangle$ in σ, $yt_1\dots t_m$ is standard in $\sigma(\Gamma_x),\gamma,\forall x_1 : S_1,\dots,\forall x_n : S_n$ and $\{x_1,\dots,x_n\}$. Moreover $dom(\gamma)\subseteq FV(t)$ and γ consists of declarations of the form $\exists z : C$ with C atomic. Conversely, if $z\in FV(t)$ is existential in $\sigma(\Gamma_x),\gamma$, then $z\in dom(\gamma)$.

Finally, let $P = \langle\Gamma\,;\, A\,;\, B\rangle$ be a matching problem for objects in $\lambda\tau$ and suppose σ is a solution for P. Then σ is called *a standard solution for P* if σ is standard in Γ and for every triple $\langle x\,;\,\gamma\,;\,\lambda x_1{:}S_1\dots\lambda x_n{:}S_n.yt_1\dots t_m\rangle\in\sigma$ and for all z in $FV(\lambda x_1{:}S_1\dots\lambda x_n{:}S_n.yt_1\dots t_m)$ such that z is universal in $\sigma(\Gamma_x)$ we have $z\in FV(B)$. (If P is a matching problem for types in $\lambda\underline{\omega}$ then the notion of standard solution for P is defined similarly.)

Theorem 4.14 *(Dowek [5]). Let $P = \langle\Gamma\,;\, A\,;\, B\rangle$ be a third-order matching problem in $\lambda\tau$ or a third-order matching problem for types in $\lambda\underline{\omega}$. From a solution σ for P one can* *construct a substitution $\sigma^\star$ for P such that $\sigma^\star$ is a standard solution for P and $x \in dom(\sigma^\star)$ iff x is a head in $\Phi(P,\sigma)$. (Note that $\sigma^\star$ depends not only on σ but also on P.)*

Moreover there exists a set, $Sol(P)$, containing standard solutions for P, such that if P has a solution σ then $\sigma^\star \in Sol(P)$ and such that for some $n \in \mathbf{N}$, computable from P only, $Sol(P)$ can be enumerated in time bounded by n.

4.3 Lifting solutions from $\lambda\tau$ to $\lambda\underline{\omega}$

In this section we describe a way to lift solutions for matching problems in $\lambda\tau$ to solutions for matching problems in $\lambda\underline{\omega}$. We first decorate $\lambda\tau$ terms in a naive way. This procedure need not yield terms that are well-typed in $\lambda\underline{\omega}$ (they will be *pre-well-typed*). In order to change them into well-typed ones, we define a third-order matching problem for types such that solutions for this matching problem, when applied to the decorated terms, yield terms of the desired type.

Definition 4.15. Let Γ be a context legal in $\lambda\underline{\omega}$. We say that a pseudo-term t is *pre-well-typed in* Γ if t satisfies the following three conditions. (i) $|\Gamma| \vdash_{\lambda\tau} |t| : S$, for some term S and $|t|$ is in LNF; (ii) for every domain D in t which is not inside another domain in t we have $\Gamma \vdash_{\lambda\underline{\omega}} D : *$; (iii) for every variable $y \in FV(t)$, not occurring in a domain in t, $\Gamma(y)$ is defined and $\Gamma \vdash_{\lambda\underline{\omega}} \Gamma(y) : *$. For example in the context $\langle \forall B : *, \forall X : * \to *, \exists A_1 : *, \forall g : XB \to XB \to XB, \exists x : XB\rangle$, which is legal in $\lambda\underline{\omega}$, the term $\lambda y{:}XA_1.gxy$ is pre-well-typed but not typable. The term becomes typable when we substitute B for A_1.

We extend the notion *standard* to pre-well-typed terms in $\lambda\underline{\omega}$. Let Γ be legal in $\lambda\underline{\omega}$. Let t be pre-well-typed in Γ. Let Z be a set of variables in Γ. Then t is called *standard in* Γ *and* Z if for every $x \in FV(t)$ such that $x \notin Z$ and $\Gamma(x) \notin \mathcal{K}_\Box$ we have that either x is universal in Γ and $\Gamma(x)$ is closed in Γ, or x is existential in Γ, $\Gamma(x)$ is atomic and x has a unique occurrence in t.

In the definition of the decorating function, we need the following auxiliary notion.

Definition 4.16. Suppose $\Gamma \vdash_{\lambda o} S : *$ and $\Gamma \vdash_{\lambda o} T : *$, S and T in HNF. Then S and T are called *twins* if $|S|_{\mathrm{T}} \equiv |T|_{\mathrm{T}}$. If we write $S \equiv S_1 \to \cdots \to S_n \to S'$ and $T \equiv T_1 \to \cdots \to T_n \to T'$, then the twinness of S and T implies the twinness of S_i and T_i $(1 \leq i \leq n)$. So we can say that *S_i is the twin of T_i in T.*

Definition 4.17. Let Γ be legal in $\lambda\underline{\omega}$. Suppose $\Delta \vdash_{\lambda\tau} \lambda x_1{:}S_1 \ldots \lambda x_n{:}S_n.yt_1 \ldots t_m : S$ and $\Gamma(z)$ and $\Delta(z)$ are twins, for all $z \in FV(\lambda x_1{:}S_1 \ldots \lambda x_n{:}S_n.yt_1 \ldots t_m) \cap dom(\Gamma)$. (Note that $\Gamma \vdash_{\lambda\underline{\omega}} \Gamma(z) : *$.) Suppose $\Gamma \vdash_{\lambda\underline{\omega}} T_1 \to \cdots \to T_n \to T : *$ and that S_i and T_i are twins, for every $1 \leq i \leq n$. Then we define: $Deco(\lambda x_1{:}S_1 \ldots \lambda x_n{:}S_n.yt_1 \ldots t_m\,;\, T_1 \to \cdots \to T_n \to T)_\Gamma = \lambda x_1{:}T_1 \ldots \lambda x_n{:}T_n.t$ and $Cont(\lambda x_1{:}S_1 \ldots \lambda x_n{:}S_n.yt_1 \ldots t_m\,;\, T_1 \to \cdots \to T_n \to T)_\Gamma = \Xi$, where (putting $\Gamma' \equiv \Gamma, \forall x_1 : T_1, \ldots, \forall x_n : T_n$)

- If for some $R_1, \ldots, R_m, R$ we have $Qy : R_1 \to \cdots \to R_m \to R \in \Gamma'$, then $t \equiv ys_1 \ldots s_m$ and $\Xi \equiv \Xi_1, \ldots, \Xi_m$. Here $s_i \equiv Deco(t_i\,;\, R_i)_{\Gamma'}$ and $\Xi_i = Cont(t_i\,;\, R_i)_{\Gamma'}$, for $1 \leq i \leq m$.
- If for no $R_1, \ldots, R_m, R$ we have that $Qy : R_1 \to \cdots \to R_m \to R \in \Gamma'$, then $t \equiv y$ and $\Xi \equiv \langle \exists y : T\rangle$.

We extend *Deco* to substitutions. Let Γ be a context legal in $\lambda\underline{\omega}$, σ well-typed in $|\Gamma|$ and $dom(\sigma) \subseteq dom(|\Gamma|)$. Let $\langle x\,;\,\gamma\,;\,t\rangle \in \sigma$. Then $Deco(\sigma)_\Gamma = \{Deco(\langle x\,;\,\gamma\,;\,t\rangle)_\Gamma \mid \langle x\,;\,\gamma\,;\,t\rangle \in \sigma\}$, where $Deco(\langle x\,;\,\gamma\,;\,t\rangle)_\Gamma = \langle x\,;\, Cont(t\,;\,\Gamma(x))_{\Gamma_x}\,;\, Deco(t\,;\,\Gamma(x))_{\Gamma_x}\rangle$.

We give an example. Let Γ be $\langle \forall B : *, \forall X : * \to *, \exists A_1 : *, \exists A_2 : *, \forall g : XB \to XB \to XB, \exists f : XA_1 \to XA_2\rangle$. Then Γ is legal in $\lambda\underline{\omega}$. Let σ be $\{\langle f\,;\,\langle \exists x : O\rangle\,;\,\lambda y{:}O.gxy\rangle\}$. Then $Deco(\sigma)_\Gamma =$

$\{\langle f\,;\,\langle\exists x : XB\rangle\,;\,\lambda y{:}XA_1.gxy\rangle\}$. Note that the term $\lambda y{:}XA_1.gxy$ is not typable in the context $\langle\forall B : *, \forall X : *\to*, \exists A_1 : *, \exists A_2 : *, \forall g : XB\to XB\to XB, \exists x : XB\rangle$, but it is pre-well-typed in this context. The term becomes typable when we substitute B for A_1. Compare this to the example in Definition 4.15. Observe that $\{\langle A_1\,;\,\langle\rangle\,;\,B\rangle, \langle A_2\,;\,\langle\rangle\,;\,B\rangle\}\circ \mathit{Deco}(\sigma)_\Gamma$ is a substitution that is well-typed in Γ.

We list a number of elementary properties of *Deco*. A trivial property is that if $\Gamma\vdash_{\lambda\underline{\omega}} t : S : *$, then $\mathit{Deco}(|t|\,;\,S)_\Gamma \equiv t$.

Lemma 4.18. *Let $P = \langle\Gamma\,;\,A\,;\,B\rangle$ be a third-order matching problem for objects in $\lambda\underline{\omega}$, σ a standard solution for $|P|$. Let $\Delta \supseteq \Gamma$ be legal such that all existential variables in Δ are of order at most n.*

1. *$\mathit{Deco}(\sigma)_\Gamma(\Delta)$ is legal in $\lambda\underline{\omega}$ and $\mathit{Deco}(\sigma)_\Gamma(\Delta) =_\square \Delta$. Moreover every existential variable in $\mathit{Deco}(\sigma)_\Gamma(\Delta)$ is of order at most n.*
2. *For each triple $\langle x\,;\,\gamma\,;\,\lambda x_1{:}S_1\ldots\lambda x_n{:}S_n.yt_1\ldots t_m\rangle \in \mathit{Deco}(\sigma)_\Gamma$ we have that $yt_1\ldots t_m$ is pre-well-typed and standard in $\mathit{Deco}(\sigma)_\Gamma(\Gamma_x),\gamma$ (and $\{x_1,\ldots,x_n\}$). Also, $\sigma = |\mathit{Deco}(\sigma)_\Gamma|$ (up to the order of declarations in the substitution contexts).*
3. *Let τ be a $\square$-substitution, well-typed in Γ. Then $|\tau\circ \mathit{Deco}(\sigma)_\Gamma| = \sigma$ (up to a permutation of declarations in the substitution contexts).*

Next, we define a function, *Targets*, which given a pre-well-typed term t, a type T and a variable x of some unspecified but atomic type, returns the set of types that x must have if t is to be of type T. Of course, we eventually want this set to be a singleton, but it is convenient not to demand that at this moment. The *Targets* function is used in Definition 4.20.

Definition 4.19. Let Γ be a context, legal in $\lambda\underline{\omega}$, and $t \equiv \lambda x_1{:}S_1\ldots\lambda x_n{:}S_n.yt_1\ldots t_m$ pre-well-typed in Γ. Suppose that $\Gamma\vdash_{\lambda\underline{\omega}} T_1\to\cdots\to T_n\to T : *$, where for each i, $1\le i\le n$, S_i and T_i are twins. Let $x \in FV(t)$ be of atomic type in Γ, $\Gamma(x)\notin \mathcal{K}_\square$. Put $I = \{i \mid x \in FV(t_i), 1\le i\le n\}$ and $\Gamma' = \Gamma, \forall x_1 : S_1,\ldots,\forall x_n : S_n$. Note that by, pre-well-typedness, either y is one of the x_i and $S_i \equiv R_1\to\cdots\to R_m\to R$ (for some $R_1,\ldots, R_m, R$), or Γ contains a declaration $Qy : R_1\to\cdots\to R_m\to R$. So we can define $\mathit{Targets}(x\,;\,t\,;\,T_1\to\cdots\to T_n\to T)_\Gamma$ as $\{T\}$ if $y\equiv x$ and $\bigcup_{i\in I} \mathit{Targets}(x\,;\,t_i\,;\,R_i)_{\Gamma'}$ otherwise. It follows from pre-well-typedness that $\mathit{Targets}(x\,;\,t_i\,;\,R_i)_{\Gamma'}$ is defined. Of course, if $\Gamma\vdash_{\lambda\underline{\omega}} t : S : *$, then $\mathit{Targets}(x\,;\,t\,;\,S)_\Gamma = \{\Gamma(x)\}$.

Next we will sketch the matching problem we need for the proof of decidability of third-order matching for objects in $\lambda\underline{\omega}$; this sketch is made precise in Definition 4.20. We call a legal term $\lambda x_1{:}S_1\ldots\lambda x_n{:}S_n.yt_1\ldots t_m$ *relevant in its ith argument* if $x_i \in FV(yt_1\ldots t_m)$.

Let $P = \langle\Gamma\,;\,A\,;\,B\rangle$ be a matching problem for objects in $\lambda\underline{\omega}$ and σ a standard solution for $|P|$. Suppose x is bound by σ. Let $t \equiv \lambda x_1{:}S_1\ldots\lambda x_n{:}S_n.yt_1\ldots t_m$ be a subterm of the decorated version of $\sigma(x)$ in Γ (w.r.t. some type) and $T \equiv T_1\to\cdots\to T_n\to T'$ a type, closed in Γ. We assume that t is standard in some $\Gamma' \supseteq \Gamma$ and $\{z_1,\ldots,z_k\}$ and we assume the decorated version of $\sigma(x)$ to be of the form $\lambda z_1{:}Z_1\ldots\lambda z_k{:}Z_k.s$. We want to define by induction on the length of t a matching problem such that every solution τ for this problem is such that $\tau(t)$ is of type T. Obviously we have to match each S_i with T_i. Now if y is free in t and not in $\{z_1,\ldots,z_k\}$ then, by standardness of t, there are two possibilities for y. Either y is universal in Γ and of closed type $R_1\to\cdots\to R_m\to R$. Hence we have to match R with T' and by induction we know how to deal with t_j and R_j, for all $1\le j\le m$. Or y is existential in Γ and of atomic type R, hence $m = 0$ and we are done if we can match R with T'. If y is a bound variable x_i, then S_i is of the form $R_1\to\cdots\to R_m\to R$ and since we have matched S_i with T_i (so we can write $T_i \equiv R'_1\to\cdots\to R'_m\to R'$) we have by induction a matching problem for t_j and R'_j and we have to match R' with T'. If $y \in \{z_1,\ldots,z_k\}$, say $y\equiv z_i$, then it is not evident which *closed* type R_j we should use as input for the matching problem for t_j and R_j, which is by induction defined. The idea is to look at every matching problem $\langle\Delta\,;\,xD_1\ldots D_n\,;\,C\rangle$ in $\Phi(|P|,\sigma)$ such that D_i is relevant in its jth argument and not a dummy variable. (In Remark 4.22 we discuss the case where such a matching problem does not exist.) Because σ is a solution for $\Phi(|P|,\sigma)$, D_i will be substituted for z_i and applied to $t_1, \ldots, t_m$. From D_i we try to read off the type t_j that should have in order that D_i can be applied to $t_1, \ldots, t_m$. D_i is a term whose type is of order ≤ 2. So either D_i is of the form $\lambda y_1{:}O\ldots\lambda y_m{:}O.y_j$ and we take $R_j \equiv T'$ or D_i is of the form $\lambda y_1{:}O\ldots\lambda y_m{:}O.es_1\ldots s_l$ ($e \not\equiv y_j$, for $1 \leq j \leq m$) and we obtain R_j by the *Targets* function applied to y_j, the decorated version of $es_1\ldots s_l$ and T'.

Definition 4.20. Let $P = \langle\Gamma\,;\,A\,;\,B\rangle$ be a third-order matching problem for objects in $\lambda\underline{\omega}$ and σ a solution for $|P|$. Let x be existential in Γ, $\Gamma(x) \notin \mathcal{K}_\Box$. Write $\Phi = \Phi(|P|,\sigma)$. Let Δ be legal, $\Delta =_\Box \Gamma$. Let $t \equiv \lambda x_1{:}S_1\ldots\lambda x_n{:}S_n.yt_1\ldots t_m$ be pre-well-typed in Δ. Suppose that $\Delta \vdash_{\lambda\underline{\omega}} T_1{\to}\cdots{\to}T_n{\to}T : *$, where T_i is a twin of S_i, for every $1 \leq i \leq n$. Put $\Delta' = \Delta, \forall x_1 : T_1, \ldots, \forall x_n : T_n$. Let Z be a set of variables that occur in Δ such that the types of the variables in Z are of order at most 2.

We define $\mathit{SubMatch}(\lambda x_1{:}S_1\ldots\lambda x_n{:}S_n.yt_1\ldots t_m\,;\,T_1{\to}\cdots{\to}T_n{\to}T)_{\Delta,Z}$. As a result of pre-well-typedness, y is one of the x_i's and $S_i \equiv R_1{\to}\cdots{\to}R_m{\to}R$ (for some $R_1,\ldots, R_m$, R) or Δ contains a declaration $Qy : R_1{\to}\cdots{\to}R_m{\to}R$. We distinguish two cases.

1. $y \notin Z$. We define $\mathit{SubMatch}(\lambda x_1{:}S_1\ldots\lambda x_n{:}S_n.yt_1\ldots t_m\,;\,T_1{\to}\cdots{\to}T_n{\to}T)_{\Delta,Z} =$ $\{\langle\Gamma\,;\,S_i\,;\,T_i\rangle \mid 1 \leq i \leq n\} \cup \{\langle\Gamma\,;\,R\,;\,T\rangle\} \cup \bigcup_{1\leq i\leq m} \mathit{SubMatch}(t_i\,;\,R_i')_{\Delta',Z}$.
   Here R_i' is R_i if $y \in \mathit{dom}(\Delta)$ and R_i' is the twin of R_i in T_j if y is x_j (for some $1 \leq j \leq n$).

2. $y \in Z$, say $y \equiv z_i$. Write $\Phi_{x,i,j}$ for
   $\{\langle\Psi\,;\,xD_1\ldots D_p\,;\,C\rangle \in \Phi \mid D_i$ relevant in its jth argument, not a dummy variable$\}$.
   Suppose $E = \langle\Psi\,;\,xD_1\ldots D_p\,;\,C\rangle$ is an element of $\Phi_{x,i,j}$. Let Ψ_1 be the $\lambda\underline{\omega}$-companion to Ψ. Because of typing reasons it is sufficient to distinguish the following two cases.

   (a) $D_i \equiv \lambda y_1{:}O\ldots\lambda y_k{:}O.y_j$.
   We define $\mathit{add}_{j,E} = \{\langle\Gamma\,;\,R_j\,;\,T\rangle\} \cup \mathit{SubMatch}(t_j\,;\,T)_{\Delta',Z}$.

   (b) $D_i \equiv \lambda y_1{:}O\ldots\lambda y_k{:}O.es_1\ldots s_l$ ($e \not\equiv y_j$, for $1 \leq j \leq k$). Put $\Psi_2 = \Psi_1, \forall y_1 : R_1, \ldots, \forall y_k : R_k$ and $\Theta = \mathit{Targets}(y_j\,;\,\mathit{Deco}(es_1\ldots s_l\,;\,T)_{\Psi_2}\,;\,T)_{\Psi_2}$. It is not difficult to check that Θ is defined. Note that because of the relevance of D_i in its jth argument, Θ is non-empty. Also note that since $\Gamma =_\Box \Psi_2 =_\Box \Delta'$ we have $\Gamma \vdash_{\lambda\underline{\omega}} U : *$ and $\Delta' \vdash_{\lambda\underline{\omega}} U : *$, for all $U \in \Theta$. We define $\mathit{add}_{j,E} = \bigcup_{U\in\Theta}(\mathit{SubMatch}(t_j\,;\,U)_{\Delta',Z} \cup \{\langle\Gamma\,;\,R_j\,;\,U\rangle\})$.

   Now define $\mathit{Add}_j = \bigcup_{E\in\Phi_{x,i,j}} \mathit{add}_{j,E}$. In Remark 4.22, the case where $\Phi_{x,i,j}$ is empty is discussed. We define $\mathit{SubMatch}(\lambda x_1{:}S_1\ldots\lambda x_n{:}S_n.yt_1\ldots t_m\,;\,T_1{\to}\cdots{\to}T_n{\to}T)_{\Delta,Z} =$ $\{\langle\Gamma\,;\,S_i\,;\,T_i\rangle \mid 1 \leq i \leq n\} \cup \{\langle\Gamma\,;\,R\,;\,T\rangle\} \cup \bigcup_{1\leq j\leq m} \mathit{Add}_j$.

Next we define *Match*. This definition is a straightforward application of *SubMatch*, except that we want to use previously obtained substitutions in the definition of *Match* and we want to restrict the use of *Submatch* to heads above a fixed depth in $\Phi(|P|,\sigma)$. Hence *Match* gets a substitution and a natural number as extra parameters. Moreover, in order to guarantee that *Match* is a matching problem rather than a unification problem, we explicitly demand that the input type of *Submatch* ($\rho(S)$, below) is closed in Γ.

Definition 4.21. Let $P = \langle\Gamma\,;\,A\,;\,B\rangle$ be a third-order matching problem for objects in $\lambda\underline{\omega}$ and σ a standard solution for $|P|$. Let ρ be a $\Box$-substitution, well-typed in Γ. Suppose the declaration $\exists x : S_1{\to}\cdots{\to}S_n{\to}S : *$ occurs in Γ, where $\rho(S)$ is closed in Γ and x is bound by σ and at depth at most d in $\Phi(|P|,\sigma)$. Write $S' \equiv S_1{\to}\cdots{\to}S_n{\to}\rho(S)$.

Let $Deco(\sigma(x)\,;\,S')_\Gamma$ be $\lambda x_1{:}S_1 \ldots \lambda x_n{:}S_n.yt_1 \ldots t_m$. Put $Z = \{x_1, \ldots, x_n\}$. Put $\Delta = Deco(\sigma)_\Gamma(\Gamma), \forall x_1 : S_1, \ldots, \forall x_n : S_n$. By Lemma 4.18 (1), Δ is legal in $\lambda\underline{\omega}$ and $\Delta =_\Box \Gamma$. By Lemma 4.18 (2), $yt_1 \ldots t_m$ is pre-well-typed in Δ. So we can define $Match(x, \rho, d)_\Gamma = SubMatch(yt_1 \ldots t_m\,;\,\rho(S))_{\Delta,Z} \cup \{\langle\Gamma\,;\,S\,;\,\rho(S)\rangle\}$. Let X be the set of variables x_i at depth at most d in $\Phi(|P|, \sigma)$ such that x_i is bound by σ and $\exists x_i : T_1 \to \cdots \to T_{n_i} \to T_i$ occurs in Γ, where $\rho(T_i)$ is closed in Γ. Put $Match(\sigma, \rho, d)_\Gamma = \bigcup_{x \in X} Match(x, \rho, d)_\Gamma$.

Remark 4.22. There is one subtlety involved in the definition above. Recall that we want a solution τ of $Match(x, \rho, d)_\Gamma$ to be such that $\tau(Deco(\sigma)_\Gamma(\Gamma)) \vdash_{\lambda\underline{\omega}} \tau(Deco(\sigma(x)\,;\,S')_\Gamma) : \tau(S')$. Suppose that in some recursive call of *SubMatch*, we are in the second case, where the head variable under consideration is an element of Z, say it is z_i and z_i is in this case the head variable of $z_i t'_1 \ldots t'_{m'}$. Write $S_i = V_1 \to \cdots \to V_{m'} \to V$. Suppose now that $\Phi_{x,i,j} = \emptyset$ (for some $1 \leq j \leq m'$). Then the further recursive calls of *Submatch* in this case do not extend *Match* with matching problems to ensure that $\tau(Deco(\sigma)_\Gamma(\Gamma)') \vdash_{\lambda\underline{\omega}} \tau(t'_j) : \tau(V_j)$, where $Deco(\sigma)_\Gamma(\Gamma)'$ is the current context extending $Deco(\sigma)_\Gamma(\Gamma)$. We explain why there is no need to extend it. Write $V_j \equiv W_1 \to \cdots \to W_l \to W$. By (the modification of) Dowek's construction of the standard term $\sigma(x)$, t'_j is in this case of the form $\lambda z_1{:}W_1 \ldots \lambda z_k{:}W_l.u$, where $\exists u : W$ occurs in $Deco(\sigma)_\Gamma(\Gamma)'$. So *every* substitution θ, well-typed in $Deco(\sigma)_\Gamma(\Gamma)'$, satisfies $\theta(Deco(\sigma)_\Gamma(\Gamma)') \vdash_{\lambda\underline{\omega}} \theta(t'_j) : \theta(V_j)$.

Lemma 4.23. *Let $P = \langle\Gamma\,;\,A\,;\,B\rangle$ be a third-order matching problem for objects in $\lambda\underline{\omega}$, σ a standard solution for $|P|$ and ρ a $\Box$-substitution, well-typed in Γ. Let d be less than or equal to the depth of $\Phi(|P|, \sigma)$. Then $Match(\sigma, \rho, d)_\Gamma$ is a finite collection of Γ-compatible third-order matching problems for types.*

The next two lemmas express soundness and completeness of *Match*.

Lemma 4.24. *Let $P = \langle\Gamma\,;\,A\,;\,B\rangle$ be a third-order matching problem for objects in $\lambda\underline{\omega}$, σ a standard solution for $|P|$ and ρ a $\Box$-substitution, well-typed in Γ. Let $\langle x\,;\,\gamma\,;\,t\rangle \in \sigma$, where $\exists x : S_1 \to \cdots \to S_n \to S$ occurs in Γ and $\rho(S)$ is closed in Γ. If τ is a solution for $Match(\sigma, \rho, d)_\Gamma$, then $\tau(Deco(\sigma)_\Gamma(\Gamma)) \vdash_{\lambda\underline{\omega}} \tau(Deco(t\,;\,S_1 \to \cdots \to S_n \to S)_\Gamma) : \tau(S_1 \to \cdots \to S_n \to S)$.*

Lemma 4.25. *Let $P = \langle\Gamma\,;\,A\,;\,B\rangle$ be a third-order matching problem for objects in $\lambda\underline{\omega}$, σ a solution for P. Let $\langle x\,;\,\gamma\,;\,t\rangle \in |\sigma|^\star$, where $\exists x : S_1 \to \cdots \to S_n \to S$ occurs in Γ and x is at depth at most d in $\Phi(P, |\sigma|^\star)$. Suppose $\sigma_\Box(S)$ is closed in Γ. Then $\sigma_\Box$ is a solution for $Match(x, \sigma_\Box, d)_\Gamma$.*

The completeness of the algorithm presented in the proof of Theorem 4.27 is obtained by showing that if P in $\lambda\underline{\omega}$ has solution σ then $\sigma_\Box$ is a solution for *Match* for *all* heads x in $\Phi(|P|, |\sigma|^\star)$. The proof of this fact proceeds by induction on the depth of heads in $\Phi(|P|, |\sigma|^\star)$. The following lemma is needed for the induction step.

Lemma 4.26. *Let $P = \langle\Gamma\,;\,A\,;\,B\rangle$ be a third-order matching problem for objects in $\lambda\underline{\omega}$ and σ a solution for P. Let x be a head in $\Phi(|P|, |\sigma|^\star)$ at depth n and of type $T_1 \to \cdots \to T_m \to T$ in Γ. Let y be a head below x at depth $n+1$ and of type $S_1 \to \cdots \to S_k \to S$ in Γ, where S is not closed in Γ. Suppose that θ is a $\Box$-substitution, well-typed in Γ, such that $\theta(T) \equiv \sigma_\Box(T)$ and $\theta(T)$ is closed in Γ. Then $Match(x, \theta, n)_\Gamma$ contains a matching problem $\langle\Gamma\,;\,S\,;\,\sigma_\Box(S)\rangle$.*

At last we have reached the point where we can state and prove our main result.

Theorem 4.27. *It is decidable whether a third-order matching problem for objects in $\lambda\underline{\omega}$ has a solution or not.*

Proof. Let $P = \langle\Gamma\,;\,A\,;\,B\rangle$ be a third-order matching problem for objects in $\lambda\underline{\omega}$. We present an algorithm which, given P as input, returns a solution for P if P has a solution and *fail* otherwise. The algorithm is as follows.

- Translate P to $|P|$. Enumerate the solutions in $Sol(|P|)$ as $\{\sigma_i \mid 1 \leq i \leq n\}$. If $Sol(|P|)$ is empty, return *fail* and stop. Else do the following.
- Let α be a meta variable ranging over substitutions and *dummy* some dummy substitution. Put $i := 1$ and $\alpha := dummy$. While $i \leq n$ do
  - Delete all triples $\langle x\,;\gamma\,;t\rangle$ in σ_i such that x is not a head in $\Phi(|P|,\sigma_i)$. (By Lemma 4.11, the resulting substitution, which we keep denoting by σ_i, is still a solution for $|P|$ and $\Phi(|P|,\sigma_i)$). If the result is the substitution $\emptyset$, put $i := n+1$ and $\alpha := \emptyset$. Else do the following.
  - Define $appr(0) := \{\emptyset\}$; $appr(k+1) := \bigcup_{\theta \in appr(k)} Sol(Match(\sigma_i, \theta, k)_\Gamma)$
  - Let d be the depth of $\Phi(|P|,\sigma_i)$. If there exist a θ_1 in $appr(d)$ such that for all x, head in $\Phi(|P|,\sigma_i)$ (with $\Gamma(x) \equiv S_1 \rightarrow \cdots \rightarrow S_n \rightarrow S$), $\theta_1(S)$ is closed in Γ and if there exists a θ_2 in $Sol(Match(\sigma_i, \theta_1, d)_\Gamma)$, put $i := n+1$ and $\alpha := \theta_2 \circ Deco(\sigma_i)_\Gamma$, else put $i := i+1$.
- If $\alpha \equiv dummy$, return *fail*, else return α. Stop.

We have to check that this algorithm is sound and complete and that it always terminates. We only check the first two points.

Soundness. Let σ be a standard solution for $|P|$ in $Sol(|P|)$ and d the depth of $\Phi(|P|,\sigma)$. If, after deletion of superfluous parts, $\sigma = \emptyset$, then we must have that $|A| \equiv |B|$, and it is easy to check that $\emptyset$ is indeed a solution for P. Next, suppose that $\sigma \neq \emptyset$. Then there must be an $x \in FV(A)$, bound by σ and of type $S_1 \rightarrow \cdots \rightarrow S_n \rightarrow S$ in Γ, with S closed in Γ. Let τ_1 be an element of $appr(d)$ such that $\tau_1(S)$ is closed in Γ for all x, head in $\Phi(|P|,\sigma)$ (with $\Gamma(x) \equiv S_1 \rightarrow \cdots \rightarrow S_n \rightarrow S$). Let τ_2 be an element of $Sol(Match(\sigma, \tau_1, d)_\Gamma)$. We show that $\tau_2 \circ Deco(\sigma)_\Gamma$ is a solution for P. By construction there exist substitutions $\rho_1, \ldots, \rho_{d-1}$ such that ρ_1 is a standard solution for $Match(\sigma, \emptyset, 0)_\Gamma$, each ρ_{i+1} is a standard solution for $Match(\sigma, \rho_i, i)_\Gamma$ and τ_1 is a standard solution for $Match(\sigma, \rho_{d-1}, d-1)_\Gamma$. By construction of *Match*, each ρ_i is well-typed in Γ. The same holds for τ_1 and τ_2. From this and Lemma 4.24 we can deduce that $\tau_2 \circ Deco(\sigma)_\Gamma$ is well-typed in Γ. By Lemma 4.18 (3), $|\tau_2 \circ Deco(\sigma)_\Gamma| = \sigma$ and hence $|\tau_2 \circ Deco(\sigma)_\Gamma|$ is a solution for $|P|$. By Lemma 4.9, $\tau_2 \circ Deco(\sigma)_\Gamma$ is a solution for P.

Completeness. Suppose P has a solution σ. We have to prove that the algorithm returns a substitution. It is no restriction to assume that $dom(\sigma) \subseteq dom(\Gamma)$. By Proposition 4.8, $|\sigma|$ is a solution for $|P|$. So $|\sigma|^\star$ is also a solution for $|P|$ and occurs in the first enumeration given by the algorithm. The interesting case is where the deleting of superfluous parts from $|\sigma|^\star$ does not yield the empty substitution. Let $d > 0$ be the depth of $\Phi(|P|,|\sigma|^\star)$. Completeness follows from the next claim, which implies that $\alpha := \theta_d \circ Deco(|\sigma|^\star)_\Gamma$ is a solution for P.

Claim. For each k, $0 \leq k \leq d$, there exist a $\zeta_k \in appr(k)$ and a $\theta_k \in appr(k+1)$ such that $\theta_k \in Sol(Match(|\sigma|^\star, \zeta_k, k)_\Gamma)$ and for all heads x at depth $l \leq k$ in $\Phi(|P|,|\sigma|^\star)$ and of type $S_1 \rightarrow \cdots \rightarrow S_n \rightarrow S$ in Γ, $\zeta_k(S) \equiv \theta_k(S) \equiv \sigma_\square(S)$ (and these types are closed in Γ).

The proof of this claim goes by induction on k, taking $\zeta_0 = \emptyset$. We only treat the induction step. So assume $k = k'+1$ and that the claim is proved for k'. We put $\zeta_{k'+1} := \theta_{k'}$. First we show that $Match(|\sigma|^\star, \theta_{k'}, k'+1)_\Gamma \subseteq Match(|\sigma|^\star, \sigma_\square, k'+1)_\Gamma$. The interesting case is where there exists a head y at depth at most $k'+1$ in $\Phi(|P|,|\sigma|^\star)$ and of type $T_1 \rightarrow \cdots \rightarrow T_m \rightarrow T$ in Γ, where T is not closed in Γ. There exists a head z above y in $\Phi(|P|,|\sigma|^\star)$ whose depth is at most k'. Let $Z_1 \rightarrow \cdots \rightarrow Z_l \rightarrow Z$ be the type of z in Γ. The induction hypothesis yields that $\zeta_{k'}(Z) \equiv \sigma_\square(Z)$ is closed in Γ. Hence $\theta_{k'}$ is a solution for $Match(z, \zeta_{k'}, k')_\Gamma$. By Lemma 4.26, there exists a matching problem $\langle \Gamma\,; T\,; \sigma_\square(T)\rangle \in Match(z, \zeta_{k'}, k')_\Gamma$. So $\theta_{k'}(T) \equiv \sigma_\square(T)$, by construction closed in Γ.

So $Match(|\sigma|^*, \theta_{k'}, k'+1)_\Gamma \subseteq Match(|\sigma|^*, \sigma_\Box, k'+1)_\Gamma$ and by Lemma 4.25, $\sigma_\Box$ is a solution for $Match(|\sigma|^*, \theta_{k'}, k'+1)_\Gamma$, hence there is a $\theta_{k'+1}$ in $Sol(Match(|\sigma|^*, \theta_{k'}, k'+1)_\Gamma)$.

We leave it to the reader to check that for all heads x at depth $l \leq k'+1$ in $\Phi(|P|, |\sigma|^*)$ and of type $S_1 \rightarrow \cdots \rightarrow S_n \rightarrow S$ in Γ, $\theta_{k'}(S) \equiv \theta_{k'+1}(S) \equiv \sigma_\Box(S)$, closed in Γ. ⊠

References

[1] H.P. Barendregt. Lambda calculi with types. In S. Abramsky, D. M. Gabbay, and T.S.E. Maibaum, editors, *Handbook of Logic in Computer Science*, volume 2, pages 117–309. Oxford Science Publications, 1992.

[2] T. Coquand and G. Huet. The calculus of constructions. *Information and Control*, 76:95–120, 1988.

[3] G. Dowek. L'indécidabilité du filtrage du troisième ordre dans les calculs avec types dépendants ou constructeurs de types. *Compte Rendu à l'Académie des Sciences*, 312, Série I:951–956, 1991. With erratum in: ibid., 318, Série I, p. 873, 1994.

[4] G. Dowek. A second-order pattern matching algorithm for the cube of typed λ-calculi. In *Mathematical Foundations of Computer Science 91*, volume 520 of *Lecture Notes in Computer Science*, pages 151–160. Springer-Verlag, 1991.

[5] G. Dowek. Third order matching is decidable. In *Proceedings* 7th *Annual Symposium on Logic in Computer Science*, Santa Cruz, California, 1992.

[6] G. Dowek. A complete proof synthesis method for the cube of type systems. *Journal of Logic and Computation*, 3:287–315, 1993.

[7] G. Dowek. The undecidability of pattern matching in calculi where primitive recursive functions are representable. *Theoretical Computer Science*, 107:349–356, 1993.

[8] J.H. Geuvers. *Logics and Type Systems.* Phd thesis, University of Nijmegen, September 1993.

[9] W.D. Goldfarb. The undecidability of the second-order unification problem. *Theoretical Computer Science*, 13:225–230, 1981.

[10] G. Huet and B. Lang. Proving and applying program transformations expressed with second-order patterns. *Acta Informatica*, 11:31–55, 1978.

[11] D. Miller. Unification under a mixed prefix. *Journal of Symbolic Computation*, 14(4):321–359, October 1992.

[12] V. Padovani. On equivalence classes of interpolation equations. This volume.

[13] V. Padovani. Fourth order dual interpolation is decidable. Manuscript, Université Paris VII - C.N.R.S, 1994.

[14] J. Springintveld. Third-order matching in the polymorphic lambda calculus. Technical Report 119, Logic Group Preprint Series, Utrecht University, September 1994.

[15] J. Springintveld. Third-order matching in the presence of type constructors. Technical Report 112, Logic Group Preprint Series, Utrecht University, May 1994.

Author Index

Springer-Verlag and the Environment

We at Springer-Verlag firmly believe that an international science publisher has a special obligation to the environment, and our corporate policies consistently reflect this conviction.

We also expect our business partners – paper mills, printers, packaging manufacturers, etc. – to commit themselves to using environmentally friendly materials and production processes.

The paper in this book is made from low- or no-chlorine pulp and is acid free, in conformance with international standards for paper permanency.

Lecture Notes in Computer Science

For information about Vols. 1–822
please contact your bookseller or Springer-Verlag

Vol. 823: R. A. Elmasri, V. Kouramajian, B. Thalheim (Eds.), Entity-Relationship Approach — ER '93. Proceedings, 1993. X, 531 pages. 1994.

Vol. 824: E. M. Schmidt, S. Skyum (Eds.), Algorithm Theory – SWAT '94. Proceedings. IX, 383 pages. 1994.

Vol. 825: J. L. Mundy, A. Zisserman, D. Forsyth (Eds.), Applications of Invariance in Computer Vision. Proceedings, 1993. IX, 510 pages. 1994.

Vol. 826: D. S. Bowers (Ed.), Directions in Databases. Proceedings, 1994. X, 234 pages. 1994.

Vol. 827: D. M. Gabbay, H. J. Ohlbach (Eds.), Temporal Logic. Proceedings, 1994. XI, 546 pages. 1994. (Subseries LNAI).

Vol. 828: L. C. Paulson, Isabelle. XVII, 321 pages. 1994.

Vol. 829: A. Chmora, S. B. Wicker (Eds.), Error Control, Cryptology, and Speech Compression. Proceedings, 1993. VIII, 121 pages. 1994.

Vol. 830: C. Castelfranchi, E. Werner (Eds.), Artificial Social Systems. Proceedings, 1992. XVIII, 337 pages. 1994. (Subseries LNAI).

Vol. 831: V. Bouchitté, M. Morvan (Eds.), Orders, Algorithms, and Applications. Proceedings, 1994. IX, 204 pages. 1994.

Vol. 832: E. Börger, Y. Gurevich, K. Meinke (Eds.), Computer Science Logic. Proceedings, 1993. VIII, 336 pages. 1994.

Vol. 833: D. Driankov, P. W. Eklund, A. Ralescu (Eds.), Fuzzy Logic and Fuzzy Control. Proceedings, 1991. XII, 157 pages. 1994. (Subseries LNAI).

Vol. 834: D.-Z. Du, X.-S. Zhang (Eds.), Algorithms and Computation. Proceedings, 1994. XIII, 687 pages. 1994.

Vol. 835: W. M. Tepfenhart, J. P. Dick, J. F. Sowa (Eds.), Conceptual Structures: Current Practices. Proceedings, 1994. VIII, 331 pages. 1994. (Subseries LNAI).

Vol. 836: B. Jonsson, J. Parrow (Eds.), CONCUR '94: Concurrency Theory. Proceedings, 1994. IX, 529 pages. 1994.

Vol. 837: S. Wess, K.-D. Althoff, M. M. Richter (Eds.), Topics in Case-Based Reasoning. Proceedings, 1993. IX, 471 pages. 1994. (Subseries LNAI).

Vol. 838: C. MacNish, D. Pearce, L. Moniz Pereira (Eds.), Logics in Artificial Intelligence. Proceedings, 1994. IX, 413 pages. 1994. (Subseries LNAI).

Vol. 839: Y. G. Desmedt (Ed.), Advances in Cryptology - CRYPTO '94. Proceedings, 1994. XII, 439 pages. 1994.

Vol. 840: G. Reinelt, The Traveling Salesman. VIII, 223 pages. 1994.

Vol. 841: I. Prívara, B. Rovan, P. Ružička (Eds.), Mathematical Foundations of Computer Science 1994. Proceedings, 1994. X, 628 pages. 1994.

Vol. 842: T. Kloks, Treewidth. IX, 209 pages. 1994.

Vol. 843: A. Szepietowski, Turing Machines with Sublogarithmic Space. VIII, 115 pages. 1994.

Vol. 844: M. Hermenegildo, J. Penjam (Eds.), Programming Language Implementation and Logic Programming. Proceedings, 1994. XII, 469 pages. 1994.

Vol. 845: J.-P. Jouannaud (Ed.), Constraints in Computational Logics. Proceedings, 1994. VIII, 367 pages. 1994.

Vol. 846: D. Shepherd, G. Blair, G. Coulson, N. Davies, F. Garcia (Eds.), Network and Operating System Support for Digital Audio and Video. Proceedings, 1993. VIII, 269 pages. 1994.

Vol. 847: A. L. Ralescu (Ed.) Fuzzy Logic in Artificial Intelligence. Proceedings, 1993. VII, 128 pages. 1994. (Subseries LNAI).

Vol. 848: A. R. Krommer, C. W. Ueberhuber, Numerical Integration on Advanced Computer Systems. XIII, 341 pages. 1994.

Vol. 849: R. W. Hartenstein, M. Z. Servít (Eds.), Field-Programmable Logic. Proceedings, 1994. XI, 434 pages. 1994.

Vol. 850: G. Levi, M. Rodríguez-Artalejo (Eds.), Algebraic and Logic Programming. Proceedings, 1994. VIII, 304 pages. 1994.

Vol. 851: H.-J. Kugler, A. Mullery, N. Niebert (Eds.), Towards a Pan-European Telecommunication Service Infrastructure. Proceedings, 1994. XIII, 582 pages. 1994.

Vol. 852: K. Echtle, D. Hammer, D. Powell (Eds.), Dependable Computing – EDCC-1. Proceedings, 1994. XVII, 618 pages. 1994.

Vol. 853: K. Bolding, L. Snyder (Eds.), Parallel Computer Routing and Communication. Proceedings, 1994. IX, 317 pages. 1994.

Vol. 854: B. Buchberger, J. Volkert (Eds.), Parallel Processing: CONPAR 94 – VAPP VI. Proceedings, 1994. XVI, 893 pages. 1994.

Vol. 855: J. van Leeuwen (Ed.), Algorithms – ESA '94. Proceedings, 1994. X, 510 pages.1994.

Vol. 856: D. Karagiannis (Ed.), Database and Expert Systems Applications. Proceedings, 1994. XVII, 807 pages. 1994.

Vol. 857: G. Tel, P. Vitányi (Eds.), Distributed Algorithms. Proceedings, 1994. X, 370 pages. 1994.

Vol. 858: E. Bertino, S. Urban (Eds.), Object-Oriented Methodologies and Systems. Proceedings, 1994. X, 386 pages. 1994.

Vol. 859: T. F. Melham, J. Camilleri (Eds.), Higher Order Logic Theorem Proving and Its Applications. Proceedings, 1994. IX, 470 pages. 1994.

Vol. 860: W. L. Zagler, G. Busby, R. R. Wagner (Eds.), Computers for Handicapped Persons. Proceedings, 1994. XX, 625 pages. 1994.

Vol: 861: B. Nebel, L. Dreschler-Fischer (Eds.), KI-94: Advances in Artificial Intelligence. Proceedings, 1994. IX, 401 pages. 1994. (Subseries LNAI).

Vol. 862: R. C. Carrasco, J. Oncina (Eds.), Grammatical Inference and Applications. Proceedings, 1994. VIII, 290 pages. 1994. (Subseries LNAI).

Vol. 863: H. Langmaack, W.-P. de Roever, J. Vytopil (Eds.), Formal Techniques in Real-Time and Fault-Tolerant Systems. Proceedings, 1994. XIV, 787 pages. 1994.

Vol. 864: B. Le Charlier (Ed.), Static Analysis. Proceedings, 1994. XII, 465 pages. 1994.

Vol. 865: T. C. Fogarty (Ed.), Evolutionary Computing. Proceedings, 1994. XII, 332 pages. 1994.

Vol. 866: Y. Davidor, H.-P. Schwefel, R. Männer (Eds.), Parallel Problem Solving from Nature - PPSN III. Proceedings, 1994. XV, 642 pages. 1994.

Vol 867: L. Steels, G. Schreiber, W. Van de Velde (Eds.), A Future for Knowledge Acquisition. Proceedings, 1994. XII, 414 pages. 1994. (Subseries LNAI).

Vol. 868: R. Steinmetz (Ed.), Multimedia: Advanced Teleservices and High-Speed Communication Architectures. Proceedings, 1994. IX, 451 pages. 1994.

Vol. 869: Z. W. Raś, Zemankova (Eds.), Methodologies for Intelligent Systems. Proceedings, 1994. X, 613 pages. 1994. (Subseries LNAI).

Vol. 870: J. S. Greenfield, Distributed Programming Paradigms with Cryptography Applications. XI, 182 pages. 1994.

Vol. 871: J. P. Lee, G. G. Grinstein (Eds.), Database Issues for Data Visualization. Proceedings, 1993. XIV, 229 pages. 1994.

Vol. 872: S Arikawa, K. P. Jantke (Eds.), Algorithmic Learning Theory. Proceedings, 1994. XIV, 575 pages. 1994.

Vol. 873: M. Naftalin, T. Denvir, M. Bertran (Eds.), FME '94: Industrial Benefit of Formal Methods. Proceedings, 1994. XI, 723 pages. 1994.

Vol. 874: A. Borning (Ed.), Principles and Practice of Constraint Programming. Proceedings, 1994. IX, 361 pages. 1994.

Vol. 875: D. Gollmann (Ed.), Computer Security – ESORICS 94. Proceedings, 1994. XI, 469 pages. 1994.

Vol. 876: B. Blumenthal, J. Gornostaev, C. Unger (Eds.), Human-Computer Interaction. Proceedings, 1994. IX, 239 pages. 1994.

Vol. 877: L. M. Adleman, M.-D. Huang (Eds.), Algorithmic Number Theory. Proceedings, 1994. IX, 323 pages. 1994.

Vol. 878: T. Ishida; Parallel, Distributed and Multiagent Production Systems. XVII, 166 pages. 1994. (Subseries (LNAI).

Vol. 879: J. Dongarra, J. Waśniewski (Eds.), Parallel Scientific Computing. Proceedings, 1994. XI, 566 pages. 1994.

Vol. 880: P. S. Thiagarajan (Ed.), Foundations of Software Technology and Theoretical Computer Science. Proceedings, 1994. XI, 451 pages. 1994.

Vol. 881: P. Loucopoulos (Ed.), Entity-Relationship Approach – ER'94. Proceedings, 1994. XIII, 579 pages. 1994.

Vol. 882: D. Hutchison, A. Danthine, H. Leopold, G. Coulson (Eds.), Multimedia Transport and Teleservices. Proceedings, 1994. XI, 380 pages. 1994.

Vol. 883: L. Fribourg, F. Turini (Eds.), Logic Program Synthesis and Transformation – Meta-Programming in Logic. Proceedings, 1994. IX, 451 pages. 1994.

Vol. 884: J. Nievergelt, T. Roos, H.-J. Schek, P. Widmayer (Eds.), IGIS '94: Geographic Information Systems. Proceedings, 1994. VIII, 292 pages. 19944.

Vol. 885: R. C. Veltkamp, Closed Objects Boundaries from Scattered Points. VIII, 144 pages. 1994.

Vol. 886: M. M. Veloso, Planning and Learning by Analogical Reasoning. XIII, 181 pages. 1994. (Subseries LNAI).

Vol. 887: M. Toussaint (Ed.), Ada in Europe. Proceedings, 1994. XII, 521 pages. 1994.

Vol. 888: S. A. Andersson (Ed.), Analysis of Dynamical and Cognitive Systems. Proceedings, 1993. VII, 260 pages. 1995.

Vol. 889: H. P. Lubich, Towards a CSCW Framework for Scientific Cooperation in Europe. X, 268 pages. 1995.

Vol. 890: M. J. Wooldridge, N. R. Jennings (Eds.), Intelligent Agents. Proceedings, 1994. VIII, 407 pages. 1995. (Subseries LNAI).

Vol. 891: C. Lewerentz, T. Lindner (Eds.), Formal Development of Reactive Systems. XI, 394 pages. 1995.

Vol. 892: K. Pingali, U. Banerjee, D. Gelernter, A. Nicolau, D. Padua (Eds.), Languages and Compilers for Parallel Computing. Proceedings, 1994. XI, 496 pages. 1995.

Vol. 893: G. Gottlob, M. Y. Vardi (Eds.), Database Theory – ICDT '95. Proceedings, 1995. XI, 454 pages. 1995.

Vol. 894: R. Tamassia, I. G. Tollis (Eds.), Graph Drawing. Proceedings, 1994. X, 471 pages. 1995.

Vol. 895: R. L. Ibrahim (Ed.), Software Engineering Education. Proceedings, 1995. XII, 449 pages. 1995.

Vol. 896: R. M. Taylor, J. Coutaz (Eds.), Software Engineering and Human-Computer Interaction. Proceedings, 1994. X, 281 pages. 1995.

Vol. 898: P. Steffens (Ed.), Machine Translation and the Lexicon. Proceedings, 1993. X, 251 pages. 1995. (Subseries LNAI).

Vol. 899: W. Banzhaf, F. H. Eeckman (Eds.), Evolution and Biocomputation. VII, 277 pages. 1995.

Vol. 900: E. W. Mayr, C. Puech (Eds.), STACS 95. Proceedings, 1995. XIII, 654 pages. 1995.

Vol. 901: R. Kumar, T. Kropf (Eds.), Theorem Provers in Circuit Design. Proceedings, 1994. VIII, 303 pages. 1995.

Vol. 902: M. Dezani-Ciancaglini, G. Plotkin (Eds.), Typed Lambda Calculi and Applications. Proceedings, 1995. VIII, 443 pages. 1995.

Vol. 903: E. W. Mayr, G. Schmidt, G. Tinhofer (Eds.), Graph-Theoretic Concepts in Computer Science. Proceedings, 1994. IX, 414 pages. 1995.